XXIV International Conference on Integrable Systems and Quantum Symmetries 2016 (ISQS-24)

Journal of Physics: Conference Series Volume 804

Prague, Czech Republic
14 – 18 June 2016

Editors:

Cestmir Burdik
Ondrej Navratil
Severin Posta

ISBN: 978-1-5108-3898-7
ISSN: 1742-6588

Printed from e-media with permission by:

Curran Associates, Inc.
57 Morehouse Lane
Red Hook, NY 12571

Some format issues inherent in the e-media version may also appear in this print version.

Printed by Curran Associates, Inc. (2017)

For permission requests, please contact the Institute of Physics
at the address below.

Institute of Physics
Dirac House, Temple Back
Bristol BS1 6BE UK

Phone: 44 1 17 929 7481
Fax: 44 1 17 920 0979

techtracking@iop.org

Additional copies of this publication are available from:

Curran Associates, Inc.
57 Morehouse Lane
Red Hook, NY 12571 USA
Phone: 845-758-0400
Fax: 845-758-2633
Email: curran@proceedings.com
Web: www.proceedings.com

Table of contents

Volume 804

XXIV International Conference on Integrable Systems and Quantum symmetries (ISQS-24)

14–18 June 2016, Prague, Czech Republic

Accepted papers received: 13 January 2017
Published online: 4 April 2017

Preface

Papers

IOP Conf. Series: Journal of Physics: Conf. Series **804** (2017) 011001 doi:10.1088/1742-6596/804/1/011001

XXIV International Conference on Integrable Systems and Quantum Symmetries (ISQS-24)

Čestmír Burdík, Ondřej Navrátil and Severin Pošta

Abstract

The XXIV International Conference on Integrable Systems and Quantum Symmetries (ISQS-24), organized by the Department of Mathematics, Faculty of Nuclear Sciences and Physical Engineering, Czech Technical University Prague and the Bogoliubov Laboratory of Theoretical Physics of the Joint Institute for Nuclear Research, belongs to the successful series of conferences held at the Czech Technical University which began in 1992 and is devoted to problems of mathematical physics related to the theory of integrable systems, quantum groups and quantum symmetries. During the last 5 years, each of the conferences gathered around 110 scientists from all over the world. 43 papers of plenary lectures and contributions presented at ISQS-24 are published in the present issue of Journal of Physics: Conference Series.

Peer review statement

All papers published in this volume of *Journal of Physics: Conference Series* have been peer reviewed through processes administered by the proceedings Editors. Reviews were conducted by expert referees to the professional and scientific standards expected of a proceedings journal published by IOP Publishing.

IOP Conf. Series: Journal of Physics: Conf. Series **804** (2017) 012001 doi:10.1088/1742-6596/804/1/012001

Direct Photon Production in Pion-Proton Collisions

A. I. Ahmadov[1], C. Aydin[2], O. Uzun[2]

[1] Department of Theoretical Physics, Baku State University, Z. Khalilov st. 23, AZ-1148, Baku, Azerbaijan,

[2] Department of Physics, Karadeniz Technical University, 61080, Trabzon, Turkey

E-mail: ahmadovazar@yahoo.com, coskun@ktu.edu.tr, oguzhan_deu@hotmail.com

Abstract. In this article we calculate the contribution of the higher-twist Feynman diagrams to the large-p_T inclusive single direct photon production cross section in πp collisions in case of the running coupling and frozen coupling approaches within holographic QCD. We determined the structure of infrared renormalon singularities of the higher-twist subprocess cross section. Also compared and analyzed the resummed higher-twist cross sections with the ones obtained in the framework of the frozen coupling approach.

1. Introduction

Quantum chromodynamics(QCD) is widely believed to be the fundamental theory of the strong interactions. One of the main goal in QCD is to describe the structure and dynamics of hadrons at the amplitude level. The hadronic distribution amplitude in terms of quark and gluon degrees of freedoms plays an important role in QCD process predictions.

The AdS/CFT model has led to important insights into the properties of QCD. AdS/CFT duality also gives accurate predictions for hadron spectroscopy and a description of the quark structure of hadrons which has scale invariance also dimensional counting at short distances and together with color confinement at large distances Based on the AdS/CFT correspondence where these are many significant theoretical advances using this theory as application, interesting and important attempts have been made to understand non-perturbative aspects of QCD under the name "holographic QCD". A holographic model based on a truncated AdS space can be used to obtain the hadronic spectrum of light $q\bar{q}$, qqq and gg bound states showed by Brodsky et al. [1]

The higher-twist effects was calculated within the frozen coupling constant approach in the References [2, 3, 4, 5, 6].

In this study, we apply the running [7] and frozen coupling approaches in order to compute the effects of the infrared renormalons on the direct photon production at πp collision in holographic QCD. This approach was also employed previously [8, 9, 10, 11, 12, 13] to calculate the inclusive meson production in pp and $\gamma\gamma$ collisions.

We are interested in the calculation and analysis of the higher-twist effects on the dependence of the pion distribution amplitude in direct single photon production at πp collision the considered approaches within holographic QCD. In this respect, the contribution of the higher-twist Feynman diagrams to a single direct photon production cross section has been computed by using various pion distribution amplitudes from holographic QCD.

The rest of the present paper is organized as follows. The formulae for the calculation of the contributions of the higher-twist and leading-twist diagrams are provided in Section 2. Some

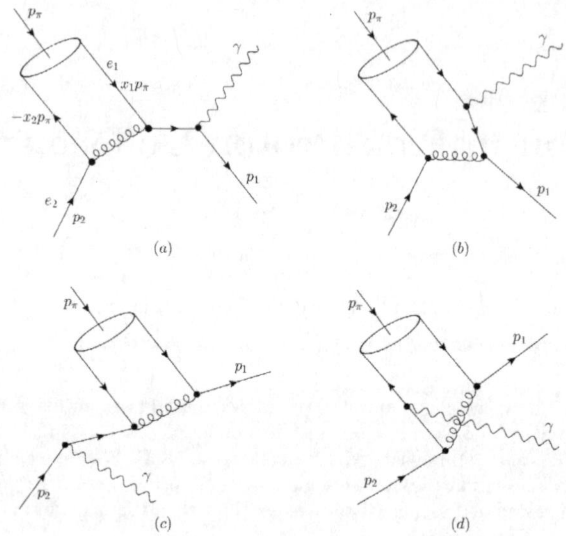

Figure 1. The higher-twist contribution to $\pi p \to \gamma X$.

formulae and analysis of the higher twist effects on the dependence of the pion distribution amplitudes by the running coupling constant approach is presented in Section 3, and the numerical results of the cross section and discussion of the dependence of the cross section on the pion distribution amplitudes and conclusions are presented in Section 4.

2. HIGHER TWIST CONTRIBUTIONS TO INCLUSIVE DIRECT PHOTON PRODUCTION

The higher-twist Feynman diagrams for the direct photon production in the pion-proton collision $\pi p \to \gamma X$ are shown in Fig.1. The amplitude for this subprocess is found by means of the Brodsky-Lepage formula [14]:

$$M(\hat{s},\hat{t}) = \int_0^1 dx_1 \int_0^1 dx_2 \delta(1-x_1-x_2)\Phi_M(x_1,x_2,Q^2)T_H(\hat{s},\hat{t};x_1,x_2), \quad (1)$$

where T_H is the sum of the graphs contributing to the hard-scattering part of the subprocess.

As higher-twist subprocess which contribute to $\pi p \to \gamma X$, we take $\pi q_N \to \gamma q$ where q_N is a constituent of the initial nucleon target. In this study, for $\pi^+ p \to \gamma X$ and for $\pi^- p \to \gamma X$ processes we take, subprocesses as $\pi^+ d_N \to \gamma u$, $\pi^- u_N \to \gamma d$, respectively.

Differential higher-twist cross section of the main process $\pi p \to \gamma X$ we can write formal analogy with deep-inelastic scattering in this form:

$$E\frac{d\sigma}{d^3p}(\pi p \to \gamma X) = \int_0^1 dx \delta(\hat{s}+\hat{t}+\hat{u})\hat{s}G_{q/p}(x,Q^2)\frac{1}{\pi}\frac{d\sigma}{d\hat{t}}(\pi q_N \to \gamma q), \tag{2}$$

where $G_{q/p}(x,Q^2)$ is the quark distribution function inside a proton.

The Mandelstam invariant variables for higher-twist subprocess $\pi q_N \to \gamma q$ are

$$\hat{s} = (p_1 + p_\gamma)^2 = (p_2 + p_\pi)^2, \quad \hat{t} = (p_\gamma - p_\pi)^2, \quad \hat{u} = (p_1 - p_\pi)^2. \tag{3}$$

The subprocess cross section for $\pi q_N \to \gamma q$ have the form [15]:

$$\frac{d\sigma}{d\hat{t}}(\hat{s},\hat{t},\hat{u}) = \frac{16\pi^2\alpha_E C_F}{9}\frac{[D(\hat{s},\hat{u})]^2}{\hat{s}^2(-\hat{t})}\left[\frac{1}{\hat{u}^2}+\frac{1}{\hat{s}^2}\right], \tag{4}$$

here $D(\hat{s},\hat{u})$ in the framework of the frozen coupling constant approach takes the form:

$$D(\hat{s},\hat{u}) = e_1\hat{u}\alpha_s(Q_1^2)\int_0^1 dx\left[\frac{\Phi_\pi(x,Q_1^2)}{x(1-x)}\right] + e_2\hat{s}\alpha_s(Q_2^2)\int_0^1 dx\left[\frac{\Phi_\pi(x,Q_2^2)}{x(1-x)}\right]. \tag{5}$$

where $Q_1^2 = (1-x)\hat{s}$ and $Q_2^2 = -x\hat{u}$ are represent the momentum squared carried by the hard gluon in Fig.1, e_1 and e_2 is the charge of quarks and $C_F = \frac{4}{3}$. For the case $\pi^- p \to \gamma X$, $e_1 = -\frac{1}{3}$ and $e_2 = \frac{2}{3}$, and for $\pi^+ p \to \gamma X$, $e_1 = \frac{2}{3}$ and $e_2 = -\frac{1}{3}$.

In this calculations we use pion distribution amplitudes predicted by AdS/CFT [16, 17, 18], the pQCD evolution [19],
are

$$\Phi^{hol}(x) = \frac{4}{\sqrt{3}\pi}f_\pi\sqrt{x(1-x)},$$

$$\Phi^{hol}_{VSBGL}(x) = \frac{A_1 k_1}{2\pi}\sqrt{x(1-x)}exp\left(-\frac{m^2}{2k_1^2 x(1-x)}\right), \quad \Phi_{asy}(x) = \sqrt{3}f_\pi x(1-x), \tag{6}$$

where f_π is the pion decay constant, respectively.

The final form for the higher-twist differential cross section for direct photon production in the process $\pi p \to \gamma X$ is written as

$$E\frac{d\sigma}{d^3p}(\pi p \to \gamma X) = \frac{s}{s+u}xG_{q/p}(x,Q^2)\frac{16\pi\alpha_E C_F}{9}\frac{[D(\hat{s},\hat{u})]^2}{\hat{s}^2(-\hat{t})}\left[\frac{1}{\hat{s}^2}+\frac{1}{\hat{u}^2}\right]. \tag{7}$$

The main problem, is extracting the higher-twist corrections to the direct photon production cross section. One can also interested consider the comparison of higher-twist corrections with leading-twist contributions. For the leading-twist subprocess in the direct photon production, we take as subprocess quark-antiquark annihilation $q\bar{q} \to g\gamma$. The corresponding cross section for subprocess $q\bar{q} \to g\gamma$ is

$$\frac{d\sigma}{d\hat{t}}(q\bar{q} \to g\gamma) = \frac{8}{9}\pi\alpha_E\alpha_s(Q^2)\frac{e_q^2}{\hat{s}^2}\left(\frac{\hat{t}}{\hat{u}}+\frac{\hat{u}}{\hat{t}}\right). \tag{8}$$

The leading-twist cross section for production of direct photon is

$$\Sigma_M^{LT} \equiv E\frac{d\sigma}{d^3p}(\pi p \to \gamma X) = \int_0^1 dx_1\int_0^1 dx_2\delta(\hat{s}+\hat{t}+\hat{u})G_{\bar{q}/M}(x_1,Q_1^2)G_{q/p}(x_2,Q_2^2)\frac{\hat{s}}{\pi}\frac{d\sigma}{d\hat{t}}(q\bar{q} \to g\gamma), \tag{9}$$

where

$$\hat{s} = x_1x_2s,\ \hat{t} = x_1t, \hat{u} = x_2u.$$

IOP Conf. Series: Journal of Physics: Conf. Series **804** (2017) 012001 doi:10.1088/1742-6596/804/1/012001

3. HIGHER TWIST EFFECTS WITHIN HOLOGRAPHIC QCD AND THE ROLE INFRARED RENORMALONS

The main aim in our study are the calculation of integral in (5) by the running coupling constant approach within holographic QCD and the extraction of pure contribution of higher-twist and renormalon effects to the cross section. The renormalization scale according to Figs.1(a) and 1(b) should be chosen equal to $\mu_{R_1}^2 = (1-x)\hat{s} = \lambda_1 \hat{s}$, whereas in Figs.1(c) and 1(d) $\mu_{R_2}^2 = -x\hat{u} = \lambda_2 \hat{u}$, which directly we obtain from Feynman diagram. The integral in Eq.(5) in the framework of the running coupling approach takes the form

$$D(\mu_R^2) = \int_0^1 \frac{\alpha_s(\mu_{R_1}^2)\Phi_M(x,\mu_F^2)dx}{x(1-x)} + \int_0^1 \frac{\alpha_s(\mu_{R_2}^2)\Phi_M(x,\mu_F^2)dx}{x(1-x)}. \tag{10}$$

The $\alpha_s(\lambda\mu_{R_0}^2)$ has the infrared singularity for $\lambda = 0$ and so the integral (10) diverges. Thus for the regularization of the integral, we express the running coupling at scaling variable $\alpha_s(\mu_R^2)$ with the aid of the renormalization group equation in terms of the fixed one $\alpha_s(Q^2)$. The solution of renormalization group equation for the running coupling $\alpha \equiv \alpha_s/\pi$ has the form

$$\frac{\alpha(\lambda)}{\alpha} = \left[1 + \alpha\frac{\beta_0}{4}\ln\lambda\right]^{-1}. \tag{11}$$

Then, for $\alpha_s(\lambda Q^2)$, we get

$$\alpha(\lambda Q^2) = \frac{\alpha_s}{1+\ln\lambda/t}, \tag{12}$$

where $t = 4\pi/\alpha_s(Q^2)\beta_0 = 4/\alpha\beta_0$.

If we insert Eq.(12) into Eq.(10), then we obtain

$$D(\hat{s},\hat{u}) = e_1\hat{u}\int_0^1 dx\frac{\alpha_s(\mu_{R_1}^2)\Phi_M(x,Q_1^2)}{x(1-x)} + e_2\hat{s}\int_0^1 dx\frac{\alpha_s(\mu_{R_2}^2)\Phi_M(x,Q_2^2)}{x(1-x)}$$

$$= e_1\hat{u}\alpha_s(\hat{s})t_1\int_0^1 dx\frac{\Phi_M(x,Q_1^2)}{x(1-x)(t_1+\ln\lambda_1)} + e_2\hat{s}\alpha_s(-\hat{u})t_2\int_0^1 dx\frac{\Phi_M(x,Q_2^2)}{x(1-x)(t_2+\ln\lambda_2)}, \tag{13}$$

where $t_1 = 4\pi/\alpha_s(\hat{s})\beta_0$ and $t_2 = 4\pi/\alpha_s(-\hat{u})\beta_0$.

Although the integral (13) is still divergent, but this expression can be transformed to more useful form as making the change of variable $z = \ln\lambda$, which gives

$$D(\hat{s},\hat{u}) = e_1\hat{u}\alpha_s(\hat{s})t_1\int_0^1 \frac{\Phi_M(x,Q_1^2)dx}{x(1-x)(t_1+z_1)} + e_2\hat{s}\alpha_s(-\hat{u})t_2\int_0^1 \frac{\Phi_M(x,Q_2^2)dx}{x(1-x)(t_2+z_2)}. \tag{14}$$

We can calculate (14) by applying the integral representation of $1/(t+z)$ [20, 21]:

$$\frac{1}{t+z} = \int_0^\infty e^{-(t+z)u}du, \tag{15}$$

gives

$$D(\hat{s},\hat{u}) = e_1\hat{u}\alpha_s(-\hat{s})t_1\int_0^1\int_0^\infty \frac{\Phi_\pi(x,Q_1^2)e^{-(t_1+z_1)u}dudx}{x(1-x)} +$$

$$+e_2\hat{s}\alpha_s(-\hat{u})t_2\int_0^1\int_0^\infty \frac{\Phi_\pi(x,Q_2^2)e^{-(t_2+z_2)u}dudx}{x(1-x)}. \tag{16}$$

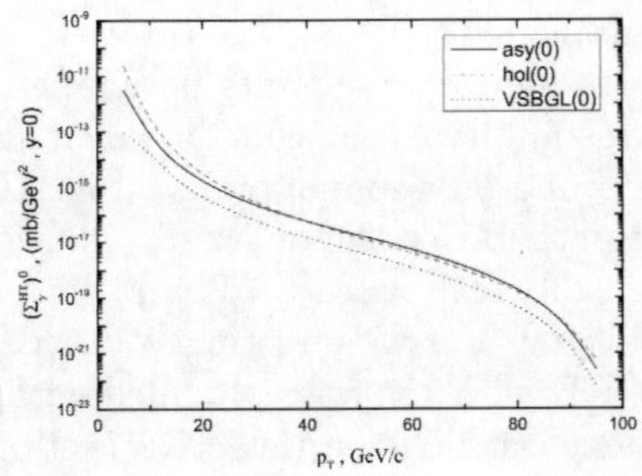

Figure 2. Higher-twist $\pi^+p \to \gamma X$ direct photon production cross section $(\Sigma_\gamma^{HT})^0$ as a function of the p_T transverse momentum of the photon at the c.m. energy $\sqrt{s} = 200\ GeV$.

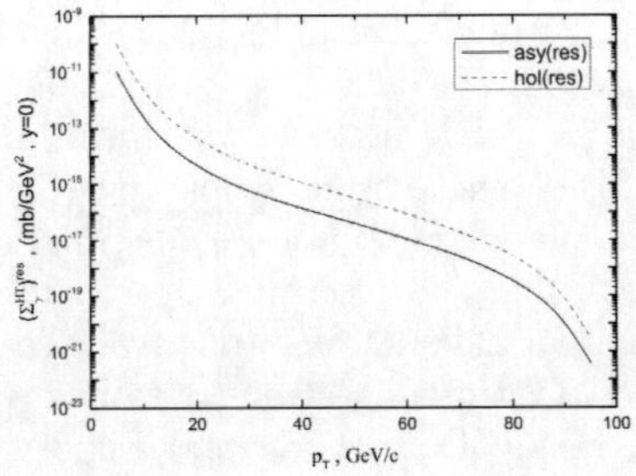

Figure 3. Higher-twist $\pi^+p \to \gamma X$ direct photon production cross section $(\Sigma_\gamma^{HT})^{res}$ as a function of the p_T transverse momentum of the photon at the c.m.energy $\sqrt{s} = 200\ GeV$.

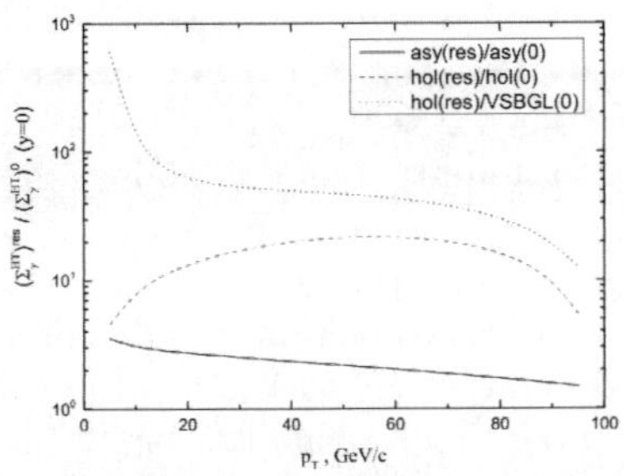

Figure 4. Ratio $(\Sigma_\gamma^{HT})^{res}/(\Sigma_\gamma^{HT})^0$, in the process $\pi^+p \to \gamma X$, where higher-twist contribution are calculated for the photon rapidity $y = 0$ at the c.m.energy $\sqrt{s} = 200\ GeV$ as a function of the photon transverse momentum, p_T.

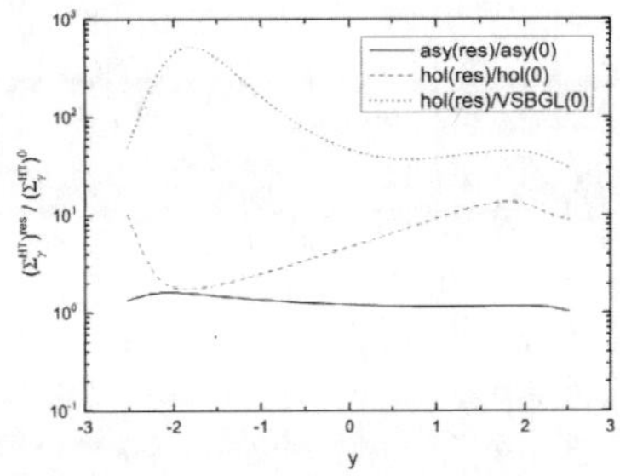

Figure 5. Ratio $(\Sigma_\gamma^{HT})^{res}/(\Sigma_\gamma^{HT})^0$,in the process $\pi^+p \to \gamma X$, as a function of the y rapidity of the photon at the transverse momentum of the photon $p_T = 15.5\ GeV/c$, at the c.m. energy $\sqrt{s} = 200\ GeV$.

For $\Phi^{hol}(x)$ the Eq.(16) is written as

$$D(\hat{s},\hat{u}) = \frac{16 f_\pi e_1 \hat{u}}{\sqrt{3}\beta_0} \int_0^\infty du e^{-t_1 u} B\left(\frac{1}{2}, \frac{1}{2} - u\right) + \frac{16 f_\pi e_2 \hat{s}}{\sqrt{3}\beta_0} \int_0^\infty du e^{-t_2 u} B\left(\frac{1}{2}, \frac{1}{2} - u\right), \tag{17}$$

and for $\Phi_{asy}(x)$ wave function

$$D(\hat{s},\hat{u}) = \frac{4\sqrt{3}\pi f_\pi e_1 \hat{u}}{\beta_0} \int_0^\infty du e^{-t_1 u} \left[\frac{1}{1-u}\right] + \frac{4\sqrt{3}\pi f_\pi e_2 \hat{s}}{\beta_0} \int_0^\infty du e^{-t_2 u} \left[\frac{1}{1-u}\right]. \tag{18}$$

where $B(\alpha, \beta)$ is Beta function and u is Borel parameter.

4. NUMERICAL RESULTS AND CONCLUSIONS

We discuss the numerical results for higher-twist and renormalon effects with higher-twist contributions calculated in the context of the running and frozen coupling approaches on the dependence of the chosen pion distribution amplitudes in the direct photon production process, $\pi p \to \gamma X$, within holographic QCD. For the numerical calculations, for $\pi^+ p \to \gamma X$ process, we take subprocess as $\pi^+ d_N \to \gamma u$. Inclusive direct photon production represents a significant test case in which higher-twist terms dominate those of leading-twist in certain kinematic domains. For the dominant leading-twist subprocess for the photon production, we take the quark-antiquark annihilation $q\bar{q} \to \gamma g$. In the numerical calculations, for the quark distribution functions inside the pion and proton used as given in [22, 23], respectively. The results of our numerical calculations are displayed in Figs.2-5. Firstly, it is very interesting comparing the higher-twist cross sections obtained within holographic QCD with the ones obtained within perturbative QCD. In Fig.2 and Fig.3 we show the dependence of higher-twist cross sections $(\Sigma_{\gamma}^{HT})^0$ (frozen cross section), $(\Sigma_{\gamma}^{HT})^{res}$ (resummed cross section) calculated in the context of the frozen and running coupling constant approaches as a function of the photon transverse momentum p_T for the pion distribution amplitudes $\Phi^{hol}(x)$, $\Phi_{asy}(x)$, $\Phi_{VSBGL}^{hol}(x)$, at $y = 0$. It is seen from Fig.2 and Fig.3 that the higher-twist cross section is monotonically decreasing with an increase in the transverse momentum of the photon. In the region $5\ GeV/c < p_T < 95\ GeV/c$ the resummed cross sections of the process $\pi^+ p \to \gamma X$ decreases in the range between $2.256 \cdot 10^{-11} \mu b/GeV^2$ to $5.331 \cdot 10^{-22} \mu b/GeV^2$. In Fig.4-Fig.5 we show the dependence of the ratios $(\Sigma_{\gamma}^{HT})^{res}/(\Sigma_{\gamma}^{HT})^0$, in the process $\pi^+ p \to \gamma X$ which are displayed as a function of p_T and y rapidity for $\Phi^{hol}(x)$, $\Phi_{asy}(x)$ and $\Phi_{VSBGL}^{hol}(x)$ pion distribution amplitudes. As shown in Fig.4, in the region $5\ GeV/c < p_T < 95\ GeV/c$, resummed cross section for $\Phi^{hol}(x)$ is suppress by about two orders of magnitude relative to the frozen cross section for $\Phi_{VSBGL}^{hol}(x)$. In Fig.5 the dependence of ratio $(\Sigma_{\gamma}^{HT})_{hol}^{res}/(\Sigma_{\gamma}^{HT})_{hol}^{0}$ for $\Phi^{hol}(x)$ distribution amplitude at the point $y = -2$ has a minimum, but for $\Phi_{VSBGL}^{hol}(x)$ pion distribution amplitude has a maximum.

We think that this feature of infrared renormalons may help the explain theoretical interpretations with future experimental data for the direct photon production cross section in the pion-proton collisions. Higher-twist cross section obtained in our study should be observable at hadron collider. The following results can be concluded from the experiments; the higher-twist contributions to single direct photon production cross section in the pion-proton collisions have important phenomenological consequences, the higher-twist photon production cross section in the pion-proton collisions depends on the form of the pion distribution amplitudes and may be used for future study. Also the contributions of renormalon effects within the framework holograpich QCD in this process is essential and may help to analyse experimental results. Further investigations are needed in order to clarify the role of higher-twist effects in QCD. In hadron-hadron collisions, real photons at high transverse momentum can serve as a short distance in the probe of the incident hadrons. Especially, the future experimental measurements will provide further tests of the dynamics of large-p_T hadron production beyond the leading twist.

Acknowledgments
One of the authors, A. I. A. thanks the organizers of the ISQS-24, in Prague especially to the Prof.Cestmir Burdik for the opportunity to speak at the conference, financial support and for their exceptional hospitality.

[1] S. J. Brodsky, G. F. de Teramond, hep-ph/0709.2072.
[2] J. A. Bagger and J. F. Gunion, Phys. Rev. **D 29**, 40 (1984).
[3] A. Bagger and J. F. Gunion, Phys. Rev. **D 25**, 2287 (1982).
[4] A. I. Ahmadov, I. Boztosun, R. Kh. Muradov, A. Soylu, and E. A. Dadashov, Int. J. Mod. Phys. **E 15**, 1209 (2006).
[5] A. I. Ahmadov, I. Boztosun, A. Soylu, and E. A. Dadashov, Int. J. Mod.Phys. **E 17**, 1041 (2008).
[6] V. N. Baier and A. Grozin, Phys. Lett. **B 96**, 181 (1980); S. Gupta, Phys. Rev. **D 24**, 1169 (1981).
[7] S. S. Agaev, Phys. Lett. **B 360**, 117 (1995); **B 369**, 379(E) (1996).

[8] A. I. Ahmadov, Coskun Aydin, Sh. M. Nagiyev, A. Hakan Yilmaz, and E. A. Dadashov, Phys. Rev. **D 80**, 016003 (2009).
[9] A. I. Ahmadov, Coskun Aydin, E. A. Dadashov and Sh. M. Nagiyev, Phys. Rev. **D 81**, 054016 (2010).
[10] A. I. Ahmadov, R. M. Burjaliyev, Int. J. Mod. Phys. **E 20**, 1243 (2011).
[11] A. I. Ahmadov, Sh. M. Nagiyev and E. A. Dadashov , Int. J. Mod. Phys. **E 21**, 1250014 (2012).
[12] A. I. Ahmadov, C. Aydin and F. Keskin, Phys. Rev. **D 85**, 034009 (2012).
[13] A. I. Ahmadov, C. Aydin and F. Keskin, Annals of Physics. **327**, 1472 (2012).
[14] G. L. Lepage and S. J. Brodsky, Phys. Rev. **D 22**, 2157 (1980).
[15] E. L. Berger, Phys. Rev. **D 26**, 105 (1982).
[16] S. J. Brodsky, arxiv: hep-ph/0707.2643.
[17] S. J. Brodsky and G. F. de Teramond, Phys. Rev. **D 77**, 056007 (2008).
[18] A. Vega, I. Schmidt, T. Branz, T.Gutsche,V. Lyubovitskij, Phys. Rev. **D 80**, 055014 (2009).
[19] G. P. Lepage and S. J. Brodsky, Phys. Lett. **B 87**, 359 (1979).
[20] J. Zinn-Justin, Phys. Rept. **70**, 109 (1981).
[21] A. Erdelyi, Higher Transcendental Functions (McGrow-Hill Book Company, New York, 1953), Vol.2.
[22] Seung-il Nam, Phys. Rev. **D 86**, 074005 (2012).
[23] A.D. Martin, W. J. Stirling, R.S. Thorne, G. Watt, arXiv:hep-ph/0901.0002.

IOP Conf. Series: Journal of Physics: Conf. Series **804** (2017) 012002 doi:10.1088/1742-6596/804/1/012002

Magnetization Plateaus and Thermal Entanglement of Spin Systems

N Ananikian[1], Č Burdík [2], L Ananikyan [1] and H Poghosyan[1]

[1] Theory Department , A.I. Alikhanyan National Science Laboratory, 0036 Yerevan, Armenia
[2] Department of Mathematics, Czech Technical University, Prague 12000, Czech Republic

E-mail: ananik@yerphi.am

Abstract. The geometrically magnetic frustrations and quantum thermal entanglement of antiferromagnetic metal-containing compounds are considered on a diamond chain. We researched the magnetic and thermal properties of the symmetric Hubbard dimers with delocalized interstitial spins and the quantum entanglement states. It is presented magnetization plateaus and negativity in spin-1 Ising-Heisenberg model using transfer matrix technique. Applying the dynamic system approach we study the magnetic curves, Lyapunov exponents and superstable point in the two-dimensional mapping for the partition function of spin-1 classical and Ising-Heisenberg models at $T \to 0$ on a diamond chain.

1. Introduction

The magnetic plateau and entanglement properties exhibit common features observable via antiferromagnetic coupling constant, external magnetic, crystal (single-ion anisotropy) fields and chemical potential. The family of known non-trivial quantum effects in the condensed matter physics is enriched with the novel phenomenonintermediate plateaus in the magnetization processes [1, 2]. The observation of a 1/3 magnetization plateau in natural azurite $Cu_3(CO_3)_2(OH)_2$ has been proposed as a realization of the exotic diamond chain of antiferromagnetically coupled spin-1/2 Cu^{+2}-monomers [3, 4]. The theoretical study of magnetization plateaus in azurite were obtained by using the density functional theory [5, 6], mean-field-like treatment based on the Gibbs-Bogoliubov inequality [7] , the density-matrix renormalization-group (DMRG) technique [8,9] and the decoration-iteration transformation [10]. On a diamond chain there were observed the plateaus of electron density as a function of chemical potential on spinless fermion, extended Hubbard and magnetization as a function of external magnetic field on distorted Ising-Hubbard models [11–13]. The magnetic behavior and magnetic susceptibility of the metal-containing complex $[Ni_3(fum)_2 - (\mu_3 - OH)_2(H_2O)_4]_n \cdot (2H_2O)_n$ was measured [14]. The spin-1 diamond-chain of the compound is characterized by both ferromagnetic and antiferromagnetic exchanges. In this article we would get magnetization and quadrupole moment plateaus at low temperatures as a function of external magnetic field and single-ion anisotropy. Quantum entanglement is considered to play a key role for understanding of strongly correlated quantum systems, quantum phase transitions and collective quantum phenomena in particular many-body spin and fermionic lattice systems [15–17] especially in antiferromagnetic models. Another important observation is the strong relationship between magnetic and entanglement properties of the antiferromagnetic system on diamond chain models

[18–25]. We also considered magnetic behavior of spin lattice models applying the dynamic system method [26–28]. The dynamical approach represents an essential tool in the theory of phase transitions and criticality and it enhanced our understanding of the phase structure and critical properties of spin models. In the case of antiferromagnetic coupling between lattice nodes of the three site interaction Ising and the Q-state Potts models ($Q < 2$) on Husimi and Bethe lattices exhibit a complex behavior, featuring doubling bifurcations, chaotic regimes, intermittency, and superstable points [29,30]. The plateaus of the maximal Lyapunov exponent coincide with magnetization plateaus on a kagome chain for multidimensional mapping [31]. The existence of magnetization plateau has been observed at one third of the saturation magnetization in the antiferromagnetic spin-1/2 IsingHeisenberg model on a diamond chain using the multidimensional mapping [32]. It has been detected that the maximal Lyapunov exponent exhibits plateau like magnetization one and found the existence of the supercritical point at $T \to 0$ in the absence of the external magnetic field corresponding to phase transition point.

In the present paper we mainly deal with the quantum entanglement and magnetic curve behavior of symmetric Hubbard dimers with delocalized interstitial electrons and spin-1 Ising - Heisenberg model using transfer matrix technique on a diamond chain. Applying the recurrence relation of the two-dimensional mapping we study magnetic plateau and their relation to maximal Lyapunov exponents, superstable point of spin-1 classical and Ising - Heisenberg models on a diamond chain. The paper is organized as follows. In Section 2 the phase behavior of the diamond chain is presented of a symmetric Hubbard-Ising model and studied the magnetic and thermal properties. We researched the quantum entanglement of electrons in Hubbard dimers and dependence of Hamiltonian parameters as well as on temperature. We give an exact solution of spin-1 Ising - Heisenberg model by means of the transfer-matrix method and study the magnetization plateaus and negativity in Section 3. In Section 4 we apply the dynamical system approach in researching magnetic plateaus and maximal Lyapunov exponent plateaus. It is observed that the superstable coincide with phase transition point at $T \to 0$ in the spin-1 classical and Ising-Heisenberg models on the diamond chain. Some concluding remarks are given in section 5.

2. Symmetric diamond chain with delocalized Hubbard interstitial spins

The Hamiltonian of symmetric Ising- Hubbard model [33] describing the diamond chain (Fig. 1)is

$\mathcal{H} = \sum_{i=1}^{N} \mathcal{H}_{i,i+1}$, where

$$\mathcal{H}_{i,i+1} = -\tilde{t}\left(a_{i\uparrow}^{+}b_{i\uparrow} + a_{i\uparrow}b_{i\uparrow}^{+} + a_{i\downarrow}^{+}b_{i\downarrow} + a_{i\downarrow}b_{i\downarrow}^{+}\right) + \widetilde{V}n_{ai}n_{bi} + \widetilde{U}\left(n_{ai\uparrow}n_{ai\downarrow} + n_{bi\uparrow}n_{bi\downarrow}\right) \qquad (1)$$
$$+\widetilde{J}_m s_{ci}s_{ci+1} + J\left(s_{ai}^{z} + s_{bi}^{z}\right)\left(s_{ci} + s_{ci+1}\right) - \widetilde{\mu}\left(n_{ai} + n_{bi}\right) - \widetilde{h}\left(s_{ai}^{z} + s_{bi}^{z} + \frac{s_{ci} + s_{ci+1}}{2}\right).$$

Here $a_{i\sigma}^{+}$ $(a_{i\sigma})$ and $b_{i\sigma}^{+}$ $(b_{i\sigma})$ are the electron creation (annihilation) operators with the spin value σ=($\uparrow, \downarrow$), respectively on sites a_i and b_i. The $n_{ai\sigma} = a_{i\sigma}^{+}a_{i\sigma}$ and $n_{bi\sigma} = b_{i\sigma}^{+}b_{i\sigma}$ are the electron number operators, respectively on sites a_i and b_i and $\tilde{t}$ is the hopping amplitude. We denote $n_{\alpha i} = n_{\alpha i\uparrow} + n_{\alpha i\downarrow}$, $s_{\alpha i}^{z} = \frac{1}{2}(n_{\alpha i\uparrow} - n_{\alpha i\downarrow})$, $(\alpha = a, b)$ and $s_{ci} = \pm\frac{1}{2}$. Here $\tilde{\mu}$ and $\widetilde{h}$ are the values of the chemical potential and the external magnetic field ($\widetilde{h} = g\mu_B B_z$) respectively. The terms which contain $\widetilde{U}$ and $\widetilde{V}$ correspond respectively to one–site and two–site Coulomb repulsions in the Hubbard dimer(interstitial spins). The terms which contain the symmetric coupling J that describe Hubbard dimer interactions with nodal Ising-type spins. The term containing $\widetilde{J}_m$ describes two nodal Ising–like spins interaction.

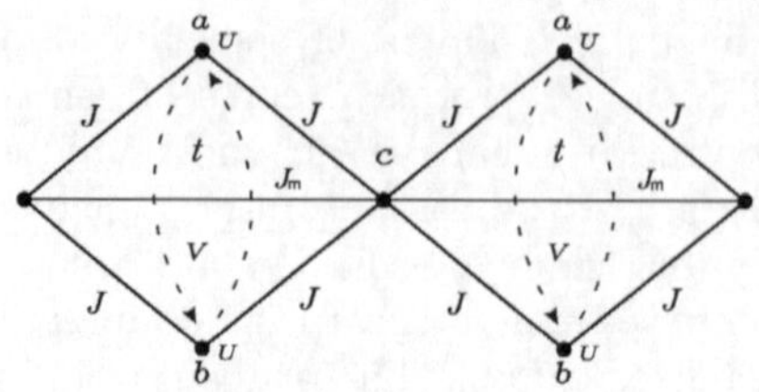

Figure 1. Two sites of the Hubbard dimer are denoted as a– and b–. There is an electron hopping between them with amplitude t. One–site (U) and two–site (V) Coulomb interactions are also present. The diamonds are connected to each other at nodal c–sites. There is an Ising–like direct interaction with J_m coupling constant between the nodal sites. The nodal sites are also connected to a– and b–sites by means of Ising–like interactions with coupling constant J.

From the identity

$$\widetilde{V} n_a n_b + \widetilde{U}\left(n_{a\uparrow} n_{a\downarrow} + n_{b\uparrow} n_{b\downarrow}\right) = \left(\widetilde{U} - \widetilde{V}\right)\left(n_{a\uparrow} n_{a\downarrow} + n_{b\uparrow} n_{b\downarrow}\right) + \frac{n\,(n-1)}{2}\widetilde{V}, \tag{2}$$

(where $n = n_{a\uparrow} + n_{a\downarrow} + n_{b\uparrow} + n_{b\downarrow}$ is the total number of electrons in the Hubbard dimer), it follows that the differences between energy levels depend on $\widetilde{U} - \widetilde{V} \equiv \widetilde{W}$ only, if n is fixed. In particular, for fixed $n \neq 2$, $(n_{a\uparrow} n_{a\downarrow} + n_{b\uparrow} n_{b\downarrow})$ is constant and the behavior of the system does not depend on $\widetilde{W}$ at all, which is physically obvious.

As all $\mathcal{H}_{i,i+1}$ commute, therefore the partition function can be written as:

$$Z_N = Tr\left(e^{-\beta\mathcal{H}}\right) = \sum_{s_{ci}=\pm\frac{1}{2}} \prod_{i=1}^{N} Tr_{a,b}\left(e^{-\beta\mathcal{H}_{i,i+1}(s_{ci},s_{ci+1})}\right) = Tr\mathbf{V}^N, \tag{3}$$

where $\mathbf{V} = \begin{pmatrix} V(+,+) & V(+,-) \\ V(-,+) & V(-,-) \end{pmatrix}$ is the transfer matrix, with $V(\pm,\pm) = \sum_j e^{-\beta E_j(\pm,\pm)}$, where $E_j(\pm,\pm) \equiv E_j\left(\pm\frac{1}{2},\pm\frac{1}{2}\right)$ are the eigenvalues of the $\mathcal{H}_{i,i+1}(\pm,\pm) \equiv \mathcal{H}_{i,i+1}\left(\pm\frac{1}{2},\pm\frac{1}{2}\right)$ Hamiltonian.

In the thermodynamic limit the free energy per diamond is

$$F = \lim_{N\to\infty} \frac{-T}{N} \ln Z_N = -\ln\lambda_1, \tag{4}$$

where $\lambda_1 = \frac{1}{2}\left(Tr\mathbf{V} + \sqrt{(Tr\mathbf{V})^2 - 4\det\mathbf{V}}\right)$ is the largest eigenvalue of the transfer matrix.

On this basis it is easy to calculate magnetization and specific heat. Analogously we use the following expression for reduced density matrix of a Hubbard dimer to study entanglement properties:

$$\begin{aligned}\rho' &= \frac{1}{\lambda_1}\left[\frac{e^{-\beta H(+,+)} + e^{-\beta H(-,-)}}{2} + e^{-\beta H(+,-)}\sin 2\theta + \right. \\ &+ \left.\frac{e^{-\beta H(+,+)} - e^{-\beta H(-,-)}}{2}\cos 2\theta\right],\end{aligned} \tag{5}$$

with $\sin 2\theta = \dfrac{2V(+,-)}{\sqrt{(Tr\mathbf{V})^2 - 4\det\mathbf{V}}}$, $\cos 2\theta = \dfrac{V(+,+) - V(-,-)}{\sqrt{(Tr\mathbf{V})^2 - 4\det\mathbf{V}}}$.

In the sequel we rescale all quantities by the J absolute value, substituting $t = \tilde{t}/|J|$, $h = \tilde{h}/|J|$, $\mu = \tilde{\mu}/|J|$, $U = \tilde{U}/|J|$, $V = \tilde{V}/|J|$, $W = \widetilde{W}/|J|$ and $J_m = \tilde{J}_m/|J|$. The magnetization curves at nonzero temperature are derived from $m = -\frac{\partial F}{\partial h}$. Here we take the number of electrons in a Hubbard dimer fixed and equals to 2. Fig. 2 shows possible forms of magnetization curves at zero and nonzero temperatures. The following types of plateaus exist here: at 0, 1/3 and 2/3 values and of classical (electrons in the Hubbard dimers are not entangled and their spin projections have definite value at each site) and non-classical (electrons in the Hubbard dimers are entangled) types.

The entanglement of formation in the special case of two spin-1/2 particles has the analytical expression [35]: $E_F = H\left(\frac{1+\sqrt{1-C^2}}{2}\right)$, where $H(x) = -x\log_2(x) - (1-x)\log_2(1-x)$, and C is the quantity called *concurrence*. For fixed values of t and J_m we have plotted the concurrence as a function of the temperature and the external magnetic field (Figs. 3). The thermal entanglement becomes significantly large when the energy level of the (non-entangled) ground state and the nearest energy level (corresponding to the entangled state) are close to each other.

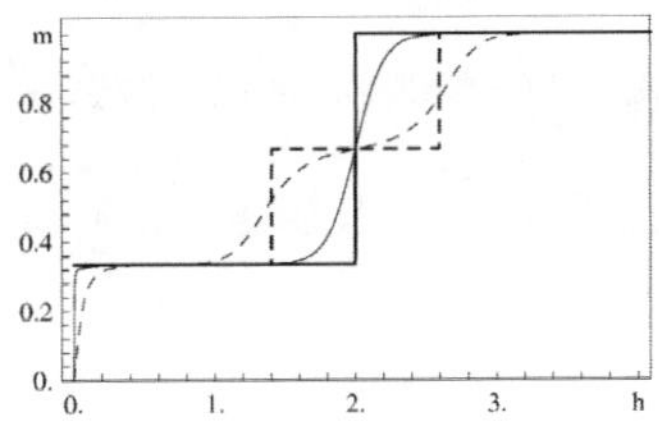

$J_m = 0$, $J_m = 0.6$

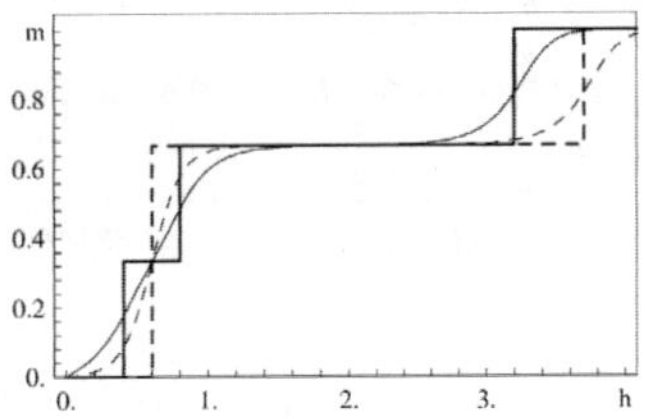

$J_m = 1.2$, $J_m = 1.7$

Figure 2. Dependence of the diamond chain magnetization (m) (per diamond in $g\mu_B$ units) on the external magnetic field (h) for different values of the internodal change interaction (J_m) at $T = 0$. The fixed chain parameters are $t = 0.3$, $W = 0$, $\mu = 0$. The case of $T = 0.1$ is also shown. Thermal fluctuations smoothen magnetization jumps. Bold lines represent at $T = 0$ and solid, dashed lines represent the different values of J_m accordingly.

3. Magnetization plateaus and negativity in spin-1 Ising - Heisenberg model

Let us consider spin-1 Ising-Heisenberg model on a diamond chain (see Fig. 6) without biquadratic coupling terms using the transfer matrix technique. The magnetic behavior of homometallic molecular ferrimagnet $[Ni_3(fum)_2-(\mu_3-OH)_2(H_2O)_4]_n\cdot(2H_2O)_n$ was measured by [14] without single-ion anisotropy parameters. We observe magnetization plateau and thermal quantum negativity in the presence of an external magnetic fields and single-ion anisotropy terms. The Hamiltonian is equal to the sum over block Hamiltonians $\mathcal{H} = \sum_i \mathcal{H}_i$ with

$$\begin{aligned}\mathcal{H}_i &= J\vec{S}_{a,i}\vec{S}_{b,i} + J_1(S^z_{a,i} + S^z_{b,i})(\sigma^z_i + \sigma^z_{i+1}) + D_H((S^z_{a,i})^2 + (S^z_{b,i})^2) + \frac{D_I}{2}((\sigma^z_i)^2 + (\sigma^z_{i+1})^2) \quad (6)\\ &- h_H g\mu_B(S^z_{a,i} + S^z_{b,i}) - \frac{h_I g\mu_B}{2}(\sigma^z_i + \sigma^z_{i+1}),\end{aligned}$$

where $S^{\alpha}_{a/b,i}$ $(\alpha = x, y, z)$ are spin-1 operators on the sites with coordinates $(a/b, i)$, while σ^z_i's are z projections of spin-1 operator resting on the sites i, J denotes the XXX interaction within the Heisenberg dimer and J_1 stands for the interaction between nodal and dimer spins. Coefficients

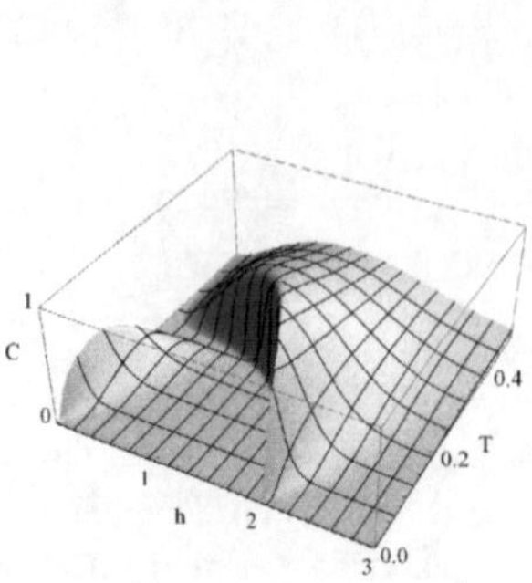

a) $t = 0.45, J_m = -2$

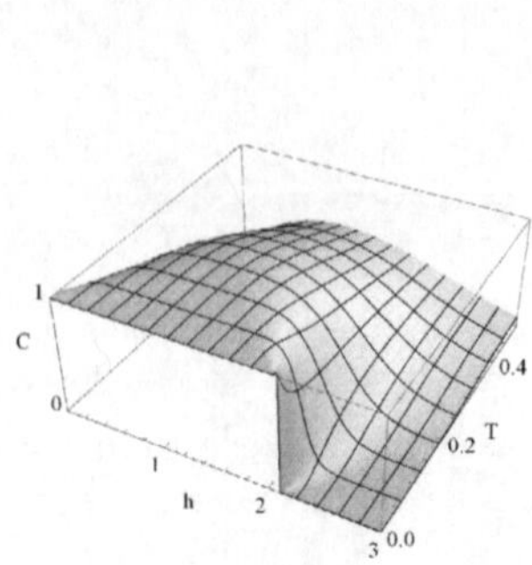

b) $t = 0.55, J_m = -2$

Figure 3. The dependence of the concurrence from temperature and external magnetic field for various values of hopping amplitude. Concurrence vanishes in high temperature limit.

D_I and D_H correspond to the longitudinal crystal(single-ion anisotropy) fields in the z direction. The gyromagnetic ratio is taken to be $g = 2.2$ in the plots drawn below, which is more or less typical for nickel containing compounds. We assume that cyclic boundary condition $\sigma_{N+1} = \sigma_1$ is applied. The block Hamiltonians($\mathcal{H}_i$) are commutative, therefore the partition function can be written like in equation (3).

After direct diagonalization of the block Hamiltonian $\mathcal{H}_i$ one gets the energy spectrum of a diamond block depending on Ising spins:

$$\begin{aligned}
E_1 &= -J - \frac{h_I g \mu_B}{2}(\sigma_i + \sigma_{i+1}) + \frac{D_I}{2}(\sigma_i^2 + \sigma_{i+1}^2) + 2D_H, \\
E_{2,3} &= \pm J - \left(J_1 + \frac{h_I g \mu_B}{2}\right)(\sigma_i + \sigma_{i+1}) + h_H g \mu_B + D_H + \frac{D_I}{2}(\sigma_i^2 + \sigma_{i+1}^2), \\
E_{4,5} &= \pm J + (J_1 - \frac{h_I g \mu_B}{2})(\sigma_i + \sigma_{i+1}) - h_H g \mu_B + D_H + \frac{D_I}{2}(\sigma_i^2 + \sigma_{i+1}^2), \\
E_{6,7} &= J + \left(\pm 2J_1 - \frac{h_I g \mu_B}{2}\right)(\sigma_i + \sigma_{i+1}) \mp 2h_H g \mu_B + \frac{D_I}{2}(\sigma_i^2 + \sigma_{i+1}^2) + 2D_H, \\
E_{8,9} &= \frac{-J \pm \Lambda}{2} - \frac{h_I g \mu_B}{2}(\sigma_i + \sigma_{i+1}) + \frac{D_I}{2}(\sigma_i^2 + \sigma_{i+1}^2) + D_H,
\end{aligned} \tag{7}$$

where $\Lambda = \sqrt{(2D_H - J)^2 + 8J^2}$. Knowledge of the spectrum allows us to perform partial trace-overs along Heisenberg degrees of freedom. Which makes representation $Z_N = \sum_{\sigma_i} \prod_{i=1}^{N} V_{\sigma_i,\sigma_{i+1}}$, with $V_{\sigma_i,\sigma_{i+1}} = Tr_i e^{-\beta \mathcal{H}_i} = \sum_{i=1}^{9} e^{-\beta E_i(\sigma_i,\sigma_{i+1})}$. $V_{\sigma_i,\sigma_{i+1}}$'s may be viewed as components of a 3×3 transfer matrix:

$$V_{\sigma_i,\sigma_{i+1}} = \begin{pmatrix} V_{-1,-1} & V_{-1,0} & V_{-1,1} \\ V_{0,-1} & V_{0,0} & V_{0,1} \\ V_{1,-1} & V_{1,0} & V_{1,1} \end{pmatrix}, \tag{8}$$

where matrix indexes are the three possible projections of Ising spins. Taking into account the cyclic boundary conditions one may write $Z_N = Tr(V^N)$ in transfer matrix notations. Hence the partition function may be expressed in terms of the transfer matrix eigenvalues:

$$Z_N = \lambda_1^N + \lambda_2^N + \lambda_3^N. \tag{9}$$

As usual, in the thermodynamic limit $N \to \infty$ only the contribution of the largest eigenvalue must be considered. Denoting the maximal eigenvalue as λ, the free energy per block for infinitely long chain takes the form

$$f = -\frac{1}{\beta} \lim_{N\to\infty} \frac{1}{N} \ln Z_N = -\frac{1}{\beta} \ln \lambda. \tag{10}$$

The exact expression of the free energy will allow us to calculate thermodynamic quantities of the system. In the present paper we will consider the sublattice magnetization and quadrupole moment of Ising spins and the sublattice magnetization of Heisenberg spins. They are given by

$$m_I = -\frac{\partial f}{\partial h_I}, \quad m_H = -\frac{1}{2}\frac{\partial f}{\partial h_H}, \quad q_I = \frac{\partial f}{\partial D_I}. \tag{11}$$

The total magnetization is equal to the average of m_I and two m_H. To illustrate the magnetization process in a few interesting cases. We show the magnetization at sufficiently low temperature ($T = 0.01K$) as a function of the magnetic field with several plateaus (Fig. 4). Thermal quantum entanglement (negativity) can be defined as [36]. We have plotted negativity

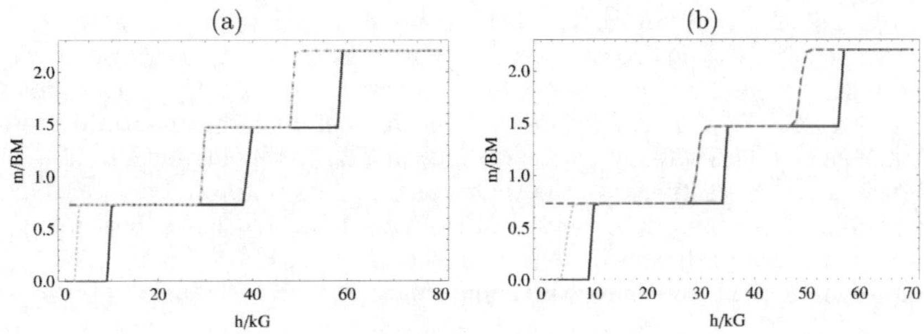

Figure 4. Low temperature ($T = 0.01K$) magnetization curves in Bohr magneton units when $J_1 = D_I = 1cm^{-1}$ and (a) $J = -6cm^{-1}$, $D_H = 1cm^{-1}$ (dashed), $D_H = 2.25cm^{-1}$ (dotted), $D_H = 4cm^{-1}$ (solid), (b) $J = 0.9cm^{-1}$, $D_H = -1cm^{-1}$ (dashed), $D_H = 0.7cm^{-1}$ (dotted), $D_H = 2cm^{-1}$ (solid).

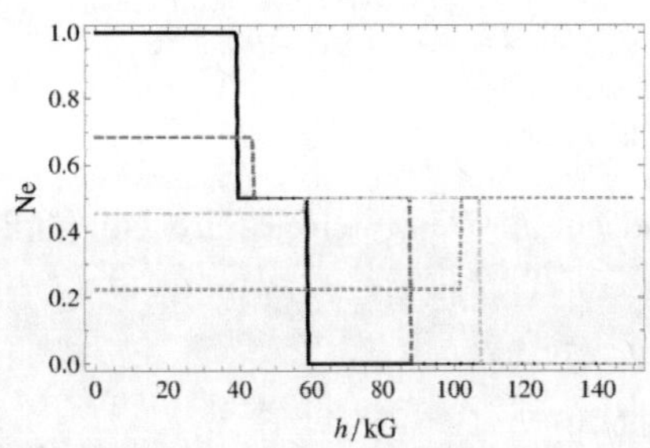

Figure 5. Negativity via magnetic field for different values of the single-ion anisotropy $D_H = 0, 3, 5, 10cm^{-1}$ (solid, dashed, dotted-dashed and dotted curves, respectively) at low temperature $T = 10^{-4}K$ and for $J = 2cm^{-1}$. Here $J_1 = D_I = 1cm^{-1}$.

vs. magnetic field Fig. 5(a) , demonstrating the change of negativity by varying D_H as well as the sharp step like behavior of the entanglement at transitions. The introduction of the single-ion anisotropy is essential for the observation of magnetic and entanglement properties for homometallic molecular ferrimagnet $[Ni_3(fum)_2 - (\mu_3 - OH)_2(H_2O)_4]_n \cdot (2H_2O)_n$.

4. Magnetic plateaus and quantum entanglement: Dynamical approach

In this section we utilise two models the spin-1 Ising and Ising-Heisenberg models. Let us first consider the spin-1 Ising model on diamond chain with free boundary conditions in the presence of an external magnetic field (see Fig. 6). The Hamiltonian operator of the model is equal to the summation of the block Hamiltonians and can be written as

$$\mathcal{H} = \sum_i^N \mathcal{H}_i = \sum_i^N (J(S_{a,i}S_{b,i}) + K(S_{a,i}S_{b,i})^2 + J_1(\mu_i + \mu_{i+1})(S_{a,i} + S_{b,i}) \tag{12}$$
$$+K_1((\mu_i)^2 + (\mu_{i+1})^2)((S_{a,i})^2 + (S_{b,i})^2) + \Delta((S_{a,i})^2 + (S_{b,i})^2) + \Delta_1 \frac{(\mu_i)^2 + (\mu_{i+1})^2}{2}$$
$$-h_H(S_{a,i} + S_{b,i}) - h_I \frac{\mu_i + \mu_{i+1}}{2})$$

Where $S_{a,i}$, $S_{b,i}$ and μ_i, μ_{i+1} are classic Ising spins and can take the values $1, 0, -1$. The parameter J stands for the linear interactions between the nearest-neighboring $S_{a,i} - S_{b,i}$ elements and J_1 is the linear interactions parameter for nearest-neighboring $S - \mu$ elements ($S_{a,i} - \mu_i$, $S_{a,i} - \mu_{i+1}$, $S_{b,i} - \mu_i$, $S_{b,i} - \mu_{i+1}$). Analogously K and K_1 stand for the quadratic interactions. Δ and Δ_1 are the single-ion anisotropy parameters, h_I and h_H are contributions of a longitudinal external magnetic field interacting with the Ising spins ($S_{a,i}$, $S_{b,i} - h_H$ and $\mu_i - h_I$).This notations are made to highlight the similarities between the spin-1 Ising and Ising-Heisenberg models. In our further calculations we will consider the case when $K = J$, $K_1 = J_1$, $\Delta = \Delta_1$, $h_H = h_I = h$. The partition function of the system with Hamiltonian (12) is

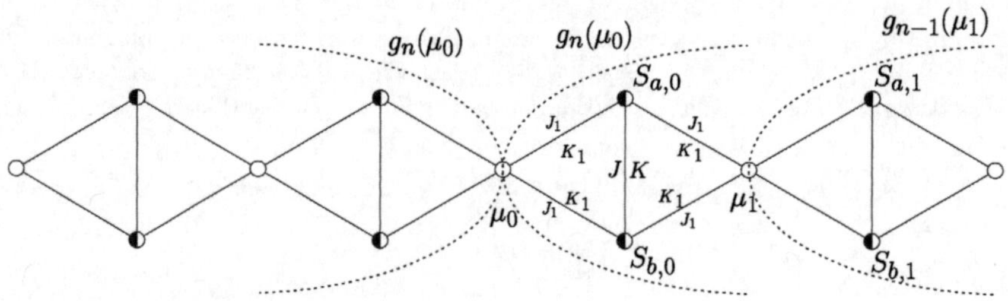

Figure 6. The procedure for derivation of the Ising (S_a and S_b are white) and Ising-Heisenberg (S_a and S_b are black) diamond chain. White circles are Ising spins and black are Heisenberg spins

$$Z = \sum_{\mu_i, S_{a,i}, S_{b,i}} \exp -\beta\mathcal{H} = \sum_{\mu_0} e^{-\frac{-h_I\mu_0 + \Delta_1\mu_0^2}{T}} g_n(\mu_0)^2 \tag{13}$$

where $\beta = (T)^{-1}$, T is the absolute temperature. By cutting diamond chain at μ_0 into two branches $g_n(\mu_0)$ (Fig. 1) the exact recursion relation for the partition function can be derived. The relation between $g_n(_0)$ and $g_{n-1}(\mu_1)$ is given by the following equation.

$$g_n(\mu_0) = \sum_{\mu_1=-1}^{1} \sum_{S_{a,0}=-1}^{1} \sum_{S_{b,0}=-1}^{1} \exp[-\beta(J(S_{a,0}S_{b,0}) + K(S_{a,0}S_{b,0})^2 + J_1(\mu_1 + \mu_0)(S_{a,0} + S_{b,0}) + \tag{14}$$
$$K_1((S_{a,0}\mu_1)^2 + (S_{a,0}\mu_0)^2 + (S_{b,0}\mu_1)^2 + (S_{b,0}\mu_0)^2) + \Delta((S_{a,0})^2 + (S_{b,0})^2) + \Delta_1\mu_1^2$$
$$-h_H(S_{a,0} + S_{b,0}) - h_I\mu_1)] * g_{n-1}(\mu_1)$$

By introducing the following notation

$$x_n = \frac{g_n(1)}{g_n(0)}, \quad y_n = \frac{g_n(-1)}{g_n(0)} \tag{15}$$

we can get two-dimensional recursion relation for the partition function

$$x_n = f1(x_{n-1}, y_{n-1}), \quad f_1(x,y) = \frac{a_1 + a_{11}x + a_{12}y}{b_0 + b_1 x + b_2 y} \tag{16}$$
$$y_n = f2(x_{n-1}, y_{n-1}), \quad f_2(x,y) = \frac{a_2 + a_{21}x + a_{22}y}{b_0 + b_1 x + b_2 y}$$

where the coefficients are easily calculated from equations (14) and (15).

Recursion relation (16) plays a crucial role in our further investigation because the order parameters magnetic and quadrupole moments can be expressed through two-dimensional rational mapping. The equations for total magnetic and quadrupole moments are expressed similar as [34].

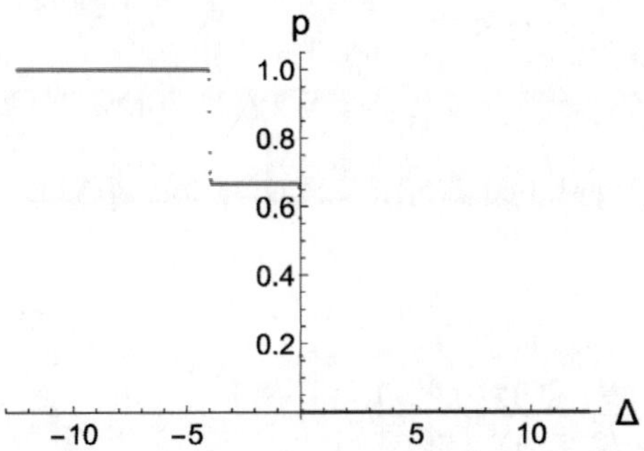

Figure 7. Classical Ising spin-1 quadrupole moment(p) for $J = 2, J_1 = 1, h = 0, T = 0.01$.

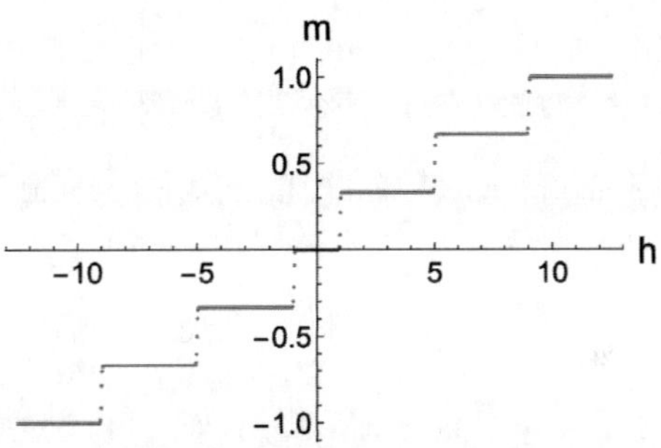

Figure 8. Classical Ising spin-1 magnetic moment(m) for $J = 2, J_1 = 1, \Delta = 1, T = 0.01$.

Now lets discus the spin-1 Ising-Heisenberg model on diamond chain with free boundary conditions in the presence of an external magnetic and crystal(single-ion anisotropy) fields (see Fig. 6). This two models are very similar therefore this time our narration will be more concise. The blocks of the diamond chain consist of two Heisenberg interstitial spins ($S^{\alpha}_{a,i}$ and $S^{\alpha}_{b,i}$). The z components of interstitial spins are coupled with nearest-neighboring nodal Ising spins (μ_i and μ_{i+1}). Like in the previous model the Hamiltonian of the system may be represented as a sum over block Hamiltonians

$$\mathcal{H} = \sum_i^N \mathcal{H}_i = \sum_i^N (J(S^x_{a,i}S^x_{b,i} + S^y_{a,i}S^y_{b,i} + DS^z_{a,i}S^z_{b,i}) + K(S^x_{a,i}S^x_{b,i} + S^y_{a,i}S^y_{b,i} + D_1 S^z_{a,i}S^z_{b,i})^2 \tag{17}$$
$$+J_1(\mu^z_i + \mu^z_{i+1})(S^z_{a,i} + S^z_{b,i}) + K_1((\mu^z_i)^2 + (\mu^z_{i+1})^2)((S^z_{a,i})^2 + (S^z_{b,i})^2) + \Delta((S^z_{a,i})^2 + (S^z_{b,i})^2)$$
$$+\Delta_1 \frac{(\mu^z_i)^2 + (\mu^z_{i+1})^2}{2} - h_H(S^z_{a,i} + S^z_{b,i}) - h_I \frac{\mu^z_i + \mu^z_{i+1}}{2})$$

In this equation, $S^{\alpha}_{a,i}$, $S^{\alpha}_{b,i}$ ($\alpha = x, y, z$) and μ^z_i represent relevant components of Heisenberg and Ising spin-1 operators. In equation (17) J is the linear Heisenberg interaction term, K is the quadratic interaction term parameter, D and D_1 are anisotropy parameters, J_1 is the interaction parameter between the nearest-neighboring Ising and Heisenberg spins, Δ (Δ_1) is the single-ion anisotropy parameter of Heisenberg (Ising) spins and the parameter K_1 is the analogue of a quadratic Ising interaction term. The last two terms in the Hamiltonian are contributions of a longitudinal external magnetic field interacting with the Heisenberg and Ising spins. It is important to notice that the commutation relation between different block Hamiltonians is zero ($[H_i, H_j] = 0$). Furthermore it can be shown that the block Hamiltonian commutes with Ising and Heisenberg spins, i.e. $[H_i, \mu_i^z] = [H_i, S^z_{a,i} + S^z_{b,i}]$ for any values of the model parameters J , K , J_1, K_1, D, D_1, Δ, Δ_1 and magnetic field, but the block Hamiltonian and quadrupole moment commute,$[H_i, (S^z_{a,i})^2 + (S^z_{b,i})^2] = 0$, only when $J = KD_1$. In the present work we consider only the case when $J = K$, $J_1 = K_1$, $D = D_1 = 1$, $\Delta = \Delta_1$, $h_H = h_I = h$. Equation (2) and (3) are conserved for this model if instead of μ_i we insert μ_i^z considering the fact that here we also cut the diamond chain at μ_0 (see Fig. 6). From them its easy to calculate the equivalent of equation (14)

$$g_n(\mu_0) = \sum_{\mu_1, S_{a,0}, S_{b,0}} \exp[-\beta(J(S^x_{a,0}S^x_{b,0} + S^y_{a,0}S^y_{b,0} + DS^z_{a,0}S^z_{b,0}) + K(S^x_{a,0}S^x_{b,0} + S^y_{a,0}S^y_{b,0} + D_1 S^z_{a,0}S^z_{b,0})^2 \tag{18}$$

$$+J_1(\mu_1^z + \mu_0^z)(S^z_{a,0} + S^z_{b,0}) + K_1((S^z_{a,0}\mu_1^z)^2 + (S^z_{a,0}\mu_0^z)^2 + (S^z_{b,0}\mu_1^z)^2 + (S^z_{b,0}\mu_0^z)^2) + \Delta((S^z_{a,0})^2 + (S^z_{b,0})^2)$$
$$+\Delta_1(\mu_1^z)^2 - h_H(S^z_{a,0} + S^z_{b,0}) - h_I\mu_1^z)] * g_{n-1}(\mu_1)$$

by inserting the eigenvalues of the operators and by using the equivalents of equations (15,16) the recursion relation has the following form

$$x_n = f1(x_{n-1}, y_{n-1}), \quad f_1(x, y) = \frac{a_1 + a_{11}x + a_{12}y}{b_0 + b_1 x + b_2 y} \tag{19}$$
$$y_n = f2(x_{n-1}, y_{n-1}), \quad f_2(x, y) = \frac{a_2 + a_{21}x + a_{22}y}{b_0 + b_1 x + b_2 y}$$

where the coefficients are easily calculated from equations (18).

The equations for total magnetic and quadrupole moments are expressed similar as [34].

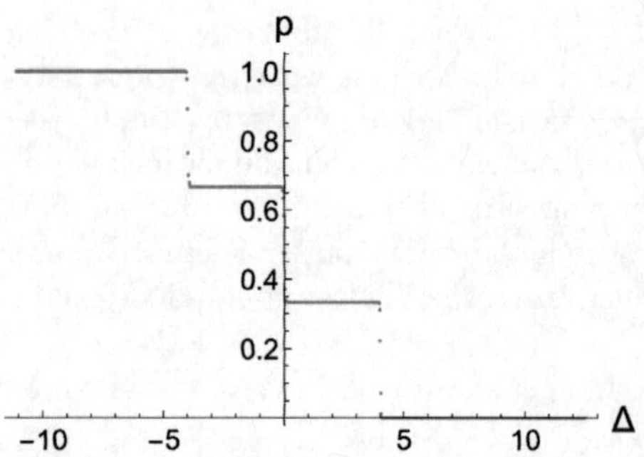

Figure 9. Quantum Ising-Heisenberg spin-1 quadrupole moment(p) for $J = 2, J_1 = 1, h = 0, T = 0.01$.

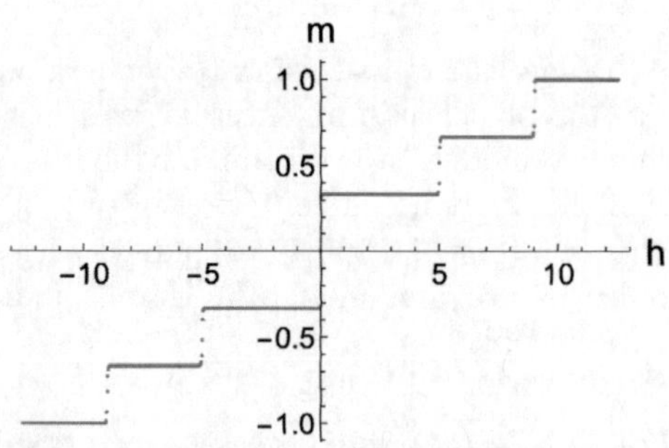

Figure 10. Quantum Ising-Heisenberg spin-1 magnetic moment(m) for $J = 2, J_1 = 1, \Delta = 1, T = 0.01$.

In the second part of this section we will focus on the thermodynamical equilibrium description of the spin-1 Ising/Ising-Heisenberg model on a diamond chain, by studying infinite-size systems. Lyapunov exponents have a crucial role in the study of "stability" or "instability"

of systems. Lyapunov exponents are the growth rate of an infinitesimal perturbation on a reference trajectory. The following values of maximal Lyapunov exponents can be observed during the investigation ($\lambda = max(\lambda_i)$).
$\lambda < 0$ negative Lyapunov exponents show that the system is dissipative or non-conservative. The systems with more negative values of Lyapunov exponent are more stable. If $\lambda = -\infty$ means that we have superstable fixed and superstable periodic points. $\lambda = 0$ corresponding to neutral fixed point. At this value of Lyapunov exponents the second-order phase transition takes place. $\lambda > 0$ corresponding to unstable and chaotic systems. In general case if we have a N dimensional map the $i-th$ Lyapunov exponent is calculated with the following equation

$$\lambda_i = \lim_{n\to\infty} \frac{\log H_i}{n}, \quad \mathbf{H} = J(x_0)J(x_1)...J(x_n) \tag{20}$$

where H_i is the $i-th$ eigenvalue of $\mathbf{H}$ and $J(x_k)$ is the Jacobian matrix of our N dimensional map after k iterations from the initial point x_0. In some cases when the eigenvalues of $\mathbf{H}$ are complex numbers and its reasonable to use $\mathbf{H}^\dagger\mathbf{H}$ ($\dagger$ denotes the transpose operator) instead of $\mathbf{H}$. In that case Lyapunov exponents will have $\lambda_i = \lim_{n\to\infty} \frac{\log H_i}{2n}$ form.

Superstable points are particularly interesting studying maximal Lyapunov exponent in classical and quantum cases. In this aspect the above mentioned models have many similarities as well as fundamental differences. There is a superstable point in the spin-1 Ising model at $h = 0$, $J = 2$, $J_1 = 1$, $T \to 0$ and $\Delta > 0$. By computational research it's shown that for many antiferromagnetic configurations there is a superstable point if $\Delta > 0$ which vanishes for $\Delta < 0$ configurations. Ising-Heisenberg models behavior is similar, many antiferromagnetic configurations have superstable points but unlike the Ising model here are two superstable points and they vanish if approximately $\Delta < -1.75$ (this point is for $h = 0$, $J = 2$, $J_1 = 1$ configuration). The fact that there is a magnetization plateau around supercritical points in the classical (see Figures 8, 11) and quantum (see Figures 10, 12) cases is of great importance.

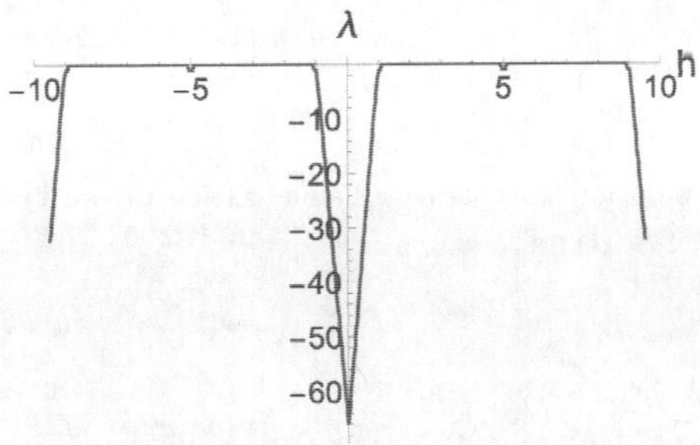

Figure 11. Classical Ising spin-1 maximal Lyapunov exponent for $J = 2, J_1 = 1, \Delta = 1, T = 0.03$.

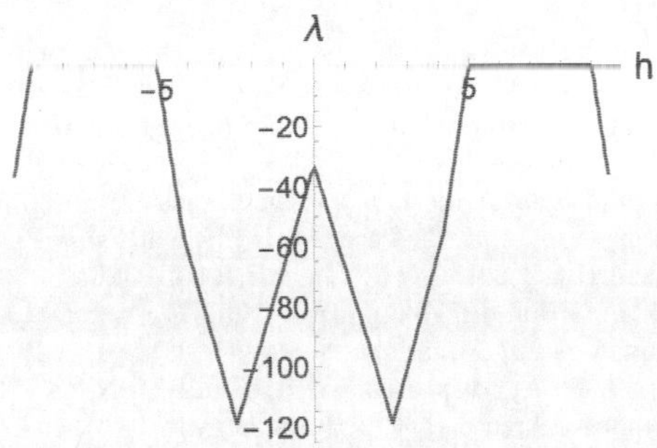

Figure 12. Quantum Ising-Heisenberg spin-1 maximal Lyapunov exponent for $J = 2, J_1 = 1, \Delta = 1, T = 0.03$.

5. Conclusions

In this paper we study symmetric IsingHubbard, spin-1 Ising and Ising-Heisenberg models on a diamond chain using transfer matrix and dynamical system methods for antiferromagnetic case. The behavior of the magnetization plateaus and thermal quantum entanglement are observed. It's detected the behavior of maximal Lyapunov exponent curves and superstable points connections with magnetization plateaus at $T \to 0$ in classical and quantum cases. It becomes exponentially more difficult to study the model with dynamical approach regarding the case of $T < 0.01$, during calculations some elements become smaller than 10^{-1000}. Similar problems arise for greater values of J, J_1, Δ and h. This type of complications may justify the use of other non-standard approaches in the future.

Acknowledgments

This work was supported by MC-IRSES No. 612707 (DIONICOS) and CS MES RA in the frame of the Research Project no. SCS 15T-1C114 grants.

References

[1] Hida K 1994 *J. Phys. Soc. Jpn.* **63** 2359
[2] Ohanyan V R and Ananikian N S 2003 *Phys. Lett.* A v**307** 76
[3] Kikuchi H, Fujii Y, Chiba M, Mitsudo S, Idehara T, Tonegawa T, Okamoto K, Sakai T, Kuwai T and Ohta H 2005 *Phys. Rev. Lett.* **94** 227201
[4] Rule K C, Wolter A U B, Süllow S, Tennant D A, Brühl A, Köhler S, Wolf B, Lang M and Schreuer J 2008 *Phys. Rev. Lett.* **100** 117202
[5] Jeschke H et al. 2011 *Phys. Rev. Lett.* **106** 217201
[6] Kang J, Lee C, Kremer R K and Whangbo M-H, 2009 *J. Phys.: Condens. Matter* **21** 392201
[7] Ananikian N, Lazaryan H, Nalbandyan M 2012 *Eur. Phys. J.* B **85** 223
[8] Gu B and Su G 2007 *Phys. Rev.* A **75** 174437
[9] Takano K, Kubo K and Sakamoto H 2003 *J. Phys.: Condens. Matter* **15** 5979
[10] Čanová L, Strečka J and Jašvur M 2006 *J. Phys.: Condens. Matter* **18** 4967
[11] Rojas O, de Souza S M 2011 *Phys.Lett.* A **375** 1295
[12] Rojas O, de Souza S M and Ananikian N S 2012 *Phys. Rev.* E **85** 061123
[13] Nalbandyan M, Lazaryan H, Rojas O, de Souza S M and Ananikian N S 2014 *J. Phys. Soc. Jpn.* **83** 074001
[14] Konar S, Mukherjee P S, Zangrando E, Lloret F and Chaudhuri N R 2002 *Angew. Chem. Int. Ed.* **41** 1561
[15] Amico L, Fazio R, Osterloh A, Vedral V 2008 *Rev. Mod. Phys.* **80** 517
[16] Horodecki R, Horodecki P, Horodecki M, Horodecki K 2009 *Rev. Mod. Phys.* **81** 865
[17] Sachdev S 1999 *Quantum Phase Transition*(Cambridge: Cambridge University Press)
[18] Torrico J, Rojas M, de Souza S M, Rojas O and Ananikian N S 2014 *EPL* **108** 50007
[19] Ananikian N S , Ananikyan L N, Chakhmakhchyan L A and Rojas O 2012 *J. Phys.: Condens. Matter* **24** 256001
[20] Chakhmakhchyan L, Ananikian N, Ananikyan L and Burdík Č 2012 *J. Phys.: Conf. Series* **343** 012022
[21] Rojas O, Rojas M, Ananikian N S, de Souza S M 2012 *Phys. Rev.* A **86** 042330
[22] Abgaryan V S, Ananikian N S, Ananikyan L N, Hovhannisyan V 2015 *Solid State Commun.* **203** 5
[23] Abgaryan V S, Ananikian N S, Ananikyan L N, Hovhannisyan V 2015 *Solid State Commun.* **224** 15
[24] Torrico J, Rojas M, de Souza S M, Rojas O 2016 Zero temperature non-plateau magnetization driven by XY-anisotropy and magnetocaloric effect in Ising-XYZ diamond chain structure *Preprint* arXiv:1602.07279
[25] Jie Q and Bin Z 2015 *Chin. Phys.* B **24** 110306
[26] Ananikian N S and Dallakian S K 1997 *Physica* D **107** 75
[27] Akheyan A Z and Ananikian N S 1996 *J. Phys.* A **29** 721
[28] Avakian A R, Ananikyan N S and Izmailyan N Sh 1990 *Phys. Lett.* A **150** 163
[29] Ananikian N, Artuso R, Chakhmakhchyan L 2014 *Commun. Nonlinear Sci. Numer. Simul.* **19** 3671
[30] Ananikian N S, Ananikyan L N and Chakhmakhchyan L A 2011 *JETP Lett.* **94** 39
[31] Ananikian N, Ananikyan L, Artuso R, Lazaryan H 2010 *Phys. Lett.* A **374** 4084
[32] Ananikian N, Hovhannisyan V 2013 *Physica* A **392** 2375
[33] Lazaryan H, Nalbandyan M and Ananikian N 2016 *Int. J. Mod. Phys.* B **30** 1650135
[34] Hovhannisyan V V, Ananikian N, Kenna R 2016 *Physica* A **453** 116
[35] Wootters W K 1998 *Phys. Rev. Lett.* **80** 2245
[36] Vidal G and Werner R F 2002 *Phys. Rev.* A **65** 032314

IOP Conf. Series: Journal of Physics: Conf. Series **804** (2017) 012003 doi:10.1088/1742-6596/804/1/012003

Quantum entropies, Schur concavity and dynamical semigroups

Paolo Aniello

Dipartimento di Fisica "Ettore Pancini", Università di Napoli "Federico II", and INFN – Sezione di Napoli, C.U. di Monte Sant'Angelo, via Cintia, I-80126 Napoli, Italy

E-mail: paolo.aniello@na.infn.it

Abstract. Entropy plays a fundamental role in several branches of physics. In the quantum setting, one usually considers the von Neumann entropy, but other useful quantities have been proposed in the literature; e.g., the Rényi and the Tsallis entropies. The evolution of an open quantum system, described by a semigroup of dynamical maps (in short, a dynamical semigroup), may decrease a quantum entropy, for some initial condition. We will discuss various characterizations of those dynamical semigroups that, for every initial condition, do not decrease a general class of quantum entropies, which is defined using the notion of Schur concavity of a function. We will not assume that such a dynamical semigroup be *completely* positive, the physical justification of this condition being controversial. Therefore, we will consider semigroups of trace-preserving, positive — but not necessarily completely positive — linear maps. We will next focus on a special class of (completely positive) dynamical semigroups, the *twirling semigroups*, having applications in quantum information science. We will argue that the whole class of dynamical semigroups that do not decrease a quantum entropy can be obtained as a suitable generalization of the twirling semigroups.

1. Introduction and main ideas

Entropy is a fundamental concept in several branches of physics; in particular, in the theory of open quantum systems [1, 2] and in quantum information science [3]. In this setting, various quantities have been considered; e.g., the von Neumann, the Tsallis and the Rényi entropies [4].

A natural problem is to study the possible temporal evolutions of an open quantum system that do not decrease a certain entropy, for every initial condition of the system [5]. Under suitable hypotheses, the evolution of an open quantum system is well described by a *dynamical semigroup* [2, 6]; i.e., by a semigroup of operators [7] consisting of dynamical maps. Given a complex Hilbert space $\mathcal{H}$ (for the sake of simplicity, we will deal with a finite-dimensional — say, with a $\mathtt{N}$-dimensional, $\mathtt{N} \geq 2$ — Hilbert space), by a *dynamical map* we mean a linear map

$$\mathfrak{Q}\colon \mathcal{B}(\mathcal{H}) \rightarrow \mathcal{B}(\mathcal{H}), \tag{1}$$

with $\mathcal{B}(\mathcal{H})$ denoting the space of linear operators in $\mathcal{H}$ (or $\mathtt{N}\times\mathtt{N}$ complex matrices), characterized by the following defining properties:

(i) $\mathfrak{Q}$ is *positive*, i.e., $\mathcal{B}(\mathcal{H}) \ni \hat{A} \geq 0 \implies \mathfrak{Q}\hat{A} \geq 0$;

(ii) $\mathfrak{Q}$ is *trace-preserving*, i.e., $\mathrm{tr}(\mathfrak{Q}\hat{A}) = \mathrm{tr}(\hat{A})$, for all $\hat{A} \in \mathcal{B}(\mathcal{H})$.

Note that, throughout the paper, we work in the Schrödinger picture.

One often assumes that $\mathfrak{Q}$ is, more specifically, *completely positive*; i.e., that

(iii) for every $n = 1, 2, \ldots$, the map

$$\mathfrak{Q} \otimes \mathrm{Id}_n \colon \mathcal{B}(\mathcal{H}) \otimes \mathbb{C}^{n\times n} \to \mathcal{B}(\mathcal{H}) \otimes \mathbb{C}^{n\times n} \tag{2}$$

— where $\mathbb{C}^{n\times n}$ is regarded as the space $\mathcal{B}(\mathbb{C}^n)$ of $n \times n$ matrices and Id_n is the identity map in this space — is positive.

This further assumption is usually justified with the possible — extremely weak, thus negligible — coupling of the relevant physical system with another system, a non-evolving n-dimensional 'ancilla' [6], $n \geq 2$. However, the validity of this and other similar arguments is controversial [8]. Moreover, the intriguing idea of describing the evolution of the universe itself, and the associated change of the entropy, by a dissipative master equation — see [9] and references therein — does not seem to entail, in any reasonable way, the notion of *complete* positivity.

In this paper, we will therefore consider the problem of characterizing the one-parameter semigroups of dynamical maps that do not decrease a suitable class of quantum entropies. To define this class of functions, we recall that the convex body of density operators in $\mathcal{H}$ — the physical states — can be endowed with a natural transitive, reflexive binary relation (a preorder) '$\prec$', usually called *majorization*; see sect. 2.

Definition 1. Let $\mathcal{S}(\mathcal{H}) \subset \mathcal{B}(\mathcal{H})$ be the convex body of density operators in $\mathcal{H}$ (the unit trace, positive operators), and let $\hat{\rho}_\star := \mathtt{N}^{-1}\hat{I} \in \mathcal{S}(\mathcal{H})$, with $\hat{I}$ denoting the identity operator in $\mathcal{H}$, be the maximally mixed state. By a *quantum entropy* we mean a map $\mathscr{E} \colon \mathcal{S}(\mathcal{H}) \to \mathbb{R}$ such that

(E1) $\mathscr{E}$ is *Schur concave* w.r.t. the natural majorization relation in $\mathcal{S}(\mathcal{H})$, namely,

$$\hat{\omega} \prec \hat{\rho} \implies \mathscr{E}(\hat{\omega}) \geq \mathscr{E}(\hat{\rho})\,; \tag{3}$$

(E2) $\mathscr{E}$ has a *strict* global maximum at $\hat{\rho}_\star$.

The axioms (E1) and (E2) are coherent, because $\hat{\rho}_\star$ is majorized by any other state. Examples of functions satisfying them — including the von Neumann entropy — will be discussed in sect. 2 (these entropies actually satisfy slightly more stringent properties). We now assign a precise meaning to the statement that a dynamical semigroup does not decrease a quantum entropy.

Definition 2. We say that a dynamical semigroup $\{\mathfrak{Q}_t \colon \mathcal{B}(\mathcal{H}) \to \mathcal{B}(\mathcal{H})\}_{t\in\mathbb{R}^+}$ *does not decrease a quantum entropy* $\mathscr{E} \colon \mathcal{S}(\mathcal{H}) \to \mathbb{R}$ if

$$\mathscr{E}(\mathfrak{Q}_t\hat{\rho}) \geq \mathscr{E}(\hat{\rho})\,, \tag{4}$$

for all $\hat{\rho} \in \mathcal{S}(\mathcal{H})$ and all $t \geq 0$. We say that $\{\mathfrak{Q}_t\}_{t\in\mathbb{R}^+}$ is *entropy-nondecreasing* if it does not decrease *every* quantum entropy (satisfying axioms (E1) and (E2)).

Note that a dynamical semigroup $\{\mathfrak{Q}_t\}_{t\in\mathbb{R}^+}$, by the semigroup property ($\mathfrak{Q}_{t+s} = \mathfrak{Q}_t\,\mathfrak{Q}_s$), does not decrease the entropy $\mathscr{E}$ if *and only if* $\mathscr{E}(\mathfrak{Q}_{t+s}\hat{\rho}) \geq \mathscr{E}(\mathfrak{Q}_t\hat{\rho})$, for all $\hat{\rho} \in \mathcal{S}(\mathcal{H})$ and all $t, s \geq 0$. It is also worth anticipating one of the results outlined in sect. 3: a dynamical semigroup does not decrease a certain quantum entropy if *and only if* it is entropy-nondecreasing. Otherwise stated, the solution of our problem does not depend on the choice of a particular type of entropy.

The class of entropy-nondecreasing dynamical semigroups turns out to contain as a remarkable subclass — see sect. 4 — the so-called *twirling semigroups* [10–16]. This type of semigroups of operators, first considered by Kossakowski [17] in the early 1970s, are characterized by an integral expression involving a representation of a locally compact group and a convolution semigroup of probability measures [18] on that group.

A related interesting fact, discussed in sect. 4, is that the whole class of entropy-nondecreasing dynamical semigroups can be obtained as a suitable generalization of the twirling semigroups. Indeed, they admit an integral expression where the semigroups of probability measures are replaced with certain (more general) families of *signed* measures.

IOP Conf. Series: Journal of Physics: Conf. Series **804** (2017) 012003 doi:10.1088/1742-6596/804/1/012003

2. Quantum entropies and concavity

Given a density operator $\hat{\rho} \in \mathcal{S}(\mathcal{H})$, the *von Neumann entropy* of $\hat{\rho}$ is defined by

$$\mathscr{S}(\hat{\rho}) := -\operatorname{tr}(\hat{\rho} \ln \hat{\rho}) = -\sum_{k=1}^{\mathrm{N}} p_k \ln p_k \,, \tag{5}$$

where $\{p_1, \ldots, p_{\mathrm{N}}\}$ is the whole set of the eigenvalues of $\hat{\rho}$ (repeated taking into account degeneracy; here, $0 \ln 0 \equiv 0$). Notice that the definition of $\mathscr{S}(\hat{\rho})$ does not depend on the way the eigenvalues $\{p_k\}$ of $\hat{\rho}$ are ordered, and accordingly in the following $\{p_k\}$ will be regarded as an unordered set. As it will be clear soon, $\mathscr{S}$ is a quantum entropy in the sense of Definition 1.

Other examples of entropies considered in the literature are [4, 5, 19]:

- The *Tsallis entropy* $\mathscr{T}_q$, labeled by the parameter q, $0 < q \neq 1$; namely,

$$\mathscr{T}_q(\hat{\rho}) := \frac{1}{1-q}\left(\operatorname{tr}(\hat{\rho}^q) - 1\right) = \tau_q(\{p_k\}) \,, \tag{6}$$

where the set $\{p_k\}$ is defined as above and

$$\tau_q(\{p_k\}) := \frac{1}{1-q}\left(\sum_{k=1}^{\mathrm{N}} p_k^q - 1\right). \tag{7}$$

Clearly, as previously noted for the von Neumann entropy, the quantity $\tau_q(\{p_k\})$ does not depend on the ordering of the elements of the probability distribution $\{p_k\}$; i.e., τ_q may be regarded as a symmetric function of the probability vector $(p_1, \ldots, p_{\mathrm{N}})$. Observe that

$$\mathscr{S}(\hat{\rho}) = \lim_{q \to 1} \mathscr{T}_q(\hat{\rho}) = \lim_{q \to 1} \tau_q(\{p_k\}) = -\sum_{k=1}^{\mathrm{N}} p_k \ln p_k =: \tau_1(\{p_k\}) \,. \tag{8}$$

It is then natural to set

$$\mathscr{T}_1(\hat{\rho}) \equiv \mathscr{S}(\hat{\rho}) \,. \tag{9}$$

For $q = 2$, the Tsallis entropy is directly related to the *purity* $\mathscr{P}(\hat{\rho}) := \operatorname{tr}(\hat{\rho}^2)$ [20], since $\mathscr{T}_2(\hat{\rho}) = 1 - \mathscr{P}(\hat{\rho})$.

- The *Rényi entropy* $\mathscr{R}_q$, $0 < q \neq 1$; namely,

$$\mathscr{R}_q(\hat{\rho}) := \frac{1}{1-q} \ln \operatorname{tr}(\hat{\rho}^q) = \frac{1}{1-q} \ln\left(\sum_{k=1}^{\mathrm{N}} p_k^q\right) =: \varrho_q(\{p_k\}) \,. \tag{10}$$

Once again we have:

$$\mathscr{R}_1(\hat{\rho}) \equiv \mathscr{S}(\hat{\rho}) = \lim_{q \to 1} \mathscr{R}_q(\hat{\rho}) = \lim_{q \to 1} \varrho_q(\{p_k\}) \,. \tag{11}$$

For $q = 2$, the Rényi entropy is related to the purity too: $\mathscr{R}_2(\hat{\rho}) = -\ln \mathscr{P}(\hat{\rho})$.

- Consider also the function

$$\mathscr{A}_{\mathsf{p}} \colon \mathcal{S}(\mathcal{H}) \ni \hat{\rho} \mapsto (1 - \|\hat{\rho}\|_{\mathsf{p}}) \in \mathbb{R}, \quad \mathsf{p} > 1 \,, \tag{12}$$

where $\|\cdot\|_{\mathsf{p}}$ is the Schatten p-norm in $\mathcal{B}(\mathcal{H})$:

$$\|\hat{A}\|_{\mathsf{p}} := \operatorname{tr}\left(|\hat{A}|^{\mathsf{p}}\right)^{1/\mathsf{p}}. \tag{13}$$

With the same notation as above, we have:

$$\mathscr{A}_{\mathsf{p}}(\hat{\rho}) = 1 - \left(\sum_{k=1}^{\mathtt{N}} p_k^{\mathsf{p}}\right)^{1/\mathsf{p}} =: \alpha_{\mathsf{p}}(\{p_k\}), \quad \mathsf{p} > 1. \tag{14}$$

Note that the following relations hold:

$$\mathtt{N}^{\frac{1-\mathsf{p}}{\mathsf{p}}} \leq \left(\sum_{k=1}^{\mathtt{N}} p_k^{\mathsf{p}}\right)^{1/\mathsf{p}} \leq 1. \tag{15}$$

These inequalities imply that

$$0 \leq \alpha_{\mathsf{p}}(\{p_k\}) \leq 1 - \mathtt{N}^{\frac{1-\mathsf{p}}{\mathsf{p}}}, \tag{16}$$

where the upper inequality is saturated by the probability distribution $\{1/\mathtt{N}, \ldots, 1/\mathtt{N}\}$ only. Thus, $\mathscr{A}_{\mathsf{p}}$ has a strict global maximum at the maximally mixed state $\hat{\rho}_\star$. To describe the points where $\mathscr{A}_{\mathsf{p}}$ attains it minimum value it is convenient to label the eigenvalues $\{p_k\}$ of $\hat{\rho}$ in decreasing order — $p_1 \geq \cdots \geq p_{\mathtt{N}} \geq 0$ — in such a way to obtain an *ordered probability vector* $\vec{p} = \vec{p}(\hat{\rho}) = (p_1, \ldots, p_{\mathtt{N}})$. Then, the lower inequality in (16) is saturated by every probability distribution $\{p_k\}$ with corresponding ordered vector of the form $\vec{p} = (1, 0, \ldots, 0)$; i.e., $\mathscr{A}_{\mathsf{p}}$ attains its minimum at every pure state. The same properties hold for the von Neumann, the Tsallis and the Rényi entropies too.

The notion of ordered vector induces a natural *majorization* relation [21, 22], a preorder. Indeed, as previously noted, with every vector $x = (x_1, \ldots, x_n)$ in $\mathbb{R}^n$ one can associate the vector $\vec{x} = (\vec{x}_1, \ldots, \vec{x}_n)$ obtained by rearranging the coordinates in decreasing order: $\vec{x}_1 \geq \cdots \geq \vec{x}_n$. Then, for $x, y \in \mathbb{R}^n$, we say that x is majorized by y — $x \prec y$ — whenever the following relations hold:

$$\sum_{k=1}^{j} \vec{x}_k \leq \sum_{k=1}^{j} \vec{y}_k, \quad 1 \leq j \leq n, \quad \text{and} \quad \sum_{k=1}^{n} \vec{x}_k = \sum_{k=1}^{n} \vec{y}_k. \tag{17}$$

If $x = (x_1, \ldots, x_n)$ is a probability vector — $x_1 \geq 0, \ldots, x_n \geq 0$ and $\sum_{k=1}^{n} x_k = 1$ — we have:

$$p_\star := (1/n, \ldots, 1/n) \prec (x_1, \ldots, x_n) \prec (1, 0, \ldots, 0). \tag{18}$$

If both $x \prec y$ and $y \prec x$ — in symbols, $x \approx y$ — then $x = \mathsf{P}\, y$, for some permutation matrix P. Clearly, $p_\star$ is *strictly* majorized by any other vector belonging to the probability simplex $\triangle_{n-1}$ in $\mathbb{R}^n$, $n \geq 2$, because $\triangle_{n-1} \ni x \prec p_\star \Longrightarrow x = p_\star$.

Note that majorization between (eigenvalue) vectors induces a corresponding majorization relation between density operators [1, 4]:

$$\hat{\omega} \prec \hat{\rho} \overset{\text{def}}{\Longleftrightarrow} \vec{p}(\hat{\omega}) \prec \vec{p}(\hat{\rho}). \tag{19}$$

The basic idea is that, for $\hat{\omega} \prec \hat{\rho}$ and $\hat{\rho} \not\prec \hat{\omega}$ ($\hat{\omega}$ strictly majorized by $\hat{\rho}$), the state $\hat{\omega}$ will be 'more mixed' or 'more chaotic' than $\hat{\rho}$ [1]. In particular, the maximally mixed state $\hat{\rho}_\star = \mathtt{N}^{-1}\hat{I} \in \mathcal{S}(\mathcal{H})$ is strictly majorized by any other state and every state is majorized by any pure state.

Therefore, a very natural property to be considered for defining a quantum entropy is the so-called *Schur concavity*, which has been indeed used in Definition 1. Recall that a function $\mathscr{F} : \mathcal{S}(\mathcal{H}) \to \mathbb{R}$ is said to be *Schur concave* if

$$\hat{\omega}, \hat{\rho} \in \mathcal{S}(\mathcal{H}), \; \hat{\omega} \prec \hat{\rho} \Longrightarrow \mathscr{F}(\hat{\omega}) \geq \mathscr{F}(\hat{\rho}); \tag{20}$$

in particular, we say that $\mathscr{F}$ is *strictly* Schur concave if $\hat{\omega} \prec \hat{\rho}$, with $\hat{\omega} \not\approx \hat{\rho}$ — i.e., $\vec{p}(\hat{\omega}) \neq \vec{p}(\hat{\rho})$ — implies that $\mathscr{F}(\hat{\rho}) > \mathscr{F}(\hat{\omega})$.

Remark 1. Since the maximally mixed state $\hat{\rho}_\star$ is strictly majorized by any other state, a strictly Schur concave function $\mathscr{E}\colon \mathcal{S}(\mathcal{H}) \to \mathbb{R}$ satisfies both the axioms (E1) and (E2) in Definition 1. It is quite natural, moreover, to assume that $\mathscr{E}(\hat{\rho}) = \mathscr{E}(\hat{\omega})$, for $\hat{\rho} \approx \hat{\omega}$ (i.e., for $\vec{p}(\hat{\rho}) = \vec{p}(\hat{\omega})$). This condition is equivalent to the unitary invariance of $\mathscr{E}$: $\mathscr{E}(\hat{\rho}) = \mathscr{E}(\hat{U}\hat{\rho}\hat{U}^*)$, for every $\hat{\rho} \in \mathcal{S}(\mathcal{H})$ and every unitary operator $\hat{U}$ in $\mathcal{H}$. We will call a unitarily invariant, strictly Schur concave real function on $\mathcal{S}(\mathcal{H})$ a *proper* quantum entropy.

The following result — see [5] — provides a convenient recipe for generating proper quantum entropies *ad libitum*.

Proposition 1. *If* h *is a real function on* $\mathbb{R}^n$*, strictly decreasing — alternatively, strictly increasing — w.r.t. each of its arguments, and* $\mathsf{g}_1, \ldots, \mathsf{g}_n$ *are strictly convex — respectively, strictly concave — continuous real functions on the interval* $[0,1]$*, then*

$$\mathcal{S}(\mathcal{H}) \ni \hat{\rho} \mapsto \mathsf{h}(\mathrm{tr}(\mathsf{g}_1(\hat{\rho})), \ldots, \mathrm{tr}(\mathsf{g}_n(\hat{\rho}))) =: \mathscr{E}(\hat{\rho}) \tag{21}$$

is a proper quantum entropy.

Proposition 2. *The families of functions* $\{\mathscr{T}_q\}_{q>0}$, $\{\mathscr{R}_q\}_{q>0}$ *— that include the von Neumann entropy* ($q = 1$) *— and* $\{\mathscr{A}_\mathsf{p}\}_{\mathsf{p}>1}$ *are proper quantum entropies.*

Proof. Apply Proposition 1. E.g., in the case of the von Neumann entropy, we have: $n = 1$, $\mathsf{g}\colon [0,1] \ni \xi \mapsto -\xi \ln \xi$ (concave) and $\mathsf{h}(x) = x$ (increasing). □

Remark 2. By Klein's inequality [4], one can show that the von Neumann entropy is also concave, in the ordinary sense: for every pair $\hat{\rho}, \hat{\omega} \in \mathcal{S}(\mathcal{H})$, $\mathscr{S}(\epsilon\hat{\rho} + (1-\epsilon)\hat{\omega}) \geq \epsilon\mathscr{S}(\hat{\rho}) + (1-\epsilon)\mathscr{S}(\hat{\omega})$, $\epsilon \in [0,1]$. The same property holds for the entropy $\mathscr{A}_\mathsf{p}$ too, for all $\mathsf{p} > 1$. For the Tsallis entropy $\mathscr{T}_q$, $q > 0$, the standard concavity holds, in general, in a slightly weaker sense: the associated function τ_q (see (7)), regarded as a symmetric function on the the probability simplex $\triangle_{\mathtt{N}-1} \subset \mathbb{R}^{\mathtt{N}}$, is concave. For the Rényi entropy $\mathscr{R}_q$, even this weaker property is no longer true, in general (i.e., for a sufficiently large $q > 1$); see Remark 9 of [5].

In connection with the previous remark, it is worth noting the following fact:

Proposition 3. *A function* $\mathscr{F}\colon \mathcal{S}(\mathcal{H}) \to \mathbb{R}$ *is unitarily invariant —* $\mathscr{F}(\hat{\rho}) = \mathscr{F}(\hat{U}\hat{\rho}\hat{U}^*)$*, for every* $\hat{\rho} \in \mathcal{S}(\mathcal{H})$ *and every unitary operator* $\hat{U}$ *in* $\mathcal{H}$ *— if and only if it is of the form*

$$\mathscr{F}(\hat{\rho}) = \varphi(\{p_k\}), \tag{22}$$

where $\{p_k\}$ *is the whole set of the eigenvalues of* $\hat{\rho}$*, repeated taking into account degeneracy, and* φ *can be regarded as a symmetric function on the probability simplex* $\triangle_{\mathtt{N}-1} \subset \mathbb{R}^{\mathtt{N}}$*. Moreover, if* $\mathscr{F}$ *is of the form (22) and the function* φ *is concave on* $\triangle_{\mathtt{N}-1}$ *— in particular, if* $\mathscr{F}$ *is concave on* $\mathcal{S}(\mathcal{H})$ *and of the form (22) — then* $\mathscr{F}$ *is Schur concave.*

Proof. Let $\lambda\colon \hat{\rho} \mapsto \{p_k\}$ be the function which assigns to each density operator in $\mathcal{H}$ the (unordered) set of its eigenvalues, repeated according to degeneracy. The relation $\lambda(\hat{\rho}) = \lambda(\hat{\omega})$ is equivalent to the fact that $\hat{\omega} = \hat{U}\hat{\rho}\hat{U}^*$, for some unitary operator $\hat{U}$ in $\mathcal{H}$. Therefore, if $\mathscr{F}$ is unitarily invariant, then $\lambda(\hat{\rho}) = \lambda(\hat{\omega}) \implies \mathscr{F}(\hat{\rho}) = \mathscr{F}(\hat{\omega})$. It follows that in this case $\mathscr{F} = \varphi \circ \lambda$, for some map φ which can be regarded as symmetric function on the probability simplex $\triangle_{\mathtt{N}-1}$. The reverse implication is clear. At this point, to complete the proof it is sufficient to observe that a concave function on a convex symmetric domain in $\mathbb{R}^{\mathtt{N}}$ — e.g., on the probability simplex $\triangle_{\mathtt{N}-1}$ — is Schur concave if (and only if) it is symmetric; see Remark 10 of [5]. □

We stress that Remark 2 and Proposition 3 provide further support to our previous claim that a very natural defining property for a quantum entropy is the Schur concavity (rather than the standard concavity).

3. Dynamical semigroups that do not decrease a quantum entropy

In this section, given a quantum entropy $\mathscr{E}$ on $\mathcal{S}(\mathcal{H})$ (satisfying the axioms (E1) and (E2) in Definition 1) — e.g., the von Neumann entropy, or some of the entropies $\mathscr{T}_q$, $\mathscr{R}_q$, $\mathscr{A}_\mathrm{p}$ defined in the previous section — we will provide various characterizations of the dynamical semigroups that do not decrease this quantity.

Like every semigroup of operators [7], a dynamical semigroup $\{\mathfrak{Q}_t\colon \mathcal{B}(\mathcal{H}) \to \mathcal{B}(\mathcal{H})\}_{t\in\mathbb{R}^+}$ — a continuous (w.r.t. any norm topology in $\mathcal{B}(\mathcal{H})$) one-parameter semigroup of trace-preserving, positive linear maps — is completely characterized by its infinitesimal generator $\mathfrak{L}$:

$$\mathfrak{L} = \lim_{t\downarrow 0} t^{-1}(\mathfrak{Q}_t - \mathrm{Id}), \quad \mathfrak{Q}_t = \exp(\mathfrak{L}\, t). \tag{23}$$

We will denote by $\langle\cdot\,,\cdot\rangle_{\mathrm{HS}}$ the Hilbert-Schmidt scalar product in $\mathcal{B}(\mathcal{H})$, and the adjoint of a linear map in $\mathcal{B}(\mathcal{H})$ will be understood as relative to this pairing.

Without assuming, in general, the complete positivity, the following result holds [5]:

Theorem 1. *Let $\{\mathfrak{Q}_t\colon \mathcal{B}(\mathcal{H}) \to \mathcal{B}(\mathcal{H})\}_{t\in\mathbb{R}^+}$ be a (continuous) semigroup of linear maps, with generator $\mathfrak{L}$. Then, the following properties are equivalent:*

- *$\{\mathfrak{Q}_t\}_{t\in\mathbb{R}^+}$ is positive, trace-preserving and does not decrease the entropy $\mathscr{E}$;*
- *$\{\mathfrak{Q}_t\}_{t\in\mathbb{R}^+}$ is positive, trace-preserving and entropy-nondecreasing;*
- *$\{\mathfrak{Q}_t\}_{t\in\mathbb{R}^+}$ is positive, trace-preserving and purity-nonincreasing: for every $t \geq 0$ and every $\hat\rho \in \mathcal{S}(\mathcal{H})$, $\mathscr{P}(\mathfrak{Q}_t\hat\rho) \leq \mathscr{P}(\hat\rho)$;*
- *$\{\mathfrak{Q}_t\}_{t\in\mathbb{R}^+}$ is positive, trace-preserving and, for every $t \geq 0$ and every $\hat\rho \in \mathcal{S}(\mathcal{H})$, $\mathfrak{Q}_t\,\hat\rho \prec \hat\rho$;*
- *$\{\mathfrak{Q}_t\}_{t\in\mathbb{R}^+}$ is positive, trace-preserving and leaves the maximally mixed state invariant: $\mathfrak{Q}_t\,\hat\rho_\star = \hat\rho_\star$, $t \geq 0$;*
- *for every set $\{\hat\psi_j\} \subset \mathcal{B}(\mathcal{H})$ of mutually orthogonal rank-one (selfadjoint) projectors forming a resolution of the identity — $\sum_j \hat\psi_j = \hat I$ —*

$$\mathrm{tr}\left(\hat\psi_j\,(\mathfrak{L}\,\hat\psi_k)\right) \geq 0, \quad \text{for } j \neq k, \tag{24}$$

and

$$\sum_{j=1}^{\mathtt{N}} \mathrm{tr}\left(\hat\psi_j\,(\mathfrak{L}\,\hat\psi_k)\right) = 0 = \sum_{k=1}^{\mathtt{N}} \mathrm{tr}\left(\hat\psi_j\,(\mathfrak{L}\,\hat\psi_k)\right); \tag{25}$$

- *for every pair of mutually orthogonal (selfadjoint) projectors $\hat P, \hat Q \in \mathcal{B}(\mathcal{H})$,*

$$\langle\hat P, \mathfrak{L}\,\hat Q\rangle_{\mathrm{HS}} \geq 0 \ \text{ and } \ \langle\hat I, \mathfrak{L}\,\hat P\rangle_{\mathrm{HS}} = 0 = \langle\hat P, \mathfrak{L}\,\hat I\rangle_{\mathrm{HS}}\,. \tag{26}$$

The reader will have noted in particular that, as anticipated in the introduction, the property of a dynamical semigroup of not decreasing a certain quantum entropy is equivalent to the property of being entropy-nondecreasing; i.e., of not decreasing *every* quantum entropy.

It is also worth observing, with regard to the last two equivalent properties in Theorem 1, that in dimension two a more explicit, but still relatively simple, characterization of the infinitesimal generator of an entropy-nondecreasing dynamical semigroup is provided by the following result [5]:

Proposition 4. *For $\dim(\mathcal{H}) = 2$, the general form of the generator of an entropy-nondecreasing dynamical semigroup is given by*

$$\mathfrak{L}\,\hat A = -\mathrm{i}\sum_{j=1}^{3} h_j\,[\hat S_j, \hat A] + \sum_{j,k=1}^{3} \varkappa_{jk}\left(\hat S_j\,\hat A\,\hat S_k - \frac{1}{2}(\hat S_j\,\hat S_k\,\hat A + \hat A\,\hat S_j\,\hat S_k)\right), \tag{27}$$

where $\hat{S}_1, \hat{S}_2, \hat{S}_3$ are the standard spin operators, $h_1, h_2, h_3 \in \mathbb{R}$, $\mathscr{K} := (\varkappa_{jk})$ is a symmetric 3×3 real matrix and — setting

$$\kappa_1 \equiv \varkappa_{22} + \varkappa_{33}\,,\ \ \kappa_2 \equiv \varkappa_{11} + \varkappa_{33}\,,\ \ \kappa_3 \equiv \varkappa_{11} + \varkappa_{22}\,,\ a \equiv -\varkappa_{23}\,,\ b \equiv -\varkappa_{13}\,,\ c \equiv -\varkappa_{12} \tag{28}$$

— the associated symmetric real matrix $\mathscr{L}$,

$$\mathscr{L} := \begin{pmatrix} \kappa_1 & c & b \\ c & \kappa_2 & a \\ b & a & \kappa_3 \end{pmatrix}, \tag{29}$$

is positive semidefinite. Thus, such a generator $\mathfrak{L}$ is selfadjoint — $\mathfrak{L} = \mathfrak{L}^*$ — if and only if $h_1 = h_2 = h_3 = 0$.

Assuming now the complete positivity, one can prove a slightly stronger result [5], w.r.t. Theorem 1. Indeed, the generators of completely positive dynamical semigroups admit a complete classification, the Gorini-Kossakowski-Lindblad-Sudarshan canonical form [23, 24]. Therefore, one can describe, within this general classification, the typical form of the generators that give rise to an entropy-nondecreasing temporal evolution.

Theorem 2. *Let $\{\mathfrak{Q}_t \colon \mathcal{B}(\mathcal{H}) \to \mathcal{B}(\mathcal{H})\}_{t\in\mathbb{R}^+}$ be a completely positive dynamical semigroup with generator $\mathfrak{L}$. Then, the following properties are equivalent:*

- *$\{\mathfrak{Q}_t\}_{t\in\mathbb{R}^+}$ does not decrease the entropy $\mathscr{E}$;*
- *$\{\mathfrak{Q}_t\}_{t\in\mathbb{R}^+}$ is entropy-nondecreasing;*
- *$\{\mathfrak{Q}_t\}_{t\in\mathbb{R}^+}$ is purity-nonincreasing;*
- *for every $t \geq 0$ and every $\hat{\rho} \in \mathcal{S}(\mathcal{H})$, $\mathfrak{Q}_t\,\hat{\rho} \prec \hat{\rho}$;*
- *$\{\mathfrak{Q}_t\}_{t\in\mathbb{R}^+}$ leaves the maximally mixed state invariant: $\mathfrak{Q}_t\,\hat{\rho}_\star = \hat{\rho}_\star$, $t \geq 0$;*
- *the adjoint semigroup of $\{\mathfrak{Q}_t\}_{t\in\mathbb{R}^+}$ is trace-preserving;*
- *$\mathfrak{L}$ is of the form*

$$\mathfrak{L}\,\hat{A} = -\mathrm{i}\big[\hat{H}, \hat{A}\big] + \mathfrak{F}\,\hat{A} - \frac{1}{2}\Big(\big(\mathfrak{F}\,\hat{I}\big)\,\hat{A} + \hat{A}\,\big(\mathfrak{F}\,\hat{I}\big)\Big), \tag{30}$$

where $\hat{H}$ is a selfadjoint operator and $\mathfrak{F}$ a completely positive map in $\mathcal{B}(\mathcal{H})$ such that

$$\mathfrak{F}\,\hat{I} = \mathfrak{F}^*\,\hat{I}; \tag{31}$$

- *$\mathfrak{L}$ is of the form*

$$\mathfrak{L}\,\hat{A} = -\mathrm{i}\big[\hat{H}, \hat{A}\big] + \sum_{k=1}^{\mathtt{N}^2-1} \gamma_k \Big(\hat{F}_k\,\hat{A}\,\hat{F}_k^* - \frac{1}{2}\big(\hat{F}_k^*\hat{F}_k\,\hat{A} + \hat{A}\,\hat{F}_k^*\hat{F}_k\big)\Big), \tag{32}$$

where $\hat{H}$ is a trace-less selfadjoint operator, $\gamma_1 \geq 0, \ldots, \gamma_{\mathtt{N}^2-1} \geq 0$, and $\hat{F}_1, \ldots, \hat{F}_{\mathtt{N}^2-1}$ are trace-less operators such that

$$\big\langle \hat{F}_j\,, \hat{F}_k \big\rangle_{\mathrm{HS}} = \delta_{jk}\,, \quad j, k = 1, \ldots, \mathtt{N}^2 - 1, \tag{33}$$

and

$$\sum_{k=1}^{\mathtt{N}^2-1} \gamma_k\,\hat{F}_k\,\hat{F}_k^* = \sum_{k=1}^{\mathtt{N}^2-1} \gamma_k\,\hat{F}_k^*\,\hat{F}_k\,. \tag{34}$$

Once again, in dimension two one has a somewhat simpler result for the characterization of the infinitesimal generators [5]:

Proposition 5. *For $\dim(\mathcal{H}) = 2$, the the general form of the infinitesimal generator of an entropy-nondecreasing, completely positive dynamical semigroup is given by (27), where $\hat{S}_1, \hat{S}_2, \hat{S}_3$ are the spin operators, h_1, h_2, h_3 are arbitrary real numbers and $\mathscr{K} := (\varkappa_{jk})$ is a positive semidefinite, symmetric real matrix.*

IOP Conf. Series: Journal of Physics: Conf. Series **804** (2017) 012003 doi:10.1088/1742-6596/804/1/012003

4. Twirling semigroups and beyond

We will now focus on a type of completely positive dynamical semigroups first considered by Kossakowski [17] in the early 1970s. Not only are these semigroups of operators interesting for applications (e.g., in quantum information [12, 14, 15]), but they also form a remarkable class of entropy-nondecreasing dynamical semigroups. Moreover, they are in some sense "prototypical", because *every* entropy-nondecreasing dynamical semigroup can be realized as a generalization of this type of completely positive dynamical semigroups, in a sense to be clarified below.

Given a locally compact (second countable, Hausdorff) topological group G, a continuous unitary representation $\mathsf{V}\colon G \to \mathrm{U}(\mathcal{H})$ of G in $\mathcal{H}$ — $\mathrm{U}(\mathcal{H})$ denoting the unitary group of $\mathcal{H}$, endowed with the strong topology — and a convolution semigroup of probability measures $\{\mu_t\}_{t\in\mathbb{R}^+}$ on G [10, 18], the family of maps $\{\mathfrak{Q}_t\colon \mathcal{B}(\mathcal{H}) \to \mathcal{B}(\mathcal{H})\}_{t\in\mathbb{R}^+}$, with

$$\mathfrak{Q}_t\,\hat{A} := \int_G \mathsf{V}(g)\hat{A}\,\mathsf{V}(g)^*\,\mathrm{d}\mu_t(g)\,, \tag{35}$$

turns out to be a completely positive dynamical semigroup; a *twirling semigroup* [10–12]. Typically, G will be a Lie group, and in fact it can be shown that the pair $(\mathsf{V}, \{\mu_t\}_{t\in\mathbb{R}^+})$ generating $\{\mathfrak{Q}_t\}_{t\in\mathbb{R}^+}$ is not unique: to obtain a *generic* twirling semigroup, one can set $G = \mathrm{SU}(\mathtt{N})$ and choose V as the defining representation (with the identification of $\mathrm{U}(\mathtt{N})$ with $\mathrm{U}(\mathcal{H})$), where $\{\mu_t\}_{t\in\mathbb{R}^+}$ spans the whole set of convolution semigroups of probability measures on $\mathrm{SU}(\mathtt{N})$ [10]. Clearly, the maximally mixed state $\hat{\rho}_\star$ is left invariant by the twirling semigroup $\{\mathfrak{Q}_t\}_{t\in\mathbb{R}^+}$, which is then entropy-nondecreasing (by Theorem 2). The generator of $\{\mathfrak{Q}_t\}_{t\in\mathbb{R}^+}$ can be written as

$$\mathfrak{L}\,\hat{A} = -\mathrm{i}\big[\hat{H}, \hat{A}\big] + \mathfrak{D}\,\hat{A} - \frac{1}{2}\Big(\big(\mathfrak{D}\,\hat{I}\big)\hat{A} + \hat{A}\,\big(\mathfrak{D}\,\hat{I}\big)\Big)\,, \quad \hat{A} \in \mathcal{B}(\mathcal{H})\,, \tag{36}$$

with $\hat{H}$ denoting a trace-less selfadjoint operator (determining the Hamiltonian component of the generator) and $\mathfrak{D}$ a completely positive map (the dissipative component of $\mathfrak{L}$) of the form

$$\mathfrak{D} = \sum_{k=1}^{\mathtt{N}^2-1} \gamma_k\,\hat{L}_k(\cdot)\hat{L}_k + \gamma_0\,\mathfrak{U}\,, \tag{37}$$

where $\gamma_0, \ldots, \gamma_{\mathtt{N}^2-1}$ are non-negative numbers, $\hat{L}_1, \ldots, \hat{L}_{\mathtt{N}^2-1}$ trace-less selfadjoint operators such that $\langle \hat{L}_j, \hat{L}_k \rangle_{\mathrm{HS}} = \delta_{jk}$ and $\mathfrak{U}$ a *random unitary map* in $\mathcal{B}(\mathcal{H})$; namely, $\mathfrak{U}$ admits a decomposition of the type

$$\mathfrak{U}\,\hat{A} = \sum_{j=1}^{\mathscr{N}} p_j\,\hat{U}_j\,\hat{A}\,\hat{U}_j^*\,, \tag{38}$$

for some set $\{\hat{U}_j\}_{j=1}^{\mathscr{N}}$ of unitary operators in $\mathcal{H}$ and some probability distribution $\{p_j\}_{j=1}^{\mathscr{N}}$. Note that the operators

$$\hat{U}_1, \ldots, \hat{U}_{\mathscr{N}}, \hat{L}_1, \ldots, \hat{L}_{\mathtt{N}^2-1} \tag{39}$$

are normal; hence, the condition $\mathfrak{D}\,\hat{I} = \mathfrak{D}^*\,\hat{I}$ is satisfied (coherently with Theorem 2). Conversely, every map $\mathfrak{L}$ of the form (36)–(37) is the generator of a twirling semigroup [10].

The class of all twirling semigroups turns out to coincide with the class of *random unitary semigroups* acting in $\mathcal{B}(\mathcal{H})$ [10, 12], namely, of those completely positive dynamical semigroups whose elements are random unitary maps. It is known [25], on the other hand, that for $\mathtt{N} = \dim(\mathcal{H}) = 2$ every completely positive dynamical map in $\mathcal{B}(\mathcal{H})$ which leaves $\hat{\rho}_\star$ invariant is actually a random unitary map (while for $\mathtt{N} > 2$ this is no longer true). By this reasoning and by Theorem 2, we conclude that, for $\mathtt{N} = 2$, the class of twirling semigroups simply *coincides* with the class of entropy-nondecreasing, completely positive dynamical semigroups. Therefore, we have the following result:

Proposition 6. *In the case where* $\dim(\mathcal{H}) = 2$, *a completely positive dynamical semigroup* $\{\mathfrak{Q}_t\colon \mathcal{B}(\mathcal{H}) \to \mathcal{B}(\mathcal{H})\}_{t\in\mathbb{R}^+}$ *is entropy-nondecreasing if and only if it is a twirling semigroup. In the integral form of such a semigroup of operators, one can set* $G = \mathrm{SU}(2)$ *and choose* $\mathsf{V}\colon G \to \mathrm{U}(\mathcal{H})$ *as the defining representation.*

Interestingly, replacing in (35) the convolution semigroup of probability measures $\{\mu_t\}_{t\in\mathbb{R}^+}$ with a suitable family of *signed* measures, one gets the whole class of entropy-nondecreasing dynamical semigroups. Precisely, the following result holds [5]:

Theorem 3. *A family of maps* $\{\mathfrak{Q}_t\colon \mathcal{B}(\mathcal{H}) \to \mathcal{B}(\mathcal{H})\}_{t\in\mathbb{R}^+}$ *is a one-parameter semigroup of entropy-nondecreasing, trace-preserving, positive linear maps if and only if it can be expressed in the form*

$$\mathfrak{Q}_t\,\hat{A} = \int_G \mathsf{V}(g)\hat{A}\,\mathsf{V}(g)^*\,\mathrm{d}\varsigma_t(g)\,, \tag{40}$$

where $(\mathsf{V}, \{\varsigma_t\}_{t\in\mathbb{R}^+})$ *is a pair formed by a continuous unitary representation* $\mathsf{V}\colon G \to \mathrm{U}(\mathcal{H})$ *of a locally compact group* G *and by a family* $\{\varsigma_t\}_{t\in\mathbb{R}^+}$ *of finite, signed Borel measures on* G *satisfying the conditions* (always verified if $\{\varsigma_t\}_{t\in\mathbb{R}^+}$ is a convolution semigroup of probability measures)*:*

(M1) ς_0 *coincides with the Dirac measure at the identity of* G *and* $\varsigma_t(G) = 1$, *for all* $t \geq 0$*;*

(M2) *for some — hence, for every — orthonormal basis* $\Psi \equiv \{\psi_k\}_{k=1}^{\mathrm{N}}$ *in* $\mathcal{H}$*,*

$$\lim_{t\downarrow 0} \int_G \mathfrak{v}_{jklm}(\Psi; g)\,\mathrm{d}\varsigma_t(g) = \delta_{jk}\,\delta_{lm} \tag{41}$$

and

$$\int_G \mathfrak{v}_{jklm}(\Psi; g)\,\mathrm{d}(\varsigma_s \circledcirc \varsigma_t)(g) = \int_G \mathfrak{v}_{jklm}(\Psi; g)\,\mathrm{d}\varsigma_{s+t}(g)\,, \quad \forall s, t > 0\,, \tag{42}$$

where $\mathfrak{v}_{jklm}(\Psi; g) := \langle\psi_j, \mathsf{V}(g)\psi_k\rangle\langle\mathsf{V}(g)\psi_l, \psi_m\rangle$ *and the measure* $\varsigma_s \circledcirc \varsigma_t$ *is the convolution of* ς_s *with* ς_t*;*

(M3) *for every orthonormal basis* $\Psi \equiv \{\psi_k\}_{k-1}^{\mathrm{N}}$ *in* $\mathcal{H}$*,*

$$\lim_{t\downarrow 0} t^{-1} \int_G \mathfrak{v}_{jkkj}(\Psi; g)\,\mathrm{d}\varsigma_t(g) \geq 0\,, \quad j \neq k\,. \tag{43}$$

(It can be shown that the limit (43) exists, provided that (M2) is verified.)

Here, one can always assume that $G = \mathrm{SU}(\mathrm{N})$ *and choose* V *as the defining representation.*

It seems natural to call a semigroup of operators of the form (40), where the family of signed measures $\{\varsigma_t\}_{t\in\mathbb{R}^+}$ satisfies conditions (M1)–(M3), a *generalized twirling semigroup*. A dynamical semigroup is entropy-nondecreasing if and only if it is a generalized twirling semigroup.

5. Conclusions

An interesting problem is the characterization of the dynamical semigroups that do not decrease a quantum entropy, for every initial state. Studying this problem one is led in a natural way to consider a general class of quantum entropies satisfying two simple axioms, relying on the notion of Schur concavity [5]; see Definition 1. The von Neumann, the Tsallis and the Rényi entropies all satisfy these axioms. They actually satisfy slightly more stringent hypotheses: they are *proper* quantum entropies; namely, unitarily invariant, *strictly* Schur concave functions on $\mathcal{S}(\mathcal{H})$. It turns out that, however, the solution of our problem does not depend neither on the fact of considering, in particular, a proper entropy, nor on any specific choice in the whole class of quantum entropies (Theorem 1). A dynamical semigroup that does not decrease a certain entropy is "entropy-nondecreasing" *tout court*. In hindsight, this is quite natural. Such

IOP Conf. Series: Journal of Physics: Conf. Series **804** (2017) 012003 doi:10.1088/1742-6596/804/1/012003

a dynamical semigroup simply maps any state into a not less mixed state; thus, the maximally mixed state $\hat{\rho}_\star$ is stationary. Conversely, if $\hat{\rho}_\star$ is left invariant by a dynamical semigroup, then this is entropy-nondecreasing. In the case of *completely positive*, entropy-nondecreasing dynamical semigroups, on can exhibit an explicit characterization of the generators, within the Gorini-Kossakowski-Lindblad-Sudarshan classification; see condition (34) in Theorem 2. Banks *et al* [9] found the selfadjointness of the operators $\hat{F}_1, \ldots, \hat{F}_{N^2-1}$ to be a condition sufficient to ensure that the associated dynamical semigroups do not decrease the von Neumann entropy. Actually this condition amounts to selecting a subclass out of the class of *twirling semigroups*; see (36)–(37). For $\dim(\mathcal{H}) = 2$, the twirling semigroups saturate the class of entropy-nondecreasing, completely positive dynamical semigroups (Proposition 6). In general — $\dim(\mathcal{H}) \geq 2$, no assumption of complete positivity — the class of entropy-nondecreasing dynamical semigroups coincides with the *generalized* twirling semigroups (Theorem 3), where the semigroups of probability measures of the standard twirling semigroups are replaced with suitable families of *signed* measures. This is somewhat reminiscent of the formalism of Wigner functions, where the classical probability distributions are replaced with (signed) quantum quasi-distributions [15, 26].

References

[1] Wehrl A 1978 General properties of entropy *Rev. Mod. Phys.* **50** 221
[2] Breuer H-P and Petruccione F 2002 *The Theory of Open Quantum Systems* (Oxford: Oxford University Press)
[3] Nielsen M A and Chuang I L 2000 *Quantum Computation and Quantum Information* (Cambridge: Cambridge University Press)
[4] Bengtsson I and Życzkowski K 2006 *Geometry of Quantum States* (Cambridge: Cambridge University Press)
[5] Aniello P and Chruściński D 2016 Characterizing the dynamical semigroups that do not decrease a quantum entropy *J. Phys. A: Math. Theor.* **49** 345301
[6] Alicki R and Lendi K 1987 *Quantum Dynamical Semigroups and Applications* (Berlin: Springer)
[7] Yosida K 1968 *Functional Analysis* (Berlin: Springer)
[8] Shaji A and Sudarshan E C G 2005 Who's afraid of not completely positive maps? *Phys. Lett. A* **341** 48
[9] Banks T, Susskind L and Peskin M E 1984 Difficulties for the evolution of pure states into mixed states *Nucl. Phys. B* **244** 125
[10] Aniello P, Kossakowski A, Marmo G and Ventriglia F 2010 Brownian motion on Lie groups and open quantum systems *J. Phys. A: Math. Theor.* **43** 265301
[11] Aniello P 2011 On a certain class of semigroups of operators *Open Syst. Inf. Dyn.* **18** 129
[12] Aniello P 2013 Quantum dynamical semigroups, group representations and convolution semigroups *Phys. Scr.* **T153** 014003
[13] Aniello P 2013 Operators versus functions: from quantum dynamical semigroups to tomographic semigroups *J. Phys.: Conf. Ser.* **474** 012005
[14] Aniello P 2014 Classical-quantum semigroups *J. Phys.: Conf. Ser.* **563** 012002
[15] Aniello P 2015 Playing with functions of positive type, classical and quantum *Phys. Scr.* **90** 074042
[16] Aniello P 2016 Functions of positive type on phase space, between classical and quantum, and beyond *J. Phys.: Conf. Ser.* **670** 012004
[17] Kossakowski A 1972 On quantum statistical mechanics of non-Hamiltonian systems *Rep. Math. Phys.* **3** 247
[18] Heyer H 1977 *Probability Measures on Locally Compact Groups* (Berlin: Springer)
[19] Naudts J 2011 *Generalized Thermostatistics* (London: Springer)
[20] Lupo C, Aniello P and Scardicchio A 2008 Bipartite quantum systems: on the realignment criterion and beyond *J. Phys. A: Math. Theor.* **41** 415301
[21] Bhatia R 1997 *Matrix Analysis* (New York: Springer)
[22] Marshall A W, Olkin I and Arnold B C 2011 *Inequalities: Theory of Majorization and Its Applications* 2nd edn (New York: Springer)
[23] Gorini V, Kossakowski A and Sudarshan E C G 1976 Completely positive dynamical semigroups of N-level systems *J. Math. Phys.* **17** 821
[24] Lindblad G 1976 On the generators of quantum dynamical semigroups *Commun. Math. Phys.* **48** 119
[25] Landau L J and Streater R F 1993 On Birkhoff's theorem for doubly stochastic completely positive maps of matrix algebras *Linear Algebra Appl.* **193** 107
[26] Aniello P, Man'ko V I and Marmo G 2008 Frame transforms, star products and quantum mechanics on phase space *J. Phys. A: Math. Theor.* **41** 285304

Lax representation and soliton solutions for the (2+1) -dimensional two-component complex modified Korteweg-de Vries equations

G T Bekova, G N Shaikhova, K R Yesmakhanova and R Myrzakulov

Eurasian International Center for Theoretical Physics and Department of General &Theoretical Physics, Eurasian National University, Astana 010008, Kazakhstan

E-mail: kryesmakhanova@gmail.com

Abstract. Under investigation in this paper is the (2+1) -dimensional two-component complex modified Korteweg-de Vries (tcmKdV) equations. Based on the Ablowitz-Kaup-Newell-Segur system, the a new Lax representation is derived of the tcmKdV equations. By Darboux transformation method soliton solutions for the tcmKdV equations are obtained, respectively.

1. Introduction

The most typical and well-studied integrable evolution equations which describe nonlinear wave phenomena for a range of dispersive physical systems are the nonlinear Schrodinger (NLS) equation , Korteweg-de Vries (KdV) equation , modified Korteweg-de Vries (mKdV) equation. The complex modified Korteweg-de Vries (cmKdV) equation is one of generalized form of the KdV equation. It admits soliton solutions. In soliton theory, (cmKdV) equation possesses all the basic characters of integrable models. From a physical point of view, cmKdV equation has been derived for, e.g. the dynamical evolution of nonlinear lattices, fluid dynamics, plasma physics, nonlinear transmission lines and so on [1]. With regard to exact solutions, many classical solving methods, such as Hirotas bilinear method [2], the IST [3, 4], Darboux transformation [5-15] has been used to solve integrable evolution equations.

In this paper, we consider the new (2+1) -dimensional two-component complex modified

Korteweg-de Vries (tcmKdV) equations as follows:

$$iq_{1t} + iq_{1yxx} + i(w_1 q_1)_x + i(f_1 q_2)_x - v_1 q_1 - u_1 q_2 = 0, \tag{1}$$
$$iq_{2t} + iq_{2yxx} + i(w_2 q_2)_x + i(f_2 q_1)_x - v_2 q_2 - u_2 q_1 = 0, \tag{2}$$
$$v_{1x} + i(2q_1^* q_{1xy} - 2q_{1xy}^* q_1 + q_2^* q_{2xy} - q_{2xy}^* q_2) + f_2 q_2^* q_1 - f_1 q_1^* q_2 = 0, \tag{3}$$
$$v_{2x} + i(2q_2^* q_{2xy} - 2q_{2xy}^* q_2 + q_1^* q_{1xy} - q_{1xy}^* q_1) + f_1 q_1^* q_2 - f_2 q_2^* q_1 = 0, \tag{4}$$
$$u_{1x} + i[(q_2^* q_{1yx} - q_{2yx}^* q_1) - f_1(|q_1|^2 + |q_2|^2) + (w_2 - w_1) q_2^* q_1] = 0, \tag{5}$$
$$u_{2x} + i[(q_1^* q_{2yx} - q_{1yx}^* q_2) - f_2(|q_1|^2 + |q_2|^2) + (w_1 - w_2) q_1^* q_2] = 0, \tag{6}$$
$$w_{1x} - (2|q_1|^2 + |q_2|^2)_y = 0, \tag{7}$$
$$w_{2x} - (|q_1|^2 + 2|q_2|^2)_y = 0, \tag{8}$$
$$f_{1x} - (q_2^* q_1)_y = 0, \tag{9}$$
$$f_{2x} - (q_1^* q_2)_y = 0, \tag{10}$$

where q_1, q_2 are complex functions, v_1, v_2, u_1, u_2, w_1, w_2, f_1, f_2 are real functions. In one-component case cmKdV equations has been studied in [16].

The paper is prepared as follows. In Section 2, the Lax pair is presented. In Section 3, we apply Darboux transformation method for the two-component complex modified Korteweg-de Vries (tcmKdV) equations. Section 4, soliton solutions are obtained by Darboux transformation, and in Section 5 conclusions are given.

2. Lax representation

In this section, we would like to present a new Lax pair of the (2+1)-dimensional tcmKdV equations. According to the Ablowitz-Kaup-Newell- Segur (AKNS) formalism, Lax pair for equations (1)-(10) is written as

$$\Psi_x = A\Psi, \tag{11}$$
$$\Psi_t = 4\lambda^2 \Psi_y + B\Psi, \tag{12}$$

with the following matrices:

$$A = i\lambda\Sigma + A_0,$$
$$B = \lambda B_1 + B_0,$$

where

$$\Sigma = \begin{pmatrix} 1 & 0 & 0 \\ 0 & -1 & 0 \\ 0 & 0 & -1 \end{pmatrix}, \quad A_0 = \begin{pmatrix} 0 & q_1 & q_2 \\ -q_1^* & 0 & 0 \\ -q_2^* & 0 & 0 \end{pmatrix}, \quad B_1 = \begin{pmatrix} a_{11} & 2iq_{1y} & 2iq_{2y} \\ 2iq_{1y}^* & a_{22} & -2if_2 \\ 2iq_{2y}^* & -2if_1 & a_{33} \end{pmatrix},$$

$$B_0 = \begin{pmatrix} b_{11} & -q_{1yx} - w_1 q_1 - f_1 q_2 & -q_{2yx} - w_2 q_2 - f_2 q_1 \\ q_{1yx}^* + w_1 q_1^* - f_2 q_2^* & b_{22} & iu_2 \\ q_{2yx}^* + w_2 q_2^* - f_1 q_1^* & iu_1 & b_{33} \end{pmatrix}.$$

Moreover, $\Psi = (\Psi_1(\lambda, x, y, t), \Psi_2(\lambda, x, y, t), \Psi_3(\lambda, x, y, t))^T$ (T mean a matrix transpose) denotes the eigenfunction of Lax pair equations (11)-(12) associated with λ. The compatible condition of equations (1)-(10) is

$$A_t - B_x + AB - BA - 4\lambda^2 A_y = 0. \tag{13}$$

In fact, by direct calculation of the compatible condition (13) we can obtain next equations

$$iq_{1t} + iq_{1yxx} - \frac{1}{2}(a_{2211}q_1)_x + i(f_1q_2)_x + ib_{2211}q_1 - u_1q_2 = 0 \tag{14}$$

$$iq_{2t} + iq_{2yxx} - \frac{1}{2}(a_{3311}q_2)_x + i(f_2q_1)_x + ib_{3311}q_2 - u_1q_1 = 0 \tag{15}$$

$$a_{11x} = 2i(|q_1|^2 + |q_2|^2)_y \tag{16}$$

$$a_{22x} = -2i(|q_1|^2)_y \tag{17}$$

$$a_{33x} = -2i(|q_2|^2)_y \tag{18}$$

$$b_{11x} = (q^*_{1yx}q_1 - q^*_1q_{1yx} + q^*_{2yx}q_2 - q^*_2q_{2yx}) \tag{19}$$

$$b_{22x} = (q^*_1q_{1yx} - q^*_{1yx}q_1) + i(f_1q^*_1q_2 - f_2q^*_2q_1) \tag{20}$$

$$b_{33x} = (q^*_2q_{2yx} - q^*_{2yx}q_2) - i(f_1q^*_1q_2 - f_2q^*_2q_1) \tag{21}$$

where

$$b_{2211} = b_{22} - b_{11} = iv_1, \tag{22}$$

$$b_{3311} = b_{33} - b_{11} = iv_2, \tag{23}$$

$$a_{2211} = a_{22} - a_{11} = -2iw_1, \tag{24}$$

$$a_{3311} = a_{33} - a_{11} = -2iw_2. \tag{25}$$

After doing some simplifications of equations (14)-(25) we can yield the (2+1)-dimensional tcmKdV equations (1)-(10).

3. Darboux transformation

Based on the Darboux transformation for AKNS system, we consider the following transformation of equations (11)-(12)

$$\Psi' = T\Psi = (\lambda I - M)\Psi, \tag{26}$$

where

$$I = \begin{pmatrix} 1 & 0 & 0 \\ 0 & 1 & 0 \\ 0 & 0 & 1 \end{pmatrix}, \quad M = \begin{pmatrix} m_{11} & m_{12} & m_{13} \\ m_{21} & m_{22} & m_{23} \\ m_{31} & m_{32} & m_{33} \end{pmatrix}. \tag{27}$$

New function Ψ' is supposed to satisfy

$$\Psi'_x = A'\Psi', \tag{28}$$

$$\Psi'_t = 4\lambda^2\Psi'_y + B'\Psi'. \tag{29}$$

where A' and B' depend on $q_1^{[1]}, q_2^{[1]}, u_1^{[1]}, u_2^{[1]}, v_1^{[1]}, v_2^{[1]}, w_1^{[1]}, w_2^{[1]}, f_1^{[1]}, f_2^{[1]}$ and λ. In order to hold equations (28)-(29) the T must satisfy

$$T_x = A'T - TA, \tag{30}$$

$$T_t = B'T - TB + 4\lambda^2 T_y. \tag{31}$$

One can choose $T = \lambda I - M$. Then the relation between $q_1, q_2, u_1, u_2, v_1, v_2, w_1, w_2, f_1, f_2$ and $q_1^{[1]}, q_2^{[1]}, u_1^{[1]}, u_2^{[1]}, v_1^{[1]}, v_2^{[1]}, w_1^{[1]}, w_2^{[1]}, f_1^{[1]}, f_2^{[1]}$ can be reduced from equations (30)-(31), which is the Darboux transformation of the (2+1)-dimensional tcmKdV equations. Similar to the case of AKNS, comparing the coefficient of $\lambda^i (i=0,1,2)$ of the two sides of equations (30), we have

$$\lambda^0 \ : \ M_x = A_0'M - MA_0, \tag{32}$$
$$\lambda^1 \ : \ A_0' = A_0 + i[\Sigma, M], \tag{33}$$
$$\lambda^2 \ : \ iI\Sigma = i\Sigma I. \tag{34}$$

From the above equations we obtain

$$q_1^{[1]} = q_1 - 2im_{12}, \tag{35}$$
$$q_2^{[1]} = q_2 - 2im_{13}, \tag{36}$$
$$q_1^{*[1]} = q_1^* - 2im_{21}, \tag{37}$$
$$q_2^{*[1]} = q_2^* - 2im_{31}. \tag{38}$$

Hence we get $m_{21} = -m_{12}^*, m_{31} = -m_{13}^*$.

Then comparing the coefficients of λ^i of the two sides of the equation (31) gives us

$$\lambda^0 \ : \ M_t = B_0'M - MB_0, \tag{39}$$
$$\lambda^1 \ : \ B_0' = B_0 - MB_1 + B_1'M, \tag{40}$$
$$\lambda^2 \ : \ B_1' = B_1 + 4M_y. \tag{41}$$

It leads to

$$B_0' = B_0 - MB_1 + (B_1 + 4M_y)M, \tag{42}$$
$$B_1' = B_1 + 4M_y. \tag{43}$$

Using the equations (42)-(43) we can found

$$v_1^{[1]} = v_1 + 2(2q_{1y}^* m_{12} + 2q_{1y}m_{12}^* + q_{2y}m_{13}^* + q_{2y}^* m_{13}) - 4i(m_{22}m_{22y} + m_{32}m_{23y} - \\ -m_{11}m_{11y} + m_{13}^* m_{13y}) - 2(f_2 m_{32} - f_1 m_{23})), \tag{44}$$

$$v_2^{[1]} = v_1 + 2(2q_{2y}^* m_{13} + 2q_{2y}m_{13}^* + q_{1y}m_{12}^* + q_{1y}^* m_{12}) - 4i(m_{33}m_{33y} + m_{23}m_{32y} - \\ -m_{11}m_{11y} + m_{12}^* m_{12y}) - 2(f_1 m_{23} - f_2 m_{32})), \tag{45}$$

$$u_1^{[1]} = u_1 + 2(q_{2y}^* m_{12} + q_{1y}m_{13}^*) - 4i(m_{22}m_{32y} + m_{32}m_{33} - m_{12}m_{13y}^*) - \\ -2f_1(m_{22} - m_{33}) + 2(w_1 - w_2)m_{32}, \tag{46}$$

$$u_2^{[1]} = u_2 + 2(q_{1y}^* m_{13} + q_{2y}m_{12}^*) - 4i(m_{33}m_{23y} + m_{23}m_{22} - m_{13}m_{12y}^*) - \\ +2f_2(m_{22} - m_{33}) - 2(w_1 - w_2)m_{23}, \tag{47}$$

$$w_1^{[1]} = w_1 + 2i(m_{22} - m_{11})_y, \tag{48}$$
$$w_1^{[1]} = w_2 + 2i(m_{33} - m_{11})_y, \tag{49}$$
$$f_1^{[1]} = f_1 + 2im_{32y}, \tag{50}$$
$$f_2^{[1]} = f_2 + 2im_{23y}. \tag{51}$$

The key step is to find the concrete form of M expressed by the column solution of equations (11)-(12). Let

$$M = H\Lambda H^{-1}, \tag{52}$$

where

$$\Lambda = \begin{pmatrix} \lambda_1 & 0 & 0 \\ 0 & \lambda_2 & 0 \\ 0 & 0 & \lambda_3 \end{pmatrix},$$

$$H = \begin{pmatrix} \Psi_1(\lambda_1, x, y, t) & \Psi_1(\lambda_2, x, y, t) & \Psi_1(\lambda_3, x, y, t) \\ \Psi_2(\lambda_1, x, y, t) & \Psi_2(\lambda_2, x, y, t) & \Psi_2(\lambda_3, x, y, t) \\ \Psi_3(\lambda_1, x, y, t) & \Psi_3(\lambda_2, x, y, t) & \Psi_3(\lambda_3, x, y, t) \end{pmatrix} := \begin{pmatrix} \Psi_{1,1} & \Psi_{1,2} & \Psi_{1,3} \\ \Psi_{2,1} & \Psi_{2,2} & \Psi_{2,3} \\ \Psi_{3,1} & \Psi_{3,2} & \Psi_{3,3} \end{pmatrix}$$

and $det(H) \neq 0$. We take

$$\lambda_3 = \lambda_2 = \lambda_1^*, \quad H = \begin{pmatrix} \Psi_1(\lambda_1, x, y, t) & \Psi_2^*(\lambda_1, x, y, t) & \Psi_3^*(\lambda_1, x, y, t) \\ \Psi_2(\lambda_1, x, y, t) & -\Psi_1^*(\lambda_1, x, y, t) & 0 \\ \Psi_3(\lambda_1, x, y, t) & 0 & -\Psi_1^*(\lambda_1, x, y, t) \end{pmatrix} \tag{53}$$

and then

$$H^{-1} = \frac{1}{|\Psi_1|^2 + |\Psi_2|^2 + |\Psi_3|^2)\Psi_1^*} \begin{pmatrix} \Psi_1^{2*} & \Psi_1^*\Psi_2^* & \Psi_1^*\Psi_3^* \\ \Psi_2\Psi_1^* & -|\Psi_1|^2 - |\Psi_3|^2 & \Psi_2\Psi_3^* \\ \Psi_3\Psi_1^* & \Psi_3\Psi_2^* & -|\Psi_1|^2 - |\Psi_2|^2 \end{pmatrix}. \tag{54}$$

So taking equation (53) and (54) back into equation (52) we get matrix M in the following form

$$M = \frac{1}{\Delta} \begin{pmatrix} \lambda_1|\Psi_1|^2 + \lambda_1^*(|\Psi_2|^2 + |\Psi_3|^2) & \Psi_1\Psi_2^*(\lambda_1 - \lambda_1^*) & \Psi_1\Psi_3^*(\lambda_1 - \lambda_1^*) \\ \Psi_2\Psi_1^*(\lambda_1 - \lambda_1^*) & \lambda_1|\Psi_2|^2 + \lambda_1^*(|\Psi_1|^2 + |\Psi_3|^2) & \Psi_2\Psi_3^*(\lambda_1 - \lambda_1^*) \\ \Psi_3\Psi_1^*(\lambda_1 - \lambda_1^*) & \Psi_3\Psi_2^*(\lambda_1 - \lambda_1^*) & \lambda_1|\Psi_3|^2 + \lambda_1^*(|\Psi_1|^2 + |\Psi_2|^2) \end{pmatrix}, \tag{55}$$

where

$$\Delta = |\Psi_1|^2 + |\Psi_2|^2 + |\Psi_3|^2.$$

4. Soliton solution

Having the explicit form of the Darboux transformation, we are ready to construct exact solutions of the (2+1)-dimensional tcmKdV equations. As an example, let us present the one-soliton solution. To get the one-soliton solutions we assume trivial seed solutions of equations (1)-(10) as

$$q_1 = 0; q_2 = 0; v_1 = 0; v_2 = 0; u_1 = 0; u_2 = 0; w_1 = 0, w_2 = 0; f_1 = 0; f_2 = 0. \tag{56}$$

From equations (11)-(12), the corresponding eigenfunctions of this seed become

$$\Psi_1 = e^{-i\lambda x + ib_1 y + 4\lambda^2 b_1 t + i\delta_1}, \tag{57}$$

$$\Psi_2 = e^{i\lambda x + ib_2 y + i4\lambda^2 b_3 t + i\delta_2}, \tag{58}$$

$$\Psi_3 = e^{i\lambda x + ib_3 y + i4\lambda^2 b_3 t + i\delta_3}, \tag{59}$$

IOP Conf. Series: Journal of Physics: Conf. Series **804** (2017) 012004 doi:10.1088/1742-6596/804/1/012004

where $\delta_1, \delta_2, \delta_3$ are real constants. If $\lambda = \alpha + i\beta$, $b_j = \nu_j + in_j$ $(j = 1, 2, 3)$ then we can rewrite equations (57)-(59) as

$$\Psi_1 = e^{\theta_1 + i\chi_1}, \tag{60}$$

$$\Psi_2 = e^{\theta_2 + i\chi_2}, \tag{61}$$

$$\Psi_3 = e^{\theta_3 + i\chi_3}, \tag{62}$$

where

$$\begin{aligned}
\theta_1 &= \beta x - n_1 y + 4(\beta^2 n_1 - \alpha^2 n_1)t - 8\alpha\beta\nu_1 t,\\
\chi_1 &= -\alpha x + \nu_1 y + 4(\alpha^2\nu_1 - \beta^2\nu_1)t - 8\alpha\beta n_1 t,\\
\theta_2 &= \beta x - n_2 y + 4(\beta^2 n_2 - \alpha^2 n_2)t - 8\alpha\beta\nu_2 t,\\
\chi_2 &= -\alpha x + \nu_2 y + 4(\alpha^2\nu_2 - \beta^2\nu_2)t - 8\alpha\beta n_2 t,\\
\theta_3 &= \beta x - n_3 y + 4(\beta^2 n_3 - \alpha^2 n_3)t - 8\alpha\beta\nu_3 t,\\
\chi_3 &= -\alpha x + \nu_3 y + 4(\alpha^2\nu_3 - \beta^2\nu_3)t - 8\alpha\beta n_3 t.
\end{aligned}$$

By taking equations (60)-(62) into equations (35),(36),(44)-(51) and doing some simplifications, we obtain the one-soliton solutions of the (2+1)-dimensional tcmKdV equations in the following form

$$\begin{aligned}
q_1^{[1]} &= -2im_{12},\\
q_2^{[1]} &= -2im_{13},\\
v_1^{[1]} &= -4i(m_{22}m_{22y} + m_{32}m_{23y} - m_{11}m_{11y} + m_{13}^{*}m_{13y}),\\
v_2^{[1]} &= -4i(m_{33}m_{33y} + m_{23}m_{32y} - m_{11}m_{11y} + m_{12}^{*}m_{12y}),\\
u_1^{[1]} &= -4i(m_{22}m_{32y} + m_{32}m_{33} - m_{12}m_{13y}^{*}),\\
u_2^{[1]} &= -4i(m_{33}m_{23y} + m_{23}m_{22} - m_{13}m_{12y}^{*}),\\
w_1^{[1]} &= 2i(m_{22} - m_{11})_y,\\
w_1^{[1]} &= 2i(m_{33} - m_{11})_y,\\
f_1^{[1]} &= 2im_{32y},\\
f_2^{[1]} &= 2im_{23y},
\end{aligned}$$

where

$$\begin{aligned}
m_{11} &= \alpha + \frac{i\beta(e^{2\theta_1} - e^{2\theta_2} - e^{2\theta_3})}{e^{2\theta_1} + e^{2\theta_2} + e^{2\theta_3}},\\
m_{22} &= \alpha + \frac{i\beta(e^{2\theta_2} - e^{2\theta_1} - e^{2\theta_3})}{e^{2\theta_1} + e^{2\theta_2} + e^{2\theta_3}},\\
m_{33} &= \alpha + \frac{i\beta(e^{2\theta_3} - e^{2\theta_1} - e^{2\theta_2})}{e^{2\theta_1} + e^{2\theta_2} + e^{2\theta_3}},\\
m_{12} &= \frac{2i\beta e^{\theta_1+\theta_2+i(\chi_1-\chi_2)}}{e^{2\theta_1} + e^{2\theta_2} + e^{2\theta_3}},\\
m_{13} &= \frac{2i\beta e^{\theta_1+\theta_3+i(\chi_1-\chi_3)}}{e^{2\theta_1} + e^{2\theta_2} + e^{2\theta_3}},\\
m_{23} &= \frac{2i\beta e^{\theta_2+\theta_3+i(\chi_2-\chi_3)}}{e^{2\theta_1} + e^{2\theta_2} + e^{2\theta_3}},\\
m_{32} &= \frac{2i\beta e^{\theta_2+\theta_3+i(\chi_3-\chi_2)}}{e^{2\theta_1} + e^{2\theta_2} + e^{2\theta_3}}.
\end{aligned}$$

5. Conclusion

In this letter, we present a new Lax pair of the (2+1)-dimensional two-component complex modified Korteweg-de Vries equations. One fold Darboux transformation is constructed. The one-soliton solutions are obtained be means of the Darboux transformation. In a recursive manner, one can obtain the multi-soliton solution from the two seeds in previous. We will study other different kind solutions of the (2+1)-dimensional tcmKdV equations in future.

Acknowledgment

The work of G.B., G.Sh., K.Y., and M.R. is supported by the Ministry of Education and Science of Republic of Kazakhstan (the base part of the state task No.0888/GF4)

References

[1] Ablowitz M J and Clarkson P A 1991 Cambridge University Press, Cambridge, NY.
[2] Hirota R 1972 *J. Phys. Soc. Jpn.* **33** 14561458.
[3] Wadati M 1973 *J. Phys. Soc. Jpn.* **34** 1289 1296.
[4] Tanaka S 1972 *Proc. Jap. Acad.* **48** 466469.
[5] Matveev V B Salle M A 1991 Berlin, Springer
[6] Li C Z, He J S and Porsezian K 2013 *Physical Review E* **87** 012913
[7] Li C Z, He J S and Porsezian K 2013 *Chin. Phys. B* **22** 4 044208
[8] Xue Y S, Tian B, etal. 2011 *Nonlinear Dynamics* **67** 2799-2806
[9] Porsezian K, Mahalingam A and Shanmugha S 2000 *Chaos, Solitons and Fractals* **11** 1261-1264
[10] Yesmakanova K R Shaikhova G N Bekova G T and Myrzakulova Zh R 2016 *Advances in Intelligent Systems and Computing* **441**, 183-198.
[11] Yesmakanova K R Shaikhova G N Bekova G T 2016 *AIP Conference Proceedings* **1759**, 020147.
[12] He J Wang L Li L Porseizian K and Erdely R 2014 *Phys. Rev. E* **89**, 062917.
[13] Li Ch. He J. 2014 *J. Sci. China Phys. Mech. Astron.* **57**, 898.
[14] Yang J. Li Ch. Li T. Cheng Zh. 2013 *Chinese Physics Letters*, **30**, 104201.
[15] Li Ch He J Porsezian K 2013 *Physical Review E* **87** 012913.
[16] Yesmakanova K R Shaikhova G N Bekova G T and Myrzakulova R 2016 *Journal of Physics: Conference Series* **738** 012018.

Self-duality in higher dimensions

A H Bilge[1], T. Dereli[2], S. Kocak[3]
[1] Kadir Has University, [2] Koc University, [3] Anadolu University
E-mail: ayse.bilge@khas.edu.tr, tdereli@ku.edu.tr, skocak@anadolu.edu.tr

Abstract. Let ω be a 2-form on a $2n$ dimensional manifold. In previous work, we called ω "strong self-dual, if the eigenvalues of its matrix with respect to an orthonormal frame are equal in absolute value. In a series of papers, we showed that strong self-duality agrees with previous definitions; in particular if ω is strong self-dual, then, in $2n$ dimensions, ω^n is proportional to its Hodge dual ω and in $4n$ dimensions, ω^n is Hodge self-dual. We also obtained a local expression of the Bonan 4-form on 8 manifolds with $Spin_7$ holonomy, as the sum of the squares of any orthonormal basis of a maximal linear subspace of strong self-dual 2-forms. In the present work we generalize the notion of strong self-duality to odd dimensional manifolds and we express the dual of the Fundamental 3-form 7 manifolds with G_2 holonomy, as a sum of the squares of an orthonormal basis of a maximal linear subspace of strong self-dual 2-forms.

1. Introduction

The mathematical framework of gauge theories consists of vector bundles over differentiable manifolds. These vector bundles are equipped with a connection that takes values in the Lie algebra of the structure group of the vector bundle. "Action integrals" are expressed in terms of the inner products of the components of the curvature 2-form. The variation of the action integral leads to the field equations of the theory. Yang-Mills theory on 4-manifolds is the prototype of this scheme. In Yang-Mills theories, the action integral is just the norm of the Lie algebra valued curvature 2-form of this vector bundle. Yang-Mills equations obtained by the variation of the action integral are satisfied provided that the curvature 2-form is self dual.

In a series of papers we defined the notion of "strong self-duality" in even dimensions and we studied the minimization of various action integrals, algebraically, by comparing inner products and wedge products of 2-forms. Strong self-dual forms turned out to be the building blocks for the curvature 2-forms that minimize various action integrals. [1],[2],[3],[4],[5] In the present work, we generalize the notion of strong self-duality to odd dimensions. This generalization relies on the results obtained in [10] pertaining skew-symmetric matrices in odd dimensions. We present a brief overview of the properties of strong self-dual 2-forms in even dimensions and we state without proof, relevant results in odd dimensions.

2. Self-duality and Strong Self-duality

The motivation of our definition of strong self-duality is the eigenvalue structure of self-dual 2-forms in 4-dimensions. We start by a description of their eigenvalues.

IOP Conf. Series: Journal of Physics: Conf. Series **804** (2017) 012005 doi:10.1088/1742-6596/804/1/012005

2.1. Self-duality in 4-dimensions

Let e^i, $i = 1, \ldots, 4$ be a local, positively oriented, orthonormal basis for the cotangent bundle T^*M of a 4-dimentional M be a 4-dimensional, oriented Riemannian manifold M. The local expression of a 2-form is given by

$$\omega = \sum_{i<j} \omega_{ij} e^{ij}, \tag{1}$$

where $e^{ij} = e^i \wedge e^j$. The self-dual ($*\omega = \omega$) and anti-self-dual ($*\omega = -\omega$) 2-forms are eigenvectors of the Hodge duality map, $*$, and in 4-dimensions, the 6-dimensional linear space of 2-forms is the orthogonal direct sum of the 3-dimensional self-dual and anti self-dual 2-forms.

The ω_{ij}'s can be viewed as a skew-symmetric matrix whose eigenvalues are pure imaginary and occur in conjugate pairs. Denoting them as $\pm i\lambda_1$ and $\pm i\lambda_2$, it can be seen that they satisfy

$$\begin{aligned} \lambda_1^2 + \lambda_2^2 &= \omega_{12}^2 + \omega_{13}^2 + \omega_{14}^2 + \omega_{23}^2 + \omega_{24}^2 + \omega_{34}^2, \\ \lambda_1^2 \lambda_2^2 &= (\omega_{12}\omega_{34} - \omega_{13}\omega_{24} + \omega_{14}\omega_{23})^2. \end{aligned}$$

Hence

$$\lambda_1 \mp \lambda_2 = \sqrt{(\omega_{12} \mp \omega_{34})^2 + (\omega_{13} \pm \omega_{24})^2 + (\omega_{14} \mp \omega_{23})^2}.$$

For self-duality $\lambda_1 = \lambda_2$, while for anti-self-duality $\lambda_1 = -\lambda_2$. These two cases are distinguished by the sign of the Pfaffian, $\mathrm{Pf}(\omega) = \omega_{12}\omega_{34} - \omega_{13}\omega_{24} + \omega_{14}\omega_{23}$. Thus in four dimensions, the equality of the absolute values of the eigenvalues gives the usual notion of self-duality in the Hodge sense.

2.2. Strong self-duality in even dimensions

In higher (even) dimensions, we proposed the equality of the eigenvalues in absolute value as a definition of the self-duality of 2-forms. Let ω be a 2-form on a $2n$ dimensional manifold M. If $\{e^1, e^2, \ldots, e^{2n}\}$ is a local orthonormal basis for the cotangent bundle of M, then ω is expressed as in (1) and the corresponding skew-symmetric matrix has eigenvalues $\{\pm i\lambda_1, \ldots, \pm i\lambda_n\}$. We dofine n ω to be strong self-dual (respectively, strong anti-self-dual) if

$$|\lambda_1| = |\lambda_2| = \cdots = |\lambda_n| \tag{2}$$

and $*\omega^n > 0$ (respectively $*\omega^n < 0$). [1] Note that this is equivalent to the statement that the distinction is based on the sign of the Pfaffian of ω with respect to a positively oriented orthonormal basis.

In terms of matrices, strong self-duality or anti self-duality can also be expressed by the minimal polynomial requirement

$$\omega^2 + \lambda^2 I = 0 \tag{3}$$

where $\lambda^2 = -\frac{1}{2n} Tr\omega^2$.

In four dimensions, the strong self-duality coincides with usual Hodge duality. More precisely, the matrices satisfying $\omega^2 + \lambda^2 I = 0$ consist of the union of the usual self-dual and anti-self-dual 2-forms (including the zero form).

2.3. Strong self-duality in odd dimensions

In odd dimensions, the right way to generalize the notion of strong self-duality is via the minimal polynomial of the corresponding skew-symmetric matrix. In odd dimension, every skew-symmetric matrix is necessarily singular. Thus, the closest analogue of (2) and (3) would be to require that the kernel be 1-dimensional and the non-zero eigenvalues be equal in absolute value. Equivalently be can require that the minimal polynomial of the corresponding matrix be

$$(\omega^2 + \lambda^2 I)\omega = 0, \tag{4}$$

IOP Conf. Series: Journal of Physics: Conf. Series **804** (2017) 012005 doi:10.1088/1742-6596/804/1/012005

with $\lambda^2 = -\frac{1}{2n}Tr\omega^2$ as before. Note that a strong self-dual 2-form in $2n+1$ dimension restricted to the complement of its kernel satisfies (3), hence it is essentially a strong self-dual 2 form in $2n$ dimensions. The complication arises when we algebraic variety defined by (3), and in particular linear subspaces lying in this variety.

The structure of matrices satisfying (4) has been analyzed in [10]. We adopt the notation of [10] and denote the rank, trace, kernel and orthogonal complement of the kernel of a matrix A by, r_A, $Tr(A)$, $Ker(A)$ and W_A. We denote the set of $(2n+1)\times(2n+1)$ matrices with minimal polynomial (4) by $\mathcal{S}_{2n+1}$. Note that the elements of $\mathcal{S}_{2n+1}$ need not have constant rank and its structure as an algebraic variety can be complicated. On the other hand, linear subspaces of $\mathcal{S}_{2n+1}$, that we denote as $\mathcal{L}$ have a more uniform structure. We briefly quote relevant results form [10]. If A and B are two matrices belonging to a linear subspace of $\mathcal{S}_{2n+1}$, then any linear combination of these matrices should satisfy (4). This gives,

$$A^2B + ABA + BA^2 + \lambda_A^2 B = 0, \qquad AB^2 + BAB + B + \lambda_B^2 A = 0. \tag{5}$$

By Remark 1, in[10], If a vector X belongs to $Ker(A)$, then $Y = ABX$ belongs to $Ker(B)$ and it follows that matrices belonging to the same linear subspace have constant rank. The structure of linear subspaces is determined by the Lemma below (Lemma 5 in [10]).

Lemma A. *Let $\{A, B\}$ be a basis for a 2-dimensional subspace in $\mathcal{S}_{2n+1}$ such that $A \perp B$.*

(i)If $Ker(A) = Ker(B)$, then $AB + BA = 0$.

(ii) If $Ker(A) \cup Ker(B) = \{0\}$, them $KerA() \perp Ker(B)$.

Linear subspaces of $\mathcal{S}_{2n+1}$ are called to be of Type 1 or Type 2, according as pairs of matrices satisfy (i) or (ii) of Lemma A. We quote below Lemma 8 in [10].

Lemma B. *Let $\mathcal{L}$ be a linear subspace of $\mathcal{S}_{2n+1}$.*

(i) If $\mathcal{L}$ is of the first type, then there exists a subspace $\mathcal{K} \subset \mathcal{S}_{2n}$ such that $dim(\mathcal{L}) = dim(\mathcal{K})$.

(ii) If $\mathcal{L}$ is of the second type, then there exists a subspace $\mathcal{M} \subset \mathcal{S}_{2n+2}$ such that $dim(\mathcal{L}) = dim(\mathcal{M})$.

3. Properties of strong self-dual 2 forms

In this section we give a brief review of the basic properties of strong self-dual 2-forms in even dimensions and we state without proof thse properties that generalize to odd dimensions.

3.1. Eigenvalue Inequalities

Well known inequalities between elementary symmetric functions of the eigenvalues of a skew symmetric matrix leads to inequalities between the norms of the powers of a 2-form ω. In this setting, strong self dual 2 forms are the ones for which the norms of ω^k are maximal. In $2n$ dimensions, all inequalities are saturated, as shown in previous work. We start by quoting a result from [9]

Lemma C. *Let s_k be the kth elementary symmetric function of the numbers $\{\alpha_1, \alpha_2, ..., \alpha_n\}$, with $\alpha_i \in R$ and let the weighted elementary symmetric polynomials, q_k's, be defined by*

$$\binom{n}{k} q_k = s_k. \tag{6}$$

Then

$$q_1 \geq q_2^{1/2} \geq q_3^{1/3} \geq \cdots \geq q_n^{1/n}, \tag{7}$$

$$q_{r-1}q_{r+1} \leq q_r^2, \qquad 1 \leq r < n. \tag{8}$$

If all α_i's are equal, then the equalities hold and if any single equality holds, then all α_i's are equal.

The norm ω^k is related to q_k as

$$(\omega^k, \omega^k) = \frac{n!k!}{(n-k)!}q_k.$$

From inequalities (7), we obtain,

$$(\omega^k, \omega^k) \leq \frac{n!k!}{n^k(n-k)!}(\omega, \omega)^k \tag{9}$$

and equalities hold if and only if ω is strong self-dual.

In $(2n+1)$ dimensions, a strong-self-dual 2-form has a 1-dimensional kernel and it its restriction to the complement of its kernel is a strong self-dual 2-form in $2n$ dimensions. Therefore the inequalities above hold for appropriate values of n and equalities hold again if and only if ω is strong self-dual.

3.2. Equivalence of strong self-duality with previous definitions of self-duality

Using the eigenvalue inequalities, we have shown that (i) a 2-form ω in $2n$ dimensions is strong self-dual if and only if ω^{n-1} is proportional to the Hodge dual of ω, and (ii) a 2-form ω in $4n$ dimensions is strong self-dual if ω^n is self-dual in the Hodge sense . The first condition has been proposed as a definition of self-duality by Trautman [11] while the second one appears in the work of Grossman et al. [8].

These properties have no direct analogues in odd dimensions. The best we can say is the following. Given the 2-form ω in $(2n+1)$-dimensions, the 1-dimensional kernel of the corresponding skew-symmetric matrix determines a 1-form α_ω. Then, if ω is strong self-dual, then

$$\omega^{n-1} \wedge \alpha = k \; *\omega, \tag{10}$$

where k is a constant.

3.3. Maximal linear subspaces of strong self-dual 2-forms in eight dimensions

Maximal linear spaces of strong self-dual 2-forms on 8 manifolds 7-dimensional. We work with the subspace with basis elements given below.

$$\begin{aligned}
\omega_1 &= -e^{12} + e^{34} + e^{56} - e^{78}, \\
\omega_2 &= -e^{13} - e^{24} + e^{57} + e^{68}, \\
\omega_3 &= -e^{14} + e^{23} + e^{58} - e^{67}, \\
\omega_4 &= -e^{15} - e^{26} - e^{37} - e^{48}, \\
\omega_5 &= -e^{16} + e^{25} - e^{38} + e^{47}, \\
\omega_6 &= -e^{17} + e^{28} + e^{35} - e^{46}, \\
\omega_7 &= -e^{18} - e^{27} + e^{36} + e^{45}.
\end{aligned} \tag{11}$$

3.4. Maximal linear subspaces of strong self-dual 2-forms in 7 dimensions

We recall that maximal linear subspaces of strong self-dual 2-forms in $(2n+1)$ dimensions are inherited from $2n$ or $2n+2$ dimensions. For $2n+1=7$, maximal linear subspaces in $2n=6$ dimensions are 1 dimensional, while the maximal linear subspaces in $2n+2=8$ dimensions are 7-dimensional, as described above. The matrices corresponding these 2-forms can be computed by solving the equations (5) and they are given in [10].

$$\begin{aligned}
\omega_1 &= -e^{12} + e^{34} + e^{56}, \\
\omega_2 &= -e^{13} - e^{24} + e^{57}, \\
\omega_3 &= -e^{14} + e^{23} - e^{67}, \\
\omega_4 &= -e^{15} - e^{26} - e^{37}, \\
\omega_5 &= -e^{16} + e^{25} + e^{47}, \\
\omega_6 &= -e^{17} + e^{35} - e^{46}, \\
\omega_7 &= -e^{27} + e^{36} + e^{45}.
\end{aligned} \tag{12}$$

4. Linear subspaces defined via $n-4$ forms

The self-duality equations given in [7] can be described as eigenspaces of the (self-adjoint) map defined via a 4-form ϕ as follows.

$$T_\Phi(\omega) = *(\Phi \wedge \omega). \tag{13}$$

This definition makes sense in any dimension, provided that such a (globally defined) 4-form exists. In 8-dimensions, the strong self-dual forms given in the previous section are eigenvectors of the map defined via the Bonan form. The restriction of the Bonan form to 7-dimensions gives the dual of the "fundamental 3-form" ϕ. In 7-dimensions ϕ defines a linear map on 2-forms.

4.1. Bonan-form in 8-dimensions

The Bonan-form Φ, which is a $Spin(7)$-invariant 4-form in 8-dimensions [6], can be constructed in terms of strong self-dual 2-forms. We consider a 7-plane $\mathcal{L}$ with an orthonormal basis $\{\eta_1, \eta_2, \ldots, \eta_7\}$ and let

$$\Phi = \eta_1^2 + \eta_2^2 + \cdots + \eta_7^2 \tag{14}$$

This expression is independent of the basis of the 7-plane. To give an explicit expression we can use the orthonormal basis $\eta_i = \frac{1}{2}\omega_i$, with ω_i given by (11) and we obtain

$$\begin{aligned}
\Phi = -\frac{3}{2}(e^{1234} + e^{1256} - e^{1278} + e^{1357} + e^{1368} + e^{1458} - e^{1467} \\
-e^{2358} + e^{2367} + e^{2457} + e^{2468} - e^{3456} + e^{3478} + e^{5678}).
\end{aligned} \tag{15}$$

Furthermore, the ω_i's are eigenvalues of the map (13) defined via the Bonan form, with eigenvalue 9/2, i.e,.

$$*(\Phi \wedge \omega_i) = \frac{9}{2}\omega_i. \tag{16}$$

4.2. The fundamental 3-form in 7-dimensions

The properties described above for the Bonan form generalize to 7-dimensions. The Fundamental 3-form ψ is G_2 invariant. We consider again a 7-plane $\mathcal{L}$ with an orthonormal basis $\{\eta_1, \eta_2, \ldots, \eta_7\}$ and let

$$*\psi = \eta_1^2 + \eta_2^2 + \cdots + \eta_7^2 \tag{17}$$

As above, this definition is independent of the basis of the 7-plane. We the orthonormal basis $\eta_i = \frac{1}{\sqrt{3}}\omega_i$, with ω_i given by (12) and we obtain

$$*\psi = -2(e^{1234} + e^{1256} + e^{1357} - e^{1467} + e^{2367} + e^{2457} - e^{3456}). \tag{18}$$

The ω_i's are eigenvalues of the map (13) defined via the the Fundamental 3-form, with eigenvalue 4, i.e,.

$$*(\psi \wedge \omega_i) = 4\omega_i. \tag{19}$$

We finally note that explicit solutions related to these structure were given for $Spin_7$ and G_2 manifolds in[12] and [13] respectively.

References

[1] Bilge A.H., Dereli T. and Koçak Ş., Self-dual Yang-Mills fields in eight dimensions, Lett. Math. Phys., 1996 , V. 36, 301.
[2] Bilge A.H., Dereli T. and Koçak Ş., The geometry of self-dual two-forms, J. Math. Phys., 1997, V. 38, N. 8, 4804.
[3] A.H.Bilge, T.Dereli,Ş.Koçak, The geometry of self-dual gauge fields, *New Symmetries in the Fundamental Interactions Theory*, edited by J. Lukierski, A.Jadczyk (PWN Polish Scientific Publishers,Warszawa,1997) p.45
[4] Bilge A.H., Dereli T. and Koçak Ş., Seiberg-Witten equations on R^8, Proceedings of 5th Gökova Geometry-Topology Conference, Edited by S.Akbulut, T.Önder, R, Stern (TUBITAK, Ankara, 1997), p.87.
[5] Bilge A.H., Dereli T. and Koçak Ş., Maximal linear subspaces of strong self-dual 2-forms and Bona 4-form, Linear Algebra and its Applications 434 (2011) 12001214
[6] Bonan E., Sur les varieties riemanniennes a groupe d'holonomie G_2 ou *Spin*(7), C.R. Acad. Sci. Paris, 1966, V.262, 127-129.
[7] Corrigan E., Devchand C., Fairlie D. and Nuyts J., First order equations for gauge fields in spaces of dimensions greater than four, Nucl. Phys. B., 1983, V. 214, 452-464.
[8] Grossman B., Kephart T.W. and Statsheff J.D., Solutions to Yang-Mills field equations in eight dimensions and the last Hopf map, Comm. Math. Phys., 1984, V. 96, 431-437.
[9] M. Marcus and H.Minc, *A Survey of Matrix Theory and Matrix Inequalities*, Allyn and Bacon Inc., Boston, (1964).
[10] Y. Ozdemir, "Spaces of Skew-Symmetric Matrices Satisfying $A^3 = \lambda A$", Linear Algebra and Its Applications, Volume 438, Page: 1365-1371, (2013).
[11] Trautman A., Solutions of the Maxwell and Yang-Mills equations associated with Hopf fiberings, Int. J. Theor. Phys., 1977, V.16, 561-565.
[12] S.Uguz and A.H. Bilge, $(3+3+2)$-warped-like manifolds with $Spin_7$ holonomy, Journal of Geometry and Physics, Volume 61, Issue 6, p. 1093-1103.
[13] S.Uguz, "Special warped-like product manifolds with (week) G_2 holonomy", Ukranian Mathematical Journal, Volume 65, No.8, pp.1126-1140, (2013).

IOP Conf. Series: Journal of Physics: Conf. Series **804** (2017) 012006 doi:10.1088/1742-6596/804/1/012006

Inequivalent representations in the functional integral framework

M Blasone[1], **P Jizba**[2] **and L Smaldone**[1]

[1] Dipartimento di Fisica, Università di Salerno, Via Giovanni Paolo II, 132 84084 Fisciano, Italy & INFN Sezione di Napoli, Gruppo collegato di Salerno, Italy

[2] FNSPE, Czech Technical University in Prague, Břehová 7, 115 19 Praha 1, Czech Republic

E-mail: blasone@sa.infn.it

E-mail: petr.jizba@fjfi.cvut.cz

E-mail: luca.smaldone@studenti.unina2.it

Abstract. An important feature of Quantum Field Theory is the existence of unitarily inequivalent representations of canonical commutation relations. When one works with the functional integral formalism, it is not clear, however, how this feature emerges. By following the seminal work of M. Swanson on canonical transformations in phase-space path integral, we generalize his approach to coherent-state functional integrals which in turn will lead to a simplified formalism which makes the appearance of the inequivalent representations more transparent.

1. Introduction

Since the early days of its formulation, it was realized that Quantum Field Theory (QFT) is not simply the extension of quantum mechanics (QM) to relativistic systems [1]. There is indeed more: QFT deals with systems with an infinite number of degrees of freedom and thus the Stone–von Neumann uniqueness theorem [2], stating the unitary equivalence of the unitary irreducible representations of the canonical commutation relations (CCR), does not apply. Consequently there exist in QFT infinitely many unitarily inequivalent representations of the field algebra or, in other words, for a given dynamics, there is an infinite number of (inequivalent) Hilbert spaces.

A related, and often not sufficiently appreciated, feature of QFT is that the interacting (Heisenberg) fields do not have a unique representation in terms of the asymptotic fields, i.e. fields which directly act on the Fock space. In fact, the functional relation between the asymptotic fields and the Heisenberg fields is known as Haag's map and represents the so called weak operatorial relation, i.e. a relation valid only for expectation values over states belonging to the Hilbert space of the asymptotic fields. All this is of course well known and indeed important phenomena like the spontaneous symmetry breaking mechanism [3, 4] or the Hawking black hole radiation [5, 6] and quantization of dissipative systems [7], are possible in QFT only because of the existence of inequivalent representations.

In the last two decades the fundamental role of inequivalent representations has been recognized in the problem of quantization of mixed particles: in such case, a simple canonical transformation (rotation) of fields with different masses has a dramatic effect on the structure of the Hilbert space, leading to the orthogonality of the vacua for fields with definite flavor

IOP Conf. Series: Journal of Physics: Conf. Series **804** (2017) 012006 doi:10.1088/1742-6596/804/1/012006

and those with definite mass. This effect has been shown to be valid for both bosonic and fermionic fields [8, 9] and lead to phenomenologically relevant corrections in flavor oscillation formulas [10]. Further recent developments include the study of dynamical generation of fermionic mixing [11, 12] with extensions to curved backgrounds [13].

Aforementioned progress has been achieved in the context of canonical (i.e., operator) formalism, where there is a clear distinction between of dynamics (i.e., the operatorial Heisenberg equations) and that of the representation which embodies the boundary conditions for the Heisenberg equations and sets up the Hilbert space. On the other hand, one should be able to achieve the same result by means of the functional integral formalism, where one deals with a generating functional for Green's functions.

Here we show what role the inequivalent representations play in the study of linear canonical transformations in functional integrals. To this end, we generalize a treatment developed by M. Swanson in Ref. [14] for path integrals, to the framework of coherent states functional integrals which are better suited for the field-theory considerations.

Plan of the paper is as follows; in the following section we present the van Hove model [15] which is a simple toy-model system exhibiting inequivalent representations [4]. In Section 3 we present a brief review of the theory of canonical transformations in phase-space path integral, following Ref. [14]. In Section 4 we present our approach to the canonical transformations in the context of coherent-state functional integrals. In Section 5 we use the general framework obtain to study van Hove model in more detail.

2. van Hove model and inequivalent representations in QFT

The van Hove model is described by the Hamiltonian [4, 15]

$$\hat{H} = \int \mathrm{d}^3k \, \left[\omega_{\mathbf{k}} \hat{a}^\dagger_{\mathbf{k}} \hat{a}_{\mathbf{k}} + \nu_{\mathbf{k}} \left(\hat{a}^\dagger_{\mathbf{k}} + \hat{a}_{\mathbf{k}}\right)\right] . \tag{1}$$

This can be diagonalized by the the so-called *dynamical map*

$$\hat{\alpha}_{\mathbf{k}} = \ddot{a}_{\mathbf{k}} + g_{\mathbf{k}} \,, \quad \ddot{\alpha}^\dagger_{\mathbf{k}} = a^\dagger_{\mathbf{k}} + g_{\mathbf{k}} \,, \tag{2}$$

with $g_{\mathbf{k}} = \frac{\nu_{\mathbf{k}}}{\omega_{\mathbf{k}}}$. The energy spectrum is defined by the dispersion relation

$$E_{\mathbf{k}} = \omega_{\mathbf{k}} \,. \tag{3}$$

The physical vacuum is defined by

$$\hat{\alpha}_{\mathbf{k}} |0(g)\rangle = 0 \,. \tag{4}$$

This is related with the "interacting" vacuum $|0\rangle$ satisfying

$$\hat{a}_{\mathbf{k}} |0\rangle \; = \; 0 \tag{5}$$

by the relation

$$|0(g)\rangle \; = \; \hat{G}^{-1} |0\rangle \,. \tag{6}$$

Here $\hat{G}$ is the generator of the canonical transformation

$$\hat{G} = \exp\left(\int \mathrm{d}^3k \, \left[g_{\mathbf{k}} \hat{a}^\dagger_{\mathbf{k}} - g^*_{\mathbf{k}} \hat{a}_{\mathbf{k}}\right]\right) . \tag{7}$$

The "scalar product" between the two vacua is

$$\langle 0|0(g)\rangle = \exp\left(\int \mathrm{d}^3k |g_{\mathbf{k}}|^2\right) . \tag{8}$$

If we now consider the case of a translationally invariant system, i.e. $\nu_{\mathbf{k}} = \nu\delta(\mathbf{k})$ and then $g_{\mathbf{k}} = c\delta(\mathbf{k})$, the foregoing product can be explicitly written as

$$\langle 0|0(g)\rangle = \exp\left(-\frac{1}{2}\frac{Vc^2}{(2\pi)^3}\right). \tag{9}$$

In the long-wave (i.e., $V \to \infty$) limit this product goes to zero, i.e. the Hilbert spaces constructed on the two vacua are orthogonal. From the second Schur's lemma [16] then follows the unitary inequivalence of the CCR representations constructed over ensuing Hilbert spaces. Actually, the Haag's map (2) can be seen as a weak relation and it is still defined. This is an example of the celebrated Haag's theorem [17].

3. Canonical transformations in phase space path integral

Let us now first briefly review the concept of canonical transformations in the framework of phase-space PIs. This subject has been treated literature by several authors [18, 19, 20, 21, 22]. In the following we will closely follow the approach introduced by M. Swanson in Ref. [14]. There the key object of interest is the *evolution kernel* in mixed phase-space representation, namely

$$W_{fi} \;=\; \langle p_f|\exp\left[-iH(p,q;t)(t_f - t_i)\right]|q_i\rangle\,. \tag{10}$$

The latter can be rewritten in the time-sliced PI form

$$W_{fi} \;=\; \frac{1}{\sqrt{2\pi}}\prod_{j=1}^{N}\int\frac{\mathrm{d}p_j\mathrm{d}q_j}{2\pi}\exp\left\{i\sum_{j=1}^{N}\left[-q_j(p_{j+1}-p_j) - \Delta t H(p_j,q_j) - iq_i p_1\right]\right\}, \tag{11}$$

where $\Delta t \equiv (t_f - t_i)/N$, $t_i \equiv t_0$, $t_f \equiv t_{N+1}$, $p(t_0 + (N+1)\Delta t) \equiv p_{N+1} = p_f$, $q(t_0) \equiv q_0 = q_i$ and the limit $\Delta t \to 0$ or equivalently $N \to \infty$ is understood to be taken at the appropriate stage of calculations.

To define a QM form of the canonical transformation, one starts from the completeness relation

$$\langle p_f|q_i\rangle = \int\frac{\mathrm{d}P_1\mathrm{d}Q_N}{\sqrt{2\pi}}\langle p_f|Q_N\rangle e^{iP_1Q_N}\langle P_1|q_i\rangle\,, \tag{12}$$

and assumes the following correspondence relations [14]

$$\langle p_f|Q_N\rangle \;=\; \frac{1}{\sqrt{2\pi}}\exp\left\{i\left[P_f(Q_f - Q_N) + F(p_f,Q_f)\right]\right\}, \tag{13}$$

$$\langle P_1|q_i\rangle \;=\; \frac{1}{\sqrt{2\pi}}\exp\left\{-i\left[P_1Q_i + F(p_i,Q_i)\right]\right\}. \tag{14}$$

In order to identify $F(p.Q)$, one substitutes these relations in Eq. (12). To save the consistence of this equation, one has to make the identifications

$$-P_f\left(Q_f - Q_i\right) \;=\; F(p_f,Q_f) - F(p_f,Q_i)\,, \tag{15}$$

$$-q_i\left(p_f - p_i\right) \;=\; F(p_f,Q_i) - F(p_i,Q_i)\,, \tag{16}$$

which are nothing but ansatz proposed by Fukutaka and Kashiwa [21]. However, these seem to lead to the correct result only for $p_i = 0$. This condition implies an over-specification of boundary conditions. In Ref. [14] Swanson deals with this problem by assuming the boundary condition $q_i = 0$, and then shifts the solution via Galilean boost.

In any case, the relations (15)-(16) define the new variables. To recover the classical result we can expand these relations:

$$P_f = -\frac{\partial F(p_f, Q_f)}{\partial Q_f} + \frac{1}{2}\frac{\partial^2 F(p_f, Q_f)}{\partial Q_f^2}(Q_f - Q_i) + \dots , \tag{17}$$

where the classical canonical transformation correspond to the first order result. Therefore, we can identify the function F with the third type classical generating function of the canonical transformation [23]. Writing (15)-(16) in the infinitesimal form

$$P_j = -\frac{F(p_j, Q_j) - F(p_j, Q_{j-1})}{\Delta Q_j}, \tag{18}$$

$$q_j = -\frac{F(p_{j+1}, Q_j) - F(p_j, Q_j)}{\Delta p_j}, \tag{19}$$

where $\Delta Q_j = Q_j - Q_{j-1}$, $\Delta p_j = p_{j+1} - p_j$ we get

$$-q_j(p_{j+1} - p_j) = P_{j+1}(Q_{j+1} - Q_j) + F(p_{j+1}, Q_{j+1}) - F(p_j, Q_j). \tag{20}$$

This relation permit us to rewrite the argument of the PI as

$$\begin{aligned}&\sum_{j=0}^{N}[-q_j(p_{j+1} - p_j) - \epsilon H(q_j, p_j)] \\ &= F(p_f, Q_f) - F(p_i, Q_i) + \sum_{j=0}^{N}[P_{j+1}(Q_{j+1} - Q_j) - \epsilon H(P_j, Q_j, \Delta P_j, \Delta Q_j)] .\end{aligned} \tag{21}$$

Now the Jacobian could give a finite contribution to the PI action because of the appearance of anomalous corrections (Liouville anomaly) [14, 19, 20, 21]:

$$J^{-1} = \prod_{j=1}^{N}\left[1 + \frac{1}{2}\left(\frac{\partial^3 F}{\partial p_j^2 \partial Q_j}\frac{\partial Q_j}{\partial q_j}\frac{\partial p_j}{\partial Q_j} + \frac{\partial^3 F}{\partial Q_j^2 \partial p_j}\frac{\partial Q_j}{\partial q_j}\right)\Delta Q_j + \frac{1}{2}\frac{\partial^3 F}{\partial p_j^2 \partial Q_j}\frac{\partial Q_j}{\partial q_j}\frac{\partial p_j}{\partial P_j}\Delta P_j\right] . \tag{22}$$

When $\Delta Q_j = O(\Delta t)$, $\Delta P_j = O(\Delta t)$, exponentiating the expression (22), we get

$$J^{-1} = e^{\prod_{j=1}^{N}\log(1 + A_j \Delta Q_j + B_j \Delta P_j)} \sim e^{\sum_{j=1}^{N}(A_j \Delta Q_j + B_j \Delta P_j)}, \tag{23}$$

where, in the last passage we have expanded the logarithm and following Ref.[14] we have introduced the short notation

$$A_j = \frac{1}{2}\left(\frac{\partial^3 F}{\partial p_j^2 \partial Q_j}\frac{\partial Q_j}{\partial q_j}\frac{\partial p_j}{\partial Q_j} + \frac{\partial^3 F}{\partial Q_j^2 \partial p_j}\frac{\partial Q_j}{\partial q_j}\right), \tag{24}$$

$$B_j = \frac{1}{2}\frac{\partial^3 F}{\partial p_j^2 \partial Q_j}\frac{\partial Q_j}{\partial q_j}\frac{\partial p_j}{\partial P_j}. \tag{25}$$

Eq. (23) can be rewritten as

$$J^{-1} = e^{\sum_{j=1}^{N}\Delta t\left(A_j \frac{\Delta Q_j}{\Delta t} + B_j \frac{\Delta P_j}{\Delta t}\right)}. \tag{26}$$

A naive approach to this subject could thus lead to various inconsistencies as discusses by many authors (cf. e.g., [22] and citations therein).

4. Canonical transformations and inequivalent representations

In this section we present our analysis of canonical transformation in the framework of coherent-state (or holomorphic-representation) functional integrals: indeed such a formalism seems to offer, in a natural way, a correspondence between canonical transformations in PIs and a change in the CCR representations.

As shown by Klauder [24], coherent states PI can be expressed in terms of a true Gaussian measure, solving convergence problems. Moreover, this measure is naturally symmetric in the integration variables, avoiding problems do to the asymmetry of the phase space path integral in the definition of canonical transformations [22]. Moreover in the holomorphic PI integral the extension to the field theoretical case is really natural. Actually, this is a context where the holomorphic functional integral is widely used [25].

4.1. QM and finite degrees of freedom systems

We are going to treat the case of quantum mechanical many body systems in the coherent states PI framework. In the next subsection we shall pass to the field theoretical case.

Let us start with some basic definitions. Let $\{z_k\}$ be a sequence of complex numbers so that

$$\sum_k z_k^* z_k < \infty \,. \tag{27}$$

A coherent states in Schrödinger picture are defined as

$$|\{z\}\rangle = e^{-\sum_k z_k \hat{a}_k^\dagger}|0\rangle \,. \tag{28}$$

These are eigenstates of annihilation operators

$$\hat{a}_k|\{z\}\rangle = z_k|\{z\}\rangle \,, \tag{29}$$

where the corresponding bra vectors are defined as

$$\langle\{z^*\}| = \langle 0|e^{-\sum_k z_k^* \hat{a}_k} \,, \tag{30}$$

and are eigenstates of creation operators

$$\langle\{z^*\}|\hat{a}_k^\dagger = \langle\{z^*\}z_k^* \,. \tag{31}$$

Moving base (or Heisenberg-picture base) vectors are defined as

$$|\{z\},t\rangle = e^{-i\hat{H}(\{\hat{a}\},\{\hat{a}^\dagger\})t}|\{z\}\rangle \,. \tag{32}$$

A resolution of the identity can be written in terms of these base states as

$$\int [\mathrm{d}\mu(z)]|\{z\},t\rangle\langle\{z^*\},t| = \mathbb{1} \,, \tag{33}$$

where

$$[\mathrm{d}\mu(z)] = \prod_k \int \mathrm{d}\mu(z_k) = \prod_k \int \frac{\mathrm{d}z_k \mathrm{d}z_k^*}{2\pi i} e^{-\sum_k z_k^* z_k} \,. \tag{34}$$

The scalar product between two coherent states is done by the relation

$$\langle\{\tilde{z}^*\}|\{z\}\rangle = e^{\sum_k \tilde{z}_k^* z_k} \,. \tag{35}$$

With these ingredients we are able to construct a functional-integral form for the evolution kernel:

$$\langle\{z^*\}_f, t_f|\{z\}_i, t_i\rangle = \int_{\{z\}(t_i)=\{z\}_i}^{\{z^*\}(t_f)=\{z^*\}_f} \prod_k \mathcal{D}z_k^* \mathcal{D}z_k e^{\sum_k z_{f\,k}^* z_{f\,k}} e^{i\sum_k \int_{t_i}^{t_f} \mathrm{d}t[iz_k^*(t)\dot{z}_k(t) - H(z_k^*, z_k)]}, \quad (36)$$

that has to be interpreted as a formal way to indicate the time sliced expression

$$\begin{aligned} \langle\{z^*\}_f, t_f|\{z\}_i, t_i\rangle &= \lim_{N\to\infty} \prod_{j=1}^{N} \int [\mathrm{d}\mu(z_j^*)\mathrm{d}\mu(z_j)] e^{z_{f\,k}^* z_{f\,k}} e^{-\sum_{j=1}^{N+1}\sum_k z_{j\,k}^*(z_{j\,k} - z_{j-1\,k})} \\ &\times\ e^{-i\sum_{j=0}^{N}\sum_k H(z_{j+1\,k}^*, z_{j\,k})\Delta t}. \end{aligned} \quad (37)$$

Here

$$H(z_{j+1\,k}^*, z_{j\,k}) \equiv \frac{\langle\{z^*\}_{j+1}, t_{j+1}|\hat{H}(\hat{a}_k, \hat{a}_k^\dagger)|\{z\}_j, t_j\rangle}{\langle\{z^*\}_{j+1}, t_{j+1}|\{z\}_j, t_j\rangle}. \quad (38)$$

Now we consider another set of coherent states $|\{\zeta\}\rangle$, constructed on the Hilbert space $\tilde{\mathcal{H}}$. Here it is very important to remark that when we consider these new states, we are dealing with another representation of the CCR. These states are defined as

$$\begin{aligned} |\{\zeta\}\rangle &= e^{\sum_k \zeta_k \hat{\alpha}_k^\dagger}|\tilde{0}\rangle, & (39)\\ \langle\{\zeta^*\}| &= \langle\tilde{0}|e^{\sum_k \zeta_k^* \hat{\alpha}_k}, & (40)\\ \hat{\alpha}_k|\{\zeta\}\rangle &= \zeta_k|\{\zeta\}\rangle, \quad \langle\{\zeta^*\}|\hat{\alpha}_k^\dagger - \zeta_k^*\langle\{\zeta^*\}|, & (41) \end{aligned}$$

where

$$\hat{\alpha}_k = \exp(-i\hat{\mathcal{K}})\hat{a}_k \exp(i\hat{\mathcal{K}}), \quad \hat{\alpha}_k^\dagger = \exp(-i\hat{\mathcal{K}})\hat{a}_k^\dagger \exp(i\hat{\mathcal{K}}). \quad (42)$$

where $\hat{\mathcal{K}} \equiv \sum_k \hat{K}(\hat{a}_k, \hat{a}_k^\dagger)\theta_k$ is a Hermitian operator. In Ref. [18, 26] it was noticed that in the phase-space framework the *correspondence relations* might be written as formal PIs

$$\begin{aligned} \langle\{z^*\}|\{\zeta\}\rangle &= \int_{\{\xi\}(0)=\{\zeta\}}^{\{\xi^*\}(\theta)=\{z^*\}} \prod_k \mathcal{D}\xi_k^*\mathcal{D}\xi_k e^{\sum_k \xi_k^*(\theta_k)\xi_k(0)} e^{i\sum_k \int_0^{\theta_k} \mathrm{d}\theta_k'[i\xi_k^*(\theta_k')\dot{\xi}_k(\theta_k') - K(\xi_k^*, \xi_k)]}, & (43)\\ \langle\{\zeta^*\}|\{z\}\rangle &= \int_{\{\xi\}(0)=\{z\}}^{\{\xi^*\}(\theta)=\{\zeta^*\}} \prod_k \mathcal{D}\xi_k^*\mathcal{D}\xi_k e^{\sum_k \xi_k^*(\theta_k)\xi_k(0)} e^{i\sum_k \int_0^{\theta_k} \mathrm{d}\theta_k'[i\xi_k^*(\theta_k')\dot{\xi}_k(\theta_k') - K(\xi_k^*, \xi_k)]}. & (44) \end{aligned}$$

These are impossible to solve in the general and we have to resort to the following ansatz

$$\begin{aligned} \langle\{z^*\}|\{\zeta\}\rangle &= A^*(\{\zeta^*\}, \{z\}) \exp[F^*(\{\zeta\}^*, \{z\})], & (45)\\ \langle\{\zeta^*\}|\{z\}\rangle &= A(\{\zeta^*\}, \{z\}) \exp[F(\{\zeta\}^*, \{z\})]. & (46) \end{aligned}$$

As shown in [27, 28] these functions can be determined, solving the formal path integrals in the stationary-phase approximation, obtaining

$$\begin{aligned} A(\{\zeta^*\}, \{z\}) &= \prod_k \left(\frac{\partial^2 F_2(\{\zeta^*\}, \{z\})}{\partial z_k \partial \zeta_k^*}\right)^{\frac{1}{2}}, & (47)\\ F(\{\zeta^*\}, \{z\}) &= F_2(\{\zeta^*\}, \{z\}) + \sum_k \frac{i}{2}\int_0^{\theta_k} \mathrm{d}\theta_k' \frac{\partial^2 K(\xi_k^c, \xi_k^{*c})}{\partial \xi_k \partial \xi_k^*}. & (48) \end{aligned}$$

where $\xi_k^c(\theta)$ and $\xi_k^{*c}(\theta)$ are stationary phase solutions and

$$F_2\left(\{\zeta\}^*,\{z\}\right) = \sum_k F_{2\,k}(\zeta_k^*, z_k)\,, \tag{49}$$

is the classical second-type generating function of the canonical transformation[1]. Let us now introduce the function $\mathcal{G}\left(\{\zeta\}^*,\{z\}\right) = F\left(\{\zeta\}^*,\{z\}\right) + \log A\left(\{\zeta\}^*,\{z\}\right)$. In QM these two representations are always unitarily equivalent, because of the Stone–von Neumann theorem. Therefore overcompleteness ensures that the expression

$$\frac{\langle\{\zeta^*\}_{j+1}|\{z\}_{j+1}\rangle}{\langle\{\zeta\}^*_{j+1}|\{z\}_j\rangle} = e^{\mathcal{G}(\{\zeta\}^*_{j+1},\{z\}_{j+1})-\mathcal{G}(\{\zeta\}^*_{j+1},\{z\}_j)}\,, \tag{50}$$

is well defined. Multiplying both sides by $\langle\{z\}^*_{j+1}|\{\zeta\}_{j+1}\rangle\langle\{\zeta\}^*_{j+1}|\{z\}_j\rangle$ and integrating over $[\mathrm{d}\mu(\zeta_{j+1})]$ we get

$$\langle\{z^*\}_{j+1}|\{z\}_{j+1}\rangle = \int [\mathrm{d}\mu(\zeta_{j+1})]\left[e^{\mathcal{G}(\{\zeta\}^*_{j+1},\{z\}_{j+1})-\mathcal{G}(\{\zeta\}^*_{j+1},\{z\}_j)}\langle\{z\}^*_{j+1}|\{\zeta\}_{j+1}\rangle\langle\{\zeta\}^*_{j+1}|\{z\}_j\rangle\right]. \tag{51}$$

With this we arrive on the consistency relation

$$\mathcal{G}(\{\zeta\}^*_{j+1},\{z\}_{j+1}) - \mathcal{G}(\{\zeta\}^*_{j+1},\{z\}_j) = \sum_k z^*_{j+1\,k}(z_{j+1\,k} - z_{j\,k})\,. \tag{52}$$

Similarly we find

$$\mathcal{G}(\{\zeta\}^*_{j+1},\{z\}_j) - \mathcal{G}(\{\zeta\}^*_j,\{z\}_j) = \sum_k \zeta_{j\,k}(\zeta^*_{j+1\,k} - \zeta^*_{j\,k})\,. \tag{53}$$

These two relations are formally analogous to Eqs. (18)-(19). Because the formalism developed by Fukutaka and Kashiwa and by Swanson works well, we could make the identification $\mathcal{G}\left(\{\zeta\}^*,\{z\}\right) = F_2\left(\{\zeta\}^*,\{z\}\right)$. As it is clear from Eq. (47), this identification is exact for linear transformations.

We find then, that the argument of the functional integral, in the new variables, reads

$$\begin{aligned}
&-\sum_{j=1}^{N+1}\sum_k z^*_{j\,k}(z_{j\,k} - z_{j-1\,k}) - i\sum_{j=0}^{N} H(z_{j\,k}, z^*_{j+1\,k})\Delta t\\
&= F_2(\{\zeta\}^*_i,\{z\}_i) - F_2(\{\zeta\}^*_f,\{z\}_f) + \sum_{j=1}^{N+1}\sum_k \zeta_{j-1\,k}(\zeta^*_{j\,k} - \zeta^*_{j-1\,k})\\
&-i\sum_{j=0}^{N} H(\zeta_{j\,k},\zeta^*_{j+1\,k},\Delta\zeta_{j\,k},\Delta\zeta^*_{j\,k})\Delta t\,.
\end{aligned} \tag{54}$$

The inverse Jacobian, with anomalous correction takes the form

$$\begin{aligned}
J^{-1} &= \prod_{j=1}^{N}\prod_k\left[1 + \frac{1}{2}\frac{\partial^3 F_{2\,k}(\zeta^*_{j\,k}.z_{j\,k})}{\partial z^2_{j\,k}\partial\zeta^*_{j\,k}}\frac{\partial\zeta^*_{j\,k}}{\partial z^*_{j\,k}}\frac{\partial z_{j\,k}}{\partial\zeta_{j\,k}}\Delta\zeta_{j\,k}\right.\\
&+ \left.\frac{1}{2}\left(\frac{\partial^3 F_2(\zeta^*_{j\,k}.z_{j\,k})}{\partial z^2_{j\,k}\partial\zeta^*_{j\,k}}\frac{\partial\zeta^*_{j\,k}}{\partial z^*_{j\,k}}\frac{\partial z_{j\,k}}{\partial\zeta^*_{j\,k}} + \frac{\partial^3 F_2(\zeta^*_{j\,k}.z_{j\,k})}{\partial\zeta^{*2}_{j\,k}\partial z_{j\,k}}\frac{\partial\zeta^*_{j\,k}}{\partial z^*_{j\,k}}\right)\Delta\zeta^*_{j\,k}\right]\,,
\end{aligned} \tag{55}$$

which very similar to the Swanson result. However, here we have not imposed particular boundary conditions for the consistency of our equations.

[1] A formulation of classical mechanics in terms of complex phase space variable was introduced by Strocchi [29]. However here we use a slightly different notation that generalizes the notation of Goldstein [23].

IOP Conf. Series: Journal of Physics: Conf. Series **804** (2017) 012006 doi:10.1088/1742-6596/804/1/012006

4.2. Field theoretical case and inequivalent representations

As seen in Section 2, in QFT the Stone–von Neumann theorem does not apply and we must deal with many unitarily inequivalent representations of the CCR. In this section, we restrict our attention to linear transformations. In this case the following correspondence relations are exact

$$\langle z^*|\zeta\rangle = \exp\left\{\int \mathrm{d}^3k \; F_2^*\left[\zeta^*(\mathbf{k}), z(\mathbf{k})\right]\right\}, \tag{56}$$

$$\langle \zeta^*|z\rangle = \exp\left\{\int \mathrm{d}^3k \; F_2\left[\zeta^*(\mathbf{k}), z(\mathbf{k})\right]\right\}. \tag{57}$$

However, $F_2[\zeta^*, z] \equiv \int \mathrm{d}^3k \; F_2\left[\zeta^*(\mathbf{k}), z(\mathbf{k})\right]$ is now a functional, defined on an infinite dimensional functional space, and as such it might be a divergent quantity. To see this we decompose F_2 as

$$F_2 = \int \mathrm{d}^3k \; \tilde{F}_2\left[\zeta^*(\mathbf{k}), z(\mathbf{k})\right] + C, \tag{58}$$

where the arbitrary constant C was singled out. But now this constant is determined by the the vacuum structure. In fact it follows easily that

$$\langle \tilde{0}|0\rangle = \exp(C). \tag{59}$$

If these two representations are unitarily inequivalent C diverges and so the generating functional as well. Therefore we cannot derive a definition of new variables as we did above in deriving the consistency relations (52)-(53). We limit ourselves to formally define the relations

$$F_2\left[\zeta^*_{j+1}(\mathbf{k}), z_{j+1}(\mathbf{k})\right] - F_2\left[\zeta^*_{j+1}(\mathbf{k}), z_j(\mathbf{k})\right] = z^*_{j+1}(\mathbf{k})[z_{j+1}(\mathbf{k}) - z_j(\mathbf{k})], \tag{60}$$

$$F_2\left[\zeta^*_{j+1}(\mathbf{k}), z_j(\mathbf{k})\right] - F_2\left[\zeta^*_{j}(\mathbf{k}), z_j(\mathbf{k})\right] = \zeta^*_{j+1}(\mathbf{k})[\zeta^*_{j+1}(\mathbf{k}) - \zeta_j(\mathbf{k})], \tag{61}$$

that there not affected by the divergence seen above. The functional integral can be now *formally* expanded in a generic representation as

$$\langle z_f^*, t_f|z_i, t_i\rangle = \prod_{\mathbf{k}} \int_{\tilde{z}(t_i)=z_i}^{\tilde{z}^*(t_f)=z_f^*} \mathcal{D}\zeta^*(\mathbf{k})\mathcal{D}\zeta^*(\mathbf{k}) e^{\int \mathrm{d}^3k \; \tilde{z}_f^*\tilde{z}_f} \\ \times \; e^{i\int \mathrm{d}^3k\left\{i\left(F_2[\zeta_f^*(\mathbf{k}),\tilde{z}_f] - F_2[\zeta_i^*(\mathbf{k}),\tilde{z}_i]\right) + \int_{t_i}^{t_f} \mathrm{d}t \; \left[-i\zeta(\mathbf{k},t)\dot{\zeta}^*(\mathbf{k},t) - H(\zeta(\mathbf{k},t),\zeta^*(\mathbf{k},t))\right]\right\}}, \tag{62}$$

where we denoted $\tilde{z} = z\left[\zeta(\mathbf{k},t), \zeta^*(\mathbf{k},t)\right]$. Let us stress that, in general, only a combination of these new variables can be fixed by boundary conditions because of the uncertainty principle. As shown by Torre [30] in the case of quadratic Hamiltonians, because of the existence of unitarily inequivalent representations, the unitary time evolution operator may not exist so that expressions such as (62) lose sense. In the following section we show a physical view of these mathematical considerations, using a close connection between QFT and statistical physics.

5. Functional integral analysis of the van Hove model

We start our analysis of the van Hove model by working in the Minkowski space. We will perform the Wick rotation to the Euclidean regime at the end of our reasoning. For the sake of simplicity we start with a single-mode van Hove model, the extension to many modes being trivial. The Feynman kernel, after an integration by parts, can be written as

$$\langle z_f^*, t_f|z_i, t_i\rangle = \lim_{N\to\infty} \prod_{j=1}^{N} \int \frac{\mathrm{d}\zeta_j^*\mathrm{d}\zeta_j}{2\pi i} e^{-F_2(\zeta_f^*, z_f) + F_2(\zeta_i^*, z_i) + \zeta_f^*\zeta_f - \zeta_i^*\zeta_i + (\zeta_f^* - g^*)(\zeta_f - g)} \\ \times \; e^{-\sum_{j=1}^{N+1}\zeta_j^*(\zeta_j - \zeta_{j-1}) - i\sum_{j=0}^{N}\omega(\zeta^*_{j+1} - g^*)(\zeta_j - g) + \nu(\zeta^*_{j+1} + \zeta_j - g - g^*)\Delta t}. \tag{63}$$

If we take $g = \nu/\omega$, we find

$$\langle z_f^*, t_f | z_i, t_i \rangle = \lim_{N\to\infty} \prod_{j=1}^{N} \int \frac{\mathrm{d}\zeta_j^* \mathrm{d}\zeta_j}{2\pi i} e^{\zeta_f^*\zeta_f - \frac{\nu}{\omega}(\zeta_f^*+\zeta_i)+\left(\frac{\nu}{\omega}\right)^2} \times e^{-\sum_{j=1}^{N+1}\zeta_j^*(\zeta_j-\zeta_{j-1}) - i\sum_{j=0}^{N}\left(\omega\zeta_{j+1}^*\zeta_j - \frac{\nu^2}{\omega}\right)\Delta t} . \quad (64)$$

Passing to the continuous limit we get

$$\langle z_f^*, t_f | z_i, t_i \rangle = e^{-\frac{\nu}{\omega}(\zeta_f^*+\zeta_i)+\left(\frac{\nu}{\omega}\right)^2} \int \mathcal{D}\zeta^*\mathcal{D}\zeta e^{\zeta_f^*\zeta_f} e^{-\int_{t_i}^{t_f}\left[\zeta^*\dot{\zeta} - i\left(\omega\zeta^*\zeta - \frac{\nu^2}{\omega}\right)\right]\mathrm{d}t} . \quad (65)$$

Apart from the constant pre-factor, fixed by the mixed boundary conditions, this is the harmonic oscillator kernel and as such it can be exactly solved in the saddle-point approximation. We thus find

$$\langle z_f^*, t_f | z_i, t_i \rangle = e^{\zeta_f^* e^{-i\omega(t_f-t_i)}\zeta_i - \frac{\nu}{\omega}(\zeta_f^*+\zeta_i)+\left(\frac{\nu}{\omega}\right)^2 - i(t_f-t_i)\frac{\nu^2}{\omega}} . \quad (66)$$

With the help of Wick's rotation we derive the Euclidean kernel

$$\langle z_f^*, \beta | z_i, 0 \rangle = e^{\zeta_f^* e^{-\beta\omega}\zeta_i - \frac{\nu}{\omega}(\zeta_f^*+\zeta_i)+\left(\frac{\nu}{\omega}\right)^2 - \beta\frac{\nu^2}{\omega}} . \quad (67)$$

The ensuing partition function is then easily obtained in the form

$$Z_{vH} = Z_{ho} e^{-\beta\frac{\nu^2}{\omega}} . \quad (68)$$

Here Z_{ho} indicates the harmonic-oscillator partition function (cf. e.g. [1, 22]). The extension of these results to systems with many degrees of freedom is now formally straightforward. The Euclidean kernel is now

$$\langle \{z^*\}_f, \beta | \{z\}_i, 0 \rangle = e^{\sum_k\left[\zeta_{f\,k}^* e^{-\beta\omega_k}\zeta_{i\,k} - \frac{\nu_k}{\omega_k}(\zeta_{f\,k}^*+\zeta_{i\,k})+\left(\frac{\nu_k}{\omega_k}\right)^2 - \beta\frac{\nu_k^2}{\omega_k}\right]} , \quad (69)$$

and the Euclidean partition function reads

$$Z_{vH} = Z_{ho} e^{-\beta\sum_k\frac{\nu_k^2}{\omega_k}} . \quad (70)$$

Reading this expression as the partition function of an interacting field-theoretical system in a box of a finite volume V we can interpret Z_{ho} as a partition function of quasi-particles. The energy gap can be obtained from the thermodynamical relation

$$\mathcal{F}_{vH} - \mathcal{F}_{ho} = \sum_k \frac{\nu_k^2}{\omega_k} , \quad (71)$$

where $\mathcal{F} = \beta \ln Z$ is the Helmholtz free energy. Taking the large volume limit, in the case of the translationally invariant van Hove model, for which $\nu(\mathbf{k})/\omega(\mathbf{k}) = c\delta(\mathbf{k})$, we find

$$\Delta\mathcal{F} = \mathcal{F}_{vH} - \mathcal{F}_{ho} = \int \mathrm{d}^3k \; c^2\omega(\mathbf{k})\delta^2(\mathbf{k}) = \frac{Vc^2\omega(\mathbf{0})}{(2\pi)^3} , \quad (72)$$

If we now perform the thermodynamic limit we see that energy gap $\Delta\mathcal{F}$ diverges. This is the physical meaning of the inequivalence between these two representations: an infinite energy is required to pass from one vacuum to another one.

6. Conclusions

In this paper we have shown how a careful study of canonical transformations in the coherent-state functional integrals makes clearer the connection between classical and quantum theory of canonical transformations. This, on one hand side, helps to bring into a new light previous results on this subject found by other authors in the context of phase-space path integrals. On the other hand, we have demonstrated how the existence of unitarily inequivalent representations of the CCR in QFT affects this study: although the (classical) generator of the canonical transformations between two representations typically do not exist, the transformation can still be (at least formally) well defined. This fact is directly reflected in the existence of an infinite energy gap among inequivalent vacuum states.

Acknowledgments

It is pleasure to acknowledge helpful conversations with G. Vitiello and H. Kleinert. PJ was supported by the GAČR Grant No. GA14-07983S.

References

[1] Itzykson M and Zuber J B 1980 *Quantum Field Theory* (New York: McGraw-Hill, Inc.)
[2] von Neumann J 1931 *Ann. Math.* **104** 570;
[3] Umezawa H, Matsumoto H and Tachiki M 1982 *Thermo Field Dynamics And Condensed States* (Amsterdam: North-holland)
[4] Blasone M, Jizba P and Vitiello G 2011 *Quantum Field Theory and its Macroscopic Manifestations* (London: World Scientific & ICP).
[5] Hawking S W 1975 *Commun.Math.Phys.* **43** 199–220;
[6] Martellini M, Sodano P, Vitiello G 1978 *Il Nuovo Cimento* **48** 341–58;
[7] Celeghini E, Rasetti M, Vitiello G 1992 *Ann.Phys.* **215** 156–70;
[8] Blasone M and Vitiello G 1995 *Ann. Phys.* **244** 283.
[9] Blasone M, Gargiulo M V and Vitiello G 2016 *Phys. Lett.* B **716** 104 ;
[10] Blasone M, Henning P A and Vitiello G 1999 *Phys. Lett.* B **451** 140;
[11] Blasone M, Jizba P, Lambiase G and Mavromatos N E 2014 *J. Phys.* **538** 012003;
[12] Blasone M, Jizba P and Smaldone L 2015 *Il Nuovo Cimento* **38 C** 201 ;
[13] Blasone M, Lambiase G and Luciano G G 2015 *J. Phys. Conf. Ser.* **631** no.1 012053.
[14] Swanson M S 1994 *Phys. Rev.* A **50** 4538;
[15] van Hove L 1952 *Physica* **18** 145–59;
[16] Tung W K 1980 *Group Theory in Physics* (Singapore: World Scientific Co.Pte.Ltd.)
[17] Streater R F and Wightman A S 1964 *PCT, Spin and Statistics and all that* (New York-Amsterdam: W.A.Benjamin,inc.)
[18] Fanelli R 1976 *J. Math. Phys.* **17** 490;
[19] Shiekh A Y 1995 *J. Math. Phys.* **36** 6681;
[20] Gervais J L and Jevicki A 1976 *Nucl. Phys.*B **110** 93–112;
[21] Fukutaka H and Kashiwa T 1988 *Ann. Phys.* **185** 301–29;
[22] Schulman L S 1982 *Techniques and Applications of Path Integration* (John Wiley and Sons, inc.)
[23] Goldstein H, Poole C and Safko J 1950 *Classical Mechanics* (Addison Wesley)
[24] Klauder J R and Skagerstam B 1985 *Coherent States-Applications in Physics and in Mathematical Physics* (Singapore: World Scientific)
[25] Faddeev L D and Slavnov A A 1980 *Gauge Fields-Introduction to Quantum Theory*(The Benjamin Cummings Publishing Company)
[26] Fanelli R 1975 *J. Math. Phys.* **16** 1729;
[27] Heller E J 1977 *Journ. Chem. Phys.* **66** 5777;
[28] Weissman Y 1982 *Journ. Chem. Phys.* **76** 4067;
[29] Strocchi F 1966 *Rev. Mod. Phys.* **38** 36–40;
[30] Torre C G 2005 *Phys. Rev.* D **71** 025004;

Mixed Nash equilibria in Eisert-Lewenstein-Wilkens (ELW) games

Katarzyna Bolonek-Lasoń[1] **and Piotr Kosiński**[2]

[1] Department of Statistical Methods, Faculty of Economics and Sociology, University of Lodz, 41/43 Rewolucji 1905 St., 90-214 Lodz, Poland.

[2] Department of Computer Science, Faculty of Physics and Applied Informatics, University of Lodz, 149/153 Pomorska St., 90-236 Lodz, Poland.

E-mail: kbolonek@uni.lodz.pl

Abstract. The classification of all mixed Nash equilibria for the original ELW game is presented. It is based on the quaternionic form of the game proposed by Landsburg (*Proc. Am. Math. Soc.* 139 (2011), 4423; Rochester Working Paper No 524 (2006); *Wiley Encyclopedia of Operations Research and Management Science* (Wiley and Sons, New York, (2011)). This approach allows to reduce the problem of finding the Nash equilibria to relatively simple analysis of the extrema of certain quadratic forms.

1. Introduction

We present here the review of results obtained in our paper [1]. Many papers concerning quantum game theory have appeared in recent decades [2]÷[53]. One of the starting points of the research was a quantum game defined by Eisert-Lewenstein-Wilkens (ELW) [2], [3]. They considered classical two-player (Alice and Bob) two-strategies noncooperative game defined by the relevant payoff matrix given in Table 1; here C and D are the possible strategies of players and X_0, X_1, X_2, X_3 are the corresponding outcomes. We will be dealing with the games which obey the conditions: (*i*) all X_α are distinct, (*ii*) all twofold sums $X_\alpha + X_\beta$ are distinct as well. The games defined by these two conditions are called generic in Landsburg's terminology.

Table 1. Payoff matrix of classical game

		Bob: C	Bob: D
Alice	C	(X_0, X_0)	(X_2, X_1)
	D	(X_1, X_2)	(X_3, X_3)

The quantization of classical game begins by ascribing to each player a two-dimensional complex Hilbert space H; then the total space of the game is the tensor product $H \otimes H$. The basic vectors in H correspond to two classical strategies

$$|C\rangle = \begin{pmatrix} 1 \\ 0 \end{pmatrix}, \qquad |D\rangle = \begin{pmatrix} 0 \\ 1 \end{pmatrix}. \tag{1}$$

IOP Conf. Series: Journal of Physics: Conf. Series **804** (2017) 012007 doi:10.1088/1742-6596/804/1/012007

Next, we define the initial state of the game

$$|\Psi_i\rangle \equiv J\left(|C\rangle \otimes |C\rangle\right) \tag{2}$$

where J is the gate operator (entangler) introducing quantum correlations. The operator J has the following form

$$J = exp\left(-\frac{i\gamma}{2}\sigma_2 \otimes \sigma_2\right) \tag{3}$$

with $\gamma \in \langle 0, \frac{\pi}{2}\rangle$ and σ_2 being the Pauli matrix. The form of gate operator is determined by two conditions: the quantum game is symmetric with respect to the exchange of players and the classical pure strategies are contained in the set of pure quantum ones.
Quantum strategies of both players are associated with unitary 2×2 operators U_A and U_B belonging to $SU(2)$. Defining the final state of the game as

$$|\Psi_f\rangle = J^+\left(U_A \otimes U_B\right) J\left(|C\rangle \otimes |C\rangle\right). \tag{4}$$

we can compute Alice and Bob payoffs

$$\begin{aligned}\$_A = X_0\left|\left(\langle C| \otimes \langle C|\right)|\Psi_f\rangle\right|^2 + X_1\left|\left(\langle D| \otimes \langle C|\right)|\Psi_f\rangle\right|^2 + \\ X_2\left|\left(\langle C| \otimes \langle D|\right)|\Psi_f\rangle\right|^2 + X_3\left|\left(\langle D| \otimes \langle D|\right)|\Psi_f\rangle\right|^2\end{aligned} \tag{5}$$

$$\begin{aligned}\$_B = X_0\left|\left(\langle C| \otimes \langle C|\right)|\Psi_f\rangle\right|^2 + X_1\left|\left(\langle C| \otimes \langle D|\right)|\Psi_f\rangle\right|^2 + \\ X_2\left|\left(\langle D| \otimes \langle C|\right)|\Psi_f\rangle\right|^2 + X_3\left|\left(\langle D| \otimes \langle D|\right)|\Psi_f\rangle\right|^2.\end{aligned} \tag{6}$$

Summarizing, the properties of the game depend on three elements:

- the choice of payoff matrix; for example, if the classical values of outcomes satisfy $X_1 > X_0 > X_3 > X_2$ we obtain the game named Prisoner Dilemma. This game provides a paradox where both players choose to protect themselves at the expense of the other player; as a result, both players find themselves in a worse state then if they had cooperated with each other.
- the value of parameter γ which determines quantum entanglement; for $\gamma = 0$ we have classical game whereas for $\gamma = \frac{\pi}{2}$ the initial state of the game is maximally entangled.
- the choice of the manifold $S \subset SU(2)$ of allowed strategies. Eisert et al. considered the set of allowed strategies belonging to the twodimensional submanifold of $SU(2)$ which itself is not a group. In our paper we assume that the set of allowed strategies is the whole $SU(2)$ group as indicated above.

One of the most important elements of the game is the notion of Nash equilibrium determining the optimal solution of the game. A pair of strategies (μ_0, ν_0) defines a Nash equilibrium iff

$$\$_A\left(\mu_0, \nu_0\right) \geq \$_A\left(\mu, \nu_0\right) \tag{7}$$

$$\$_B\left(\mu_0, \nu_0\right) \geq \$_B\left(\mu_0, \nu\right) \tag{8}$$

for all strategies μ and ν. Landsburg described the general structure of possible, in general mixed, Nash equilibria for maximally entangled ELW game in the papers [25], [37]÷[40]. In the present paper, we classify all Nash equilibria for the original ELW game. We use the method proposed by Landsburg with some modifications.

Landsburg proved that all Nash equilibria have a particularly simple form. He showed that any mixed strategy is equivalent to the one supported on at most four orthogonal quaternions and what's more the analysis of possible mixed strategies reduces to that of degeneracies of highest eigenvalues of the matrices determining the players payoffs. If we combine the symmetries of the game with this principle the method of finding and classifying Nash equilibria becomes transparent.

The paper is organized as follows. In Sec. 2 we present ELW game in terms of quaternion algebra. Then in Sec. 3 Landsburg's method of classifying Nash equilibria is described. The classification of Nash equilibria for the original ELW game is considered in Sec. 4. The last Section contains conclusions.

2. ELW game in terms of quaternion algebra

The $SU(2)$ group is isomorphic to the group of unit quaternions; therefore, we can rewrite the ELW game using quaternions. We start with the 4D real vector space spanned by the orthonormal vectors e_α, $\alpha = 0, 1, 2, 3$, $(e_\alpha, e_\beta) = \delta_{\alpha,\beta}$. We have the following multiplication law:

$$e_0 = \mathbb{1}, \quad e_i^2 = -\mathbb{1}, \quad e_i e_j = \varepsilon_{ijk} e_k, \quad i, j, k = 1, 2, 3. \tag{9}$$

The product of two quaternions $p = p_\alpha e_\alpha$, $q = q_\alpha e_\alpha$ has form

$$\begin{aligned}(pq)_0 &= p_0 q_0 - p_i q_i \\ (pq)_i &= p_0 q_i + q_0 p_i + \varepsilon_{ijk} p_j q_k\end{aligned} \tag{10}$$

(summations over latin indices run from 1 to 3).

The group isomorphism is defined by the mapping

$$U \longleftrightarrow p = p_\alpha e_\alpha \tag{11}$$

and any matrix $U \in SU(2)$ can be written as

$$U = p_0 \mathbb{1} - i p_k \sigma_k, \quad p_0^2 + \sum_{k=1}^{3} p_k^2 = 1. \tag{12}$$

In what follows we consider the particular case $\gamma = \frac{\pi}{2}$. Then the initial state of the game is maximally entangled and the role of quantum correlations is the most significant. The strategies of Alice and Bob in terms of $SU(2)$ matrices have the form

$$U_A = \begin{pmatrix} a & b \\ -\bar{b} & \bar{a} \end{pmatrix} \qquad |a|^2 + |b|^2 = 1 \tag{13}$$

$$U_B = \begin{pmatrix} \alpha & \beta \\ -\bar{\beta} & \bar{\alpha} \end{pmatrix} \qquad |\alpha|^2 + |\beta|^2 = 1. \tag{14}$$

Using the isomorphism (11) the Alice strategy is represented by the quaternion

$$p = \mathrm{Re}(a)\, e_0 - \mathrm{Im}(b)\, e_1 - \mathrm{Re}(b)\, e_2 - \mathrm{Im}(a)\, e_3 \tag{15}$$

The quaternion corresponding to Bob strategy reads

$$q = \mathrm{Re}(\alpha)\, e_0 + \mathrm{Re}(\beta)\, e_1 + \mathrm{Im}(\beta)\, e_2 + \mathrm{Im}(\alpha)\, e_3 \tag{16}$$

which corresponds to the alternative isomorphism of $SU(2)$ and unit quaternions

$$e_1 \longleftrightarrow i\sigma_2, \quad e_2 \longleftrightarrow i\sigma_1, \quad e_3 \longleftrightarrow i\sigma_3. \tag{17}$$

The gate operator, defined by eq. (3), takes now the following form

$$J = \cos\left(\frac{\gamma}{2}\right) e_0 \otimes e_0 + i \sin\left(\frac{\gamma}{2}\right) e_2 \otimes e_2 = \frac{1}{\sqrt{2}} \left(e_0 \otimes e_0 + ie_2 \otimes e_2\right). \tag{18}$$

Finally, we can check that the payoff functions of both players, eqs. (5) and (6), take the form

$$\begin{aligned} \$_A(p,q) &= X_0 \left(pq^{-1}\right)_0^2 + X_1 \left(pq^{-1}\right)_1^2 + X_2 \left(pq^{-1}\right)_2^2 + X_3 \left(pq^{-1}\right)_3^2 \equiv \\ &\equiv \sum_{\alpha=0}^{3} X_\alpha \left(pq^{-1}\right)_\alpha^2 \end{aligned} \tag{19}$$

$$\begin{aligned} \$_B(p,q) &= X_0 \left(pq^{-1}\right)_0^2 + X_2 \left(pq^{-1}\right)_1^2 + X_1 \left(pq^{-1}\right)_2^2 + X_3 \left(pq^{-1}\right)_3^2 \equiv \\ &\equiv \sum_{\alpha=0}^{3} \widetilde{X}_\alpha \left(pq^{-1}\right)_\alpha^2 \end{aligned} \tag{20}$$

with $\widetilde{X}_{0,3} = X_{0,3}$, $\widetilde{X}_{1,2} = X_{2,1}$.
We see that the payoff functions depend only on the product pq^{-1}. The reason for such a simple form of payoff functions is the form of the stability subgroup which is maximal for the maximally entangled initial state. The stability subgroup G_s is a set of elements g belonging to $SU(2) \times SU(2)$ which do not change the initial state. Any element $g \in G_s$ has the form $g = \left(U, \overline{U}\right)$, $U \in SU(2)$. Therefore, two pairs of strategies, (U_A, U_B) and (U'_A, U'_B), differing only by element g belonging to the stability subgroup

$$(U_A, U_B) = \left(U'_A, U'_B\right) \cdot g \tag{21}$$

provide to the same final result. The maximal entanglement allows us to write the following decomposition

$$(U_A, U_B) = \left(U_A U_B^T, I\right) \cdot \left(\overline{U}_B, U_B\right). \tag{22}$$

Eqs. (21) and (22) lead us to the conclusion that the outcomes of both players depend only on the product $U_A U_B^T$.

3. Landsburg's method of classifying Nash equilibria

A mixed strategy is a probability distribution on the space of unit quaternions representing pure strategies. If both players play mixed strategies the expected payoffs read

$$\$_A(\mu,\nu) = \iint\limits_{S^3 \times S^3} \$_A(p,q)\, \mathrm{d}\mu(p)\, \mathrm{d}\nu(q) \tag{23}$$

$$\$_B(\mu,\nu) = \iint\limits_{S^3 \times S^3} \$_B(p,q)\, \mathrm{d}\mu(p)\, \mathrm{d}\nu(q). \tag{24}$$

Landsburg's method starts from the observation that any mixed strategy is equivalent to the one supported on at most four orthogonal quaternions. To see this we write two definitions concerning equivalence. Two mixed strategies ν and ν' are left equivalent iff

$$\int\limits_{S^3} (pq)_\alpha^2\, \mathrm{d}\nu(q) = \int\limits_{S^3} (pq)_\alpha^2\, \mathrm{d}\nu'(q) \tag{25}$$

for any unit quaternion p and $\alpha = 0, 1, 2, 3$. Analogously, two mixed strategies μ and μ', are right equivalent iff

$$\int\limits_{S^3} (pq)_\alpha^2 \, \mathrm{d}\mu(p) = \int\limits_{S^3} (pq)_\alpha^2 \, \mathrm{d}\mu'(p) \tag{26}$$

for any unit quaternion q and $\alpha = 0, 1, 2, 3$. One can show that (see [1]) any mixed strategy ν is left equivalent to the one supported on at most four orthonoraml quaternions. The analogous principle is true for right equivalence.
The second important element of Landsburg's analysis concerns the search for Nash equilibria. It reduces to finding the strategies corresponding to maximal eigenvalues of quadratic forms defined by payoff functions. In more detail, the payoff functions can be always written in the form

$$\$_{A,B}(\mu, \nu) = \sum_{a,b=1}^{4} \sigma_a \rho_b \$_{A,B}\left(p^{(a)}, q^{(b)}\right) \tag{27}$$

with $\rho_a \geq 0$, $\sigma_a \geq 0$, $\sum_{a=1}^{4} \rho_a = \sum_{a=1}^{4} \sigma_a = 1$. Our aim is to find all sets $\left(\sigma_a, p^{(a)}\right)$ and $\left(\rho_b, q^{(b)}\right)$ such that $\$_A$ as a function of first strategy maximizes on $\left(\sigma_a, p^{(a)}\right)$ while $\$_B$ as a function of second strategy maximizes on $\left(\rho_b, q^{(b)}\right)$. Let $\Lambda \subset \{1, 2, 3, 4\}$ be the set of indices a for which $\sigma_a > 0$. Then the payoff function of Alice is a convex combination of the quantities $\sum_{b=1}^{4} \rho_b \$_A\left(p^{(a)}, q^{(b)}\right)$ and acquires a maximal value iff these quantities take the same maximal value for all $a \in \Lambda$. Considering the Alice payoff

$$\$_A(p, \nu) \equiv \sum_{b=1}^{4} \rho_b \$_A\left(p, q^{(b)}\right). \tag{28}$$

which is a quadratic form in p, we conclude that the vectors $p^{(a)}$, $a \in \Lambda$, span the eigenspace corresponding to the maximal eigenvalue; therefore, the highest eigenvalue has $|\Lambda|$-fold degeneracy. The same conclusion concerns Bob payoff function. If $\Sigma \subset \{1, 2, 3, 4\}$ is the set of inidces b such that $\rho_b > 0$ then the eigenspace of the matrix defined by the quadratic form in q

$$\$_B(\mu, q) = \sum_{a=1}^{4} \sigma_a \$_B\left(p^{(a)}, q\right) \tag{29}$$

which corresponds to the maximal eigenvalue is spanned by the vectors $q^{(b)}$, $b \in \Sigma$.

Summarizing, in order to determine Nash equilibria we have to find two sets, $\mu = \left\{\sigma_a, p^{(a)}\right\}$ and $\nu = \left\{\rho_b, q^{(b)}\right\}$, such that $p^{(a)}$, $a \in \Lambda$ span the maximal eigenspace of the quadratic form $\$_A(p, \nu)$, eq. (28), while $q^{(b)}$, $b \in \Sigma$ span the maximal eigenspace of quadratic form $\$_B(\mu, q)$, eq. (29). As a result we can introduce the following:

Definition 1. *The pair* $\left\{\left\{\sigma_a, p^{(a)}\right\}, \left\{\rho_b, q^{(b)}\right\}\right\}$ *is called a Nash equilibrium of the* (M, N) *type if* $|\Lambda| = M$, $|\Sigma| = N$, $1 \leq M, N \leq 4$ (cf. [37]).

The problem of finding Nash equilibria becames easier if we use symmetries of game. In our consideration two symmetries are useful:

- the structure of the game does not change, except the actual values of the payoffs, if one makes the substitution

$$X_\alpha \to \lambda X_\alpha + \mu, \quad \mu \in \mathbb{R}, \quad \lambda \in \mathbb{R}_+. \tag{30}$$

IOP Conf. Series: Journal of Physics: Conf. Series **804** (2017) 012007 doi:10.1088/1742-6596/804/1/012007

- for any unit quaternion r eqs. (19) and (20) imply (with q replaced by q^{-1} which simplifies notation)
$$\$_{A,B}\left(p,q\right) = \$_{A,B}\left(pr^{-1}, rq\right). \tag{31}$$
- both classical and quantum games are symmetric with respect to the exchange of players. In order to see this on the quaternionic level, let
$$r \equiv \frac{1}{\sqrt{2}}\left(e_0 + e_3\right); \tag{32}$$
then
$$re_0r^{-1} = e_0, \quad re_1r^{-1} = e_2, \quad re_2r^{-1} = -e_1, \quad re_3r^{-1} = e_3. \tag{33}$$
Therefore, the payoff functions of both players can be written as
$$\$_A\left(p,q\right) \equiv \$\left(p,q\right) \tag{34}$$
$$\$_B\left(p,q\right) = \$\left(r\,\overline{q}\,r^{-1}, r\,\overline{p}\,r^{-1}\right) \tag{35}$$
which express symmetric role of both players.

4. Classification of Nash equilibria for the original ELW game

The original ELW game [2] is defined by the payoffs $X_0 = 3$, $X_1 = 5$, $X_2 = 0$, $X_3 = 1$. We will classify the Nash equilibria according to Definition 1. Due to the symmetry of the game we can assume $M \geq N$. The procedure can be described as follows. Bob chooses the strategy q. We insert this strategy to the payoff function of Alice and find its maximum. Then the strategy maximizing Alice's payoff is inserted back to the payoff function of Bob and it is checked whether its maximum is achieved for the same quaternion q which we have chosen at the begining. Considering all possible pairs (M, N) we found [1] (cf. [37]÷[40]) that the Nash equilibria exist only for $N = 2$; therefore, we present only this case in detail.

Due to the symmetry $p \to pr$, $q \to r^{-1}q$ one can assume $q^{(1)} = \mathbb{1} - e_0$, $q^{(2)} - q$, $q = q_1e_1 + q_2e_2 + q_3e_3$, $q^2 = -e_0$. Denote $\rho_1 = \rho$, $\rho_2 = 1 - \rho$, $0 < \rho < 1$. We find
$$\$_A\left(p,\nu\right) = \rho\sum_{\alpha=0}^{3} X_\alpha p_\alpha^2 + (1-\rho)\sum_{\alpha=0}^{3}\left(pq\right)_\alpha^2. \tag{36}$$
Let $X = \mathrm{diag}\left(X_0, X_1, X_2, X_3\right)$; moreover, one can write
$$\left(pq\right)_\alpha \equiv \widetilde{m}_{\alpha\beta}\left(q\right)p_\beta \tag{37}$$
where
$$\widetilde{m}\left(q\right) = \begin{pmatrix} q_0 & -q_1 & -q_2 & -q_3 \\ q_1 & q_0 & q_3 & -q_2 \\ q_2 & -q_3 & q_0 & q_1 \\ q_3 & q_2 & -q_1 & q_0 \end{pmatrix}; \tag{38}$$
$q^2 = -e_0$, i.e. $q_0 = 0$, implies $\widetilde{m}^T\left(q\right) = -\widetilde{m}\left(q\right)$ and $\widetilde{m}^2\left(q\right) = -I$. The payoff function of Alice (36) can be rewritten as
$$\$_A\left(p,\nu\right) = p^T\left(\rho X + (1-\rho)\,\widetilde{m}^T\left(q\right)X\widetilde{m}\left(q\right)\right)p. \tag{39}$$
The matrix
$$Y\left(q,\rho,X\right) \equiv \rho X + (1-\rho)\,\widetilde{m}^T\left(q\right)X\widetilde{m}\left(q\right) \tag{40}$$

should posses M-fold ($M \geq 2$) degenerate highest eigenvalue. $Y(q,\rho,X)$ is a real symmetric matrix; explicitly

$$Y(q,\rho,X) = \begin{pmatrix} Y_0 & (1-\rho)X_{32}q_2q_3 & (1-\rho)X_{13}q_1q_3 & (1-\rho)X_{21}q_1q_2 \\ (1-\rho)X_{32}q_2q_3 & Y_1 & (1-\rho)X_{03}q_1q_2 & (1-\rho)X_{02}q_1q_3 \\ (1-\rho)X_{13}q_1q_3 & (1-\rho)X_{03}q_1q_2 & Y_2 & (1-\rho)X_{01}q_2q_3 \\ (1-\rho)X_{21}q_1q_2 & (1-\rho)X_{02}q_1q_3 & (1-\rho)X_{01}q_2q_3 & Y_3 \end{pmatrix} \quad (41)$$

where $X_{\alpha\beta} \equiv X_\alpha - X_\beta$ and

$$Y_0 = \rho X_0 + (1-\rho)\left(X_1q_1^2 + X_2q_2^2 + X_3q_3^2\right) \quad (42)$$

$$Y_1 = \rho X_1 + (1-\rho)\left(X_0q_1^2 + X_2q_3^2 + X_3q_2^2\right) \quad (43)$$

$$Y_2 = \rho X_2 + (1-\rho)\left(X_0q_2^2 + X_1q_3^2 + X_3q_1^2\right) \quad (44)$$

$$Y_3 = \rho X_3 + (1-\rho)\left(X_0q_3^2 + X_1q_2^2 + X_2q_1^2\right). \quad (45)$$

Assume first that only one component of q is nonzero. Then the matrix $Y(q,\rho,X)$ is diagonal

$$Y(q,\rho,X) = \begin{pmatrix} \rho X_0 + (1-\rho)X_3 & 0 & 0 & 0 \\ 0 & \rho X_1 + (1-\rho)X_2 & 0 & 0 \\ 0 & 0 & \rho X_2 + (1-\rho)X_1 & 0 \\ 0 & 0 & 0 & \rho X_3 + (1-\rho)X_0 \end{pmatrix}. \quad (46)$$

Using the actual values of the payoffs X_α we find easily that the highest eigenvalue is degenerate only provided $\rho = \frac{1}{2}$. Then the eigenspace corresponding to the maximal eigenvalues is spanned by e_1 and e_2. Therefore, according to the general discussion presented above, Alice plays the strategies

$$\begin{aligned} p^{(1)} &= e_1\cos\theta + e_2\sin\theta \\ p^{(2)} &= -e_1\sin\theta + e_2\cos\theta \end{aligned} \quad (47)$$

with the probabilities σ and $(1-\sigma)$, respectively. Then the quadratic matrix defining the Bob payoff reads

$$Z(p,\sigma,X) = \begin{pmatrix} Z_0 & 0 & 0 & (2\sigma-1)cs\left(\widetilde{X}_1 - \widetilde{X}_2\right) \\ 0 & Z_1 & (2\sigma-1)cs\left(\widetilde{X}_0 - \widetilde{X}_3\right) & 0 \\ 0 & (2\sigma-1)cs\left(\widetilde{X}_0 - \widetilde{X}_3\right) & Z_2 & 0 \\ (2\sigma-1)cs\left(\widetilde{X}_1 - \widetilde{X}_2\right) & 0 & 0 & Z_3 \end{pmatrix} \quad (48)$$

where $c \equiv \cos\theta$, $s \equiv \sin\theta$ and

$$Z_0 = \widetilde{X}_1\left(s^2 + \sigma\left(c^2 - s^2\right)\right) + \widetilde{X}_2\left(c^2 - \sigma\left(c^2 - s^2\right)\right) \quad (49)$$

$$Z_1 = \widetilde{X}_0\left(s^2 + \sigma\left(c^2 - s^2\right)\right) + \widetilde{X}_3\left(c^2 - \sigma\left(c^2 - s^2\right)\right) \quad (50)$$

$$Z_2 = \widetilde{X}_0\left(c^2 - \sigma\left(c^2 - s^2\right)\right) + \widetilde{X}_3\left(s^2 + \sigma\left(c^2 - s^2\right)\right) \quad (51)$$

$$Z_3 = \widetilde{X}_1\left(c^2 - \sigma\left(c^2 - s^2\right)\right) + \widetilde{X}_2\left(s^2 + \sigma\left(c^2 - s^2\right)\right). \quad (52)$$

Now, $q = e_0$ should be an eigenvector corresponding to the highest eigenvalue. This implies

$$(2\sigma - 1)\cos\theta\sin\theta = 0. \quad (53)$$

If $\sigma = \frac{1}{2}$ we find double degeneracy of highest eigenvalue corresponding to $q^{(1)} = e_0$, $q^{(3)} = \pm e_3$. Therefore, we find the whole family of Nash equilibria parametrized by the angle θ:

Alice:

The strategies

$$\begin{aligned} p^{(1)} &= e_1 \cos\theta + e_2 \sin\theta \\ p^{(2)} &= -e_1 \sin\theta + e_2 \cos\theta \end{aligned} \tag{54}$$

played with the probabilities $\frac{1}{2}$;

Bob:

The strategies

$$\begin{aligned} q^{(1)} &= e_0 \\ q^{(2)} &= \pm e_3 \end{aligned} \tag{55}$$

played with the probabilities $\frac{1}{2}$.

Another solution to eq. (53) is $\sin 2\theta = 0$; however, it does not lead to Nash equilibrium. In the case $M = N = 2$, one can check using MATHEMATICA that no solution exists unless two components of q vanish. We have also checked [1] that for the case $M = 3, 4$ there are no Nash equilibria. We conclude that all Nash equilibria for $M = N = 2$ are given by eqs. (31), (54) and (55).

5. Conclusion

We have sketched the way to classify all Nash equilibria for original ELW game. To this end we used,with some modifications, the method described by Landsburg. Landsburg's method is based on two important pillars. First, any mixed strategy is equivalent to the one supported on at most four orthogonal quaternions; second, the condition that the strategy is supported on a given number of quaternions implies that the highest eigenvalues of the matrices determining the players payoffs must be degenerate with the multiplicities equal to the number of quaternions enterings the supports of the corresponding measures. Additionally, we used the symmetries of the game what makes the classification easier.

For the case $M = N = 2$, where the strategies of both players are supported on two quaternions, we gave the complete description of Nash equilibria (for maximally entangled game).

Acknowledgment
Research of Katarzyna Bolonek-Lasoń was supported by NCN Grant no. DEC-2012/05/D/ST2/00754.

References

[1] Bolonek-Lasoń K, Kosiński P 2015, *International Journal of Quantum Information* **13**(8), 1550066
[2] Eisert J, Wilkens M, Lewenstein M 1999, *Phys. Rev. Lett.* **83**, 3077
[3] Eisert J, Wilkens M 2000, *J. Mod. Opt.* **47**, 2543
[4] Meyer D 1999, *Phys. Rev. Lett.* **82**, 1052
[5] Marinatto L, Weber T 2000, *Phys. Lett*, **A272**, 291
[6] Benjamin S 2000, *Phys. Lett.*, **A277**, 180
[7] Flitney A, Abbott D 2000, *Fluct. Noise Lett.* **2**, R175
[8] Marinatto L, Weber T 2000, *Phys. Lett.* **A 277**, 183
[9] Benjamin S, Hayden P 2001, *Phys. Rev. Lett.* **87(6)**, 069801
[10] Iqbal A, Toor A 2001, *Phys. Lett* **A280**, 249
[11] Du J, Li H, Xu X, Zhou X, Han R 2001, *Phys. Lett* **A289**, 9
[12] Eisert J, Wilkens M, Lewenstein M 2001, *Phys. Rev.Lett.* **87**, 069802
[13] Iqbal A, Toor A 2002, *Phys. Rev.* **A65**, 052328

[14] Du J, Li H, Xu X, Shi M, Wu J, Zhou X, Han R 2002, *Phys. Rev. Lett.* **88**, 137902
[15] van Enk S.J, Pike R 2002, *Phys. Rev* **A66**, 024306
[16] Du J, Xu X, Li H, Zhou X, Han R 2002, *Fluct. Noise Lett.* **2**, R189
[17] Guinea F, Martin-Delgado M.A 2003, *Journ. Phys.* **A36**, L197
[18] Flitney A, Abbott D 2003, *Proc. R. Soc. Lond.* **A459**, 2463
[19] Piotrowski E, Sladkowski J 2003, *Int. Journ. Theor. Phys.* **42**, 1089
[20] Du J, Li H, Xu X, Zhou X, Han R 2003, *Journ. Phys.* **A36**, 6551
[21] Zhou L, Kuang L 2003, *Phys. Lett* **A315**, 426
[22] Chen L, Ang H, Kiang D, Kwek L, Lo C 2003, *Phys. Lett.* **A316**, 317
[23] Lee C.F, Johnson N.F 2003, *Phys. Rev.* **A67**, 022311
[24] Shimamura J, Ozdemir S, Morikoshi F, Imoto N 2004, *Int. Journ. Quant. Inf.* **2**, 79
[25] Landsburg S 2004, *Notices of the Am. Math. Soc.* **51**, 394
[26] Rosero A, "Classification of Quantum Symmetric Non-zero Sum 2×2 Games in the Eisert Scheme", arXiv: quant-phys/0402117
[27] Nawaz A, Toor A 2004, *Journ. Phys.* **A37**, 11457
[28] Nawaz A, Toor A 2004, *Journ. Phys.* **A37**, 4437
[29] Iqbal A, "Studies in the theory of quantum games", arXiv: quant-phys/0503176
[30] Flitney A, Abbott D 2005, *Journ. Phys.* **A38**, 449
[31] Ichikawa T, Tsutsui I 2007, *Ann. Phys.* **322**, 531
[32] Cheon T, Tsutsui I 2006, *Phys. Lett.* **A348**, 147
[33] Patel N 2007, *Nature* **445**, 144
[34] Ichikawa T, Tsutsui I, Cheon T 2008 *Journ. Phys. A: Math. and Theor.* **41**, 135303
[35] Flitney A, Hollenberg L 2007, *Phys. Lett.* **A363**, 381
[36] Nawaz A, "The generalized quantization schemes for games and its application to quantum information", arXiv:1012.1933
[37] Landsburg S 2006, *Rochester Working Paper* No 524
[38] Landsburg S 2011, *Proc. Am. Math. Soc.* **139**, 4423
[39] Landsburg S 2011, *Wiley Encyclopedia of Operations Research and Management Science* (Wiley and Sons, New York)
[40] Landsburg S, Dahl G.B, "Quantum Strategies", arXiv:1110.4678
[41] Schneider D 2011, *Journ. Phys.* **A44**, 095301
[42] Schneider D 2012, *Journ. Phys.* **A45**, 085303
[43] Avishai Y, "Some Topics in Quantum Games", arXiv:1306.0284
[44] Bolonek-Lasoń K, Kosiński P 2013, *Prog. Theor. Exp. Phys.* (7), 073A02
[45] Ramzan M 2013, *Quant. Inf. Process.* **12**, 577
[46] M. Ramzan, Khan M.K 2013, *Fluctuation and Noise Letters* **12**, 1350025
[47] Nawaz A 2013, *Chin. Phys. Lett.* **30(5)**, 050302
[48] Frackiewicz P 2013, *Acta Phys. Polonica B* **44**, 29
[49] Nawaz A, "Werner-like States and Strategies form of Quantum games", arXiv:1307.5508
[50] Nawaz A 2012, *J. Phys. A: Math. Theor. J. Phys.* **45**, 195304
[51] Iqbal A, Chappell J.M, Li Q, Pearce C.E.M, Abbott D 2014, *Quantum Information Processing* **13**, 2783
[52] Iqbal A, Chappell J.M, Abbott D 2015, *Physica A: Statistical Mechanics and its Applications* **436**, 798
[53] Bolonek-Lasoń K 2015, *Prog. Theor. Exp. Phys.* **2**, 023A03
[54] Nash J 1950, *Proceedings of the National Academy of Sciences* **36(1)**, 48
[55] Glicksberg I.L 1952, *Proceedings of the American Mathematical Society*, **Vol. 3, No. 1**, 170
[56] Myerson R.B 1997, *Game Theory Analysis of Conflict*, Harvard University Press, Cambridge

IOP Conf. Series: Journal of Physics: Conf. Series **804** (2017) 012008 doi:10.1088/1742-6596/804/1/012008

Do the Kontsevich tetrahedral flows preserve or destroy the space of Poisson bi-vectors?

Anass Bouisaghouane and Arthemy V Kiselev

Johann Bernoulli Institute for Mathematics and Computer Science, University of Groningen, P.O.Box 407, 9700 AK Groningen, The Netherlands

E-mail: A.V.Kiselev@rug.nl

Abstract.
From the paper "Formality Conjecture" (Ascona 1996):

I am aware of only one such a class, it corresponds to simplest good graph, the complete graph with 4 *vertices* (*and* 6 *edges*). *This class gives a remarkable vector field on the space of bi-vector fields on* $\mathbb{R}^d$. *The evolution with respect to the time* t *is described by the following non-linear partial differential equation:* ..., *where* $\alpha = \sum_{i,j} \alpha_{ij}\partial/\partial x_i \wedge \partial/\partial x_j$ *is a bi-vector field on* $\mathbb{R}^d$.

It follows from general properties of cohomology that 1) ***this evolution preserves the class of*** (***real-analytic***) ***Poisson structures***, ...

In fact, I cheated a little bit. In the formula for the vector field on the space of bivector fields which one get from the tetrahedron graph, an additional term is present. ... *It is possible to prove formally that* ***if*** α ***is a Poisson bracket, i.e. if*** $[\alpha, \alpha] = 0 \in T^2(\mathbb{R}^d)$, ***then the additional term shown above vanishes.***

By using twelve Poisson structures with high-degree polynomial coefficients as explicit counterexamples, we show that both the above claims are false: neither does the first flow preserve the property of bi-vectors to be Poisson nor does the second flow vanish identically at Poisson bi-vectors. The counterexamples at hand suggest a correction to the formula for the "exotic" flow on the space of Poisson bi-vectors; in fact, this flow is encoded by the balanced sum involving both the Kontsevich tetrahedral graphs (that give rise to the flows mentioned above). We reveal that it is only the balance 1 : 6 for which the flow does preserve the space of Poisson bi-vectors.

Introduction. The Kontsevich graph complex is the language of deformation quantisation on finite-dimensional Poisson manifolds [1, 2]. We consider the class of oriented graphs on two sinks and $k \geqslant 1$ internal vertices (of which, each is the tail of two edges and carries a copy of the Poisson bi-vector $\mathcal{P}$). Encoding bi-differential operators, such graphs determine the flows on the space of bi-vectors on a Poisson manifold at hand. The two flows with $k = 4$ internal vertices in the graphs are provided by the two tetrahedra [1], see Fig. 1 on the next page. By producing 12 counterexamples, we prove that the claim [1, 2] of preservation of the Poisson property is false as stated. Simultaneously, we reveal that the flow which is determined by the second graph is not always vanishing by virtue of the skew-symmetry and Jacobi identity for Poisson bi-vectors $\mathcal{P}$.

This paper is structured as follows. First we recall the correspondence between graphs and polydifferential operators [3, 4] and we indicate the mechanism for such an operator to vanish, cf. [5, 6]. In section 2 we recall three constructions of Poisson brackets with polynomial coefficients of arbitrarily high degree (see [7, 8, 9]). In Tables 1–4 on pp. 7–8 we then summarise the properties of all structures from our 12 counterexamples to the claim [1] that

(i) the flow $\dot{\mathcal{P}} = \Gamma_1(\mathcal{P})$ which the first graph in Fig. 1 encodes on the space of bi-vectors $\mathcal{P}$ would preserve their property to be Poisson (in fact, it does not), and that

(ii) the flow $\dot{\mathcal{P}} = \Gamma_2(\mathcal{P})$ would always be trivial whenever the bi-vector $\mathcal{P}$ is Poisson (in fact, this is not true).

In particular, the twelfth counterexample pertains to the infinite-dimensional jet-space geometry of variational Poisson structures [11]. (Quoted from [12], the Hamiltonian differential operator for that variational Poisson bi-vector $\boldsymbol{\mathcal{P}}$ is processed by using the techniques from [13, 14, 15]).

Finally, we examine at which balance the linear combination of the Kontsevich tetrahedral flows preserves the space of Poisson structures on finite-dimensional manifolds. We argue that the ratio 1 : 6 does the job; this claim has been proved in [6].

1. The graphs and operators

Let us formalise a way to encode polydifferential operators using oriented graphs. Consider the space $\mathbb{R}^n$ with Cartesian coordinates $\boldsymbol{x} = (x_1, \ldots, x_n)$, here $2 \leqslant n < \infty$; for typographical reasons we use the lower indices to enumerate the variables, so that $x_1^2 = (x_1)^2$, etc. By definition, the decorated edge $\bullet \xrightarrow{i} \bullet$ denotes at once the derivation $\partial/\partial x_i \equiv \partial_i$ (that acts on the content of the arrowhead vertex) and the summation $\sum_{i=1}^n$ (over the index i in the object which is contained in the arrowtail vertex). For example, the graph $\bullet \xleftarrow{i} \mathcal{P}^{ij}(\boldsymbol{x}) \xrightarrow{j} \bullet$ encodes the bi-differential operator $\sum_{i=1}^n (\cdot)\overleftarrow{\partial_i}\, \mathcal{P}^{ij}(\boldsymbol{x})\, \overrightarrow{\partial_j}(\cdot)$. If its coefficients $\mathcal{P}^{ij}$ are antisymmetric, then the graph $\bullet \xleftarrow{i} \bullet \xrightarrow{j} \bullet$ encodes the bi-vector $\mathcal{P} = \mathcal{P}^{ij}\, \partial_i \wedge \partial_j$, where $\partial_i \wedge \partial_j = \frac{1}{2}(\partial_i \otimes \partial_j - \partial_j \otimes \partial_i)$. It then specifies the Poisson bracket $\{\cdot,\cdot\}_{\mathcal{P}}$ if the $\frac{n(n-1)}{2}$-tuple of coefficients solves the system of equations

$$(\mathcal{P}^{ij})\overleftarrow{\partial_\ell} \cdot \mathcal{P}^{\ell k} + (\mathcal{P}^{jk})\overleftarrow{\partial_\ell} \cdot \mathcal{P}^{\ell i} + (\mathcal{P}^{ki})\overleftarrow{\partial_\ell} \cdot \mathcal{P}^{\ell j} = 0, \qquad (1)$$

hence the bracket $\bullet \underset{L}{\xleftarrow{i}} \mathcal{P}^{ij} \underset{R}{\xrightarrow{j}} \bullet$ satisfies the Jacobi identity. Clearly, $\mathcal{P}^{ij}(\boldsymbol{x}) = \{x_i, x_j\}_{\mathcal{P}}$.

From now on, let us consider only the oriented graphs whose vertices are either sinks, with no issued edges, or tails for an ordered pair of arrows, each decorated with its own index (see Fig. 1). Allowing the only exception in footnote 1, we shall always assume that there are neither tadpoles, nor double oriented edges, nor two-edge loops.

We also postulate that every vertex which is not a sink carries a copy of a given Poisson bi-vector $\mathcal{P} = \mathcal{P}^{ij}(\boldsymbol{x})\, \partial_i \wedge \partial_j$; the ordering of decorated out-going edges coincides with the ordering "first $\prec$ second" of the indexes in the coefficients of $\mathcal{P}$.

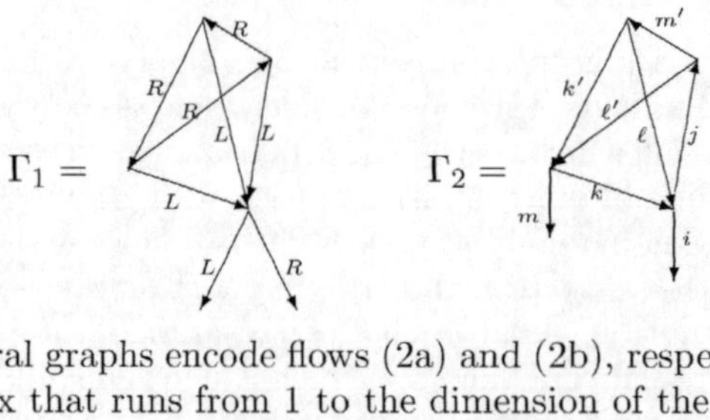

Figure 1. These tetraheral graphs encode flows (2a) and (2b), respectively. Each oriented edge carries a summation index that runs from 1 to the dimension of the Poisson manifold at hand. For each internal vertex (where a copy of the Poisson bi-vector $\mathcal{P}$ is stored), the pair of out-going edges is ordered, L $\prec$ R: the left edge (L) carries the first index and the other edge (R) carries the second index in the bi-vector coefficients. (In retrospect, the ordering and labelling of the indexed oriented edges can be guessed from formulas (2) on p. 3.)

IOP Conf. Series: Journal of Physics: Conf. Series **804** (2017) 012008 doi:10.1088/1742-6596/804/1/012008

Example 1. Under all these assumptions, the two tetrahedra which are portrayed in Fig. 1 are, up to a symmetry, the only admissible graphs with $k = 4$ internal vertices, $2k = 6 + 2$ edges, and two sinks. The first graph in Fig. 1 encodes the bi-vector

$$\Gamma_1(\mathcal{P}) = \sum_{i,j=1}^{n} \Bigg(\sum_{k,\ell,m,k',\ell',m'=1}^{n} \frac{\partial^3 \mathcal{P}^{ij}}{\partial x_k \partial x_\ell \partial x_m} \frac{\partial \mathcal{P}^{kk'}}{\partial x_{\ell'}} \frac{\partial \mathcal{P}^{\ell\ell'}}{\partial x_{m'}} \frac{\partial \mathcal{P}^{mm'}}{\partial x_{k'}} \Bigg) \frac{\partial}{\partial x_i} \wedge \frac{\partial}{\partial x_j}. \tag{2a}$$

Likewise, the second graph in Fig. 1 yields the bi-vector

$$\Gamma_2(\mathcal{P}) = \sum_{i,m=1}^{n} \Bigg(\sum_{j,k,\ell,k',\ell',m'=1}^{n} \frac{\partial^2 \mathcal{P}^{ij}}{\partial x_k \partial x_\ell} \frac{\partial^2 \mathcal{P}^{km}}{\partial x_{k'} \partial x_{\ell'}} \frac{\partial \mathcal{P}^{k'\ell}}{\partial x_{m'}} \frac{\partial \mathcal{P}^{m'\ell'}}{\partial x_j} \Bigg) \frac{\partial}{\partial x_i} \wedge \frac{\partial}{\partial x_m}. \tag{2b}$$

In this paper we examine

(i) whether the respective flows $\frac{\mathrm{d}}{\mathrm{d}\varepsilon}(\mathcal{P}) = \Gamma_\alpha(\mathcal{P})$ at $\alpha = 1, 2$ preserve or, in fact, destroy the property of bi-vectors $\mathcal{P}(\varepsilon)$ to be Poisson, provided that the Cauchy datum $\mathcal{P}\big|_{\varepsilon=0}$ is such;

(ii) we also inspect whether the second flow is (actually, it is not) vanishing identically at all ε, provided that the Cauchy datum is a Poisson bi-vector.

Remark 1. Whenever the bi-vector $\mathcal{P}$ in every internal vertex of a non-empty graph Γ is Poisson, the bi-differential operator which is encoded by Γ can vanish identically. First, this occurs due to the skew-symmetry of coefficients of the bi-vector.[1] Second, the operators encoded using graphs (with a copy of the Poisson bi-vector $\mathcal{P}$ at every internal vertex) can vanish by virtue of the Jacobi identity, see (1), or its differential consequences. This mechanism has been illustrated in [5]; making a part of our present argument (see [6]), it is a key to the proof of the fact that the balanced flow $\frac{\mathrm{d}}{\mathrm{d}\varepsilon}(\mathcal{P}) = \Gamma_1(\mathcal{P}) + 6\,\Gamma_2(\mathcal{P})$ does preserve the property of bi-vectors $\mathcal{P}(\varepsilon)$ to be (infinitesimally) Poisson whenever the Cauchy datum $\mathcal{P}\big|_{\varepsilon=0}$ is such.

So, each of the two claims (i–ii) is false if it does not hold for at least one Poisson structure (itself already known to have skew-symmetric coefficients and turn the left-hand side of the Jacobi identity into zero for any triple of arguments of the Jacobiator). To examine both claims, we need a store of Poisson structures such that the coefficients $\mathcal{P}^{ij}(\boldsymbol{x})$ are not mapped to zero by the third or second order derivatives in (2a) and (2b), respectively. For that, a regular generator of Poisson structures with polynomial coefficients of arbitrarily high degree would suffice.

2. The generators

Let us recall three regular ways to generate the Poisson brackets or modify a given one, thus obtaining a new such structure. These generators will be used in section 3 to produce the counterexamples to both claims from [1].

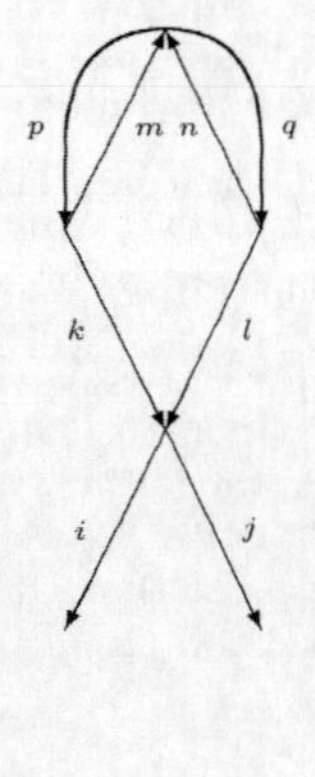

[1] For example, consider this oriented graph with ordered pairs of indexed edges ($i \prec j$, $k \prec \ell$, $m \prec n$, $p \prec q$). We claim that due to the antisymmetry of $\mathcal{P}$ which is contained in each of the four internal vertices, the operator (which this graph encodes) vanishes identically. Indeed, it equals minus itself:

$$\begin{aligned} &\partial_m\partial_n(\mathcal{P}^{pq})\partial_p(\mathcal{P}^{km})\partial_q(\mathcal{P}^{\ell n})\partial_k\partial_\ell(\mathcal{P}^{ij})\,\partial_i \wedge \partial_j \\ &= -\partial_m\partial_n(\mathcal{P}^{qp})\partial_p(\mathcal{P}^{km})\partial_q(\mathcal{P}^{\ell n})\partial_k\partial_\ell(\mathcal{P}^{ij})\,\partial_i \wedge \partial_j \\ &= -\partial_n\partial_m(\mathcal{P}^{pq})\partial_q(\mathcal{P}^{\ell n})\partial_p(\mathcal{P}^{km})\partial_\ell\partial_k(\mathcal{P}^{ij})\,\partial_i \wedge \partial_j = 0. \end{aligned}$$

To establish the second equality, we interchanged the labelling of indices ($p \rightleftarrows q$, $k \rightleftarrows \ell$, and $m \rightleftarrows n$) and we recalled that the partial derivatives commute.

IOP Conf. Series: Journal of Physics: Conf. Series **804** (2017) 012008 doi:10.1088/1742-6596/804/1/012008

2.1. The determinant construction

This generator of Poisson bi-vectors is described in [7], cf. [16] and references therein. The construction goes as follows. Let $x_1, \ldots, x_n$ be the Cartesian coordinates on $\mathbb{R}^{n\geqslant 3}$. Let $\vec{g} = (g_1, \ldots, g_{n-2})$ be a fixed tuple of smooth functions in these variables. For any a, $b \in C^\infty(\mathbb{R}^n)$, put

$$\{a, b\}_{\vec{g}} = \det\big(\mathbf{J}(g_1, \ldots, g_{n-2}, a, b)\big)$$

where $\mathbf{J}(\cdot, \ldots, \cdot)$ is the Jacobian matrix. Clearly, the bracket $\{\cdot,\cdot\}_{\vec{g}}$ is bi-linear and skew-symmetric. Moreover, it is readily seen to be a derivation in each of its arguments: $\{a, b\cdot c\}_{\vec{g}} = \{a, b\}_{\vec{g}} \cdot c + b \cdot \{a, c\}_{\vec{g}}$. For the validity mechanism of the Jacobi identity for this particular instance of the Nambu bracket we refer to [16] again (see also [17]).

Example 2 (see entry 3 in Table 2 on p. 7). Fix the functions $g_1 = x_2^3 x_3^2 x_4$ and $g_2 = x_1 x_3^4 x_4$, and insert them in the determinant generator of Poisson bi-vectors. We thus obtain the bi-vector $\mathcal{P}_0$,

$$\mathcal{P}_0^{ij} = \begin{pmatrix} 0 & -2\,x_1x_2^3x_3^5x_4 & -3\,x_1x_2^2x_3^6x_4 & 12\,x_1x_2^2x_3^5x_4^2 \\ 2\,x_1x_2^3x_3^5x_4 & 0 & -x_2^3x_3^6x_4 & 2\,x_2^3x_3^5x_4^2 \\ 3\,x_1x_2^2x_3^6x_4 & x_2^3x_3^6x_4 & 0 & -3\,x_2^2x_3^6x_4^2 \\ -12\,x_1x_2^2x_3^5x_4^2 & -2\,x_2^3x_3^5x_4^2 & 3\,x_2^2x_3^6x_4^2 & 0 \end{pmatrix}.$$

By construction, the above matrix is skew-symmetric. The validity of Jacobi identity (1) is straightforward: indexed by i, j, k, all the components $[\![\mathcal{P},\mathcal{P}]\!]^{ijk}$ of the tri-vector vanish.[2] This Poisson bi-vector $\mathcal{P}$ is used in section 3 in the list of counterexamples to the claims under study.

2.2. Pre-multiplication in the 3-dimensional case

Let x, y, z be the Cartesian coordinates on the vector space $\mathbb{R}^3$. For every bi-vector $\mathcal{P} = \mathcal{P}^{ij}\,\partial_i \wedge \partial_j$, introduce the differential one-form $\mathrm{P} = \mathrm{P}_1\,dx + \mathrm{P}_2\,dy + \mathrm{P}_3\,dz$ by setting $\mathrm{P} := -\mathcal{P} \lrcorner\, dx \wedge dy \wedge dz$, so that $\mathrm{P}_1 = -\mathcal{P}^{23}$, $\mathrm{P}_2 = \mathcal{P}^{13}$, and $\mathrm{P}_3 = -\mathcal{P}^{12}$. It is readily seen [8] that the original Jacobi identity for the bi-vector $\mathcal{P}$ now reads[3] $\mathrm{dP} \wedge \mathrm{P} = 0$ for the respective one-form P. But let us note that the pre-multiplication $\mathrm{P} \mapsto f \cdot \mathrm{P}$ of the form P by a smooth function f preserves this reading of the Jacobi identity: $\mathrm{d}(f\mathrm{P}) \wedge (f\mathrm{P}) = f \cdot \big[\mathrm{d}f \wedge \mathrm{P} \wedge \mathrm{P} + f \cdot \mathrm{dP} \wedge \mathrm{P}\big] = f^2 \cdot \mathrm{dP} \wedge \mathrm{P} = 0$. This shows that the bi-vector $f\mathcal{P}$ which the form $f\mathrm{P}$ yields on $\mathbb{R}^3$ is also Poisson.

This pre-multiplication trick provides the examples of Poisson structures of arbitrarily high polynomial degree coefficients (in a manifestly non-symplectic three-dimensional set-up).[4]

2.3. The Vanhaecke construction

In [9], Vanhaecke created another construction of high polynomial degree Poisson bi-vectors. Let u be a monic degree d polynomial in λ and v be a polynomial of degree $d-1$ in λ:

$$u(\lambda) = \lambda^d + u_1\lambda^{d-1} + \ldots + u_{d-1}\lambda + u_d, \qquad v(\lambda) = v_1\lambda^{d-1} + \ldots + v_{d-1}\lambda + v_d.$$

[2] Indeed, there are four tuples of distinct values of the indices i, j, and k up to permutations; we let $1 \leqslant i < j < k \leqslant n = 4$ so that the check runs over the set of triples $\{(1,2,3), (1,2,4), (1,3,4), (2,3,4)\}$. For example, $[\![\mathcal{P},\mathcal{P}]\!]^{123} = 6x_1x_2^5x_3^{11}x_4^2 - 6x_1x_2^5x_3^{11}x_4^2 - 6x_1x_2^5x_3^{11}x_4^2 + 6x_1x_2^5x_3^{11}x_4^2 - 18x_1x_2^5x_3^{11}x_4^2 + 18x_1x_2^5x_3^{11}x_4^2 + 12x_1x_2^5x_3^{11}x_4^2 - 6x_1x_2^5x_3^{11}x_4^2 - 6x_1x_2^5x_3^{11}x_4^2 = 0$. Therefore, $[\![\mathcal{P},\mathcal{P}]\!] = \sum_{1\leqslant i<j<k\leqslant 4} [\![\mathcal{P},\mathcal{P}]\!]^{ijk}(\boldsymbol{x})\,\partial_i \wedge \partial_j \wedge \partial_k = 0$.

[3] The exterior differential dP is equal to $\mathrm{dP} = (\partial_x\mathcal{P}^{13} + \partial_y\mathcal{P}^{23})\,dx \wedge \mathrm{d}y + (-\partial_x\mathcal{P}^{12} + \partial_z\mathcal{P}^{23})\,dx \wedge \mathrm{d}z + (-\partial_y\mathcal{P}^{11} - \partial_z\mathcal{P}^{13})\,\mathrm{d}y \wedge \mathrm{d}z$. The wedge product is $\mathrm{dP} \wedge \mathrm{P} = \big(\partial_x\mathcal{P}^{31}\,\mathcal{P}^{12} + \partial_y\mathcal{P}^{23}\,\mathrm{P}^{21} + \partial_x\mathcal{P}^{12}\,\mathcal{P}^{13} + \partial_z\mathcal{P}^{23}\,\mathcal{P}^{31} + \partial_y\mathcal{P}^{12}\,\mathcal{P}^{23} + \partial_z\mathcal{P}^{31}\,\mathcal{P}^{32}\big)\,dx \wedge \mathrm{d}y \wedge \mathrm{d}z = (-[\![\mathcal{P},\mathcal{P}]\!] \lrcorner\, dx \wedge \mathrm{d}y \wedge \mathrm{d}z)\,dx \wedge \mathrm{d}y \wedge \mathrm{d}z$.

[4] In dimension three, this pre-multiplication procedure also provides the examples of Poisson bi-vectors at which the second flow (2b) does not vanish identically.

Consider the space $\Bbbk^{2d}$ (e.g., set $\Bbbk := \mathbb{R}$) with Cartesian coordinates $u_1, \ldots, u_n, v_1, \ldots, v_d$. To define the Poisson bracket, fix a bivariate polynomial $\phi(\cdot,\cdot)$ and for all $1 \leqslant i,j \leqslant d$ set

$$\{u_i,u_j\} = \{v_i,v_j\} = 0, \qquad \{u_i,v_j\} = \text{coeff. of } \lambda^j \text{ in } \phi\big(\lambda, v(\lambda)\big) \cdot \left[\frac{u(\lambda)}{\lambda^{d-i+1}}\right]_+ \quad \operatorname{mod} u(\lambda), \quad (3)$$

where we denote by $[\ldots]_+$ the argument's polynomial part and where the remainder modulo the degree d polynomial $u(\lambda)$ is obtained using the Euclidean division algorithm.

Let us emphasise that these Poisson bi-vector are defined on the even-dimensional spaces. Indeed, the coefficients of Poisson bracket (3) are arranged in the block matrix $\left(\begin{smallmatrix} 0 & U \\ -U & 0 \end{smallmatrix}\right)$, where the components of the matrix U are $U^{ij} = \{u_i, v_j\}$.

2.4. The Hamiltonian differential operators on jet spaces

The variational Poisson brackets $\{\cdot,\cdot\}_{\boldsymbol{\mathcal{P}}}$ for functionals of sections of affine bundles generalise the notion of Poisson brackets $\{\cdot,\cdot\}_{\mathcal{P}}$ for functions on finite-dimensional Poisson manifolds $(N^n, \{\cdot,\cdot\}_{\mathcal{P}})$. Namely, let us consider the space $J^\infty(\pi)$ of infinite jets of sections for a given bundle π over a manifold M^n of positive dimension m. The variational Poisson brackets $\{\cdot,\cdot\}_{\boldsymbol{\mathcal{P}}}$ on $J^\infty(\pi)$ are then specified by using the Hamiltonian differential operators (which we shall denote by A and the order of which is typically positive).[5] The formalism of variational Poisson bi-vectors $\boldsymbol{\mathcal{P}} = \frac{1}{2}\langle \boldsymbol{\xi} \cdot \vec{A}(\boldsymbol{\xi})\rangle$ and the variational Schouten bracket $[\![\cdot,\cdot]\!]$ is standard (see [11, 19]). The geometry of iterated variations is revealed in [13]; the correspondence between the Kontsevich graphs and local variational polydifferential operators is explained in [14].

Example 3. To inspect whether either of the two claims (which we quote from [1] on the title page) would hold in the variational set-up, it is enough to consider a Hamiltonian differential operator with (differential-)polynomial coefficients of degree $\geqslant 3$. Let us take the Hamiltonian operator[6] $A = u^2 \circ \mathrm{d}/\mathrm{d}x \circ u^2$ for the Harry Dym equation (see [12]); here u is the fibre coordinate in the trivial bundle $\pi\colon \mathbb{R}\times\mathbb{R} \to \mathbb{R}$ and x is the base variable. This operator is obviously skew-adjoint, whence the variational Poisson bracket $\{\cdot,\cdot\}_{\boldsymbol{\mathcal{P}}}$ is skew-symmetric. The Jacobi identity for $\{\cdot,\cdot\}_{\boldsymbol{\mathcal{P}}}$ is also easy to check: the variational master-equation $[\![\boldsymbol{\mathcal{P}},\boldsymbol{\mathcal{P}}]\!] \cong 0$ does hold for the variational bi-vector $\boldsymbol{\mathcal{P}} = \frac{1}{2}\langle \xi \cdot \vec{A}(\xi)\rangle$.

3. The counterexamples

We now examine the properties of both tetrahedral flows (2) whenever each of them is evaluated at a given Poisson bi-vector. (Examples of such bi-vectors are produced by using the techniques from section 2.) To motivate the composition of Tables 1–4 and clarify the meaning of their content, let us consider an example: namely, we first take the Poisson bi-vector which was obtained in section 2.1 (see p. 4).

Example 4 (continued). Rewriting the Poisson bi-vector $\mathcal{P}_0 \in \Gamma\big(\bigwedge^2 TN^4\big)$ in terms of the parity-odd variables $\boldsymbol{\xi}$, we obtain that under the isomorphism $\Gamma\big(\bigwedge^\bullet TN^n\big) \simeq C^\infty(\Pi T^*N^n)$ the bi-vector $\mathcal{P}_0^{ij}(\boldsymbol{x})\,\partial_i \wedge \partial_j$ becomes $\frac{1}{2}\mathcal{P}_0^{ij}(\boldsymbol{x})\,\xi_i\xi_j$; we have that $\mathcal{P}_0 =$

$$-2\,x_1x_2^3x_3^5x_4\xi_1\xi_2 - 3\,x_1x_2^2x_3^6x_4\xi_1\xi_3 + 12\,x_1x_2^2x_3^5x_4^2\xi_1\xi_4 - x_2^3x_3^6x_4\xi_2\xi_3 + 2\,x_2^3x_3^5x_4^2\xi_2\xi_4 - 3\,x_2^2x_3^6x_4^2\xi_3\xi_4.$$

[5] In fact, the Poisson geometry of finite-dimensional affine manifolds $(N^n, \{\cdot,\cdot\}_{\mathcal{P}})$ is a zero differential order sub-theory in the variational Poisson geometry of infinite jet spaces $J^\infty(\pi)$. Indeed, let the fibres in the bundle π be N^n and proclaim that only *constant* sections are allowed.

[6] More examples of variational Poisson structures, which are relevant for our present purpose, can be found in [20] or, e.g., in [21] (see also the references contained therein).

Now, we calculate the right-hand sides $\mathcal{P}_1 := \Gamma_1(\mathcal{P}_0)$ and $\mathcal{P}_2 := \Gamma_2(\mathcal{P}_0)$ of tetrahedral flows (2). The coefficient matrix of the bi-vector $\mathcal{P}_1$ is

$$\mathcal{P}_1^{ij} = \begin{pmatrix} 0 & -24480\,x_1x_2^9x_3^{20}x_4^4 & -51840\,x_1x_2^8x_3^{21}x_4^4 & 12960\,x_1x_2^8x_3^{20}x_4^5 \\ 24480\,x_1x_2^9x_3^{20}x_4^4 & 0 & -15480\,x_2^9x_3^{21}x_4^4 & 2448\,x_2^9x_3^{20}x_4^5 \\ 51840\,x_1x_2^8x_3^{21}x_4^4 & 15480\,x_2^9x_3^{21}x_4^4 & 0 & -18144\,x_2^8x_3^{21}x_4^5 \\ -12960\,x_1x_2^8x_3^{20}x_4^5 & -2448\,x_2^9x_3^{20}x_4^5 & 18144\,x_2^8x_3^{21}x_4^5 & 0 \end{pmatrix}.$$

In a similar way, the polydifferential operator Γ_2 (encoded by the second tetrahedral graph in Fig. 1) yields the matrix

$$\mathcal{P}_2^{ij} = \begin{pmatrix} 16920x_1^2x_2^8x_3^{20}x_4^4 & -12060\,x_1x_2^9x_3^{20}x_4^4 & -16380\,x_1x_2^8x_3^{21}x_4^4 & 42840\,x_1x_2^8x_3^{20}x_4^5 \\ 2700\,x_1x_2^9x_3^{20}x_4^4 & -7200\,x_2^{10}x_3^{20}x_4^4 & 4680\,x_2^9x_3^{21}x_4^4 & -252\,x_2^9x_3^{20}x_4^5 \\ -13140\,x_1x_2^8x_3^{21}x_4^4 & 5040\,x_2^9x_3^{21}x_4^4 & -12060\,x_2^8x_3^{22}x_4^4 & 13716\,x_2^8x_3^{21}x_4^5 \\ -80280\,x_1x_2^8x_3^{20}x_4^5 & -18036\,x_2^9x_3^{20}x_4^5 & 21708\,x_2^8x_3^{21}x_4^5 & -58104\,x_2^8x_3^{20}x_4^6 \end{pmatrix}.$$

Notice that this coefficient matrix is not yet antisymmetric, but its *symmetric* counterpart is skipped out in the construction of the bi-vector $\mathcal{P}_2$ and its transcription by using the anticommuting variables $\boldsymbol{\xi}$. Therefore, we antisymmetrise the above matrix at once, the output to be used in what follows. We obtain that the bi-vector is

$$\begin{aligned}\mathcal{P}_2 = &-7380x_1x_2^9x_3^{20}x_4^4\xi_1\xi_2 - 1620x_1x_2^8x_3^{21}x_4^4\xi_1\xi_3 + 61560x_1x_2^8x_3^{20}x_4^5\xi_1\xi_4 \\ &- 180x_2^9x_3^{21}x_4^4\xi_2\xi_3 + 8892x_2^9x_3^{20}x_4^5\xi_2\xi_4 - 3996x_2^8x_3^{21}x_4^5\xi_3\xi_4.\end{aligned}$$

We now see that for the Poisson bi-vector $\mathcal{P}_0$ from Example 2 on p. 4, ***the bi-vector $\mathcal{P}_2$ does not vanish***, thereby disavowing the second claim from [1].

To check the compatibility of the original Poisson bi-vector $\mathcal{P}_0$ with the newly obtained bi-vector $\mathcal{P}_1$, we calculate their Schouten bracket:

$$\begin{aligned}[\![\mathcal{P}_0, \mathcal{P}_1]\!] = &\ 46008\,x_1x_2^{11}x_3^{26}x_4^5\xi_1\xi_2\xi_3 + 852768\,x_1x_2^{11}x_3^{25}x_4^6\xi_1\xi_2\xi_4 \\ &+ 1246752\,x_1x_2^{10}x_3^{26}x_4^6\xi_1\xi_3\xi_4 + 340200\,x_2^{11}x_3^{26}x_4^6\xi_2\xi_3\xi_4 \neq 0.\end{aligned}$$

The above expression is not identically zero. Therefore, ***the leading term $\mathcal{P}_1$ in the deformation $\mathcal{P}_0 \mapsto \mathcal{P}(\varepsilon) = \mathcal{P}_0 + \varepsilon\mathcal{P}_1 + \bar{o}(\varepsilon)$ destroys the property of bi-vector $\mathcal{P}(\varepsilon)$ to be Poisson*** at $\varepsilon \neq 0$ on all of $\mathbb{R}^4$.

The same compatibility test for $\mathcal{P}_0$ and its second flow (2b) yields that

$$\begin{aligned}[\![\mathcal{P}_0, \mathcal{P}_2]\!] = &-7668\,x_1x_2^{11}x_3^{26}x_4^5\xi_1\xi_2\xi_3 - 142128\,x_1x_2^{11}x_3^{25}x_4^6\xi_1\xi_2\xi_4 \\ &- 207792\,x_1x_2^{10}x_3^{26}x_4^6\xi_1\xi_3\xi_4 - 56700\,x_2^{11}x_3^{26}x_4^6\xi_2\xi_3\xi_4.\end{aligned}$$

Again, this expression does not vanish identically on all of the Poisson manifold $\big(\mathbb{R}^4, \{\cdot,\cdot\}_{\mathcal{P}_0}\big)$. We conclude that neither of two flows (2) preserve the property of bi-vector $\mathcal{P}(\varepsilon)$ to stay (infinitesimally) Poisson at $\varepsilon \neq 0$ for this example of Poisson bi-vector.[7]

[7] Let us also inspect whether the Jacobi identity holds for any of the bi-vectors $\mathcal{P}_1$ and $\mathcal{P}_2$. For $\mathcal{P}_1$ we have that the left-hand side of the Jacobi identity is equal to

$$[\![\mathcal{P}_1, \mathcal{P}_1]\!] = -2963589120 \cdot \big(x_1x_2^{17}x_3^{41}x_4^8\xi_1\xi_2\xi_3 + 5\,x_1x_2^{17}x_3^{40}x_4^9\xi_1\xi_2\xi_4 - 2\,x_1x_2^{16}x_3^{41}x_4^9\xi_1\xi_3\xi_4\big),$$

which does not vanish. For $\mathcal{P}_2$ the left-hand side of the Jacobi identity equals

$$[\![\mathcal{P}_2, \mathcal{P}_2]\!] = -262517760 \cdot \big(x_1x_2^{17}x_3^{41}x_4^8\xi_1\xi_2\xi_3 + 5\,x_1x_2^{17}x_3^{40}x_4^9\xi_1\xi_2\xi_4 - 2\,x_1x_2^{16}x_3^{41}x_4^9\xi_1\xi_3\xi_4\big).$$

This expression also does not vanish, so that neither $\mathcal{P}_1$ nor $\mathcal{P}_2$ are Poisson bi-vectors.

Remark 2. In the above example, the Schouten brackets $[\![\mathcal{P}_0, \mathcal{P}_1]\!]$ and $[\![\mathcal{P}_0, \mathcal{P}_2]\!]$ are determined by the same polynomials in the variables $\boldsymbol{x}$ and $\boldsymbol{\xi}$: we see that $[\![\mathcal{P}_0, \mathcal{P}_1]\!] = -6 \cdot [\![\mathcal{P}_0, \mathcal{P}_2]\!]$. This implies that for this example of Poisson bi-vector $\mathcal{P}_0$, the leading term $\mathcal{Q} := \mathcal{P}_1 + 6\mathcal{P}_2$ does (infinitesimally) preserve the property of $\mathcal{P}(\varepsilon)$ to be Poisson in the course of deformation $\mathcal{P}_0 \mapsto \mathcal{P}_0 + \varepsilon\mathcal{Q} + \bar{o}(\varepsilon)$.

Moreover, it is readily seen that the ratio 1 : 6 is the *only* way to balance the two flows, (2a) vs (2b), such that their nontrivial linear combination $\mathcal{Q}$ is compatible with the Poisson bi-vector $\mathcal{P}_0$ from Example 2.[8]

Remark 3. In Example 4 the linear combination $\mathcal{Q} = \mathcal{P}_1 + 6\mathcal{P}_2 \neq 0$ of two flows (2) is not identically equal to zero. (For other examples this may happen incidentally.) The leading term $\mathcal{Q}$ in the infinitesimal deformation $\mathcal{P}_0 \mapsto \mathcal{P}_0 + \varepsilon\mathcal{Q} + \bar{o}(\varepsilon)$ is trivial in the Poisson cohomology with respect to $\boldsymbol{\partial}_{\mathcal{P}_0}$, i. e. $\mathcal{Q} = [\![\mathcal{P}_0, \mathcal{X}]\!]$ for some vector $\mathcal{X}$ on the four-dimensional space.[9] Hence this $\mathcal{Q}$ is trivially compatible with the Poisson bi-vector $\mathcal{P}_0$: namely, $[\![\mathcal{P}_0, \mathcal{Q}]\!] \equiv 0$, see p. 8 below.

In the three tables below we summarise the results about the flows $\mathcal{P}_1$ and $\mathcal{P}_2$, which we evaluate at the examples of Poisson bi-vectors $\mathcal{P}_0$. Special attention is paid to the leading deformation term $\mathcal{Q} = \mathcal{P}_1 + 6\mathcal{P}_2$ in each case: we inspect whether this bi-vector incidentally vanishes and whether it is (indeed, always) compatible with the original Poisson structure $\mathcal{P}_0$.

Table 1. The Poisson bi-vectors $\mathcal{P}_0$ are generated using the determinant method from section 2.1 (the dimension is equal to 3, so we specify the fixed argument g_1); that generator is combined with the pre-multiplication $(f\cdot)$ as explained in section 2.2.

№	dim	Argument & pre-factor	$[\![\mathcal{P}_0, \mathcal{P}_1]\!] = 0$?	$\mathcal{P}_2 \overset{?}{=} 0$	$[\![\mathcal{P}_0, \mathcal{P}_2]\!] = 0$?	$\mathcal{Q} \overset{?}{=} 0$	$[\![\mathcal{P}_0, \mathcal{Q}]\!] = 0$?
1.	3	$[x_1^5 x_2^3 x_3^4 + x_1^2 x_3^5 + x_1 x_2^5 x_3]$ $x_1^3 + x_2^2$	✗	✗	✗	✗	✓
2.	3	$[x_1 x_2 + x_1 x_3 + x_2 x_3]$ $x_1^2 + x_2$	✗	✗	✗	✗	✓

For both examples in Table 1 we have that neither does $\mathcal{P}_1$ preserve the property of $\mathcal{P}_0 + \varepsilon\mathcal{P}_1 + \bar{o}(\varepsilon)$ to be (infinitesimally) Poisson nor does $\mathcal{P}_2$ vanish identically — which is in contrast with both the claims from [1].

Table 2. In dimensions higher than 3, we generate the Poisson bi-vectors $\mathcal{P}_0$ by using the determinant method from section 2.1: the auxiliary arguments $g_1, \ldots, g_{n-2}$ are specified.

№	dim	Arguments	$[\![\mathcal{P}_0, \mathcal{P}_1]\!] = 0$?	$\mathcal{P}_2 \overset{?}{=} 0$	$[\![\mathcal{P}_0, \mathcal{P}_2]\!] = 0$?	$\mathcal{Q} \overset{?}{=} 0$	$[\![\mathcal{P}_0, \mathcal{Q}]\!] = 0$?
3.	4	$[x_2^3 x_3^2 x_4, x_1 x_3^4 x_4]$	✗	✗	✗	✗	✓
4.	4	$[x_1^2 x_2^3 x_3^4 x_4^5, x_1 x_2 x_3 x_4]$	✗	✗	✗	✓	✓
5.	4	$[x_2^2 x_3^2 x_4^2, x_1^2 x_3^2 x_4^2]$	✗	✗	✗	✓	✓
6.	5	$[x_2^3 x_3^2 x_4, x_1 x_3^4 x_4, x_3^3 x_4^2 x_5^4]$	✗	✗	✗	✗	✓

[8] The balance $1 : \frac{4}{3}$ was considered in [22, §5.2] for the linear combination of flows (2a) and (2b), respectively.

[9] In all the two-dimensional Poisson geometries, the first flow $\mathcal{P}_1$ is always cohomologically trivial, i.e. it is of the form $\mathcal{P}_1 = [\![\mathcal{P}_0, \mathcal{X}]\!]$ for some one-vector $\mathcal{X}$, see [1].

IOP Conf. Series: Journal of Physics: Conf. Series **804** (2017) 012008 doi:10.1088/1742-6596/804/1/012008

In Table 2 we again have that neither is the property to be (infinitesimally) Poisson preserved for $\mathcal{P}_0 + \varepsilon\mathcal{P}_1 + \bar{o}(\varepsilon)$ nor is the bi-vector $\mathcal{P}_2$ vanishing identically.

Table 3. The results for the Vanhaecke method from section 2.3: we here specify the bivariate polynomials ϕ.

№	dim	$\phi(x,y)$	$[\![\mathcal{P}_0,\mathcal{P}_1]\!] \overset{?}{=} 0$	$\mathcal{P}_2 \overset{?}{=} 0$	$[\![\mathcal{P}_0,\mathcal{P}_2]\!] \overset{?}{=} 0$	$\mathcal{Q} \overset{?}{=} 0$	$[\![\mathcal{P}_0,\mathcal{Q}]\!] \overset{?}{=} 0$
7.	4	$[x^2y^2]$	✗	✗	✗	✗	✓
8.	4	$[x^2y]$	✗	✗	✗	✗	✓
9.	4	$[x^3y^2]$	✗	✗	✗	✗	✓
10.	4	$[x^3y^3]$	✗	✗	✗	✗	✓
11.	6	$[x^2y^2]$	✗	✗	✗	✗	✓

The entries in Table 3 report on the use of the generator from section 2.3: experimentally established, the properties of these Poisson bi-vectors do not match both the claims from [1].

Table 4. The results for the infinite-dimensional case.

№	dim	Operator	$[\![\boldsymbol{\mathcal{P}}_0,\boldsymbol{\mathcal{P}}_1]\!] \overset{?}{\cong} 0$	$\boldsymbol{\mathcal{P}}_2 \overset{?}{\cong} 0$
12.	∞	$u^2 \circ \mathrm{d}/\mathrm{d}x \circ u^2$	✗	✓

The variational bi-vector $\boldsymbol{\mathcal{P}}_1 = \frac{1}{2}\langle \xi \cdot \vec{A}_1(\xi)\rangle$, which we construct from the variational Poisson bi-vector $\boldsymbol{\mathcal{P}}_0 = \frac{1}{2}\langle \xi \cdot u^2\, \vec{\mathrm{d}}/\mathrm{d}x(u^2\,\xi)\rangle$ by using the geometric technique from [13] (see also [14]), is determined by the (skew-adjoint part of the) first-order differential operator $A_1 = 192\,(9u^8 u_x u_{xx} - u^9 u_{xxx})\,\mathrm{d}/\mathrm{d}x$ in total derivatives. Again (see Table 4), the two variational bi-vectors are *not* compatible: we check that $[\![\boldsymbol{\mathcal{P}}_0,\boldsymbol{\mathcal{P}}_1]\!] \not\cong 0$ under the variational Schouten bracket. Remarkably, the variational bi-vector $\boldsymbol{\mathcal{P}}_2$ is specified by the second-order total differential operator whose skew-adjoint component vanishes, whence the respective variational bi-vector *is* equal to zero (modulo exact terms within its horizontal cohomology class [11]).

Conclusion

The linear combination $\mathcal{Q} = \mathcal{P}_1 + 6\mathcal{P}_2$ of the Kontsevich tetrahedral flows preserves the space of Poisson bi-vectors $\mathcal{P}_0$ under the infinitesimal deformations $\mathcal{P}_0 \mapsto \mathcal{P}_0 + \varepsilon\mathcal{Q} + \bar{o}(\varepsilon)$. This is manifestly true for all the examples of Poisson bi-vectors on finite-dimensional (vector or affine) spaces $\mathbb{R}^n$ which we have considered so far. We conjectured that the leading deformation term $\mathcal{Q} = \mathcal{Q}(\mathcal{P}_0)$ always has this property, that is, ***the bi-vector*** $\mathcal{Q}$ ***marks a*** $\boldsymbol{\partial}_{\mathcal{P}_0}$***-cohomology class for every Poisson bi-vector*** $\mathcal{P}_0$ on a finite-dimensional affine manifold. (Recall that such class can be $\boldsymbol{\partial}_{\mathcal{P}_0}$-trivial; moreover, the bi-vector $\mathcal{Q}$ can vanish identically — yet the above examples confirm the existence of Poisson geometries where neither of the two options is realised.)

Let us conclude that every claim of an object's vanishing by virtue of the skew-symmetry and Jacobi identity for a given Poisson bi-vector, which that object depends on by construction, must be accompanied with an explicit description of that factorisation mechanism (e.g., see [5]) or at least, with a proof of that mechanism's existence. Apart from the trivial case (here, $\mathcal{Q} = 0$ so that $[\![\mathcal{P}_0,\mathcal{Q}]\!] \equiv 0$), such factorisation through the master-equation $[\![\mathcal{P}_0,\mathcal{P}_0]\!] = 0$ can be immediate: here, we have that $[\![\mathcal{P}_0,\mathcal{Q}]\!] = [\![\mathcal{P}_0,[\![\mathcal{P}_0,\mathcal{X}]\!]]\!] = \frac{1}{2}[\![[\![\mathcal{P}_0,\mathcal{P}_0]\!],\mathcal{X}]\!] = (\frac{1}{2}[\![\cdot,\mathcal{X}]\!])\,([\![\mathcal{P}_0,\mathcal{P}_0]\!])$ for all $\boldsymbol{\partial}_{\mathcal{P}_0}$-exact infinitesimal deformations $\mathcal{Q} = \boldsymbol{\partial}_{\mathcal{P}_0}(\mathcal{X})$ of the Poisson bi-vectors $\mathcal{P}_0$. Elaborated in [5], the Poisson cohomology estimate mechanism of the vanishing $[\![\mathcal{P}_0,\mathcal{Q}]\!] \doteq 0$ via $[\![\mathcal{P}_0,\mathcal{P}_0]\!] = 0$

works – for the nontrivial cocycles $\mathcal{Q} \notin \operatorname{im} \boldsymbol{\partial}_{\mathcal{P}_0}$ in the $\boldsymbol{\partial}_{\mathcal{P}_0}$-cohomology – due to much more refined principles. That vanishing mechanism is applied to the factorisation problem at hand in the paper [6] (joint with R. Buring), where we prove the above conjecture.

Acknowledgments
The second author thanks the Organizing committee of XXIV International conference 'Integrable systems & quantum symmetries' (13 – 19 June 2016; CVUT Prague, Czech Republic) for a warm atmosphere during the meeting. The research of A. B. was supported by JBI RUG project 190.135.021; A. K. was supported by NWO grant VENI 639.031.623 (The Netherlands) and JBI RUG project 106552 (Groningen). A. B. thanks R. Buring for fruitful cooperation; A. K. thanks M. Kontsevich for posing the problem and stimulating discussion.

Appendix A. The mechanism of vanishing for $[\![\mathcal{P}, \mathcal{Q}_{1:6}(\mathcal{P})]\!] = 0$: an example

We wish to recognize the differential consequences of the Jacobi identity in the compatibility equation $[\![\mathcal{P}, \mathcal{Q}_{1:6}(\mathcal{P})]\!] = 0$, to understand why it holds. By a straightforward calculation we learn that $[\![\mathcal{P}, \mathcal{Q}_{1:6}(\mathcal{P})]\!] = 0$ for all Poisson bi-vectors on $\mathbb{R}^3$. But as soon as the differential consequences of the Jacobi identity are recognized, they can be translated into graphs. Independent of dimension, the language of graphs then answers the question which we started out with. This answer is found in [6].

Let us now illustrate a more analytic approach to the factorization problem for $[\![\mathcal{P}, \mathcal{Q}_{1:6}]\!] = 0$ via $[\![\mathcal{P}, \mathcal{P}]\!] = 0$ (see [6, App. D] for details). The compatibility equation is a vanishing expression, which is impossible to factorize through the Jacobi identity, which itself is also zero. To make both visible, we perturb a given Poisson bi-vector $\mathcal{P}$ using $\tilde{\mathcal{P}} = \mathcal{P} + \epsilon \cdot \Delta$ for a bi-vector Δ, in such a way that $\tilde{\mathcal{P}}$ is no longer Poisson, thereby $[\![\tilde{\mathcal{P}}, \tilde{\mathcal{P}}]\!] \neq 0$. The goal is to perturb the bi-vector $\mathcal{P}$ such that the left-hand side $[\![\tilde{\mathcal{P}}, \tilde{\mathcal{Q}}_{1:6}]\!]$ becomes non-zero as well. Now the Jacobi identity's non-zero differential consequences becomes recognizable in the non-zero expression $[\![\tilde{\mathcal{P}}, \tilde{\mathcal{Q}}_{1:6}]\!]$.

Example 5. Consider the Poisson bi-vector obtained on $\mathbb{R}^3$ from the determinant construction using two functions $g(z)$ and $f(x)$ as argument and pre-multiplication factor, respectively. Let the perturbation Δ be given component-wise by $\Delta^{12} = f_1(y,z)$, $\Delta^{13} = f_2(y,z)$ and $\Delta^{23} = 0$. The perturbed bi-vector then equals

$$\tilde{\mathcal{P}} = \begin{bmatrix} 0 & f \cdot \mathrm{d}g/\mathrm{d}z & 0 \\ -f \cdot \mathrm{d}g/\mathrm{d}z & 0 & 0 \\ 0 & 0 & 0 \end{bmatrix} + \epsilon \cdot \begin{bmatrix} 0 & f_1 & f_2 \\ -f_1 & 0 & 0 \\ -f_2 & 0 & 0 \end{bmatrix}.$$

The left-hand sides of the Jacobi identity and of the compatibility condition are evaluated to

$$[\![\tilde{\mathcal{P}}, \tilde{\mathcal{P}}]\!]^{123} = \epsilon f_2 \cdot \frac{\mathrm{d}f}{\mathrm{d}x}\frac{\mathrm{d}g}{\mathrm{d}z} + \bar{o}(\epsilon), \qquad [\![\tilde{\mathcal{P}}, \tilde{\mathcal{Q}}]\!]^{123} = -\epsilon \cdot \frac{\partial^3 f_2}{\partial y^3}\left(\frac{\mathrm{d}f}{\mathrm{d}x}\right)^4 \left(\frac{\mathrm{d}g}{\mathrm{d}z}\right)^4 + \bar{o}(\epsilon).$$

There is only one way to recognize a differential consequence of the Jacobiator inside $[\![\tilde{\mathcal{P}}, \tilde{\mathcal{Q}}_{1:6}]\!]^{123}$. Namely, the Jacobi identity contains a product of f_2 and derivatives of f and g. The same is true for its non-zero differential consequences. Let us extract this product from $[\![\tilde{\mathcal{P}}, \tilde{\mathcal{Q}}_{1:6}]\!]^{123}$. The only differential consequences of f_2, $\mathrm{d}f/\mathrm{d}x$, and $\mathrm{d}g/\mathrm{d}y$ in $[\![\mathcal{P}, \mathcal{Q}_{1:6}]\!]^{123}$ are $\partial^3 f_2/\partial y^3$, $\mathrm{d}f/\mathrm{d}x$ and $\mathrm{d}g/\mathrm{d}z$, respectively. This hints that we have the differential consequence $[\![\mathcal{P}, \mathcal{P}]\!]^{123}_{yyy}$. To understand what its coefficient is, we note that the remaining co-factors in $[\![\tilde{\mathcal{P}}, \tilde{\mathcal{Q}}_{1:6}]\!]^{123}$ form $(\mathcal{P}^{12}_x)^3$. We conclude that the left-hand side of the compatibility equation factorizes through the Jacobi identity as follows

$$[\![\mathcal{P}, \mathcal{Q}_{1:6}]\!]^{123} = \mathcal{P}^{12}_x \mathcal{P}^{12}_x \mathcal{P}^{12}_x [\![\mathcal{P}, \mathcal{P}]\!]^{123}_{yyy} + \cdots.$$

Looking at this expression, we construct a list of graphs that can encode it. Such a list fully formed, it is subtracted from $[\![\mathcal{P}, \mathcal{Q}_{1:6}]\!]$ and resolved with respect to the coefficients of every proposed graph. We keep subtracting the already found graphs from any non-zero perturbations of $[\![\mathcal{P}, \mathcal{Q}_{1:6}]\!]$ in the future, once the coefficients are known. The example under study gave us the tripod graph, which is the first entry in [6, Eq. (6)]. Proceeding in the same way, we also recognized the 'elephant' graph, which is the sixth entry in that solution (cf. [6, Remarks 10–11]).

References

[1] Kontsevich M (1997) Formality conjecture *Deformation theory and symplectic geometry (Ascona 1996)* Math. Phys. Stud. **20** (Dordrecht: Kluwer Acad. Publ.) 139–156

[2] Kontsevich M (2003) Deformation quantization of Poisson manifolds *Lett. Math. Phys.* **66**:3 157–216 (*Preprint* `q-alg/9709040`)

[3] Kontsevich M (1995) Homological algebra of mirror symmetry *Proc. Intern. Congr. Math.* **1** (Basel: Birkhäuser) 120–139

[4] Kontsevich M (1994) Feynman diagrams and low-dimensional topology *First Europ. Congr. of Math.* **2** (Paris, 1992) Progr. Math. **120** (Basel: Birkhäuser) 97–121

[5] Buring R and Kiselev A V (2017) On the Kontsevich $\star$-product associativity mechanism *PEPAN Letters* **14**:2 accepted (*Preprint* `arXiv:1602.09036` [q-alg])

[6] Bouisaghouane A, Buring R and Kiselev A V (2016) The Kontsevich tetrahedral flows revisited *Preprint* `arXiv:1608.01710` (v2) [q-alg] 1–20

[7] Donin J (1998) On the quantization of quadratic Poisson brackets on a polynomial algebra of four variables *Lie Groups and Lie Algebras. Their representations, generalisations and applications* Math. Appl. **433** (Dordrecht: Kluwer Acad. Publ.) 17–25

[8] Grabowski J, Marmo G and Perelomov A M (1993) Poisson structures: towards a classification *Mod. Phys. Lett.* **A8**:18 1719–1733

[9] Vanhaecke P (1996) *Integrable systems in the realm of algebraic geometry* Lect. Notes Math. **1638** (Berlin: Springer–Verlag)

[10] Gerstenhaber M (1964) On the deformation of rings and algebras *Ann. Math.* **79** 59–103

[11] Olver P J (1993) *Applications of Lie groups to differential equations* Grad. Texts in Math. **107** (2nd ed.) (NY: Springer–Verlag)

[12] Wang J P (2002) A list of $1+1$ dimensional integrable equations and their properties *J. Nonlin. Math. Phys.* **9** suppl. 1 213–233

[13] Kiselev A V (2013) The geometry of variations in Batalin–Vilkovisky formalism *J. Phys.: Conf. Ser.* **474** 012024 1–51 (*Preprint* `1312.1262 [math-ph]`)

[14] Kiselev A V (2015) Deformation approach to quantisation of field models *Preprint* IHÉS/M/15/13 (Bures-sur-Yvette, France) 1–37

[15] Kiselev A V (2016) The right-hand side of the Jacobi identity: to be naught or not to be? *J. Phys.: Conf. Ser.* **670** 012030 1–17 (*Preprint* `arXiv:1410.0173` [math-ph])

[16] Vinogradov A and Vinogradov M (1998) On multiple generalizations of Lie algebras and Poisson manifolds *Secondary calculus and cohomological physics* Contemp. Math. **219** (Providence RI: AMS) 273–287

[17] Nambu Y (1973) Generalized Hamiltonian dynamics *Phys. Rev. D* **7** 2405–2412

[18] Omori H, Maeda Y and Yoshioka A (1993) A construction of a deformation quantization of a Poisson algebra *Geometry and its applications* (River Edge NJ: World Sci. Publ.) 201–218

[19] Kiselev A V (2015) The calculus of multivectors on noncommutative jet spaces *Preprint* IHÉS/M/14/39 (Bures-sur-Yvette, France) `arXiv:1210.0726` (v3) [math.DG] 1–41

[20] Fokas A S, Olver P J and Rosenau P (1997) A plethora of integrable bi-Hamiltonian equations *Algebraic aspects of integrable systems* Progr. Nonlin. Diff. Equations and Their Appl. **26** (Boston MA: Birkhäuser) 93–101

[21] Vodová J (2013) Low-order Hamiltonian operators having momentum *J. Math. Anal. Appl.* **401**:2 724–732 (*Preprint* `arXiv:1111.6434` [math-ph])

[22] Merkulov S A (2010) Exotic automorphisms of the Schouten algebra of polyvector fields *Preprint* `arXiv:0809.2385` (v6) [q-alg] 1–37

IOP Conf. Series: Journal of Physics: Conf. Series **804** (2017) 012009 doi:10.1088/1742-6596/804/1/012009

Supergroups in Critical Dimensions and Division Algebras

Čestmir Burdik[1,2,†], **Sultan Catto**[3,4,††], **Yasemin Gürcan**[5], **Amish Khalfan**[6,†††], **Levent Kurt**[5] **and V. Kato La**[7]

[1]Bogoliubov Laboratory of Theoretical Physics, JINR, 141980 Dubna, Russia
[2]Department of Mathematics, Faculty of Nuclear Sciences and Physical Engineering, Czech Technical University, Prague, Trojanova 13, CZ-120 00 Czech Republic
[3]Physics Department, The Graduate School, City University of New York, New York, NY 10016-4309
[4]Theoretical Physics Group, Rockefeller University, 1230 York Avenue, New York, NY 10021-6399
[5] Department of Science, Borough of Manhattan CC, The City University of NY, New York, NY 10007
[6] Physics Department, LaGuardia CC, The City University of New York, LIC, NY 11101
[7] Columbia University, New York, NY 10027

[†]Grant No. SG15/215/OHK4/3T/14 Czech Technical University in Prague
[††] Work supported in part by the PSC-CUNY Research Awards, and DOE contracts No. DE-AC-0276 ER 03074 and 03075; and NSF Grant No. DMS-8917754.
[†††] Work supported in part by the PSC-CUNY Research Award No:69379-00 47

Abstract.

We establish a link between classical heterotic strings and the groups of the magic square associated with Jordan algebras, allowing for a uniform treatment of the bosonic and superstring sectors of the heterotic string.

1. Division and Jordan Algebras, Magic Squares

This will be an introduction to the construction of the super-Poincaré group in special dimensions $d = 3, 4, 6, 10$, called critical dimensions in which it is possible to write supersymmetric classical Lagrangians for Yang-Mills theories and superstrings. It is known that the conventional quantization of superstrings[1] is only possible in $d = 10$. However, parastatistical quantization appears to be possible in the other critical dimensions[2]. In critical dimensions special algebraic structures occur. They are associated with the four division algebras of the Hurwitz's theorem[3], i.e. $\mathcal{R}$ (the real numbers), $\mathcal{C}$ (complex numbers), $\mathcal{H}$ (quaternions) and $\mathcal{O}$ (octonions or Cayley numbers). Their dimensions are respectively 1, 2, 4 and 8 or $d-2$ if d is the critical dimension. They have positive semi-definite quadratic multiplicative norms. The first two are commutative and associative. Quaternions are associative but not commutative and hence can be represented by 2×2 complex or 4×4 real matrices. Octonions are neither commutative nor associative and as a result have no direct matrix representations. However, their associator is totally antisymmetric, so that the octonion algebra is "alternative".

We shall make use of the norm groups and the automorphism groups of the Hurwitz algebras. The norm groups are formed by the linear transformations of the components of an element of the algebra that preserve the quadratic norm. For real numbers it it the discrete group Z_2. For complex numbers it is $O(2) \sim U(1)$. For quaternions it is $O(4) \sim SU(2) \times SU(2)$ and for octonions it is $O(8)$. The automorphism groups are the groups that leave invariant the multiplication table of the imaginary units of the algebra. It is Z_2 $(i \to -i)$ for complex numbers, $SU(2)$ for quaternions and the exceptional group G_2 for octonions. The automorphism groups, unlike the norm groups, leave the real (or scalar) part of an algebra element invariant. All Hurwitz algebras are power associative, so that if $N = n_1 + n_2 + \ldots + n_m$, we can write

$$x^N = x^{n_1} . \, x^{n_2} \, \ldots x^{n_m} \tag{1}$$

unambiguously and independently of the brackets in the nonassociative case. Besides alternative algebras Jordan algebras are also power associative. We now give a brief description of Jordan algebras.

Jordan algebras have elements that have a commutative product

$$a.b = b.a \tag{2}$$

which is not associative but is subject to the Jordan identity

$$[a \, b \, a^2] = 0 \,, \tag{3}$$

where the square bracket denotes the associator defined by

$$[a \, b \, c] = (a.b).c \; - \; a.(b.c) \,. \tag{4}$$

An example of a Jordan algebra is the algebra of observables with elements that are $n \times n$ hermitian complex matrices with the composition law

$$a.b = \frac{1}{2}\{a \, b\} = \frac{1}{2}(a \, b + b \, a) \,. \tag{5}$$

The identity (3) is satisfied. There is also the exceptional Jordan algebra J_3^8 of 3×3 octonionic matrices that are hermitian with respect to octonionic conjugation. Note that in a conjugation the sign of imaginary units is reversed. The commutative Jordan product in this case is still defined by (5) and the Jordan identity still holds.

The classical Lie groups can be regarded as the automorphism groups of Jordan algebras. If J_n^1, J_n^2 and J_n^4 denote the Jordan algebras of $n \times n$ hermitian matrices over $\mathcal{R}$, $\mathcal{C}$ and $\mathcal{H}$ respectively they have the corresponding automorphism groups $O(n)$, $SU(n)$ and $Sp(n)$. The automorphism group of the exceptional Jordan algebra is the exceptional group F_4. The automorphism groups leave the trace of the Jordan matrices invariant. Hence traceless hermitian matrices are modules for their automorphism groups. For 3×3 matrices we find the following groups and dimensions:

Division algebras	$\mathcal{R}$	$\mathcal{C}$	$\mathcal{H}$	$\mathcal{O}$
Jordan algebras	J_3^1	J_3^2	J_3^4	J_3^8
Dim. of traceless elements	5	8	14	26
Automorphism groups	$O(3)$	$SU(3)$	$Sp(3)$	F_4
Dim. of automorphism groups	3	8	21	52

These groups form the first line of the celebrated magic square of Rozenfeld, Freudenthal and Tits[5]. To proceed further we must introduce the Freudenthal algebra[6]. Consider a 3×3 complex octonionic matrix J that is hermitian with respect to octonionic conjugation. It is a 27-dimensional module (27) for the exceptional group E_6, which is a complex group like $SU(3)$. Since $27 \times \bar{27}$ contains the identity representation and 27×27 contains the $(\bar{27})$ representation of E_6 we can define a scalar product and a symmetric product associated with the mappings

$$J \times J^* \to \mathcal{C} \quad \text{and} \quad J \times J \to J^*, \qquad J^* \times J^* \to J\,. \tag{6}$$

The scalar product is defined by

$$(J\,, K^*) = \text{Tr}\,(J\,.K^*) \tag{7}$$

while the "Freudenthal product" of J and K, both belonging to the (27) representation of E_6 is obtained by first introducing

$$J \times J = J^{-1}\,\text{Det}\,J = J^2 - J\,\text{Tr}\,J - \frac{1}{2}\,I\,\text{Tr}\,(J^2 - J\,\text{Tr}\,J)\,, \tag{8}$$

and then defining

$$J \times K = \frac{1}{2}\,(J + K) \times (J + K) \; - \; \frac{1}{2}\,J \times J \; - \; \frac{1}{2}\,K \times K\,, \tag{9}$$

we find

$$J \times K = J.K - \frac{1}{2}\,J\,\text{Tr}\,K - \frac{1}{2}\,K\,\text{Tr}\,J - \frac{1}{2}\,I\,(\text{Tr}\,J.K - \text{Tr}\,J\,\text{Tr}\,K)\,. \tag{10}$$

If J and K transform like (27), then $J \times K$ transforms like $\bar{27}$ of E_6. Similarly $J^* \times K^*$ transforms like a (27). The group E_6 is an automorphism group for the combination of the scalar product (7) and the Freudenthal product (10).

If we consider automorphism groups of 3×3 matrices with entries that are combinations of complex numbers with the four division algebras, requiring hermiticity with respect to the latter only, and then define scalar and Freudenthal products as for the octonionic matrices, we find

	$\mathcal{R}$	$\mathcal{C}$	$\mathcal{H}$	$\mathcal{O}$
Automorphism group $\mathcal{C}$:	$SU(3)$	$SU(3) \times SU(3)$	$SU(6)$	E_6
Dimension of group	8	16	35	78
Complex Dim. of module	6	9	15	27

This is the second line of the magic square. Note that the traces of the 3×3 matrices are not invariant under the automorphism groups of the Freudenthal algebra, but Det J, Tr $(J.K^*)$ and Tr $(J^*.K)$ are invariant.

The combination of the scalar product and the Freudenthal product also allows for the definition of a triple product that is E_6 covariant. Given F and K that belong to the (27) representation we define

$$M = \{F\;K^*\;F\} = F\;\text{Tr}\,(K^*.F) - 2\,(F \times F) \times K^*\,. \tag{11}$$

Then M also transforms like a (27). In the case of ordinary matrices it can be expressed in terms of the usual matrix product as

$$M = FK^*F\,, \tag{12}$$

so that

$$\text{Det}\;M = (\text{Det}F)^2\;\text{Det}\;K^*\,, \tag{13}$$

a property that also holds for the exceptional Jordan algebra. Given three elements of the Freudenthal algebra F, K, L, we have the triple product:

IOP Conf. Series: Journal of Physics: Conf. Series **804** (2017) 012009 doi:10.1088/1742-6596/804/1/012009

$$N = \{F\ K^*\ L\} = \frac{1}{2}\{(F+L)\ K^*\ (F+L)\} - \frac{1}{2}\{F\ K^*\ F\} - \frac{1}{2}\{L\ K^*\ L\}\,. \tag{14}$$

In terms of the Jordan product it can be written in the form[4]

$$N = \{F\ K^*\ L\} = (F.K^*).L + F.(K^*.L) - K^*.(F.L)\,. \tag{15}$$

E_6 is also the automorphism group of this ternary algebra.

In the case of the exceptional group E_7 there is a 56-dimensional fundamental representation X consisting of two complex scalars α and β^* and two Jordan matrices J and K^* belonging respectively to the (27) and ($\bar{27}$) representations of the E_6 subgroup. Now the direct product of three 56 dimensional representations also contains a (56). Hence there is a ternary algebra of the 56-dimensional modules of E_7. There is also a scalar product of a (56) and its complex conjugate. E_7 can be regarded as the automorphism group of the combined algebra of ternary products and scalar products.

Ranging over the automorphism groups of the ternary algebra when the module is constructed out of the four division algebras we find the third line of the magic square

	$\mathcal{R}$	$\mathcal{C}$	$\mathcal{H}$	$\mathcal{O}$
Aut. group of the ternary prod.	$Sp(3)$	$SU(6)$	$SO(12)$	E_7
Dimension of group	21	35	66	133
Dimension of the module X	14	20	32	56

Finally we can generate groups by considering fractional linear representations of these modules[7], giving nonlinear (coset space) realizations of the exceptional groups F_4, E_6, E_7 and E_8 that form the last line of the magic square. They have the respective dimensions 52, 78, 133 and 248.

The Magic square groups also have many non compact forms. A typical example is the $SL(3,R)$ real form associated with $SU(3)$ and the non compact groups $SL(3,R)\times SL(3,R)$ and $SL(3,C)$ associated with the compact group $SU(3)\times SU(3)$. In general they are subgroups of the complexified forms of the compact groups. They appear in the couplings of scalar fields in supergravity[8],[9] and also in the algebras associated with non simply laced lattices[10]. We shall see that they are also connected with the representations of the supersymmetric kinematical groups in critical dimensions.

Here is one set of a magic square of non compact groups that are subgroups of the non compact group $E_{8(-24)}$. Note that the number in brackets represents the number of non compact generators minus that of compact generators.

$\mathcal{R}$	$\mathcal{C}$	$\mathcal{H}$	$\mathcal{O}$
$O(2,1)$	$SU(2,1)$	$Sp(2,1)$	$F_{4(-20)}$
$SL(3,R)$	$SL(3,C)$	$SU^*(6)$	$E_{6(-26)}$
$Sp(6,R)$	$SU(3,3)$	$SO^*(12)$	$E_{7(-25)}$
$F_{4(+4)}$	$E_{6(+2)}$	$E_{7(-5)}$	$E_{8(-24)}$

The first line can be replaced by the compact groups $O(3)$, $SU(3)$, $Sp(3)$ and F_4 which are also subgroups of the non-compact groups of the second line. The starred groups are quaternionic groups. $SU^*(6)$ is isomorphic to the linear group over quaternions $SL(3,H)$. All of these non compact groups occur in supergravity theories.

Another remarkable series of critical dimensions is given by J_2^D, the Jordan algebra of 2×2 hermitian matrices over the division algebras, with D= 1, 2, 4 and 8. They form subalgebras of J_3^8. Here are the dimensions of the algebras and their automorphism groups.

	$\mathcal{R}$	$\mathcal{C}$	$\mathcal{H}$	$\mathcal{O}$
Algebra	J_2^1	J_2^2	J_2^4	J_2^8
Real Dim. of alg.	3	4	6	10
Aut. group	$O(2)$	$O(3)$	$O(5)$	$O(9)$
[covering group]:	$[U(1)]$	$[SU(2)]$	$[Sp(2)]$	$[spin9]$

These groups are seen to be subgroups of the first lines of the magic squares for compact and non compact groups given above. To look for the corresponding subgroups of the second line of the non-compact magic square, we start from the well known case of J_2^2, the algebra of 2×2 hermitian complex matrices. An example is the matrix associated with the momentum p that can be written

$$p = p_0 + \vec{\sigma} \cdot \vec{p} = \begin{pmatrix} p_o + p_3 & p_1 - ip_2 \\ p_1 + ip_2 & p_o - p_3 \end{pmatrix} \tag{16}$$

in terms of the Pauli matrices. The unitary representations of the Poincaré group are built through the use of the little groups that leave the standard form of p invariant. In the time-like and space-like cases p can be brought along its time component or one space-like components respectively, being left invariant by the corresponding groups $SO(3)$ and $SO(2,1)$. These are just the automorphism groups of the first lines of the magic squares given above. Similarly, the momentum vectors in critical dimensions are represented by hermitian 2×2 matrices over the division algebras. They are $(D+2)$ dimensional vectors in the Minkowski spaces $(D+1,1)$. The transverse part of the vector $(p_\perp)$, relevant in the massless, light-like case is represented by an element of the division algebra of dimension D[11]. The Lorentz group in critical dimensions is given by linear, homogeneous transformations that preserve the hermiticity of p and its Minkowski norm given by

$$\text{Det}\, p = p\,\bar{p}, \qquad \bar{p} = -p + I\,\text{Tr}p\,, \tag{17}$$

with I being 2×2 identity matrix.

In $d = 4$ $(D = 2)$ dimensions it is given by

$$p' = L\,p\,L^\dagger, \qquad (\text{Det}L = 1)\,, \tag{18}$$

where L is a unimodular complex 2×2 matrix. Then we have

$$L\,\varepsilon\,SL(2,C) \sim SO(3,1)\,. \tag{19}$$

For $d = 3$ it is a real unimodular matrix representing $SL(2,R) \sim SO(2,1)$, while for $d = 6$ it is a quaternionic unimodular matrix $SL(2,H) \sim SO(5,1)$. For $d = 8$ the matrix is octonionic and the linear action on p must also involve associators[12]. The action with associators is contained in the automorphism group G_2, the 14 parameters of which must be added to the $4 \times 8 - 1 = 31$ parameters of a 2×2 octonionic matrix, giving a linear group of dimension 45, namely $O(9,1)$. Thus we can construct the following table:

Algebras	J_2^1	J_2^2	J_2^4	J_2^8
Minkowski dim.	3	4	6	10
Norm preserving group [Lorentz group]	$SL(2,R)$ $[\sim SO(2,1)]$	$SL(2,C)$ $[\sim SO(3,1)]$	$SL(2,H)$ $[\sim SO(5,1)]$	$Spin(9,1)$ $[\sim SO(9,1)]$
Little group for massless particle	T_1	$IO(2)$	$IO(4)$	$IO(8)$
Helicity group	$\cdots$	$O(2)$	$O(4)$	$O(8)$

In this table I means the inhomogeneous group, T_n is the translation group in n dimension and the helicity group is the homogeneous part of the little group for massless particles which preserves p_0+p_{d-1}. The helicity groups are seen to be the norm groups of the division algebras associated with the critical Minkowski dimensions. The Lorentz groups are seen to be subgroups of the second line of the magic square of the non compact groups, including $O(9,1)$ which is a subgroup of $E_{6(-26)}$. We have already seen that the little groups are subgroups of the first line of the same table.

2. Spinors in Critical Dimensions

For a massless particle the momentum is light-like, so that

$$p\,\bar{p} = \text{Det}\,p = 0 \tag{20}$$

and the hermitian matrix p factorizes in the form

$$p = \psi\,\psi^\dagger\,,\ \ \psi = \begin{pmatrix}\psi_1\\ \psi_2\end{pmatrix}\,,\ \ \psi^\dagger = (\bar{\psi}_1\ \ \bar{\psi}_2)\,. \tag{21}$$

Here the components ψ_1 and ψ_2 of the spinor ψ are elements of the division algebras. The bar denotes division algebra conjugation. Under a Lorentz transformation

$$\psi' = L\,\psi\,. \tag{22}$$

In the octonionic case ($d = 10$) it is understood that the automorphism group G_2 of octonions is adjoined to the left multiplication by L. The real dimensions of these spinors is $2D = 2(d-2)$ where D is the dimension of the division algebra. Let us also display the dimensions of the vector combined with a spinor, a combination that appears for representing a point in superspace.

Minkowski (vector dimension d)	3	4	6	10
Spinor dimension	2	4	8	16
Superspace dimension	5	8	14	26

The spinor dimensions correspond to those of a real spinor in $d = 3$, a Weyl or Majorana spinor in $d = 4$, a Weyl spinor in $d = 6$ and finally a spinor that is both Weyl and Majorana in $d = 10$. The superspace dimensions are the same as those for traceless 3×3 Jordan algebras of the first line of the magic square. This suggest a correspondence between a superspace point z (with vector component x and spinor component θ) and a matrix F of Jordan form

$$F = \begin{pmatrix} x & \theta \\ \theta^\dagger & a \end{pmatrix} \tag{23}$$

where the constant a is determined by a Lorentz invariant condition in terms of x and θ, such as

$$\text{Det}\, F = 1\,. \tag{24}$$

At this stage the correspondence is suggestive but not exact, as the components of θ for a superspace point are Grassmann numbers while they are real or complex numbers for an element of a Jordan algebra. To make the correspondence more precise we shall have to generalize the Jordan algebra to a graded Jordan algebra[13]. On the other hand, the 26-dimensional bosonic string vector can be represented by an element of the usual exceptional Jordan algebra of the form

$$J = \begin{pmatrix} y & u \\ u^\dagger & b \end{pmatrix}, \quad \text{Det}\, J \;=\; 1\,, \tag{25}$$

where y is a 2×2 hermitian matrix, b is a scalar and u a 2×1 matrix over octonions. When the octonions are real we have seen that J is a (27) module for $E_{6(-26)}$. J^{-1} and hence $J \times J$ is a $(\bar{27})$ module. It is known that the determinant of J is a multiplicative cubic norm for the exceptional Jordan algebra. We have

$$I\ \text{Det}\, J = (J \times J).\, J, \qquad \text{Det}\, J = \frac{1}{3}\ \text{Tr}\ [(J\times J).\, J]\,. \tag{26}$$

With the vector-spinor decomposition of J given in Eq. (25), using Eq. (8), one finds

$$J \times J \;=\; \begin{pmatrix} b\bar{y} - \overline{uu}^\dagger & -\bar{y}u \\ -u^\dagger\bar{y} & y\bar{y} \end{pmatrix} \tag{27}$$

and

$$\text{Det}\, J = b\,\bar{y}y - \text{Sc}\,(u^\dagger \bar{y} u)\,. \tag{28}$$

Note that the last formula is unambiguous because of the identity

$$\text{Sc}\,[u^\dagger\,(\bar{y}u)] \;=\; \text{Sc}\,[(u^\dagger\bar{y})\,u] \;=\; \frac{1}{2}\,(u^\dagger\bar{y})\,u \;+\; \frac{1}{2}\,u^\dagger\,(\bar{y}u)\,. \tag{29}$$

The bar notation over a 2×2 hermitian matrix has been defined as in eq.(17) by

$$y = \begin{pmatrix} y_+ & y_\perp \\ \bar{y}_\perp & y_- \end{pmatrix}, \qquad \bar{y} = \begin{pmatrix} y_- & -y_\perp \\ -\bar{y}_\perp & y_+ \end{pmatrix}, \tag{30}$$

and $u^\dagger$ has the same meaning as in eq.(21).

Let us emphasize that when J is a hermitian matrix over real octonions it corresponds to the (27) representation of $E_{6(-26)}$ and Det J is invariant under this noncompact group. Hence, it is also invariant with respect to its $O(9,1)$ subgroup under which y transforms like a vector, u like a Weyl-Majorana spinor and b like a scalar. There is an independent real $(\bar{27})$ representation of $E_{6(-26)}$ that has the same vector-spinor-scalar decomposition in 10-dimensional Minkowski space. Since the $(\bar{27})$ involves a right handed spinor instead of the left handed spinor u, these two representations will be assigned different chiralities.

We will leave out some further discussion on the Poincaré group and superPoincaré group in (D+1,1), the covariant superstring action, bosonic string formalism, and compactification and discretization for another publication due to lack of space. For further works on octonions we refer our reader to references at the end of this paper.

IOP Conf. Series: Journal of Physics: Conf. Series **804** (2017) 012009 doi:10.1088/1742-6596/804/1/012009

3. Remarks

We hope to have shown that there is a one to one correspondence between a graded matrix representation of the super Poincaré group and division algebras for Minkowski dimensions $(D+1, 1)$ in which classical superstring theories and classical super Yang-Mills theories exist, D denoting the dimension of the division algebra. On the other hand classical bosonic strings can exist in any dimension. However, if closed bosonic strings are associated with superstrings to form classical heterotic strings, then the dimension of the bosonic string must be $3D + 2$. The internal symmetry group of the classical heterotic string becomes a group of rank $2D$. Quantization only works for $D = 8$ (octonionic case) but paraquantization might be possible in the other cases[2]. The internal symmetry group was obtained by discretizing $2D$ elements of a certain subgroup of $O(3D + 1, 1)$. If a bosonized ghost dimension is added as in the case of covariant treatment, then the Lorentz group is extended to $O(3D + 1, 2)$. For $D = 8$, the complex Lorentz group admits as subgroup a real form of E_6 that has the same Lorentz subgroup $O(9, 1)$ as $O(25, 2)$. The free bosonic Lagrangian than becomes invariant under both $O(25, 2)$ and $E_{6(-26)}$. This approach establishes a link between classical heterotic strings and the groups of the magic square associated with Jordan algebras, and also allows a uniform treatment of the bosonic and superstring sectors of the heterotic string.

References

[1] For a review of Superstrings see M.B. Green, J.H. Schwarz and E.Witten, "Superstring Theory" (Cambridqe U. Press, 1987).

[2] F. Ardalan and F. Mansouri, Phys. Rev. D9, 3341 (1974); Phys. Rev. Lett. 56, 2456 (1986); Phys. Lett. B176, 99 (1986).

[3] For a review see R.D. Schafer, An Introduction to Nonassociative Algebras (Academic Press, New York 1966).

[4] For a review see N. Jacobson, "Structure and Representations of Jordan Algebras" (Am. Math. Soc. Providence 1968) and "Exceptional Lie Algebras" (M. Dekker, New York 1971).

[5] H. Freudenthal, Advances in Mathematics, vol. I., p. 145 (1962), J. Tits, Proc. Colloq. Utrecht, p. 175 (1962), B.A. Rozenfeld, Proc.Colloq. Utrecht, p. 135 (1962).

[6] See H. Freudenthal in Ref. 5 and Ch. 10 of second book in Ref. 4.

[7] M. Koecher, Am. J. Math. 89, 787 (1967) and Invent. Math. 3, 136 (1967).

[8] M. Günaydin, G. Sierra and P.K. Townsend, Nucl. Phys. B242, 244 (1984). For other approaches to the occurrence of the Magic Square in Supergravity see B. Julia in 5th John Hopkins Workshop on particle Theory, Eds. G. Domokos and S. Kövesi-Domokos, p. 23 (Baltimore, 1981), and also P. Truini, G. Olivieri and L.C. Biedenharn, Lett. in Math. Phys. 9, 255 (1985).

[9] S. Ferrara, P. Fre and L. Girardello, Nucl. Phys. B274, 600 (1986), see also C. A. Savoy, 1987 Capri Symposium, F. Buccella, ed.

[10] P. Goddard, W. Nahm, D. Olive and A. Schwimmer, Comm. Math. Phys. 107, 179 (1986). For the emergence of the octonion algebra see P. Goddard, W. Nahm, D.I. Olive, H. Ruegg and A. Schwimmer, "Fermions and Octonions", (Imp. Coll. reprint circa (1987)).

[11] T. Kugo and P. Townsend, Nucl. Phys B221, 357, (1983), A. Sudbery, Math. Gen. 17, 939 (1984).

[12] R. Dündarer, F. Gürsey and H.C. Tze, Nucl. Phys. B266, 440 (1986).

[13] F. Gürsey and L. Marchildon, J. Math. Phys. 19,942 (1978), and Phys. Rev. D17, 2038 (1978). F. Gürsey, "Chiral Structures in Particle Physics" Proc. Tokyo Symposium 1987 in honor of Prof. K. Nishijima, Eds. A. Ukawa and K. Kawarabayashi.

[14] S. Catto, Y. Gürcan, A. Khalfan and L. Kurt, *Journal of Physics,***411** (2013) 012009

[15] S. Catto, Y. Gürcan, A. Khalfan and L. Kurt, *Journal of Physics,***563** (2014) 0120096

[16] S. Catto, Y. Gürcan, A. Khalfan and L. Kurt, *Journal of Physics,***670** (2016) 012016

[17] S. Catto and F. Gürsey, *Nuovo Cimento* **A 86** *(1985) 201*

[18] S. Catto and F. Gürsey, *Nuovo Cimento* **A99** *(1988) 685*

[19] C. Burdik, S.Catto, Y. Gürcan, A. Khalfan and L. Kurt, *Physics of Elementary Particles and Atomic Nuclei, Theory* in print 2016.

Acknowledgments:

One of us (SC) would like to thank Professor Čestmir Burdik for the invited talk given in Prague and for his kind hospitality. We would like to thank Professors Vladimir Akulov, Itzhak Bars and Francesco Iachello for enlightening conversations.

D2-brane within coset approach

Č. Burdík[1] **and S. Krivonos**[2]
[1]Department of Mathematics, Faculty of Nuclear Sciences and Physical Engineering, Czech Technical University, Prague
[2]Bogoliubov Laboratory of Theoretical Physics, JINR, 141980 Dubna, Russia

E-mail: burdices@kmlinux.fjfi.cvut.cz, krivonos@theor.jinr.ru

Abstract. In this notes, we derive the component on-shell action of the space-filling D2-brane, i.e. $N = 1$, $d = 3$ supersymmetric Born-Infeld action within the nonlinear realization approach. The Bianchi identity admits direct covariantization with respect to the broken $N = 1, d = 3$ supersymmetry. In the corresponding on-shell component action all fermionic terms combine into covariant, with respect to broken supersymmetry, objects: covariant derivatives and fierbein and therefore the D2-brane component action has very simple form. Similarly to the cases of p-branes, it mimics the bosonic Born-Infeld action.

1. Introduction

The main property of the D-branes system considered within the nonlinear realization approach is non Goldstone nature of the field strengths. In other words, they never show up as the parameters of the coset space. Instead, they are hidden inside the higher in θ's-components of the superfields - fermionic or bosonic which select the given supermultiplet. Thus, the main problem in the description of such systems is to find the proper covariant Bianchi identities generalizing the linear ones for the cases with additional spontaneously broken supersymmetries. Among different D-branes the simplest one is the space filling $D2$ brane in $D = 3$. The superfield action of the $D2$ brane has been constructed in [1] within linear realization approach. Then, the component action within the nonlinear realization approach has been obtained by the duality transformation from the action of super membrane in $D = 4$ [2]. However, while performing the duality transformation we have lost the information about transformation properties of the fields and, therefore, one cannot check the invariance of the action explicitly. Now we are going to re-construct the component action of $D2$ brane from first principles.

2. D2 brane

2.1. Kinematics

The $D2$ brane action provides a $N = 2$ superextension of $d = 3$ Born-Infeld one and it corresponds to the partial breaking of $N = 2$, $d = 3$ supersymmetry to $N = 1$, $d = 3$ one. Thus, we are starting from $N = 2, d = 3$ Poincaré superalgebra

$$\{Q_a, Q_b\} = P_{ab}, \qquad \{S_a, S_b\} = P_{ab}, \tag{1}$$

$a, b = 1, 2$ being the $d = 3$ sl_2 spinor indices[1] The generators P_{ab}, Q_a and S_a transform in a standard way under $d = 3$ Lorentz group consisting of the generators M_{ab}. All these symmetries

[1] The indices are raised and lowered as follows: $V^a = \epsilon^{ab} V_b, V_a = \epsilon_{ab} V^b$, $\epsilon_{ab}\epsilon^{bc} = \delta_a^c$.

can be realized by the left multiplication of the coset element g defined as

$$g = e^{x^{ab}P_{ab}} e^{\theta^a Q_a} e^{\psi^a S_a}. \tag{2}$$

Here, x^{ab}, θ^a are the $N=1$, $d=3$ superspace coordinates, while the $\psi^a \equiv \psi^a(x,\theta)$ is a Goldstone superfield. Thus, Q supersymmetry remains unbroken, while the S-supersymmetry is spontaneously broken. Acting on the coset element (2) from the left by different elements of the $N=2, d=3$ Poincaré supergroup, one may find the transformation properties of the coset coordinates:

- Q-supersymmetry

$$\delta x^{ab} = -\frac{1}{4}\epsilon^a\theta^b - \frac{1}{4}\epsilon^b\theta^a \equiv -\frac{1}{2}\epsilon^{(a}\theta^{b)}, \qquad \delta\theta^a = \epsilon^a, \tag{3}$$

- S-supersymmetry

$$\delta x^{ab} = -\frac{1}{2}\varepsilon^{(a}\psi^{b)}, \qquad \delta\psi^a = \varepsilon^a. \tag{4}$$

The Cartan forms which specify the local geometric properties of the system are defined in a standard way as

$$g^{-1}dg = \mathrm{i}\omega_P^{ab}P_{ab} + \mathrm{i}\omega_Q^a Q_a + \mathrm{i}\omega_S^a S_a \tag{5}$$

and have a very simple form:

$$\omega_P^{ab} = dx^{ab} + \frac{1}{2}\theta^{(a}d\theta^{b)} + \frac{1}{2}\psi^{(a}d\psi^{b)}, \quad \omega_Q^a = d\theta^a, \ \omega_S^a = d\psi^a. \tag{6}$$

Using the covariant differentials $\omega_P^{ab}, d\theta^a$ (6), one may construct the covariant derivatives

$$\nabla_{ab} = \left(E^{-1}\right)_{ab}^{cd}\partial_{cd}, \quad \nabla_a = D_a + \frac{1}{2}\psi^b D_a\psi^c \nabla_{bc} = D_a + \frac{1}{2}\psi^b\nabla_a\psi^c\partial_{bc}, \tag{7}$$

where

$$D_a = \frac{\partial}{\partial\theta^a} + \frac{1}{2}\theta^b\partial_{ab}, \ \{D_a, D_b\} = \partial_{ab}, \qquad E_{ab}^{cd} = \delta_a^{(c}\delta_b^{d)} + \frac{1}{2}\psi^{(c}\partial_{ab}\psi^{d)}. \tag{8}$$

They obey the following algebra

$$\begin{aligned} &\{\nabla_a, \nabla_b\} = \nabla_{ab} + \nabla_a\psi^m\nabla_b\psi^n\nabla_{mn}, \\ &[\nabla_{ab}, \nabla_c] = \nabla_{ab}\psi^m\nabla_c\psi^n\nabla_{mn}, \\ &[\nabla_{ab}, \nabla_{cd}] = -\nabla_{ab}\psi^m\nabla_{cd}\psi^n\nabla_{mn}. \end{aligned} \tag{9}$$

2.2. Bianchi identities

To select a irreducible supermultiplet one has to impose on the superfield ψ^a the covariant version of the standard constraint selecting $N=1, d=3$ vector supermultiplet

$$D_a\psi^a = 0. \tag{10}$$

Keeping in the mind that the Cartan forms (6) are invariant with respect to $N=2, d=3$ super Poincaré group, one may immediately conclude that the proper version of the irreducibility constraint is

$$\nabla_a\psi^a = 0. \tag{11}$$

IOP Conf. Series: Journal of Physics: Conf. Series **804** (2017) 012010 doi:10.1088/1742-6596/804/1/012010

Similarly to the flat case (10), the constraint (11) leaves in the superfield $\boldsymbol{\psi}^a$ the following components

$$\psi^a = \boldsymbol{\psi}^a|_{\theta=0}, \qquad \lambda_{ab} = -\frac{1}{2}\nabla_{(a}\boldsymbol{\psi}_{b)}|_{\theta=0}. \tag{12}$$

In addition, the same constraint (11) has to impose, similarly to the flat case, the covariant version of Bianchi identities to be able to treat the component λ_{ab} as the field strength. To get this additional restriction one has to hit the basic constraint (11) by $\nabla^b\nabla_b$ which will result in the equation

$$\nabla^b\{\nabla_b,\nabla_a\}\boldsymbol{\psi}^a + \frac{1}{2}\{\nabla_a,\nabla_b\}\nabla^a\boldsymbol{\psi}^b = 0. \tag{13}$$

Using the explicit form of the covariant derivatives (7) and their commutators (9), after some rather straightforward calculations one may represent the equation (13) in the form

$$\nabla_{ab}\left[\frac{\nabla^a\boldsymbol{\psi}^b}{1-\frac{1}{2}(\nabla\boldsymbol{\psi})^2}\right] + \frac{\nabla^a\boldsymbol{\psi}^b}{1-\frac{1}{2}(\nabla\boldsymbol{\psi})^2}\nabla_{ab}\boldsymbol{\psi}^m\,\nabla_{mn}\boldsymbol{\psi}^n = 0, \tag{14}$$

where $(\nabla\boldsymbol{\psi})^2 = \nabla^a\boldsymbol{\psi}^b\,\nabla_a\boldsymbol{\psi}_b$. Now, from the definition of the covariant derivatives ∇_{ab} (7) and the explicit form of the commutator $[\nabla_{ab},\nabla_{cd}]$ (9) it follows that

$$\nabla_{ab}\left(E^{-1}\right)_{cd}^{kl} E_{kl}^{mn} - \nabla_{cd}\left(E^{-1}\right)_{ab}^{kl} E_{kl}^{mn} = -\nabla_{ab}\boldsymbol{\psi}^{(m}\,\nabla_{cd}\boldsymbol{\psi}^{n)} \tag{15}$$

and, therefore

$$E_{kl}^{mn}\nabla_{ab}\left(E^{-1}\right)_{mn}^{kl} - \partial_{mn}\left(E^{-1}\right)_{ab}^{mn} = -\nabla_{ab}\boldsymbol{\psi}^m\,\nabla_{mn}\boldsymbol{\psi}^n. \tag{16}$$

Keeping in the mind that

$$\begin{aligned}\partial_{cd}\left[\det E\,\left(E^{-1}\right)_{ab}^{cd}\right] &= \det E\left[\left(E^{-1}\right)_{mn}^{kl}\partial_{cd}E_{kl}^{mn}\left(E^{-1}\right)_{ab}^{cd} + \partial_{cd}\left(E^{-1}\right)_{ab}^{cd}\right] = \\ &= \det E\left[\left(E^{-1}\right)_{mn}^{kl}\nabla_{ab}E_{kl}^{mn} + \partial_{cd}\left(E^{-1}\right)_{ab}^{cd}\right] = \\ &= \det E\left[\nabla_{ab}\boldsymbol{\psi}^m\,\nabla_{mn}\boldsymbol{\psi}^n\right],\end{aligned} \tag{17}$$

where the last equality follows from (16), we may rewrite the equation (14) as

$$\begin{aligned}\det E\left(E^{-1}\right)_{ab}^{cd}\partial_{cd}\left[\frac{\nabla^a\boldsymbol{\psi}^b}{1-\frac{1}{2}(\nabla\boldsymbol{\psi})^2}\right] &+ \frac{\nabla^a\boldsymbol{\psi}^b}{1-\frac{1}{2}(\nabla\boldsymbol{\psi})^2}\partial_{cd}\left[\det E\,\left(E^{-1}\right)_{ab}^{cd}\right] = \\ &= \partial_{cd}\left[\det E\frac{\left(E^{-1}\right)_{ab}^{cd}\,\nabla^a\boldsymbol{\psi}^b}{1-\frac{1}{2}(\nabla\boldsymbol{\psi})^2}\right] = 0.\end{aligned} \tag{18}$$

Thus, the proper covariantization of the field strength is given by the expression (we choose the coefficient $-\frac{1}{2}$ for further convenience)

$$F^{ab} = -\frac{1}{2}\left[\det E\frac{\left(E^{-1}\right)_{cd}^{ab}\,\nabla^c\boldsymbol{\psi}^d}{1-\frac{1}{2}(\nabla\boldsymbol{\psi})^2}\right], \qquad \partial_{ab}F^{ab} = 0, \tag{19}$$

and obeys standard Bianchi identities.

2.3. Component action

To construct the component action of $D2$ brane we will follow the procedure developed in [2, 3]. In the application to the present case, the basic steps of this method can be summarized as follows:

- Firstly, our supermultiplet contains the physical components ψ^a, λ^{ab} (12), where the components λ^{ab} are subjected to Bianchi identity
$$\partial_{ab}\left[\det\mathcal{E}\,\frac{(\mathcal{E}^{-1})_{cd}^{ab}\,\lambda^{cd}}{1-2\lambda^2}\right]=0, \tag{20}$$
where
$$\mathcal{E}_{ab}^{cd}=E_{ab}^{cd}\Big|_{\theta=0}=\delta_a^{(c}\delta_b^{d)}+\frac{1}{2}\psi^{(c}\partial_{ab}\psi^{d)}. \tag{21}$$
- Secondly, with respect to broken S supersymmetry $\delta\theta^a=0$ (4). This means, that the transformation properties of the ψ^a can be extracted from the reduced coset
$$g|_{\theta=0}=e^{x^{ab}P_{ab}}e^{\psi^a S_a}, \tag{22}$$
while λ^{ab} do not transform $\delta_S\lambda^{ab}=0$.
- Finally, due to nontrivial transformations of x^{ab} and ψ^a under S supersymmetry (3)
$$\delta_S x^{ab}=-\frac{1}{2}\varepsilon^{(a}\psi^{b)},\qquad \delta\psi^a=\varepsilon^a, \tag{23}$$
the Goldstone fermion ψ^a may enter the action through the determinant of the dreibein $\mathcal{E}$ (21) or through the covariant derivatives $\mathcal{D}_{ab}$ of matter fields
$$\mathcal{D}_{ab}=\left(\mathcal{E}^{-1}\right)_{ab}^{cd}\partial_{cd}. \tag{24}$$
Thus, the minimal component action, invariant with respect to spontaneously broken S supersymmetry reads
$$S=\int d^3x\,\det\mathcal{E}\,\mathcal{F}[\lambda^2], \tag{25}$$
where $\mathcal{F}$ is an arbitrary, for the time being, function.

The last step is to select from the family of the actions (25) the action which corresponds to $D2$ brane, i.e. the one which possesses the invariance with respect to unbroken Q supersymmetry. Let us start from the transformations of ψ^a
$$\delta_Q\psi^a\equiv-\epsilon^b D_b\psi^a|_{\theta=0}=2\epsilon^b\lambda_b^a-\epsilon^b\psi^m\lambda_b^n\partial_{mn}\psi^a. \tag{26}$$
Repeating these calculations for all needed ingredients entering the action (25) we will get
$$\begin{aligned}\delta_Q\det\mathcal{E}&=2\det\mathcal{E}\,\epsilon^a\lambda_a^b\mathcal{D}_{bc}\psi^n-\partial_{bc}\left(\det\mathcal{E}\,\epsilon^a\lambda_a^b\psi^c\right),\\ \delta_Q\lambda^2&=\frac{1}{2}\left(1-2\lambda^2\right)\epsilon^a\lambda^{bc}\mathcal{D}_{bc}\psi_a-\left(1-2\lambda^2\right)\epsilon^a\lambda_a^b\mathcal{D}_{bc}\psi^c-\epsilon^a\lambda_a^b\psi^c\partial_{bc}\lambda^2.\end{aligned} \tag{27}$$
Now we can calculate the variation of the action (25) with respect to unbroken supersymmetry:
$$\begin{aligned}\delta_Q S&=\int d^3x\left[\frac{1}{2}\det\mathcal{E}\,\left(1-2\lambda^2\right)\epsilon^a\lambda^{bc}\mathcal{D}_{bc}\psi_a\,\mathcal{F}'+\det\mathcal{E}\,\epsilon^a\lambda_a^b\mathcal{D}_{bc}\psi^c\left(2\mathcal{F}-\left(1-2\lambda^2\right)\mathcal{F}'\right)-\right.\\ &-\left.\partial_{ab}\left(\det\mathcal{E}\,\epsilon^c\lambda_c^a\psi^b\right)\mathcal{F}-\left(\det\mathcal{E}\,\epsilon^c\lambda_c^a\psi^b\right)\partial_{ab}\mathcal{F}\right].\end{aligned} \tag{28}$$

The last two terms in (28) combine into a full divergence and thus they are eliminated after integration over d^3x. The cancellation of the second term gives

$$2\mathcal{F} - \left(1 - 2\lambda^2\right)\mathcal{F}' = 0 \quad \Rightarrow \quad \mathcal{F} = \frac{\gamma}{1 - 2\lambda^2}. \tag{29}$$

Thus, the variation of the action (28) with the explicit form of the function $\mathcal{F}$ given in (29) reads

$$\delta_Q S = \gamma \int d^3x \, \frac{\det \mathcal{E} \; \epsilon^a \lambda^{bc}}{1 - 2\lambda^2} \mathcal{D}_{bc}\psi_a = -\gamma \int d^3x \epsilon^a \partial_{cd} \left[\frac{\det \mathcal{E} \; \left(\mathcal{E}^{-1}\right)_{mn}^{cd} \lambda^{mn}}{1 - 2\lambda^2} \right] \psi_a = 0 \tag{30}$$

in virtue of (20).

So, the component action of $D2$ brane reads

$$S_{D2} = \gamma \int d^3x \det \mathcal{E} \, \frac{1}{1 - 2\lambda^2}, \tag{31}$$

where the bosonic component λ_{ab} obeys the covariant version of Bianchi identities (20). Being rewritten in terms of the new variable $\widetilde{F}_{ab}$ it acquires the Born-Infeld like form

$$\widetilde{F}_{ab} \equiv \frac{\lambda_{ab}}{1 - 2\lambda^2} \quad \rightarrow \quad S_{D2} = \frac{\gamma}{2} \int d^3x \det \mathcal{E} \left(1 + \sqrt{1 + 8\widetilde{F}^2}\right). \tag{32}$$

3. Conclusion

In this paper, following the approach developed in [2, 3], we derive the component on-shell action of the space-filling D2-brane, i.e. $N = 1, d = 3$ supersymmetric Born-Infeld action. The main property of this D2-brane is that in full analogy with the cases of p-branes where irreducibility constraints can immediately covariantized, the Bianchi identity admits direct covariantization too. Of course, the corresponding on-shell component action for $N = 1, d = 3$ Born-Infeld theory can be found from the superfield one [1]. However, the action obtaining in such way will contain a long tail of fermionic terms without any visible symmetry while in the present approach all fermionic terms combine into covariant, with respect to broken supersymmetry, objects: covariant derivatives and fierbein. The D2-brane component action, being written in terms of these covariant objects, has very simple form. Similarly to the cases of p-branes, it mimics the bosonic Born-Infeld action.

Acknowledgments

The work S.K. was partially supported by the RSCF, grant 14-11-00598 and partially by the RFBR, grant 15-52-05022 Arm-a.

References

[1] E. Ivanov, S. Krivonos, *N=1 D=4 supermembrane in the coset approach*, Phys.Lett. **B453** (1999) 237; Erratum-ibid. **B657** (2007) 269; `arXiv:hep-th/9901003`.

[2] S. Bellucci, S. Krivonos, A. Sutulin, *Supersymmetric component actions via coset approach*, Phys.Lett. **B726** (2013) 497; `arXiv:1306.1115[hep-th]`.

[3] S. Bellucci, S. Krivonos, A. Sutulin, *Coset approach to the partial breaking of global supersymmetry*, Springer Proc.Phys. **153** (2014) 205; `arXiv:1401.2613[hep-th]`.

IOP Conf. Series: Journal of Physics: Conf. Series **804** (2017) 012011 doi:10.1088/1742-6596/804/1/012011

The tetrahexahedric angular Calogero model

F Correa[1] and O Lechtenfeld[2]

[1] Instituto de Ciencias Fisicas y Matemáticas, Universidad Austral de Chile, Casilla 567, Valdivia, Chile
[2] Institut für Theoretische Physik and Riemann Center for Geometry and Physics, Leibniz Universität Hannover

E-mail: francisco.correa@uach.cl, lechtenf@itp.uni-hannover.de

Abstract. We investigate the spherical reduction of the rational A_{n-1} Calogero model. It defines a quantum superintegrable model of a particle trapped in one of $n!$ spherical $(n-2)$-simplices on S^{n-2} in a special S_n-symmetric potential. The energy levels (including degeneracy) and eigenstates are given, and the construction of conserved charges and Hamiltonian intertwiners outlined. We describe superintegrable complex $\mathcal{PT}$ deformations which remove the trapping walls and in some cases create a $\mathbb{Z}_2$-graded ("supersymmetric") extension of the spectrum. Details are worked out for the cases of $n{=}3$ and $n{=}4$.

1. Introduction – some history

The Calogero model [1] has been the workhorse for integrable systems for 45 years. It is therefore mildly surprising that new aspects of it can still be uncovered. This talk describes the superintegrable spherical reduction of the rational A_{n-1} quantum Calogero model and some of its complex $\mathcal{PT}$ deformations. The emphasis is on the energy spectrum including degeneracy and eigenstates, and on the conserved charges and intertwiners, in particular for a coupling strength $g(g-1)$ with $g \in \mathbb{Z}$. We discuss all features in some detail for the cases $n{=}3$ (the hexagonal or Pöschl-Teller model) and for $n{=}4$ (the tetrahexahedric model), the latter being the simplest non-separable case. We close with a number of open issues.

It is worth recalling the relevant part (for this talk) of the Calogero model's long history:

- 1971 Calogero [2]:
  Solution of the one-dimensional N-body problem with . . . inversely quadratic pair potentials
- 1981 Olshanetsky & Perelomov [3, 4]:
  Classical integrable finite-dimensional systems related to Lie algebras (1983: quantum)
- 1983 Wojciechowski [5]:
  Superintegrability of the Calogero–Moser system
- 1989 Dunkl [6]:
  Differential-difference operators associated to reflection groups
- 1990 Veselov & Chalykh [7]:
  Commutative rings of partial differential operators and Lie algebras, supercompleteness
- 1991 Heckman [8]:
  Elementary construction for commuting charges and intertwiners (shift operators)
- 2003 M. Feigin [9]:
  Intertwining relations for the spherical parts of generalized Calogero operators

- 2008 A. Fring, M. Znojil [10, 11]:
  $\mathcal{PT}$-symmetric deformations of Calogero models
- 2008 Hakobyan, Nersessian, Yeghikyan [12]:
  The cuboctahedric Higgs oscillator from the rational Calogero model (classical)
- 2013 M. Feigin, Lechtenfeld, Polychronakos [13]:
  The quantum angular Calogero–Moser model (spectra, eigenstates)
- 2013 Correa, Lechtenfeld, Plyushchay [14]:
  Nonlinear supersymmetry in the quantum Calogero model
- 2014 M. Feigin, Hakobyan [15]:
  On the algebra of Dunkl angular momentum operators
- 2015 Correa, Lechtenfeld [16]:
  The tetrahexahedric angular Calogero model

This talk is based on [13] and [16] but contains yet unpublished material concerning $\mathcal{PT}$ deformations.

2. The angular (relative) Calogero model

The Hamiltonian for the A_{n-1} rational Calogero system reads

$$H \;=\; \sum_{\mu<\nu}^{n} \Big\{ \tfrac{1}{2n}(p_\mu - p_\nu)^2 + \frac{g(g-1)}{(x^\mu - x^\nu)^2} \Big\} , \tag{1}$$

where x^μ and p_μ with $\mu = 1,2,\ldots,n$ denote position and momentum in $\mathbb{R}$ for the μth particle, and g is a constant. Even though $H^{(1-g)} = H^{(g)}$, it is useful to admit any real value for g. The quantum theory demands that $[x^\mu, p_\nu] = \mathrm{i}\,\delta^\mu_\nu$. Note that the center-of-mass degree of freedom has already been removed, so we introduce *radial* variables in the remaining $(n-1)$-dimensional coordinate and momentum spaces,

$$\tfrac{1}{n}\sum_{\mu<\nu}(x^\mu - x^\nu)^2 \;=\; r^2 \qquad\text{and}\qquad \tfrac{1}{n}\sum_{\mu<\nu}(p_\mu - p_\nu)^2 \;=\; p_r^2 + \tfrac{1}{r^2}L^2 + \tfrac{(n-2)(n-4)}{4\,r^2} \,. \tag{2}$$

Furthermore, it is convenient to pass to $n-1$ *relative* coordinates and momenta,

$$r^2 = \sum_{i=1}^{n-1}(y^i)^2 \quad,\quad p_i \equiv p_{y^i} \quad,\quad L_{ij} \;=\; -\mathrm{i}(y^i p_j - y^j p_i) \quad,\quad L^2 \;=\; -\sum_{i<j} L_{ij}^2 \,, \tag{3}$$

and to define the corresponding angular momenta L_{ij}. The $sl(2,\mathbb{R})$ conformal algebra is generated by

$$H \;=\; \tfrac12 p_r^2 + \tfrac{(n-2)(n-4)}{8\,r^2} + \tfrac{1}{r^2}H_\Omega \quad,\qquad D \;=\; \tfrac12(r\,p_r + p_r r) \quad,\qquad K \;=\; \tfrac12 r^2 \,, \tag{4}$$

where all the interactions are hiding in the angular Calogero Hamiltonian H_Ω (or the $sl(2,\mathbb{R})$ Casimir C),

$$H_\Omega \;=\; \tfrac12 L^2 + U(\vec\theta) \;=\; C - \tfrac18(n-1)(n-5) \qquad\text{with}\qquad C \;=\; K\,H + H\,K - \tfrac12 D^2 \,. \tag{5}$$

It defines a model for a single particle moving on the $(n-2)$-sphere (angular coordinates $\vec\theta$) under the influence of a special potential

$$U(\vec\theta) \;=\; r^2 \sum_{\mu<\nu} \frac{g(g-1)}{(x^\mu - x^\nu)^2} \;=\; r^2 \sum_{\alpha\in\mathcal{R}_+} \frac{g(g-1)}{(\alpha\cdot y)^2} \;=\; \tfrac{g(g-1)}{2} \sum_{\alpha\in\mathcal{R}_+} \cos^{-2}\theta_\alpha \tag{6}$$

which is a superposition of "Higgs oscillators" $\cos^{-2}\theta$ [17, 18] centered at the directions of the $\frac{1}{2}n(n-1)$ positive roots α for A_{n-1}. It is singular at the $(n-3)$-dimensional intersections of the Weyl chamber walls with the unit $(n-2)$-sphere, so the particle is trapped in one of $n!$ spherical $(n-2)$-simplices tessalating the sphere. This potential breaks the SO$(n-1)$ invariance of the free particle on S^{n-2} to the permutation group S_n, which is the Weyl group of A_{n-1}.

From now on, we pass to the position representation,

$$p_i \mapsto -\mathrm{i}\partial_i \qquad \Longrightarrow \qquad p_r \mapsto -\mathrm{i}\big(\partial_r + \tfrac{n-2}{2r}\big)\ , \tag{7}$$

$$\begin{aligned} H \ &\mapsto\ -\tfrac12\big(\partial_r^2 + \tfrac{n-2}{r}\partial_r\big) + \tfrac{1}{r^2}H_\Omega \ =\ w^{-1}\Big[-\tfrac12\big(\partial_r^2 - \tfrac{(n-2)(n-4)}{4r^2}\big) + \tfrac{1}{r^2}H_\Omega\Big]\,w\ , \\ H_\Omega &\mapsto -\tfrac12\sum_{i<j}\big(y^i\partial_j - y^j\partial_i\big)^2 + r^2\sum_{\alpha\in\mathcal{R}_+}\frac{g(g-1)}{(\alpha\cdot y)^2} \qquad \text{with} \quad w = r^{\frac{n-2}{2}}\ . \end{aligned} \tag{8}$$

The energy spectrum and radial eigenfunctions of the full Calogero model are well known,

$$\begin{aligned} H\,\Psi_{E,q} &= E\,\Psi_{E,q} \qquad \text{with} \qquad E\in\mathbb{R}_{\geq 0}\ , \\ \Psi_{E,q}(r,\vec\theta) &= r^{-\frac{n-3}{2}}\,J_{q+(n-3)/2}\big(\sqrt{2E}\,r\big)\,v_q(\vec\theta)\ , \end{aligned} \tag{9}$$

but the angular part of these wave functions is part of the spectral problem for the angular model,

$$\begin{aligned} H_\Omega\,v_q &= \varepsilon_q\,v_q \qquad \text{with} \qquad \varepsilon_q = \tfrac12\,q\,(q+n-3) \qquad \text{and} \\ q &= \tfrac12 n(n-1)\,g + \ell \qquad \text{where} \qquad \ell = 3\ell_3 + 4\ell_4 + \ldots + n\ell_n \in \mathbb{N}_0\ . \end{aligned} \tag{10}$$

Naturally, the spectrum $\{\varepsilon_q\}$ is discrete, for a fixed g governed by an integer-valued angular momentum quantum number ℓ. For a free particle on S^{n-2}, the degeneracy at level ℓ is $\binom{\ell+n-2}{n-2} - \binom{\ell+n-4}{n-2}$. With the potential U turned on, the $SO(n-1)$ representation content has to be decomposed with respect to the S_n subgroup. The full A_{n-1} Calogero system describes n *identical* particles, whose wave function picks up just a phase factor of $\mathrm{e}^{\mathrm{i}\pi g}$ under a particle exchange, so – apart from this – only S_n singlet representations are kept. Here we keep this heritage for the angular model and, therefore, admit only permutation invariant states. This greatly reduces the degeneracy at level ℓ to the number of all $(n-2)$-tuples $(\ell_3, \ell_4, \ldots, \ell_n)$ subject to (10). It can be expressed as

$$\deg_n(\varepsilon_q) \;=\; p_n(\ell) - p_n(\ell-1) - p_n(\ell-2) + p_n(\ell-3)\ , \tag{11}$$

where $p_n(\ell)$ is the number of partitions of ℓ into integers not larger than n, which can be generated via

$$p_n(t) := \sum_{\ell=0}^{\infty} p_n(\ell)\,t^\ell \;=\; \prod_{m=1}^{n}\big(1-t^m\big)^{-1}\ . \tag{12}$$

For small values of n, there are explicit formulæ [19]:

$$\begin{aligned} \deg_3(\ell) &= \begin{cases} 0 & \text{for} \quad \ell = 1,2 \bmod 3 \\ 1 & \text{for} \quad \ell = 0 \bmod 3 \end{cases}\ , \\ \deg_4(\ell) &= \Big\lfloor \frac{\ell}{12} \Big\rfloor + \begin{cases} 0 & \text{for} \quad \ell = 1,2,5 \bmod 12 \\ 1 & \text{for} \quad \ell = \text{else} \bmod 12 \end{cases}\ , \\ \deg_5(\ell) &= \Big\lfloor \frac{6\ell^2 + 72\ell - 89}{720} \Big\rfloor + \begin{cases} 0 & \text{for} \quad \ell = 2,22,26,46 \bmod 60 \\ 2 & \text{for} \quad \ell = 0,48 \bmod 60 \\ 1 & \text{for} \quad \ell = \text{else} \bmod 60 \end{cases}\ . \end{aligned} \tag{13}$$

The main task lies in the construction of the angular eigenfunctions, which are given by

$$v_q(\vec{\theta}) \equiv v_\ell^{(g)}(\vec{\theta}) \sim r^{n-3+q}\Big(\prod_{\mu=3}^{n} \sigma_\mu(\{\mathcal{D}_i\})^{\ell_\mu}\Big)\,\Delta^g\, r^{3-n-n(n-1)g}\,. \tag{14}$$

The key ingredients are the Vandermonde factor and the Dunkl operators,

$$\Delta = \prod_{\alpha\in\mathcal{R}_+} \alpha\cdot y \qquad \text{and} \qquad \mathcal{D}_i = \partial_i - g\sum_{\alpha\in\mathcal{R}_+}\frac{\alpha_i}{\alpha\cdot y}\,s_\alpha\,, \tag{15}$$

where s_α denotes the reflection about the hyperplane orthogonal to the root α. The latter are combined into the elementary Weyl-symmetric polynomials $\sigma_\mu(y)$ of degree μ, with $\sigma_2 = \sum_i (y^i)^2$ being absent because we freeze the radial excitations. The vanishing locus of the Vandermonde Δ produces singularities for $g < 0$, rendering these states non-normalizable. Splitting off a universal factor, the angular wave functions become Dunkl-deformed Weyl-symmetric harmonic polynomials of degree ℓ,,

$$v_\ell^{(g)}(\vec{\theta}) = r^{-q}\,\Delta^g\, h_\ell^{(g)} \qquad \text{with} \qquad H\left(\Delta^g\, h_\ell^{(g)}\right) = 0\,. \tag{16}$$

Just like 'Dunkl-deforming' the momenta, i.e. $\partial_i \to \mathcal{D}_i$, generates the full Calogero model from the free particle, the angular submodel can be obtained from free motion on S^{n-2} by an analogous deformation of the angular momenta,

$$L_{ij} \mapsto -(y^i\partial_j - y^j\partial_i) \qquad \Longrightarrow \qquad \mathcal{L}_{ij} = -(y^i\mathcal{D}_j - y^j\mathcal{D}_i)\,. \tag{17}$$

In both cases, their square yields a differential-difference operator ('pre-Hamiltonian') whose restriction to Weyl symmetric functions is basically the Hamiltonian,

$$\mathcal{H} = -\tfrac12\sum_i \mathcal{D}_i^2 \qquad \text{and} \qquad \mathcal{H}_\Omega = -\tfrac12\sum_{i<j}\mathcal{L}_{ij}^2 + \tfrac12\, g{\textstyle\sum}_\alpha s_\alpha\,(g{\textstyle\sum}_\alpha s_\alpha + n-3)\,, \tag{18}$$

$$H = \operatorname{res}(\mathcal{H}) \qquad \text{and} \qquad H_\Omega = \operatorname{res}(\mathcal{H}_\Omega) = \tfrac12\operatorname{res}\big(\mathcal{L}^2\big) + \varepsilon_q(\ell{=}0)\,. \tag{19}$$

In fact, any Weyl-invariant polynomial in the $\mathcal{L}_{ij}$ of some degree t provides a conserved quantity,

$$\mathcal{C}_t(\mathcal{L}_{ij}) \ \text{Weyl-invariant} \qquad \Longrightarrow \qquad C_t = \operatorname{res}(\mathcal{C}_t) \ \text{ commutes with } H_\Omega\,. \tag{20}$$

In contrast to the full model, these are not in involution and hence do not constitute Liouville charges. The reason is that, while $[\mathcal{D}_i, \mathcal{D}_j] = 0$, angular momenta (deformed or not) do not commute, rather [15]

$$[\mathcal{L}_{ij}, \mathcal{L}_{k\ell}] = \mathcal{L}_{i\ell}\mathcal{S}_{jk} - \mathcal{L}_{ik}\mathcal{S}_{j\ell} - \mathcal{L}_{j\ell}\mathcal{S}_{ik} + \mathcal{L}_{jk}\mathcal{S}_{i\ell} \tag{21}$$

$$\text{with} \qquad \mathcal{S}_{ij} = \begin{cases} -g\, s_{ij} & \text{for } i\neq j \\ 1 + g\sum_{k(\neq i)} s_{ik} & \text{for } i = j \end{cases}\,, \tag{22}$$

$$[\mathcal{S}_{ij}, \mathcal{L}_{k\ell}] = 0 \quad , \qquad \{\mathcal{S}_{ij}, \mathcal{L}_{ij}\} = 0 \quad , \qquad \mathcal{S}_{ij}\mathcal{L}_{ik} = \mathcal{L}_{jk}\mathcal{S}_{ij}\,. \tag{23}$$

This algebra is a 'Dunkl deformation' of $so(n{-}1)$, with H_Ω being the Casimir invariant. More technically, it is a subalgebra of the rational Cherednik algebra generated by $\{\mathcal{D}_i, y^j\}$ and the Weyl reflections.

It is equally interesting to consider Weyl-*anti*invariant polynomials $\mathcal{M}_s(\mathcal{L}_{ij})$ of some degree s, since

$$\mathcal{M}_s(\mathcal{L}_{ij}) \ \text{Weyl-antiinvariant} \qquad \Longrightarrow \qquad M_s = \operatorname{res}(\mathcal{M}_s) \ \text{ is an intertwiner for } H \text{ and } H_\Omega\,. \tag{24}$$

Indeed, since $[\mathcal{L}_{ij},\mathcal{H}]=0$ we have

$$[\mathcal{M}_s,\mathcal{H}]=0 \quad\Longrightarrow\quad M_s^{(g)}H^{(g)} = H^{(-g)}M_s^{(g)} = H^{(g+1)}M_s^{(g)} \quad\text{and}\quad M_s^{(g)}\Psi_{E,q}^{(g)} \sim \Psi_{E,q}^{(g+1)}\,, \tag{25}$$

$$[\mathcal{M}_s,\mathcal{H}_\Omega]=0 \quad\Longrightarrow\quad M_s^{(g)}H_\Omega^{(g)} = H_\Omega^{(-g)}M_s^{(g)} = H_\Omega^{(g+1)}M_s^{(g)} \quad\text{and}\quad M_s^{(g)}v_\ell^{(g)} \sim v_{\ell-n(n-1)/2}^{(g+1)}\,. \tag{26}$$

This allows us to connect the states for integer increments of the coupling g and allows one to generate the wave functions for $g\in\mathbb{N}$ directly from the free ones (g=0 or g=1). Therefore, we restrict ourselves to $g\in\mathbb{Z}$ from now on (it will prove beneficial to formally admit $g<0$), keeping only Bose (g even) and Fermi (g odd) statistics. The intertwiners M_s clearly exchange bosons with fermions. Since $H_\Omega^{(g)} = H_\Omega^{(1-g)}$, at each $g\in\mathbb{N}$ the spectrum of $H_\Omega^{(g)}$ formally contains a bosonic and a fermionic tower of states, but one of them is unphysical because it is constructed with a non-positive g-value and thus not normalizable. The only exception occurs for $g=1$ because the $g=0$ states are non-singular, but this is obvious since for vanishing potential the only restriction arises from Weyl invariance.

3. $\mathcal{PT}$ deformation

It has been known for a long time that hermiticity is not an esential feature of a Hamiltonian for its spectrum to be real. For instance, it suffices that the Hamiltonian commutes with an antilinear involution (called $\mathcal{PT}$) which also leaves the eigenfunctions invariant ("unbroken $\mathcal{PT}$ symmetry") [20]. Such a non-hermitian Hamiltonian is related to a hermitian one by a (non-unitary) similarity transformation, which may be impossibly complicated. Often, however, there exists a family H_ϵ of non-hermitian $\mathcal{PT}$-invariant Hamiltonians representing a smooth deformation of a hermitian H_0. In this case we speak of a "$\mathcal{PT}$ deformation", with the parameter ϵ measuring the deviation from hermiticity. For rational Calogero models, a particularly nice set of $\mathcal{PT}$ deformations can be generated by a specific *complex* orthogonal deformation of the coordinates x^μ (or y^i) in the expression for the Hamiltonian. If such a $\mathcal{PT}$ deformation is in accordance with the Weyl symmetry of the system, its integrability will be preserved.

For our case of the rational A_{n-1} Calogero model, the following SU(1,1)×SU(1,1)×··· ⊂ SO(n−1, $\mathbb{C}$) coordinate deformations $y^i\mapsto y^i_\epsilon$ with $\epsilon_a\in\mathbb{R}$ are admissible:

$$\begin{pmatrix} y^1\\ y^2\\ y^3\\ y^4\\ \vdots \end{pmatrix} \longmapsto \begin{pmatrix} \cosh\epsilon_1 & -\mathrm{i}\sinh\epsilon_1 & 0 & 0 & \cdots\\ \mathrm{i}\sinh\epsilon_1 & \cosh\epsilon_1 & 0 & 0 & \cdots\\ 0 & 0 & \cosh\epsilon_2 & -\mathrm{i}\sinh\epsilon_2 & \cdots\\ 0 & 0 & \mathrm{i}\sinh\epsilon_2 & \cosh\epsilon_2 & \cdots\\ \vdots & \vdots & \vdots & \vdots & \ddots \end{pmatrix} \begin{pmatrix} y^1\\ y^2\\ y^3\\ y^4\\ \vdots \end{pmatrix}. \tag{27}$$

This is equivalent to a complex deformations of angles, $\phi_a\mapsto\phi_a+\mathrm{i}\epsilon_a$, which parametrize mutually orthogonal planes. The complex coordinate deformation induces a non-unitary similarity transformation in the Hilbert space, sending

$$L^2 \mapsto L^2 \qquad\text{and}\qquad U(\vec\theta) \mapsto U(\vec\theta_\epsilon) =: U_\epsilon(\vec\theta) \tag{28}$$

thus leaving the energy spectrum unchanged. Moreover, the singular Higgs oscillator potential gets regularized,

$$U_\epsilon(\vec\theta) = r^2\sum_{\mu<\nu}\frac{g(g-1)}{(x^\mu_\epsilon-x^\nu_\epsilon)^2} = r^2\sum_{\alpha\in\mathcal{R}_+}\frac{g(g-1)}{(\alpha\cdot y_\epsilon)^2} = \tfrac{g(g-1)}{2}\sum_{\alpha\in\mathcal{R}_+}\cos^{-2}\theta_\alpha(\epsilon)\,, \tag{29}$$

IOP Conf. Series: Journal of Physics: Conf. Series **804** (2017) 012011 doi:10.1088/1742-6596/804/1/012011

and the Weyl-chamber wall singularities disappear due to

$$\frac{1}{\cos^2(\theta_\alpha + \mathrm{i}\eta_\alpha(\vec{\theta},\epsilon))} = \frac{\cosh^2\eta_\alpha\cos^2\theta_\alpha - \sinh^2\eta_\alpha\sin^2\theta_\alpha + \frac{\mathrm{i}}{2}\sinh 2\eta_\alpha \sin 2\theta_\alpha}{(\cosh^2\eta_\alpha\cos^2\theta_\alpha + \sinh^2\eta_\alpha\sin^2\theta_\alpha)^2}, \tag{30}$$

where the azimuthal angle for the Higgs oscillator pertaining to the root α is $\theta_\alpha(\epsilon) = \theta_\alpha + \mathrm{i}\eta_\alpha(\vec{\theta})$. The $\mathcal{PT}$ deformation frees the particle up to escape from its Weyl chamber and move anywhere on the $(n{-}2)$-sphere. Singularities of codimension two remain at the vanishing loci of η_α. We shall see later that such a change of boundary conditions for the spectral problem may bring back previously unphysical singular wave functions to the physical sprectrum, essentially doubling the degeneracy of the energy levels.

4. Warmup: the hexagonal or Pöschl-Teller model

The first interesting case is $n{=}3$, leading to a one-dimensional integrable system known as the Pöschl-Teller model for a particle on a circle. The Jacobi relative coordinates on the two-plane perpendicular to the center-of-mass X are

$$\begin{aligned} x^1 &= X + \tfrac{1}{\sqrt{2}}y^1 + \tfrac{1}{\sqrt{6}}y^2 , & \partial_{x^1} &= \tfrac13\partial_X + \tfrac{1}{\sqrt{2}}\partial_{y^1} + \tfrac{1}{\sqrt{6}}\partial_{y^2} , \\ x^2 &= X - \tfrac{1}{\sqrt{2}}y^1 + \tfrac{1}{\sqrt{6}}y^2 , & \partial_{x^2} &= \tfrac13\partial_X - \tfrac{1}{\sqrt{2}}\partial_{y^1} + \tfrac{1}{\sqrt{6}}\partial_{y^2} , \\ x^3 &= X - \tfrac{2}{\sqrt{6}}y^2 , & \partial_{x^3} &= \tfrac13\partial_X - \tfrac{2}{\sqrt{6}}\partial_{y^2} , \end{aligned} \tag{31}$$

and we also use polar and complex coordinates in that plane,

$$y^1 = r\cos\phi \quad\text{and}\quad y^2 = r\sin\phi \qquad\Longrightarrow\qquad w := y^1 + \mathrm{i}y^2 = r\,\mathrm{e}^{\mathrm{i}\phi} . \tag{32}$$

The angular Hamiltonian reads

$$H_\Omega = \tfrac12\big(w\partial_w - \bar{w}\partial_{\bar{w}}\big)^2 + g(g{-}1)\frac{18\,(w\bar{w})^3}{(w^3+\bar{w}^3)^2}, \tag{33}$$

with the potential coming from

$$U(\phi) = \tfrac{g(g-1)}{2}\sum_{k=0,1,2}\cos^{-2}(\phi{+}k\tfrac{2\pi}{3}) = \tfrac92 g(g{-}1)\cos^{-2}(3\phi) . \tag{34}$$

Specializing our general formulæ for the energy eigenvalues and -functions from the previous section and denoting by (s_0, s_+, s_-) the three Coxeter reflections, we get $\ell = 3\,\ell_3$ and

$$\varepsilon_q = \tfrac12 q^2 \qquad\text{with}\qquad q = 3g + \ell = 3(g+\ell_3) \qquad\text{and}\qquad \deg(\varepsilon_q) = 1 , \tag{35}$$

$$\Psi_{E,q}(r,\phi) = J_q(\sqrt{2E}\,r)\,v_q(\phi) , \tag{36}$$

$$v_q(\phi) \equiv v_\ell^{(g)}(\phi) \sim r^q\big(\mathcal{D}_w^3 - \mathcal{D}_{\bar{w}}^3\big)^{\ell_3}\Delta^g r^{-6g} = r^{-q}\Delta^g h_\ell^{(g)}(w^3,\bar{w}^3) , \tag{37}$$

$$\Delta \sim w^3 + \bar{w}^3 \sim r^3\cos(3\phi) \qquad \text{vanishing at } \phi = \pm\tfrac{\pi}{6}, \pm\tfrac{\pi}{2}, \pm\tfrac{5\pi}{6} , \tag{38}$$

$$\mathcal{D}_w = \partial_w - g\Big\{\frac{1}{w+\bar{w}}\,s_0 + \frac{\rho}{\rho w + \bar{\rho}\bar{w}}\,s_+ + \frac{\bar{\rho}}{\bar{\rho}w + \rho\bar{w}}\,s_-\Big\} \quad\text{with}\quad \rho = \mathrm{e}^{2\pi\mathrm{i}/3} , \tag{39}$$

$$h_\ell^{(g)}(w^3,\bar{w}^3) = \sum_{k=0}^{\ell_3}(-1)^k\,\frac{\Gamma(1{+}\ell_3)\,\Gamma(g{+}k)\,\Gamma(g{+}\ell_3{-}k)}{\Gamma(2g{+}\ell_3)\,\Gamma(g)\,\Gamma(1{+}k)\,\Gamma(1{+}\ell_3{-}k)}\,w^{\ell-3k}\bar{w}^{3k} , \tag{40}$$

where the final equation is the result of a nontrivial computation. We display the homogeneous polynomials $h_\ell^{(g)}$ defining the wave functions (un-normalized, see above) for $g = 0, 1, 2$ and the lowest five energy levels, using the notation $(m\,\bar{m}) := w^{3m}\bar{w}^{3\bar{m}} = (y_1 + \mathrm{i}y_2)^{3m}(y_1 - \mathrm{i}y_2)^{3\bar{m}}$,

ℓ	$h_\ell^{(0)}$	$h_\ell^{(1)}$	$h_\ell^{(2)}$	$\cdots$
0	(00)	(00)	(00)	$\cdots$
3	(10) − (01)	(10) − (01)	(10) − (01)	$\cdots$
6	(20) + (02)	(20) − (11) + (02)	3(20) − 4(11) + 3(02)	$\cdots$
9	(30) − (03)	(30) − (21) + (12) − (03)	4(30) − 6(21) + 6(12) − 4(03)	$\cdots$
12	(40) + (04)	(40) − (31) + (22) − (13) + (04)	5(40) − 8(31) + 9(22) − 8(13) + 5(04)	$\cdots$
$\vdots$	$\vdots$	$\vdots$	$\vdots$	

These states can also be obtained by applying the angular intertwiner $M_1 = \mathrm{res}(\mathcal{M}_1)$,

$$\begin{aligned} \mathcal{M}_1^{(g)} &\sim \mathrm{i}\big(w\mathcal{D}_w - \bar{w}\mathcal{D}_{\bar{w}}\big) \\ &\sim \mathrm{i}\big(w\partial_w - \bar{w}\partial_{\bar{w}}\big) - \mathrm{i}\,g\Big\{\frac{w-\bar{w}}{w+\bar{w}}\,s_0 + \frac{\rho w - \bar{\rho}\bar{w}}{\rho w + \bar{\rho}\bar{w}}\,s_+ + \frac{\bar{\rho}w - \rho\bar{w}}{\bar{\rho}w + \rho\bar{w}}\,s_-\Big\}\,, \end{aligned} \tag{41}$$

$$M_1^{(g)} \sim \mathrm{i}\big(w\partial_w - \bar{w}\partial_{\bar{w}}\big) - 3\mathrm{i}\,g\,\frac{w^3-\bar{w}^3}{w^3+\bar{w}^3} = \mathrm{i}\,\Delta^g\big(w\partial_w - \bar{w}\partial_{\bar{w}}\big)\Delta^{-g} = \partial_\phi + 3\,g\,\tan 3\phi\,. \tag{42}$$

Since $M_1^{(g)}v_\ell^{(g)} \sim v_{\ell-3}^{(g+1)}$, we learn that

$$h_\ell^{(g+1)} \sim \mathrm{i}\,\Delta^{-1}\big(w\partial_w - \bar{w}\partial_{\bar{w}}\big)\,h_{\ell+3}^{(g)}\,, \tag{43}$$

which may be iterated to arrive at (40). This intertwiner does not yield additional conserved charges, because

$$\big(M_1^\dagger M_1\big)^{(g)} = -2\,H_\Omega^{(g)} + 9\,g^2 = -\mathrm{res}(\mathcal{L}^2) = -C_2^{(g)}\,. \tag{44}$$

What are the possible $\mathcal{PT}$ deformations of the Pöschl-Teller system? Without loss of generality, we may fix $\mathcal{P}$ to be the Coxeter reflection about the line perpendicular to $\alpha \sim \left(\begin{smallmatrix}1\\0\end{smallmatrix}\right)$, hence

$$\mathcal{P}:\ s_0 = \left(\begin{smallmatrix}-1&0\\0&1\end{smallmatrix}\right) \qquad \text{and} \qquad \mathcal{T}:\ \text{complex conjugation} \quad (\text{but } w\mapsto w \text{ and } \bar{w}\mapsto\bar{w}\,!)\,. \tag{45}$$

The most general $\mathcal{PT}$ deformation compatible with this involution is an SU(1,1) ⊂ SO(2,ℂ) boost,

$$\begin{pmatrix} y^1 \\ y^2 \end{pmatrix} \longmapsto \begin{pmatrix} \cosh\epsilon & -\mathrm{i}\sinh\epsilon \\ \mathrm{i}\sinh\epsilon & \cosh\epsilon \end{pmatrix}\begin{pmatrix} y^1 \\ y^2 \end{pmatrix} = r\begin{pmatrix} \cos(\phi{+}\mathrm{i}\epsilon) \\ \sin(\phi{+}\mathrm{i}\epsilon) \end{pmatrix}, \tag{46}$$

which in the original particle coordinates acquires the less transparent form

$$\begin{pmatrix} x^1 \\ x^2 \\ x^3 \end{pmatrix} \longmapsto \frac{1}{3}\begin{pmatrix} 1{+}2\cosh\epsilon & 1{-}\cosh\epsilon{-}\mathrm{i}\sqrt{3}\sinh\epsilon & 1{-}\cosh\epsilon{+}\mathrm{i}\sqrt{3}\sinh\epsilon \\ 1{-}\cosh\epsilon{+}\mathrm{i}\sqrt{3}\sinh\epsilon & 1{+}2\cosh\epsilon & 1{-}\cosh\epsilon{-}\mathrm{i}\sqrt{3}\sinh\epsilon \\ 1{-}\cosh\epsilon{-}\mathrm{i}\sqrt{3}\sinh\epsilon & 1{-}\cosh\epsilon{+}\mathrm{i}\sqrt{3}\sinh\epsilon & 1{+}2\cosh\epsilon \end{pmatrix}\begin{pmatrix} x^1 \\ x^2 \\ x^3 \end{pmatrix}. \tag{47}$$

In polar and complex coordinates, it is simply

$$\phi \mapsto \phi + \mathrm{i}\epsilon \qquad \Longleftrightarrow \qquad (w,\bar{w}) \mapsto (\mathrm{e}^{-\epsilon}w, \mathrm{e}^{\epsilon}\bar{w})\,, \tag{48}$$

leading to the complex and completely regular potential

$$U_\epsilon(\phi) \;=\; U(\phi{+}\mathrm{i}\epsilon) \;=\; 9\,g(g{-}1)\,\frac{(1+\cosh 6\epsilon\cos 6\phi)+2\mathrm{i}\sinh 6\epsilon\sin 6\phi}{(\cosh 6\epsilon+\cos 6\phi)^2}\;. \tag{49}$$

The energy levels are independent of ϵ, but previously singular states for $g{<}0$ now have become normalizable and thus appear in the physical spectrum (but $q<0$ yields nothing new!),

$$\varepsilon_q \;=\; \tfrac12 q^2 \qquad \text{with} \qquad q \;=\; 3g+\ell \;=\; 3(g+\ell_3) \qquad \text{and} \qquad \ell_3 \ge \max(-g,0)\;, \tag{50}$$

$$\Delta_\epsilon \;\sim\; \mathrm{e}^{-3\epsilon}w^3+\mathrm{e}^{3\epsilon}\bar w^3 \;\sim\; r^3\big(\cosh(3\epsilon)\cos(3\phi)-\mathrm{i}\sinh(3\epsilon)\sin(3\phi)\big)\;\neq 0\;, \tag{51}$$

$$h_\ell^{\epsilon(g)}(w^3,\bar w^3) \;=\; \sum_{k=0}^{\ell_3}(-1)^k\,\frac{\Gamma(1{+}\ell_3)\,\Gamma(g{+}k)\,\Gamma(g{+}\ell_3{-}k)}{\Gamma(2g{+}\ell_3)\,\Gamma(g)\,\Gamma(1{+}k)\,\Gamma(1{+}\ell_3{-}k)}\,(\mathrm{e}^{-\epsilon}w)^{\ell-3k}(\mathrm{e}^{\epsilon}\bar w)^{3k}\;, \tag{52}$$

where the latter expression also holds for $g<0$ with proper $\frac{\infty}{\infty}$ regularization. Here is a list of the low-lying deformed polynomials including the $g{=}{-}1$ tower as a negative-g example, with the adapted notation $(m\,\bar m) \;:=\; \mathrm{e}^{-3(m-\bar m)\epsilon}\,w^{3m}\bar w^{3\bar m}$:

ℓ	$h_\ell^{\epsilon(-1)}$	$h_\ell^{\epsilon(0)}$	$h_\ell^{\epsilon(1)}$	$h_\ell^{\epsilon(2)}$
0		(00)	(00)	(00)
3	(10) − (01)	(10) − (01)	(10) − (01)	(10) − (01)
6	(00)	(20) + (02)	(20) − (11) + (02)	3(20) − 4(11) + 3(02)
9	(30) + 3(21) − 3(12) − (03)	(30) − (03)	(30) − (21) + (12) − (03)	4(30) − 6(21) + 6(12) − 4(03)
⋮	⋮	⋮	⋮	⋮

Since the Hamiltonian depends only on the combination $g(g{-}1)$, we should restrict to $g\ge 1$ but join to the energy eigenstates of any positive g-value the new tower of states at coupling $1{-}g$. This enhances the level degeneracy to

$$\deg(\varepsilon_q) \;\overset{g\ge 0}{=}\; \begin{cases} 1 & \text{for } q<3g \\ 2 & \text{for } q\ge 3g \end{cases} \tag{53}$$

and yields a totally Weyl symmetric and a totally Weyl antisymmetric state at each but the lowest energy levels. These states are related by a new conserved charge,

$$Q^{(g)} \;=\; M_1^{(g-1)}M_1^{(g-2)}\cdots M_1^{(1-g)} \quad\Longrightarrow\quad Q^{(g)}H_\Omega^{(g)} \;=\; Q^{(g)}H_\Omega^{(1-g)} \;=\; H_\Omega^{(g)}Q^{(g)}\;. \tag{54}$$

It is algebraically independent of the Hamiltonian but squares to

$$\big(Q_\epsilon^{(g)}\big)^2 \;=\; \prod_{j=1-g}^{g-1}\big(-2\,H_\epsilon^{(g)}+9j^2\big)\;, \tag{55}$$

which provides a kind of "nonlinear supersymmetry" [21].

5. Tetrahexahedric model: the spectrum

The A_2 model is atypical since it is completely separable. It is much more exciting to study the next more complicated system, which is based on $A_3 \simeq D_3$ and has been termed the "tetrahexahedric model", due to the A_3 Coxeter system and its tetrahedral finite reflection group. For convenience, we rotate the Jacobi relative coordinates adapted to the A_3 root system to a D_3 basis, called Walsh-Hadamard coordinates:

$$\begin{aligned} x^1 &= X+\tfrac12(+x+y+z) \quad , \qquad \partial_{x^1} = \tfrac14\partial_X+\tfrac12(+\partial_x+\partial_y+\partial_z)\,, \\ x^2 &= X+\tfrac12(+x-y-z) \quad , \qquad \partial_{x^2} = \tfrac14\partial_X+\tfrac12(+\partial_x-\partial_y-\partial_z)\,, \\ x^3 &= X+\tfrac12(-x+y-z) \quad , \qquad \partial_{x^3} = \tfrac14\partial_X+\tfrac12(-\partial_x+\partial_y-\partial_z)\,, \\ x^4 &= X+\tfrac12(-x-y+z) \quad , \qquad \partial_{x^4} = \tfrac14\partial_X+\tfrac12(-\partial_x-\partial_y+\partial_z)\,, \end{aligned} \tag{56}$$

and introduce polar coordinates

$$x = r\sin\theta\cos\phi \quad , \qquad y = r\sin\theta\sin\phi \quad , \qquad z = r\cos\theta\,. \tag{57}$$

The angular momenta

$$L_x = -(y\partial_z - z\partial_y)\,, \qquad L_y = -(z\partial_x - x\partial_z)\,, \qquad L_z = -(x\partial_y - y\partial_x) \tag{58}$$

are the building blocks of the S^2 Laplacian

$$L^2 = -(L_x^2+L_y^2+L_z^2) = -\tfrac{1}{\sin\theta}\partial_\theta\sin\theta\,\partial_\theta - \tfrac{1}{\sin^2\theta}\partial_\phi^2\,, \tag{59}$$

and the full and angular Hamiltonians read

$$H = -\tfrac12(\partial_x^2+\partial_y^2+\partial_z^2) + 2g(g{-}1)\Big(\frac{x^2+y^2}{(x^2-y^2)^2}+\frac{y^2+z^2}{(y^2-z^2)^2}+\frac{z^2+x^2}{(z^2-x^2)^2}\Big) \tag{60}$$

and $H_\Omega = \tfrac12 L^2 + U(\vec\theta)$, respectively, with

$$U(\theta,\phi) = 2g(g{-}1)\Big\{\frac{1}{\sin^2\theta\,\cos^2 2\phi}+\frac{\cos^2\theta+\sin^2\theta\cos^2\phi}{(\cos^2\theta-\sin^2\theta\cos^2\phi)^2}+\frac{\cos^2\theta+\sin^2\theta\sin^2\phi}{(\cos^2\theta-\sin^2\theta\sin^2\phi)^2}\Big\}\,, \tag{61}$$

which blows up on 6 great circles of S^2. The potential is invariant under the S_4 Weyl group action, whose elementary reflections are realized as

$$\begin{aligned} s_{x+y}&: (x,y,z)\mapsto(-y,-x,+z) \quad , \qquad s_{x-y}: (x,y,z)\mapsto(+y,+x,+z)\,, \\ s_{y+z}&: (x,y,z)\mapsto(+x,-z,-y) \quad , \qquad s_{y-z}: (x,y,z)\mapsto(+x,+z,+y)\,, \\ s_{z+x}&: (x,y,z)\mapsto(-z,+y,-x) \quad , \qquad s_{z-x}: (x,y,z)\mapsto(+z,+y,+x)\,. \end{aligned} \tag{62}$$

Specializing our general formulæ to $n{=}4$ with $\sigma_3 = x\,y\,z$ and $\sigma_4 = x^4+y^4+z^4$, we get

$$\varepsilon_q = \tfrac12 q\,(q{+}1) \qquad \text{with} \qquad q = 6g+\ell = 6g+3\ell_3{+}4\ell_4\,, \tag{63}$$

$$\Psi_{E,q}(r,\theta,\phi) = j_q(\sqrt{2E}\,r)\,v_q(\theta,\phi)\,, \tag{64}$$

$$v_\ell^{(g)}(\theta,\phi) \sim r^{q+1}\,(\mathcal{D}_x\mathcal{D}_y\mathcal{D}_z)^{\ell_3}\,(\mathcal{D}_x^4{+}\mathcal{D}_y^4{+}\mathcal{D}_z^4)^{\ell_4}\,\Delta^g\,r^{1-12g} = r^{-q}\,\Delta^g\,h_\ell^{(g)}(x,y,z)\,, \tag{65}$$

$$\Delta \sim (x^2-y^2)(y^2-z^2)(x^2-z^2)\,, \tag{66}$$

with the linear Dunkl operators

$$\begin{aligned} \mathcal{D}_x &= \partial_x - \frac{g}{x+y}\,s_{x+y} - \frac{g}{x-y}\,s_{x-y} - \frac{g}{z+x}\,s_{x+z} - \frac{g}{x-z}\,s_{z-x}\,, \\ \mathcal{D}_y &= \partial_y - \frac{g}{y+x}\,s_{x+y} - \frac{g}{y-x}\,s_{x-y} - \frac{g}{y+z}\,s_{y+z} - \frac{g}{y-z}\,s_{y-z}\,, \\ \mathcal{D}_z &= \partial_z - \frac{g}{z+x}\,s_{z+x} - \frac{g}{z-x}\,s_{z-x} - \frac{g}{z+y}\,s_{y+z} - \frac{g}{z-y}\,s_{y-z}\,. \end{aligned} \tag{67}$$

With the short-hand notation $\{rst\} := x^ry^sz^t + x^ry^tz^s + x^sy^tz^r + x^sy^rz^t + x^ty^rz^s + x^ty^sz^r$, we display the eigenstate polynomials for ten lowest energy levels at $g{=}0$ and $g{=}1$,

ℓ	ℓ_3	ℓ_4	$h_\ell^{(0)}$
0	0	0	$\{000\}$
3	1	0	$\{111\}$
4	0	1	$\{400\}-3\{220\}$
6	2	0	$\{600\}-15\{420\}+30\{222\}$
7	1	1	$3\{511\}-5\{331\}$
8	0	2	$\{800\}-28\{620\}+35\{440\}$
9	3	0	$9\{711\}-63\{531\}+70\{333\}$
10	2	1	$\{1000\}-45\{820\}+42\{640\}+504\{622\}-630\{442\}$
11	1	2	$5\{911\}-60\{731\}+63\{551\}$
12	4	0	$36\{1200\}-2376\{1020\}+2445\{840\}+46125\{822\}+4893\{660\}-215250\{642\}+179375\{444\}$
12	0	3	$101\{1200\}-6666\{1020\}+47100\{840\}+8685\{822\}-42609\{660\}-40530\{642\}+33775\{444\}$

ℓ	ℓ_3	ℓ_4	$h_\ell^{(1)}$
0	0	0	$\{000\}$
3	1	0	$\{111\}$
4	0	1	$3\{400\}-11\{220\}$
6	2	0	$3\{600\}-39\{420\}+196\{222\}$
7	1	1	$5\{511\}-13\{331\}$
8	0	2	$\{800\}-20\{620\}+23\{440\}+12\{422\}$
9	3	0	$3\{711\}-27\{531\}+56\{333\}$
10	2	1	$15\{1000\}-425\{820\}+576\{640\}+7568\{622\}-14454\{442\}$
11	1	2	$35\{911\}-476\{731\}+477\{551\}+204\{533\}$
12	4	0	$12\{1200\}-456\{1020\}+657\{840\}+13581\{822\}+1137\{660\}-88842\{642\}+114007\{444\}$
12	0	3	$813\{1200\}-30894\{1020\}+165652\{840\}+72131\{822\}-147943\{660\}-169702\{642\}+57527\{444\}$

The first degeneracy occurs at $\ell = 12 = 4\cdot 3 = 3\cdot 4$, and it increases linearly in steps of 12.

What kind of $\mathcal{PT}$ deformation is admissible in this model? Without loss of generality, we may take $\mathcal{P} = s_{x-y}$. Compatibility with this $\mathcal{PT}$ involution restricts the coset SO$(3,\mathbb{C})/$SO$(3,\mathbb{R})$ of complex orthogonal deformations to a two-parameter family, which contains e.g. the SU(1,1) boost in the xy plane,

$$\begin{pmatrix} y^1 \\ y^2 \\ y^3 \end{pmatrix} \longmapsto \begin{pmatrix} \cosh\epsilon & -\mathrm{i}\sinh\epsilon & 0 \\ \mathrm{i}\sinh\epsilon & \cosh\epsilon & 0 \\ 0 & 0 & 1 \end{pmatrix} \begin{pmatrix} y^1 \\ y^2 \\ y^3 \end{pmatrix} = r \begin{pmatrix} \sin\theta\cos(\phi{+}\mathrm{i}\epsilon) \\ \sin\theta\sin(\phi{+}\mathrm{i}\epsilon) \\ \cos\theta \end{pmatrix} \tag{68}$$

equivalent to

$$\begin{pmatrix} x^1 \\ x^2 \\ x^3 \\ x^4 \end{pmatrix} \longmapsto \begin{pmatrix} 1{+}\cosh\epsilon & \mathrm{i}\sinh\epsilon & -\mathrm{i}\sinh\epsilon & 1{-}\cosh\epsilon \\ -\mathrm{i}\sinh\epsilon & 1{+}\cosh\epsilon & 1{-}\cosh\epsilon & \mathrm{i}\sinh\epsilon \\ \mathrm{i}\sinh\epsilon & 1{-}\cosh\epsilon & 1{+}\cosh\epsilon & -\mathrm{i}\sinh\epsilon \\ 1{-}\cosh\epsilon & -\mathrm{i}\sinh\epsilon & \mathrm{i}\sinh\epsilon & 1{+}\cosh\epsilon \end{pmatrix} \begin{pmatrix} x^1 \\ x^2 \\ x^3 \\ x^4 \end{pmatrix} \tag{69}$$

but more simply given by

$$\phi \mapsto \phi + \mathrm{i}\epsilon \qquad \text{or} \qquad (x \pm \mathrm{i}y, z) \mapsto \big(\mathrm{e}^{\mp\epsilon}(x \pm \mathrm{i}y), z\big)\ . \tag{70}$$

The ensuing complex potential

$$\frac{U_\epsilon(\theta,\phi)}{2g(g{-}1)} = \frac{1}{\sin^2\theta\,\cos^2 2(\phi{+}\mathrm{i}\epsilon)} + \frac{\cos^2\theta + \sin^2\theta\cos^2(\phi{+}\mathrm{i}\epsilon)}{(\cos^2\theta - \sin^2\theta\cos^2(\phi{+}\mathrm{i}\epsilon))^2} + \frac{\cos^2\theta + \sin^2\theta\sin^2(\phi{+}\mathrm{i}\epsilon)}{(\cos^2\theta - \sin^2\theta\sin^2(\phi{+}\mathrm{i}\epsilon))^2} \tag{71}$$

is free of singular lines but still blows up at 5 pairs of antipodal points. Because these singularities are also carried by

$$\Delta_\epsilon \sim \sin^2\theta\,\cos 2(\phi{+}\mathrm{i}\epsilon)\,(\cos^2\theta - \sin^2\theta\cos^2(\phi{+}\mathrm{i}\epsilon))\,(\cos^2\theta - \sin^2\theta\sin^2(\phi{+}\mathrm{i}\epsilon))\ , \tag{72}$$

the formal $g{<}0$ wave functions $v_\ell^{\epsilon(g)} \sim \Delta_\epsilon^g$, obtained by simply applying the deformation to the argument of $v_\ell^{(g)}$, remain non-normalizable. Hence, except for the free system ($g{=}1$), the degeneracy remains to be given by $\deg_4(\ell)$ for any $g \in \mathbb{N}$.

If we were able to invent a (nonlinear) $\mathcal{PT}$ deformation (for a suitable involution) which gets rid of all zeros of Δ (and hence of the potential), then the $g{<}0$ states would be resurrected, extending the eigenspace for any given $g{\ge}2$ by the new states with angular momenta $\tilde\ell$ for a coupling $\tilde g = 1{-}g \le -1$. For small values of $\tilde\ell$ this requires $q < 0$, so the second branch of (63) would get occupied, and the $g{>}0$ spectrum would be modified to

$$\varepsilon_q = \tfrac12 q\,(q{+}1) \qquad \text{with} \qquad q = 6g + \ell = 6(1{-}g) + \tilde\ell \quad \text{or} \quad -1{-}q = 6(1{-}g) + \tilde\ell\ , \tag{73}$$

where we restricted $q \ge 0$ for uniqueness, and both ℓ and $\tilde\ell$ are composed of 3's and 4's. This would enhance the degeneracy to

$$\begin{aligned}\deg(\varepsilon_q) &= \deg_4(\ell) + \deg_4(\tilde\ell) = \deg_4(q{-}6g) + \deg_4(q{+}6g{-}6) + \deg_4(-q{+}6g{-}7)\\ &= \begin{cases} g{-}1 + \begin{Bmatrix} 0 & \text{for} & q+6g = 0,3,4,7,8,11 \bmod 12 \\ 1 & \text{for} & q+6g = 1,2,5,6,9,10 \bmod 12 \end{Bmatrix} & \text{if} \quad q < 6g{-}6 \\[1ex] \left\lfloor \frac{q}{6} \right\rfloor + \begin{Bmatrix} 0 & \text{for} & q = 1,2,5 \bmod 6 \\ 1 & \text{for} & q = 0,3,4 \bmod 6 \end{Bmatrix} & \text{if} \quad q \ge 6g{-}6 \end{cases}\end{aligned} \tag{74}$$

which shows that its linear growth with q would become g-independent at high energy.

6. Tetrahexahedric model: intertwiner & integrability

As can already be glanced from the tetrahexahedric case, the angular Calogero models possess a rich structure of intertwiners. For their construction, we need the angular Dunkl operators:

$$\begin{aligned}\mathcal{L}_x &= L_x + g\big\{\tfrac{z}{x-y}s_{x-y} - \tfrac{z}{x+y}s_{x+y} - \tfrac{y}{x-z}s_{z-x} + \tfrac{y}{z+x}s_{z+x} - \tfrac{y+z}{y-z}s_{y-z} + \tfrac{y-z}{y+z}s_{y+z}\big\}\ ,\\ \mathcal{L}_y &= L_y + g\big\{\tfrac{x}{y-z}s_{y-z} - \tfrac{x}{y+z}s_{y+z} - \tfrac{z}{y-x}s_{x-y} + \tfrac{z}{y+x}s_{x+y} - \tfrac{z+x}{z-x}s_{z-x} + \tfrac{z-x}{z+x}s_{z+x}\big\}\ ,\\ \mathcal{L}_z &= L_z + g\big\{\tfrac{y}{z-x}s_{z-x} - \tfrac{y}{z+x}s_{z+x} - \tfrac{x}{z-y}s_{y-z} + \tfrac{x}{z+y}s_{y+z} - \tfrac{x+y}{x-y}s_{x-y} + \tfrac{x-y}{x+y}s_{x+y}\big\}\ .\end{aligned} \tag{75}$$

Although we cannot (yet) prove it, we have ample evidence that there exist exactly two algebraically independent angular intertwiners. The first one, of order three, derives from the Weyl antiinvariant

$$\mathcal{M}_3 \sim \tfrac16\big(\mathcal{L}_x\mathcal{L}_y\mathcal{L}_z + \mathcal{L}_x\mathcal{L}_z\mathcal{L}_y + \mathcal{L}_y\mathcal{L}_z\mathcal{L}_x + \mathcal{L}_y\mathcal{L}_x\mathcal{L}_z + \mathcal{L}_z\mathcal{L}_x\mathcal{L}_y + \mathcal{L}_z\mathcal{L}_y\mathcal{L}_x\big)\ , \tag{76}$$

whose restriction to symmetric functions yields

$$
\begin{aligned}
M_3 \sim\ & y^2 z\partial_{zxx} - yz^2\partial_{xxy} + \tfrac12(y^2-z^2)\,\partial_{xx} + 4g\,\tfrac{yz}{y^2-z^2}\big(yz\partial_{xx} + x^2\partial_{yz} - zx\partial_{xy}\big) \\
& + g\Big[2g\,y^2z^2\big(\tfrac{8g}{(x^2-y^2)(z^2-x^2)} + \tfrac{16g}{(z^2-x^2)(y^2-z^2)} - \tfrac{2g-1}{(x^2-y^2)^2} + \tfrac{2g-1}{(z^2-x^2)^2}\big) \\
& \quad - \tfrac{2x^2y^2}{(z^2-x^2)^2} + \tfrac{2x^2z^2}{(x^2-y^2)^2} - \tfrac{2y^2}{x^2-y^2} - \tfrac{2z^2}{z^2-x^2} - 2\tfrac{y^2+z^2}{y^2-z^2}\Big]\,x\partial_x \\
& + 2g(g-1)(g+2)\,x^2\Big[\tfrac{y^2+z^2}{(y^2-z^2)^2} + z\big(\tfrac{1}{(y-z)^3} - \tfrac{1}{(y+z)^3}\big)\Big] + g\,(2g^2+8g-1)\,\tfrac{y^2+z^2}{y^2-z^2} \\
& + 2g^2(8+9g)\tfrac{x^2y^2z^2}{(x^2-y^2)(x^2-z^2)(y^2-z^2)} - \tfrac23 g^3\,\tfrac{x^6+y^6+z^6}{(x^2-y^2)(x^2-z^2)(y^2-z^2)} + \text{cyclic}(x\,y\,z)\,.
\end{aligned}
\tag{77}
$$

It somewhat simplifies in the "potential-free frame",

$$
\begin{aligned}
\Delta^{-g} M_3\,\Delta^{g} \sim\ & y^2 z\partial_{zxx} - yz^2\partial_{xxy} + \tfrac12(y^2-z^2)\,\partial_{xx} + 2g\,\tfrac{y^2z^2(y^2-z^2)}{(x^2-y^2)(x^2-z^2)}\,\partial_{xx} \\
& + 4g\,\tfrac{xy^2z}{x^2-z^2}\,\partial_{xz} + 2g\,x\Big[\tfrac{y^2(x^2+3z^2)}{(x^2-z^2)^2} - \tfrac{z^2(x^2+3y^2)}{(x^2-y^2)^2}\Big]\,\partial_x + \text{cyclic}(x\,y\,z)\,.
\end{aligned}
\tag{78}
$$

The second one, of order six, emerges from

$$
\mathcal{M}_6 \sim \{\mathcal{L}_x^4,\mathcal{L}_y^2\} - \{\mathcal{L}_y^4,\mathcal{L}_x^2\} + \{\mathcal{L}_y^4,\mathcal{L}_z^2\} - \{\mathcal{L}_z^4,\mathcal{L}_y^2\} + \{\mathcal{L}_z^4,\mathcal{L}_x^2\} - \{\mathcal{L}_x^4,\mathcal{L}_z^2\}\,, \tag{79}
$$

whose symmetric restriction $M_6 = \text{res}(\mathcal{M}_6)$ is a rather lengthy expression not displayed here. (Hopefully $\Delta^{-g} M_6\,\Delta^{g}$ is a bit shorter.)

Higher angular intertwiners can be reduced to M_3 and M_6 in combination with the conserved charges

$$
J_k := \text{res}\big(\mathcal{L}_x^k + \mathcal{L}_y^k + \mathcal{L}_z^k\big) \qquad \text{for} \quad k = (0,)2,4,6 \tag{80}
$$

$$
\text{with} \qquad J_0 = C_0 = 1 \qquad \text{and} \qquad J_2 = -C_2 = -2\,H_\Omega + 6g(6g+1)\,. \tag{81}
$$

In fact, any word in the letters $\{J_2, J_4, J_6\}$ is conserved, but they are not in involution because $[J_4, J_6]$ and $\{J_4, J_6\}$ both are nontrivial new words, and the center of the algebra of conserved charges is spanned by the Casimir J_2 (and J_0 of course). This is different from the situation in the full A_3 model, since we deal with a Dunkl-deformed nonabelian so(3) algebra. However, like in the full model, higher conserved charges are algebraically dependent, e.g.

$$
\begin{aligned}
6J_8 =\ & 8J_6J_2 + 3J_4J_4 - 6J_4J_2J_2 + J_2J_2J_2J_2 \\
& - 12(8+5g+12g^2)J_6 + 4(34+23g+30g^2)J_4J_2 - 8(5+3g+3g^2)J_2J_2J_2 \\
& + 24(13+15g-102g^2-72g^3)J_4 - 4(43+70g-252g^2-144g^3)J_2J_2 \\
& - 48(1+3g)(1+4g)(1-12g)J_2\,.
\end{aligned}
\tag{82}
$$

The intertwining relations for the conserved charges beyond the Hamiltonian are also more involved than in the full model. In particular, they are no longer diagonal in either set. The basic relations for M_3 are

$$
\begin{aligned}
M_3^{(g)} J_2^{(g)} =\ & \big(J_2^{(g+1)} - 6(7+12g)\big)\,M_3^{(g)}\,, \\
M_3^{(g)} J_4^{(g)} =\ & \big(J_4^{(g+1)} - 4(11+12g)J_2^{(g+1)} + 48(26+73g+48g^2)\big)\,M_3^{(g)} \\
& + 2\,M_6^{(g)}\,, \\
M_3^{(g)} J_6^{(g)} =\ & \big(J_6^{(g+1)} - (35+36g)J_4^{(g+1)} - 3(7+4g)J_2^{(g+1)}J_2^{(g+1)} \\
& + 2(1111+2668g+1392g^2)J_2^{(g+1)} \\
& + 96(457+1933g+2717g^2+1368g^3+144g^4)\big)\,M_3^{(g)} \\
& + \big(3J_2^{(g+1)} - (115+200g+48g^2)\big)\,M_6^{(g)}\,,
\end{aligned}
\tag{83}
$$

and those for M_6 look similar (but more lengthy).

Finally, for $g > 0$ the intertwining ladder $1{-}g \to 2{-}g \to \ldots \to g{-}2 \to g{-}1 \to g$ extends the algebra of conserved charges to a $\mathbb{Z}_2$-graded one by adjoining the new "odd" charges ($* = 3$ or 6)

$$Q^{(g)} = M_*^{(g-1)} M_*^{(g-2)} \cdots M_*^{(1-g)} \tag{84}$$

relating bosonic and fermionic states at any given energy level and positive integer coupling g. However, these operators are only well defined if a suitable $\mathcal{PT}$ deformation can fully regularize the potential (as is the case for the Pöschl-Teller model) and so allows one to double the state space by combining the states at $1{-}g$ and g. It is clear that such fermionic charges are independent of the J_k because they are of odd order, but they square to a polynomial in the J_k. We conjecture that the various above choices for Q (depending on the details of '$*$') are all related by multiplication with appropriate even charges, and so the $\mathbb{Z}_2$-graded nonlinear algebra would be generated by $\{Q, J_2, J_4, J_6\}$.

7. Summary and outlook

The angular rational A_{n-1} Calogero model for rank $n{-}1 \geq 3$ is superintegrable but not separable. It describes the Weyl-symmetric states of a particle on S^{n-2} subject to a particular S_n-symmetric potential. We have formulated the spectral problem and given a constructive solution, including a discussion of the conserved charges and the Hamiltonian intertwiners shifting the coupling $g \mapsto g{+}1$. A complex $\mathcal{PT}$ deformation reduces the dimension of the singular loci of the potential. If all singularities can be removed (as we demonstrate for $n{=}3$), then the degeneracy of the energy spectrum will essentially double, and for $g \in \mathbb{Z}$ the algebra of conserved charges gets enhanced to a $\mathbb{Z}_2$-graded one ("supersymmetrized") by the appearance of an additional, odd, charge. We have analyzed in some detail the cases of $n{=}3$ (the hexagonal or Pöschl-Teller model) and $n{=}4$ (the tetrahexahedric model).

Various issues remain unresolved. First, the question of Liouville charges (the first test appears at $n{=}5$) and, more generally, the algebra of the conserved charges depend on the structure of the deformed so($n{-}1$) algebra of Dunkl-deformed angular momenta $\mathcal{L}_{ij}$, about which little is known. Second, a classification of all independent Hamiltonian intertwiners, via Weyl antiinvariants built from the $\mathcal{L}_{ij}$, is open. Third, a systematic study of $\mathcal{PT}$ deformations compatible with the Weyl symmetry is warranted, in order to decide if and when the state space can be doubled and a nonlinear supersymmetry can be realized for integral values of g. Finally, there is always the potential generalization to trigonometric, hyperbolic or elliptic Calogero systems.

References

[1] A.P. Polychronakos A P 2006 *J. Phys.* A: *Math. Gen.* **39** 12793–846
[2] Calogero F 1971 *J. Math. Phys.* **12** 419–436
[3] Olshanetsky M A and Perelomov A M 1981 *Phys. Rept.* **71** 313–400
[4] Olshanetsky M A and Perelomov A M 1983 *Phys. Rept.* **94** 313–404
[5] Wojciechowski S 1983 *Phys. Lett.* **95**A 279–281
[6] Dunkl C F 1989 *Trans. Amer. Math. Soc.* **311** 167–183
[7] Chalykh O A and Veselov A P 1990 *Commun. Math. Phys.* **126** 597–611
[8] Heckman G J 1991 *Progr. Math.* **101** 181–191
[9] Feigin M V 2003 *Theor. Math. Phys.* **135** 497–509
[10] Fring A and Znojil M 2008 *J. Phys.* A **41** 194010
[11] Fring A 2013 *Phil. Trans. Roy. Soc. Lond.* A **371** 20120046
[12] Hakobyan T, Nersessian A and Yeghikyan V 2009 *J. Phys.* A: *Math. Theor.* **42** 205206
[13] Feigin M, Lechtenfeld O and Polychronakos A 2013 *J. High Energy Phys.* JHEP07(2013)162
[14] Correa F, Lechtenfeld O and Plyushchay M 2014 *J. High Energy Phys.* JHEP04(2014)151
[15] Feigin M and Hakobyan T 2015 *J. High Energy Phys.* JHEP11(2015)107
[16] Correa C and Lechtenfeld O 2015 *J. High Energy Phys.* JHEP10(2015)191
[17] Higgs P W 1979 *J. Phys.* A: *Math. Gen.* **12** 309–323
[18] Leemon H I 1979 *J. Phys.* A: *Math. Gen.* **12** 489–501
[19] Gupta H, Gwyther C E and Miller J C P 1958 *Tables of Partitions* (Royal Society Mathematical Tables Volume 4) (Cambridge: Cambridge University Press)
[20] Bender C M and Boettcher S 1989 *Phys. Rev. Lett.* **80** 5243–46
[21] Klishevich S M and Plyushchay M S 2001 *Nucl. Phys.* B **606** 583–612

ISQS IOP Publishing
IOP Conf. Series: Journal of Physics: Conf. Series **804** (2017) 012012 doi:10.1088/1742-6596/804/1/012012

New spinor fields classes and applications

Roldao da Rocha
CMCC, Universidade Federal do ABC,
09210-580, Santo André, Brazil
E-mail: roldao.rocha@ufabc.edu.br

Abstract. After revisiting the well-known Lounesto classification of spinor fields, wherefrom the bilinear covariants are considered as the landmark beyond Wigner classification, relevant features of regular and singular spinor fields are presented. Hence, non-standard spinor fields are scrutinised, together with their associated dynamics, paving recently found applications in Physics. The case of the new classes of spinors on 7-manifolds is revisited to provide concrete examples.

1. Introduction

The Lounesto classification of spinor fields [1] is much more recent than previous classifications, like the Wigner and the Cartan classifications [2]. The well-known Weyl, Dirac, and Majorana spinor fields occupy, in the Lounesto classification, a restricted – but not less important – place, among the whole panorama of possibilities comprised by the six disjoint classes of regular and singular spinors, in the Lounesto classification [1]. To fix the denomination of spinors in the six spinor field classes proposed by Lounesto, the standard – textbook – Dirac spinor is an eigenspinor of the parity operator and occupied a subset in the first class of regular spinors. The Weyl spinor is a very particular dipole spinor in the sixth class in Lounesto classification [1]. Possibly the most astonishing class of spinors is the fifth one, consisting of flagpole spinor fields. This class encompasses Majorana spinors, spinors that are solutions of the Dirac equation in a Kerr black hole background [3, 4, 5], and, moreover, also encompasses Elko (dark) spinor field – that are eigenspinors of the charge conjugation operator with dual helicity, having mass dimension one – as a flagpole spinor [6]. Hence, the fifth class in Lounesto classification has a subset of spinors that are neutral and another subset of spinors that are charged, whereas the fifth class of spinors has a subset of fermions of spin 1/2 that has mass dimension one and another subset of fermions that have canonical mass dimension one. To find the complementary subsets in this fifth class is still an open problem in the literature. In spite of those known examples in the field theoretical literature [6, 7, 8, 9, 10, 11, 12, 13, 14, 15, 16, 17, 20, 21], the dynamics ruling the spinor fields in this class still lacks, despite of the previously mentioned known spinor fields [22, 23, 24]. A bottom-up method consisting of searching for sharp examples. The methods designated in Refs. [7, 8] are successful efforts towards the derivation of the dynamics that govern each subset in each class of spinor fields, in the Lounesto spinor fields classification. Applications gravitation and field theory are abundant [9, 11, 10, 12]. Elko spinor fields were the first example of a (singular) spinor field that is beyond the Dirac, Majorana and Weyl spinor fields. Without going further in the second quantised fields, the 4-component Elko spinors are very similar to the Majorana spinors, since both are eigenspinors of the charge conjugation operator, in a subset of neutral spinors in the class of flagpole spinors, in Lounesto classification. However, Majorana's choice did not originally include both self-conjugate and anti-self-conjugate spinors. From first principles, Elko spinors consist of a four spinors family that, besides the charge conjugation operator, needs the helicity operator to act on them. Therefore Elko present dual helicity. Recently, flag-dipoles spinors, that are singular spinors in the fourth class of Lounesto classification, have been found to be solutions of the Dirac equation in a $f(R)$-background with torsion, in the context of Einstein-Sciama-Kibble (ESK) gravity. These kind of singular spinors work appropriately as a source of matter in ESK gravity, being the first explicit construction in Physics of flag-dipole spinors [8, 11]. Furthermore, solutions of the Dirac equation in a Kerr black hole background have been more recently shown to generate flag-dipoles and flagpoles spinor fields rather than the expected Dirac spinor field [3, 5]. Flagpoles comprise realisations of mass dimension one Elko spinors, which are candidates for solving the dark matter problem [9, 10, 12]. Far beyond the standard Dirac spinors paradigm, regular and singular spinors range new applications that encompass dark spinors Hawking radiation across black holes and black strings [17, 18], cosmology, gravity on extra-dimensional thick branes [29, 30], particle physics and condensed matter [9, 10]. In fact, this matter has been further explored in the context of black hole thermodynamics, where tunnelling methods were studied for the eigenspinors of the charge conjugation operator having dual helicity [15, 16], as special type of flagpoles [17, 19, 23]. Experimental signatures of type-5, flagpole, dark spinors in Lounesto classification are proposed as being a byproduct of the Higgs bosonic field at LHC [20]. In fact, the existence of spin $1/2$ fermions with canonical mass dimension one, in the Lounesto classification, demands the Weinberg no go theorem must be eloped [15].

New classes of spinors and the Lounesto classification were primordially defined in four-dimensional, Minkowski, spacetime. On 2015, new surprises occurred, now consisting of the discovery of additional classes of spinors on higher dimensional spacetimes. Some of the spinors fields in this context have defined dynamics, and can be obtained from the analysis of the geometric Fierz identities, that constrain the bilinear covariants. This procedure turns possible to classify spinors fields on manifolds of arbitrary dimensions and signatures. Constructions on specific dimensions and signatures have been derived. In fact, the classification of spinor fields on seven-dimensional Riemannian [25] and Lorentzian [26] manifolds has been constructed, showing that the Lounesto classification in Minkowski spacetime can also exist in higher dimensional spacetimes [25, 26]. Particular interest on these new solutions has arisen in $\mathrm{AdS}_4 \times S^7$ compactified setups, since the new classes of spinor fields solutions can represent new solutions in string theory. Besides, new classes of spinor fields have been also shown to exist on compactifications of type $\mathrm{AdS}_5 \times S^5$, being further important to the analysis of fermion localisation on brane-worlds. Moreover, new fermionic field in a bulk correspond to the new spinor field classes that are evinced from the geometric Fierz identities, in five-dimensional Lorentzian spacetimes. Our main goal here is to present this framework and their applications, discussing the utility of considering spinors that are quite distinct of the usual Dirac, Majorana and Weyl constructions.

This paper is organised as follows: the bilinear covariants are the stage to revisit the classification of spinor fields in Minkowski spacetime, according to the Lounesto classification prescription, and the Fierz aggregate and its related boomerang are defined. The geometric Fierz identities are employed and, from the admissible pairings between spinor fields, the number of classes in the spinor field classification is constrained, by the geometric Fierz identities [27]. We study the case of spinor fields on Lorentzian 5-manifolds and conclude that five-dimensional spinors pertain to solely one class, through a similar approach to the one previously acquired in seven-dimensional [25, 26, 27]. Spinor fields on 5-manifolds can be then classified in eight classes, from the classification of quaternionic spinor fields. One of the new classes encompasses the standard five-dimensional spinor fields, and six others non-trivial classes provide new candidates for physical solutions, for instance, in thick brane models and gravity on AdS_5.

2. Geometric Fierz Identities and Bilinear Covariants

In order to fix the notation, let an oriented manifolds (M,g) endowed with the metric g, of signature $p-q$, has a spacetime structure [22], having an exterior bundle $\Omega(M)=\oplus_{i=0}^{\infty}\Omega_i(M)$ whereupon the endomorphisms can be usually defined, for a k-form $a\in\Omega^k(M)$, namely, the grade involution, $\hat{a}=(-1)^k a$, that is an automorphism; the reversion, $\tilde{a}=(-1)^{[k/2]}a$, where $[k]$ is the integer part of k, that is an anti-automorphism; and the conjugation anti-automorphism, defined by the commutative composition between the grade involution and the reversion, denoted hereon by $\bar{a}\equiv\hat{\tilde{a}}=\tilde{\hat{a}}$. Those three morphism can be extended to the whole exterior bundle, by linearity and, further, to the Clifford bundle, likewise. By endowing the exterior bundle $\Omega(M)$ with the universal Clifford product $u\circ a=u\wedge a+u\lrcorner a$, for all form fields $u\in\sec\Omega^1(M)$, where $\lrcorner$ is the left contraction, the Clifford bundle can be, therefore, defined.

The spinor bundle in four-dimensional Minkowski spacetime corresponds to the bundle $\mathcal{P}_{\mathrm{Spin}^e_{1,3}}(M)\times_\sigma\mathbb{C}^4$. Classical spinor fields $\psi\in\sec\mathcal{P}_{\mathrm{Spin}^e_{1,3}}(M)\times_\sigma\mathbb{C}^4$ carry the $\left(\frac{1}{2},0\right)\oplus\left(0,\frac{1}{2}\right)$ representations σ of the Lorentz group. The bundle $\mathcal{P}_{\mathrm{Spin}^e_{1,3}}(M)$ has fibers, whereupon the group $\mathrm{Spin}^e_{1,3}$ acts. The bilinear covariants are sections of the exterior bundle $\Omega(M)$, whose basis is given by $\{e^\mu\}$ [1] (hereupon $\mu,\nu=0,1,2,3$ are Minkowski spacetime indexes):

$$\sigma = \bar{\psi}\psi\in\sec\Omega^0(M)\,, \tag{1a}$$
$$\mathbf{J} = \mathrm{J}_\mu e^\mu\in\sec\Omega^1(M)\,, \tag{1b}$$
$$\mathbf{S} = S_{\mu\nu}e^\mu\wedge e^\nu\in\sec\Omega^2(M)\,, \tag{1c}$$
$$\mathbf{K} = K_\mu e^\nu\in\sec\Omega^3(M)\,, \tag{1d}$$
$$\omega = \bar{\psi}\gamma_5\psi\in\sec\Omega^4(M)\,, \tag{1e}$$

where

$$\mathrm{J}_\mu = \bar{\psi}\gamma_\mu\psi\,, \tag{2}$$
$$S_{\mu\nu} = \tfrac{1}{2}i\bar{\psi}\gamma_{\mu\nu}\psi\,, \tag{3}$$
$$K_\mu = i\bar{\psi}\gamma_5\gamma_\mu\psi\,, \tag{4}$$

are the respective components in Eqs. (1b) – (1d) and $\gamma_5:=i\gamma_0\gamma_1\gamma_2\gamma_3$, where the spinor conjugation is denoted by $\bar{\psi}=\psi^\dagger\gamma_0$ [2, 22]. The gamma matrices satisfy the usual Clifford relation $\gamma_\mu\gamma_\nu+\gamma_\nu\gamma_\mu=2g_{\mu\nu}\mathbf{1}$, where $g_{\mu\nu}$ denotes the Minkowski spacetime metric components. The physical observables, exclusively for the Dirac electron theory, are realised by bilinear covariants. In fact, the current density $\mathbf{J}$, the spin density $\mathbf{S}$, and the chiral current $\mathbf{K}$, satisfy the Fierz identities [1]

$$-(\omega+\sigma\gamma_5)\mathbf{S}=\mathbf{J}\wedge\mathbf{K},\qquad \mathbf{K}\lrcorner\mathbf{K}+\mathbf{J}\lrcorner\mathbf{J}=0=\mathbf{J}\llcorner\mathbf{K},\qquad \mathbf{J}\lrcorner\mathbf{J}=\omega^2+\sigma^2\,. \tag{5}$$

Lounesto derived, from the bilinear covariants, a classification of spinor fields [1]. The expression $\mathbf{J}\neq 0$ holds for all spinors in Lounesto class. However, this condition that was firstly motivated by the Dirac electron theory can be circumvented in three additional classes that were recently derived in Minkowski spacetime [24], conjectured to consist of ghost spinors:

1) $\omega\neq 0,\quad \sigma\neq 0,\quad \mathbf{K}\neq 0,\quad \mathbf{S}\neq 0$ (regular spinors) (6a)
2) $\omega=0,\quad \sigma\neq 0,\quad \mathbf{K}\neq 0,\quad \mathbf{S}\neq 0$ (regular spinors) (6b)
3) $\omega\neq 0,\quad \sigma=0,\quad \mathbf{K}\neq 0\quad \mathbf{S}\neq 0$ (regular spinors) (6c)
4) $\omega=\sigma=0,\quad \mathbf{K}\neq 0,\quad \mathbf{S}\neq 0$ (flag-dipole, singular, spinors) (6d)
5) $\omega=\sigma=0,\quad \mathbf{K}=0,\quad \mathbf{S}\neq 0$ (flagpole, singular, spinors) (6e)
6) $\omega=\sigma=0,\quad \mathbf{K}\neq 0,\quad \mathbf{S}=0$ (dipole, singular, spinors) (6f)

Singular spinors consist of flag-dipole, flagpole, and dipole spinors, respectively in the fourth, fifth, and sixth classes in the just mentioned six classes. The standard Dirac spinor is an element of the set of regular spinors in class 1. Moreover, Majorana spinors – that have mass dimension 3/2 – and Elko spinor fields – presenting canonical mass dimension 1 – are neutral spinors that embrace particular realisations of flagpole spinors, that further includes a set of spinors that are solutions of the Dirac equation in a Kerr black hole background. The chiral Weyl spinors consist of a tiny subset of dipole spinors. Nevertheless, regular, flagpole, flag-dipole, and dipole spinor fields provide a plethora of opportunities that have not been explored in the literature whatsoever. Although dozen of explicit constructions, consisting of non-standard spinors, with defined dynamics, were derived in the last decade, like the flag-dipole spinors in Einstein-Sciama-Kibble theory of gravity [11], that was the first concrete example of flag-dipole spinor field in the literature, largely searched for by Lounesto itself [1]. than those realisations.

The Fierz-Pauli-Kofink (FPK) identities (5) are well known not to be valid for the case of singular spinors. In this case, based upon a Fierz aggregate,

$$\mathbf{Z} = \gamma_5(\omega + \mathbf{K}) + i\mathbf{S} + \mathbf{J} + \sigma\,, \tag{7}$$

the FPK identities (5) can be replaced by the the most general equations

$$4\mathbf{Z} = \frac{i}{\omega}\mathbf{Z}\gamma_5\mathbf{Z} = \frac{1}{\mathrm{J}_\mu}\mathbf{Z}\gamma_\mu\mathbf{Z} = \frac{i}{S_{\mu\nu}}\mathbf{Z}\gamma_\mu\gamma_\nu\mathbf{Z} = -\frac{1}{K_\mu}\mathbf{Z}\gamma_5\gamma_\mu\mathbf{Z}\,, \tag{8}$$

that can be led to (5) when not both σ and ω equal zero, namely, for the case of a regular spinor field.

The geometric underlying content of the Fierz aggregate can be encompassed by physical considerations. The irreducible representation of the Lorentz group can be split into homogeneous components of the Fierz aggregate, as follows:

$$[(0,\tfrac{1}{2})\oplus(\tfrac{1}{2},0)]\otimes[(0,\tfrac{1}{2})\oplus(\tfrac{1}{2},0)] = \underbrace{\overbrace{(0,0)}}_{\sigma}\oplus\underbrace{\overbrace{\left(\frac{1}{2},\frac{1}{2}\right)}}_{\mathbf{J}}\oplus\underbrace{\overbrace{(1,0)\oplus(0,1)}}_{\mathbf{S}}\oplus\underbrace{\overbrace{\left(\frac{1}{2},\frac{1}{2}\right)}}_{\mathbf{K}}\oplus\underbrace{\overbrace{(0,0)}}_{\omega} \tag{9}$$

It is worth to emphasise that in classes 4 and 5, in Lounesto classification, the form fields $\{\mathbf{J},\mathbf{K}\}$ are not elements of a basis for the dual Minkowski spacetime cotangent bundle, since for these cases the linear span of the set $\{\mathbf{J},\mathbf{K}\}$ collapses into a null line with direction provided by $\mathbf{J}$. The 1-form field $\mathbf{J}$ is interpreted as being a pole, and flagpoles are consequently elements of the class 5 in Lounesto classification. In fact, for this case one finds by definition that $\mathbf{K}=0$ and $\mathbf{S}\neq 0$. The 2-form field $\mathbf{S}$ can be realised as being a plaquette, namely, a flag-pole. Regarding now type-4 spinor fields, both the 2-form field $\mathbf{S}$ and the 1-form field $\mathbf{K}$ do not equal zero, thus constituting a flag-dipole structure. Both concepts encompass Penrose flagpoles [2]. Moreover, flagpole and dipoles are, by definition, limiting cases of flag-dipole spinor fields.

Generalising the FPK relations beyond the Minkowski spacetime setup has brought unexpectedly prominent panoramas for higher dimensional theories in Physics. Interesting features of newly found non-standard spinor fields, with respect to the textbook ones, have also unexpected applications. In this context, the standard FPK identities were extended and introduced for form-valued spinor bilinear covariants, for arbitrary spacetime dimensions and metric signatures [27, 28]. Indeed, Refs. [27, 28] show that the geometric Fierz identities strongly obstruct the number, and even the existence, of classes of pinor and spinor fields. We have recently found how to circumvent these constructions, to provide two new classes of spinors (on the spin bundle of a Riemannian 7-manifold) [25], three new classes of spinor fields (on the spin bundle of a Lorentzian 7-manifold) [26], and some additional spinor classes (on the spin bundle of a Lorentzian 5-manifold, that serves as a 5-dimensional bulk) [5].

On manifolds of arbitrary dimensions having spin structure, the Kähler-Atiyah bundle $(\sec\Omega(M), \circ)$ is then employed, whose associated spin bundle S a module setup that is defined by the morphism $\gamma : (\Omega(M), \circ) \to (\mathrm{End}(S), \bullet)$ [2, 27, 28]. In addition, there is a splitting $S = S_{(0)} \oplus S_{(1)}$ for an idempotent $\mathrm{R} \in \Gamma(\mathrm{End}(S))$ in the space of the endomorphisms $\sec \mathrm{End}(S))$ [27]. The sets $S_{(0)}$ and $S_{(1)}$ are respectively determined by the eigenvalues ± 1 of R [2]. Spin projectors are defined by $\Pi_\pm = \frac{1}{2}(I \pm \mathrm{R})$, providing the splitting $S = \Pi_+(S) \oplus \Pi_-(S) \equiv S^+ \oplus S^-$. Sections of $S^\pm$ are called Majorana spinors when $p-q \equiv 7 \mod 8$ [25, 26]. The complex structure $J \in \Gamma(\mathrm{End}(S))$ further equips the Kähler-Atiyah bundle [27], and an endomorphism D on the spin bundle, that commutes with the grade involution, is defined, [27], by $D \bullet D = (-1)^{\frac{p-q+1}{4}} I$. A non-degenerate bilinear pairing $B : S \times S \to \mathbb{R}$ induces a more general bilinear form [27, 28], by the complexification procedure [25, 26]. In fact, smooth sections $\psi, \psi' \in \Gamma(S)$ are employed to define a bilinear form $\beta_0 : S \times S \to \mathbb{R}$, as [27, 28, 25, 26].

$$\beta_0(\psi,\psi') = B\big({}_{(\mathrm{Re})}\psi, {}_{(\mathrm{Re})}\psi'\big) - B\big({}_{(\mathrm{Im})}\psi, {}_{(\mathrm{Im})}\psi'\big) + i\big[B\big({}_{(\mathrm{Re})}\psi, {}_{(\mathrm{Im})}\psi'\big) + B\big({}_{(\mathrm{Im})}\psi, {}_{(\mathrm{Re})}\psi'\big)\big], \tag{10}$$

where ${}_{(\mathrm{Re})}\psi = \frac{1}{2}(\psi + D(\psi))$ $[{}_{(\mathrm{Im})}\psi = \frac{1}{2}(\psi - D(\psi))]$ is the real [imaginary] parts of ψ [27].

For arbitrary dimensions, a spin-invariant product induces a spinor conjugation, $\bar{\psi} = \psi^\dagger b^{-1}$, where b is some appropriate invertible element, uniquely defined for each spacetime dimension/signature, in the Clifford bundle, yielding the corresponding spin bilinear form on S:

$$\beta_k(\psi,\psi') = B(\psi, \gamma_{\rho_1\ldots\rho_k}\psi') = \bar{\psi}\gamma_{\rho_1\ldots\rho_k}\psi'. \tag{11}$$

The particular case of Minkowski spacetime reads the well-known spinor conjugation $\bar{\psi} = \psi^\dagger\gamma^0$ [2].

The important case studied in [27, 25, 26] consist of the signature $p - q \equiv 7 \mod 8$ [27]. The real chiral spinor bundles $S^\pm \equiv P_\pm(S)$ define an element ψ to be a Majorana spinor. Bilinear pairings can be restricted by the condition $B(\psi, \gamma^{\rho_1\cdots\rho_k}\psi) \neq 0$ if k is odd [27]. The endomorphisms $\mathcal{A}_{\psi|\psi'}$ of the spin bundle S were defined [27] as:

$$\mathcal{A}_{\psi_1|\psi_2}(\psi) := B(\psi,\psi_2)\psi_1, \quad \text{for all} \quad \psi,\psi_1,\psi_2 \in \Gamma(S). \tag{12}$$

It determines the geometric Fierz identities [27]

$$\mathcal{A}_{\psi_1|\psi_2} \bullet \mathcal{A}_{\psi_3|\psi_4} = B(\psi_3,\psi_2)\mathcal{A}_{\psi_1|\psi_4}. \tag{13}$$

Consider now the completeness relation $\mathcal{A}_{\psi|\psi'} \sim \sum_k \frac{1}{k!}(-1)^k B(\psi, \gamma_{\rho_1\ldots\rho_k}\psi')e^{\rho_1\cdots\rho_k}$, where the symbol "$\sim$" denotes an equivalence modulus a constant that equals either a suitable power of the constant 2 or two times this constant, that is spacetime signature dependent [27]. Moreover, every element in the space $\Gamma(\mathrm{End}(S))$, in particular the $\mathcal{A}_{\psi|\psi'}$, can be split uniquely as $\mathcal{A}_{\psi|\psi'} = D \bullet \mathcal{A}^1_{\psi|\psi'} + \mathcal{A}^0_{\psi|\psi'}$ [27], where

$$\mathcal{A}^0_{\psi|\psi'} \sim \sum_k \frac{(-1)^k}{k!} B(\psi, \gamma_{\rho_1\ldots\rho_k}\psi')e^{\rho_1\cdots\rho_k}, \tag{14a}$$

$$\mathcal{A}^1_{\psi|\psi'} \sim \sum_k \frac{1}{k!}(-1)^{\left(k+\frac{1+p-q}{4}\right)} B(\psi, D \bullet \gamma_{\rho_1\ldots\rho_k}\psi')e^{\rho_1\cdots\rho_k}. \tag{14b}$$

The FPK identities (5) read, for generalised for arbitrary dimensions/signatures, hence [27]:

$$\widehat{\mathcal{A}^0_{\psi_1|\psi_2}} \circ \mathcal{A}^1_{\psi_3|\psi_4} + \mathcal{A}^1_{\psi_1|\psi_2} \circ \mathcal{A}^0_{\psi_3|\psi_4} = B(\psi_3,\psi_2)\mathcal{A}^1_{\psi_1|\psi_4}, \tag{15a}$$

$$\mathcal{A}^0_{\psi_1|\psi_2} \circ \mathcal{A}^0_{\psi_3|\psi_4} + (-1)^{\frac{1+p-q}{4}} \widehat{\mathcal{A}^1_{\psi_1|\psi_2}} \circ \mathcal{A}^1_{\psi_3|\psi_4} = B(\psi_3,\psi_2)\mathcal{A}^0_{\psi_1|\psi_4}, \tag{15b}$$

being reduced to Eqs. (5) for the Minkowski spacetime. As we are interested in determining the nature of Majorana spinor fields according to bilinear covariants in 7-manifolds, we focus in the particular, however important case of $n = p + q = 7 + 0$. Moreover, the expression

$$\varphi_k = \frac{1}{k!} B(\psi, \gamma_{\rho_1 \cdots \rho_k} \psi) e^{\rho_1 \cdots \rho_k} \tag{16}$$

equals zero except if k is even. Together with the symmetry of B, this shows that the element $A^{0,k} \equiv A^{0,k}_{\psi|\psi}$ vanishes except if $k = 0, 3, 4, 7$. Combining this with (16), it implies that, for ψ a Majorana spinor, the forms φ_k equal zero except when $k = 0$ or $k = 4$. By regarding ψ normalized such that $B(\psi, \psi) = 1$, it follows that $\mathcal{A}^{0,0} = 1$ and the following bilinear can be defined [27]:

$$\varphi_4 = \frac{1}{4!} B(\psi, \gamma_{\rho_1\rho_2\rho_3\rho_4} \psi) e^{\rho_1\rho_2\rho_3\rho_4} \,, \tag{17}$$

which are the components of the first generator $\mathcal{A}^0 = \frac{1}{16}(1 + \varphi_4)$ of the so called Fierz algebra represented in (14a, 14b) [27], where $\mathcal{A}^\lambda \equiv \mathcal{A}^\lambda_{\psi|\psi}$. The product $\Delta_k : \Omega(M) \times \Omega(M) \to \Omega(M)$ is iteratively defined by

$$\theta_1 \, \Delta_{k+1} \, \theta_2 = \frac{1}{k+1} g_{ab} (e^a \lrcorner \theta_1) \, \Delta_k \, (e^b \lrcorner \theta_2) \,, \qquad \chi, \vartheta \in \sec \Omega(M) \,, \tag{18}$$

where the symbol g_{ab} stands for the metric tensor coefficients. One fixes $\Delta_0 = \wedge$ as being the exterior product.

In the case here to be analyzed $n = p + q = 7$, we already know that $\varphi_k = 0$ except for the values $k \in \{0, 3, 4, 7\}$. In addition, due to the restriction that $B(\psi, J \circ \gamma_{\rho_1 \ldots \rho_k} \psi) = -B(\psi, J \circ D \circ \gamma_{\rho_1 \ldots \rho_k} \psi) = 0$, unless if k is odd, one obtains the restriction $B'(\psi, \gamma_{\rho_1 \ldots \rho_k} \psi) = 0$, unless $k = 3$ or $k = 7$. It was shown in [27, 26, 25] that merely the forms $\varphi_0 = B(\psi, \psi)$ and $\varphi_4 = \frac{1}{4!} B(\psi, \gamma_{\rho_1\rho_2\rho_3\rho_4} \psi) e^{\rho_1\rho_2\rho_3\rho_4}$ are different of zero. The Fierz identities imply, in particular, that $\varphi_4 \neq 0$, corresponding to a single class of – Majorana – spinors:

$$\varphi_0 \neq 0, \quad \varphi_1 = 0, \quad \varphi_2 = 0, \quad \varphi_3 = 0, \quad \varphi_4 \neq 0, \quad \varphi_5 = 0, \quad \varphi_6 = 0, \quad \varphi_7 = 0 \,. \tag{19}$$

Eq. (11) can be thus generalized, in order to encompass the complex case, providing the higher degree generalization of (10):

$$\begin{aligned} \beta_k(\psi, \gamma_{\rho_1 \ldots \rho_k} \psi') &= -B\left({}_{(\mathrm{Im})}\psi, \gamma_{\rho_1 \ldots \rho_k \, (\mathrm{Im})}\psi'\right) + B\left({}_{(\mathrm{Re})}\psi, \gamma_{\rho_1 \ldots \rho_k \, (\mathrm{Re})}\psi'\right) \\ &\quad + i\left[+B\left({}_{(\mathrm{Im})}\psi, \gamma_{\rho_1 \ldots \rho_k \, (\mathrm{Re})}\psi'\right) + B\left({}_{(\mathrm{Re})}\psi, \gamma_{\rho_1 \ldots \rho_k \, (\mathrm{Im})}\psi'\right)\right] . \end{aligned} \tag{20}$$

Hence, by defining $\underline{\phi}_k := \frac{1}{k!} \beta_k(\psi, \gamma_{\rho_1 \ldots \rho_k} \psi) e^{\rho_1 \cdots \rho_k}$, the non-trivial classes of spinor fields $\psi \in \Gamma(S)$ read [25]:

$$1) \quad \underline{\phi}_0 = 0, \quad \underline{\phi}_4 \neq 0, \tag{21a}$$

$$2) \quad \underline{\phi}_0 \neq 0, \quad \underline{\phi}_4 = 0, \tag{21b}$$

$$3) \quad \underline{\phi}_0 \neq 0, \quad \underline{\phi}_4 \neq 0 \,. \tag{21c}$$

All other $\underline{\phi}_k = 0$ for $k \neq 4$. Hence, novel classes of spinors, under the geometric Fierz identities, was here illustrated, for the case of Riemmanian 7-manifolds [25, 26]. Other cases, regarding new spinors in a five-dimensional bulk and in Lorentzian 7-manifolds [26] were found. Other few signatures can be, further, explored [27, 28].

2.1. Acknowledgments

RdR thanks FAPESP (grant No. 2015/10270-0) and CNPq (grant No. 303293/2015-2; No. 473326/2013-2), for partial financial support.

References

[1] Lounesto P 2002 *Clifford Algebras and Spinors* (Cambridge: Cambridge Univ. Press)
[2] Vaz J and da Rocha R 2016 *An Introduction to Clifford Algebras and Spinors* (Oxford: Oxford Univ. Press)
[3] Mei J 2012 Spinor Fields and Symmetries of the spacetime *Gen. Rel. Grav.* **44** 2191
[4] Brito K P S and da Rocha R 2016 New spinors in the bulk, *J. Phys. A*, accepted.
[5] da Rocha R and Cavalcanti R T 2016 Flag-dipole and flagpole spinors fluid flows in Kerr spacetimes, (*Preprint* arXiv:1602.02441 [hep-th]
[6] da Rocha R and Rodrigues W A 2006 Where are ELKO spinor fields in Lounesto spinor field classification? *Mod. Phys. Lett. A* **21** 65
[7] Cavalcanti R T 2014 Looking for the Classification of Singular Spinor Fields Dynamics and other Mass Dimension One Fermions: Characterization of Spinor Fields *Int. J. Mod. Phys. D* **23** 1444002
[8] Fabbri L 2016 A generally-relativistic gauge classification of the Dirac fields *Int. J. Geom. Meth. Mod. Phys.* **13** 1650078
[9] Hoff da Silva J M and da Rocha R 2013 Unfolding Physics from the Algebraic Classification of Spinor Fields *Phys. Lett. B* **718** 1519
[10] Rocha R, Bernardini A E and Hoff da Silva J M 2011 Exotic Dark Spinor Fields *J. High Energy Phys.* JHEP04(2011)110
[11] da Rocha R, Fabbri L, Hoff da Silva J M, Cavalcanti R T and Silva-Neto J A 2013 Flag-Dipole Spinor Fields in ESK Gravities *J. Math. Phys.* **54** 102505
[12] Lee C Y 2016 A Lagrangian for mass dimension one fermionic dark matter *Phys. Lett. B* **760** 164
[13] da Rocha R and Hoff da Silva J M 2010 ELKO, flagpole and flag-dipole spinor fields, and the instanton Hopf fibration *Adv. Appl. Clifford Algebras* **20** 847
[14] Bernardini A E and da Rocha R 2012 Dynamical dispersion relation for ELKO dark spinor fields *Phys. Lett. B* **717** 238
[15] Ahluwalia D V, Lee C Y, Schritt D and Watson T F 2010 Elko as self-interacting fermionic dark matter with axis of locality *Phys. Lett. B* **687** 248
[16] Ahluwalia D V and Grumiller D 2005 Spin half fermions with mass dimension one: Theory, phenomenology, and dark matter *J. Cosm. Astrop. Phys.* JCAP07(2005)012
[17] da Rocha R and Hoff da Silva J M 2014 Hawking Radiation from Elko Particles Tunnelling across Black Strings Horizon *Europhys. Lett.* **107** 50001
[18] da Rocha R and Hoff da Silva J M 2014 Hawking Radiation from Elko Particles Tunnelling across Black Strings Horizon *Europhys. Lett.* **107** 50001
[19] A. P. S. S., S. H. Pereira and J. F. Jesus, Eur. Phys. J. C **75** (2015) no.1, 36
[20] Dias M, de Campos F and Hoff da Silva J M 2012 Exploring Elko typical signature *Phys. Lett. B* **706** 352
[21] Cavalcanti R T, Hoff da Silva J M and da Rocha R 2014 VSR symmetries in the DKP algebra: the interplay between Dirac and Elko spinor fields *Eur. Phys. J. Plus* **129** 246
[22] Rodrigues W A, da Rocha R and Vaz J 2005 Hidden consequence of active local Lorentz invariance *Int. J. Geom. Meth. Mod. Phys.* **2** 305
[23] Cavalcanti R T and da Rocha R 2016 Dark Spinors Hawking Radiation in String Theory Black Holes *Adv. High Energy Phys.* **2016** Article ID 4681902
[24] Coronado Villalobos C H, Hoff da Silva J M and da Rocha R 2015 Questing mass dimension 1 spinor fields *Eur. Phys. J. C* **75** 266
[25] Bonora L, de Brito K P S and da Rocha R 2015, Spinor fields classification in arbitrary dimensions and new classes of spinor fields on 7-manifolds *J. High Energy Phys.* JHEP02(2015)069
[26] Bonora L and da Rocha R 2016 New spinor fields on Lorentzian 7-manifolds *J. High Energy Phys.* JHEP01(2016)133
[27] Lazaroiu C I, Babalic E M and Comam I A 2013 The geometric algebra of Fierz identities in arbitrary dimensions and signatures *J. High Energy Phys.* JHEP09(2013)156
[28] Babalic E M and Lazaroiu C I 2013 Revisiting eight-manifold flux compactifications of M-theory using geometric algebra techniques *Rom. Journ. Phys.* **58** 414
[29] Y. X. Liu, X. N. Zhou, K. Yang and F. W. Chen, Phys. Rev. D **86** (2012) 064012
[30] Jardim I C, Alencar G, Landim R R and Costa Filho R N 2015 Solutions to the problem of ELKO spinor localization in brane models *Phys. Rev. D* **91** 085008

On a Microscopic Representation of Space-Time V

R Dahm[1]

[1]beratung für Informationssysteme und Systemintegration, Gärtnergasse 1, D-55116 Mainz, Germany

E-mail: dahm@bf-IS.de

Abstract. In previous parts of this publication series, starting from the Dirac algebra and SU∗(4), the 'dual' compact rank-3 group SU(4) and Lie theory, we have developed some arguments and the reasoning to use (real) projective and (line) Complex geometry directly. Here, we want to extend this approach further in terms of line and Complex geometry and give some analytical examples. As such, we start from quadratic Complexe which we've identified in parts III and IV already as yielding naturally the 'light cone' $x_1^2 + x_2^2 + x_3^2 - x_0^2 = 0$ when being related to (homogeneous) point coordinates x_α^2 and infinitesimal dynamics by tetrahedral Complexe (or line elements). This introduces naturally projective transformations by preserving anharmonic ratios. We summarize some old work of Plücker relating quadratic Complexe to optics and discuss briefly their relation to spherical (and Schrödinger-type) equations as well as an obvious interpretation based on homogeneous coordinates and relations to conics and second order surfaces. Discussing (linear) symplectic symmetry and line coordinates, the main purpose and thread within this paper, however, is the identification and discussion of special relativity as direct invariance properties of line/Complex coordinates as well as their relation to 'quantum field theory' by complexification of point coordinates or Complexe. This can be established by the Lie mapping[1] which relates lines/Complexe to sphere geometry so that SU(2), SU(2)×U(1), SU(2)×SU(2) and the Dirac spinor description emerge without additional assumptions. We give a short outlook in that quadratic Complexe are related to dynamics e.g. power expressions in terms of six-vector products of Complexe, and action principles may be applied. (Quadratic) products like $F^{\mu\nu}F_{\mu\nu}$ or $F^{a\,\mu\nu}F^a_{\mu\nu}$, $1 \leq a \leq 3$ are natural quadratic Complex expressions which may be extended by line constraints $\lambda k \cdot \epsilon = 0$ with respect to an 'action principle' so that we identify 'quantum field theory' with projective or line/Complex geometry having applied the Lie mapping.

1. Introduction

With this fifth part[2] of our series, we've reached a certain milestone to pause and make up the balance in departing from SU(4) vs. SU∗(4), the various derived symmetry breaking patterns [2], [3] and the identification of photons in SSB patterns. So starting from the (physical!) interpretation of the Lie algebra su(4) (in terms of spin, isospin and chiral symmetry) and its various real forms, the main path – also in the context of noncompact su∗(4) and the Dirac algebra – concentrated on a 10⊕5 reductive decomposition and its interpretation (see [3]

[1] We use 'Lie mapping' to denote the line-sphere mapping [14] which needs to be distinguished from 'Lie transformations' used in phase or function space discussions by means of Lie series resp. power series expansions.

[2] This paper, being limited as conference proceedings, yields mainly the direct arguments and calculations related to the abstract. A more detailed paper containing additional background and details is going to appear as part VI soon.

and references). In collecting various algebraic and physical aspects, we were led to projective geometry, transfer principles and especially line and Complex geometry [4], [5].

Here, we want to present briefly some additional details and identifications with respect to typical/basic 'ingredients' of dynamical theories like special relativity, Dirac's 'square root' of the Klein-Gordon equation, and – most important – identification of states resp. interpretation of the mathematical description. Whereas a 'light cone', Lorentz transformations and special relativity can be related to (real) projective geometry, we need the Lie mapping [14] to introduce Pauli matrices (respectively quaternions) which relates to 'quantum theory' and 'quantum' notion.

The central theme is[3] that this transformation relates *line* representations[4] (NOT *point* reps!) to spheres (and elliptic geometry). So first of all we have to use and respect a priori *two* distinct spaces (and not one and the same space-time) respectively their related individual 'physical' identification(s) and, more important, our fundamental geometrical space element is a (nonlocal) 4-dim *line* and afterwards a set of lines, the linear Complex, not as usual the point or 'vector' rep. So points emerge only as line incidences which must be treated by projective geometry. However, dynamics being concerned with tangents, momenta and conic sections benefits ernormously according to this different choice of the fundamental geometrical space element due to (linear) Complexe being related to null systems, the Hamilton formalism and differential equations [19], [15]. This comprises a very long history where lines, various compositions of lines (i.e. Complexe, congruences, etc.) and null systems have been applied with great use and success to physics. So here, we start right from the beginning with line and Complex geometry[5].

2. Quadratic Complexe and the 'Light Cone'

Why do we want to see the 'light cone' $x_1^2 + x_2^2 + x_3^2 - x_0^2 = 0$ (in point coordinates) emerging directly from quadratic Complexe like in [4]? Well, the reason is simple: As soon as we interpret this 'light cone' (as usual) in terms of a 'metric' on point spaces and/or in metric coordinates, we have to introduce additional physical identification or at least an additional dimension (i.e. we have to use *five* homogeneous coordinates, see e.g. [13], appendix §5) in order to treat absolute elements ('infinities'). From our viewpoint, it is much easier and much more consistent (and we'll give some additional arguments below) to understand the 'light cone' as an absolute element (or 'gauge surface') when switching from quadratic line/Complex reps to point reps. This allows to introduce and relate velocities via a 'common' (system) time parameter t to point coordinates x_α, i.e. the geometrical line picture to a dynamic point picture and as such to relations/ratios of velocities in terms of $\beta_i = v_i/c$. So the velocity of light c may serve to parametrize the absolute coordinate x_0 by its correspondence to physical observations and identifications, using the velocity picture $x_i \sim v_i t$, of course (see [4]), and the 'gauge surface' in the geometrical line picture to apply the Cayley-Klein mechanism towards metric coordinates and spaces. We have mentioned already the introduction of equivalence classes by this velocity concept. Although it is known from classical mechanics how to work with (euclidean) coordinate projections and 'vectors', it is also known[6] that in order to describe kinematics and dynamics, the naive (3-dim euclidean) 'vector' picture of velocities, momenta and forces is not sufficient but one has to work with force and null systems which were generalized by Plücker to Complexe and 6-dim

[3] Here, we discuss 'simple' geometrical issues and identifications according to [14] and postpone the viewpoint of 'advanced geometry' in terms of the Plücker-Klein quadric in $\mathbb{R}^5$, Study's points of view [23], or even higher dimensional transfers to upcoming publications/discussions.

[4] As before, subsequently we use the shorthand notation 'rep'.

[5] For introductory details from physical identifications to projective geometry, please see [4] and [5], some basic definitions have been repeated in [4] section 1.2ff which originate mainly from Plücker's work [19] on line Complexe.

[6] As an example, we cite [21] which, starting from Complex description of rigid bodies, gives insight into some deeper aspects of 'advanced geometry' by addressing null systems and the tetrahedral Complex, 3rd and higher order elements, etc.

'Dynamen' [18]. This means, we use *four homogeneous* coordinates x_α, and the 'metric'/the 'light cone' is a point representation related to the line representation of a quadratic Complex. So this picture yields a natural relation in the context of projective geometry of 3-space, and natural representations of lines in terms of points and planes (see [4], p. 9, and references).

Here, we skip some background on quadratic Complexe and postpone the details to part VI. In brief, having mentioned ([5], III C) Plücker's work on relating ellipsoids and lines/Complexe already, some more aspects can be found in Klein's book[7] [12] §§4-7 with respect to potential theory, confocal surfaces and elliptic coordinates. Physical applications in optics [17], [19] should be mentioned, too, in that double refraction at a crystal surface yields a triple (line) vertex or in generalizing the variational action principle [18], [11].

In addition, we have learned from [7] a basic set of requirements to express relativity, and we have to state that – being in charge to introduce those requirements in point reps and differential geometry – it is Complex geometry (and especially in conjunction with quadratic Complexe) which yields the requirements with respect to necessary conformal, projective and affine structures automatically, especially such non-local requirements like second-order cones at each point of the curve or geodesic families with certain behavior. So using quadratic Complexe, we control a superset *to derive* those features – there is no need to introduce them by hand, but we have to use Complex geometry. So the unifying space element is a linear Complex, and we have to relate our reasoning in 3-dim space to higher order Complexe and calculation patterns in order to compare to physics and extract principles.

2.1. Infinitesimal Dynamics

Recalling tetrahedral Complexe, those quadratic Complexe (while preserving by definition anharmonic ratios, see [5] and references there) emphasize projective transformation, i.e. transformations leaving the anharmonic ratio invariant. This provides the most 'natural link' via Klein's Erlanger program to introduce and justify group theory in terms of linear reps. So departing here, we may study projective transformations in various reps e.g. only on appropriate point sets, their transformation groups or even only infinitesimal transformations related to them[8]. With respect to more general geometric considerations, however, by changing the underlying space element (e.g. from points to lines or spheres), we want to make first steps into linear and quadratic Complex geometry[9]. Additional details can be found in [12] §§21-24.

Now the shorthand approach to dynamics can be based on line incidence and in a second step on Complex involutions [11]. Whereas the incidence relation of two lines with coordinates $p_{\alpha\beta}$ and $p'_{\alpha\beta}$ reads as ([12] §20)

$$p_{12}p'_{34} + p_{13}p'_{42} + p_{14}p'_{23} + p_{34}p'_{12} + p_{42}p'_{13} + p_{23}p'_{14} = 0\,, \tag{1}$$

choosing the linear Complex as basic element introduces new notations and invariant structures [12] §§21-24. The incidence equation (1) above, with line coordinates $p_{\alpha\beta}$ being replaced by Complex coordinates, declare involution of Complexe, and in analyzing the quadratic form Ω, Klein introduced [10], [11] six fundamental Complexe and determined their invariants to be ± 1.

[7] For the sake of simplicity, we use this book and the references given there as a kind of dictionary in order to avoid lengthy details or historical background.

[8] We want to remember that projective geometry of $\mathbb{R}^3$ not only involves 'collineations' but also 'correlations' and dual elements, and it allows for application of higher transfer principles! In the context of transfer principles, it is necessary to recall the identification of certain *point* coordinates in higher dimensional spaces with (even extended) geometrical elements in ordinary real 3-dim space, e.g. points with respect to the Plücker-Klein quadric or the Lie mapping later in section 4.

[9] Of course, we may choose other elements and transfer them to (higher-dim) point spaces. Here, we stay in the regime of 3-dim space, the Complex as basic element, the Lie mapping, sphere and Laguerre geometry.

IOP Conf. Series: Journal of Physics: Conf. Series **804** (2017) 012013 doi:10.1088/1742-6596/804/1/012013

Because of their relation to handedness, the six fundamental Complexe[10] decompose into $3 \oplus 3$ left-/right-handed, the simultaneous invariant of each pair being 0.

If – as before [4] – we identify $F^{\mu\nu}F_{\mu\nu}$ as a Complex invariant, on the one hand we can use Klein's work to postulate an action (see section 4.2), on the other hand one has to recall (see e.g. [16], [18], [10] or [11]) the discussion of up to six acting forces. So with respect to $F^{\mu\nu}F_{\mu\nu}$ and $F^{a\,\mu\nu}F^{a}_{\mu\nu}$, $1 \leq a \leq 3$, emerging in actions, we have settled the basic environment here from within the framework of Complex geometry and involutions.

Last not least, it is noteworthy that the (infinitesimal) action (and as such the very foundation of the action principle) may be written as a quadratic Complex (see e.g. Klein [13] or Dirac [6] eq. (5.1)) so that projective transformations are automatically transformation groups of the action (which of course allows to apply Klein's Erlanger program in all details with respect to (linear) transformation groups and subgroups). We've had related $F^{\mu\nu}$ already to a (linear) Complex [4], so starting from the known rep $F^{\mu\nu}$, Ω in [6], eq. (5.1) suggests several interpretations and generalizations. Thus Ω can be interpreted in terms of incidence/conjugation of lines, involution of linear Complexe or generalized in terms of squares of two (general linear) Complexe respectively the most general possible form of a quadratic Complex. Moreover, this is not unique in that [6], eq. (5.1) allows to add other Complex reps as well as additional (incidence/conjugation) terms equating to 0 as long as the quadratic character of Ω is preserved[11].

3. Complexe and Symplectic Symmetry

The notion 'symplectic symmetry' emerged because Weyl renamed the symmetry of the 'Komplexgruppe' to greek notion in order to avoid misconceptions and/or misunderstanding of complex numbers and Complexe[12]. We use linear Complexe and their point rep as given e.g. by Hamermesh (see [8], ch. 10, eq. (10-73)).

3.1. (Linear) Symplectic Symmetry

If we do *not* split $n = 2\nu$ and the given decomposition into $(n-2)$-dim subspaces [8], but if instead we use general n and fix $n = 4$ in a linear enumeration of the coordinate reps x and y, we may write the bilinear form according to $x^T Jy = x^T_\alpha J_{\alpha\beta} y_\beta$, $1 \leq \alpha, \beta \leq 4$, which (in this special case) leads to $x_1 y_2 - x_2 y_1 + x_3 y_4 - x_4 y_3$. Interpreting the coordinates as homogeneous coordinates, we may replace them by line coordinate reps (see e.g. [4], ch. 1.2), i.e. the invariance requirement of $x^T Jy$ using point coordinates reads as the invariance of $p_{12} + p_{34}$ in terms of (homogeneous!) line coordinates $p_{\alpha\beta}$. Now, $p_{12}+p_{34}$ is a special case of a general linear Complex $A_{\alpha\beta}p_{\alpha\beta}$, and due to simplicity we'll use the expression $p_{12} \pm kp_{34} = 0$ for subsequent discussions, $k \in \mathbb{R}$ (for the moment) being nothing but a real parameter.

3.2. Special Relativity as Invariance Property

Having established the (symplectic) invariance requirement of $x^T Jy = p_{12}+p_{34}$ and generalized it to $p_{12} \pm kp_{34} = 0$, we may ask for (point coordinate) transformations keeping such objects

[10] With respect to SU(2)×SU(2) structures, we still have to introduce the Lie mapping, see section 4.

[11] In order to preserve the grade of a Complex under projective transformations, one has to use Plücker's 5-dim (inhomogeneous) coordinates r, s, ρ, σ, and $\eta = r\sigma - s\rho$, or six homogeneous coordinates, *not* only the original 4-dim set of line coordinates r, s, ρ and σ! This introduces the (quadratic) Plücker condition which may be used as an additional term while introducing an additional parameter into the action. Please note here for later use, that if we introduce a matrix notation $M := \begin{pmatrix} r & \rho \\ s & \sigma \end{pmatrix}$ then $\eta = \det M$. So at least for linear transformations of line coordinates this matrix rep provides a calculus (or symbolism) to transform line coordinates by performing matrix transformations.

[12] We hope to avoid such misery by strictly using capitalized 'C' and old german plural form in conjunction with the treatment of line Complexe.

invariant. In selecting the two most trivial cases[13], it is obvious that at a first glance, we can keep $p_{\alpha\beta}$ invariant by itself when keeping the coordinates (and as such also the area element) $x_\alpha, y_\alpha \longrightarrow x'_\alpha = x_\alpha, y'_\alpha = y_\alpha$ invariant.

A second, almost trivial possibility is to apply transformations according to

$$\begin{array}{ll} x_1 \longrightarrow x'_1 = x_1\,, \quad x_2 \longrightarrow x'_2 = x_2\,, & y_1 \longrightarrow y'_1 = y_1\,, \quad y_2 \longrightarrow y'_2 = y_2\,, \\ x_3 \longrightarrow x'_3 = x_3 \cosh\eta - x_4 \sinh\eta\,, & y_3 \longrightarrow y'_3 = y_3 \cosh\eta - y_4 \sinh\eta\,, \\ x_4 \longrightarrow x'_4 = x_4 \cosh\eta - x_3 \sinh\eta\,, & y_4 \longrightarrow x'_4 = y_4 \cosh\eta - y_3 \sinh\eta\,. \end{array} \tag{2}$$

Direct calculation shows $(x'_3)^2 - (x'_4)^2 = x_3^2 - x_4^2$, or even $(x'_1)^2 + (x'_2)^2 + (x'_3)^2 - (x'_4)^2 = x_1^2 + x_2^2 + x_3^2 - x_4^2$, so the transformation (2) itself can be identified with a Lorentz transformation (see also [22], eq. (2.1.7)), with $\eta = \alpha$ denoting the rapidity and relating the 4- (or 0-) coordinate to the standard 'time' coordinate. So $\cosh\eta = \gamma$, $\sinh\eta = \gamma\beta$ and $\tanh\eta = \beta$ with $\gamma = (1-\beta)^{-\frac{1}{2}}, \beta = \frac{v}{c}$. The same holds, of course, for the second point y.

Now the interesting and important fact is that p_{34} (or $x_3y_4 - x_4y_3$) by itself is invariant[14], too, i.e.

$$\begin{aligned} p'_{34} &= x'_3y'_4 - x'_4y'_3 \\ &\overset{(2)}{=} x_3y_4\,(\cosh^2\eta - \sinh^2\eta) + x_4y_3\,(\sinh^2\eta - \cosh^2\eta) \\ &= x_3y_4 - x_4y_3 \\ &= p_{34}\,. \end{aligned} \tag{3}$$

Without branching into the details of (Lorentz, Poincaré or general) group theory as usual at this point (see e.g. [22] or [9]), for us it is sufficient having derived this property from a (linear) Complex like $p_{12} \pm kp_{34} = 0$. Here, we do not have room to start the geometrical discussion of this Complex but we want to mention that this linear Complex incorporates already special choices (see e.g. [17], nrs. 28ff or [12], §§16, 17). Based mainly on Möbius's and Plücker's work (e.g. [16], [17], [18], [19] and references therein), Klein recalls point-plane mappings, null systems and line conjugation, and closes the gap to dynamics and second order surfaces (i.e. polar theory).

Defining the axis of the null system as conjugate (line) polar of an absolute line and using a special choice of an orthogonal coordinate system, Klein discussed in detail how to obtain the rep $p_{12} \pm kp_{34} = 0 \longrightarrow (xy' - x'y) + k(z - z') = 0$, k denoting the 'parameter of the null system' via $k = \frac{a_{12}a_{34} + a_{13}a_{42} + a_{14}a_{23}}{a_{12}^2 + a_{13}^2 + a_{14}^2}$. This notion is used as departure in various other contexts (see e.g. [1] or [24]) discussing aspects of (linear) line Complexe, sometimes using different notations. For our purpose, we use Lie's approach [14] departing from the interpretation of imaginary elements in projective geometry and using them to establish a mapping of two complexified planar real coordinates onto lines, i.e. a 4-dim planar space to 4-dim line space.

4. Lie Mapping (or Lie Transfer)

While having the equivalent point/plane/line coordinate reps from within [4], section 1.4, in mind, we choose the line rep for further discussion. In the general case, we thus have to discuss the (conjugation/reciprocity) relation of two linear Complexe (see e.g. [19], [10], [11], [20] II 18. Vortrag, and especially [14], §7ff) versus occurences and use of quadratic Complexe. So

[13] We can forget all background and think with respect to $p_{\alpha\beta}$ just in terms of 2×2 determinants of some point coordinates.

[14] In addition to direct calculation, one can use the 2×2-determinant representation and standard rules of determinant calculus. This yields some more insight into standard rep theory if we 'transform' (or map) the determinant 'back' to the bilinear *point coordinate* 'invariant' x^Ty and use 'matrix' reps of group (or algebra) transformations, i.e. $y' = Gy, x'^T = (Gx)^T = x^TG^T$ with respect to the two columns. Also note that if we complexify the coordinates, i.e. $x_3, x_4, y_3, y_4 \in \mathbb{R} \longrightarrow \psi_1, \psi_2, \psi'_1, \psi'_2 \in \mathbb{C}$, we obtain expressions $\psi_1\psi'_2 - \psi_2\psi'_1$ to be compared with (spinorial) singlet structures using the same mapping onto matrices and e.g. $\psi = \psi'^+$ with hermitean conjugation. This, however, will be addresses in section 4 in more detail.

IOP Conf. Series: Journal of Physics: Conf. Series **804** (2017) 012013 doi:10.1088/1742-6596/804/1/012013

Dirac's quest of finding a linear representation in p (or sloppy speaking 'the root' of p^2) can be cast onto various related and interwoven topics:

- Given a quadratic Complex $\mathfrak{C}^2$, how can we determine an appropriate linear Complex $\mathfrak{C}$, its geometrical and its dynamical and transformation/covariance properties?
- Given a quadratic Complex $\mathfrak{C}^2$, can we find other geometrical elements and/or representations $\mathfrak{C}$ which square to a quadratic Complex, or what is the most general form of $\mathfrak{C}$, respectively?
- Determining the parameters of (planar) line pencils or the intersection points of lines e.g. with the absolute quadric, we obtain quadratic relations in terms of line coordinates. How can we relate them to physical observations and principles related to linear and quadratic line Complexe?

4.1. Complexification of 'Point' Coordinates

Based on [14], §4, number 11, one can consider general reciprocity mappings between two real 3-dim spaces r and R. For our purpose the specialization given in §7 eq. (1), based on choosing two line Complexe, is substantial. So points in r map to lines in R and vice versa. Following §§8 and 9, Lie obtained the special mapping[15] $2\rho = X + iY, 2s = X - iY, 2\sigma = Z \pm H, 2r = -(Z \mp H) = -Z \pm H$ which constrains the linear Complex $r + \sigma$ in r mapped to minimal lines[16] in R with $dX^2 + dY^2 + dZ^2 = 0$. Besides a lot of other features and results (see [14] or the overview in [12] §§25-27), this mapping accordingly transforms *lines* of the 3-dim real space r uniquely into *spheres* of R. Vice versa, a sphere of R described by (X, Y, Z, H^2), $\pm H = \sigma + r$, is transformed into only *two* lines of r! Both lines $(X, Y, Z, +H)$ and $(X, Y, Z, -H)$ are polar with respect to the linear Complex $\pm H = \sigma + r = 0$ [14]. The uniqueness may be established by introducing the notion of orientation and Laguerre geometry, however, according to our earlier footnote by introducing the matrix M, we can represent M in terms of coordinates of R by

$$M = \begin{pmatrix} r & \rho \\ s & \sigma \end{pmatrix} \longrightarrow \begin{pmatrix} \frac{1}{2}(-Z \pm H) & \frac{1}{2}(X + iY) \\ \frac{1}{2}(X - iY) & \frac{1}{2}(Z + H) \end{pmatrix} = \frac{1}{2}\left(\pm H\sigma_0 + X\sigma_1 - Y\sigma_2 - Z\sigma_3\right). \tag{4}$$

i.e. we obtain a mapping of lines (or line Complexe or even higher elements) of the space r into the space R. Lines of r are mapped onto Pauli matrices σ_α with real coefficients, or simply SU(2). No need to recall the possibility to introduce quaternions here. However, it is apparent that higher line and Complex geometry have to find their appropriate counterparts in R space if lines are mapped to Pauli matrices (or quaternions). So a priori we expect appropriate 'reflections' of higher line and Complex geometry in Clifford algebras.

It is however noteworthy, by starting from Plücker's line equations $x = rz + \rho, y = sz + \sigma$ to mention one more fundamental issue where x and y denote (euclidean) *projections* of a line in space (parametrized by z) onto the xz and yz planes, respectively. $\eta \sim (ry - sz) = r\sigma - s\rho = \det M$ denotes the projection onto the xy-plane[17]. If we rewrite those equations in matrix rep, we obtain

$$\begin{pmatrix} x \\ y \end{pmatrix} = \begin{pmatrix} r & \rho \\ s & \sigma \end{pmatrix}\begin{pmatrix} z \\ 1 \end{pmatrix} \longrightarrow \begin{pmatrix} x_1 \\ x_2 \end{pmatrix} = \begin{pmatrix} r & \rho \\ s & \sigma \end{pmatrix}\begin{pmatrix} x_3 \\ x_4 \end{pmatrix}$$

[15] Standard small letters denote elements of the space r, capital letters those of R. Although r is used parallel with Plücker's line coordinate r, the respective context should misunderstandings.

[16] Which later have to be expressed in Pauli or spinorial counterparts.

[17] Please note once more the *alternative* rep of the line by $x(t)$, $y(t)$ and $z(t)$ or by a 3-dim euclidean 'vector' $\vec{x}(t)$ in terms of a time parameter and a related velocity 'vector', time-dependent or not. Both pictures can be 'completed' by appropriate (additional) homogeneous coordinates, however, it is necessary to care about the respective interpretations.

IOP Conf. Series: Journal of Physics: Conf. Series **804** (2017) 012013 doi:10.1088/1742-6596/804/1/012013

after having introduced homogeneous (point) coordinates x_α. This requires to switch to quaternary matrix reps in order to describe transformations of the full set of homogeneous (point) coordinates x_α. It is obvious, that 2×2 inverse and/or conjugate matrices have to be considered when applying the Lie mapping, i.e. we expect 2×2 block structures in matrix reps of R transformations. Moreover, the 2×2 calculus of the Lie mapping lightens some background of expressions like $\vec{\sigma} \cdot \vec{v}\vec{\sigma} \cdot \vec{v} = \vec{v}^2$ on the context of line coordinates and Complex conjugation and/or quadratic Complexe.

At the time of writing, we tend to identify the space r with physical observable (projective) space whereas eq. (4) represents 'quantum' reps respectively R represents 'quantum space', e.g. in terms of 2×2 Pauli or 4×4 Dirac matrices considering two lines (or Complexe). However, it is important to note that this relation yields some assumptions related to linear (line) Complexe in r [14], §7 and §8, and it is far from a general description especially if we relax the restrictions in r and proceed to general dynamics and higher geometry. Vice versa, we can establish a mapping to a Pauli/quaternionic or Clifford calculus in order to relate physical observations and observables.

In [14], Lie presents various further very interesting results, also with respect to Complex cones, differential geometry and mappings of tangent to curvature properties which we suppress here. However, there are two more important facts which Lie used to introduce contact interactions and his theory of partial differential equations but which for us is important to relate to physics. So he mentioned Klein's work on six Complexes in involution[18] as a superset, in addition in [14], §9, number 28, he showed that two incident lines (being reciprocal/conjugate polars with respect to the special Complex $H = 0$) are mapped to two spheres being in contact, i.e. fulfilling $(X_1 - X_2)^2 + (Y_1 - Y_2)^2 + (Z_1 - Z_2)^2 = (H_1 \pm H_2)^2$. If we rewrite the (line) incidence relation like above (see also [4] section 2.1) and respect [19], [10], [11], the action can be treated by quadratic Complex theory. There is an interesting side effect when restricting observations to planar problems in that we may investigate (planar) Complex curves and still discuss (energy) conic sections in the planes as well as two pencils of lines related to two linear Complexe (and their congruence). We'll address this elsewhere.

4.2. Energy and Second Order Surfaces

Last not least it is noteworthy to focus on the description of physical actions which typically starts from an action principle and variations. From above, we are already equipped with an action principle; from the background of projective geometry, we are equipped with a coordinate tetrahedron, with transformation groups and (linear) representation theory, and last not least, we may use transfer principles.

So from above, we may include (invariant) products like $F^{\mu\nu}F_{\mu\nu}$ or $F^{a\,\mu\nu}F^a_{\mu\nu}$, $1 \leq a \leq 3$, and Complex geometry. However, we may as well try to find appropriate representations of line Complexe on other rep spaces. As such, we can observe that the product of two (real) quaternions a^+, b, $^+$ denoting quaternionic conjugation, yields Complex coefficients if we neglect the trace for the moment. So the mapping $a^+ \cdot b \longrightarrow A_{\alpha\beta}\, q^+_\alpha q_\beta \longrightarrow A_{\alpha\beta}\, q_\alpha \otimes q_\beta$ allows to extract the relevant (real) coefficients and relate them to SL(2,$\mathbb{H}$) and its real forms. On the other hand, placing two quaternions q^+, q into a 'spinor', we find of course a rotated 'spinor' set $1/2(q^+ \pm q)$ which can be related to Dirac theory and Clifford algebra. In the background, it is of course the Lie mapping which is active in relating lines and Complexe to spheres and sphere Complexe, etc. For our purpose here, however, it is sufficient to justify the 'spinorial rep' of Complexe in an action principle which is based on the Lie mapping. For all calculations, the standard mechanisms can be applied to calculate energies and states, however, the interpretation has to be changed at a couple of places. We are going to investigate such topics in upcoming work.

[18] Recall the handedness of the null systems!

References

[1] Ball R S 1876. *The Theory of Screws: A Study in the Dynamics of a Rigid Body.* (Dublin: Hodges, Foster, and Co., Grafton-Street)

[2] Dahm R 2010. *A Symmetry Reduction Scheme of the Dirac Algebra without Dimensional Defects. Yad. Fis.* **73** 297; *Phys. Atom. Nuclei* **73** 276

[3] Dahm R 2012. *On A Microscopic Representation of Space-Time. Yad. Fis.* **75** 1244; *Phys. Atom. Nuclei* **75** 1173; *http://arxiv.org/abs/1102.0027*

[4] Dahm R 2015. *On A Microscopic Representation of Space-Time III. ICGT 30, Ghent and ICCA 10, Tartu, http://arxiv.org/abs/1508.06872, submitted to AACA*

[5] Dahm R 2015. *On A Microscopic Representation of Space-Time IV. QTS 9, Yerevan, http://arxiv.org/abs/1512.07119, submitted to* Yad. Fis./Phys. Atom. Nuclei

[6] Dirac, P A M 1973. *Long range forces and broken symmetries. Proceedings of the Royal Society of London. Series A, Mathematical and Physical Sciences* **333** 403

[7] Ehlers J, Pirani F and Schild A 1972. *The Geometry of Free Fall and Light Propagation.* In: *Studies in Relativity.* ed L O'Raifeartaigh (Oxford: Clarendon Press) pp 63–84

[8] Hamermesh M 1962. *Group Theory.* (London-Paris: Pergamon Press)

[9] Jackson J D 1983. *Klassische Elektrodynamik.* 2nd edition (1975, German translation) (Berlin: de Gruyter)

[10] Klein F 1869. *Zur Theorie der Linien-Complexe des ersten und zweiten Grades. Math. Ann.* **2** 198; *Die allgemeine lineare Transformation der Linien-Co-ordinaten. Math. Ann.* **2** 366

[11] Klein F 1871. *Notiz betreffend den Zusammenhang der Linien-Geometrie mit der Mechanik starrer Körper. Math. Ann.* **4** 403

[12] Klein F 1926. *Vorlesungen über höhere Geometrie.* (Die Grundlehren der mathematischen Wissenschaft XXII) (Berlin: Springer)

[13] Klein F 1928. *Vorlesungen über nicht-euklidische Geometrie.* (Die Grundlehren der mathematischen Wissenschaft XXVI) (Berlin: Springer)

[14] Lie S 1872. *Ueber Complexe, insbesondere Linien- und Kugel-Complexe, mit Anwendung auf die Theorie partieller Differential-Gleichungen. Math. Ann.* **5** 145; *Ueber einige partielle Differential-Gleichungen zweiter Ordnung. Math. Ann.* **5** 209

[15] Lie S 1896. *Geometrie der Berührungstransformationen.* (Leipzig: Teubner)

[16] Möbius A F 1837. *Lehrbuch der Statik.* (Leipzig: Göschen)

[17] Plücker J 1865. *On a new Geometry of Space. Phil. Trans. R. Soc. London* **155** 725

[18] Plücker J 1866. *Fundamental Views regarding Mechanics. Phil. Trans. R. Soc. London* **156** 361

[19] Plücker J 1868/1869. *Neue Geometrie des Raumes.* ed A Clebsch and F Klein (Leipzig: B. G. Teubner)

[20] Reye Th 1899. *Die Geometrie der Lage I III.* (4th edition, Leipzig: Baumgärtners Buchhandlung) (1866/1868)

[21] Schoenflies A 1884. *Zur Theorie der Bewegung starrer räumlicher Systeme. Journal für die reine und angewandte Mathematik* **98** 265

[22] Sexl R U, Urbantke H K 1992. *Relativität, Gruppen, Teilchen.* (3rd edition, Wien New York: Springer)

[23] Study E 1903. *Geometrie der Dynamen.* (Leipzig: Teubner)

[24] Zindler K 1902/1906. *Liniengeometrie mit Anwendungen.* (Sammlung Schubert XXXIV/LI) (Leipzig: G. J. Göschensche Verlagshandlung)

IOP Conf. Series: Journal of Physics: Conf. Series **804** (2017) 012014 doi:10.1088/1742-6596/804/1/012014

T-dualization of a weakly curved background

Lj. Davidović, B. Nikolić and B. Sazdović

Institute of Physics, University of Belgrade, Belgrade, Serbia

Abstract. We consider a string moving in a weakly curved background, composed of a constant metric and a linearly coordinate dependent Kalb-Ramond field with an infinitesimal strength. We discuss the T-dualization procedure which we developed for a closed bosonic string moving in a weakly curved background. The procedure is a generalization of a Buscher T-dualization procedure and enables the T-dualization of the nonisometry directions. The same procedure is used to investigate the T-duals of an open bosonic string as well. The generalized T-dualizations give insight to the connection between the geometrical properties of the T-dual spaces.

1. Introduction

In string theory there exists a symmetry, T-duality, which allows the physical equivalence of the string living on the different geometrical structures of the compactified dimensions. The string living in a space with one dimension compactified on a radius R, has the same physical features as a string leaving in a space with one dimension compactified on a radius $\frac{\alpha'}{R}$, where α' is a Regge slope parameter. T-duality was first described in the context of toroidal compactification in [1] (thoroughly explained in [2]), it can be generalized to the arbitrary toroidal compactification [3], and extended to the non-flat conformal backgrounds [4]. The origin of T-duality is seen in a possibility that, unlike a point particle, the string can wrap around compactified dimensions.

The first T-dualization procedure, the prescription for obtaining a theory which is T-dual of a given theory, was defined by Buscher [5]. The procedure was done for a string sigma model, describing a string moving in a background composed of a metric $G_{\mu\nu}$, an antisymmetric field $B_{\mu\nu}$ and a dilaton field Φ. It is required that the metric admits at least one continuous abelian isometry which leaves the action invariant. The procedure is founded in gauging the isometry by introducing the gauge fields v^μ_α. In order to preserve the physical content of the original theory, one requires that the new fields v^μ_α are nonphysical, which is achieved by the requirement that the gauge fields have a vanishing field strength $F^\mu_{\alpha\beta} = \partial_\alpha v^\mu_\beta - \partial_\beta v^\mu_\alpha$. This requirement is included in the theory by adding the Lagrange multiplier term $y_\mu F^\mu_{01}$ into the Lagrangian. Fixing the gauge one obtains the gauge fixed Lagrangian which carries the information on both initial and a T-dual theory. The integration over the Lagrange multipliers y_μ, simply recovers the original theory. The integration over the gauge fields v^μ_α, produces the T-dual theory.

The standard T-dualization procedure is applicable along directions which do not appear as the background field arguments. The generalized T-dualization procedure which is applicable along an arbitrary coordinate was done in Refs. [6, 7, 8]. The procedure is founded in the standard procedure and keeps the main rules of the standard procedure. In order to gauge the global isometry, one introduces the gauge fields v^μ_α, as usual. The replacement of the derivatives $\partial_\alpha x^\mu$ with the covariant ones $D_\alpha x^\mu$, does not as before make the whole action invariant. The obstacle is the background field $B_{\mu\nu}$ depending on x^μ, which is not locally gauge invariant. So, as a new rule we substitute the argument of the background fields by an invariant argument Δx^μ_{inv}, defined as the line integral of the covariant derivatives of the original argument. As before, in order to obtain the theory physically equivalent to the original one, we add the Lagrange multiplier term. Using the local gauge freedom we fix the gauge taking $x^\mu(\xi) = x^\mu(\xi_0)$. The obtained gauge fixed action reduces to to the original action for the equations of motion for the Lagrange multiplier. The T-dual theory is obtained for the equations of motion for the gauge fields v^μ_α.

The generalized T-dualization procedure was investigated for a string moving in a weakly curved background composed of a constant metric, a linearly coordinate dependent Kalb-Ramond field with an infinitesimal filed strength and a constant dilaton field. It was first applied to all space-time coordinates in Ref. [6], and a T-dual was obtained. In Ref. [8], the procedure was applied to an arbitrary set of the initial coordinates. Choosing d arbitrary directions, we denote $\mathcal{T}^a = \circ_{n=1}^d T^{\mu_n}$, $\mathcal{T}^i = \circ_{n=d+1}^D T^{\mu_n}$, and $\mathcal{T} = \circ_{n=1}^D T^{\mu_n}$, where T^μ stands for a T-dualization along direction x^μ and $\mathcal{T}_a = \circ_{n=1}^d T_{\mu_n}$, $\mathcal{T}_i = \circ_{n=d+1}^D T_{\mu_n}$, $\tilde{\mathcal{T}} = \circ_{n=1}^D T_{\mu_n}$, where T_μ stands for the T-dualization along a dual direction y_μ. Performing the generalized procedure we proved the following composition laws:

$$\mathcal{T}^i \circ \mathcal{T}^a = \mathcal{T}, \quad \mathcal{T}_i \circ \mathcal{T}_a = \tilde{\mathcal{T}}, \quad \mathcal{T}_a \circ \mathcal{T}^a = 1, \tag{1}$$

where 1 denotes the identical transformation (T-dualization not performed). We found the explicit forms of the resulting theories and the corresponding T-dual coordinate transformation laws. These results complete the T-dualization diagram connecting all the theories T-dual to the initial theory.

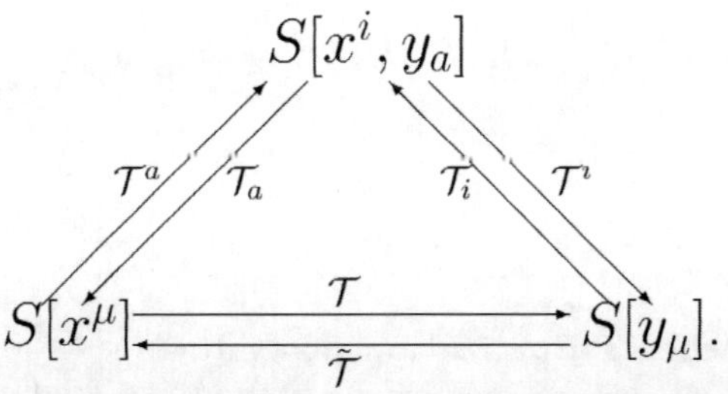

The initial theory, describing the bosonic string moving in the weakly curved background is defined on the geometrical space. All its T-dual theories are non-geometric and non-local because they depend on variable V^μ, which is a line integral of the derivatives of the dual coordinates. To all of these theories there corresponds a flux which is of the same type as the R flux unlike the non-geometric theories with Q flux, which have a local geometric description.

2. Bosonic string action

Let us consider the action [9, 10] describing the propagation of the bosonic string in a background composed of a space-time metric $G_{\mu\nu}$, a Kalb-Ramond field $B_{\mu\nu}$ and a dilaton field Φ

$$S[x] = \kappa \int_\Sigma d^2\xi \sqrt{-g}\Big[\Big(\frac{1}{2}g^{\alpha\beta}G_{\mu\nu}(x) + \frac{\varepsilon^{\alpha\beta}}{\sqrt{-g}}B_{\mu\nu}(x)\Big)\partial_\alpha x^\mu \partial_\beta x^\nu + \frac{1}{4\pi\kappa}\Phi(x)R^{(2)}\Big]. \tag{2}$$

The integration goes over two-dimensional world-sheet Σ parametrized by ξ^α ($\xi^0 = \tau$, $\xi^1 = \sigma$), $g_{\alpha\beta}$ is intrinsic world-sheet metric, $R^{(2)}$ corresponding 2-dimensional scalar curvature, $x^\mu(\xi)$, $\mu = 0, 1, ..., D-1$ are the coordinates of the D-dimensional space-time, $\kappa = \frac{1}{2\pi\alpha'}$ and $\varepsilon^{01} = -1$.

In order to have a world-sheet conformal invariance on the quantum level, the background fields have to obey the space-time equations of motion which in the lowest order in slope parameter α', have the following form

$$\begin{aligned} &R_{\mu\nu} - \frac{1}{4}B_{\mu\rho\sigma}B_{\nu}{}^{\rho\sigma} + 2D_\mu\partial_\nu\Phi = 0, \\ &D_\rho B^\rho{}_{\mu\nu} - 2\partial_\rho\Phi B^\rho{}_{\mu\nu} = 0, \\ &4(\partial\Phi)^2 - 4D_\mu\partial^\mu\Phi + \frac{1}{12}B_{\mu\nu\rho}B^{\mu\nu\rho} - R + 4\pi\kappa\frac{D-26}{3} = 0\,, \end{aligned} \tag{3}$$

where $B_{\mu\nu\rho} = \partial_\mu B_{\nu\rho} + \partial_\nu B_{\rho\mu} + \partial_\rho B_{\mu\nu}$ is the field strength of the field $B_{\mu\nu}$, and $R_{\mu\nu}$ and D_μ are Ricci tensor and covariant derivative with respect to space-time metric. We consider the weakly curved background, defined by

$$G_{\mu\nu} = const, \quad B_{\mu\nu}(x) = b_{\mu\nu} + \frac{1}{3}B_{\mu\nu\rho}x^\rho \equiv b_{\mu\nu} + h_{\mu\nu}(x), \quad \Phi = const. \tag{4}$$

The Kalb-Ramond field strength $B_{\mu\nu\rho}$ is taken to be infinitesimal. All the calculations are done in the first order in $B_{\mu\nu\rho}$. In this approximation the weakly curved background is the solution of the space-time equations of motion (3).

Introducing the light-cone coordinates

$$\xi^{\pm} = \frac{1}{2}(\tau \pm \sigma)$$

and their derivatives $\partial_{\pm} = \partial_\tau \pm \partial_\sigma$, taking a conformal gauge $g_{\alpha\beta} = e^{2F}\eta_{\alpha\beta}$, the action (2) can be written as

$$S[x] = \kappa \int_\Sigma d^2\xi\ \partial_+ x^\mu \Pi_{+\mu\nu}(x)\partial_- x^\nu, \tag{5}$$

where

$$\Pi_{\pm\mu\nu}(x) = B_{\mu\nu}(x) \pm \frac{1}{2}G_{\mu\nu}(x). \tag{6}$$

3. The Generalized Buscher T-dualization procedure

The standard T-dualization procedure, enables one to find a T-dual of a given theory, applying the procedure to the coordinate directions which do not appear as the background field arguments. The generalized T-dualization procedure does not have this limitation. Both procedures are grounded in a localization of a global coordinate shift symmetry $\delta x^\mu = \lambda^\mu = const$. The first rule of the procedures is the introduction of the gauge fields v^μ_α and the substitution of the ordinary derivatives with the covariant derivatives, defined by

$$\partial_\alpha x^\mu \to D_\alpha x^\mu = \partial_\alpha x^\mu + v^\mu_\alpha. \tag{7}$$

If one imposes the following transformation law for the gauge fields

$$\delta v^\mu_\alpha = -\partial_\alpha \lambda^\mu, \quad (\lambda^\mu = \lambda^\mu(\tau,\sigma)) \tag{8}$$

one obtains $\delta D_\alpha x^\mu = 0$. In the case when the background does not depend on the coordinates, along which the T-dualization is performed, the first step is sufficient to obtain the gauge invariant action. However if the background depends on all the coordinates, an additional rule must be introduced. The new rule reads: *Substitute the background field argument (the* *coordinate x^μ), by the invariant argument (invariant coordinate)*, defined as a line integral of the covariant derivatives of the original coordinate

$$\Delta x^\mu_{inv} \equiv \int_P d\xi^\alpha D_\alpha x^\mu = x^\mu - x^\mu(\xi_0) + \Delta V^\mu, \quad \Delta V^\mu \equiv \int_P d\xi^\alpha v^\mu_\alpha. \tag{9}$$

The invariant coordinate is by definition nonlocal. The consequence of this will be a nonlocal T-dual theory, defined on the doubled geometrical space composed of the dual coordinate y_μ and its double $\tilde{y}_\mu$.

The common rule of the procedures is the addition of the Lagrange multiplier term which makes the introduced gauge fields nonphysical, by requiring that there field strength

$$F^\mu_{\alpha\beta} \equiv \partial_\alpha v^\mu_\beta - \partial_\beta v^\mu_\alpha \tag{10}$$

must be zero. This enables the physical equivalence of the theories. Following these rules we built the gauge invariant action.

The main object and the main crossway of the procedure are the gauge fixed action and their equations of motion, because for the equation of motion obtained varying the action over the Lagrange multipliers, one returns to the initial action. On the other hand for the equation of motion obtained varying the gauge fixed action over the gauge fields one obtains the T-dual theory. Comparing the solutions for the gauge fields in these two directions, one obtains the T-dual coordinate transformation laws. These laws are used in investigation of the relations between the non-commutativity characteristics of the spaces connected by T-duality.

The generalized procedure, can be generalized once more in order to allow the T-dualization of the backgrounds which do not have a global symmetry. The generalization was made in Ref. [7] for a bosonic string moving in a weakly curved background of the second order, which consists of the coordinate dependent metric and Kalb-Ramond field. One postulates the auxiliary action which inherits two important features of the gauge fixed action. It reduces to the initial theory for the equations of motion for the Lagrange multipliers and to the T-dual action for the equations of motion for the auxiliary fields.

3.1. Complete T-dualization

If one applies the T-dualization procedure to all coordinates, one obtains a following gauge invariant action

$$S_{inv} = \kappa \int d^2\xi \Big[D_+ x^\mu \Pi_{+\mu\nu}(\Delta x_{inv}) D_- x^\nu + \frac{1}{2}(v^\mu_+ \partial_- y_\mu - v^\mu_- \partial_+ y_\mu)\Big], \tag{11}$$

which is physically equivalent to the initial action. Fixing the gauge by $x^\mu(\xi) = x^\mu(\xi_0)$, one obtains the gauge fixed action

$$S_{fix}[y, v_\pm] = \kappa \int d^2\xi \Big[v^\mu_+ \Pi_{+\mu\nu}(\Delta V) v^\nu_- + \frac{1}{2}(v^\mu_+ \partial_- y_\mu - v^\mu_- \partial_+ y_\mu)\Big]. \tag{12}$$

In order to find a T-dual action one has to integrate out the gauge fields from (12).

The equations of motion with respect to the gauge fields $v^\mu_\pm$ are

$$\Pi_{\mp\mu\nu}(\Delta V) v^\nu_\pm + \frac{1}{2}\partial_\pm y_\mu = \mp\beta^\mp_\mu(V), \tag{13}$$

with the right hand side coming from the variation of the background fields argument, with $\beta^\pm_\mu(x) = \mp\frac{1}{2}h_{\mu\nu}[x]\partial_\mp x^\nu$. The equation of motion can be rewritten as

$$v^\mu_\pm(y) \;=\; -\kappa\,\Theta^{\mu\nu}_\pm[\Delta V(y)]\Big[\partial_\pm y_\nu \pm 2\beta^\mp_\nu[V(y)]\Big], \tag{14}$$

where

$$\Theta_{\pm}^{\mu\nu}[\Delta V] = -\frac{2}{\kappa}(G_E^{-1}\Pi_{\pm}G^{-1})^{\mu\nu} = \theta^{\mu\nu}[\Delta V] \mp \frac{1}{\kappa}(G_E^{-1})^{\mu\nu}[\Delta V], \tag{15}$$

and $G_{\mu\nu}^E \equiv [G - 4BG^{-1}B]_{\mu\nu}$, $\theta^{\mu\nu} \equiv -\frac{2}{\kappa}(G_E^{-1}BG^{-1})^{\mu\nu}$ are the open string background fields: the effective metric and the non-commutativity parameter respectively. They are defined in analogy with the flat space-time open string background fields introduced in [11]. Tensors $\Pi_{\mp\mu\nu}$ and $\Theta_{\pm}^{\mu\nu}$ are connected by $\Theta_{\pm}^{\mu\nu}\Pi_{\mp\nu\rho} = \frac{1}{2\kappa}\delta_\rho^\mu$. Substituting (14) into the action (12), we obtain T-dual action

$${}^\star S[y] \equiv S_{fix}[y] = \frac{\kappa^2}{2}\int d^2\xi\ \partial_+ y_\mu \Theta_-^{\mu\nu}[\Delta V^{(0)}(y)]\partial_- y_\nu, \tag{16}$$

where we neglected the term $\beta_\mu^- \beta_\nu^+$ as the infinitesimal of the second order, and the argument is given by

$$\Delta V^{(0)\mu}(y) = -\kappa\theta_0^{\mu\nu}\Delta y_\nu^{(0)} + (g^{-1})^{\mu\nu}\Delta\tilde{y}_\nu^{(0)}. \tag{17}$$

Comparing the initial action (5) with the T-dual action (16), we see that they are equal under following transformations $\partial_\pm x^\mu \to \partial_\pm y_\mu$ and $\Pi_{+\mu\nu}[x] \to \frac{\kappa}{2}\Theta_-^{\mu\nu}[\Delta V^{(0)}]$, which implies

$$\begin{aligned} G_{\mu\nu} &\to {}^\star G^{\mu\nu} = (G_E^{-1})^{\mu\nu}[\Delta V^{(0)}], \\ B_{\mu\nu}[x] &\to {}^\star B^{\mu\nu} = \frac{\kappa}{2}\theta^{\mu\nu}[\Delta V^{(0)}], \end{aligned} \tag{18}$$

where $(G_E^{-1})^{\mu\nu}$ and $\theta^{\mu\nu}$ are introduced in (15).

The initial background consisted of a constant metric and a linearly coordinate dependent Kalb-Ramond field with an infinitesimal field strength. The T-dual background consists of coordinate dependent metric and Kalb-Ramond field, with the argument ΔV^μ, which is the linear combination of y_μ and its double $\tilde{y}_\mu$. Note that the variable V^μ and consequently T-dual action is not defined on the geometrical space (defined by the coordinate y_μ) but on the so called doubled target space [12] composed of both y_μ and $\tilde{y}_\mu$.

3.2. Partial T-dualization

If one choses only a subset of the initial coordinates, say d coordinates x^a, and performs T-dualization procedure along these coordinates, one obtains the following gauge invariant action

$$\begin{aligned} S_{inv}[x^\mu, x^a_{inv}, y_a] &= \kappa\int d^2\xi\Big[\partial_+ x^i\Pi_{+ij}(x^i, \Delta x^a_{inv})\partial_- x^j \\ &+ \partial_+ x^i\Pi_{+ia}(x^i, \Delta x^a_{inv})D_- x^a + D_+ x^a\Pi_{+ai}(x^i, \Delta x^a_{inv})\partial_- x^i \\ &+ D_+ x^a\Pi_{+ab}(x^i, \Delta x^a_{inv})D_- x^b + \frac{1}{2}(v_+^a\partial_- y_a - v_-^a\partial_+ y_a)\Big]. \end{aligned} \tag{19}$$

This action is obtained localizing the global shift symmetry only for the coordinates x^a, by introducing the gauge fields v_α^a. The ordinary derivatives $\partial_\alpha x^a$ were substituted by the covariant derivatives $D_\alpha x^a = \partial_\alpha x^a + v_\alpha^a$. The covariant derivatives are invariant under the standard gauge transformations $\delta v_\alpha^a = -\partial_\alpha\lambda^a$. The coordinates x^a in the argument of the background fields were substituted by their invariant extension, defined by $\Delta x^a_{inv} \equiv \int_P d\xi^\alpha\, D_\alpha x^a = x^a - x^a(\xi_0) + \Delta V^a$, where $\Delta V^a \equiv \int_P d\xi^\alpha v_\alpha^a$. The physical equivalence is preserved by adding the Lagrange multiplier term (the last term in the action). Fixing the gauge by $x^a(\xi) = x^a(\xi_0)$ one obtains the gauge fixed action

$$\begin{aligned} S_{fix}[x^i, v^a_\pm, y_a] &= \kappa \int d^2\xi \Big[\partial_+ x^i \Pi_{+ij}(x^i, \Delta V^a)\partial_- x^j \\ &+ \partial_+ x^i \Pi_{+ia}(x^i, \Delta V^a) v^a_- + v^a_+ \Pi_{+ai}(x^i, \Delta V^a)\partial_- x^i \\ &+ v^a_+ \Pi_{+ab}(x^i, \Delta V^a) v^b_- + \frac{1}{2}(v^a_+ \partial_- y_a - v^a_- \partial_+ y_a)\Big]. \end{aligned} \tag{20}$$

This action reduces to the initial one for the equations of motion obtained varying over the Lagrange multipliers. The T-dual action is obtained for the equations of motion for the gauge fields. It reads

$$\begin{aligned} S[x^i, y_a] &= \kappa \int d^2\xi \Big[\partial_+ x^i \overline{\Pi}_{+ij}(x^i, \Delta V^a(x^i, y_a))\partial_- x^j \\ &-\kappa\, \partial_+ x^i \Pi_{+ia}(x^i, \Delta V^a(x^i, y_a))\tilde{\Theta}^{ab}_-(x^i, \Delta V^a(x^i, y_a))\partial_- y_b \\ &+\kappa\, \partial_+ y_a \tilde{\Theta}^{ab}_-(x^i, \Delta V^a(x^i, y_a))\Pi_{+bi}(x^i, \Delta V^a(x^i, y_a))\partial_- x^i \\ &+\frac{\kappa}{2}\, \partial_+ y_a \tilde{\Theta}^{ab}_-(x^i, \Delta V^a(x^i, y_a))\partial_- y_b\Big]. \end{aligned} \tag{21}$$

The T-dual background fields compositions are the inverses of the already known background compositions, divided into two coordinate subspaces, the subspace formed by the coordinates we T-dualize and the subspace formed by the rest of the coordinates. The background field compositions $\overline{\Pi}_{\pm ij}$ and $\tilde{\Theta}^{ab}_\pm$ are defined as the inverses of the background field compositions $\Theta^{jk}_\mp$ and $\Pi_{\mp bc}$, which are the parts of $\Theta^{\mu\nu}_\mp$ and $\Pi_{\mp\mu\nu}$ in an appropriate subspace

$$\begin{aligned} \overline{\Pi}_{\pm ij}\Theta^{jk}_\mp &= \Theta^{kj}_\mp \overline{\Pi}_{\pm ji} = \frac{1}{2\kappa}\delta^k_i, \\ \tilde{\Theta}^{ab}_\pm \Pi_{\mp bc} &= \Pi_{\mp cb}\tilde{\Theta}^{ba}_\pm = \frac{1}{2\kappa}\delta^a_c. \end{aligned} \tag{22}$$

It can be shown that

$$\overline{\Pi}_{+ij} \equiv \Pi_{+ij} - 2\kappa \Pi_{+ia}\tilde{\Theta}^{ab}_- \Pi_{+bj}. \tag{23}$$

The argument of the background fields is

$$\begin{aligned} \Delta V^{(0)a}(x^i, y_a) &= -\kappa\Big[\tilde{\Theta}^{ab}_{0+}\Pi_{0-bi} + \tilde{\Theta}^{ab}_{0-}\Pi_{0+bi}\Big]\Delta x^{(0)i} \\ &- \kappa\Big[\tilde{\Theta}^{ab}_{0+}\Pi_{0-bi} - \tilde{\Theta}^{ab}_{0-}\Pi_{0+bi}\Big]\Delta \tilde{x}^{(0)i} \\ &- \frac{\kappa}{2}\Big[\tilde{\Theta}^{ab}_{0+} + \tilde{\Theta}^{ab}_{0-}\Big]\Delta y^{(0)}_b - \frac{\kappa}{2}\Big[\tilde{\Theta}^{ab}_{0+} - \tilde{\Theta}^{ab}_{0-}\Big]\Delta \tilde{y}^{(0)}_b. \end{aligned} \tag{24}$$

Calculating the symmetric and antisymmetric part of the background fields we obtain a T-dual metric and a T-dual Kalb-Ramond field

$$\begin{aligned} &{}^\bullet G_{ij} = \overline{G}_{ij} = G_{ij} - G_{ia}(\tilde{G}^{-1}_E)^{ab}G_{bj} \\ &\quad -2\kappa\Big(B_{ia}\tilde{\theta}^{ab}G_{bj} + G_{ia}\tilde{\theta}^{ab}B_{bj}\Big) - 4B_{ia}(\tilde{G}^{-1}_E)^{ab}B_{bj}, \\ &{}^\bullet B_{ij} = \overline{B}_{ij} = B_{ij} - \frac{\kappa}{2}G_{ia}\tilde{\theta}^{ab}G_{bj} - B_{ia}(\tilde{G}^{-1}_E)^{ab}G_{bj} \\ &\quad -G_{ia}(\tilde{G}^{-1}_E)^{ab}B_{bj} - 2\kappa B_{ia}\tilde{\theta}^{ab}B_{bj}, \\ &{}^\bullet G^{ab} = (\tilde{G}^{-1}_E)^{ab}, \\ &{}^\bullet B^{ab} = \frac{\kappa}{2}\tilde{\theta}^{ab}, \\ &{}^\bullet G^a{}_i = \kappa\tilde{\theta}^{ab}G_{bi} + 2(\tilde{G}^{-1}_E)^{ab}B_{bi}, \\ &{}^\bullet B^a{}_i = \kappa\tilde{\theta}^{ab}B_{bi} + \frac{1}{2}(\tilde{G}^{-1}_E)^{ab}G_{bi}. \end{aligned} \tag{25}$$

As the constituents of the T-dual background field there appear the effective metric in the subspace a, defined by $\tilde{G}_{Eab} \equiv G_{ab} - 4B_{ac}(\tilde{G}^{-1})^{cd}B_{db}$, the non-commutativity parameter in the same subspace $\tilde{\theta}^{ab} \equiv -\frac{2}{\kappa}(\tilde{G}_E^{-1})^{ac}B_{cd}(\tilde{G}^{-1})^{db}$, which combined give the new theta function $\tilde{\Theta}^{ab}_{\pm} = \tilde{\theta}^{ab} \mp \frac{1}{\kappa}(\tilde{G}_E^{-1})^{ab}$.

4. Open string T-dualization

In paper [13] we investigated a T-duality of an open string moving in a weakly curved background. The open string moving in a weakly curved background was a subject of investigation in our papers [14, 15, 16]. Solving the boundary conditions at the open string end-points, one obtains the effective closed string described by the effective closed string theory S^{eff}, defined on the doubled space $(q^\mu, \tilde{q}^\mu)$. As the effective theory is closed string theory, one can try to apply the generalized T-dualization procedure to this theory. The effective theory is defined on the doubled theory, just as the T-duals of the closed string theory moving in the weakly curved background. So, the application in this case resembles the application of the T-dualization procedure to the T-dual theories.

The effective theory of the open string moving in the weakly curved background, obtained for the solution of the boundary conditions equals

$$S^{eff} = \kappa \int d\tau \int_{-\pi}^{\pi} d\sigma\, \partial_+ q^\mu\, \Pi^{eff}_{+\mu\nu}(q, 2b\tilde{q})\, \partial_- q^\nu, \tag{26}$$

where

$$\Pi^{eff}_{\pm\mu\nu}(q, 2b\tilde{q}) \equiv B^{eff}_{\mu\nu}(2b\tilde{q}) \pm \frac{1}{2} G^{eff}_{\mu\nu}(q). \tag{27}$$

The effective variable is $q^\mu(\sigma)$, an even part of the initial coordinate. The effective metric and the Kalb-Ramond field are explicitly given by

$$\begin{aligned} G^{eff}_{\mu\nu}(q) &= G^E_{\mu\nu}(q) := (G - 4B^2(q))_{\mu\nu}, \\ B^{eff}_{\mu\nu}(2b\tilde{q}) &= -\frac{\kappa}{2}\Big(g_E \Delta\theta(2b\tilde{q}) g_E\Big)_{\mu\nu}, \end{aligned} \tag{28}$$

where $\Delta\theta^{\mu\nu}$ is the infinitesimal part of the non-commutativity parameter $\theta^{\mu\nu} = -\frac{2}{\kappa}\Big[G_E^{-1}BG^{-1}\Big]^{\mu\nu} = \theta_0^{\mu\nu} - \frac{2}{\kappa}\Big[g_E^{-1}(h + 4bhb)g_E^{-1}\Big]^{\mu\nu}$. In paper [13] we applied the generalized Buscher T-dualization procedure, to the effective theory along all effective directions q^μ. Following the procedure we find the gauge fixed action

$$\mathcal{S}_{fix} = \kappa \int d^2\xi \Big[v^\mu_+ \Pi^{eff}_{+\mu\nu}(\Delta V, 2b\Delta\tilde{V}) v^\nu_- + \frac{1}{2}(v^\mu_+ \partial_- \varrho_\mu - v^\mu_- \partial_+ \varrho_\mu)\Big], \tag{29}$$

obtained from the effective action (26), by substituting the light-cone derivatives $\partial_\pm q^\mu$ with the covariant derivatives $D_\pm q^\mu = \partial_\pm q^\mu + v^\mu_\pm$, where $v^\mu_\pm$ are the gauge fields, which transform as $\delta v^\mu_\pm = -\partial_\pm \lambda^\mu$. The argument of the background fields is substituted with an invariant argument, which is obtained substituting the effective coordinate q^μ and its double $\tilde{q}^\mu$ with an invariant effective coordinate and its double, defined by the following line integrals of the gauge fields $\Delta V^\mu = \int_P (d\xi^+ v^\mu_+ + d\xi^- v^\mu_-)$, and $\Delta\tilde{V}^\mu = \int_P (d\xi^+ v^\mu_+ - d\xi^- v^\mu_-)$. The physical equivalence was achieved by adding the Lagrange multiplier term $\frac{1}{2}(v^\mu_+ \partial_- \varrho_\mu - v^\mu_- \partial_+ \varrho_\mu)$ and the gauge is fixed with $q^\mu(\xi) = q^\mu(\xi_0)$.

The T-dual theory was obtained for the equation of motion for the gauge fields. The T-dual action reads

$${}^\star\mathcal{S} = \kappa \int d^2\xi \partial_+ \varrho_\mu \frac{\kappa}{2} (\Theta^{eff}_-)^{\mu\nu}(\Delta V(\varrho), 2b\Delta\tilde{V}(\varrho)) \partial_- \varrho_\nu, \tag{30}$$

where

$$(\Theta_{\pm}^{eff})^{\mu\nu}(x,y) \equiv \Theta_{\pm}^{\mu\nu}(G_{eff}(x), B_{eff}(y)) = \theta_{eff}^{\mu\nu}(y) \mp \frac{1}{\kappa}(G_E^{-1})^{\mu\nu}(x), \tag{31}$$

$\theta_{eff}^{\mu\nu} := \theta^{\mu\nu}(G_{eff}(x), B_{eff}(y)) = -\frac{2}{\kappa}\Big(G_E^{-1}(G_{eff}(x), B_{eff}(y))B_{eff}(y)G_{eff}^{-1}(x)\Big)^{\mu\nu}$ and the argument is

$$\begin{aligned} V_0^\mu(\varrho) &= (g_E^{-1})^{\mu\nu}(G^{eff}, B^{eff})\tilde{\varrho}_\nu = (g_E^{-1})^{\mu\nu}\tilde{\varrho}_\nu, \\ \tilde{V}_0^\mu(\varrho) &= (g_E^{-1})^{\mu\nu}(G^{eff}, B^{eff})\varrho_\nu = (g_E^{-1})^{\mu\nu}\varrho_\nu. \end{aligned} \tag{32}$$

The T-dual metric ${}^\star G^{\mu\nu}$ which depends on the first variable ΔV^μ and the T-dual Kalb-Ramond field ${}^\star B^{\mu\nu}$, which depends on the second variable $2b^\mu{}_\nu \Delta\tilde{V}^\nu$ are

$$\begin{aligned} {}^\star G^{\mu\nu} &= (G_E^{-1})^{\mu\nu}(\Delta V), \\ {}^\star B^{\mu\nu} &= \frac{\kappa}{2}(\theta^{eff})^{\mu\nu}(2b\Delta\tilde{V}) = \frac{\kappa}{2}\Delta\theta^{\mu\nu}(2b\Delta\tilde{V}). \end{aligned} \tag{33}$$

We see, that the effective metric has transformed to its inverse and that the Kalb-Ramond field has transformed to the infinitesimal part of the non-commutativity parameter.

Finally, we searched for the open string theory $\tilde{S}$ such that its effective theory is ${}^\star S^{eff}$ exactly. We found

$$\tilde{S}[y] = \kappa \int_\Sigma d^2\xi\ \partial_+ y_\mu \tilde{\Pi}_+^{\mu\nu}(y)\partial_- y_\nu, \tag{34}$$

with

$$\begin{aligned} \tilde{G} &= -(C^T)^{-1}GC^{-1}, \\ \tilde{B}(y) &= \pm(C^T)^{-1}(b - h(C^{-1}y))C^{-1}, \end{aligned} \tag{35}$$

where C makes a connection between the variables of the effective theory of $\tilde{S}$ and the T-dual theory (30)

$$\begin{aligned} q_\mu(y) &= C_{\mu\nu}(g_E^{-1})^{\nu\rho}\tilde{\varrho}_\rho, \\ \bar{q}_\mu(y) &= \mp C_{\mu\nu}2(G^{-1}bg_E^{-1})^{\nu\rho}\varrho_\rho. \end{aligned} \tag{36}$$

In the closed string moving in the weakly curved background case, the T-duality transforms the geometrical background into a doubled non-geometrical background. It transforms a constant metric to a coordinate dependent effective metric inverse, while the linearly coordinate dependent Kalb-Ramond field is transformed into a coordinate dependent non-commutativity parameter. In the open string case, the T-dual theory remains geometric. T-duality transforms the constant metric of the weakly curved background to a constant T-dual metric, while the coordinate dependent Kalb-Ramond field transforms again to the coordinate dependent field.

In paper [17] a generalization of the standard analysis of the open bosonic string moving in a flat background is addressed. The T-dualization was performed in two ways, first in terms of non-constant vector fields in which case the Buscher T-dualization procedure can not be applied and second in terms of the field strengths of the gauge fields. The role of the gauge fields, which live on the string boundary, is to restore the symmetries of the closed string: the local gauge symmetry of the Kalb-Ramond field and the general coordinate transformations, at the string end-points. The investigation lead to a discovery of the geometrical features of the non-geometry.

Conclusion

The generalized T-dualization procedure, enabled T-dualization over the non isometry directions. It gives the new insights into a connection between the spaces connected by T-duality. It enabled further investigations of the closed string non-commutativity [18]. Comparing the solutions for the gauge fields which transform the gauge fixed actions into the initial or the T-dual actions, one obtains the T-dual coordinate transformation laws. Using these laws one can find how does for example a standard Poisson bracket transform. It is obtained that the original theory which is commutative is equivalent to the non-commutative T-dual theory, whose Poisson brackets are proportional to the background fluxes times winding and momentum numbers. The obtained results add novelty to the form and the origin of different non-commutative structures.

Acknowledgments

Research is supported in part by the Serbian Ministry of Education, Science and Technological Development (project No. 171031) and by The National Scholarship L'Oreal-UNESCO "For Women in Science" which Lj.D. obtained in 2015.

References

[1] L.Brink, M.B. Green, J.H. Schwarz, *Nucl. Phys.* **B 198** (1982) 474; K. Kikkawa and M.Yamasaki, *Phys. Lett.* **B 149** (1984) 357; N. Sakai and I. Senda, *Prog. Theor. Phys.* **75** (1984) 692.
[2] E. Alvarez, L. Alvarez-Gaume and Y. Lozano, *Nucl. Phys. Proc. Suppl.* **41** (1995) 1; A. Giveon, M.Parrati and E. Rabinovici, *Phys. Rep.* **244** (1994) 77.
[3] K.Narain, H. Sarmadi and E. Witten, *Nucl. Phys.* **B 279** (1987) 369.
[4] T. Buscher, *Phys. Lett.* **B 194** (1987) 51; **201** (1988) 466.
[5] T. Buscher, Phys. Lett. **B 194** (1987) 59; M. Roček and E.P. Verlinde, Nucl. Phys. **B 373** (1992) 630.
[6] Lj. Davidović and B. Sazdović, EPJ**C 74** No. 1 (2014) 2683.
[7] Lj. Davidović and B. Sazdović, EPJ**C 75** No. 12 (2015) 576.
[8] Lj. Davidović, B. Nikolić and B. Sazdović, JHEP **11** (2015) 119.
[9] B. Zwiebach, *A First Course in String Theory*, (Cambridge University Press, 2004).
[10] K. Backer, M. Backer and J. Schwarz, *String Theory and M-theory*, (Cambridge University Press, 2007).
[11] N. Seiberg, E. Witten, J. High Energy Phys. **09** (1999) 032.
[12] C. Hull, *JHEP* **10** (2005) 065; A. Dabholkar and C. Hull, *JHEP* **05** (2006) 009; C. Hull, *JHEP* **07** (2007) 080; J. Shelton, W. Taylor and B. Wecht, *JHEP* **10** (2005) 085; A. Dabholkar and C. Hull, *JHEP* **09** (2003) 054; A. Flournoy and B. Williams, *JHEP* **01** (2006) 166; T. Kugo, B. Zwiebach, *Prog. Theor. Phys.* **87** (1992) 801; O. Hohm, C. Hull and B. Zwiebach, *JHEP* **08** (2010) 008; O. Hohm, C. Hull and B. Zwiebach, *JHEP* **07** (2010) 016; B. Wecht, *Class.Quant.Grav.* **24 21** (2007) S773; D. Lust, *JHEP* **12** (2010) 084; D.Mylonas, P. Schupp and R. Szabo, arXiv:1207.0926 [hep-th].
[13] Lj. Davidović, EPJ**C 76** (2016) 660.
[14] Lj. Davidović and B. Sazdović, Phys. Rev. **D 83** (2011) 066014.
[15] Lj. Davidović and B. Sazdović, JHEP **08** (2011) 112.
[16] Lj. Davidović and B. Sazdović, EPJ**C 72** No. 11 (2012) 2199.
[17] B. Sazdović, arXiv:1606.01938 [hep-th].
[18] Lj. Davidović, B. Nikolić and B. Sazdović, EPJ**C 74** (2014) 2734.

Positive Energy Unitary Irreducible Representations of the Superalgebras $osp(1|2n, I\!R)$ and Character Formulae for $n = 3$

V.K. Dobrev

Institute of Nuclear Research and Nuclear Energy, Bulgarian Academy of Sciences, 72 Tsarigradsko Chaussee, 1784 Sofia, Bulgaria

I. Salom

Institute of Physics, University of Belgrade, Pregrevica 118, 11080 Zemun, Belgrade, Serbia

Abstract. We overview our study of the positive energy (lowest weight) unitary irreducible representations of the superalgebras $osp(1|2n, I\!R)$. We give more explicitly character formulae for these representations in the case $n = 3$.

1. Introduction

Recently, superconformal field theories in various dimensions are attracting more interest, in particular, due to their duality to AdS supergravities. Until recently only those for $D \leqslant 6$ were studied since in these cases the relevant superconformal algebras satisfy [1] the Haag-Lopuszanski-Sohnius theorem [2]. Thus, such classification was known only for the $D = 4$ superconformal algebras $su(2, 2/N)$ [3] (for $N = 1$), [4, 5, 6, 7] (for arbitrary N). More recently, the classification for $D = 3$ (for even N), $D = 5$, and $D = 6$ (for $N = 1, 2$) was given in [8] (some results are conjectural), and then the $D = 6$ case (for arbitrary N) was finalized in [9].

On the other hand the applications in string theory require the knowledge of the UIRs of the conformal superalgebras for $D > 6$. Most prominent role play the superalgebras $osp(1|2n)$. Initially, the superalgebra $osp(1|32)$ was put forward for $D = 10$ [10]. Later it was realized that $osp(1|2n)$ would fit any dimension, though they are minimal only for $D = 3, 9, 10, 11$ (for $n = 2, 16, 16, 32$, resp.) [11]. In all cases we need to find first the UIRs of $osp(1|2n, I\!R)$ which study was started in [12] and [13]. Later, in [14] we finalized the UIR classification of [12] as Dobrev-Zhang-Salom (DZS) Theorem. There we also proved the DZS Theorem for $osp(1|6)$, while the case $osp(1|8)$ was proved in [15].

In the present paper we present more explicitly the character formulae for $osp(1|6)$. For the lack of space we refer for extensive literature on the subject in [12, 14].

2. Preliminaries on representations

Our basic references for Lie superalgebras are [16, 17], although in this exposition we follow [12].

The even subalgebra of $\mathcal{G} = osp(1|2n, I\!R)$ is the algebra $sp(2n, I\!R)$ with maximal compact subalgebra $\mathcal{K} = u(n) \cong su(n) \oplus u(1)$.

We label the relevant representations of $\mathcal{G}$ by the signature:

$$\chi = [\,d\,;\, a_1\,, ..., a_{n-1}\,] \tag{1}$$

where d is the conformal weight, and $a_1, ..., a_{n-1}$ are non-negative integers which are Dynkin labels of the finite-dimensional UIRs of the subalgebra $su(n)$ (the simple part of $\mathcal{K}$).

We present the classification of the positive energy (lowest weight) UIRs of $\mathcal{G}$ following [12, 14] where were used the methods used for the $D = 4, 6$ conformal superalgebras, cf. [4, 5, 6, 7, 9]. The main tool is an adaptation of the Shapovalov form [18] on the Verma modules V^χ over the complexification $\mathcal{G}^{\mathbb{C}} = osp(1|2n)$ of $\mathcal{G}$.

The root system of $\mathcal{G}^{\mathbb{C}}$ are given in terms of $\delta_1 \ldots, \delta_n$, $(\delta_i, \delta_j) = \delta_{ij}$, $i, j = 1, ..., n$. The even and odd roots systems are [16]:

$$\begin{aligned} \Delta_{\bar{0}} &= \{\pm\delta_i \pm \delta_j\ ,\ 1 \leqslant i < j \leqslant n\ ,\ \ \pm 2\delta_i\ ,\ 1 \leqslant i \leqslant n\}\ , \\ \Delta_{\bar{1}} &= \{\pm\delta_i\ ,\ 1 \leqslant i \leqslant n\} \end{aligned} \tag{2}$$

(we remind that the signs $\pm$ are not correlated). We shall use the following distinguished simple root system [16]:

$$\Pi = \{\,\delta_1 - \delta_2\,,\,, \ldots, \delta_{n-1} - \delta_n\,,\, \delta_n\,\}\ , \tag{3}$$

or introducing standard notation for the simple roots:

$$\begin{aligned} \Pi &= \{\alpha_1, ..., \alpha_n\}, \\ \alpha_j &= \delta_j - \delta_{j+1}, \quad j = 1, ..., n-1, \quad \alpha_n = \delta_n . \end{aligned} \tag{4}$$

The root $\alpha_n = \delta_n$ is odd, the other simple roots are even. The Dynkin diagram is:

$$\underset{1}{\circ} — \cdots — \underset{n-1}{\circ} \Longrightarrow \underset{n}{\bullet} \tag{5}$$

The black dot is used to signify that the simple odd root is not nilpotent. In fact, the superalgebras $B(0,n) = osp(1|2n)$ have no nilpotent generators unlike all other types of basic classical Lie superalgebras [16].

The corresponding to Π positive root system is:

$$\Delta^+_{\bar{0}} = \{\delta_i \pm \delta_j, \ 1 \leqslant i < j \leqslant n, \ 2\delta_i, \ 1 \leqslant i \leqslant n\}, \qquad \Delta^+_{\bar{1}} = \{\delta_i, \ 1 \leqslant i \leqslant n\} \tag{6}$$

Conversely, we give the elementary functionals through the simple roots:

$$\delta_k = \alpha_k + \cdots + \alpha_n . \tag{7}$$

From the point of view of representation theory more relevant is the restricted root system, such that:

$$\begin{aligned} \bar{\Delta}^+ &= \bar{\Delta}^+_{\bar{0}} \cup \Delta^!_{\bar{1}} , \\ \bar{\Delta}^+_{\bar{0}} &\equiv \{\alpha \in \Delta^+_{\bar{0}} \mid \tfrac{1}{2}\alpha \notin \Delta^+_{\bar{1}}\} = \{\delta_i \pm \delta_j, \ 1 \leqslant i < j \leqslant n\} \end{aligned} \tag{8}$$

The superalgebra $\mathcal{G} = osp(1|2n, I\!R)$ is a split real form of $osp(1|2n)$ and has the same root system.

The above simple root system is also the simple root system of the complex simple Lie algebra B_n (dropping the distinction between even and odd roots) with Dynkin diagram:

$$\underset{1}{\circ} — \cdots — \underset{n-1}{\circ} \Longrightarrow \underset{n}{\circ} \tag{9}$$

and root system:

$$\Delta^+_{\mathrm{B_n}} = \{\delta_i \pm \delta_j, \ 1 \leqslant i < j \leqslant n, \ \delta_i, \ 1 \leqslant i \leqslant n\} \cong \bar{\Delta}^+ \tag{10}$$

This shall be used essentially below.

We need explicitly the lowest weight $\Lambda \in \mathcal{H}^*$ (where $\mathcal{H}$ is the Cartan subalgebra of $\mathcal{G}^{\mathbb{C}}$) the values of which should be related to the signature (1):

$$(\Lambda, \alpha_k^\vee) \ = - \, a_k \ , \quad 1 \leqslant k \leqslant n \ , \tag{11}$$

where $\alpha_k^\vee \equiv 2\alpha_k/(\alpha_k, \alpha_k)$, and the minus signs anticipate the fact that we shall use lowest weight Verma modules (instead of the highest weight modules used in [17]) and to Verma module reducibility w.r.t. the roots α_k (this is explained in detail in [6, 12]).

Obviously, a_n must be related to the conformal weight d which is a matter of normalization so as to correspond to some known cases. Thus, our choice is:

$$a_n \ = \ - \, 2d - a_1 - \cdots - a_{n-1} \ . \tag{12}$$

The actual Dynkin labelling is given by:

$$m_k \ = \ (\rho - \Lambda, \alpha_k^\vee) \tag{13}$$

where $\rho \in \mathcal{H}^*$ is given by the difference of the half-sums $\rho_{\bar{0}}\,, \rho_{\bar{1}}$ of the even, odd, resp., positive roots (cf. (6):

$$\begin{aligned} \rho \ \doteq \ \rho_{\bar{0}} - \rho_{\bar{1}} \ &= \ (n - \tfrac{1}{2})\delta_1 + (n - \tfrac{3}{2})\delta_2 + \cdots + \tfrac{3}{2}\delta_{n-1} + \tfrac{1}{2}\delta_n \ , \\ \rho_{\bar{0}} \ &= \ n\delta_1 + (n-1)\delta_2 + \cdots + 2\delta_{n-1} + \delta_n \ , \\ \rho_{\bar{1}} \ &= \ \tfrac{1}{2}(\delta_1 + \cdots + \delta_n) \ . \end{aligned} \tag{14}$$

Naturally, the value of ρ on the simple roots is 1: $(\rho, \alpha_i^\vee) = 1$, $i = 1, ..., n$.

Unlike $a_k \in \mathbb{Z}_+$ for $k < n$ the value of a_n is arbitrary. In the cases when a_n is also a non-negative integer, and then $m_k \in \mathbb{N}$ $(\forall k)$ the corresponding irreps are the finite-dimensional irreps of $\mathcal{G}$.

To introduce Verma modules we use the standard decomposition:

$$\mathcal{G}^{\mathbb{C}} \ = \ \mathcal{G}^+ \oplus \mathcal{H} \oplus \mathcal{G}^- \tag{15}$$

where $\mathcal{G}^+$, $\mathcal{G}^-$, resp., are the subalgebras corresponding to the positive, negative, roots, resp., and $\mathcal{H}$ denotes the Cartan subalgebra.

We consider lowest weight Verma modules, so that $V^\Lambda \cong U(\mathcal{G}^+) \otimes v_0$, where $U(\mathcal{G}^+)$ is the universal enveloping algebra of $\mathcal{G}^+$, and v_0 is a lowest weight vector v_0 such that:

$$\begin{aligned} Z\, v_0 \ &= \ 0 \ , \quad Z \in \mathcal{G}^- \\ H\, v_0 \ &= \ \Lambda(H)\, v_0 \ , \quad H \in \mathcal{H} \ . \end{aligned} \tag{16}$$

Further, for simplicity we omit the sign $\otimes$, i.e., we write $p\,v_0 \in V^\Lambda$ with $p \in U(\mathcal{G}^+)$.

Adapting the criterion of [17] to lowest weight modules, one finds that a Verma module V^Λ is reducible w.r.t. the positive root β iff the following holds [12]:

$$(\rho - \Lambda, \beta^\vee) = m_\beta \;, \qquad \beta \in \Delta^+ \;, \quad m_\beta \in \mathbb{N} \;. \tag{17}$$

If a condition from (17) is fulfilled then V^Λ contains a submodule which is a Verma module $V^{\Lambda'}$ with shifted weight given by the pair m, β : $\Lambda' = \Lambda + m\beta$. The embedding of $V^{\Lambda'}$ in V^Λ is provided by mapping the lowest weight vector v_0' of $V^{\Lambda'}$ to the **singular vector** $v_s^{m,\beta}$ in V^Λ which is completely determined by the conditions [19]:

$$\begin{aligned} X\, v_s^{m,\beta} &= 0 \;, \quad X \in \mathcal{G}^- \;, \\ H\, v_s^{m,\beta} &= \Lambda'(H)\, v_0 \;, \quad H \in \mathcal{H} \;, \quad \Lambda' = \Lambda + m\beta \;. \end{aligned} \tag{18}$$

Explicitly, $v_s^{m,\beta}$ is given by a polynomial in the positive root generators [20, 21]:

$$v_s^{m,\beta} = P^{m,\beta}\, v_0 \;, \quad P^{m,\beta} \in U(\mathcal{G}^+) \;. \tag{19}$$

Thus, the submodule I^β of V^Λ which is isomorphic to $V^{\Lambda'}$ is given by $U(\mathcal{G}^+)\, P^{m,\beta}\, v_0$.

Certainly, (17) may be fulfilled for several positive roots (even for all of them). Let Δ_Λ denote the set of all positive roots for which (17) is fulfilled, and let us denote: $\tilde{I}^\Lambda \equiv \cup_{\beta\in\Delta_\Lambda} I^\beta$. Clearly, $\tilde{I}^\Lambda$ is a proper submodule of V^Λ. Let us also denote $F^\Lambda \equiv V^\Lambda/\tilde{I}^\Lambda$.

The Verma module V^Λ contains a unique proper maximal submodule I^Λ ($\supseteq \tilde{I}^\Lambda$) [17, 22]. Among the lowest weight modules with lowest weight Λ there is a unique irreducible one, denoted by L_Λ, i.e., $L_\Lambda = V^\Lambda/I^\Lambda$.

It may happen that the maximal submodule I^Λ coincides with the submodule $\tilde{I}^\Lambda$ generated by all singular vectors. This is, e.g., the case for all Verma modules if rank $\mathcal{G} \leqslant 2$, or when (17) is fulfilled for all simple roots (and, as a consequence for all positive roots). Here we are interested in the cases when $\tilde{I}^\Lambda$ is a proper submodule of I^Λ. We need the following notion.

Definition: [22, 23, 24] *Let V^Λ be a reducible Verma module. A vector $v_{\rm ssv} \in V^\Lambda$ is called a* **subsingular vector** *if $v_{\rm su} \notin \tilde{I}^\Lambda$ and the following holds:*

$$X\, v_{\rm su} \in \tilde{I}^\Lambda \;, \quad \forall X \in \mathcal{G}^- \tag{20}$$

Going from the above more general definitions to $\mathcal{G}$ we recall that in [12] it was established that from (17) follows that the Verma module $V^{\Lambda(\chi)}$ is reducible if one of the following relations holds:

$$\mathbb{N} \ni m^-_{ij} = j - i + a_i + \cdots + a_{j-1} \quad (21a)$$

$$\mathbb{N} \ni m^+_{ij} = 2n - i - j + 1 + a_j + \cdots + a_{n-1} - a_1 - \cdots - a_{i-1} - 2d \quad (21b)$$

$$\mathbb{N} \ni m_i = 2n - 2i + 1 + a_i + \cdots + a_{n-1} - a_1 + \cdots - a_{i-1} - 2d \quad (21c)$$

$$\mathbb{N} \ni m_{ii} = n - i + \tfrac{1}{2}(1 + a_i + \cdots + a_{n-1} - a_1 + \cdots - a_{i-1}) - d \quad (21d)$$

corresponding to the roots $\delta_i - \delta_j$, $\delta_i + \delta_j$, $(i < j)$, δ_i, $2\delta_i$, resp. Further we shall use the fact from [12] that we may eliminate the reducibilities and embeddings related to the roots $2\delta_i$. Indeed, since $m_i = 2m_{ii}$, whenever (21d) is fulfilled also (21c) is fulfilled.

For further use we introduce notation for the root vector $X^+_j \in \mathcal{G}^+$, $j = 1, \ldots, n$, corresponding to the simple root α_j.

Further, we notice that all reducibility conditions in (21a) are fulfilled. In particular, for the simple roots from those condition (21a) is fulfilled with $\beta \to \alpha_i = \delta_i - \delta_{i+1}$, $i = 1, ..., n-1$ and $m^-_i \equiv m^-_{i,i+1} = 1 + a_i$. The corresponding submodules $I^{\alpha_i} = U(\mathcal{G}^+)\, v^i_s$, where $\Lambda_i = \Lambda + m^-_i \alpha_i$ and $v^i_s = (X^+_i)^{1+a_i}\, v_0$. These submodules generate an invariant submodule which we denote by $I^\Lambda_c \subset \tilde{I}^\Lambda$. Since these submodules are nontrivial for all our signatures in the question of unitarity instead of V^Λ we shall consider also the factor-modules:

$$F^\Lambda_c = V^\Lambda / I^\Lambda_c \supset F^\Lambda . \quad (22)$$

We shall denote the lowest weight vector of F^Λ_c by $|\Lambda_c\rangle$ and the singular vectors above become null conditions in F^Λ_c :

$$(X^+_i)^{1+a_i}\, |\Lambda_c\rangle = 0 , \quad i = 1, ..., n-1. \quad (23)$$

If the Verma module V^Λ is not reducible w.r.t. the other roots, i.e., (21b,c,d) are not fulfilled, then $F^\Lambda_c = F^\Lambda$ is irreducible and is isomorphic to the irrep L_Λ with this weight.

In fact, for the factor-modules reducibility is controlled by the value of d, or in more detail:

The maximal d coming from the different possibilities in (21b) are obtained for $m^+_{ij} = 1$ and they are:

$$d_{ij} \equiv n + \tfrac{1}{2}(a_j + \cdots + a_{n-1} - a_1 - \cdots - a_{i-1} - i - j) , \quad i < j, \quad (24)$$

the corresponding root being $\delta_i + \delta_j$.

IOP Conf. Series: Journal of Physics: Conf. Series **804** (2017) 012015 doi:10.1088/1742-6596/804/1/012015

The maximal d coming from the different possibilities in (21c) are obtained for $m_i = 1$ and they are:

$$d_i \equiv n - i + \tfrac{1}{2}(a_i + \cdots + a_{n-1} - a_1 - \cdots - a_{i-1}) , \qquad (25)$$

the corresponding roots being δ_i.

There are some orderings between these maximal reduction points [12]:

$$\begin{aligned} d_1 &> d_2 > \cdots > d_n , \\ d_{i,i+1} &> d_{i,i+2} > \cdots > d_{in} , \\ d_{1,j} &> d_{2,j} > \cdots > d_{j-1,j} , \\ d_i &> d_{jk} > d_\ell , \qquad i \leqslant j < k \leqslant \ell . \end{aligned} \qquad (26)$$

Obviously the first reduction point is:

$$d_1 = n - 1 + \tfrac{1}{2}(a_1 + \cdots + a_{n-1}) . \qquad (27)$$

Below we shall use the following notion. The singular vector v_1 is called **descendant** of the singular vector $v_2 \notin \mathbb{C}v_1$ if there exists a homogeneous polynomial P_{12} in $U(\mathcal{G}^+)$ so that $v_1 = P_{12}\, v_2$. Clearly, in this case we have: $I^1 \subset I^2$, where I^k is the submodule generated by v_k. Thus, when we factor the submodule I_2 this means factoring also the submodule I_1.

3. Unitarity

The first results on the unitarity were given in [12], and then improved in [14]. Thus, the statement below should be called *Dobrev-Zhang-Salom Theorem*:

Theorem DZS: All positive energy unitary irreducible representations of the superalgebras $osp(1|\,2n, I\!R)$ characterized by the signature χ in (1) are obtained for real d and are given as follows:

$$\begin{aligned}
d &\geqslant n-1+\tfrac{1}{2}(a_1+\cdots+a_{n-1}) = d_1\ , \quad a_1 \neq 0\ , && (28)\\
d &\geqslant n-\tfrac{3}{2}+\tfrac{1}{2}(a_2+\cdots+a_{n-1}) = d_{12}\ , \quad a_1=0,\ a_2\neq 0\ ,\\
d &= n-2+\tfrac{1}{2}(a_2+\cdots+a_{n-1}) = d_2 > d_{13}\ , \quad a_1=0,\ a_2\neq 0\ ,\\
d &\geqslant n-2+\tfrac{1}{2}(a_3+\cdots+a_{n-1}) = d_2 = d_{13}\ , \quad a_1=a_2=0,\ a_3\neq 0\ ,\\
d &= n-\tfrac{5}{2}+\tfrac{1}{2}(a_3+\cdots+a_{n-1}) = d_{23} > d_{14}\ , \quad a_1=a_2=0,\ a_3\neq 0\ ,\\
d &= n-3+\tfrac{1}{2}(a_3+\cdots+a_{n-1}) = d_3 = d_{24} > d_{15}\ , \quad a_1=a_2=0,\ a_3\neq 0\ , && (29)\\
&\ldots\\
d &\geqslant n-1-\kappa+\tfrac{1}{2}(a_{2\kappa+1}+\cdots+a_{n-1})\ , \quad a_1=\ldots=a_{2\kappa}=0,\ a_{2\kappa+1}\neq 0\ ,\\
&\qquad \kappa=\tfrac{1}{2},1,\ldots,\tfrac{1}{2}(n-1)\ ,\\
d &= n-\tfrac{3}{2}-\kappa+\tfrac{1}{2}(a_{2\kappa+1}+\cdots+a_{n-1})\ , \quad a_1=\ldots=a_{2\kappa}=0,\ a_{2\kappa+1}\neq 0\ ,\\
&\ldots\\
d &= n-1-2\kappa+\tfrac{1}{2}(a_{2\kappa+1}+\cdots+a_{n-1})\ , \quad a_1=\ldots=a_{2\kappa}=0,\ a_{2\kappa+1}\neq 0\ ,\\
&\ldots\\
d &\geqslant \tfrac{1}{2}(n-1)\ , \quad a_1=\ldots=a_{n-1}=0\\
d &= \tfrac{1}{2}(n-2)\ , \quad a_1=\ldots=a_{n-1}=0\\
&\ldots\\
d &= \tfrac{1}{2}\ , \quad a_1=\ldots=a_{n-1}=0\\
d &= 0\ , \quad a_1=\ldots=a_{n-1}=0
\end{aligned}$$

Parts of the *Proof* were given in [12], while in [14] was given a detailed sketch of the Proof. In [14] was given also the Proof for the case $n=3$, while the proof for $n=4$ was given in [15].

4. Character formulae

Let $\hat{\mathcal{G}}$ be a simple Lie algebra of rank ℓ with Cartan subalgebra $\hat{\mathcal{H}}$, root system $\hat{\Delta}$, simple root system $\hat{\pi}$. Let Γ, (resp. Γ_+), be the set of all integral, (resp. integral dominant), elements of $\hat{\mathcal{H}}^*$, i.e., $\lambda\in\hat{\mathcal{H}}^*$ such that $(\lambda,\alpha_i^\vee)\in\mathbb{Z}$, (resp. $\mathbb{Z}_+$), for all simple roots α_i, $(\alpha_i^\vee\equiv 2\alpha_i/(\alpha_i,\alpha_i))$. Let V be a lowest weight module with lowest weight Λ and lowest weight vector v_0. It has the following decomposition:

$$\begin{aligned}
V &= \underset{\mu\in\Gamma_+}{\oplus} V_\mu\ , && (30)\\
V_\mu &= \{u\in V \mid Hu=(\Lambda+\mu)(H)u,\ \forall\ H\in\mathcal{H}\}
\end{aligned}$$

(Note that $V_0 = \mathbb{C}v_0$.) Let $E(\mathcal{H}^*)$ be the associative abelian algebra consisting of the series $\sum_{\mu\in\mathcal{H}^*} c_\mu e(\mu)$, where $c_\mu \in \mathbb{C}$, $c_\mu = 0$ for μ outside the union of a finite number of sets of the form $D(\lambda) = \{\mu \in \mathcal{H}^* | \mu \geqslant \lambda\}$, using some ordering of $\mathcal{H}^*$, e.g., the lexicographic one; the formal exponents $e(\mu)$ have the properties: $e(0) = 1$, $e(\mu)e(\nu) = e(\mu+\nu)$.

Then the (formal) character of V is defined by:

$$\begin{aligned} ch_0\ V &= \sum_{\mu\in\Gamma_+} (\dim\ V_\mu)\ e(\Lambda+\mu) = \\ &= e(\Lambda) \sum_{\mu\in\Gamma_+} (\dim\ V_\mu)\ e(\mu) \end{aligned} \tag{31}$$

(We shall use subscript '0' for the even case.)

For a Verma module, i.e., $V = V^\Lambda$ one has $\dim V_\mu = P(\mu)$, where $P(\mu)$ is a generalized partition function, $P(\mu) = \#$ of ways μ can be presented as a sum of positive roots β, each root taken with its multiplicity $\dim \mathcal{G}_\beta$ ($= 1$ here), $P(0) \equiv 1$. Thus, the character formula for Verma modules is:

$$\begin{aligned} ch_0\ V^\Lambda &= e(\Lambda) \sum_{\mu\in\Gamma_+} P(\mu) e(\mu) = \\ &= e(\Lambda) \prod_{\alpha\in\Delta^+} (1-e(\alpha))^{-1} \end{aligned} \tag{32}$$

Further we recall the standard reflections in $\hat{\mathcal{H}}^*$:

$$s_\alpha(\lambda) = \lambda - (\lambda, \alpha^\vee)\alpha, \quad \lambda \in \hat{\mathcal{H}}^*, \quad \alpha \in \hat{\Delta}. \tag{33}$$

The Weyl group W is generated by the simple reflections $s_i \equiv s_{\alpha_i}$, $\alpha_i \in \hat{\pi}$. Thus every element $w \in W$ can be written as the product of simple reflections. It is said that w is written in a reduced form if it is written with the minimal possible number of simple reflections; the number of reflections of a reduced form of w is called the length of w, denoted by $\ell(w)$.

The Weyl character formula for the finite-dimensional irreducible LWM L_Λ over $\hat{\mathcal{G}}$, i.e., when $\Lambda \in -\Gamma_+$, has the form:

$$ch_0\ L_\Lambda = \sum_{w\in W} (-1)^{\ell(w)}\ ch_0\ V^{w\cdot\Lambda}, \quad \Lambda \in -\Gamma_+ \tag{34}$$

where the dot $\cdot$ action is defined by $w \cdot \lambda = w(\lambda - \rho) + \rho$. For future reference we note:

$$s_\alpha \cdot \Lambda = \Lambda + n_\alpha \alpha \tag{35}$$

where

$$n_\alpha = n_\alpha(\Lambda) \doteq (\rho - \Lambda, \alpha^\vee) = (\rho - \Lambda)(H_\alpha), \qquad \alpha \in \Delta^+. \tag{36}$$

In the case of basic classical Lie superalgebras the first character formulae were given by Kac. They are more complicated than the bosonic case, except for the algebras we consider. Actually, for $osp(1/2n)$ the Verma module character formula is the same as (32):

$$ch\ V^\Lambda = e(\Lambda) \prod_{\alpha\in\bar\Delta^+} \frac{1}{1-e(\alpha)} \tag{37}$$

using the restricted root system $\bar\Delta^+$. Naturally, the character formula for the finite-dimensional irreducible LWM L_Λ is again (34) using the Weyl group W_n of B_n.

Multiplets

A Verma module V^Λ may be reducible w.r.t. to many positive roots, and thus there maybe many Verma modules isomorphic to its submodules. They themselves may be reducible, and so on.

One main ingredient of the approach of [20] is as follows. We group the (reducible) Verma modules with the same Casimirs in sets called *multiplets* [20]. The multiplet corresponding to fixed values of the Casimirs may be depicted as a connected graph, the vertices of which correspond to the reducible Verma modules and the lines between the vertices correspond to embeddings between them. The explicit parametrization of the multiplets and of their Verma modules is important for understanding of the situation.

If a Verma module V^Λ is reducible w.r.t. to all simple roots (and thus w.r.t. all positive roots), i.e., $m_k \in \mathbb{N}$ for all k, then the irreducible submodules are isomorphic to the finite-dimensional irreps of $\mathcal{G}^\mathbb{C}$ [20]. (Actually, this is a condition only for m_n since $m_k \in \mathbb{N}$ for $k = 1, \ldots, n-1$.) In these cases we have the *main multiplets* which are isomorphic to the Weyl group of $\mathcal{G}^\mathbb{C}$ [20].

In the cases of non-dominant weight Λ the character formula for the irreducible LWM is [25] :

$$ch\ L_\Lambda = \sum_{\substack{w\in W \\ w\leqslant w_\Lambda}} (-1)^{\ell(w_\Lambda w)}\ P_{w,w_\Lambda}(1)\ ch\ V^{w\cdot(w_\Lambda^{-1}\cdot\Lambda)}, \quad \Lambda\in\Gamma \tag{38}$$

where $P_{y,w}(u)$ are the Kazhdan–Lusztig polynomials $y, w \in W$ [25] (for an easier exposition see [24]), w_Λ is a unique element of W with minimal

IOP Conf. Series: Journal of Physics: Conf. Series **804** (2017) 012015 doi:10.1088/1742-6596/804/1/012015

length such that the signature of $\Lambda_0 = w_\Lambda^{-1} \cdot \Lambda$ is anti-dominant or semi-anti-dominant:

$$\chi_0 = (m'_1, \ldots, m'_n), \qquad m'_k = 1 - \Lambda_0(H_k) \in \mathbb{Z}_- . \tag{39}$$

Note that $P_{y,w}(1) \in \mathbb{N}$ for $y \leqslant w$.

When Λ_0 is semi-anti-dominant, i.e., at least one $m'_k = 0$, then in fact W is replaced by a reduced Weyl group W_R.

Most often the value of $P_{y,w}(1)$ is equal to 1 (as in the character formula for the finite-dimensional irreps), while the cases $P_{y,w}(1) > 1$ are related to the appearance of subsingular vectors, though the situation is more subtle, see [24].

It is interesting to see how the reducible points relevant for unitarity fit in the multiplets. In the case of d_{ij} using (24) we have:

$$m_n(d_{ij}) = 1 - 2m_j - \cdots - 2m_{n-1} - m_i - \cdots - m_{j-1} . \tag{40}$$

In the case of d_i (25) we have:

$$m_n(d_i) = 1 - 2m_i - \cdots - 2m_{n-1} . \tag{41}$$

As expected the weights related to positive energy d are not dominant $(m_n(d_{ij}) \in \mathbb{Z}_-,\ m_n(d_i) \in -\mathbb{N},\ (i < n))$, since the positive energy UIRs are infinite-dimensional. (Naturally, $m_n(d_n) = 1$ falls out of the picture since $d_n < 0$.)

Thus, the Verma modules with weights related to positive energy would be somewhere in the main multiplet (or in a reduction of the main multiplet), and the first task for calculating the character is to find the w_Λ in the character formula (38). This we do in the next subsection in the case $n = 3$.

5. The case of osp(1|6)

For $n = 3$ formula (26) simplifies to:

$$d_1 > d_{12} > d_2 > d_{23} > d_3$$
$$\hookrightarrow > d_{13} > \nearrow$$

The Theorem now reads:

$$\begin{aligned}
d &\geqslant 2+\tfrac{1}{2}(a_1+a_2) \;=\; d_1 \,, \quad a_1 \neq 0 \,, \\
d &\geqslant \tfrac{3}{2}+\tfrac{1}{2}a_2 \;=\; d_{12} \,, \quad a_1 = 0,\; a_2 \neq 0 \,, \\
d &= 1+\tfrac{1}{2}a_2 \;=\; d_2 > d_{13} \,, \quad a_1 = 0,\; a_2 \neq 0 \,, \\
d &\geqslant 1 \;=\; d_2 = d_{13} \,, \quad a_1 = a_2 = 0 \,, \\
d &= \tfrac{1}{2} \;=\; d_{23} \,, \quad a_1 = a_2 = 0 \,, \\
d &= 0 \;=\; d_3 \,, \quad a_1 = a_2 = 0 \,.
\end{aligned} \tag{42}$$

The Weyl group W_n of B_n has $2^n n!$ elements, i.e., 48 for B_3. Let $S = (s_1, s_2, s_3)$, $s_i \equiv s_{\alpha_i}$, be the simple reflections. They fulfill the following relations:

$$s_1^2 = s_2^2 = s_3^2 = e, \;\; (s_1 s_2)^3 = e, \;\; (s_2 s_3)^4 = e, \;\; s_1 s_3 \;=\; s_3 s_1 \,, \tag{43}$$

e being the identity of W_3. The 48 elements may be listed as:

$$\begin{aligned}
& e\,,\; s_1\,,\; s_2\,,\; s_3 \\
& s_1s_2\,,\; s_1s_3\,,\; s_2s_1\,,\; s_2s_3\,,\; s_3s_2\,, \\
& s_1s_2s_1\,,\; s_1s_2s_3\,,\; s_1s_3s_2\,,\; s_2s_1s_3\,,\; s_2s_3s_2\,, \\
& s_3s_2s_1\,,\; s_3s_2s_3\,, \\
& s_1s_2s_1s_3\,,\; s_1s_2s_3s_2\,,\; s_1s_3s_2s_1\,,\; s_1s_3s_2s_3\,, \\
& s_2s_3s_2s_1\,,\; s_2s_1s_3s_2\,,\; s_3s_2s_3s_1\,,\; s_3s_2s_3s_2\,, \\
& s_1s_2s_3s_2s_1\,,\; s_1s_3s_2s_1s_3\,,\; s_1s_2s_1s_3s_2\,, \\
& s_1s_3s_2s_3s_2\,,\, s_2s_1s_3s_2s_1\,,\; s_2s_1s_3s_2s_3\,, \\
& s_3s_2s_3s_1s_2\,,\; s_3s_2s_3s_2s_1\,, \\
& s_1s_3s_2s_3s_2s_1\,,\; s_1s_3s_2s_1s_3s_2\,,\; s_1s_2s_1s_3s_2s_1\,, \\
& s_2s_1s_3s_2s_1s_3\,,\, s_2s_1s_3s_2s_3s_2\,,\; s_3s_2s_3s_1s_2s_1\,, \\
& s_3s_2s_3s_1s_2s_3\,,\, s_2s_1s_3s_2s_3s_2s_1\,, \\
& s_2s_1s_3s_2s_3s_1s_2\,,\, s_3s_2s_1s_2s_3s_2s_1\,, \\
& s_3s_2s_3s_1s_2s_1s_3\,,\; s_3s_2s_3s_1s_2s_3s_2\,, \\
& s_2s_3s_2s_1s_2s_3s_2s_1\,,\; s_3s_2s_1s_3s_2s_3s_2s_1\,, \\
& s_3s_2s_1s_3s_2s_3s_1s_2\,,\, s_2s_3s_2s_1s_3s_2s_3s_2s_1\;.
\end{aligned} \tag{44}$$

The character formula for the Verma modules in our case is given explicitly by:

$$ch\, V^{\Lambda} = \frac{e(\Lambda)}{(1-t_1)(1-t_2)(1-t_1t_2)} \times$$
$$\times \frac{1}{(1-t_3)(1-t_2t_3)(1-t_1t_2t_3)} \times$$
$$\times \frac{1}{(1-t_2t_3^2)(1-t_1t_2t_3^2)(1-t_1t_2^2t_3^2)} \tag{45}$$

where $t_j \equiv e(\alpha_j)$.

Now we give the character formulae of the five boundary or isolated unitarity cases. Below we shall denote the signature of the dominant weight Λ_0 which determines the main multiplet by (m'_1, m'_2, m'_3), $m'_k \in \mathbb{N}$, using primes to distinguish from the signatures of the weights we are interested. We shall use also reductions of the main multiplet when the weights are semi-dominant, i.e., when some $m'_k = 0$.

- In the case of $d = d_1 = 2 + \frac{1}{2}(a_1 + a_2)$ there are twelve members of the multiplet which is a submiltiplet of a main multiplet. (Remember that that $m_1 > 1$ since $a_1 \neq 0$.) They are grouped into two standard $sl(3)$ submultiplets of six members. The first submultiplet starts from $V^{\Lambda_0^{d_1}}$, where $\Lambda_0^{d_1} = w \cdot \Lambda_0$, $w = w_{\Lambda_0^{d_1}} = s_2 s_1 s_3 s_2 s_3$, with signature:

$$\Lambda_0^{d_1} \quad : \quad (m_1, m_2, m'_3 = 1 - 2m_{12}) \,, \tag{46}$$
$$m_1, m_2 \in \mathbb{N} \,, \quad m_{12} \equiv m_1 + m_2 \quad .$$

The other submultiplet starts from $V^{\Lambda'_0}$ with $\Lambda'_0 = \Lambda_0^{d_1} + \delta_1 = \Lambda_0^{d_1} + \alpha_1 + \alpha_2 + \alpha_3$, with signature: $\Lambda'_0 \,:\, (m_1 - 1, m_2, m'_3 = 1 - 2m_{12})$, $m_1 > 1$. The character formula is (38) with $w_\Lambda = w_{\Lambda_0^{d_1}}$:

$$\mathrm{ch}\,\Lambda_0^{d_1} = \frac{e(\Lambda_0^{d_1})}{(1-t_3)(1-t_2t_3)(1-t_1t_2t_3)} \times \tag{47}$$
$$\times \frac{1}{(1-t_2t_3^2)(1-t_1t_2t_3^2)(1-t_1t_2^2t_3^2)} \times$$
$$\times \{\, \mathrm{ch}\,\Lambda_{m_1,m_2}(t_1,t_2) \;-\; t_1t_2t_3\,\mathrm{ch}\,\Lambda_{m_1-1,m_2}(t_1,t_2) \,\} \,, \; m_1 > 1$$

where $\mathrm{ch}\,\Lambda_{m_1,m_2}(t_1,t_2)$ is the normalized character of the finite-dimensional $sl(3)$ irrep with Dynkin labels (m_1, m_2) (and dimension $m_1m_2(m_1 + m_2)/2$):

$$\mathrm{ch}\,\Lambda_{m_1,m_2}(t_1,t_2) \;=\; \frac{1 - t_1^{m_1} - t_2^{m_2} + t_1^{m_1}t_2^{m_{12}} + t_1^{m_{12}}t_2^{m_2} - t_1^{m_{12}}t_2^{m_{12}}}{(1-t_1)(1-t_2)(1-t_1t_2)} \tag{48}$$

Naturally, the latter formula is a polynomial in t_1, t_2, e.g., $\mathrm{ch}\,\Lambda_{1,1}(t_1,t_2) = 1$, $\mathrm{ch}\,\Lambda_{2,1}(t_1,t_2) = 1+t_1+t_1t_2$.

In the case $m_1 = 2, m_2 = 1$ the character formula (47) simplifies to:

$$\mathrm{ch}\,\Lambda_0^{d_1} = \frac{e(\Lambda_0^{d_1})}{(1-t_3)(1-t_2t_3)(1-t_2t_3^2)(1-t_1t_2t_3^2)(1-t_1t_2^2t_3^2)} \times \\ \times \left(1 + \frac{t_1(1+t_2)}{1-t_1t_2t_3}\right), \qquad m_1 = 2, m_2 = 1 \tag{49}$$

• In the case of $d = d_{12} = \frac{1}{2}(3+a_2)$ which is relevant for unitarity, i.e., $m_1 = 1$, there are again twelve members of the multiplet. Omitting the details [14] the character f-la is:

$$\mathrm{ch}\,\Lambda_0^{d_{12}} = \frac{e(\Lambda_0^{d_{12}})}{(1-t_3)(1-t_2t_3)(1-t_1t_2t_3)} \times \tag{50}$$
$$\times \frac{1}{(1-t_2t_3^2)(1-t_1t_2t_3^2)(1-t_1t_2^2t_3^2)}$$
$$\times\{\,\mathrm{ch}\,\Lambda_{1,m_2}(t_1,t_2) - (t_1t_2^2t_3^2)^{m_2}\,\mathrm{ch}\,\Lambda_{1,m_2-1}(t_1,t_2)\,\}, \quad m_2 > 1$$

In the case $m_2 = 2$ it simplifies to:

$$\mathrm{ch}\,\Lambda_0^{d_{12}} = \frac{e(\Lambda_0^{d_{12}})}{(1-t_3)(1-t_2t_3)(1-t_1t_2t_3)(1-t_2t_3^2)(1-t_1t_2t_3^2)} \times \\ \times\{\,1+t_1t_2^2t_3^2 + \frac{t_2(1+t_1)}{1-t_1t_2^2t_3^2}\,\} \tag{51}$$

• In the case $d = d_2 = d_{13} = 1$ and $a_1 = a_2 = 0$, $m_1 = m_2 = 1$, the signature is:

$$\Lambda_0^{d_2=d_{13}} : (1,1,-1)\,. \tag{52}$$

Again there are twelve members of the multiplet which has two $sl(3)$ submultiplets. First there is a $sl(3)$ sextet starting from $\Lambda_0^{d_2=d_{13}}$ with parameters $(1,1)$. Then there is a $sl(3)$ sextet starting from $\Lambda_0^{d_2=d_{13}} + \alpha_1 + 2\alpha_2 + 3\alpha_3$ with parameters $(1,1)$. Note that that $\alpha_1+2\alpha_2+3\alpha_3 = \delta_1+\delta_2+\delta_3$ is the weight of a subsingular vector [14], yet the corresponding KL polynomial $P_{y,w}(1)$ is equal to 1. Thus, the character formula is [14]:

$$\mathrm{ch}\,\Lambda_0^{d_2=d_{13}} = \tag{53}$$
$$= \frac{e(\Lambda_0^{d_2=d_{13}})\,(\,1-t_1t_2^2t_3^3\,)}{(1-t_3)(1-t_2t_3)(1-t_1t_2t_3)(1-t_2t_3^2)(1-t_1t_2t_3^2)(1-t_1t_2^2t_3^2)}$$

IOP Conf. Series: Journal of Physics: Conf. Series **804** (2017) 012015 doi:10.1088/1742-6596/804/1/012015

Note that the above formula may be rewritten as:

$$\mathrm{ch}\,\Lambda_0^{d_2=d_{13}} = \frac{e(\Lambda_0^{d_2=d_{13}})}{(1-t_2t_3)(1-t_1t_2t_3)(1-t_2t_3^2)(1-t_1t_2t_3^2)} \times \times \left(\frac{1}{1-t_1t_2^2t_3^2} + \frac{t_3}{1-t_3} \right) \tag{54}$$

• In the case of $d = d_2 = 1+\frac{1}{2}a_2 > d_{13} = 1$, i.e., $m_1 = 1$, $m_2 = 1+a_2 > 1$. The multiplet has 24 members for $m_2 > 2$. Omitting the details [14] the character f-la is:

$$\begin{aligned} \mathrm{ch}\,\Lambda_0'^{d_2} = & \frac{e(\Lambda_0'^{d_2})}{(1-t_3)(1-t_2t_3)(1-t_1t_2t_3)} \times \\ & \times \frac{1}{(1-t_2t_3^2)(1-t_1t_2t_3^2)(1-t_1t_2^2t_3^2)} \times \\ & \times \{ \mathrm{ch}\,\Lambda_{1,m_2}(t_1,t_2) - t_2t_3\,\mathrm{ch}\,\Lambda_{2,m_2-1}(t_1,t_2) + \\ & + t_1t_2^3t_3^3\,\mathrm{ch}\,\Lambda_{2,m_2-2}(t_1,t_2) - t_1^2t_2^4t_3^4\,\mathrm{ch}\,\Lambda_{1,m_2-2}(t_1,t_2) \} \end{aligned} \tag{55}$$

When $m_2 = 2$ $(a_2 = 1)$ the multiplet reduces to only 12 members, and the character formula simplifies to:

$$\begin{aligned} \mathrm{ch}\,\Lambda_0'^{d_2} = & \frac{e(\Lambda_0'^{d_2})}{(1-t_2t_3^2)(1-t_1t_2t_3^2)(1-t_1t_2^2t_3^2)} \times \\ & \times \left\{ \frac{1}{(1-t_3)(1-t_1t_2t_3)} + \frac{t_2}{(1-t_3)(1-t_2t_3)} \right. \\ & \left. + \frac{t_1t_2}{(1-t_2t_3)(1-t_1t_2t_3)} \right\} \end{aligned} \tag{56}$$

• In the case of $d = d_{23} = \frac{1}{2}$, $a_1 = a_2 = 0$, i.e., $m_1 = m_2 = 1$, again we have a multiplet with 24 members. Omitting the details [14] the character formula is:

$$\begin{aligned}
\mathrm{ch}\,\Lambda_0^{d_{23}} &= \frac{e(\Lambda_0^{d_{23}})}{(1-t_3)(1-t_2t_3)(1-t_1t_2t_3)} \times \\
&\times \frac{1}{(1-t_2t_3^2)(1-t_1t_2t_3^2)(1-t_1t_2^2t_3^2)} \times \\
&\times \{\, 1 \;-\; t_2t_3^2\,\mathrm{ch}\,\Lambda_{2,1}(t_1,t_2) \;+\; t_1t_2^2t_3^4\,\mathrm{ch}\,\Lambda_{1,2}(t_1,t_2) \;-\; t_1^2t_2^4t_3^6 \,\} = \\
&= \frac{e(\Lambda_0^{d_{23}})}{(1-t_3)(1-t_2t_3)(1-t_1t_2t_3)\;(1-t_2t_3^2)(1-t_1t_2t_3^2)(1-t_1t_2^2t_3^2)} \times \\
&\times \{\, 1 \;-\; t_2t_3^2\,(1+t_1+t_1t_2) \;+\; t_1t_2^2t_3^4\,(1+t_2+t_1t_2) \;-\; t_1^2t_2^4t_3^6 \,\}
\end{aligned} \tag{57}$$

Note that the above formula may be rewritten as:

$$\mathrm{ch}\,\Lambda_0^{d_{23}} = \frac{e(\Lambda_0^{d_{23}})}{(1-t_3)(1-t_2t_3)(1-t_1t_2t_3)} \tag{58}$$

Note that formulae (49),(51),(54),(56),(58) are new w.r.t. [14].

Acknowledgments

V.K.D. thanks the Organizers for the kind invitation to give a plenary talk at the 24-th International Conference on Integrable Systems and Quantum Symmetries, Prague, June 2016. V.K.D. was supported by COST Actions MP1210 and MP1405, and by Bulgarian NSF Grant DFNI T02/6.

I.S. was supported by COST Action MP1405 and Serbian Ministry of Science and Technological Development, grant OI 171031.

References

[1] W. Nahm, *Supersymmetries and their representations*, Nucl. Phys. **B135** (1978) 149-166.

[2] R. Haag, J.T. Lopuszanski and M. Sohnius, *All possible generators of supersymmetries of the S-matrix*, Nucl. Phys. **B88** (1975) 257-274.

[3] M. Flato and C. Fronsdal, *Representations of conformal supersymmetry*, Lett. Math. Phys. **8** (1984) 159-162.

[4] V.K. Dobrev and V.B. Petkova, *On the group-theoretical approach to extended conformal supersymmetry : classification of multiplets*, Lett. Math. Phys. **9** (1985) 287-298.

[5] V.K. Dobrev and V.B. Petkova, *All positive energy unitary irreducible representations of extended conformal supersymmetry*, Phys. Lett. **162B** (1985) 127-132.

[6] V.K. Dobrev and V.B. Petkova, *On the group-theoretical approach to extended conformal supersymmetry : function space realizations and invariant differential operators*, Fortschr. d. Phys. **35** (1987) 537-572.

[7] V.K. Dobrev and V.B. Petkova, *All positive energy unitary irreducible representations of the extended conformal superalgebra*, in: A.O. Barut and H.D. Doebner (eds.), *Conformal Groups and Structures*, Lecture Notes in Physics, Vol. 261 (Springer-Verlag, Berlin, 1986) pp. 300-308.

[8] S. Minwalla, *Restrictions imposed by superconformal invariance on quantum field theories*, Adv. Theor. Math. Phys. **2** (1998) 781-846.

[9] V.K. Dobrev, *Positive energy unitary irreducible representations of D=6 conformal supersymmetry*, J. Phys. **A35** (2002) 7079-7100.

[10] P.K. Townsend, *M(embrane) theory on T(9)*, Nucl. Phys. Proc. Suppl. **68** (1998) 11-16; J.P. Gauntlett, G.W. Gibbons, C.M. Hull and P.K. Townsend, *BPS states of D=4 N=1 supersymmetry*, Commun. Math. Phys. **216** (2001) 431-459.

[11] R. D'Auria, S. Ferrara, M.A. Lledo and V.S. Varadarajan, *Spinor algebras*, J. Geom. Phys. **40** , (2001) 101-128; R. D'Auria, S. Ferrara and M.A. Lledo, *On the embedding of space-time symmetries into simple superalgebras*, Lett. Math. Phys. **57** (2001) 123-133; S. Ferrara and M.A. Lledo, *Considerations on super Poincare algebras and their extensions to simple superalgebras*, Rev. Math. Phys. **14** (2002) 519-530.

[12] V.K. Dobrev and R.B. Zhang, *Positive Energy Unitary Irreducible Representations of the Superalgebras* $osp(1|2n, R)$, Phys. Atom. Nuclei, **68** (2005) 1660-1669.

[13] V.K. Dobrev, A.M. Miteva, R.B. Zhang and B.S. Zlatev, *On the unitarity of D=9,10,11 conformal supersymmetry*, Czech. J. Phys. **54** (2004) 1249-1256.

[14] V.K. Dobrev and I. Salom, *Positive Energy Unitary Irreducible Representations of the Superalgebras* $osp(1|2n, R)$ *and Character Formulae*, Proceedings of the VIII Mathematical Physics Meeting, (Belgrade, 24-31 August 2014) SFIN XXVIII (A1), eds. B. Dragovich et al, (Belgrade Inst. Phys. 2015) [ISBN 978-86-82441-43-4], pp. 59-81.

[15] V.K. Dobrev and I. Salom, *Positive Energy Unitary Irreducible Representations of the Superalgebras* $osp(1|8, R)$, Publications de l'Institut Mathematique, Belgrade, to appear (2016), (IF 0.195), arXiv:1607.03008.

[16] V.G. Kac, *Lie superalgebras*, Adv. Math. **26** (1977) 8-96; *A sketch of Lie superalgebra theory*, Commun. Math. Phys. **53** (1977) 31-64; the second paper is an adaptation for physicists of the first paper.

[17] V.G. Kac, *Representations of classical Lie superalgebras*, Lect. Notes in Math. **676** (Springer-Verlag, Berlin, 1978) pp. 597-626.

[18] N.N. Shapovalov, *On a bilinear form on the universal enveloping algebra of a complex semisimple Lie algebra*, Funkts. Anal. Prilozh. **6** (4) 65 (1972); English translation: Funkt. Anal. Appl. **6**, 307 (1972).

[19] J. Dixmier, *Enveloping Algebras*, (North Holland, New York, 1977).

[20] V.K. Dobrev, *Canonical construction of intertwining differential operators associated with representations of real semisimple Lie groups*, Rept. Math. Phys. **25** (1988) 159-181.

[21] V.K. Dobrev, *Singular vectors of quantum groups representations for straight Lie algebra roots*, Lett. Math. Phys. **22** (1991) 251-266.

[22] I.N. Bernstein, I.M. Gel'fand and S.I. Gel'fand, *Structure of representations generated by highest weight vectors*, Funkts. Anal. Prilozh. **5** (1) (1971) 1; English translation: Funct. Anal. Appl. **5** (1971) 1.

[23] V.K. Dobrev, *Subsingular vectors and conditionally invariant (q-deformed) equations*, J. Phys. **A28** (1995) 7135-7155.

[24] V.K. Dobrev, *Kazhdan-Lusztig polynomials, subsingular vectors, and conditionally invariant (q-deformed) equations*, Invited talk at the Symposium "Symmetries in Science IX", Bregenz, Austria, (August 1996), Proceedings, eds. B. Gruber and M. Ramek, (Plenum Press, New York and London, 1997) pp. 47-80.

[25] D. Kazhdan and G. Lusztig, *Representations of Coxeter Groups and Hecke Algebras*, Inv. Math. **53** (1979) 165-184.

IOP Conf. Series: Journal of Physics: Conf. Series **804** (2017) 012016 doi:10.1088/1742-6596/804/1/012016

Determining quantum correlations in bipartite systems - from qubit to qutrit and beyond

Andrzej Frydryszak[1], **Lech Jakóbczyk, Piotr Ługiewicz**

Institute of Theoretical Physics, University of Wroclaw, Poland

E-mail: `andrzej.frydryszak@ift.uni.wroc.pl`

Abstract. We advocate the step change in properties of discrete d-level quantum systems, between $d = 2$ and $d \geq 3$. Qubit systems, or multipartite systems containing qubit subsystem, are exceptional in their relative simplicity. One faces a step in complexity in valuating measures of quantum correlations for qutrits and then other higher dimensional qudits. There is a growing number of arguments leading to such conclusion: recently found no-go theorem for generalization of the Peres-Horodecki's PPT criterion [1], change in geometry of state spaces of qubit and higher degree qudits (the so called 'generalized Bloch ball' is not a ball anymore), restricted possibilities for diagonalization of correlation matrices for bipartite systems, more difficult way for handling the set of relevant families of orthogonal projectors.

1. Introduction

Quantum correlations in finite dimensional quantum systems are new resources which can fuel quantum information and computing. Their quantification is crucial, but difficult task. Its complexity depends on the chosen correlation, definition of its measure and dimension d of the bipartite d-level system. Generic statement valid for all known quantum correlations is: pure states can be uncorrelated or entangled. Historically first and mostly studied correlation is quantum entanglement of qubits, extended also to qutrits [2] and then for qudits [3], but only for limited types of states. Results known for qubit systems can be generalized to some extend to a qubit-qudit case. In that case a lower dimensional system sets the level of the complexity of the problem. An interesting aspect of finding the value of correlation measure is a possibility of relating it to the mean values of a selected observables of a system. One example of such relation is given for qubits, where entanglement measure can be expressed in terms of the mean value of spin [4].

The correlation we want to focus here on is the quantum discord. The general definition of its measure, while it distinguishes the quantum and classical character of correlations in compound systems, is hardly operational, even for qubit systems. That is why we restrict our considerations to the so called measurement-induced one sided quantum geometric discord (MIQGD). It is a version of the geometric measure of quantum correlations related to the distance used in definition. To have measure contractive under completely positive trace preserving maps we chose trace norm (Schatten 1-norm). Such distance used in definition produces proper quantum correlation measure in contrast to the Hilbert-Schmidt distance. We shall discuss the MIQGD for various d-level systems, $d = 2,3,\ldots$. However, properties of bipartite systems with $d \geq 3$ change with restpect to $d = 2$. For two-qutrit states there is no finite set of criteria of separability [1], what means that there is no extension of the Peres-Horodecki's

[1] Talk given by A.F.

IOP Conf. Series: Journal of Physics: Conf. Series **804** (2017) 012016 doi:10.1088/1742-6596/804/1/012016

necessary and sufficient PPT-criterion valid for qubit-qubit and qubit-qutrit systems[5, 6] to higher level systems.

2. Single qudit system quantum state space

In the commonly used notation for $su(d)$ Lie algebras we can describe the one-partite states by the set of generators (we use here uniform notation, where usually for $su(2)$, $\lambda_j = \sigma_j$)

$$\mathrm{tr}\,\lambda_j = 0, \quad \mathrm{tr}\,(\lambda_j\lambda_k) = 2\delta_{jk}, \quad \text{and} \quad \lambda_j\lambda_k = \frac{2}{d}\delta_{jk}\,\mathbb{1}_d + \sum_l (d_{jkl} + i f_{jkl})\lambda_l \tag{1}$$

where $j,k = 1,\ldots,d^2-1$; the totally symmetric and antisymmetric, respectively, structure constants d_{jkl} and f_{jkl} are given by $d_{jkl} = \frac{1}{4}\mathrm{tr}\,([\lambda_j,\lambda_k]_+\lambda_l)$ and $f_{jkl} = \frac{1}{4i}\mathrm{tr}\,([\lambda_j,\lambda_k]\lambda_l)$.

- $su(2)$:

$$d_{ijk} = 0, \quad f_{ijk} = \varepsilon_{ijk}$$

- $su(3)$:

$$d_{ijk} = \begin{cases} \frac{1}{2} & \text{for } (ijk) = (146),\ (157),\ (256),\ (344),\ (355), \\ -\frac{1}{2} & \text{for } (ijk) = (247),\ (366),\ (377), \\ \frac{1}{\sqrt{3}} & \text{for } (ijk) = (118),\ (228),\ (338), \\ -\frac{1}{\sqrt{3}} & \text{for } (ijk) = (888), \\ -\frac{1}{2\sqrt{3}} & \text{for } (ijk) = (448),\ (558),\ (668),\ (778) \end{cases}$$

and

$$f_{ijk} = \begin{cases} 1 & \text{for } (ijk) = (123), \\ \frac{1}{2} & \text{for } (ijk) = (147),\ (246),\ (257),\ (345),\ (516),\ (637), \\ \frac{\sqrt{3}}{2} & \text{for } (ijk) = (458),\ (678). \end{cases}$$

Using these structure constants one can introduce the $\star$-product and $\wedge$-products of $\mathbb{R}^{d^2-1}$ vectors. For $n, m \in \mathbb{R}^{d^2-1}$ we define

$$(n\star m)_j = \sqrt{\frac{d(d-1)}{2}}\frac{1}{d-2}\sum_{k,l} d_{jkl}n_k m_l \quad \text{and} \quad (n\wedge m)_j = \sqrt{\frac{d(d-1)}{2}}\frac{1}{d-2}\sum_{k,l} f_{jkl}n_k m_l \tag{2}$$

Let $\lambda = (\lambda_1,\ldots,\lambda_{d^2-1})$ and

$$\langle n, \lambda\rangle = \sum_j n_j\lambda_j \tag{3}$$

The set of states ρ of d - level system i.e. the set of hermitian of a unit trace matrices which are positive definite is customarily parameterized in the following way

$$\rho = \frac{1}{d}\left(\mathbb{1}_d + \sqrt{\frac{d(d-1)}{2}}\langle n, \lambda\rangle\right), \quad n \in \mathbb{R}^{d^2-1}, \tag{4}$$

where the components of the vector n are

$$n_j = \frac{d}{\sqrt{2d(d-1)}}\mathrm{tr}\,(\rho\lambda_j), \quad j = 1,\ldots,d^2-1.$$

For arbitrary d-level system, pure states get simple characterization

$$\langle n, n\rangle = 1 \quad \text{and} \quad n\star n = n. \tag{5}$$

i.e. for solutions of above conditions relevant ρ is a projector. Only in the case of qubits the $\star$-product is trivial and above condition select the boundary of the Bloch ball with maximally mixed state laying in its center. Dimension of the quantum state space Q_d is d^2-1 and only for the $d=2$ one gets direct description of the set of states as the ball embeded in $\mathbb{R}^3$. For higher d this set has much richer structure and is highly difficult to characterize explicitly, what is related also to the fact, that characterization of the positive-definiteness of ρ in terms of vector n for arbitrary d-level system gets complicated and, in fact, is not known for generic case. Nevertheless, for the $d>2$ the frequently adopted name in literature for this set is "generalized Bloch ball" (GBB), what shouldn't be taken literary. A generic property of the GBB is that as a convex set it has boundary placed in between two spheres, the outsphere of radius R_d and the insphere of the radius r_d, where in the Hilbert-Schmidt distance [7]

$$R_d = \sqrt{(d-1)/2d}, \qquad r_d = \frac{1}{\sqrt{2d(d-1)}}, \tag{6}$$

what means that

$$R_d = (d-1)r_d.$$

As stated in [7] the Q_d has a constant height, in the sense that

$$\frac{Ar_d}{V} = d^2-1, \tag{7}$$

where A denotes an area of the boundary of the Q_d and V is its volume. More precisely, it was shown in [8] that such property holds for the convex set of separable two qubits states and the set of positive partial transpose states of an arbitrary bipartite system. Let us comment on the generalized Bloch ball Q_d for $d=2,3$.

Bloch ball

The Q_2 is the simplest quantum state space and its geometry is the best known. Here $R_d = r_d$ and boundary of Q_2, ∂Q_2 is identical with the boundary of outer/inner sphere. Pure states are characterized by $<n,n>=1$. The condition $n\star n = n$ in this case is not present.

Single qutrit state space

Explicit description of the qutrit quantum state space Q_3 and its visualization has been intriguing for some time. As for all Q_d this is a question of describing a convex set with boundary placed between inner and outer spheres, here $R_3 = 2r_3$. In the Ref. [7] one can find what objects cannot serve as a model of Q_3 and then what information one gets using two-dimensional projections and cross-sections. The question of the 'visualization' of the Q_3 was firstly discussed in Ref. [9] and recently published [10]. Tree dimensional 'visualizations' can be obtained by means of the other parametrizations of the Q_3. As shown in Ref. [11], instead of the vector n one can produce a graphical representation of a qutrit using a three dimensional vector a and a metric tensor η with distribution of eight independent parameters into 3+5 respectively. Qutrit states are described by a and η such that, $\eta \cdot a \leq 1$.

Using the description of states by means of the vector n, we know, that pure states should satisfy conditions (5), what means that such states are scattered on the outer sphere and discrete rotation is needed to map them geometrically. This is result of the $n\star n = n$ condition and non-triviality of the d_{ijk} constants for the $su(3)$ algebra (for qubits arbitrary rotations are allowed).

Single qudit state space

Despite some generic properties, there is very little known about generalized Bloch ball for $d\geq 4$. As described in the Ref. ([7]) it is easier to enumerate what characteristics that convex set does not have. Let us quote here selected properties o the GBB for qudits from the list given in the Ref. ([7]):

IOP Conf. Series: Journal of Physics: Conf. Series **804** (2017) 012016 doi:10.1088/1742-6596/804/1/012016

- Q_d has the 'no hair' property i.e. it is a d^2-1 dimensional convex set topologically equivalent to to a ball and it does not have parts of lower dimension.
- $B_{r_d} \subset Q_d \subset B_{R_d}$, where B_r denotes a ball with the radius r.
- ∂Q_d is d^2-2 dimensional and contains all states with rank smaller then maximal
- The set of pure states is $2d-2$ dimensional and connected. It has zero measure with respect to ∂Q_d.
- Q_d has constant hight, cf. Eq.(7).

3. Bipartite qudit system state space

Two qubits

The simplest bipartite quantum system in which we can discuss quantum correlations is a system of two qubits.The total space of states is equal to Q_4 which detailed geometrical structure is not known. To get some insight into the geometry of Q_4, one can consider some lower dimensional sections of this set. In [12] two - dimensional sections of the corresponding space of generalized Bloch vectors were considered. On the other hand, the problem of discrimination between separable and entangled states is in this case completely solved. The separability condition based on the notion of partial transposition [5, 6] is simple and effective. Applied to the above mentioned sections of the set of states, this condition gives some information about geometrical shapes of the sets of separable and entangled states [12].

Two qutrits and beyond

In the case of two qutrits, the geometry of the space Q_9 is yet more complicated. Some aspects of this geometry was studied in the class of so called Bell - diagonal states which form a simplex living in the nine - dimensional real linear space [13]. This analysis was then extended to the case of general qudits [14]. What mainly differs qutrits from qubits is that in the system of two qutrits separability condition based on partial transposition is not sufficient. It only shows that the states which are not positive after this operation (NPPT states) are entangled. It turns out that all entangled states can be divided into two classes: free entangled states that can be distilled using local operations and classical communication (LOCC); bound entangled states for which no LOCC strategy can be used to extract pure state entanglement [15]. Last but not least recourse, from the pragmatic point of view, might be the Monte Carlo sampling of the quantum state space [16]. It allows obtaining high-quality random samples of quantum states from higher dimensional Q_d, respecting the relevant target distributions and allowing to evaluate global extremum of a given correlation measure function. This approach is still to be applied to the geometric quantum discord measures.

New difficulties for higher dimensional systems

Let us point out some difficulties emerging in systems with higher d:

(i) Considerably more complex structure of the set of separable bipartite states, quantum-classical states etc.

(ii) Change in structure of universal enveloping algebra for $su(d)$. Vanishing d_{ijk} symmetric structure constants for $su(2)$ became nontrivial for $d \geq 3$ and modify the geometry of state space via $\star$-product.

(iii) Relatively 'shrinked' orthogonal subgroup $R(G)$ originating as the adjoint representation from the unitary transformations $G = SU(d)$, $dim\, SU(d) = d^2-1$ compared to $dim\, O(d^2-1)$ (**Table 1**). Only for qubits we have the same dimension of these groups and $SU(2)$ is just universal double cover of $SO(3)$.

(iv) No possibility to diagonalize all correlation matrices ie. emerging new sectors in comparison to the qubit intuition.

(v) Steep curve of growing complexity; hopeless perspective to perform effectively minimization and produce analytical formulas for correlation measures for arbitrary state.

IOP Conf. Series: Journal of Physics: Conf. Series **804** (2017) 012016 doi:10.1088/1742-6596/804/1/012016

Table 1. Dimensions of orthogonal subgroups.

G	$\dim G$	$\dim R(G)$	$\dim O(d^2-1)$
SU(2)	3	3	3
SU(3)	8	8	28
SU(4)	15	15	105
...	...	...	...
SU(d)	d^2-1	d^2-1	$\frac{1}{2}(d^2-1)(d^2-2)$

(vi) For qubits various correlation measures are equivalent, but split for $d > 3$.

4. Quantifying quantum correlations in bipartite systems - from qubits to qutrits and beyond

Consider now two qudits $\mathcal{A}$ and $\mathcal{B}$. It is convenient to parametrize the set of states of composite system as follows

$$\rho = \frac{1}{d^2}\left(\mathbb{1}_d \otimes \mathbb{1}_d + \sqrt{\frac{d(d-1)}{2}}\,\langle x, \lambda\rangle \otimes \mathbb{1}_d + \mathbb{1}_d \otimes \sqrt{\frac{d(d-1)}{2}}\,\langle y, \lambda\rangle + \sum_{k=1}^{d^2-1} \langle \mathcal{K} e_k, \lambda\rangle \otimes \langle e_k, \lambda_k\rangle\right) \tag{8}$$

where $x, y \in \mathbb{R}^{d^2-1}$ and $\{e_k\}_{k=1}^{d^2-1}$ are the vectors of canonical orthonormal basis of $\mathbb{R}^{d^2-1}$. Notice that

$$x_j = \frac{d}{\sqrt{2d(d-1)}}\,\mathrm{tr}\,(\rho\lambda_j \otimes \mathbb{1}_d), \quad y_j = \frac{d}{\sqrt{2d(d-1)}}\,\mathrm{tr}\,(\rho\,\mathbb{1}_d \otimes \lambda_j)$$

and the correlation matrix $\mathcal{K}$ has elements

$$\mathcal{K}_{jk} = \frac{d^2}{4}\,\mathrm{tr}\,(\rho\lambda_j \otimes \lambda_k).$$

The parametrization (8) is chosen is such a way, that the marginals $\mathrm{tr}_{\mathcal{A}}\rho$ and $\mathrm{tr}_{\mathcal{B}}\rho$ are given by the vectors x and y as in (4).

Measurement-induced geometric discord

Let us assume that bipartite system $\mathcal{AB}$ is prepared in a state ρ. Any local measurement on the subsystem $\mathcal{A}$ will disturb almost all states ρ. This observation yields the definition of a measure of quantum discord. The one-sided measurement induced geometric discord is defined as the minimal disturbance induced by any projective measurement $\mathbb{P}_{\mathcal{A}}$ on subsystem $\mathcal{A}$ [17, 18, 19]. A distance in the set of states is given by the trace norm. Namely,

$$D_1(\rho) = \min_{\mathbb{P}_{\mathcal{A}}} ||\rho - \mathbb{P}_{\mathcal{A}}(\rho)||_1, \quad ||\sigma||_1 = \mathrm{tr}\,|\sigma|. \tag{9}$$

Local projective measurement $\mathbb{P}_{\mathcal{A}}$ is given by the one-dimensional projectors $P_1, P_2, \ldots, P_d$ on $\mathbb{C}^d$, such that

$$P_1 + P_2 + \cdots + P_d = \mathbb{1}_d, \quad P_j P_k = \delta_{jk} P_k,$$
$$\mathbb{P}_{\mathcal{A}} = \mathbb{P} \otimes id,$$

where

$$\mathbb{P}(\sigma) = P_1\sigma P_1 + P_2\sigma P_2 + \cdots + P_d\sigma P_d.$$

By P_k^0 we shall denote canonical projections on standard orthonormal basis in $\mathbb{C}^d$ and respective projections on orthogonal complement spaces by $\mathcal{M}_0 = \mathbb{1} - P_0$, $\mathcal{M} = \mathbb{1} - P$. The disturbance of the state after the measurement $\mathbb{P}_{\mathcal{A}}$ can be written as

$$S(\mathcal{M}) \equiv \rho - \mathbb{P}_{\mathcal{A}}(\rho) = \frac{1}{d^2}\Big[\sqrt{\frac{d(d-1)}{2}}\langle \mathcal{M}x, \lambda\rangle \otimes \mathbb{1}_d + \sum_k \langle \mathcal{M}\mathcal{K}e_k, \lambda\rangle \otimes \langle e_k, \lambda\rangle\Big] \tag{10}$$

$$D_1(\rho) = \frac{d}{2(d-1)} \min_{\mathcal{M}} \operatorname{tr} \sqrt{Q(\mathcal{M})}, \qquad Q(\mathcal{M}) \equiv S(\mathcal{M})S(\mathcal{M})^*, \tag{11}$$

where minimum is taken over all M corresponding to measurements on subsystem $\mathcal{A}$.

Simplifications

Even such defined measure of quantum discord, more operational then general one, is still difficult to calculate and some simplifying assumptions are necessary. Let us consider class of locally maximally mixed states ρ

$$\operatorname{tr}_{\mathcal{A}}\rho = \frac{\mathbb{1}_d}{d}, \quad \operatorname{tr}_{\mathcal{B}}\rho = \frac{\mathbb{1}_d}{d}. \tag{12}$$

In the chosen parametrization this corresponds to $x = y = 0$ what results in simplified states of the form

$$\rho = \frac{1}{d^2}\left(\mathbb{1}_d \otimes \mathbb{1}_d + \sum_{j=1}^{d^2-1} \langle \mathcal{K}e_j, \lambda\rangle \otimes \langle e_j, \lambda\rangle\right) \tag{13}$$

Note that the set of correlation matrices defining above states is convex and is contained in the ball

$$\mathbb{B}_2 = \left\{A \in \mathbb{M}_{d^2-1}(\mathbb{R}) \;:\; ||A||_2 \leq \frac{d}{2}\sqrt{d^2-1}\right\} \tag{14}$$

Maximally entangled pure states of this class are defined by correlation matrices lying on the boundary of the ball $\mathbb{B}_2$. However, not every matrix lying on the border corresponds to some state. Detailed characterization of the set of correlation matrices is not known.

For the states under consideration the disturbance after measurement has the form

$$S(\mathcal{M}) = \frac{1}{d^2}\sum_{j=1}^{d^2-1} \langle \mathcal{M}\mathcal{K}e_j, \lambda\rangle \otimes \langle e_j, \lambda\rangle, \tag{15}$$

and

$$\begin{aligned} Q(\mathcal{M}) = \frac{1}{d^4}\Bigg[& \frac{4}{d^2}\sum_j \langle \mathcal{M}\mathcal{K}e_j, \mathcal{M}\mathcal{K}e_j\rangle \mathbb{1}_d \otimes \mathbb{1}_d + \frac{2}{dd'}\sum_j \langle \mathcal{M}\mathcal{K}e_j \star \mathcal{M}\mathcal{K}e_j, \lambda\rangle \otimes \mathbb{1}_d \\ & + \frac{2}{dd'}\sum_{j,k} \langle \mathcal{M}\mathcal{K}e_j, \mathcal{M}\mathcal{K}e_k\rangle \mathbb{1}_d \otimes \langle e_j \star e_k, \lambda\rangle + \frac{1}{d'^2}\sum_{j,k} \langle \mathcal{M}\mathcal{K}e_j \star \mathcal{M}\mathcal{K}e_k, \lambda\rangle \otimes \langle e_j \star e_k, \lambda\rangle \\ & - \frac{1}{d'^2}\sum_{j,k} \langle \mathcal{M}\mathcal{K}e_j \wedge \mathcal{M}\mathcal{K}e_k, \lambda\rangle \otimes \langle e_j \wedge e_k, \lambda\rangle \Bigg] \end{aligned} \tag{16}$$

The spectrum of $Q(\mathcal{M})$ for arbitrary correlation matrix is rather difficult to obtain, but interestingly enough, one can find a universal lower bound for D_1 and for analogous quantum discord measure based on the Hilbert-Schmidt distance D_2 [20]. Let $\mathcal{K}$ defines the corresponding locally maximally mixed state ρ, then

$$D_2(\rho) \geq \frac{4}{d^3(d-1)}\,\Xi(\mathcal{K}) \quad \text{and} \quad D_1(\rho) \geq \frac{1}{d(d-1)}\sqrt{\Xi(\mathcal{K})}, \quad \text{where} \quad \Xi(\mathcal{K}) = \sum_{j=d}^{d^2-1} \eta_j^{\downarrow}.$$

By the $\{\eta_j^{\downarrow}\}$ we denote the collection of eigenvalues of $\mathcal{K}\mathcal{K}^T$ taken in non- increasing order. Let us observe that states related to correlation matrices with *rank* $\mathcal{K} \geq d$ have non-zero quantum discord.

However, to get more specific information one has to admit further restrictions on the form of the correlation matrices. For $\mathcal{K} = t\,V_0$, and $V_0 \in O(d^2-1)$ one can proof that [20]

$$Q(\mathcal{M}) = \frac{t^2}{d^4}\left[\frac{4(d-1)}{d}\,\mathbb{1}_d \otimes \mathbb{1}_d + \frac{2}{d}\,\mathbb{1}_d \otimes \sum_k X_k \lambda_k + \sum_{j,k} Y_{jk}\lambda_j \otimes \lambda_k\right] \tag{17}$$

where

$$X_k = \mathrm{tr}\,(\mathcal{M}\,V_0 \Delta_k V_0^T), \quad Y_{jk} = \mathrm{tr}\,(V_0^T \mathcal{M} \Delta_j \mathcal{M} V_0 \Delta_k + V_0^T \mathcal{M} F_j \mathcal{M} V_0 F_k) \tag{18}$$

and $(\Delta_k)_{jl} = d_{jkl}$, $(F_k)_{jl} = f_{jkl}$. Let us observe that terms involving Y_{jk} are related to $\star$ and $\wedge$ - products in $\mathbb{R}^{(d^2-1)}$.

Above form of the $Q(\mathcal{M})$ can be starting point to obtain analytical expression for trace - norm quantum discord. The additional conditions on the correlation matrices one has to impose come from the case study of the system of two qutrits [19, 20]. Final answer valid for two-qudit systems can be formulated as follows: there two families of correlation matrices defining states for which analytical formula for MIQGD can be obtained. They are given by correlation matrices of the form

$$\mathcal{K}^a = t\,V, \quad V \in R(SU(d)) \tag{19}$$

and

$$\mathcal{K}^{aa} = t\,\mathcal{T}, \quad \mathcal{T} = V_1 I_0 V_2^T,\ V_1, V_2 \in R(SU(d)), \tag{20}$$

where

$$(I_0)_{kk} = \frac{1}{2}\mathrm{tr}(\lambda_k^T \lambda_k), \quad k = 1, \ldots, d^2-1$$

For the first family we obtain

$$Q^a(\mathcal{M}) = \frac{t^2}{d^4}\left[\left(\frac{2}{d}\right)^2 d(d-1)\,\mathbb{1}_d \otimes \mathbb{1}_d - 2\sum_{k=2}^{d} U\lambda_{k^2-1}U^* \otimes \tau_V(U)\tau_{\mathcal{T}}(\lambda_{k^2-1})\tau_V(U^*)\right] \tag{21}$$

and for the second one

$$Q^{aa}(\mathcal{M}) = \frac{t^2}{d^4}\Bigg[\left(\frac{2}{d}\right)^2 d(d-1)\,\mathbb{1}_d \otimes \mathbb{1}_d + 2\Bigg((d-2)\sum_k \lambda_k \otimes \tau_{\mathcal{T}}(\lambda_k) + \sum_{k=2}^{d} U\lambda_{k^2-1}U^* \otimes \tau_{\mathcal{T}}(U^*)\tau_{\mathcal{T}}(\lambda_{k^2-1})\tau_{\mathcal{T}}(U)\Bigg)\Bigg], \tag{22}$$

where τ_V and $\tau_{\mathcal{T}}$ are Jordan automorphisms and antiautomorphism of the $\mathbb{M}_d(\mathbb{C})$ (for details cf. Ref. [20]).

For above classes of states one gets explicit formulas:

(i) for $\rho \in \mathcal{E}^a$

$$D_1(\rho) = |t|, \quad -\frac{d}{2(d-1)} \leq t \leq \frac{d}{2(d+1)} \tag{23}$$

(ii) for $\rho \in \mathcal{E}^{aa}$

$$D_1(\rho) = \frac{2}{d}|t|, \quad -\frac{d}{2(d^2-1)} \leq t \leq \frac{d}{2} \tag{24}$$

It is interesting to compare above families to the two known distinguished classes of states: Werner states and isotropic states. Condition defining Werner states is

$$\rho = U \otimes U \rho U^* \otimes U^*, \quad U \in SU(d) \tag{25}$$

In turns out that these states fall into the class $\mathcal{E}^a$ and property (4) means that

$$Q^a(\mathcal{M}) = U \otimes U\, Q^a(\mathcal{M}_0)\, U^* \otimes U^*$$

Condition defining the so called isotropic states is

$$\rho = U \otimes \overline{U} \rho U^* \otimes U^T, \quad U \in SU(d) \tag{26}$$

These states fall into the class $\mathcal{E}^{aa}$ and for them

$$Q^{aa}(\mathcal{M}) = U \otimes \overline{U}\, Q^{aa}(\mathcal{M}_0)\, U^* \otimes U^T.$$

Hence, in both cases a minimization procedure is not needed.

Conclusions

In our work we have presented some aspects of the complex problem of finding values of quantum correlation measures. As an illustration we have discussed the measurement-induced one sided quantum geometric discord based on the trace distance. While, on the one hand we enlist and comment types of difficulties arising for the higher d-level systems and stress the step change between $d = 2$ and $d \geq 3$ systems, on the other hand we show that there are important instances, where one can effectively avoid troublesome minimization procedure and obtain strict results for the MIQGD.

References

[1] Skowronek Ł 2016 *There is no direct generalization of positive partial transpose criterion to the three-by-three case* `arXiv:1605.05254`
[2] Caves CM, Milburn GJ 2000 *Optics Communications* **179** 439
[3] P. Rungta, W.J. Munro, K. Nemoto, P. Deuar, G. J. Milburn, and C. M. Caves 2001 in Lecture Notes in Physics vol. **561** Springer, Berlin 149
[4] Frydryszak A M, Tkachuk V M 2012 *Geometric measure of entanglement for pure states and mean value of spin* `arXiv:1211.6472`
[5] Peres A 1996 *Phys. Rev. Lett.* **77** 1413
[6] Horodecki M, Horodecki P and Horodecki R 1996 *Phys Lett.* A**223** 1
[7] Bengtsson I, Weis S and Życzkowski K 2011 in *Geometric Methods in Physics* Springer, Basel 175-197
[8] Szarek S, Bengtsson I and Życzkowski K 2006 *J. Phys.* **A** L119
[9] Goyal S K, Simon B N, Singh R and Simon S 2011 *Geometry of the generalized Bloch sphere for qutrit*, arXiv:1111.4427
[10] Goyal S K, Simon S N, Singh R and Simon S 2016 *J. Phys. A: Math. Theor.***49** 165203
[11] Kurzynski P, Kolodziejski A, Laskowski W and Markiewicz M 2016 *Phys. Rev.* A **93** 062126
[12] Jakóbczyk L and Siennicki M 2001 *Phys. Lett.* A**286** 383
[13] Baumgartner B, Hiesmayr B C and Narnhofer H 2006 *Phys. Rev.* A **74** 032327
[14] Baumgartner B, Hiesmayr B C and Narnhofer H 2007 *J. Phys. A: Math. Theor.* **40** 7919
[15] Horodecki M, Horodecki P and Horodecki R 1998 *Phys. Rev. Lett.* **80** 5239
[16] Shang J, Seah Y-L, Ng H K, Nott D J and Englert B-G 2015 *New J. Phys.* **17** 043017
[17] Dakić B, Vedral V and Brukner C 2011 *Phys. Rev. Lett.* **107** 170502
[18] Paula F M, de Oliveira TR and Sarandy M S 2013 *Phys. Rev.* A**87** 064101
[19] Jakóbczyk L, Frydryszak A M and Ługiewicz P 2016 *Phys. Lett* **A 380** 1535
[20] Ługiewicz P, Frydryszak A M and Jakóbczyk L 2016 *Measurement - induced qudit geometric discord* `arXiv:1607.08753`

IOP Conf. Series: Journal of Physics: Conf. Series **804** (2017) 012017 doi:10.1088/1742-6596/804/1/012017

Deformations of the standard model: $SU_q(3)$ flavor symmetry

Niels G. Gresnigt

Department of Mathematical Sciences, Xi'an Jiaotong-Liverpool University, 111 Ren'ai Road, Suzhou SIP, Jiangsu, China 215123

E-mail: niels.gresnigt@xjtlu.edu.cn

Abstract. The quantum group $SU_q(3) = U_q(su(3))$ is taken as a baryon flavor symmetry and exceptionally accurate octet and decuplet baryons mass relations are obtained by accounting for the electromagnetic contribution to baryon masses (to zeroth order). The resulting charge specific q-deformed octet and decuplet baryon mass sum formulas are accurate to 0.02% and 0.08% respectively; a factor of 20 reduction in error compared to the standard Gell-Mann-Okubo mass formulas. An explicit formula for the Cabibbo angle, taken to be $\frac{\pi}{14}$, in terms of the deformation parameter q and spin parity J^P of the baryons is obtained.

1. Introduction

In this short paper we report on a recent result by the author where the flavor symmetry is deformed from a classical group to a quantum group and the electromagnetic contributions to baryon masses taken into account to obtain exceptionally accurate octet and decuplet baryon mass sum formulas [1]. That result is part of a larger program to study the applications of algebraic deformations to the generalization of both spacetime and internal symmetries.

Deformations of algebras provide a systematic approach to generalizing both spacetime and internal symmetries. Deformations of Lie algebras in particular have played a role in generalizing the isometries of spacetime [2, 3, 4, 5]. For example, both the quantum and relativistic revolutions of the 20th century can be considered as deformations of the Lie-algebras of classical mechanics and Galilean relativity into those of quantum mechanics and special relativity [6]. Much of the formal theory relating to deformations of Lie algebras dates back to the works of Gerstenhaber [7], Nijenhuis and Richardson [8] in the 1960s.

A second class of deformations are q-deformations (q for quantization). These are deformations of the universal enveloping algebra of a semi-simple Lie algebra into quantum groups. That is, these are deformations in the category of Hopf algebras. Quantum groups (which are algebras rather than groups) provide a generalization of familiar symmetry concepts through the deformation of Lie groups and algebras. These deformations of classical Lie groups (algebras) into quantum groups (algebras) extends the domain of classical group theory. Such quantum groups are deformations on Hopf algebras and depend on a deformation parameter q with the value $q = 1$ returning the undeformed universal enveloping algebra. First formalized by Jimbo [9] and Drinfeld [10] as a class of Hopf algebras, quantum groups have found many applications in theoretical physics, see [11, 12, 13, 14, 15, 16, 17] and references therein.

In the present paper we formulate octet and decuplet baryon mass relations of greatly improved accuracy based on a deformed flavor symmetry. We do this by first replacing the classical flavor symmetries by their quantum group counterparts (as in [18, 19, 20]) and then account for the electromagnetic contributions to baryons masses (as in [21, 22, 23]). The resulting octet and decuplet baryon mass sum formulas are accurate to 0.02% and 0.08% respectively.

The standard Gell-Mann-Okubo mass formula [24, 25], a result of $SU(3)$ flavor symmetry, is

$$M = \alpha_0 + \alpha_1 Y + \alpha_2 \left[I(I+1) - \frac{1}{4} Y^2 \right], \tag{1}$$

where M is the mass of a hadron within a specific multiplet, Y and I are the hypercharge and isospin respectively, and $\alpha_0, \alpha_1, \alpha_2$ are free parameters. By eliminating the free parameters $\alpha_0, \alpha_1, \alpha_2$, one obtains mass relations between the different baryons within a given multiplet. For the case of octet baryons $(1/2)^+$ one obtains the standard relation

$$N + \Xi = \frac{3}{2}\Lambda + \frac{1}{2}\Sigma, \tag{2}$$

whereas for the decuplet baryons $(3/2)^+$ one obtains the equal spacing rule

$$\Delta - \Sigma^* = \Sigma^* - \Xi^* = \Xi^* - \Omega \tag{3}$$

These relations hold to first order in flavor symmetry breaking only. Using the most recent data formula (2) is accurate to about 0.6%. The equal spacing rule (3) is less accurate ($\Delta - \Sigma^* = -152.6$MeV, $\Sigma^* - \Xi^* = -148.8$MeV, $\Xi^* - \Omega = -139.1$MeV), however a modified relation $\Omega - \Delta = 3(\Xi^* - \Sigma^*)$ due to Okubo [26] is accurate to about 1.4%.

Adopting instead q-deformed flavor symmetries obtained by replacing the unitary flavor symmetry $SU(3)$ by its quantum group counterpart $SU_q(3) \equiv U_q(su(3))$, Gavrilik was able to derive deformed mass relations for octet and decuplet baryons of increased accuracy [20, 19, 27]. By fixing a definite value for the deformation parameter q (through fitting the data), the modified baryon mass sum rules are accurate to an impressive 0.06% for the octet baryons and 0.32% for the decuplet baryons.

The masses used in both the standard as well as the q-deformed mass relations are the averages of the isospin multiplets (isoplets). At the level of accuracy of the q-deformed mass relations, the mass splittings within isoplets become significant however and should be accounted for. For example $\Sigma^- - \Sigma^+ = 8$MeV, $\Xi^{*-} - \Xi^{*0} = 3.2$MeV and $\Sigma^{*-} - \Sigma^{*+} = 4.4$MeV, representing about $\sim 0.2 - 0.4\%$ of the average isoplet mass. The impressive accuracy of the deformed mass relations lose their significance when these mass splittings (due to electromagnetic contributions to the masses of baryons) are ignored because the errors in these relations are less than the variation of masses within the isoplets.

Our starting point is the q-deformed mass relations derived in [18, 19, 20]. We then make use of the QCD general parametrization scheme of Morpurgo [28] in which the electromagnetic contributions (to zeroth order flavor symmetry breaking) to octet and decuplet baryon masses are expressed in terms of four parameters. Applying this parametrization to the deformed mass relations allows us to obtain relations that have equal electromagnetic mass contributions on both sides.

The values of the deformation parameter q that give these exceptionally accurate octet and decuplet baryons, together with the assertion made in [27, 29] that there should be a direct connection between q and the Cabibbo angle conspire to suggest a particularly simple formula for the Cabibbo angle in terms of q and spin parity J^P.

In section 2 a very brief overview of quantum groups is given and their application as flavor symmetries discussed. Section 3 focuses on electromagnetic mass splittings within isoplets and how to account for these electromagnetic contributions to baryon masses before charge specific and q-deformed mass relations for octet and decuplet baryons are presented in section 4. Finally in section 5 we look at the connection between q and the Cabibbo angle.

2. q-Deformed baryon mass relations

2.1. *The quantum groups* $SU_q(n) \equiv U_q(su(n))$

Only a very brief introduction to the relevant aspects of quantum groups is provided here. The literature on quantum groups and algebras is extensive. For an excellent introduction the reader is directed to [30]. Various applications of quantum groups to physics are discussed in [17].

The quantum (enveloping) algebra $SU_q(n) \equiv U_q(su(n))$ corresponding to a one-parameter deformation of the universal enveloping algebra of $su(n)$, is a Hopf algebra with unit $\mathbf{1}$ and generators H_i, $X_i^{\pm}$, $i = 1, 2, ..., n-1$, defined through the commutation relations in the Cartan-Chevalley basis as

$$[H_i, H_j] = 0 \quad (4)$$

$$\left[H_i, X_j^{\pm}\right] = a_{ij} X_j^{\pm} \quad (5)$$

$$\left[X_i^{+}, X_j^{-}\right] = \delta_{ij}[H_i]_q \equiv \delta_{ij}\frac{q^{H_i} - q^{-H_i}}{q - q^{-1}}, \quad (6)$$

together with the quadratic and cubic deformed q-Serre relations

$$\left[X_i^{\pm}, X_j^{\pm}\right] = 0, \quad j \neq i \pm 1, \quad 1 \leq i, j \leq n-1 \quad (7)$$

and

$$(X_i^{\pm})^2 X_j^{\pm} - [2]_q X_i^{\pm} X_j^{\pm} X_i^{\pm} + X_j^{\pm}(X_i^{\pm})^2 = 0, \quad j = i \pm 1, \quad 1 \leq i, j \leq n-1 \quad (8)$$

respectively [9, 31]. Here a_{ij} is an element of the Cartan matrix

$$a_{ij} = \begin{cases} 2 & j = i \\ -1 & j = i \pm 1 \\ 0 & \text{otherwise.} \end{cases}$$

The q-number

$$[N]_q = \frac{q^N - q^{-N}}{q - q^{-1}} \quad (9)$$

is defined for both operators (as in equation (6)) and real numbers[1] (as in equation (8)). The definition of the algebra is completed by the Hermiticity properties

$$(H_i)^{\dagger} = H_i, \qquad (X_i^{\pm})^{\dagger} = X_i^{\mp}. \quad (10)$$

The quantum algebra $SU_q(n)$ has the structure of a Hopf algebra admitting a coproduct, counit and antipode. These are not used here and so we do not define them (see [31]). In the limit $q = 1$ the above relations approach the relations for the universal enveloping algebra $U(su(n))$ but for general q they represent a deformation of the universal enveloping algebra of $su(n)$.

[1] In this paper we will only have to deal with integer values of N. The definition however holds for real numbers.

2.2. Quantum groups as flavor symmetries

In [18, 20, 19, 27], the q-analogues $SU_q(n) \equiv U_q(su(n))$ instead of the groups $SU(n)$ are taken as hadronic flavor symmetries and improved baryon mass relations are derived. The basic approach of the construction is the representation theory of $U_q(su(n))$ [32, 27]. q-Deformed mass relations are computed from the expectation value of the mass operator which is defined in terms of the generators of the dynamical algebras (quantum groups) $U_q(u(n+1))$ or $U_q(u(n,1))$. The expectation values are computed from the matrix elements of these generators. Utilizing the q-algebras $U_q(u(n+1))$ or $U_q(u(n,1))$ of dynamical symmetry, breaking of n-flavor symmetries up to exact isospin symmetry $SU_q(2)$ are realized and the q-analogues of mass sum rules for baryon multiplets are derived.

The general procedure (see [32, 20, 18] for details) is as follows:

(i) Use the Gelfand-Tsetlin basis vectors for baryon states of n-flavor $U_q(u(n))$ embedded into dynamical $U_q(u(n+1))$,

(ii) Construct a mass operator $\hat{M}_n$ invariant under the isospin+hypercharge deformed $U_q(u(2))$ from the generators of the dynamical algebra $U_q(u(n+1))$,

(iii) Calculate expressions for masses $M_{B_i} =< B_i|\hat{M}_n|B_i >$ (with $|B_i >$ suitably defined) involving M_0 and flavor symmetry breaking parameters α and β as well as the deformation parameter q,

(iv) Exclude the undetermined parameters (except q) from final expressions for M_{B_i} to obtain q-deformed baryon mass relations.

2.3. Octet baryons

With the restriction that $q = q_n = e^{i\pi/n}$ for integer n, the q-deformed mass relation for octet baryons obtained in [20] is

$$N + \frac{1}{[2]_{q_n} - 1}\Xi = \frac{[3]_{q_n}}{[2]_{q_n}}\Lambda + \left(\frac{[2]_{q_n}}{[2]_{q_n} - 1} - \frac{[3]_{q_n}}{[2]_{q_n}}\right)\Sigma. \tag{11}$$

The deformed mass relations depend on the deformation parameter q and the set of integers n produces an infinite set of mass relations. The value on n that leads to the closest agreement with experimental data [33] is $n = 7$ where it is the average mass of isoplets that has been used[2]. When $q = e^{i\pi/7}$, $[3]_{q_7} = \frac{[2]_{q_7}}{[2]_{q_7} - 1}$ and the mass relation (11) simplifies to

$$N + \frac{\Xi}{[2]_{q_7} - 1} = \frac{\Lambda}{[2]_{q_7} - 1} + \Sigma, \quad q_7 = e^{i\pi/7}, \tag{12}$$

which is accurate to 0.06%! this represents a very significant tenfold increase in accuracy compared to the GMO formula (2) which is accurate to about 0.6%. The accuracy to which the formula (12) holds is surprising given that the masses within an isoplet can differ by around $\sim 0.5\%$ whereas the masses used above are the averages of the isoplets.

2.4. Decuplet baryons

For the decuplet baryons, the equal spacing rule (3) is deformed to[32]

$$(\Sigma^* - \Delta + \Omega - \Xi^*) = [2]_q(\Xi^* - \Sigma^*). \tag{13}$$

Rearranging, (13) may be written as

$$\Omega - \Delta = (1 + [2]_q)(\Xi^* - \Sigma^*), \tag{14}$$

[2] For a table where the accuracy of the mass relations is evaluated for different values of n, see Gavrilik [20].

which is reminiscent of the relations obtained by Okubo [26]

$$\Omega - \Delta = 3(\Xi^* - \Sigma^*), \tag{15}$$

that, unlike the equal spacing rule holds for second order flavor symmetry breaking.

Taking $q = e^{i\pi/n}$ one finds excellent agreement with data for $n = 14$. We note however that $n = 14$ does not provide the best fit to the experimental data. Solving for integer n suggests $n = 16$ should provide the best fit. Experimental uncertainties of the masses are however ignored in this analysis and a range of values for n provide an excellent fit to the data. The choice of $n = 14$ in [20, 19] is motivated by the observation that $[2]_{q_{14}}$ is readily related to $[2]_{q_7}$ (the best fit for the q-deformed octet formula)

$$([2]_{q_{14}})^2 = q_{14}^2 + 2 + q_{14}^{-2} = q_7 + q_7^{-1} + 2 = [2]_{q_7} + 2. \tag{16}$$

Solving both (12) and (13) for $[2]_{q_n}$ and using the above relation, one obtains a new octet-decuplet mass relation [20]

$$\frac{\Omega - \Xi^* + \Sigma^* - \Delta}{\Xi^* - \Sigma^*} = \left(3 + \frac{\Xi - \Lambda}{\Sigma - N}\right)^{1/2}. \tag{17}$$

Using the latest data and averaging the isoplet masses, this relation holds within $\sim 1.5\%$.

3. Electromagnetic contributions

The electromagnetic contributions to baryon masses may be determined within the QCD general parametrization scheme in the spin-flavor space considered by Morpurgo [28]. To zeroth order symmetry breaking the electromagnetic contributions to the octet baryon masses are given in terms of four parameters as [22]

$$\begin{aligned}
\delta_0 p &= \mu + \frac{5}{9}\nu + \eta + \rho \qquad \delta_0 n = \frac{2}{3}\mu \\
\delta_0 \Lambda &= \frac{2}{3}\mu + \frac{1}{9}\nu \qquad \delta_0 \Sigma^+ = \mu + \frac{5}{9}\nu + \eta + \rho \\
\delta_0 \Sigma^- &= \frac{1}{3}\mu + \frac{1}{9}\nu + \eta + \frac{1}{3}\rho \qquad \delta_0 \Sigma^0 = \frac{2}{3}\mu + \frac{1}{3}\nu \\
\delta_0 \Xi^0 &= \frac{2}{3}\mu \qquad \delta_0 \Xi^- = \frac{1}{3}\mu + \frac{1}{9}\nu + \eta + \frac{1}{3}\rho.
\end{aligned}$$

It is easily checked that

$$\delta_0 p + \delta_0 \Xi^0 = \frac{3}{2}\delta_0 \Lambda^0 + \frac{1}{2}(2\delta_0 \Sigma^+ - \delta_0 \Sigma^0). \tag{18}$$

Accounting for the electromagnetic mass contributions the standard Gell-Mann-Okubo octet formula (2) becomes [23]

$$\frac{1}{2}(p + \Xi^0) + T = \frac{1}{4}(3\Lambda + 2\Sigma^+ - \Sigma^0), \tag{19}$$

where $T = \Xi^{*-} - \frac{1}{2}(\Omega^- + \Sigma^{*-}) = 5.18\text{MeV}$ is a decuplet correction. This equation has equal electromagnetic mass contributions on both sides of the equation (the electromagnetic contribution of the decuplet correction is zero, $\delta_0 T = 0$). This equation holds to within $\sim 0.13\%$, a significant improvement over the standard GMO octet relation.

Applying the same parametrization procedure to the decuplet baryons, one finds that[3]

$$\delta_0\Omega^- = \delta_0\Sigma^{-*} = \delta_0\Xi^{-*} = \delta_0\Delta^-. \tag{20}$$

Consequently, the Okubo relation becomes the charge specific decuplet mass relation

$$\Omega^- - \Delta^- = 3(\Xi^{*-} - \Sigma^{*-}), \tag{21}$$

which is accurate to $\sim 0.67\%$, a factor of two improvement over Okubo's relation.

4. Charge specific q-deformed baryon mass relations

As discussed in the introduction, the impressive accuracy of the q-deformed baryon mass relations means that mass splitting within isoplets can no longer be ignored. To account for this mass splitting specific charged baryons must be inserted into the mass relations in such a way that the electromagnetic contribution to masses is the same on both sides of the equation. In this section we apply the work discussed in section 3 to the q-deformed octet and decuplet mass relations (12) and (13). What we find is remarkably accurate charge specific q-deformed mass relations for both octet and decuplet baryons.

4.1. New octet baryon mass relation

Our starting point is the q-deformed octet baryon mass relations, eqn. (11). We now apply the parametrization of Morpurgo [22] in order to rewrite the deformed mass relations in such a way that the electromagnetic mass contributions cancel. Although somewhat more tedious than the case described in the previous section, one may check that

$$p + \frac{2[3]_q}{3[2]_q}\Xi^0 + \left(\frac{1}{[2]_q - 1} - \frac{2[3]_q}{3[2]_q}\right)\Xi^- = \frac{[3]_q}{[2]_q}\Lambda - \frac{[3]_q}{3[2]_q}\Sigma^0 + \left(\frac{1}{[2]_q - 1} - \frac{2[3]_q}{3[2]_q}\right)\Sigma^- + \Sigma^+ \tag{22}$$

is unaffected by the electromagnetic corrections (to zeroth order in flavor breaking).

Substituting the experimental masses for the baryons, we may determine the value of $q = q_n = e^{i\pi/n}$ for integer n that minimizes the error in the above mass relation. One finds that $n = 7$ continues to provide the best fit to the experimental data. For this value of q, equation (22) simplifies to

$$p + \frac{(2\Xi^0 + \Xi^-)}{3([2]_{q_7} - 1)} = \frac{\Lambda}{[2]_{q_7} - 1} + \frac{(\Sigma^- - \Sigma^0)}{3([2]_{q_7} - 1)} + \Sigma^+, \tag{23}$$

which has an error of only 0.02%, a threefold reduction in error compared to the q-deformed mass relation (11) which ignores electromagnetic contributions to mass.

4.2. New decuplet baryon mass relation

Similarly to the octet case, accounting for the electromagnetic mass contribution to the $SU_q(3)$ deformed decuplet mass relation (13) leads to the new decuplet formula

$$\Omega^- - \Delta^- = ([2]_q + 1)(\Xi^{*-} - \Sigma^{*-}). \tag{24}$$

In [20] it was suggested that choosing $q = e^{i\pi/14}$ provides a good fit to data. Although this remains true in the present case, there are other values of q for which the error is smaller. In

[3] It is worth noting that the second order flavor symmetry breaking affects only baryons with a strangeness of 2 or 3, namely Ξ, Ξ^* and Ω.

particular, we consider the case where $q = e^{i\pi/21}$. Although this is not the absolute best choice for q [4], because 21 is a multiple of 7, it allows for an elegant new relation between octet and decuplet baryons (this is also the reason why $n = 14$ is chosen in [20, 19]). Taking $n = 21$ gives LHS=440.45MeV and RHS=440.10MeV, an error of only 0.08%.

4.3. Octet-decuplet mass relation
Both formulas (23) and (24) provide excellent fits to experimental data. The octet formula is valid only for $q = e^{i\pi/7}$ whereas the decuplet formula is valid for all any $q_n = e^{i\pi/n}$, where n is an integer. In particular we chose $n = 21$ as it provides (with the exception of $n = 22$) the best fit to data. Because 21 is a multiple of 7, we can find a relation between octet and decuplet baryon masses by solving both mass relations for $[2]_{q_7}$ and $[2]_{q_{21}}$ respectively and using the fact that $[2]^3_{q_{21}} = (q_{21} + q_{21}^{-1})^3 = [2]_{q_7} + 3[2]_{q_{21}}$ to obtain

$$\left(\frac{\Omega^- - \Delta^-}{\Xi^{*-} - \Sigma^{*-}} - 1\right)^3 - 3\left(\frac{\Omega^- - \Delta^-}{\Xi^{*-} - \Sigma^{*-}} - 1\right) = \left(1 + \frac{\Lambda + \frac{1}{3}\left(\Sigma^- - \Sigma^0 - 2\Xi^0 - \Xi^-\right)}{(p - \Sigma^+)}\right). \tag{25}$$

This formula has an error of $\sim 1.2\%$; a slight improvement over the formula in [20, 19] with the choice $n = 14$ (and error $\sim 1.5\%$).

5. The Cabibbo angle as a function of q and J^P

It is asserted in [27, 29] that there should be a direct connection between the deformation parameter q and the Cabibbo angle θ_C.

With the choice of $\theta_{\mathbf{8}} = \frac{\pi}{7}$, such that $q = e^{i\theta_{\mathbf{8}}}$ for octet baryons and $\theta_{\mathbf{10}} = \frac{\pi}{14}$, such that $q = e^{i\theta_{\mathbf{10}}}$ for decuplet baryons, the suggested explicit form between these angles (which are related to q) and the Cabibbo angle is

$$\theta_{\mathbf{10}} = \theta_C, \qquad \theta_{\mathbf{8}} = 2\theta_C. \tag{26}$$

The Cabibbo angle here takes the exact value $\theta_C = \frac{\pi}{14}$.

These choices for $\theta_{\mathbf{8}}$ and $\theta_{\mathbf{10}}$ are a result of fitting data to the q-deformed octet and decuplet mass sum rules in [20, 19]. Once the electromagnetic contributions to mass are taken into account however, the choice $q = e^{i\theta_{\mathbf{10}}} = e^{i\pi/21}$ is superior, providing a significantly better fit to data than $\theta_{\mathbf{10}} = \frac{\pi}{14}$. For these improved formulas, the explicit form (26) no longer holds true. Instead we write

$$\theta_C = \frac{1}{2}\theta_{\mathbf{8}} = \frac{3}{2}\theta_{\mathbf{10}} = \frac{\pi}{14}, \tag{27}$$

Making the observation that for octet baryons $J^p = (1/2)^+$ and for decuplet baryons $J^P = (3/2)^+$, the Cabibbo angle can then be written as a function of the spin parity J^P and deformation parameter q

$$\theta_C = -iJ^P \ln q. \tag{28}$$

[4] Putting the experimental data into (24), substituting $q = e^{i\pi/n}$, and solving for n gives $n = 22$ (for integer n). The error in relation (24) when $n = 22$ is just 0.01%!

6. Discussion and conclusion

In earlier works, the quantum group $SU_q(3)$ has been taken as a flavor symmetry and mass formulas of improved accuracy derived for octet and decuplet baryons. In [1] we built on these earlier works by taking into account (via Morpurgo's parametrization scheme) the electromagnetic contributions to octet and decuplet baryons masses.

The deformed mass relations derived in [18, 19, 20] use the average of the isospin multiplet masses. The mass differences within the isoplets should not be ignored at the level of accuracy afforded by these deformed mass relations because they exceed the error in these formulas (for the best fit value of the deformation parameter q).

Accounting for the electromagnetic contributions to mass by selecting charge specific baryon masses (in a way that the electromagnetic mass contributions are balanced) rather than using isoplet mass averages leads to new octet and decuplet baryon mass relations accurate to 0.02% and 0.08% respectively. This represents more than a factor of 20 reduction in error compared to the standard GMO formulas.

A fixed value for the deformation parameter q can be determined from the experimental data. The choice of $q = e^{i\pi/7}$ for octet baryons and $q = e^{i\pi/21}$ for decuplet baryons leads to a new relation between octet and decuplet baryons with an error of around 1.2%, a slight improvement over an earlier formula (see equation (17). The choice of $n = 21$ for decuplet baryons suggests a simple formula for the Cabibbo angle in terms of q and spin parity J^P. As a result, q may not be a free parameter but rather be fixed from the baryons under consideration (either $J^P = (1/2)^+$ or $J^P = (3/2)^+$).

Quantum groups are the result of one-parameter deformation in the universal enveloping algebra. These are just one type of algebraic deformation that can be considered [7, 8]. Lie-type deformations (those that deform a Lie algebra) have proven very useful in generalizing spacetime symmetries [2, 4, 3, 5]. q-Deformation on the other hand seem to have particular applications in generalized descriptions of internal and gauge symmetries [12, 34, 35]. Considered together therefore, it seems that Lie-type and q-deformations offer a consistent framework within which to develop physics in the 21st century [36, 37, 38, 39].

Acknowledgments

This work is supported in part by XJTLU research grant RDF-14-03-13 and Natural Science Foundation of China grant RR0116.

References

[1] N. G. Gresnigt. Charge specific baryon mass formulas with deformed SUq(3) flavor symmetry. *(preprint arXiv:1607.01632)*, 2016.

[2] R. V. Mendes. Deformations, stable theories and fundamental constants. *Journal of Physics A: Mathematical and General*, 27(24):8091–8104, 1994.

[3] C Chryssomalakos and E Okon. Generalized Quantum Relativistic Kinematics: a Stability Point of View. *International Journal of Modern Physics D*, 13(10):2003–2034, 2004. [arXiv preprint hep-th/0407080].

[4] D. V. Ahluwalia, N. G. Gresnigt, A. B. Nielsen, D. Schritt, and T. F. Watson. Possible polarization and spin-dependent aspects of quantum gravity. *International Journal of Modern Physics D*, 17(3/4):495, 2008.

[5] N. G. Gresnigt, P. F. Renaud, and P. H. Butler. The Stabilized Poincare-Heisenberg Algebra: a Clifford Algebra Viewpoint. *Int. J. Mod. Phys. D*, 16(09):1519–1529, 2007. [arXiv:hep-th/0611034].

[6] L. D. Faddeev. *Mathematician's view on the development of physics, Frontiers in Physics: High technology and mathematics*. Singapore: Word Scientific, 1989.

[7] M. Gerstenhaber. On the Deformation of Rings and Algebras. *The Annals of Mathematics*, 79(1):59–103, 1964.

[8] A. Nijenhuis and R. W. Richardson. Deformation of Lie algebra structures. *J. Math. Mech*, 17:89, 1967.

[9] Michio Jimbo. A q-difference analogue of U(g) and the Yang-Baxter equation. *Letters in Mathematical Physics*, 10(1):63–69, 1985.

[10] VG Drinfeld. Soviet math. dokl. 32 (1985) 254; m. jimbo. *Lett. Math. Phys*, 10:63, 1985.

IOP Conf. Series: Journal of Physics: Conf. Series **804** (2017) 012017 doi:10.1088/1742-6596/804/1/012017

[11] R. J. Finkelstein. On q-Electroweak. *(preprint hep-th/0110075)*, 2001.

[12] Robert J Finkelstein. An SLq(2) Extension of the Standard Model. *(preprint arXiv:1205.1026)*, 2012.

[13] Harold Steinacker. Finite dimensional unitary representations of quantum anti–de sitter groups at roots of unity. *Communications in mathematical physics*, 192(3):687–706, 1998.

[14] Daniel Sternheimer. The geometry of space-time and its deformations from a physical perspective. In *From geometry to quantum mechanics*, pages 287–301. Springer, 2007.

[15] Shahn Majid and Henri Ruegg. Bicrossproduct Structure of κ-Poincare Group and Non-Commutative Geometry. *Physics Letters B*, 334(3):348–354, 1994.

[16] Jerzy Lukierski and Anatol Nowicki. Doubly Special Relativity Versus κ-Deformation of Relativistic Kinematics. *International Journal of Modern Physics A*, 18(01):7–18, 2003.

[17] Leonardo Castellani and Julius Wess. *Quantum groups and their applications in physics*, volume 127. IOS Press, 1996.

[18] AM Gavrilik and NZ Iorgov. Quantum groups as flavor symmetries: account of nonpolynomial su (3)-breaking effects in baryon masses. *arXiv preprint hep-ph/9807559*, 1998.

[19] AM Gavrilik. Quantum algebras in phenomenological description of particle properties. *Nuclear Physics B-Proceedings Supplements*, 102:298–305, 2001.

[20] AM Gavrilik. Quantum groups in hadron phenomenology. *(preprint hep-ph/9712411)*, 1997.

[21] G Morpurgo. Electromagnetic mass differences of the octet and decuplet baryons. *Physical Review D*, 45(5):1686, 1992.

[22] G Morpurgo. New mass formula for octet and decuplet baryons. *Physical review letters*, 68(2):139, 1992.

[23] G Dillon and G Morpurgo. On the miracle of the Coleman–Glashow and other baryon mass formulas. *Physics Letters B*, 481(2):239–244, 2000.

[24] Murray Gell-Mann. The eightfold way: A theory of strong interaction symmetry. Technical report, California Inst. of Tech., Pasadena. Synchrotron Lab., 1961.

[25] Susumu Okubo. Note on unitary symmetry in strong interactions. *Progress of Theoretical Physics*, 27(5):949–966, 1962.

[26] S Okubo. Some consequences of unitary symmetry model. *Physics Letters*, 4(1):14–16, 1963.

[27] AM Gavrilik. Quantum algebras, particle phenomenology, and (quasi) supersymmetry. *(preprint hep-ph/0402082)*, 2004.

[28] G Morpurgo. Field theory and the nonrelativistic quark model: A parametrization of the baryon magnetic moments and masses. *Physical Review D*, 40(9):2997, 1989.

[29] Alexandre Gavrilik. Can the Cabibbo mixing originate from noncommutative extra dimensions? In *Noncommutative Structures in Mathematics and Physics*, pages 343–355. Springer, 2001.

[30] R Jaganathan. Some introductory notes on quantum groups, quantum algebras, and their applications. *(preprint math-ph/0105002)*, 2001.

[31] Christiane Quesne. Complementarity of suq(3) and uq(2) and q-boson realization of the suq(3) irreducible representations. *Journal of Physics A: Mathematical and General*, 25(22):5977, 1992.

[32] AM Gavrilik, II Kachurik, and AV Tertychnyj. Representations of the Uq(u(4,1)) and a q-polynomial that determines baryon mass sum rules. *(preprint hep-ph/9504233)*, 1995.

[33] Keith A Olive, Particle Data Group, et al. Review of particle physics. *Chinese Physics C*, 38(9):090001, 2014.

[34] Robert J Finkelstein. A knot model suggested by the standard electroweak theory. *International Journal of Modern Physics A*, 20(28):6487–6494, 2005.

[35] Robert J Finkelstein. The elementary particles as quantum knots in electroweak theory. *International Journal of Modern Physics A*, 22(24):4467–4480, 2007.

[36] N. G. Gresnigt and A. B. Gillard. Electroweak symmetries from the topology of deformed spacetime with minimal length scale. *(preprint arXiv:1512.04339)*, 2015.

[37] Daniel Sternheimer. "the important thing is not to stop questioning", including the symmetries on which is based the standard model. In *Geometric Methods in Physics*, pages 7–37. Springer, 2014.

[38] Moshe Flato. Deformation View of Physical Theories. *Czechoslovak Journal of Physics*, 32(4):472–475, 1982.

[39] Philippe Bonneau and Daniel Sternheimer. Topological Hopf algebras, quantum groups and deformation quantization. *(arXiv preprint math/0307277)*, 2003.

IOP Conf. Series: Journal of Physics: Conf. Series **804** (2017) 012018 doi:10.1088/1742-6596/804/1/012018

Symmetrized Vibrational-Rotational Basis for Collective Nuclear Models

A.A. Gusev, S.I. Vinitsky
Joint Institute for Nuclear Research, Dubna, Russia

A. Góźdź, A. Dobrowolski, A. Szulerecka
Institute of Physics, Department of Mathematical Physics, University of Maria Curie-Skłodowska, Lublin, Poland

A. Pędrak
National Centre for Nuclear Research, Warsaw, Poland

E-mail: vinitsky@theor.jinr.ru

Abstract. The generalized projection operators for the intrinsic group acting in the space $L_2(SO(3))$ and in the space spanned by the eigenfunctions of a multidimensional harmonic oscillator are constructed. New symbolic-numerical algorithm implemented in computer algebra system for generating irreducible representations of the point symmetry groups in the rotor + shape vibrational space of a nuclear collective model in the intrinsic frame is presented. The efficiency of the algoritm is investigated by calculating the bases of irreducible representations subgroup $\overline{\mathrm{D}}_{4y}$ of octahedral group in the intrinsic frame of a quadrupole-octupole collective nuclear model. The discrete variable representation algorithm proposed for solving eigenvalue problem, describing vibrational-rotational motion of collective nuclear model in intrinsic frame.

1. Introduction

The most efficient way of construction of collective models in nuclear and molecular physics is using of the so called intrinsic frame [1] The most useful is the rotating intrinsic frame. However, to construct vibrational-rotational basis of the collective models in an intrinsic frame one needs to notice some important features of this notion [2] and the technique of generalized projection operators [3].

In this paper we propose a simple and effective algorithm for numerical construction of such basis for the vibrational + rotational collective motion based on algorithms elaborated in refs. [4, 5]. We consider application of the algorithm on the example of construction of the rotation-vibrational quadrupole-octupole basis of the irreducible representation (irr.) $A1 = \Gamma_1$ of the point group $\overline{\mathrm{D}}_{4y}$ (in classification of Ref. [6]) in the intrinsic frame considered in the zero and one-phonon approximation, as in ref.[7]. Structure of the paper is following. In Section 2, a setting of intrinsic frame and *symmetrization groups* are presented In Section 3, a setting of quadrupole-octupole harmonic oscillator in rotating frame is formulated. In Section 4, parametric vibration basis functions of irr. $A1 = \Gamma_1$ of intrinsic dihedral $\overline{\mathrm{D}}_{4y}$ point group are derived. In Section 5, an algorithm for construction of orthonormal vibrational-rotational basis is presented. In Section 6, solving an eigenvalue problem in the intrinsic frame using discrete variable representation (DVR) is described.

2. Intrinsic frame and symmetrization group

The classical rotation is well understood phenomenon in which the orientation of a body is changing with time. Contrary, the quantum rotation allows to determine only the probability of a given orientation and there is no time variable in the wave function.

The notion of the quantum rotational motion allows to define the rotating intrinsic frame, for the collective variables $\{\alpha_{\lambda\mu}^{(lab)}\}$ defined by the equation of surface [1]:

$$R(\alpha^{(lab)};\theta,\phi) = R_0\left(1+\sum\nolimits_{\lambda\mu}\alpha_{\lambda\mu}^{(lab)\star}Y_{\lambda\mu}(\theta,\phi)\right). \tag{1}$$

The corresponding collective variables in the intrinsic frame we denote by $\{\alpha_{\lambda\mu}\}$. They can be obtained by the quantum rotation of the laboratory collective variables $\{\alpha_{\lambda\mu}^{(lab)}\}$ with the rotation operator $\hat{R}(\Omega)$

$$\alpha_{\lambda}^{(lab)} = \hat{R}(\Omega^{-1})\alpha_{\lambda} \tag{2}$$

assuming, in addition, that the rotation group $SO(3) \ni \hat{R}(\Omega)$ parameters, represented by the Euler angles $\Omega = (\Omega_1,\Omega_2,\Omega_3)$, are considered as a part of intrinsic variables. The intrinsic variables α_λ are invariant in respect to the laboratory rotations $\hat{R}(\Omega)$. It is important to notice that inclusion of the Euler angles into the set of intrinsic variables makes this set of variables redundant, 3 variable more than needed. It implies that, the definition of the intrinsic frame requires three additional conditions which leads the same number of variables in both frames

$$F_k(\alpha,\Omega) = 0, \quad \text{where } k = 1,2,3. \tag{3}$$

In this way one can obtain a new description of a physical system, e.g. a nucleus, in which the rotational motion can be directly described by the Euler angles, i.e. we start to work in the rotating frame.

An interesting question is how to investigate symmetries of a nucleus in the intrinsic frame. It is clear, that the transformations furnishing an intrinsic symmetry group have to be defined in the intrinsic frame.

A convenient definition of intrinsic groups was formulated in [2] in the following form: for each element g of the group G, one can define a corresponding operator $\hat{\bar{g}}$ in the group linear space $\mathcal{L}_G$ as:

$$\hat{\bar{g}}|S\rangle = |Sg\rangle, \qquad \text{for all } |S\rangle \in \mathcal{L}_G, \tag{4}$$

where all elements inside the ket vectors $S = \sum_{g\in \mathrm{G}} c_g g$, here c_g are the complex numbers, form a group algebra of the group G. In this definition the notion of the group linear space $\mathcal{L}_G$ is used. This space is defined as the linear space spanned by all possible formal linear combinations of the elements of the group G

$$\mathcal{L}_G = \left\{|S\rangle : |S\rangle = \sum\nolimits_{g\in \mathrm{G}} c_g g, \text{ where } c_g \in \mathbf{C}\right\}. \tag{5}$$

It looks like the group algebra mentioned above, but, it is important that the elements of $\mathcal{L}_G$ have to be considered only as vectors, not as the elements of the group algebra. The group formed by the collection of the operators $\hat{\bar{g}}$ is called the intrinsic group $\overline{\mathrm{G}}$ related to the group G.

One of the most important property of the intrinsic group $\overline{\mathrm{G}}$ is that this group commutes with its partner group G (this can be called a laboratory group, because it acts in the laboratory frame):

$$[\mathrm{G}, \overline{\mathrm{G}}] = 0. \tag{6}$$

The groups G and $\overline{\mathrm{G}}$ are antyisomorphic. The required anti-isomorphism between the partner groups G and $\overline{\mathrm{G}}$ is given by

$$\phi_G : \overline{\mathrm{G}} \to \mathrm{G}, \text{ where } \quad \phi_G(\bar{g}) = g \text{ and } \phi_G(\bar{g}\bar{g}') = \phi_G(\bar{g}')\phi_G(\bar{g}). \tag{7}$$

This property suggests that the partner groups G and $\overline{\mathrm{G}}$ have a lot of common properties as e.g. similar structure of representations, decompositions of the Kronecker products, the Clebsch-Gordan coefficients and many others.

As an example let us consider a relation among representations of both groups. Because the partner groups commute one can find common basis $|\Gamma mk\rangle$ for representations of the group G and the group $\overline{\mathrm{G}}$. The representations are defined as:

$$g|\Gamma mk\rangle = \sum\nolimits_{m'} \Delta^{(\Gamma)}_{m'm}(g)|\Gamma m'k\rangle, \quad \bar{g}|\Gamma mk\rangle = \sum\nolimits_{k'} \bar{\Delta}^{(\Gamma)}_{k'k}(\bar{g})|\Gamma mk'\rangle. \tag{8}$$

To compare both representation one can use as the basis in the form of the generalized projection operators (elements of the group linear space $\mathcal{L}_G$) [3]

$$|\Gamma mk\rangle = \frac{\dim[\Gamma]}{\mathrm{card(G)}} \sum_{g\in \mathrm{G}} \Delta^{(\Gamma)}_{mk}(g)^\star g, \tag{9}$$

where $\dim[\Gamma]$ denotes the dimension of the representation Γ and card(G) is the number of elements in the group G. This allows to calculate (8)

$$\bar{g}|\Gamma mk\rangle = \frac{\dim[\Gamma]}{\mathrm{card(G)}} \sum_{g'\in \mathrm{G}} \Delta^{(\Gamma)}_{mk}(g')^\star g'g = \frac{\dim[\Gamma]}{\mathrm{card(G)}} \sum_{g'\in \mathrm{G}} \Delta^{(\Gamma)}_{mk}(g'g^{-1})^\star g'$$

$$= \frac{\dim[\Gamma]}{\mathrm{card(G)}} \sum_{g'\in \mathrm{G}} \sum_{k'} \Delta^{(\Gamma)}_{mk'}(g')^\star \Delta^{(\Gamma)}_{kk'}(g) g' = \sum_{k'} \Delta^{(\Gamma)}_{kk'}(g)|\Gamma mk'\rangle, \tag{10}$$

where $\Delta^{(\Gamma)}_{mm'}(g)$ are matrix elements of the representation Γ of the group G. Comparing both expression one can see that the matrices of both representations are related. The representations of the intrinsic group are transposed representations of the partner group

$$\bar{\Delta}^{(\Gamma)}_{mk}(\bar{g}) = \Delta^{(\Gamma)}_{km}(g). \tag{11}$$

A bit different are definitions of irreducible tensors in respect to the laboratory group G and the intrinsic group $\overline{\mathrm{G}}$. By definition the irreducible tensors in respect to the laboratory group G transform as

$$\hat{g} T^{(\Gamma)}_m \hat{g}^{-1} = \sum\nolimits_l \Delta^{(\Gamma)}_{lm}(g) T^{(\Gamma)}_l. \tag{12}$$

The tensors in respect to the intrinsic group $\overline{\mathrm{G}}$, due to the anti-isomorphism between both groups, have to be defined in the following way

$$\hat{\bar{g}} \overline{T}^{(\Gamma)}_k \hat{\bar{g}}^{-1} = \sum\nolimits_l \Delta^{(\Gamma)}_{lk}(g^{-1}) \overline{T}^{(\Gamma)}_l. \tag{13}$$

As an example, let us consider the action of the intrinsic group in the collective space consisted of the square integrable functions of the deformation parameters and the Euler angles. The intrinsic rotation operators $\hat{R}(\bar{g}_1, \bar{g}_2) \in \overline{\mathrm{SO}(3)}_\alpha \times \overline{\mathrm{SO}(3)}_\Omega$ (the indices α and Ω show the variables which are affected by the corresponding group) are defined as follows

$$\hat{R}(\bar{g}_1, \bar{g}_2) f(\alpha, \Omega) = f(\{\hat{\bar{g}}_1 \alpha\}, \Omega \phi_G(\bar{g}_2)^{-1}), \tag{14}$$

where $\bar{g}_1 \in \overline{\mathrm{SO}(3)}_\alpha$ and $\bar{g}_2 \in \overline{\mathrm{SO}(3)}_\Omega$. The action of the group $\overline{\mathrm{SO}(3)}_\alpha$ onto the deformation variables is a bit non-standard and is given by the following equation

$$\hat{\bar{g}}_1 \alpha_{\lambda\mu} = \sum_{\mu'} D^\lambda_{\mu'\mu}(\phi_G(\bar{g}_1)^{-1}) \alpha_{\lambda\mu'}, \tag{15}$$

where $D^\lambda_{\mu'\mu}$ are Wigner functions [8]. The intrinsic group $\overline{\mathrm{SO}(3)}$ corresponding to the laboratory rotation group SO(3) defined in the laboratory frame consists of all rotations $\hat{R}(\bar{g}, \bar{g})$ for which the deformation parameters and the Euler angles are rotated with the same angles. The required anti-isomorphism between the partner groups SO(3) and $\overline{\mathrm{SO}(3)}$ is given by (7). It is important to notice that, in general, not all transformations $(\bar{g}_1, \bar{g}_2) \in \overline{\mathrm{SO}(3)}_\alpha \times \overline{\mathrm{SO}(3)}_\Omega$

$$(\bar{g}_1, \bar{g}_2)\colon\ (\alpha, \Omega) \to (\alpha', \Omega') \tag{16}$$

are allowed in the intrinsic frame. They are allowed if they do not break the conditions which define the intrinsic frame (3)

$$(\hat{\bar{g}}_1, \hat{\bar{g}}_2) F_k(\alpha, \Omega) = F_k(\hat{\bar{g}}_1 \alpha, \Omega \bar{g}_2^{-1}) = 0, \text{ where } k = 1, 2, 3. \tag{17}$$

For example, in the case of the quadrupole collective variables α_2 with the standard Bohr condition which define the intrinsic frame: $\alpha_{2,\pm1} = 0$ and $\alpha_{22} = \alpha_{2-2}$, the allowed intrinsic rotations $\hat{R}(\bar{g}_1, \bar{g}_2) \in \overline{\mathrm{SO}(3)}_\alpha \times \overline{\mathrm{SO}(3)}_\Omega$ have to fulfil the following conditions

$$\hat{R}(\bar{g}_1, e_G)\alpha_{2\pm1} = 0 \text{ and } \hat{R}(\bar{g}_1, e_G)\alpha_{22} = \hat{R}(\bar{g}_1, e_G)\alpha_{2-2}, \tag{18}$$

where the second argument represents the unit element of the group $\overline{\mathrm{SO}(3)}_\Omega$. The Bohr conditions allow for the arbitrary rotations $\bar{g}_2 \in \overline{\mathrm{SO}(3)}_\Omega$.

Using the conditions (18) the allowed rotations of the deformation parameters α have to satisfy the following equations

$$D^2_{0,\pm1}(\bar{g}_1^{-1}){=}0, \quad D^2_{-2\pm1}(\bar{g}_1^{-1}) + D^2_{2\pm1}(\bar{g}_1^{-1}){=}0, \quad D^2_{02}(\bar{g}_1^{-1}) - D^2_{0,-2}(\bar{g}_1^{-1}){=}0,$$
$$D^2_{-2,-2}(\bar{g}_1^{-1}) + D^2_{2,-2}(\bar{g}_1^{-1}){=}D^2_{-22}(\bar{g}_1^{-1}) + D^2_{22}(\bar{g}_1^{-1}). \tag{19}$$

In this case, the octahedral point group $\overline{\mathrm{O}}_\alpha \subset \overline{\mathrm{SO}(3)}_\alpha$ acting only on the variables α provide the solution of the set of equations (19).

In practice, the transformation from the laboratory frame to the intrinsic frame is not a one-to-one function. In principle, there are two possibilities to achieve uniqueness of transformation from the laboratory to the intrinsic frame:

- first, one can define the appropriate region of the intrinsic collective variables in which the transformation from the laboratory to intrinsic frame is a one-to-one function,
- second, one can allow for the whole range of collective variables but then one needs to fulfil some special conditions for physical states.

However, narrowing of the domain of transformations is not a physical solution. For example, narrowing the ranges of Euler angles Ω destroys the angular momentum operators – the operators become ill defined. To keep the appropriate physical properties in space of state one have to accept that the second possibility.

For further purpose it is useful to define the special group of intrinsic transformations $\bar{h} \in \overline{\mathrm{G}}_S$:

$$(\alpha, \Omega) \to \bar{h}(\alpha', \Omega'), \tag{20}$$

where $\alpha = \{\alpha_{\lambda\mu}\}$ and which leave invariant the laboratory variables as function of the intrinsic variables:

$$\alpha^{(lab)}(\alpha', \Omega') = \alpha^{(lab)}(\alpha, \Omega), \qquad F_k(\alpha', \Omega') = F_k(\alpha, \Omega) = 0, \text{ for } k = 1, 2, 3, \tag{21}$$

where $\alpha^{(lab)}(\alpha, \Omega) = \hat{R}(\Omega^{-1})\alpha$, see (2). The group $\overline{\mathrm{G}}_S$ we call the *symmetrization group.* The symmetrization group decomposes the collective manifold into orbits of physically equivalent points.

Let the function $\Psi^{(lab)}$ denotes a state vector of a nucleus in the laboratory frame. The corresponding state vector in the intrinsic frame has to fulfil the obvious equation

$$\Psi(\alpha, \Omega) = \Psi^{(lab)}(\alpha^{(lab)}) \tag{22}$$

which represents the fact that the wave function of the physical system written in the laboratory frame has to be well and uniquely defined function. However, after transformation of Eq. (22) with the elements of the symmetrization group we potentially change Ψ and do not change the laboratory state vector:

$$\Psi(\alpha', \Omega') = \Psi^{(lab)}(\alpha^{(lab)}). \tag{23}$$

To omit the obvious contradiction, we have to define the uniqueness condition for the states in the intrinsic frame

$$\Psi(\alpha', \Omega') = \Psi(\alpha, \Omega). \tag{24}$$

This condition we will call the symmetrization condition. It can be expressed as invariance of the intrinsic state vectors in respect to all transformations $\bar{h} \in \overline{\mathrm{G}}_S$,

$$\bar{h}\Psi(\alpha, \Omega) = \Psi(\alpha, \Omega), \tag{25}$$

where the group $\overline{\mathrm{G}}_S$ is the symmetrization group for a given intrinsic frame.

In quantum mechanics the collective states in the laboratory/intrinsic frame belong the spaces of square integrable functions. In the laboratory frame the quantum state space is usually represented by the whole space $\mathrm{L}_2(\{\alpha^{(lab)}_{\lambda\mu}\})$, however, the quantum state space in the intrinsic frame has to be restricted to the subspace of functions satisfying the symmetrization condition (25): $\mathcal{K}_S \subset \mathrm{L}_2(\{\alpha_{\lambda\mu}\})$. This implies that in calculations we cannot use any standard basis, but we have to construct the basis dedicated to the subspace $\mathcal{K}_S$.

3. Shape vibrational basis in the intrinsic frame

To define the rotations of intrinsic spherical tensors with respect to intrinsic rotation group, we rewrite the expansion (1) in terms of spherical harmonics $Y_{lm}(\theta', \phi')$ in the intrinsic frame:

$$R(\theta, \phi) = R(\theta', \phi') = R_0\Big(1 + \sum_l a_{l0} Y_{l0}(\theta', \phi') + \sum_{lm}\Big(a_{lm} Y^{(+)}_{lm}(\theta', \phi') + b_{lm} Y^{(-)}_{lm}(\theta', \phi')\Big)\Big), \tag{26}$$

$$Y^{(\pm)}_{lm}(\theta', \phi') = \imath^{(1\mp1)/2}(Y_{lm}(\theta', \phi') \pm (-1)^m Y_{l-m}(\theta', \phi'))/\sqrt{2}.$$

Note, that these spherical harmonics are tensors in respect to intrinsic rotations. We rewrite the transformation $\hat{\bar{g}}\alpha_{\lambda\mu}=\bar{\alpha}_{\lambda\mu}$, $\hat{\bar{g}}\in\bar{G}$, between two intrinsic coordinate frames

$$\bar{\alpha}_{\lambda\mu} = \sum_{\mu'=-\lambda}^{\lambda} D^{\lambda}_{\mu\mu'}(g)^{\star}\alpha_{\lambda\mu'}, \quad Y_{\lambda\mu}(\bar{\theta}',\bar{\varphi}') = \sum_{\mu'=-\lambda}^{\lambda} D^{\lambda}_{\mu\mu'}(g)^{\star}Y_{\lambda\mu'}(\theta',\varphi'). \tag{27}$$

We use the relations between the complex-valued coefficients $\bar{\alpha}_{\lambda\mu}$, $\alpha_{\lambda\mu}$ and real-valued ones $\bar{a}_{\lambda\mu}$, $\bar{b}_{\lambda\mu}$, $a_{\lambda\mu}$, $b_{\lambda\mu}$

$$\bar{\alpha}_{\lambda 0}=\bar{a}_{\lambda 0}, \quad \bar{\alpha}_{\lambda\mu}=(\bar{a}_{\lambda\mu}-\imath\bar{b}_{\lambda\mu})/\sqrt{2}, \quad \bar{\alpha}_{\lambda-\mu}=(-1)^{\mu}(\bar{a}_{\lambda\mu}-\imath\bar{b}_{\lambda\mu})/\sqrt{2}. \tag{28}$$

The transformation between both intrinsic frames $\bar{\mathbf{q}} = \bar{\mathbf{M}}(g)\mathbf{q}$ is represented by a block diagonal matrix $\lambda = 2,3,...,\lambda^{\max}$, with block matrices of the dimension $(2\lambda+1)\times(2\lambda+1)$:

$$\bar{\mathbf{q}}=\bar{\mathbf{M}}(g)\mathbf{q}, \ \bar{\mathbf{q}}=\begin{pmatrix} \bar{\mathbf{q}}_2(g) \\ \bar{\mathbf{q}}_3(g) \\ \vdots \end{pmatrix}, \ \mathbf{q}=\begin{pmatrix} \mathbf{q}_2(g) \\ \mathbf{q}_3(g) \\ \vdots \end{pmatrix}, \ \bar{\mathbf{M}}(g)=\begin{pmatrix} \bar{\mathbf{M}}_2(g) & \mathbf{0} & \dots \\ \mathbf{0} & \bar{\mathbf{M}}_3(g) & \dots \\ \vdots & \vdots & \ddots \end{pmatrix}, \tag{29}$$

The transformation of pairs of components $\bar{\mathbf{q}}_{\lambda} = (\ \bar{a}_{\lambda\lambda} \ \dots \ \bar{a}_{\lambda 1} \ \bar{a}_{\lambda 0} \ \bar{b}_{\lambda 1} \ \dots \ \bar{b}_{\lambda\lambda}\ )^T$ and $\mathbf{q}_{\lambda} = (\ a_{\lambda\lambda}, \ \dots \ a_{\lambda 1}, \ a_{\lambda 0} \ b_{\lambda 1} \ \dots \ b_{\lambda\lambda}\ )^T$ are connected by the transformations $\bar{\mathbf{q}}_{\lambda} = \bar{\mathbf{M}}_{\lambda}(g)\mathbf{q}_{\lambda}$, with the blocks matrices $\bar{\mathbf{M}}_{\lambda}\equiv\bar{\mathbf{M}}_{\lambda}(g)$ of the following form

$$\bar{\mathbf{M}}_{\lambda}=\begin{pmatrix} & \vdots & \vdots & \vdots & \vdots & \vdots & \\ \dots & C^{\lambda}_{22}+C^{\lambda}_{2-2} & C^{\lambda}_{21}-C^{\lambda}_{2-1} & \sqrt{2}C^{\lambda}_{20} & S^{\lambda}_{21}+S^{\lambda}_{2-1} & S^{\lambda}_{22}-S^{\lambda}_{2-2} & \dots \\ \dots & C^{\lambda}_{12}+C^{\lambda}_{1-2} & C^{\lambda}_{11}-C^{\lambda}_{1-1} & \sqrt{2}C^{\lambda}_{10} & S^{\lambda}_{11}+S^{\lambda}_{1-1} & S^{\lambda}_{12}-S^{\lambda}_{1-2} & \dots \\ \dots & \sqrt{2}C^{\lambda}_{02} & \sqrt{2}C^{\lambda}_{01} & C^{\lambda}_{00} & \sqrt{2}S^{\lambda}_{01} & \sqrt{2}S^{\lambda}_{02} & \dots \\ \dots & -S^{\lambda}_{12}-S^{\lambda}_{1-2} & -S^{\lambda}_{11}+S^{\lambda}_{1-1} & -\sqrt{2}S^{\lambda}_{10} & C^{\lambda}_{11}+C^{\lambda}_{1-1} & C^{\lambda}_{12}-C^{\lambda}_{1-2} & \dots \\ \dots & -S^{\lambda}_{22}-S^{\lambda}_{2-2} & -S^{\lambda}_{21}+S^{\lambda}_{2-1} & -\sqrt{2}S^{\lambda}_{20} & C^{\lambda}_{21}+C^{\lambda}_{2-1} & C^{\lambda}_{22}-C^{\lambda}_{2-2} & \dots \\ & \vdots & \vdots & \vdots & \vdots & \vdots & \end{pmatrix}. \tag{30}$$

where $C^{\lambda}_{\mu\mu'} = \cos(\mu\alpha+\mu'\gamma)d^{\lambda}_{\mu\mu'}(\beta)$, $S^{\lambda}_{\mu\mu'} = \sin(\mu\alpha+\mu'\gamma)d^{\lambda}_{\mu\mu'}(\beta)$, and $d^{\lambda}_{\mu\mu'}(\beta)$ are "reduced" Wigner functions [8]. The inverse transformation is implemented using the transposed matrix $\bar{\mathbf{M}}^{-1} = \bar{\mathbf{M}}^T$ because $\bar{\mathbf{M}}$ is the orthogonal matrix, i.e. $\bar{\mathbf{M}}\bar{\mathbf{M}}^T = \bar{\mathbf{M}}^T\bar{\mathbf{M}} = \mathbf{I}$.

The quadrupole-octupole vibrational Hamiltonian $\hat{H}_v$, having the form of that of a harmonic oscillator, can be written in the intrinsic frame as:

$$\hat{H}^{(0)}_v=\left[-\frac{\hbar}{2B_2}\left(\frac{\partial^2}{\partial(a_{20})^2}+\frac{\partial^2}{\partial(a_{22})^2}\right)+\frac{B_2\omega_2^2}{2}\left((a_{20})^2+(a_{22})^2\right)\right] \tag{31}$$

$$+\left[-\frac{\hbar}{2B_3}\left(\sum_{\mu=0}^{3}\frac{\partial^2}{\partial(a_{3\mu})^2}+\sum_{\mu=1}^{3}\frac{\partial^2}{\partial(b_{3\mu})^2}\right)+\frac{B_3\omega_3^2}{2}\left(\sum_{\mu=0}^{3}(a_{3\mu})^2+\sum_{\mu=1}^{3}(b_{3\mu})^2\right)\right].$$

$$E^{N_{23}}_{\nu}=\hbar\omega_2(n_1+n_2+1)+\hbar\omega_3(n_3+...+n_9+7/2), \quad N_{23}=N_2+N_3=n_1+...+n_9.$$

The corresponding eigenfunctions characterized by a set of quantum numbers $\nu=\{n_1,...,n_9\}$ of 9D harmonic oscillator:

$$\psi^{\tilde{N}}_{\nu}(\mathbf{q}) = \psi^{N_2}_{n_1,n_2}(\mathbf{q}_2)\psi^{N_3}_{n_3,...,n_9}(\mathbf{q}_3)\delta_{N_{23},n_1+...+n_9}, \tag{32}$$

$$\psi^{N_2}_{n_1,n_2}(\mathbf{q}_2) = \psi_{n_1}(\eta_2 a_{20})\psi_{n_2}(\eta_2 a_{22}), \quad \psi^{N_3}_{n_3,...,n_9}(\mathbf{q}_3) = \psi^{N_3}_{n_3n_4n_5n_6}(\mathbf{q}^a_3)\psi^{N_3}_{n_7n_8n_9}(\mathbf{q}^b_3),$$

$$\psi^{N_3}_{n_3n_4n_5n_6}(\mathbf{q}^a_3) = \psi_{n_3}(\eta_3 a_{33})\psi_{n_4}(\eta_3 a_{32})\psi_{n_5}(\eta_3 a_{31})\psi_{n_6}(\eta_3 a_{30}),$$

$$\psi^{N_3}_{n_7n_8n_9}(\mathbf{q}^b_3) = \psi_{n_7}(\eta_3 b_{31})\psi_{n_8}(\eta_3 b_{32})\psi_{n_9}(\eta_3 b_{33}), \quad \eta_{\lambda} = B_{\lambda}\omega_{\lambda}/\hbar,$$

where $\check{N}=(N_2,N_3,N_{23})$. They have to satisfy the orthonormalization conditions

$$\int \psi_{\nu}^{\check{N}}(\mathbf{q})\psi_{\nu'}^{\check{K}'}(\mathbf{q})d\mathbf{q}=\delta_{\check{N},\check{K}'}\delta_{n_1n_1'}...\delta_{n_9n_9'}. \tag{33}$$

3.1. Generalized Projection Operators

To construct the appropriate bases in the space of vibrational functions (32) for the intrinsic point groups $\overline{\mathrm{G}}$ we have to use the action of the generalized projection operators $\hat{\bar{P}}_{ab}^{\Gamma}$ corresponding to the *intrinsic* group:

$$\Psi_{\Gamma ab_0\nu}^{\check{N}}(\mathbf{q})=\hat{\bar{P}}_{ab}^{\Gamma}\psi_{\nu}^{\check{N}}(\mathbf{q})=\frac{\dim(\Gamma)}{\mathrm{card}(G)}\sum_{g\in G}\Delta^{\Gamma}(g)_{ab_0}{}^{\star}\hat{\bar{g}}\psi_{\nu}^{\check{N}}(\mathbf{q})=\frac{\dim(\Gamma)}{\mathrm{card}(G)}\sum_{g\in G}\Delta^{\Gamma}(g)_{ab_0}{}^{\star}\psi_{\nu}^{\check{N}}(\bar{\mathbf{M}}(g)\mathbf{q})$$
$$=\frac{\dim(\Gamma)}{\mathrm{card}(G)}\sum_{g\in G}\Delta^{\Gamma}(g)_{ab_0}{}^{\star}\psi_{n_1,n_2}^{N_2}(\bar{\mathbf{M}}_2(g)\mathbf{q}_2)\psi_{n_3,\dots,n_9}^{N_3=N_{23}-N_2}(\bar{\mathbf{M}}_3(g)\mathbf{q}_3), \tag{34}$$

where the matrices of the representations $\Delta^{\Gamma}(g)_{ab_0}$ obtained from the action of $g\in G$ in the space of basis functions of the Cartesian variables are used.

In the above expressions the labels $N_{23}=n_1+...+n_9$, or $N_{23}=N_2+N_3$, $N_2=n_1+n_2$, $N_3=n_3+...+n_9$, $n_1,n_2=N_2-n_1$ run over the full range $n_1,n_2=0,...,N_2$ and $n_3,n_4,...,n_9$ run over the full range $n_3,...,n_9=0,...,N_3$ that provides the decomposition of the vibration space for a given N_2 and N_3 into the irreducible representations Γ.

Note, that the matrices $\bar{\mathbf{M}}_2(g)$ act on the vector $\mathbf{q}_2=(a_{22},a_{21},a_{20},b_{21},b_{22})^T$ under the condition $a_{21}=b_{21}=b_{22}=0$, that follows from Bohr constrains (18). They transform pairs of components a_{22} and a_{20} into combination of $\bar{a}_{22}$ and $\bar{a}_{20}$. In this case we have

$$\begin{pmatrix}\bar{a}_{22}\\ \bar{a}_{20}\end{pmatrix}=\bar{\mathbf{M}}_2(g)\begin{pmatrix}a_{22}\\ a_{20}\end{pmatrix},\ \bar{\mathbf{M}}_2(C_{4y})=\begin{bmatrix}\frac{1}{2} & \frac{\sqrt{3}}{2}\\ \frac{\sqrt{3}}{2} & -\frac{1}{2}\end{bmatrix}. \tag{35}$$

In turn, for the group $\overline{\mathrm{D}}_{4y}$ the matrices $\bar{\mathbf{M}}_3(g)$ act on the vector $\mathbf{q}_3=(a_{33},a_{32},a_{31},a_{30},b_{31},b_{32},b_{33})^T$ and transform the pairs a_{32}, a_{30} into a combination of $\bar{a}_{32}$ and $\bar{a}_{30}$, a_{33}, a_{31} into a combination of $\bar{a}_{33}$ and $\bar{a}_{31}$ and b_{33}, b_{31} into a combination of $\bar{b}_{33}$ and $\bar{b}_{31}$. An exception is the variable b_{32} which transforms into itself because it is an invariant of the tetrahedral symmetry. In this case we have

$$\begin{pmatrix}\bar{a}_{33}\\ \bar{a}_{32}\\ \bar{a}_{31}\\ \bar{a}_{30}\end{pmatrix}=\bar{\mathbf{M}}_3(g)\begin{pmatrix}a_{33}\\ a_{32}\\ a_{31}\\ a_{30}\end{pmatrix},\ \bar{\mathbf{M}}_3(C_{4y})=\begin{bmatrix}0 & -\frac{\sqrt{6}}{4} & 0 & -\frac{\sqrt{10}}{4}\\ \frac{\sqrt{6}}{4} & 0 & \frac{\sqrt{10}}{4} & 0\\ 0 & -\frac{\sqrt{10}}{4} & 0 & \frac{\sqrt{6}}{4}\\ \frac{\sqrt{10}}{4} & 0 & -\frac{\sqrt{6}}{4} & 0\end{bmatrix}. \tag{36}$$

The vibration basis functions with parameters $a_{20}^o,...,a_{33}^o$ of irr. $A1{=}\Gamma_1$ of $\overline{\mathrm{D}}_{4y}$ have the form

$$\Psi_{[n]}^{N}(a_{20},...,a_{33};a_{20}^o,...,a_{33}^o)=\left[\sum_{i=1}^{8}\Psi_{[n]}^{gi}\right];\quad \Psi_{[n]}^{gi=1,...,4}=\frac{c_i}{8}\psi_{n_1}(a_{22}-a_{22}^o)\psi_{n_2}(a_{20}-a_{20}^o)$$
$$\times\psi_{n_3}(a_{33}+d_{3i}a_{33}^o)\psi_{n_4}(a_{32}+d_{2i}a_{32}^o)\psi_{n_5}(a_{31}+d_{1i}a_{31}^o)\psi_{n_6}(a_{30}+d_{0i}a_{30}^o),$$
$$\Psi_{[n]}^{gi=5,...,8}=\frac{c_i}{8}\psi_{n_1}(\frac{\sqrt{3}}{2}a_{20}+\frac{1}{2}a_{22}-a_{22}^o)\psi_{n_2}(\frac{1}{2}a_{20}-\frac{\sqrt{3}}{2}a_{22}+a_{20}^o)\psi_{n_3}(\frac{\sqrt{10}}{4}a_{30}+\frac{\sqrt{6}}{4}a_{32}+d_{3i}a_{33}^o)$$
$$\times\psi_{n_4}(\frac{\sqrt{10}}{4}a_{31}+\frac{\sqrt{6}}{4}a_{33}+d_{2i}a_{32}^o)\psi_{n_5}(\frac{\sqrt{6}}{4}a_{30}-\frac{\sqrt{10}}{4}a_{32}+d_{1i}a_{31}^o)\psi_{n_6}(\frac{\sqrt{6}}{4}a_{31}-\frac{\sqrt{10}}{4}a_{33}+d_{0i}a_{30}^o),$$

c_1=1, c_2=$(-1)^{n_4+n_6}$, c_3=$(-1)^{n_3+n_5}$, c_4=$(-1)^{n_3+n_4+n_5+n_6}$, c_5=$(-1)^{n_2+n_5+n_6}$, c_6=$(-1)^{n_2+n_4+n_5}$, c_7=$(-1)^{n_2+n_3+n_6}$, c_8=$(-1)^{n_2+n_3+n_4}$, d_{3i}=$(-1)^{[(i+1)/2]}$, d_{2i}=d_{0i}=$(-1)^i$, d_{1i}=$(-1)^{[(i+5)/4]}$, where "$[x]$" means a greatest integer less than or equal to a number x, and $\psi_n(a)$ are normalized 1D oscillator functions.

4. Algorithm for construction of the orthonormal vibrational-rotational basis

Calculation of the parametric vibrational-rotational basis functions of irr. $A1$=Γ_1 of $\overline{\mathrm{D}}_{4y}$.

Step 1. To calculate overlaps,

$$\mathcal{B}^{JN}_{n_1,\ldots,n_6,K,n'_1,\ldots,n'_6,K'} = \langle \Psi^{JN}_{n_1,\ldots,n_6,K} | \Psi^{JN}_{n'_1,\ldots,n'_6,K'} \rangle, \tag{37}$$

$$\Psi^{JN}_{n_1,\ldots,n_6,K}(a_{20},\ldots,a_{33},\Omega) = \sum_{K'=-J}^{J} \left[\sum_{i=1}^{8} \Psi^{g_i}_{n_1,\ldots,n_6}(a_{20},\ldots,a_{33}) D^J_{KK'}(g_i) \right] r^J_{MK'}(\Omega),$$

where $r^J_{MK'}(\Omega) = \sqrt{2J+1} {D^J_{MK'}}^*(\Omega)$, form the finite $(2J+1)S\times(2J+1)S$ dimensional hermitian matrix called the Gramm matrix, which we denote here by $\mathcal{B}$.

Step 2. Solving the eigenvalue problem of the matrix $\mathcal{B}$ allows to find the orthonormal basis in which this matrix is diagonal:

$$\mathcal{B}\mathcal{X}_t(s,K) = \lambda_t \mathcal{X}_t(s,K). \tag{38}$$

For $\lambda_t \neq 0$ one gets the following basis of orthonormal states:

$$|u_t\rangle = C_t \sum_{K'=-J}^{J} \left[\sum_{K=-J}^{J} \sum_{s=1}^{S} \mathcal{X}_t(s,K) \left[\sum_{i=1}^{8} \Psi^{g_i}_s(a_{20},\ldots,a_{33}) D^J_{KK'}(g_i) \right] \right] r^J_{MK'}(\Omega), \tag{39}$$

where J, M, Γ and b fixed, and C_t is the normalization coefficient.

Step 3. To calculate the scalar products:

$$\begin{aligned} \langle u_{t'}|u_t\rangle &= C_{t'}C_t \textstyle\sum_{s',K',s,K} \mathcal{X}_{t'}(s',K')^\star \mathcal{X}_t(s,K) \langle \Psi^{JN}_{n'_1,\ldots,n'_6,K'} | \Psi^{JN}_{n_1,\ldots,n_6,K}\rangle = \\ &= C_{t'}C_t \textstyle\sum_{s,K} \mathcal{X}_{t'}(s,K)^\star \lambda_t \mathcal{X}_t(s,K) = \delta_{t't} C_{t'}C_t\lambda_t \sum_{s,K} |\mathcal{X}_t(s,K)|^2. \end{aligned} \tag{40}$$

It means that the states $|u_t\rangle$ for which the eigenvalue $\lambda_t = 0$ are the zero vectors, and for the cases when $\lambda_t \neq 0$ the normalization coefficient is equal: $C_t = \left(\lambda_t \sum_{s,K} |\mathcal{X}_t(s,K)|^2\right)^{-1/2}$.

Note that because the eigenequation (38) is, in fact, the eigenequation for the matrix representation of the projection operator $\hat{\mathrm{P}}^{\Gamma}_{bb}$ the eigenvalues are equal to $\lambda_t > 0$. In this way we have obtained the required basis in the projected subspace $\mathcal{K}^{\Gamma}_{Jb}$, e.g. K_S from section 2.

Taking into account relation $\langle r^J_{MK'}|r^J_{MK}\rangle = \delta_{KK'}$ we have expression for the matrix $(2J+1)S \times (2J+1)S$ at each $J>0$:

$$B^{JN}_{n_1,\ldots,n_6,K_1;n'_1,\ldots,n'_6,K_2}(a^o_{20},\ldots,a^o_{33}) = \eta_2^{-2}\eta_3^{-4} \sum_{i_1=1}^{\nu} \ldots \sum_{i_6=1}^{\nu} \Bigg(\sum_{s,t=1}^{8} \sum_{K'_1=-J}^{J} D^J_{K_2K'_1}(g_s)^* D^J_{K_1K'_1}(g_t)$$

$$\times \Psi^{g_s}_{111[n_1,\ldots,n_6]}(a_{20}=\frac{x_{i_1}}{\eta_2},\ldots,a_{33}=\frac{x_{i_6}}{\eta_3}) \Psi^{g_t}_{111[n'_1,\ldots,n'_6]}(a_{20}=\frac{x_{i_1}}{\eta_2},\ldots,a_{33}=\frac{x_{i_6}}{\eta_3}) \prod_{k=1}^{6} W_{i_k}\Bigg). \tag{41}$$

During the integration we use Gauss-Hermite quadratures of νth order at $\nu=10$:

$$\int_{-\infty}^{+\infty} e^{-x^2} f(x)dx = \sum_{i=1}^{\nu} w_i f(x_i), \;\Leftrightarrow\; \int_{-\infty}^{+\infty} g(x)dx = \sum_{i=1}^{\nu} W_i g(x_i),\; W_i = w_i e^{x_i^2}. \tag{42}$$

Table 1. Output for $v^{JM}_{1tpq_0}(\Omega)$: $\Gamma = \Gamma_1 = A1$, $J = 0, 2, 4, 5$; t distinguishes equiv.i.r. of D_{4y} point group of intrinsic frame. Here $r^{J\pm}_{MK}(\Omega)$=$(r^J_{MK}(\Omega) \pm r^J_{M-K}(\Omega))/\sqrt{2}$.

J	Γ	p	q_0	t	$v^{JM}_{1tpq_0}(\Omega)$
0	$\Gamma_1 = A1$	1	1	1	$v^{0M}_{1111}(\Omega) = r^0_{M0}(\Omega)$
2	$\Gamma_1 = A1$	1	1	1	$v^{2M}_{1111}(\Omega) = (\sqrt{3}r^2_{M0}(\Omega) + 3r^{2+}_{M2}(\Omega))/(2\sqrt{3})$
4	$\Gamma_1 = A1$	1	1	1	$v^{4M}_{1111}(\Omega) = (\sqrt{35}r^4_{M0}(\Omega) + 2\sqrt{7}r^{4+}_{M2}(\Omega) + 9r^{4+}_{M4}(\Omega))/12$
4	$\Gamma_1 = A1$	1	1	2	$v^{4M}_{1211}(\Omega) = (-2\sqrt{5}r^4_{M0}(\Omega) + 5r^{4+}_{M2}(\Omega))/(3\sqrt{5})$
5	$\Gamma_1 = A1$	1	1	1	$v^{5M}_{1111}(\Omega) = (\sqrt{3}r^{5-}_{M2}(\Omega) - r^{5-}_{M4}(\Omega))/2$

Here x_i are abscissas (zeros of Hermite polynomials) and w_i are weight factors [9].

Example of calculations of pure rotational basis functions of irr. $A1 = \Gamma_1$ of $\overline{\mathrm{D}}_{4y}$ symmetrization point group in zero phonon approximation at $N = 0$ is given in table 1. At $J = 0$ and $a^o_{20} = 7/10, a^o_{30} = 1/4, a^o_{22} = a^o_{31} = a^o_{32} = a^o_{33} = 0$, $\eta_2 = \eta_3 = 1$, Gram matrix for $N = 0, 1$, $s = 1, ..., S$, $S = 7$ have the following eigenvalues $\{\lambda_t\}^7_{t=1} = \{0.9897, 0.2521, 0.0293, 0.8204, 0.0151, -3. \cdot 10^{-17}, -7. \cdot 10^{-19}\}$. From $\lambda_6 = -3. \cdot 10^{-17} \equiv 0$, $\lambda_7 = -7. \cdot 10^{-19} \equiv 0$ follows $\{\mathcal{X}_{s,6}\}^7_{s=1} \equiv 0$, $\{\mathcal{X}_{s,7}\}^7_{s=1} \equiv 0$.

5. DVR algorithm for solving eigenvalue problems in intrinsic frame

One can solve eigenvalue problem describing rotational-vibrational motion of collective nuclear model with potential energy $V(a_{20}, ..., a_{33})$ in intrinsic frame, for example at $J = 0$:

$$(\hat{H}_v - E_v)\psi_v(a_{20}, ..., a_{33}) = 0, \quad \hat{H}_v = \hat{H}^{(0)} + V(a_{20}, ..., a_{33}),$$

using rotational-vibrational basis $u^{(0)}_s$ of irr $A1 = \Gamma_1$ of $\overline{\mathrm{D}}_{4y}$ group from Eq. (39) for the quadrupole-octupole harmonic oscillator

$$(\hat{H}^{(0)} - E^{(0)}_s)u^{(0)}_s = 0, \quad u^{(0)}_s = C_s \sum_{s'=1}^{S} \mathcal{X}_s(s')\psi^{(0)}_{s'}(a_{20}, ..., a_{33}),$$

where scalar product is calculated similarly to Eq. (41) using Gauss-Hermite quadratures

$$\langle u^{(0)}_s | u^{(0)}_{s'} \rangle = \sum_{t=[i_1,...,i_6]} A^T_{st} A_{s't}, \quad A_{s't} = W^{1/2}_{i_1} \times ... \times W^{1/2}_{i_6} \sum_{s''=1}^{S} C_{s''}\psi^{(0)}_{s''}(x_1, ..., x_6)\mathcal{X}_{s'}(s''). \quad (43)$$

We seek solution in the form of the Galerkin expansion over the orthogonal basis functions $u^{(0)}_s(a_{20}, ..., a_{33})$ with unknown coefficients X_{sv}

$$\psi_v(a_{20}, ..., a_{33}) = \sum_{s'=1}^{S'_{max}} u^{(0)}_{s'}(a_{20}, ..., a_{33})X_{s'v}. \quad (44)$$

Averaging over the basis with scalar product (43) leads to the algebraic eigenvalue problem

$$\sum_{s'=1}^{S'_{max}} \left[\langle u^{(0)}_s | \hat{H}^{(0)} | u^{(0)}_{s'} \rangle + \langle u^{(0)}_s | V | u^{(0)}_{s'} \rangle - E_v \delta_{ss'}\right] X_{s'v} = 0. \quad (45)$$

We will solve this problem using transformation to the discrete variable representation (DVR)

$$X_{s'v} = \sum_{t=1}^{S'_{max}} A_{s't} Y_{tv}, \quad \sum_{t=1}^{S'_{max}} A^T_{st} A_{s't} = \delta_{ss'}, \quad \sum_{s=1}^{S'_{max}} A_{st} A^T_{st'} = \delta_{tt'},$$

that was successfully applied for solving eigenvalue problems describing large amplitude molecular vibrations [10]. In the DVR the problem (45) is reduced to the following one with respect to the new unknown eigenvectors $Y_{t'v}$

$$\sum_{t'=1}^{S'_{max}} \left[\left[\sum_{s=1}^{S'_{max}} A^T_{st} E^{(0)}_s A_{st'} \right] + V\left(a_{20} = \frac{x_{i_1}}{\eta_2}, ..., a_{33} = \frac{x_{i_3}}{\eta_3}\right) \delta_{tt'} - E_v \delta_{tt'} \right] Y_{t'v} = 0.$$

Here matrix of potential energy $V\left(a_{20} = \frac{x_{i_1}}{\eta_2}, ..., a_{33} = \frac{x_{i_3}}{\eta_3}\right)$ is diagonal matrix which elements is calculated in Gauss-Hermitian nodes (42), while dense real-symmetric matrix of kinetic energy is calculated as product of known eigenvalues $E^{(0)}_s$ and known symmetric matrixes A_{st} consisting of basis eigenfunctions in Gauss-Hermitian nodes from Eq. (43). Thus, in the DVR there is no problem of cumbersome numerical calculations of the multiple integrals $\langle u^{(0)}_s | V | u^{(0)}_{s'} \rangle$. However, it is need to solve the interpolation problem of the potential energy $V\left(a_{20} = \frac{x_{i_1}}{\eta_2}, ..., a_{33} = \frac{x_{i_3}}{\eta_3}\right)$ in Gauss-Hermitian nodes if it is given in a tabular form. Generalization of the above algorithm in the case of $J > 0$ is given by direct application of the expression Eq. (39) of the rotational-vibrational basis $u^{(0)}_{sK}$ at $-J < K < J$ with corresponding modification of the scalar product Eq. (43) similar to Eq. (41).

6. Resume

New symbolic algorithm implemented in computer algebra system for generating irreducible representations of the point symmetry groups in the rotor + shape vibrational space of a nuclear collective model in the intrinsic frame was presented. The efficiency of the algorithm was investigated by calculating the symmetrized bases of irreducible representations *symmetrization group* $\bar{G}_S$*:* subgroup $\overline{\mathrm{D}}_{4y}$ of octahedral group in the intrinsic frame of a quadrupole-octupole nuclear collective model. The DVR algorithm was proposed for solving eigenvalue problems, describing vibrational-rotational motion of the collective nuclear model in intrinsic frame, that may be solved using the fast eigensolver for dense real-symmetric matrices[11] and MPI technology.

7. Acknowledgment

This work has bee partially supported by the Polish–French COPIN collaboration under the project 04-113, the Bogoliubov-Infeld JINR program and the grant RFBR 17–01–00298.

References

[1] Eisenberg J M and Greiner W 1970 *Nuclear Theory* Vol. 1 (Amsterdam: North-Holland)
[2] Chen J Q, Ping J and Wang F 2002 *Group representation theory for physicists* (Singapore: World Sci.)
[3] Barut A and Rączka R 1977 *Theory of Group Representations and Applications* (Warszawa: PWN)
[4] Gusev A A et al 2015 *Lecture Notes Comp. Sci.* **9301** 166
[5] Gusev A A et al 2016 *Lecture Notes Comp. Sci.* **9890** 228
[6] Cornwell J F 1984 *Group theory in physics* (New York: Academic Press)
[7] Szulerecka A, Dobrowolski A and Góźdź A 2014 *Phys. Scr.* **89** 054033
[8] Varshalovich D A, Moskalev A N and Khersonskii V K 1989 *Quantum theory of angular momentum* (Singapore: World Sci.)
[9] Abramowitz M and Stegun I A 1965 *Handbook of Mathematical Functions* (New York: Dover)
[10] Bacic Z, Whitnell R M, Brown D and Light J C 1988 *Comput. Phys. Commun.* **51** 35
[11] Bunge C F 2001 *Comput. Phys. Commun.* **138** 92

IOP Conf. Series: Journal of Physics: Conf. Series **804** (2017) 012019 doi:10.1088/1742-6596/804/1/012019

l–conformal Galilei symmetry in the general relativistic context

D. Chernyavsky and A. Galajinsky

Laboratory of Mathematical Physics, Tomsk Polytechnic University, 634050 Tomsk, Lenin Ave. 30, Russian Federation

E-mail: chernyavsky@tpu.ru, galajin@tpu.ru

Abstract. The method of constructing Ricci–flat metrics with l–conformal Galilei symmetry is discussed.

1. Introduction

In recent years there has been extensive investigation of the nonrelativistic conformal algebras [1]–[9]. The conformal extension of the Galilei algebra is parametrized by a (half)integer parameter l which gives rise to the name l–conformal Galilei algebra [10]. So far most of the studies focused on the construction of various dynamical realizations with a particular emphasis on the issues of the presence of higher derivative terms and functional independence of the acceleration generators in the algebra. Applications of the l–conformal Galilei symmetry within the general relativistic context are unknown. The goal of this note is to adjust the conventional group–theoretic construction so as to build Ricci–flat metrics with the l–conformal Galilei isometry group.

In Sect. 2 we consider Maurer–Cartan one–forms associated with the l–conformal Galilei algebra. Sect. 3 is devoted to the construction of a Ricci–flat spacetime of the ultrahyperbolic signature which enjoys the l–conformal Galilei isometry group. We summarize our results in Sect. 4.

2. l–conformal Galilei algebra and Maurer–Cartan one–forms

The l–conformal Galilei algebra involves the generators of time translation H, dilatation D, special conformal transformation K, spatial rotations M_{ij} (with $i = 1, \ldots, d$), spatial translations $C_i^{(0)}$, Galilei boosts $C_i^{(1)}$ and accelerations $C_i^{(\alpha)}$ with $\alpha = 2, \ldots, 2l$. The structure relations of the algebra read

$$
\begin{aligned}
&[H, D] = iH, \qquad [H, K] = 2iD, \qquad [D, K] = iK, \\
&[H, C_i^{(n)}] = inC_i^{(n-1)}, \qquad [D, C_i^{(n)}] = i(n-l)C_i^{(n)}, \qquad [K, C_i^{(n)}] = i(n-2l)C_i^{(n+1)}, \\
&[M_{ij}, C_k^{(n)}] = -i(\delta_{ik}C_j^{(n)} - \delta_{jk}C_i^{(n)}), \; [M_{ij}, M_{kl}] = -i(\delta_{ik}M_{jl} + \delta_{jl}M_{ik} - \delta_{il}M_{jk} - \delta_{jk}M_{il}),
\end{aligned} \tag{1}
$$

where $n = 0, \ldots, 2l$.

Let us choose a subgroup L generated by D and M_{ij} and consider the coset space

$$\tilde{G} = e^{itH} e^{irK} e^{ix_i^{(n)} C_i^{(n)}} \times L \tag{2}$$

parametrized by the coordinates t, r and $x_i^{(n)}$. Left multiplication by the group element $g = e^{iaH} e^{ibK} e^{icD} e^{i\lambda_i^{(n)} C_i^{(n)}} e^{\frac{i}{2}\omega_{ij} M_{ij}}$ determines the action of the group on the coset

$$\begin{aligned} t' &= t + a + bt^2 + ct, \qquad r' = r + b(1 - 2tr) - cr, \\ x_i'^{(n)} &= x_i^{(n)} - 2bt(n-l)x_i^{(n)} - c(n-l)x_i^{(n)} - \sigma_{ij} x_j^{(n)} + \\ &+ \sum_{s=0}^{n} \sum_{m=s}^{2l} \frac{(-1)^{n-s} m!(2l-s)!}{s!(m-s)!(n-s)!(2l-n)!} t^{m-s} r^{n-s} \lambda_i^{(m)}, \end{aligned} \tag{3}$$

where $a, b, c, \lambda_i^{(n)}$ and $\sigma_{ij} = -\sigma_{ji}$ are infinitesimal parameters corresponding to the time translations, special conformal transformations, dilatations, vector generators in the algebra, and spatial rotations, respectively.

The Maurer-Cartan one–forms $\tilde{G}^{-1} d\tilde{G} = i(\omega_H H + \omega_K K + \omega_D D + \omega_i^{(n)} C_i^{(n)})$

$$\begin{aligned} \omega_i^{(n)} &= dx_i^{(n)} + 2r(n-l)x_i^{(n)} dt - (n+1)x_i^{(n+1)} dt - (n-2l-1)x_i^{(n-1)}(r^2 dt + dr), \\ \omega_H &= dt, \qquad \omega_K = r^2 dt + dr, \qquad \omega_D = -2r dt, \end{aligned} \tag{4}$$

in which

$$x_i^{(-1)} = x_i^{(2l+1)} = 0 \tag{5}$$

are the building blocks for constructing a metric with the l–conformal Galilei isometry group. For what follows it proves useful to introduce a new temporal coordinate

$$t = \frac{1}{2}\left(\tilde{t} + \frac{1}{r}\right) \tag{6}$$

which yields

$$\begin{aligned} \tilde{\omega}_i^{(n)} &= dx_i^{(n)} + \left(r(n-l)x_i^{(n)} - \frac{1}{2}(n+1)x_i^{(n+1)}\right)\left(d\tilde{t} - \frac{dr}{r^2}\right) - \\ &- \frac{1}{2}(n-2l-1)x_i^{(n-1)} r^2 \left(d\tilde{t} + \frac{dr}{r^2}\right). \end{aligned} \tag{7}$$

3. l–conformal Galilei symmetry in the general relativistic context

In order to construct a metric which holds invariant under the action of the l–conformal Galilei group, we first note that the one–forms $\tilde{\omega}_i^{(n)}$ are invariant under the time translation and $\lambda_i^{(n)}$–transformations, while with respect to the dilatations, special conformal transformations and rotations they transform homogeneously

$$\begin{aligned} &\tilde{\omega}_i'^{(n)} = (1 - c(n-l))\tilde{\omega}_i^{(n)}, \qquad \tilde{\omega}_i'^{(n)} = \left(1 - b(n-l)\left(\tilde{t} + \frac{1}{r}\right)\right)\tilde{\omega}_i^{(n)}, \\ &\tilde{\omega}_i'^{(n)} = (\delta_{ij} - \sigma_{ij})\tilde{\omega}_j^{(n)}. \end{aligned} \tag{8}$$

As a result, the quadratic form

$$ds^2 = \left(r^2 d\tilde{t}^2 - \frac{dr^2}{r^2}\right) + S_{n,m}\tilde{\omega}_i^{(n)}\tilde{\omega}_i^{(m)}, \qquad S_{n,m} = S_{m,n} = k_n \delta_{n+m,2l}, \tag{9}$$

with k_n to be fixed below, has the l–conformal Galilei isometry group.

In order to promote (9) to a Ricci–flat metric, we minimally extend it by an extra coordinate y

$$ds^2 = \alpha(y)\left(r^2 d\tilde{t}^2 - \frac{dr^2}{r^2}\right) + S_{n,m}\tilde{\omega}_i^{(n)}\tilde{\omega}_i^{(m)} + \epsilon dy^2, \tag{10}$$

where $\alpha(y)$ is a function to be fixed below and $\epsilon = \pm 1$. It is assumed that y remains intact under the l–conformal Galilei transformations so that the metric maintains the symmetry of its predecessor (9). The vacuum Einstein equations then fix $\alpha(y)$

$$\alpha(y) = c_1 y^2, \quad c_1 = \epsilon + \frac{l(l+1)(2l+1)d\epsilon}{6} - \frac{d\epsilon}{4}\sum_{p=0}^{2l-1}\sum_{q=1}^{2l}(p+1)(q-2l-1)\tilde{S}^{p+1,q-1}S_{p,q}, \tag{11}$$

and impose the recurrence relations on the form of the components $S_{n,m}$

$$\begin{aligned} & n^2 S_{n-1,2l-n+1} + (2l-n)^2 S_{n+1,2l-n-1} \\ & - \left[(n+1)^2\tilde{S}^{n+1,2l-n-1} + (n-2l-1)^2\tilde{S}^{n-1,2l-n+1}\right](S_{n,2l-n})^2 = 0, \end{aligned} \tag{12}$$

where $n = 0,\ldots,2l$. In Eq. (12) $\tilde{S}^{n,m} = \tilde{S}^{m,n} = \frac{1}{S_{n,m}} = \frac{1}{k_n}\delta_{n+m,2l}$ (no sum in n) stands for the inverse of $S_{n,m}$ and it is assumed that

$$S_{2l+1,n} = S_{-1,n} = \tilde{S}^{2l+1,n} = \tilde{S}^{-1,n} = 0. \tag{13}$$

It is straightforward to verify that Eqs. (12) algebraically relate all the components $S_{n,m}$ to $S_{0,2l}$, the latter being unspecified.

4. Conclusion

Let us discuss the status of the metrics constructed above. Given the spatial dimension d in which the original l–conformal Galilei algebra (1) is realized and the value of the (half)integer parameter l, the AdS_2–spacetime in (10) is extended by $(2l+1)d$ extra dimensions parametrized by the coordinates $x_i^{(0)},\ldots,x_i^{(2l)}$, $i = 1,\ldots,d$. For half–integer l, these are split into $\frac{(2l+1)d}{2}$ spatial and $\frac{(2l+1)d}{2}$ temporal dimensions, while for integer l one reveals ld temporal and $(l+1)d$ spatial dimensions or vice versa depending on which sign is chosen for the components $S_{n,m}$ linked to $S_{0,2l}$. Depending on whether $\epsilon = 1$ or $\epsilon = -1$ is chosen in the last term in (10), the remaining coordinate y brings about one more temporal or spatial dimension. The resulting Ricci–flat spacetime is thus $[(2l+1)d+3]$–dimensional and of the ultrahyperbolic signature.

Acknowledgements

This work was supported by the MSE program Nauka under the project 3.825.2014/K.

References

[1] Fedoruk S, Ivanov E and Lukierski J 2011 *Phys. Rev.* D **83** 085013 (*Preprint* hep-th/1101.1658)
[2] Duval C, Horváthy P 2011 *J. Phys.* A **44** 335203 (*Preprint* hep-th/1104.1502)
[3] Galajinsky A, Masterov I 2011 *Phys. Lett.* B **702** 265 (*Preprint* hep-th/1104.5115)
[4] Gomis J, Kamimura K 2012 *Phys. Rev.* D **85** 045023 (*Preprint* hep-th/1109.3773)
[5] Galajinsky A, Masterov I 2013 *Nucl. Phys.* B **866** 212 (*Preprint* hep-th/1208.1403)
[6] Galajinsky A, Masterov I 2013 *Phys. Lett.* B **723** 190 (*Preprint* hep-th/1303.3419)
[7] Andrzejewski K, Gonera J 2013 *Phys. Lett.* B **721** (2013) 319
[8] Andrzejewski K, Gonera J, Kosinski P and Maslanka P 2013 *Nucl. Phys.* B **876** 309 (*Preprint* hep-th/1305.6805)
[9] Andrzejewski K, Galajinsky A, Gonera J and Masterov I 2014 *Nucl. Phys.* B **885** 150 (*Preprint* hep-th/1402.1297)
[10] Negro J, del Olmo M and Rodriguez-Marco A 1997 *J. Math. Phys.* **38** 3786.

IOP Conf. Series: Journal of Physics: Conf. Series **804** (2017) 012020 doi:10.1088/1742-6596/804/1/012020

Off- and on-shell harmonic superspaces for $6D$ SYM theories

Evgeny Ivanov

Bogoliubov Laboratory of Theoretical Physics, JINR, 141980, Dubna, Moscow Region, Russia

E-mail: eivanov@theor.jinr.ru

Abstract. It is a brief account of the harmonic superspace formulations of $\mathcal{N} = (1,0)$ and $\mathcal{N} = (1,1)$ SYM theories in six dimensions. The on-shell $\mathcal{N} = (1,1)$ harmonic superspace is argued to provide an efficient tool of constructing candidate counterterms and other invariants of $\mathcal{N} = (1,1)$ SYM. It allows one, e.g., to find out an essential difference between the single- and double-trace dimension $\underline{d = 10}$ invariants, which could be capable to explain the absence of the three-loop double-trace (non-planar) counterterms in this theory. The defining superspace constraints of $\mathcal{N} = (1,1)$ SYM are solved in terms of $\mathcal{N} = (1,0)$ harmonic superfields.

1. Motivations

During recent years, much attention was paid to the maximally extended (with 16 supercharges) supersymmetric gauge theories in diverse dimensions (see, e.g., [1]),

$$\mathcal{N} = 4\,,\ 4D \quad \Longrightarrow \quad \mathcal{N} = (1,1)\,,\ 6D \quad \Longrightarrow \quad \mathcal{N} = (1,0)\,,\ 10D\,.$$

The renowned $\mathcal{N} = 4, 4D$ SYM theory was the first example of an UV *finite* theory. Perhaps, it is also *completely integrable* [2]. The $\mathcal{N} = (1,1), 6D$ SYM is not renormalizable by formal counting (the coupling constant is dimensionful) but it is also expected to feature unique properties. In particular, in the perturbation theory it exhibits "dual conformal symmetry", like its $4D$ counterpart which respects "dual superconformal symmetry" [3]. It can be viewed as an effective theory for description of some particular low energy sectors of string theory, such as D5-brane dynamics. The quantum effective action of D5-brane as a generalization of the microscopic $\mathcal{N} = (1,1)$ SYM action was conjectured to be of non-abelian Born-Infeld type [4, 5]. The $\mathcal{N} = (1,1)$ SYM is anomaly free, as opposed to $\mathcal{N} = (1,0)$ SYM theory.

The $\mathcal{N} = (1,1)$ and $\mathcal{N} = (1,0)$ SYM theories provide an appropriate "laboratory" for studying $\mathcal{N} = 8$ supergravity and its some lower $\mathcal{N}$ descendants, which are also non-renormalizable according by the standard counting.

The latest perturbative calculations in $\mathcal{N} = (1,1)$ SYM showed plenty of unexpected cancelations of the UV divergencies. The theory is UV finite up to 2 loops, while at 3 loops only a single-trace (planar) counterterm of canonical dim 10 is needed. The permissible double-trace (non-planar) counterterms do not appear [6] - [8]. Various arguments to explain this

IOP Conf. Series: Journal of Physics: Conf. Series **804** (2017) 012020 doi:10.1088/1742-6596/804/1/012020

phenomenon were suggested [9] - [12], though the complete understanding of it is still lacking. One could expect the existence of some new non-renormalization theorems in this connection.

The maximal off-shell supersymmetry that one can achieve in $6D$ is $\mathcal{N} = (1,0)$ supersymmetry. The natural off-shell formulation of $\mathcal{N} = (1,0)$ SYM theory is given in harmonic $\mathcal{N} = (1,0), 6D$ superspace [13, 14], generalizing the harmonic $\mathcal{N} = 2, 4D$ one [15, 16]. The harmonic $6D$ formulations were further worked out in [17] - [20] and [21]. The $\mathcal{N} = (1,1)$ SYM theory in the harmonic formalism is a hybrid of two $\mathcal{N} = (1,0)$ theories, [$\mathcal{N} = (1,1)$ SYM] = [$\mathcal{N} = (1,0)$ SYM] + [$6D$ hypermultiplets], with the second hidden $\mathcal{N} = (0,1)$ supersymmetry. The natural question is as to how to construct higher-dimension $\mathcal{N} = (1,1)$ invariants in the $\mathcal{N} = (1,0)$ superfield approach.

One way to approach this issue is the "brute-force" method. One starts with the appropriate dimension $\mathcal{N} = (1,0)$ SYM invariant and then completes it to $\mathcal{N} = (1,1)$ invariant step by step, adding the proper hypermultiplet terms. This approach is rather cumbersome.

Some simplifications arise due to the fact that for finding superfield counterterms it suffices to stay on the mass shell. In a recent paper [21] there was suggested a new approach to constructing higher-dimension $\mathcal{N} = (1,1)$ invariants. It makes use of the concept of the *on-shell* $\mathcal{N} = (1,1)$ harmonic superspace with the double set of the harmonic variables $u^{\pm}_i, u^{\hat{\pm}}_A, i = 1,2; A = 1,2$ [22]. The novel point of the consideration in [21] is solving the $\mathcal{N} = (1,1)$ SYM constraints [23, 24] in terms of $\mathcal{N} = (1,0)$ superfields. The dimension $\underline{d = 8}$ and $\underline{d = 10}$ invariants were constructed in a simple way and an essential difference between the single- and double-trace dimension $\underline{d = 10}$ invariants was observed. The present contribution is a brief survey of the $6D$ harmonic methods, with the main emphasis on their recent applications in [21].

2. $6D$ superspaces and superfields

2.1. $6D$ superspaces

- The customary $\mathcal{N} = (1,0), 6D$ superspace is parametrized by the following coordinate set:

$$z = (x^M, \theta^a_i), \quad M = 0, \ldots, 5\,,\ a = 1, \ldots, 4\,,\ i = 1,2\,, \tag{1}$$

- The harmonic $\mathcal{N} = (1,0)$ superspace amounts to adding $SU(2)$ harmonics to (1):

$$Z := (z, u) = (x^M, \theta^a_i, u^{\pm i})\,, \quad u^-_i = (u^+_i)^*, u^{+i}u^-_i = 1\,,\ u^{\pm i} \in SU(2)_R/U(1)\,. \tag{2}$$

- The *analytic* $\mathcal{N} = (1,0)$ superspace forms an invariant subspace in (2):

$$\zeta := (x^M_{(\mathrm{an})}, \theta^{+a}, u^{\pm i}) \subset Z\,, \quad x^M_{(\mathrm{an})} = x^M + \frac{i}{2}\theta^a_k \gamma^M_{ab} \theta^b_l u^{+k} u^{-l}, \quad \theta^{\pm a} = \theta^a_i u^{\pm i}\,. \tag{3}$$

The differential operators in the analytic basis $Z_A := (x^M_{(\mathrm{an})}, \theta^{+a}, u^{\pm i}, \theta^{-a})$ are defined according to

$$\begin{aligned}
&D^+_a = \partial_{-a}\,,\ D^-_a = -\partial_{+a} - 2i\theta^{-b}\partial_{ab}\,,\ D^0 = u^{+i}\frac{\partial}{\partial u^{+i}} - u^{-i}\frac{\partial}{\partial u^{-i}} + \theta^{+a}\partial_{+a} - \theta^{-a}\partial_{-a}\\
&D^{++} = \partial^{++} + i\theta^{+a}\theta^{+b}\partial_{ab} + \theta^{+a}\partial_{-a}\,,\ D^{--} = \partial^{--} + i\theta^{-a}\theta^{-b}\partial_{ab} + \theta^{-a}\partial_{+a}\,,\\
&\partial_{\pm a}\theta^{\pm b} = \delta^b_a\,, \quad \partial^{++} = u^{+i}\frac{\partial}{\partial u^{-i}}, \quad \partial^{--} = u^{-i}\frac{\partial}{\partial u^{+i}}\,.
\end{aligned} \tag{4}$$

2.2. $6D$ superfields

The fundamental quantity of $\mathcal{N} = (1,0)$ SYM theory is the analytic gauge connection $V^{++}(\zeta)$

$$\nabla^{++} = D^{++} + V^{++}\,, \quad \delta V^{++} = -\nabla^{++}\Lambda\,, \quad \Lambda = \Lambda(\zeta)\,. \tag{5}$$

The second harmonic (non-analytic) connection $V^{--}(Z)$ entering the covariantized harmonic derivative ∇^{--},

$$\nabla^{--} = D^{--} + V^{--}\,, \quad \delta V^{--} = -\nabla^{--}\Lambda\,,$$

is expressed through V^{++} from the harmonic "flatness" condition

$$\begin{aligned}&[\nabla^{++}, \nabla^{--}] = D^0 \Leftrightarrow D^{++}V^{--} - D^{--}V^{++} + [V^{++}, V^{--}] = 0\\ &\Rightarrow V^{--} = V^{--}(V^{++}, u^{\pm})\,.\end{aligned} \tag{6}$$

The off-shell contents of $\mathcal{N} = (1,0)$ SYM theory is revealed in the Wess-Zumino gauge for the analytic gauge potential:

$$V^{++} \;=\; \theta^{+a}\theta^{+b}A_{ab} + 2(\theta^+)^3_a\lambda^{ai}u^-_i - 3(\theta^+)^4\mathcal{D}^{ik}u^-_i u^-_k\,. \tag{7}$$

Here A_{ab} is the gauge field, λ^{ai} is the gaugino and $\mathcal{D}^{ik} = \mathcal{D}^{ki}$ are the auxiliary fields.

The $\mathcal{N} = (1,0)$ SYM covariant derivatives are defined as

$$\begin{aligned}&\nabla^-_a = [\nabla^{--}, D^+_a] = D^-_a + \mathcal{A}^-_a\,, \qquad \nabla_{ab} = \frac{1}{2i}[D^+_a, \nabla^-_b] = \partial_{ab} + \mathcal{A}_{ab}\,,\\ &\mathcal{A}^-_a(V) = -D^+_a V^{--}\,, \quad \mathcal{A}_{ab}(V) = \frac{i}{2}D^+_a D^+_b V^{--}\,,\\ &[\nabla^{++}, \nabla^-_a] = D^+_a\,, \quad [\nabla^{++}, D^+_a] = [\nabla^{--}, \nabla^-_a] = [\nabla^{\pm\pm}, \nabla_{ab}] = 0\,.\end{aligned} \tag{8}$$

The covariant superfield strengths are constructed by the appropriate connections

$$\begin{aligned}&[D^+_a, \nabla_{bc}] \;=\; \frac{i}{2}\varepsilon_{abcd}W^{+d}\,, \qquad [\nabla^-_a, \nabla_{bc}] \;=\; \frac{i}{2}\varepsilon_{abcd}W^{-d}\,,\\ &W^{+a} = -\frac{1}{6}\varepsilon^{abcd}D^+_b D^+_c D^+_d V^{--}\,, \quad W^{-a} := \nabla^{--}W^{+a}\,,\\ &\nabla^{++}W^{+a} = \nabla^{--}W^{-a} \;=\; 0\,, \quad \nabla^{++}W^{-a} = W^{+a}\,,\\ &D^+_b W^{+a} = \delta^a_b F^{++}\,, \quad F^{++} = \frac{1}{4}D^+_a W^{+a} = (D^+)^4 V^{--}\,,\\ &\nabla^{++}F^{++} = 0\,, \quad D^+_a F^{++} = 0\,.\end{aligned} \tag{9}$$

The hypermultiplet is described off shell by an analytic superfield $q^{+A}(\zeta)\,, (A = 1, 2),$:

$$q^{+A}(\zeta) = q^{iA}(x)u^+_i - \theta^{+a}\psi^A_a(x) + \text{ An infinite tail of auxiliary fields}\,. \tag{10}$$

2.3. $\mathcal{N} = (1,0)$ *superfield actions*

The $\mathcal{N} = (1,0)$ SYM action was constructed by Boris Zupnik [14]:

$$\begin{aligned}&S^{SYM} = \frac{1}{f^2}\sum_{n=1}^{\infty}\frac{(-1)^{n+1}}{n}\mathrm{Tr}\int d^6x\, d^8\theta\, du_1 \ldots du_n \frac{V^{++}(z, u_1)\ldots V^{++}(z, u_n)}{(u^+_1 u^+_2)\ldots(u^+_n u^+_1)}\,,\\ &\delta S^{SYM} = 0 \Rightarrow F^{++} = 0\,.\end{aligned} \tag{11}$$

Here, $(u^+_1 u^+_2)^{-1}, \ldots (u^+_n u^+_1)^{-1}$ are harmonic distributions [16].

The hypermultiplet action, with q^{+A} in adjoint representation of the gauge group, is written down as

$$\begin{aligned}&S^q = -\frac{1}{2f^2}\mathrm{Tr}\int d\zeta^{-4} q^{+A}\nabla^{++}q^+_A\,, \quad \nabla^{++}q^+_A = D^{++}q^+_A + [V^{++}, q^+_A]\,,\\ &\delta S^q = 0 \Rightarrow \nabla^{++}q^{+A} = 0\,.\end{aligned} \tag{12}$$

The $\mathcal{N}=(1,0)$ superfield form of the $\mathcal{N}=(1,1)$ SYM action is a sum of the two superfield actions given above:

$$S^{(V+q)} = S^{SYM} + S^{q} = \frac{1}{f^2}\Big(\int dZ\mathcal{L}^{\rm SYM} - \frac{1}{2}{\rm Tr}\int d\zeta^{-4} q^{+A}\nabla^{++}q^{+}_{A}\Big),$$
$$\delta S^{(V+q)} = 0 \;\Rightarrow\; F^{++} + \frac{1}{2}[q^{+A}, q^{+}_{A}] = 0\,, \quad \nabla^{++}q^{+A} = 0\,. \tag{13}$$

It is invariant under the second hidden $\mathcal{N}=(0,1)$ supersymmetry acting as:

$$\delta V^{++} = \epsilon^{+A}q^{+}_{A}\,, \quad \delta q^{+A} = -(D^{+})^4(\epsilon^{-}_{A}V^{--})\,, \quad \epsilon^{\pm}_{A} = \epsilon_{aA}\theta^{\pm a}\,. \tag{14}$$

These transformations have the correct closure among themselves and with the manifest $\mathcal{N}=(1,0)$ supersymmetry only on shell.

3. Higher-dimensional invariants

3.1. *Dimension* $\underline{d=6}$

In the pure $\mathcal{N}=(1,0)$ SYM theory the $\underline{d=6}$ invariant is defined uniquely [17]:

$$S^{(6)}_{SYM} = \frac{1}{2g^2}{\rm Tr}\int d\zeta^{-4}du\,\left(F^{++}\right)^2 \;\sim\; {\rm Tr}\int d^6x[(\nabla^M F_{ML})^2 + \ldots]. \tag{1}$$

It vanishes on shell, when $F^{++}=0$. Using the results of [18], its $\mathcal{N}=(1,1)$ completion is defined up to a real parameter

$$\mathcal{L}^{d=6} = \frac{1}{2g^2}{\rm Tr}\int dud\zeta^{-4}\left(F^{++} + \frac{1}{2}[q^{+A}, q^{+}_{A}]\right)\left(F^{++} + 2\beta[q^{+A}, q^{+}_{A}]\right). \tag{2}$$

But it vanishes by itself on the full $\mathcal{N}=(1,1)$ SYM mass shell! This proves the *one-loop finiteness* of $\mathcal{N}=(1,1)$ SYM theory.

3.2. *Dimension* $\underline{d=8}$

All superfield operators of the canonical dimension $\underline{d=8}$ in the $\mathcal{N}=(1,0)$ SYM theory vanish on shell, in accord with the statement of ref. [24]. Can this conclusion be changed upon adding the hypermultiplet terms? We have checked that there exist no $\mathcal{N}=(1,0)$ off-shell invariants of the dimension $\underline{d=8}$ which would respect the on-shell $\mathcal{N}=(1,1)$ invariance.

Surprisingly, the $\underline{d=8}$ superfield expression which is non-vanishing on shell and respects the on-shell $\mathcal{N}=(1,1)$ supersymmetry can be constructed by *giving up* the demand of *off-shell* $\mathcal{N}=(1,0)$ supersymmetry.

An example of such an invariant in $\mathcal{N}=(1,0)$ SYM theory is very simple

$$\tilde{S}^{(8)}_1 \sim {\rm Tr}\int d\zeta^{-4}\,\varepsilon_{abcd}W^{+a}W^{+b}W^{+c}W^{+d}\,. \tag{3}$$

Indeed, $D^{+}_{a}W^{+b} = \delta^{b}_{a}F^{++}$, which vanishes on shell, with $F^{++}=0$. Thus, W^{+a} is on-shell analytic, for which reason the above action respects $\mathcal{N}=(1,0)$ supersymmetry on shell. Also, an analogous double-trace on-shell invariant exists.

These invariants possess $\mathcal{N}=(1,1)$ completions. For (3) such a completion reads

$$\begin{aligned}\mathcal{L}^{+4}_{(1,1)} &= {\rm Tr}_{(S)}\Big\{\frac{1}{4}\varepsilon_{abcd}W^{+a}W^{+b}W^{+c}W^{+d} + 3iq^{+A}\nabla_{ab}q^{+}_{A}W^{+a}W^{+b}\\ &- q^{+A}\nabla_{ab}q^{+}_{A}\,q^{+B}\nabla^{ab}q^{+}_{B} - W^{+a}[D^{+}_{a}q^{-}_{A}, q^{+}_{B}]q^{+A}q^{+B}\\ &- \frac{1}{2}[q^{+C}, q^{+}_{C}][q^{-}_{A}, q^{+}_{B}]q^{+A}q^{+B}\Big\}.\end{aligned} \tag{4}$$

Here, $\mathrm{Tr}_{(S)}$ stands for the *symmetrized* trace. This Lagrangian is analytic, $D^+_a \mathcal{L}^{+4}_{(1,1)} = 0$, on the total mass shell $F^{++} + \frac{1}{2}[q^{+A}, q^+_A] = 0$, $\nabla^{++} q^{+A} = 0$, hence it is on-shell $\mathcal{N} = (1,1)$ supersymmetric.

Though the nontrivial on-shell $\underline{d=8}$ invariants exist, the perturbative expansion for the amplitudes in the $\mathcal{N} = (1,1)$ SYM theory does not involve divergences at the two-loop level. The reason is that these $\underline{d=8}$ invariants do *not* possess the full off-shell $\mathcal{N} = (1,0)$ supersymmetry which the physically relevant counterterms should obey.

4. $\mathcal{N} = (1,1)$ on-shell harmonic superspace

Apart from the fact that the $\underline{d=8}$ terms mentioned above cannot come out as counterterms in $\mathcal{N} = (1,1)$ SYM theory, they can arise, e.g., as quantum corrections to the effective Wilsonian action. For the pure $\mathcal{N} = (1,0)$ SYM theory this was recently observed in [20]. It was desirable to have some systematic way of constructing such higher-order $\mathcal{N} = (1,1)$ invariants. This proves to be possible within the on-shell harmonic $\mathcal{N} = (1,1)$ superspace.

Let us start by extending the $\mathcal{N} = (1,0)$ superspace to the $\mathcal{N} = (1,1)$ one,

$$z = (x^{ab}, \theta^a_i) \;\Rightarrow\; \hat{z} = (x^{ab}, \theta^a_i, \hat{\theta}^A_a)\,. \tag{1}$$

Then we introduce the gauge-covariantized spinor derivatives,

$$\nabla^i_a = \frac{\partial}{\partial \theta^a_i} - i\theta^{bi}\partial_{ab} + \mathcal{A}^i_a\,, \quad \hat{\nabla}^{aA} = \frac{\partial}{\partial \hat{\theta}_{Aa}} - i\hat{\theta}^A_b \partial^{ab} + \hat{\mathcal{A}}^{aA}\,. \tag{2}$$

The superspace constraints defining the $\mathcal{N} = (1,1)$ SYM theory can be then written as follows [23, 24]:

$$\begin{aligned} &\{\nabla^{(i}_a, \nabla^{j)}_b\} = \{\hat{\nabla}^{a(A}, \hat{\nabla}^{bB)}\} = 0\,, \quad \{\nabla^i_a, \hat{\nabla}^{bA}\} = \delta^b_a \phi^{iA} \\ &\Rightarrow \quad \nabla^{(i}_a \phi^{j)A} = \hat{\nabla}^{a(A}\phi^{B)i} = 0 \quad (\text{By Bianchis})\,. \end{aligned} \tag{3}$$

As the next step, we introduce the $\mathcal{N} = (1,1)$ harmonic superspace [22],

$$Z = (x^{ab}, \theta^a_i, u^\pm_k) \;\Rightarrow\; \hat{Z} = (x^{ab}, \theta^a_i, \hat{\theta}^A_b, u^\pm_k, u^{\hat{\pm}}_A)\,, \tag{4}$$

pass to the analytic basis in it and choose the "hatted" spinor derivatives short, $\nabla^{\hat{+}a} = D^{\hat{+}a} = \frac{\partial}{\partial \theta^{\hat{-}}_a}$. The set of constraints (3) is equivalently rewritten as

$$\begin{aligned} &\{\nabla^+_a, \nabla^+_b\} = 0\,, \quad \{D^{\hat{+}a}, D^{\hat{+}b}\} = 0\,, \quad \{\nabla^+_a, D^{\hat{+}b}\} = \delta^b_a \phi^{+\hat{+}}\,, \\ &[\nabla^{\hat{+}\hat{+}}, \nabla^+_a] = 0\,,\ [\tilde{\nabla}^{++}, \nabla^+_a] = 0\,,\ [\nabla^{\hat{+}\hat{+}}, D^{a\hat{+}}] = 0\,,\ [\tilde{\nabla}^{++}, D^{a\hat{+}}] = 0\,, \\ &[\tilde{\nabla}^{++}, \nabla^{\hat{+}\hat{+}}] = 0\,, \end{aligned} \tag{5}$$

$$\begin{aligned} &\nabla^+_a = D^+_a + \mathcal{A}^+_a(\hat{Z})\,,\ \tilde{\nabla}^{++} = D^{++} + \tilde{V}^{++}(\hat{\zeta})\,, \quad \nabla^{\hat{+}\hat{+}} = D^{\hat{+}\hat{+}} + V^{\hat{+}\hat{+}}(\hat{\zeta})\,, \\ &\hat{\zeta} = (x^{ab}_{\rm an}, \theta^{\pm a}, \theta^{\hat{+}}_c, u^\pm_i, u^{\hat{\pm}}_A)\,. \end{aligned} \tag{6}$$

5. Solving $\mathcal{N} = (1,1)$ SYM constraints through $\mathcal{N} = (1,0)$ superfields

The starting point of our analysis in [21] was the WZ gauge for the extra connection $V^{\hat{+}\hat{+}}(\hat{\zeta})$

$$V^{\hat{+}\hat{+}} = i\theta^{\hat{+}}_a \theta^{\hat{+}}_b \hat{\mathcal{A}}^{ab} + \varepsilon^{abcd}\theta^{\hat{+}}_a \theta^{\hat{+}}_b \theta^{\hat{+}}_c \varphi^A_d u^{\hat{-}}_A + \varepsilon^{abcd}\theta^{\hat{+}}_a \theta^{\hat{+}}_b \theta^{\hat{+}}_c \theta^{\hat{+}}_d \mathcal{D}^{AB} u^{\hat{-}}_A u^{\hat{-}}_B\,, \tag{1}$$

where $\hat{\mathcal{A}}^{ab}$, φ^A_d and $\mathcal{D}^{(AB)}$ are some $\mathcal{N} = (1,0)$ harmonic superfields, still arbitrary at this step.

Then the above constraints are reduced to some sets of harmonic equations. We have solved them and, as the eventual result, found that the first harmonic connection V^{++} coincides with the previous $\mathcal{N} = (1,0)$ one, $V^{++} = V^{++}(\zeta)$, while the dependence of all other geometric $\mathcal{N} = (1,1)$ objects on the "hatted" variables is fixed as

$$\begin{aligned} V^{\hat{+}\hat{+}} &= i\theta^{\hat{+}}_a\theta^{\hat{+}}_b\mathcal{A}^{ab} - \frac{1}{3}\epsilon^{abcd}\theta^{\hat{+}}_a\theta^{\hat{+}}_b\theta^{\hat{+}}_c D^{+}_d q^{-\hat{-}} + \frac{1}{8}\epsilon^{abcd}\theta^{\hat{+}}_a\theta^{\hat{+}}_b\theta^{\hat{+}}_c\theta^{\hat{+}}_d[q^{+\hat{-}}, q^{-\hat{-}}] \\ \phi^{+\hat{+}} &= q^{+\hat{+}} - \theta^{\hat{+}}_a W^{+a} - i\theta^{\hat{+}}_a\theta^{\hat{+}}_b\nabla^{ab}q^{+\hat{-}} + \frac{1}{6}\varepsilon^{abcd}\theta^{\hat{+}}_a\theta^{\hat{+}}_b\theta^{\hat{+}}_c[D^{+}_d q^{-\hat{-}}, q^{+\hat{-}}] \\ &+ \frac{1}{24}\varepsilon^{abcd}\theta^{\hat{+}}_a\theta^{\hat{+}}_b\theta^{\hat{+}}_c\theta^{\hat{+}}_d[q^{+\hat{-}}, [q^{+\hat{-}}, q^{-\hat{-}}]]\,. \end{aligned} \tag{2}$$

Here, $q^{+\hat{\pm}} = q^{+A}(\zeta)u^{\hat{\pm}}_A$, $q^{-\hat{\pm}} = q^{-A}(\zeta)u^{\hat{\pm}}_A$ and $W^{+a}, q^{\pm A}$ are just the $\mathcal{N} = (1,0)$ superfields we dealt with previously. In the course of solving the constraints, there naturally appear the superfield equations of motion

$$\nabla^{++}q^{+A} = 0\,, \quad F^{++} = \frac{1}{4}D^{+}_a W^{+a} = -\frac{1}{2}[q^{+A}, q^{+}_A]\,. \tag{3}$$

Also, the structure of the spinor covariant derivatives is completely fixed

$$\begin{aligned} \nabla^{\hat{+}}_a &= D^{\hat{+}}_a - \theta^{\hat{+}}_a q^{+\hat{-}} + \theta^{\hat{-}}_a\phi^{+\hat{+}}\,, \\ \nabla^{-}_a &= D^{-}_a - D^{+}_a V^{--} - \theta^{\hat{+}}_a q^{-\hat{-}} + \theta^{\hat{-}}_a\phi^{-\hat{+}}\,, \quad \phi^{-\hat{+}} = \nabla^{--}\phi^{+\hat{+}}\,. \end{aligned} \tag{4}$$

The crucial point of our analysis was the requirement that the vector $6D$ connections in the sectors of hatted and unhatted variables are identical to each other.

The basic advantage of using the constrained $\mathcal{N} = (1,1)$ strengths $\phi^{\pm\hat{+}}$ for constructing invariants is the very simple transformation rules of $\phi^{\pm\hat{+}}$ under the hidden $\mathcal{N} = (0,1)$ supersymmetry

$$\delta\phi^{\pm\hat{+}} = -\epsilon^{\hat{+}}_a\frac{\partial}{\partial\theta^{\hat{+}}_a}\phi^{\pm\hat{+}} - 2i\epsilon^{\hat{-}}_a\theta^{\hat{+}}_b\partial^{ab}\phi^{\pm\hat{+}} - [\Lambda^{(comp)}, \phi^{\pm\hat{+}}]\,, \tag{5}$$

where $\Lambda^{(comp)}$ is some composite gauge parameter which makes no contribution under the trace.

6. Invariants in $\mathcal{N} = (1,1)$ superspace

The single-trace dimension <u>$d = 8$</u> invariant (4) can be readily rewritten in $\mathcal{N} = (1,1)$ superspace

$$S_{(1,1)} = \int d\zeta^{-4}\mathcal{L}^{+4}_{(1,1)}\,, \; \mathcal{L}^{+4}_{(1,1)} = -\mathrm{Tr}\,\frac{1}{4}\int d\hat{\zeta}^{-4}d\hat{u}\,(\phi^{+\hat{+}})^4, \quad d\hat{\zeta}^{-4} \sim (D^{\hat{-}})^4 \tag{1}$$

$$\delta\mathcal{L}^{+4}_{(1,1)} = -2i\partial^{ab}\mathrm{Tr}\int d\hat{\zeta}^{-4}d\hat{u}\,\Big[\epsilon^{\hat{-}}_a\theta^{\hat{+}}_b\frac{1}{4}(\phi^{+\hat{+}})^4\Big].$$

Analogously, the double-trace <u>$d = 8$</u> invariant is given by

$$\hat{\mathcal{L}}^{+4}_{(1,1)} = -\frac{1}{4}\int d\hat{\zeta}^{-4}d\hat{u}\,\mathrm{Tr}\,(\phi^{+\hat{+}})^2\,\mathrm{Tr}\,(\phi^{+\hat{+}})^2. \tag{2}$$

Now it is easy to construct the single- and double-trace <u>$d = 10$</u> invariants

$$\begin{aligned} S^{(10)}_1 &= \mathrm{Tr}\int dZ d\hat{\zeta}^{-4}d\hat{u}\,(\phi^{+\hat{+}})^2(\phi^{-\hat{+}})^2, \quad \phi^{-\hat{+}} = \nabla^{--}\phi^{+\hat{+}}\,, \\ S^{(10)}_2 &= -\int dZ d\hat{\zeta}^{-4}d\hat{u}\,\mathrm{Tr}\big(\phi^{+\hat{+}}\phi^{-\hat{+}}\big)\,\mathrm{Tr}\big(\phi^{+\hat{+}}\phi^{-\hat{+}}\big)\,. \end{aligned} \tag{3}$$

It is notable that the single-trace $\underline{d=10}$ invariant admits a representation as an integral over the *full* $\mathcal{N}=(1,1)$ superspace

$$S_1^{(10)} \sim \mathrm{Tr} \int dZ d\hat{Z} d\hat{u}\, \phi^{++\hat{+}} \phi^{--\hat{-}}\,, \quad \phi^{--\hat{-}} = \nabla^{\hat{-}\hat{-}} \phi^{-\hat{+}}\,, \tag{4}$$

with $d\hat{Z} \sim (D^{\hat{-}})^4 (D^{\hat{+}})^4$. On the other hand, the double-trace $\underline{d=10}$ invariant *cannot* be written as the total integral and so looks as being UV *protected*.

This could explain why in the perturbative calculations of the amplitudes in the $\mathcal{N}=(1,1)$ SYM theory single-trace 3-loop divergence is seen, while no double-trace structures at the same order were observed [6], [7], [8]. However, this does not seem to be like the standard non-renormalization theorems because the quantum calculation of $\mathcal{N}=(1,0)$ supergraphs should give invariants in the *off-shell* $\mathcal{N}=(1,0)$ superspace, not in the *on-shell* $\mathcal{N}=(1,1)$ superspace. So the above property seems not enough to explain the absence of the double-trace divergences and some additional piece of reasoning is needed.

7. Summary and outlook

Based on refs. [14], [17], [18] and [21], the off-shell $\mathcal{N}=(1,0)$ and on-shell $\mathcal{N}=(1,1)$ harmonic superfield approaches were accounted for. It was argued that they are very efficient for constructing higher-dimensional invariants in the $\mathcal{N}=(1,0)$ and $\mathcal{N}=(1,1)$ SYM theories. The novel solution of the $\mathcal{N}=(1,1)$ SYM constraints in terms of harmonic $\mathcal{N}=(1,0)$ superfields was given. This allowed us to explicitly construct the full set of the dimensions $\underline{d=8}$ and $\underline{d=10}$ superfield invariants revealing $\mathcal{N}=(1,1)$ on-shell supersymmetry.

All possible $\underline{d=6}$ $\mathcal{N}=(1,1)$ invariants were shown to be on-shell vanishing, thus proving the UV finiteness of $\mathcal{N}=(1,1)$ SYM at one loop.

The off-shell $\underline{d=8}$ invariants which would be non-vanishing on shell, are absent. Assuming that the $\mathcal{N}=(1,0)$ supergraphs yield integrals over the full $\mathcal{N}=(1,0)$ harmonic superspace, this means the absence of two-loop counterterms as well.

Two dimension $\underline{d=10}$ invariants were constructed as integrals over the whole $\mathcal{N}=(1,0)$ harmonic superspace. The single-trace invariant can be rewritten as an integral over $\mathcal{N}=(1,1)$ superspace, while the double-trace one cannot. This property combined with an additional reasoning (e.g., based on the algebraic renormalization scheme [25]) could explain why the double-trace invariant is UV protected.

Some further lines of development:

(a). It would be tempting and instructive to construct the $\underline{d \geq 12}$ invariants in the $\mathcal{N}=(1,1)$ SYM theory using the on-shell $\mathcal{N}=(1,1)$ harmonic superspace techniques and to see whether they exhibit the properties similar to the $\underline{d=10}$ invariants. It is probable that the proper corrections to the hidden supersymmetry transformations will be of need, while tackling this issue.
(b). It is worth to apply the same method for constructing the Born-Infeld action with the manifest off-shell $\mathcal{N}=(1,0)$ and hidden on-shell $\mathcal{N}=(0,1)$ supersymmetries.
(c). The closely related problem is to recover the higher dimension invariants listed above from the quantum $\mathcal{N}=(1,0)$ supergraph techniques. The first steps in this direction were undertaken in a recent paper [30].
(d). An interesting task is to develop an analogous on-shell bi-harmonic $\mathcal{N}=4, 4D$ superspace approach to the $\mathcal{N}=4, 4D$ SYM theory in the $\mathcal{N}=2$ superfield formulation (by solving the $\mathcal{N}=4$ SYM constraints in terms of $\mathcal{N}=2$ superfields) and apply it to the problem of constructing the $\mathcal{N}=4$ SYM effective action. It is curious that such a formulation has not been constructed so far in full generality, despite the existence of various more sophisticated superspace formulations (see [31] for a recent review).

(e). As was mentioned in sect. 5, the crucial last step in solving the constraints of $\mathcal{N} = (1,1)$ SYM theory was identifying the vector connections in the sectors with the standard and "hatted" harmonics. Only after this identification, the constraints in the bi-harmonic superspace get fully equivalent to those in the standard setting and imply the equations of motion for the involved superfields. It would be interesting to develop a superspace formulation with two *independent* vector connections and to inquire whether it could give rise to an off-shell description of $\mathcal{N} = (1,1), 6D$ (and, perhaps, of $\mathcal{N} = 4, 4D$) SYM theories. The introduction of two vector connections seems to imply doubling of the x-coordinate, in an obvious parallel with the recent eight-dimensional reformulation of $\mathcal{N} = 4, 4D$ SYM theory in [32].
(f). Applications in supergravity? Lacking the double-trace divergent structures in the 3-loop amplitude in $\mathcal{N} = (1,1)$ SYM theory is similar to the absence of 3-loop and 4-loop divergences for the four-graviton amplitudes in $\mathcal{N} = 4, 4D$ and $\mathcal{N} = 5, 4D$ supergravities [26], [27], [28], [29]. All these UV divergence cancelations could find a common explanation within the harmonic superspace approach [1].

Acknowledgments
I thank the organizers of the conference ISQS'2016 for the kind hospitality in Prague. I am grateful to my co-authors Guillaume Bossard, Andrei Smilga and Boris Zupnik. A partial support from the RFBR grant no. 15-02-06670, grant of Russian Science Foundation no. 16-12-10306 and a grant of Heisenberg-Landau program is acknowledged.

References
[1] Seiberg N 1998 Notes on theories with 16 supercharges *Nucl. Phys. Proc. Suppl.* **67** 158 (*Preprint* hep-th/9705117)
[2] Beisert N 2010 On Yangian Symmetry in Planar $\mathcal{N} = 4$ SYM *In *Trieste 2010, Gribov-80 Memorial Volume** 413-438 (*Preprint* 1004.5423 [hep-th])
[3] Dennen T and Huand Yu-tin 2011 Dual Conformal Properties of Six-Dimensional Maximal Super Yang-Mills Amplitudes *JHEP* **1101** 140 (*Preprint* 1010.5874 [hep-th])
[4] Tseytlin A A 1997 On non-abelian generalization of Born-Infeld action in string theory *Nucl. Phys.* B **501** 41-52 (*Preprint* hep-th/9701125)
[5] Drummond J M, Heslop P J, Howe P S and Kerstan S F 2003 Integral invariants in $\mathcal{N} = 4$ SYM and the effective action for coincident D-branes *JHEP* **0308** 016 (*Preprint* hep-th/0305202)
[6] Bern Z, Dixon L J and Smirnov V A 2005 Iteration of planar amplitudes in maximally supersymmetric Yang-Mills theory at three loops and beyond *Phys. Rev.* D **72** 085001 (*Preprint* hep-th/0505205)
[7] Bern Z, Carrasco J J M, Dixon L J, Johansson H and Roiban R 2010 The Complete Four-Loop Four-Point Amplitude in N=4 Super-Yang-Mills Theory *Phys. Rev.* D **82** 125040 (*Prteprint* 1008.3327 [hep-th])
[8] Bern Z, Carrasco J J M, Dixon L J, Johansson H and Roiban R 2012 Simplifying Multiloop Integrands and Ultraviolet Divergences of Gauge Theory and Gravity Amplitudes *Phys. Rev.* D **85** 105014, (*Preprint* 1201.5366 [hep-th])
[9] Bossard G, Howe P S, Lindström U, Stelle K S and Wulff L 2011 Integral invariants in maximally supersymmetric Yang-Mills theories *JHEP* **1105** 021 (*Preprint* 1012.3142 [hep-th])
[10] Berkovits N, Green M B, Russo J G and Vanhove P 2009 Non-renormalization conditions for four-gluon scattering in supersymmetric string and field theory *JHEP* **0911** 063 (*Preprint* 0908.1923 [hep-th])
[11] Bjornsson J and Green M B 2010 5 loops in 24/5 dimensions *JHEP* **1008** 132 (*Preprint* 1004.2692 [hep-th])
[12] Bjornsson J 2011 Multi-loop amplitudes in maximally supersymmetric pure spinor field theory *JHEP* **1101** 002 (*Preprint* 1009.5906 [hep-th])
[13] Howe P S, Stelle K S and West P C 1985 $\mathcal{N} = 1, d = 6$ harmonic superspace *Class. Quant. Grav.* **2** 815
[14] Zupnik B M 1986 Six-dimensional Supergauge Theories in the Harmonic Superspace *Sov. J. Nucl. Phys.* **44** 512 [*Yad. Fiz.* **44** 794-802]
Zupnik B M 1987 The Action of the Supersymmetric $\mathcal{N} = 2$ Gauge Theory in Harmonic Superspace *Phys. Lett.* B **183** 175-176
[15] Galperin A S, Ivanov E A, Kalitzin S, Ogievetsky V I and Sokatchev E S 1984 Unconstrained $\mathcal{N} = 2$ matter, Yang-Mills and supergravity theories in harmonic superspace *Class. Quant. Grav.* **1** 469-498

[1] For a recent relevant discussion see [33].

[16] Galperin A S, Ivanov E A, Ogievetsky V I and Sokatchev E S 2001 Harmonic Superspace *Cambridge University Press* 306 p.
[17] Ivanov E A, Smilga A V and Zupnik B M 2005 Renormalizable supersymmetric gauge theory in six dimensions *Nucl. Phys.* B **726** 131-148 (*Preprint* hep-th/0505082)
[18] Ivanov E A and Smilga A V 2006 Conformal properties of hypermultiplet actions in six dimensions *Phys. Lett.* B **637** 374-381 (*Preprint* hep-th/0510273)
[19] Buchbinder I L and Pletnev N G 2015 Construction of $6D$ supersymmetric field models in $\mathcal{N} = (1,0)$ harmonic superspace *Nucl. Phys.* B **892** 21-48 (*Preprint* 1411.1848 [hep-th])
[20] Buchbinder I L and Pletnev N G 2015 Leading low-energy effective action in the $6D$ hypermultiplet theory on a vector/tensor background *Phys. Lett.* B **744** 125-130 (*Preprint* 1502.03257 [hep-th])
[21] Bossard G, Ivanov E and Smilga A 2015 Ultraviolet behavior of 6D supersymmetric Yang-Mills theories and harmonic superspace *JHEP* **1512** 085 (*Preprint* 1509.08027 [hep-th])
[22] Bossard G, Howe P S and Stelle K S 2009 The Ultra-violet question in maximally supersymmetric field theories *Gen. Rel. Grav.* **41** 919 (*Preprint* 0901.4661 [hep-th])
[23] Howe P S, Sierra G and Townsend P K 1983 Supersymmetry in six dimensions *Nucl. Phys.* B **221** 331-348
[24] Howe P S and Stelle K S 1984 Ultraviolet Divergences in Higher Dimensional Supersymmetric Yang-Mills Theories *Phys. Lett.* B **137** 175-180
[25] Piguet O and Sorella S P 1995 Algebraic renormalization: Perturbative renormalization, symmetries and anomalies *Lect. Notes Phys.* M **28** 1
[26] Bern Z, Davies S, Dennen T and Huang Yu-tin 2012 Absence of three-loop four-point divergences in $\mathcal{N} = 4$ supergravity *Phys. Rev. Lett.* **108** 201301 (*Preprint* 1202.3423 [hep-th])
[27] Tourkine P and Vanhove P 2012 An R^4 non-renormalization theorem in $\mathcal{N} = 4$ supergravity *Class. Quant. Grav.* **29** 115006 (*Preprint* 1202.3692 [hep-th])
[28] Bern Z, Davies S, Dennen T and Huang Yu-tin 2012 Ultraviolet cancellations in half-maximal supergravity as a consequence of the double-copy structure *Phys. Rev.* D **86** 105014 (*Preprint* 1209.2472 [hep-th])
[29] Bern Z, Davies S and Dennen T 2014 Enhanced ultraviolet cancellations in $\mathcal{N} = 5$ supergravity at four loops *Phys. Rev.* D **90** 105011 (*Preprint* 1409.3089 [hep-th])
[30] Buchbinder I L, Ivanov E A, Merzlikin B S and Stepanyantz K V 2016 One-loop divergences in the $6D$, $\mathcal{N} = (1,0)$ abelian gauge theory (*Preprint* 1609.00975 [hep-th])
[31] Buchbinder I L, Ivanov E A and Samsonov I B 2016 The Low-Energy $\mathcal{N} = 4$ SYM Effective Action in Diverse Harmonic Superspaces (*Preprint* 1603.02768 [hep-th])
[32] Chicherin D and Sokatchev E 2016 $\mathcal{N} = 4$ super-Yang-Mills in LHC superspace. Part I: Classical and quantum theory (*Preprint* CERN-TH-2016-018, 1601.06803 [hep-th])
[33] Smilga A 2016 Ultraviolet divergences in non-renormalizale supersymmetric theories (*Preprint* 1603.06811 [hep-th])

Long $\mathcal{N} = 2,\, 4$ multiplets in supersymmetric mechanics

Evgeny Ivanov[1], **Antonio Rivasplata Mendoza**[1,2], **Stepan Sidorov**[1]

[1] Bogoliubov Laboratory of Theoretical Physics, JINR, 141980 Dubna, Moscow Region, Russia
[2] Departamento de Física Universidad Nacional de Trujillo, Trujillo, Perú

E-mail: eivanov@theor.jinr.ru, antrivas@unitru.edu.pe, sidorovstepan88@gmail.com

Abstract. We define $SU(2|1)$ supermultiplets described by chiral superfields having non-zero external spins with respect to $SU(2) \subset SU(2|1)$ and show that their splitting into $\mathcal{N} = 2$, $d = 1$ multiplets contains the so called "long" indecomposable $\mathcal{N} = 2$, $d = 1$ multiplets $(\mathbf{2}, \mathbf{4}, \mathbf{2})_l$. We give superfield formulation for this type of $\mathcal{N} = 2$ long multiplets and construct their most general superfield action. A simple example of long $\mathcal{N} = 4$, $d = 1$ multiplet is also considered, both in the superfield and the component formulations.

1. Introduction

In [1], $SU(2|1)$ supersymmetric mechanics was proposed as a deformation of the standard $\mathcal{N} = 4$ mechanics by a mass parameter m. Superfield approach based on the deformed $SU(2|1)$ superspaces allowed to reproduce many previously known models [2, 3, 4, 5] and to construct new ones [1, 6, 7, 8]. In the paper [9], $SU(2|1)$ supersymmetric quantum mechanics was obtained via dimensional reduction from the superconformal model on the four-dimensional curved space-time $S^3 \times \mathbb{R}$ and applied to compute vacuum energy of the model. For simplicity, the authors considered supersymmetric mechanics in the framework of $\mathcal{N} = 2$, $d = 1$ supersymmetry and revealed a new type of supermultiplets, the so-called "long multiplets". As was shown in [13], the long $\mathcal{N} = 2$ multiplet can be embedded into a generalized $SU(2|1)$ chiral multiplet described by a chiral superfield Φ_A carrying some external index A with respect to the subgroup $SU(2)$ of the supergroup $SU(2|1)$.

Generalizations to $\mathcal{N} = 4$ supersymmetry with various extended sets of component fields were considered in [10, 11, 12]. The main distinguishing feature of long (non-minimal) multiplets is that they accommodate extended sets of component fields. The long $\mathcal{N} = 2$ multiplet [9] can be interpreted as a deformation of the pair of chiral multiplets $(\mathbf{2}, \mathbf{2}, \mathbf{0})$ and $(\mathbf{0}, \mathbf{2}, \mathbf{2})$ by a mass-dimension parameter, *i.e.* it has an extended set of component fields $(\mathbf{2}, \mathbf{4}, \mathbf{2})_l$. The long multiplet $(\mathbf{4}, \mathbf{8}, \mathbf{4})_l$ considered in [12] joins two $\mathcal{N} = 4$ chiral multiplets $(\mathbf{2}, \mathbf{4}, \mathbf{2})$ through a dimensionless parameter.

In this contribution we give a brief account of the long $\mathcal{N} = 2$ multiplet, as it was discussed in [13], and present some new results for the long $\mathcal{N} = 4$ multiplet suggested in [12]. To be more precise, we give the superfield description for the long $\mathcal{N} = 2,\, 4$ multiplets which were studied at the component level in [9, 12].

IOP Conf. Series: Journal of Physics: Conf. Series **804** (2017) 012021 doi:10.1088/1742-6596/804/1/012021

2. $SU(2|1)$ supersymmetric mechanics

We proceed from the centrally-extended superalgebra $\hat{su}(2|1)$ with the following non-vanishing (anti)commutators:

$$\{Q^i, \bar{Q}_j\} = 2m\left(I^i_j - \delta^i_j F\right) + 2\delta^i_j H, \qquad \left[I^i_j, I^k_l\right] = \delta^k_j I^i_l - \delta^i_l I^k_j,$$
$$\left[I^i_j, \bar{Q}_l\right] = \frac{1}{2}\,\delta^i_j \bar{Q}_l - \delta^i_l \bar{Q}_j, \qquad \left[I^i_j, Q^k\right] = \delta^k_j Q^i - \frac{1}{2}\,\delta^i_j Q^k,$$
$$\left[F, \bar{Q}_l\right] = -\frac{1}{2}\,\bar{Q}_l, \qquad \left[F, Q^k\right] = \frac{1}{2}\,Q^k. \tag{1}$$

Its bosonic sector contains the central charge generator H (commuting with all other generators) and the $U(2)_{\text{int}}$ generators I^i_j and F. In the limit $m = 0$, this superalgebra becomes the standard $\mathcal{N} = 4$, $d = 1$ Poincaré superalgebra.

The supersymmetric $SU(2|1)$ transformations of the superspace coordinates $\zeta := \left\{t, \theta_i, \bar{\theta}^k\right\}$, $\bar{\theta}^i = \overline{(\theta_i)}$, are given by

$$\delta\theta_i = \epsilon_i + 2m\,\bar{\epsilon}^k\theta_k\theta_i, \quad \delta\bar{\theta}^i = \bar{\epsilon}^i - 2m\,\epsilon_k\bar{\theta}^k\bar{\theta}^i, \qquad \delta t = i\left(\bar{\epsilon}^k\theta_k + \epsilon_k\bar{\theta}^k\right). \tag{2}$$

The $SU(2|1)$ measure invariant under these transformations is

$$d\zeta = dt\, d^2\theta\, d^2\bar{\theta}\left(1 + 2m\,\bar{\theta}^k\theta_k\right), \quad \delta\left(d\zeta\right) = 0. \tag{3}$$

The left chiral subspace $\zeta_L = \{t_L, \theta_i\}$, where t_L is defined as

$$t_L = t + i\bar{\theta}^k\theta_k - \frac{i}{2}\,m\left(\theta\right)^2\left(\bar{\theta}\right)^2, \tag{4}$$

is closed under the $SU(2|1)$ transformations

$$\delta\theta_i = \epsilon_i + 2m\,\bar{\epsilon}^k\theta_k\theta_i, \qquad \delta t_L = 2i\bar{\epsilon}^k\theta_k\,. \tag{5}$$

Conjugating the coordinates of the subspace ζ_L, one obtains the right-chiral subspace ζ_R.

The $SU(2|1)$ covariant derivatives are defined as

$$\begin{aligned}
\mathcal{D}^i \;&=\; \left[1 + m\,\bar{\theta}^k\theta_k - \frac{3m^2}{8}\left(\theta\right)^2\left(\bar{\theta}\right)^2\right]\frac{\partial}{\partial\theta_i} - m\,\bar{\theta}^i\theta_j\frac{\partial}{\partial\theta_j} - i\bar{\theta}^i\partial_t \\
&\quad + m\,\bar{\theta}^i\tilde{F} - m\,\bar{\theta}^j\left(1 - m\,\bar{\theta}^k\theta_k\right)\tilde{I}^i_j, \\
\bar{\mathcal{D}}_j \;&=\; -\left[1 + m\,\bar{\theta}^k\theta_k - \frac{3m^2}{8}\left(\theta\right)^2\left(\bar{\theta}\right)^2\right]\frac{\partial}{\partial\bar{\theta}^j} + m\,\bar{\theta}^k\theta_j\frac{\partial}{\partial\bar{\theta}^k} + i\theta_j\partial_t \\
&\quad - m\,\theta_j\tilde{F} + m\,\theta_l\left(1 - m\,\bar{\theta}^k\theta_k\right)\tilde{I}^l_j,
\end{aligned} \tag{6}$$

where $\tilde{F}$ and $\tilde{I}^i_k$ are the "matrix" parts of the generators F and I^i_k. The latter non-trivially act on the covariant derivatives:

$$\tilde{I}^i_j\bar{\mathcal{D}}_l = \delta^i_l\bar{\mathcal{D}}_j - \frac{1}{2}\,\delta^i_j\bar{\mathcal{D}}_l\,, \qquad \tilde{I}^i_j\mathcal{D}^k = \frac{1}{2}\,\delta^i_j\mathcal{D}^k - \delta^k_j\mathcal{D}^i,$$
$$\tilde{F}\bar{\mathcal{D}}_l = \frac{1}{2}\,\bar{\mathcal{D}}_l\,, \qquad \tilde{F}\mathcal{D}^k = -\frac{1}{2}\,\mathcal{D}^k. \tag{7}$$

An $SU(2|1)$ superfield Φ_A can carry an external $U(2)$ representation corresponding to these matrix parts and it transforms according to this representations as

$$\begin{aligned}
&\delta\Phi_A = \left(i\delta\hat{h}\tilde{F} - i\delta h_{ij}\tilde{I}^{ij}\right)\Phi_A\,, \\
&\delta\hat{h} = -\,im\left(\epsilon_k\bar{\theta}^k + \bar{\epsilon}^k\theta_k\right), \quad \delta h_{ij} = im\left(\epsilon_{(i}\bar{\theta}_{j)} + \bar{\epsilon}_{(i}\theta_{j)}\right)\left(1 - m\,\bar{\theta}^k\theta_k\right).
\end{aligned} \tag{8}$$

2.1. Chiral superfields

Chiral $SU(2|1)$ superfields can carry non-zero external spins s with respect to $SU(2) \subset SU(2|1)$. The simplest chiral superfield with $s=0$ has the field contents $(\mathbf{2},\mathbf{4},\mathbf{2})$. As compared to the $SU(2)$ singlet chiral superfields, the number of component fields in the superfield Φ_A carrying non-zero external spins $s = 1/2, 1, \ldots$ increases according to

$$(\mathbf{2[2s+1]}, \mathbf{4[2s+1]}, \mathbf{2[2s+1]}). \tag{9}$$

The decomposition of the $s=0$ chiral supermultiplet $(\mathbf{2},\mathbf{4},\mathbf{2})$ into $\mathcal{N}=2$ multiplets is given by a direct sum of two chiral multiplets, $(\mathbf{2},\mathbf{2},\mathbf{0})$ and $(\mathbf{0},\mathbf{2},\mathbf{2})$. Below, we will consider the analogous decompositions of the $s \neq 0$ chiral multiplets.

The singlet ($s=0$) chiral superfield Φ in the $U(2)$ representation $(0, 2\kappa)$ satisfies the chirality condition

$$\bar{\mathcal{D}}_i\Phi = 0, \qquad \tilde{F}\Phi = 2\kappa\,\Phi, \qquad \tilde{I}^k_l\Phi = 0. \tag{10}$$

In the case of $s = 1/2, 1, 3/2\ldots$, the chiral superfield $\Phi_{(i_1\ldots i_{2s})}$ belongs to the $U(2)$ representation $(s, 2\kappa)$ and is defined by the constraints

$$\begin{aligned} &\bar{\mathcal{D}}_j\Phi_{(i_1\ldots i_{2s})} = 0, \qquad \tilde{F}\Phi_{(i_1\ldots i_{2s})} = 2\kappa\,\Phi_{(i_1\ldots i_{2s})}\,, \\ &\tilde{I}^k_l\Phi_{(i_1\ldots i_{2s})} = \sum_{n=1}^{2s}\left[\delta^k_{i_n}\Phi_{(i_1\ldots i_{n-1}\,l\,i_{n+1}\ldots i_{2s})} - \frac{1}{2}\,\delta^k_l\Phi_{(i_1\ldots i_{2s})}\right]. \end{aligned} \tag{11}$$

$SU(2|1)$ transformations of chiral superfields can be found from (8).

2.2. The case $s = 1/2$

The $SU(2|1)$ chiral superfield Φ_i $(i=1,2)$ in the $U(2)$ representation $(1/2, 2\kappa)$ is defined by the constraints

$$\bar{\mathcal{D}}_j\Phi_i = 0, \qquad \tilde{I}^k_l\Phi_i = \delta^k_i\Phi_l - \frac{1}{2}\,\delta^k_l\Phi_i, \qquad \tilde{F}\Phi_i = 2\kappa\,\Phi_i. \tag{12}$$

The chirality condition is solved by

$$\begin{aligned} \Phi_i\left(t_L,\theta_i,\bar{\theta}_k\right) &= \left(1+2m\,\bar{\theta}^l\theta_l\right)^{-\kappa}\left[1-\frac{3m^2}{16}\,(\theta)^2\left(\bar{\theta}\right)^2\right]\phi_i\left(t_L,\theta_i\right) \\ &\quad - m\left(\frac{1}{2}\,\delta^j_i\,\bar{\theta}^k\theta_k - \bar{\theta}^j\theta_i\right)\phi_j\left(t_L,\theta_i\right), \\ \phi_i\left(t_L,\theta_i\right) &= z_i + \theta_i\psi - \sqrt{2}\,\theta^k\psi_{(ik)} + \theta_k\theta^k B_i\,. \end{aligned} \tag{13}$$

The superfields Φ_i and ϕ_i transform as

$$\begin{aligned} \delta\Phi_i &= m\left(1-m\,\bar{\theta}^l\theta_l\right)\left[\frac{1}{2}\,\delta^j_i\left(\epsilon_k\bar{\theta}^k + \bar{\epsilon}^k\theta_k\right) - \left(\epsilon_i\bar{\theta}^j + \bar{\epsilon}^j\theta_i\right)\right]\Phi_j + 2\kappa m\left(\epsilon_k\bar{\theta}^k + \bar{\epsilon}^k\theta_k\right)\Phi_i, \\ \delta\phi_i &= 4\kappa m\left(\bar{\epsilon}^k\theta_k\right)\phi_i + 2m\left(\frac{1}{2}\,\delta^j_i\,\bar{\epsilon}^k\theta_k - \bar{\epsilon}^j\theta_i\right)\phi_j\,. \end{aligned} \tag{14}$$

The relevant $SU(2|1)$ transformations of the component fields read

$$\begin{aligned} &\delta z^i = -\,\epsilon^i\psi - \sqrt{2}\,\epsilon_k\psi^{(ik)}, \qquad \delta B^i = -\,\bar{\epsilon}^i\left(i\,\nabla_t\psi - \frac{m}{2}\,\psi\right) - \sqrt{2}\,\bar{\epsilon}_k\left(i\,\nabla_t\psi^{(ik)} + \frac{3m}{2}\,\psi^{(ik)}\right), \\ &\delta\psi = \bar{\epsilon}^k\left(i\,\nabla_t z_k + \frac{3m}{2}\,z_k\right) - \epsilon^k B_k, \qquad \delta\psi^{(ik)} = \sqrt{2}\,\bar{\epsilon}^{(k}\left[i\,\nabla_t z^{i)} - \frac{m}{2}\,z^{i)}\right] - \sqrt{2}\,\epsilon^{(i}B^{k)}, \end{aligned} \tag{15}$$

where

$$\nabla_t = \partial_t + 2i\kappa m\,, \qquad \bar{\nabla}_t = \partial_t - 2i\kappa m\,. \tag{16}$$

2.3. Decomposition into $\mathcal{N}=2$ multiplets

Singling out the subset of $\mathcal{N}=2$ transformations associated with the parameter $\epsilon_1 \equiv \epsilon$ in (15), we can identify the component fields $\left(z_i, \psi^{(ik)}, \psi, B^i\right)$ with the system of three $\mathcal{N}=2$ multiplets:

$$\text{one long multiplet } (\mathbf{2},\mathbf{4},\mathbf{2})_{\text{l}} \quad \oplus \quad \text{one short multiplet } (\mathbf{2},\mathbf{2},\mathbf{0}) \quad \oplus \quad \text{one short multiplet } (\mathbf{0},\mathbf{2},\mathbf{2}).$$

The $\mathcal{N}=2$-irreducible multiplets $(\mathbf{2},\mathbf{2},\mathbf{0})$ and $(\mathbf{0},\mathbf{2},\mathbf{2})$ are composed of the fields $(z_2, \psi^{(11)})$ and $(\psi^{(22)}, B^2)$, while the rest of component fields $\left(z_1, \psi^{(12)}, \psi, B^1\right)$ forms a multiplet with the field contents $(\mathbf{2},\mathbf{4},\mathbf{2})_{\text{l}}$ that was called "long" multiplet [9]. In the limit $m=0$, the indecomposable long $\mathcal{N}=2$ multiplet $(\mathbf{2},\mathbf{4},\mathbf{2})_{\text{l}}$ splits into the direct sum of two "short" irreducible $\mathcal{N}=2$ multiplets $(\mathbf{2},\mathbf{2},\mathbf{0})$ and $(\mathbf{0},\mathbf{2},\mathbf{2})$. At $m \neq 0$, such a splitting cannot be accomplished by any field redefinition.

In the general case $s>0$, we have the following sum of $\mathcal{N}=2$ multiplets:

$$2s \text{ long multiplets} \quad \oplus \quad \text{one short multiplet } (\mathbf{2},\mathbf{2},\mathbf{0}) \quad \oplus \quad \text{one short multiplet } (\mathbf{0},\mathbf{2},\mathbf{2}).$$

From this decomposition one can figure out that all long multiplets have mass-dimension parameters proportional to m.

The relevant $\mathcal{N}=2$ subalgebra of (1) can be identified with

$$\{Q,\bar{Q}\} = 2\,(H-\Sigma)\,, \quad Q^2 = \bar{Q}^2 = 0\,, \tag{17}$$

where

$$Q \equiv Q^1, \qquad \bar{Q} \equiv \bar{Q}_1, \qquad \Sigma \equiv m\left(F - I^1_1\right). \tag{18}$$

If we forget about the $su(2|1)$ origin of this superalgebra, the presence of the central Σ is not necessary since it can always be removed by a field redefinition, and $H-\Sigma$ can be chosen as the Hamiltonian:

$$H - \Sigma \;\rightarrow\; H \equiv i\partial_t\,. \tag{19}$$

For the long multiplet $(\mathbf{2},\mathbf{4},\mathbf{2})_{\text{l}}$, this shift can be performed through the redefinitions

$$\begin{aligned} z &= z_1 e^{i(2\kappa-1/2)mt}, \qquad \xi = \left(\frac{\psi}{\sqrt{2}} - \psi_{(12)}\right) e^{i(2\kappa-1/2)mt}, \\ B &= -B^1 e^{i(2\kappa-1/2)mt}, \qquad \pi = \left(\frac{\psi}{\sqrt{2}} + \psi_{(12)}\right) e^{i(2\kappa-1/2)mt}. \end{aligned} \tag{20}$$

Then, the corresponding $\mathcal{N}=2$ supersymmetry transformations,

$$\begin{aligned} &\delta z = -\sqrt{2}\,\epsilon\,\xi, \qquad \delta\xi = \sqrt{2}\,i\bar{\epsilon}\dot{z}, \\ &\delta\pi = -\sqrt{2}\,\epsilon B + \underline{\sqrt{2}\,m\,\bar{\epsilon}z}, \qquad \delta B = \sqrt{2}\,i\bar{\epsilon}\,\dot{\pi} - \underline{\sqrt{2}\,m\,\bar{\epsilon}\,\xi}, \end{aligned} \tag{21}$$

close on the standard $\mathcal{N}=2$ supersymmetry algebra

$$\left\{Q,\bar{Q}\right\} = 2H, \qquad \{Q,Q\} = \left\{\bar{Q},\bar{Q}\right\} = 0, \quad H = i\partial_t\,. \tag{22}$$

The free $SU(2|1)$ Lagrangian is given by

$$\mathcal{L}^{\text{free}} = \frac{1}{4}\int d^2\theta\, d^2\bar{\theta}\left(1 + 2m\,\bar{\theta}^k\theta_k\right)\Phi_i\bar{\Phi}^i. \tag{23}$$

IOP Conf. Series: Journal of Physics: Conf. Series **804** (2017) 012021 doi:10.1088/1742-6596/804/1/012021

Rewriting it in terms of the component fields,

$$\begin{aligned}\mathcal{L}^{\text{free}} &= \bar{\nabla}_t\bar{z}^i\,\nabla_t z_i + \frac{i}{2}\left(\bar{\psi}\,\nabla_t\psi + \psi\,\bar{\nabla}_t\bar{\psi}\right) + \frac{i}{2}\left(\bar{\psi}_{(ik)}\,\nabla_t\psi^{(ik)} + \psi^{(ik)}\,\bar{\nabla}_t\bar{\psi}_{(ik)}\right) + B^i\bar{B}_i \\ &\quad - \frac{i}{2}m\left(z_i\bar{\nabla}_t\bar{z}^i - \bar{z}^i\nabla_t z_i\right) - \frac{3m^2}{4}\,z_i\bar{z}^i + \frac{m}{2}\left(\psi\bar{\psi} - 3\psi^{(ik)}\bar{\psi}_{(ik)}\right), \end{aligned} \tag{24}$$

one can check that it splits into a sum of the three free $\mathcal{N}=2$ Lagrangians:

$$\mathcal{L}^{\text{free}} = \mathcal{L}^{\text{free}}_{(\mathbf{2},\mathbf{4},\mathbf{2})_1} + \mathcal{L}^{\text{free}}_{(\mathbf{0},\mathbf{2},\mathbf{2})} + \mathcal{L}^{\text{free}}_{(\mathbf{2},\mathbf{2},\mathbf{0})}\,. \tag{25}$$

After the redefinition (20), the component Lagrangian of the long multiplet in (25) reads

$$\mathcal{L}^{\text{free}}_{(\mathbf{2},\mathbf{4},\mathbf{2})_1} = \dot{\bar{z}}\dot{z} + \frac{i}{2}\left(\bar{\xi}\dot{\xi} - \dot{\bar{\xi}}\xi\right) + \frac{i}{2}\left(\bar{\pi}\dot{\pi} - \dot{\bar{\pi}}\pi\right) + B\bar{B} - \underline{m\left(\xi\bar{\pi} + \pi\bar{\xi}\right)} - \underline{m^2 z\bar{z}}. \tag{26}$$

3. The long $\mathcal{N}=2$ multiplet

Now we consider a superfield description for the long $\mathcal{N}=2$ supermultiplet defined by the transformations (21). First we define $\mathcal{N}=2$ covariant derivatives D, $\bar{D}$,

$$D = \frac{\partial}{\partial\theta} - i\bar{\theta}\partial_t\,, \qquad \bar{D} = -\frac{\partial}{\partial\bar{\theta}} + i\theta\partial_t\,, \qquad \left\{D,\bar{D}\right\} = 2i\partial_t\,. \tag{27}$$

The $\mathcal{N}=2$, $d=1$ superspace coordinates $\left\{t,\theta,\bar{\theta}\right\}$ transform in the familiar way:

$$\delta\theta = \epsilon, \qquad \delta\bar{\theta} = \bar{\epsilon}, \qquad \delta t = i\left(\epsilon\,\bar{\theta} + \bar{\epsilon}\,\theta\right). \tag{28}$$

The chiral superfields are defined by the standard conditions

$$\bar{D}Z = 0, \qquad \bar{D}\Pi = 0. \tag{29}$$

The bosonic superfield Z describes an irreducible multiplet $(\mathbf{2},\mathbf{2},\mathbf{0})$, while the fermionic superfield Π has the field contents $(\mathbf{0},\mathbf{2},\mathbf{2})$. The component expansion of Π and Z reads

$$Z = z + \sqrt{2}\,\theta\,\xi - i\theta\bar{\theta}\dot{z}, \qquad \Pi = \pi + \sqrt{2}\,\theta B - i\theta\bar{\theta}\,\dot{\pi}. \tag{30}$$

The "passive" superfield transformations $\delta\Pi = \delta Z = 0$ amount to the two independent sets of transformations for the component fields:

$$\delta z = -\sqrt{2}\,\epsilon\,\xi, \quad \delta\xi = \sqrt{2}\,i\bar{\epsilon}\dot{z}, \qquad \delta\pi = -\sqrt{2}\,\epsilon B, \quad \delta B = \sqrt{2}\,i\bar{\epsilon}\,\dot{\pi}. \tag{31}$$

The long multiplet is described by the pair of fermionic and bosonic $\mathcal{N}=2$ superfields Ψ and Z (of the same dimension as Π and Z above) which are subjected to the following conditions, with m being a deformation parameter:

$$\bar{D}\Psi = \sqrt{2}\,mZ, \qquad \bar{D}Z = 0. \tag{32}$$

As a solution of (32) (non-singular at $m=0$), the superfield Ψ can be represented as

$$\Psi = \Pi - \sqrt{2}\,m\,\bar{\theta}Z, \qquad \bar{D}\Pi = 0. \tag{33}$$

The first condition in (32) expresses some components of Ψ through the components of Z and so forces the superfunction Π to transform through Z.

IOP Conf. Series: Journal of Physics: Conf. Series **804** (2017) 012021 doi:10.1088/1742-6596/804/1/012021

The transformations $\delta\Psi = \delta Z = 0$ give rise to the following transformation law for Π:

$$\delta\Pi = \sqrt{2}\, m\, \bar{\epsilon} Z. \tag{34}$$

It generates deformed $\mathcal{N} = 2$ supersymmetry transformations which coincide with the transformations (21). Thus the considered multiplet involves an irreducible chiral multiplet $(\mathbf{2},\mathbf{2},\mathbf{0})$ and a set of fields $(\mathbf{0},\mathbf{2},\mathbf{2})$ which are described by the chiral $d = 1$ superfunctions Z and Π, respectively. The quantity Z is a chiral superfield, while Π has the non-standard transformation law (34) $\sim m$. Thus the parameter m is a deformation parameter responsible for unifying the two former chiral "short" multiplets into a single "long" multiplet.

3.1. Invariant Lagrangians

The most general Lagrangian of the long multiplet can be written down as

$$\mathcal{L}_{(\Psi,Z)} = \int d\bar{\theta}\, d\theta \left[\bar{D}\bar{Z}\, DZ\, h_0\left(Z,\bar{Z}\right) + \Psi\bar{\Psi}\, h_1\left(Z,\bar{Z}\right) + \mu\, h_{(\mu)}\left(Z,\bar{Z}\right)\right], \tag{35}$$

where h_0, h_1, $h_{(\mu)}$ are arbitrary functions and μ is a mass-dimension parameter. The free Lagrangian of the long multiplet is given by a sum of the three superfield invariants:

$$\mathcal{L}^{\text{free}}_{(\mathbf{2},\mathbf{4},\mathbf{2})_1} = \frac{1}{4}\int d\bar{\theta}\, d\theta \left(\bar{D}\bar{Z}\, DZ + 2\Psi\bar{\Psi} - 2\mu Z\bar{Z}\right). \tag{36}$$

In the component form it reads

$$\begin{aligned}\mathcal{L}^{\text{free}}_{(\mathbf{2},\mathbf{4},\mathbf{2})_1} &= \dot{\bar{z}}\dot{z} + \frac{i}{2}\left(\bar{\xi}\dot{\xi} - \dot{\bar{\xi}}\xi\right) + \frac{i}{2}\left(\bar{\pi}\dot{\pi} - \dot{\bar{\pi}}\pi\right) + B\bar{B} - m\left(\xi\bar{\pi} + \pi\bar{\xi}\right) - m^2 z\bar{z}\\ &\quad - \mu\left[\frac{i}{2}\left(z\dot{\bar{z}} - \bar{z}\dot{z}\right) + \xi\bar{\xi}\right]. \end{aligned}\tag{37}$$

The free Lagrangian (26) corresponds to the choice $\mu = 0$.

4. The long $\mathcal{N} = 4$ multiplet

In this section, we give a superfield description of the indecomposable long $\mathcal{N} = 4$ supermultiplet suggested in [12]. The standard $\mathcal{N} = 4$ supersymmetry superalgebra is formed by the (anti)commutators

$$\left\{Q^i, \bar{Q}_j\right\} = 2\delta^i_j H, \qquad \left\{Q^i, Q^j\right\} = \left\{\bar{Q}_i, \bar{Q}_j\right\} = 0, \quad H = i\partial_t\,. \tag{38}$$

The covariant $\mathcal{N} = 4$, $d = 1$ derivatives are defined in the standard way as

$$D^i = \frac{\partial}{\partial\theta_i} - i\bar{\theta}^i\partial_t\,, \quad \bar{D}_j = -\frac{\partial}{\partial\bar{\theta}^j} + i\theta_j\partial_t\,, \qquad \left\{D^i, \bar{D}_j\right\} = 2\delta^i_j H. \tag{39}$$

The $\mathcal{N} = 4$, $d = 1$ superspace coordinates $\left\{t, \theta_i, \bar{\theta}^j\right\}$ undergo the transformations:

$$\delta\theta_i = \epsilon_i, \qquad \delta\bar{\theta}^j = \bar{\epsilon}^j, \qquad \delta t = i\left(\epsilon_i\bar{\theta}^i + \bar{\epsilon}^i\theta_i\right). \tag{40}$$

The indecomposable long $\mathcal{N} = 4$ supermultiplet is parametrized by a real dimensionless parameter α and is described by the system of complex $\mathcal{N} = 4$ superfields V and W obeying the constraints

$$\bar{D}_i V = i\alpha D_i W, \qquad \bar{D}_i W = 0. \tag{41}$$

In the limit $\alpha = 0$, these constraints are reduced to those defining two ordinary $(\mathbf{2},\mathbf{4},\mathbf{2})$ chiral multiplets. The constraints (41) are solved by

$$V\left(t,\theta_i,\bar\theta^j\right) = V_0\left(t,\theta_i,\bar\theta^j\right) + i\alpha\,\bar\theta_k\frac{\partial}{\partial\theta_k}W\left(t,\theta_i,\bar\theta^j\right), \qquad \bar D_k V_0 = 0, \tag{42}$$

implying the following transformation properties for the involved objects

$$\delta V_0\left(t_L,\theta_i\right) = -\,i\alpha\,\bar\epsilon_k\frac{\partial}{\partial\theta_k}W\left(t_L,\theta_i\right), \qquad \delta W = \delta V = 0. \tag{43}$$

In components, the solution (42) reads

$$\begin{aligned} V &= y + \sqrt{2}\,\theta_i\xi^i + \theta_i\theta^i A + i\bar\theta^i\theta_i\dot y + \sqrt{2}\,i\bar\theta^j\theta_j\theta_i\dot\xi^i - \frac14\bar\theta^j\bar\theta_j\theta_i\theta^i\ddot y \\ &\quad + i\alpha\,\bar\theta_i\left(\sqrt{2}\,\psi^i + 2\theta^i C - i\bar\theta^i\dot x + \sqrt{2}\,i\theta^i\bar\theta_j\dot\psi^j\right), \\ W &= x + \sqrt{2}\,\theta_i\psi^i + \theta_i\theta^i C + i\bar\theta^i\theta_i\dot x + \sqrt{2}\,i\bar\theta^j\theta_j\theta_i\dot\psi^i - \frac14\bar\theta^j\bar\theta_j\theta_i\theta^i\ddot x, \end{aligned} \tag{44}$$

with

$$\begin{aligned} &\delta y = -\sqrt{2}\,\epsilon_i\xi^i - \sqrt{2}\,i\alpha\,\bar\epsilon_i\psi^i, \qquad \delta\xi^i = \sqrt{2}\,i\bar\epsilon^i\left(\dot y - \alpha C\right) - \sqrt{2}\,\epsilon^i A, \qquad \delta A = -\sqrt{2}\,i\bar\epsilon_i\dot\xi^i, \\ &\delta x = -\sqrt{2}\,\epsilon_i\psi^i, \qquad \delta\psi^i = \sqrt{2}\,i\bar\epsilon^i\dot x - \sqrt{2}\,\epsilon^i C, \qquad \delta C = -\sqrt{2}\,i\bar\epsilon_i\dot\psi^i. \end{aligned} \tag{45}$$

The components of W have the standard transformations inherent to the chiral multiplet $(\mathbf{2},\mathbf{4},\mathbf{2})$, while the transformations of the remaining fields y, ξ^i, A acquire additional pieces involving the components of W (they are proportional to α).

4.1. Lagrangian

The general kinetic Lagrangian is written as

$$\mathcal{L}^{\rm kin} = \frac14\int d^2\theta\, d^2\bar\theta\, f\left(V,\bar V,W,\bar W\right), \tag{46}$$

where f is just an arbitrary real function of superfields. Like in [12], we can define six bilinear invariant kinetic Lagrangians:

$$V\bar V, \quad W\bar W, \quad V\bar W + W\bar V, \quad i\left(V\bar W - W\bar V\right), \quad V^2 + \bar V^2, \quad i\left(V^2 - \bar V^2\right). \tag{47}$$

Possible terms VW and $\bar W\bar V$ do not contribute. Dependence on α remains only in the superfield V, so only five out of six bilinear kinetic Lagrangians involve the parameter α.

One can write a superpotential Lagrangian for the chiral superfield W as

$$\mathcal{L}_1^{\rm pot} = \int d^2\theta\,\mathcal{F}\left(W\right) + \int d^2\bar\theta\,\bar{\mathcal{F}}\left(\bar W\right). \tag{48}$$

Another option is to write the following superpotential Lagrangian:

$$\mathcal{L}_2^{\rm pot} = \int d^2\theta\, h'\left(W\right)V_0 + \int d^2\bar\theta\,\bar h'\left(\bar W\right)\bar V_0\,. \tag{49}$$

The transformation property (43) of V_0 allows to represent transformations of this term as

$$\delta\mathcal{L}_2^{\rm pot} = -\,i\alpha\bar\epsilon_k\int d^2\theta\,\frac{\partial}{\partial\theta_k}h\left(W\right) + \text{c.c.} = 0. \tag{50}$$

Above superpotential Lagrangians have no dependence on α since the chiral superfunction V_0 corresponds just to the limit $\alpha = 0$ of V.

To make comparison with [12], we can consider two bilinear superpotential terms

$$\gamma_1 \int d^2\theta\, V_0 W + \bar{\gamma}_1 \int d^2\bar{\theta}\, \bar{V}_0 \bar{W} \quad \text{and} \quad \gamma_2 \int d^2\theta\, W^2 + \bar{\gamma}_2 \int d^2\bar{\theta}\, \bar{W}^2. \tag{51}$$

Here, γ_1 and γ_2 are complex parameters of mass dimension. These bilinear superpotential terms in components generate the so called "Super-Zeeman" invariant terms of ref.[12], one of them containing expressions proportional to α and corresponding to a coupling to an external magnetic field. In [12], such terms were also referred to as Wess-Zumino type terms. This Wess-Zumino term can in fact be eliminated by redefining the fields (in the notations of ref. [12]) as

$$F_4 \to F_4 - \alpha\dot{\phi}_6\,, \qquad \psi_1 \to \psi_1 + \alpha\psi_8\,, \qquad \psi_2 \to \psi_2 + \alpha\psi_7\,. \tag{52}$$

Then all "Super-Zeeman" invariant terms become independent of α, and the same is true for our general superpotential Lagrangians.

In our notations, Wess-Zumino terms appear only after the elimination of auxiliary fields[1]. An example of such a term is given in the next subsection.

4.2. Free model

As an instructive example, let us consider the simple free Lagrangian given by

$$\begin{aligned} \mathcal{L}^{\text{free}} &= \frac{1}{4}\int d^2\theta\, d^2\bar{\theta} \left[V\bar{V} + \left(1-\alpha^2\right) W\bar{W}\right] + \frac{1}{4}\int d^2\theta \left(2\mu_1 V_0 + \mu_2 W\right) W \\ &\quad + \frac{1}{4}\int d^2\bar{\theta} \left(2\mu_1 \bar{V}_0 + \mu_2 \bar{W}\right) \bar{W}, \end{aligned} \tag{53}$$

where μ_1 and μ_2 are real parameters of the mass dimension. The coefficient $(1-\alpha^2)$ in front of $W\bar{W}$ was chosen to gain the correctly normalized kinetic terms in the off-shell Lagrangian:

$$\begin{aligned} \mathcal{L}^{\text{free}} &= \dot{y}\dot{\bar{y}} + \dot{x}\dot{\bar{x}} + \frac{i}{2}\left(\bar{\xi}_i\dot{\xi}^i - \dot{\bar{\xi}}_i\xi^i\right) + \frac{i}{2}\left(\bar{\psi}_i\dot{\psi}^i - \dot{\bar{\psi}}_i\psi^i\right) + A\bar{A} + \left(1+\alpha^2\right)C\bar{C} - \alpha\left(C\dot{\bar{y}} + \bar{C}\dot{y}\right) \\ &\quad + \mu_1\left(Cy + Ax + \bar{C}\bar{y} + \bar{A}\bar{x} + \xi^i\psi_i + \bar{\xi}_i\bar{\psi}^i\right) + \mu_2\left(Cx + \bar{C}\bar{x} + \frac{1}{2}\psi^i\psi_i + \frac{1}{2}\bar{\psi}_i\bar{\psi}^i\right). \end{aligned} \tag{54}$$

After eliminating the auxiliary fields A and C by their equations of motion,

$$\left(1+\alpha^2\right) C = \alpha\dot{y} - \mu_1\bar{y} - \mu_2\bar{x}, \qquad A = -\,\mu_1\bar{x}, \tag{55}$$

and neglecting a total time-derivative, we obtain the on-shell Lagrangian

$$\begin{aligned} \mathcal{L}^{\text{free}} &= \frac{\dot{y}\dot{\bar{y}}}{1+\alpha^2} + \dot{x}\dot{\bar{x}} + \frac{i}{2}\left(\bar{\xi}_i\dot{\xi}^i - \dot{\bar{\xi}}_i\xi^i\right) + \frac{i}{2}\left(\bar{\psi}_i\dot{\psi}^i - \dot{\bar{\psi}}_i\psi^i\right) + \frac{\alpha\mu_2\left(x\dot{y} + \bar{x}\dot{\bar{y}}\right)}{1+\alpha^2} - (\mu_1)^2 x\bar{x} \\ &\quad - \frac{\left(\mu_1\bar{y} + \mu_2\bar{x}\right)\left(\mu_1 y + \mu_2 x\right)}{1+\alpha^2} + \mu_1\left(\xi^i\psi_i + \bar{\xi}_i\bar{\psi}^i\right) + \frac{\mu_2}{2}\left(\psi^i\psi_i + \bar{\psi}_i\bar{\psi}^i\right). \end{aligned} \tag{56}$$

The relevant on-shell transformations are given by

$$\begin{aligned} &\delta y = -\sqrt{2}\,\epsilon_i\xi^i - \sqrt{2}\,i\alpha\,\bar{\epsilon}_i\psi^i, \quad \delta\xi^i = \frac{\sqrt{2}\,i\bar{\epsilon}^i}{1+\alpha^2}\left[\dot{\bar{y}} + \alpha\left(\mu_1\bar{y} + \mu_2\bar{x}\right)\right] + \sqrt{2}\,\mu_1\,\epsilon^i\bar{x}, \\ &\delta x = -\sqrt{2}\,\epsilon_i\psi^i, \qquad \delta\psi^i = \sqrt{2}\,i\bar{\epsilon}^i\dot{x} - \frac{\sqrt{2}\,\epsilon^i}{1+\alpha^2}\left(\alpha\dot{y} - \mu_1\bar{y} - \mu_2\bar{x}\right). \end{aligned} \tag{57}$$

[1] Such a possibility was also mentioned in [12].

The on-shell Lagrangian (56) contains the Wess-Zumino type term describing an interaction between two chiral $\mathcal{N}=4$ multiplets $(\mathbf{2},\mathbf{4},\mathbf{2})$:

$$\sim \frac{\alpha\mu_2\left(x\dot{y}+\bar{x}\dot{\bar{y}}\right)}{1+\alpha^2}. \tag{58}$$

This term matches with the statement of [12] that the elimination of auxiliary fields induces additional terms which can be treated as a coupling to an external magnetic field.

When $\mu_1=\mu_2=0$, we can make rescaling $y\rightarrow\sqrt{1+\alpha^2}\,y$ in the Lagrangian (56) and obtain the α-independent Lagrangian

$$\mathcal{L}^{\rm free}\,|_{\mu_1=\mu_2=0}=\dot{y}\dot{\bar{y}}+\dot{x}\dot{\bar{x}}+\frac{i}{2}\left(\bar{\xi}_i\dot{\xi}^i-\dot{\bar{\xi}}_i\xi^i\right)+\frac{i}{2}\left(\bar{\psi}_i\dot{\psi}^i-\dot{\bar{\psi}}_i\psi^i\right). \tag{59}$$

The transformations (57) become

$$\delta y=\frac{1}{\sqrt{1+\alpha^2}}\left[-\sqrt{2}\,\epsilon_i\xi^i-\sqrt{2}\,i\alpha\,\bar{\epsilon}_i\psi^i\right],\qquad \delta\xi^i=\frac{\sqrt{2}\,i\bar{\epsilon}^i\dot{y}}{\sqrt{1+\alpha^2}},$$
$$\delta x=-\sqrt{2}\,\epsilon_i\psi^i,\qquad \delta\psi^i=\sqrt{2}\,i\bar{\epsilon}^i\dot{x}-\frac{\sqrt{2}\,\alpha\,\epsilon^i\dot{y}}{\sqrt{1+\alpha^2}}. \tag{60}$$

The Lagrangian (59) is thus invariant under supersymmetry transformations with various parameters α, since it has no dependence on α. For instance, it is invariant under the undeformed $\alpha=0$ transformations

$$\delta y=-\sqrt{2}\,\eta_i\xi^i,\qquad \delta\xi^i=\sqrt{2}\,i\bar{\eta}^i\dot{y},\qquad \delta x=-\sqrt{2}\,\eta_i\psi^i,\qquad \delta\psi^i=\sqrt{2}\,i\bar{\eta}^i\dot{x}. \tag{61}$$

Their closure with (60) yields additional bosonic transformations

$$\delta y=a\dot{x},\qquad \delta x=\bar{a}\dot{y}, \tag{62}$$

which leave the Lagrangian (59) invariant and commute (on-shell) with the supersymmetric transformations (60) for any α. It would be interesting to see whether a similar phenomenon takes place in the interaction case too.

5. Summary and outlook

We have shown how long $\mathcal{N}=2$, $d=1$ multiplets can be embedded into $SU(2|1)$ chiral multiplets in the framework of $SU(2|1)$ supersymmetric mechanics. They naturally appear in $SU(2|1)$ mechanics of chiral multiplets, when the chiral superfield Φ_A carries some external index A with respect to the subgroup $SU(2)$ of the supergroup $SU(2|1)$. We studied this multiplet in the framework of $\mathcal{N}=2$ superspace and constructed its general superfield action.

We considered the long $\mathcal{N}=4$ multiplet [12] within the standard $\mathcal{N}=4$ superspace. Defining and solving the constraint (41), we obtained the superfields V and W describing the long multiplet $(\mathbf{4},\mathbf{8},\mathbf{4})_1$ and constructed general superfield Lagrangians consisting of the kinetic (sigma-model type) and superpotential Lagrangians. We considered the free Lagrangian (53), where the superpotential term $\sim\mu_2$ is responsible for appearing Wess-Zumino type term in the on-shell Lagrangian (56).

In conclusion, we outline some further possible lines of investigation.

- Quantization of the model (56) and construction of the Hilbert space of wave functions.
- Study of some other generalizations of long multiplets to the standard flat $\mathcal{N}=4$, $d=1$ supersymmetry [10, 11].

- Generalizing the constraints (41) to the $SU(2|1)$ covariantized constraints:

$$\bar{\mathcal{D}}_k V = i\alpha \mathcal{D}_k W, \qquad \bar{\mathcal{D}}_k W = 0. \tag{63}$$

  Such a generalization is possible for the second type of the $SU(2|1)$ chirality [6], when the spinor derivatives are inert under the induced $U(1)$ transformations.
- Answering the question whether it is possible to find out $d > 1$ analogs of long multiplets.

Acknowledgments
E.I. and S.S. thank the organizers of the conference ISQS24 for the kind hospitality in Prague. S.S. acknowledges support from the Votruba-Blokhintsev program. The work of E.I. and S.S. was partially supported by the Russian Science Foundation, Grant No. 16-12-10306. A.R.M. thanks the Directorate of BLTP and JINR for the opportunity to work in a scientific atmosphere.

References

[1] Ivanov E and Sidorov S 2014 Deformed supersymmetric mechanics *Class. Quant. Grav.* **31** 075013 (*Preprint* 1307.7690 [hep-th])
[2] Smilga A V 2004 Weak supersymmetry *Phys. Lett.* B **585** 173 (*Preprint* hep-th/0311023)
[3] Bellucci S and Nersessian A 2004 *Supersymmetric Kähler oscillator in a constant magnetic field* (*Preprint* hep-th/0401232)
[4] Römelsberger C 2006 Counting chiral primaries in $\mathcal{N} = 1$, $d = 4$ superconformal field theories *Nucl. Phys.* B **747** 329-353 (*Preprint* hep-th/0510060)
[5] Römelsberger C 2007 Calculating the superconformal index and Seiberg duality (*Preprint* 0707.3702 [hep-th])
[6] Ivanov E and Sidorov S 2014 Super Kähler oscillator from $SU(2|1)$ superspace *J. Phys.* A **47** 292002 (*Preprint* 1312.6821 [hep-th]).
[7] Ivanov E, Sidorov S and Toppan F 2015 Superconformal mechanics in $SU(2|1)$ superspace *Phys. Rev.* D **91** 085032 (*Preprint* 1501.05622 [hep-th])
[8] Ivanov E and Sidorov S 2016 $SU(2|1)$ mechanics and harmonic superspace *Class. Quant. Grav.* **33** 055001 (*Preprint* 1507.00987 [hep-th])
[9] Assel B, Cassani D, Di Pietro L, Komargodski Z, Lorenzen J and Martelli D 2015 The Casimir energy in curved space and its supersymmetric counterpart *J. High Energy Phys.* JHEP07(2015)043 (*Preprint* 1503.05537 [hep-th])
[10] Gonzales M, Khodaee S and Toppan F 2011 On non-minimal N=4 supermultiplets in 1D and their associated sigma-models *J. Math. Phys.* **52** 013514 (*Preprint* 1006.4678 [hep-th])
[11] Gonzales M, Iga K, Khodaee S and Toppan F 2012 Pure and entangled $\mathcal{N} = 4$ linear supermultiplets and their one-dimensional sigma-models *J. Math. Phys.* **53** 103513 (*Preprint* 1204.5506 [hep-th])
[12] Hubsch T and Katona G A 2016 A Q-continuum of off-shell supermultiplets *Adv. High Energy Phys.* **2016** 7350892 (*Preprint* 1310.3256 [hep-th])
[13] Ivanov E and Sidorov S 2016 Long multiplets in supersymmetric mechanics *Phys. Rev.* D **93** 065052 (*Preprint* 1509.05561 [hep-th])

IOP Conf. Series: Journal of Physics: Conf. Series **804** (2017) 012022 doi:10.1088/1742-6596/804/1/012022

Orthogonal and symplectic Yangians: oscillator realization

D Karakhanyan[1] **and R Kirschner**[2]

[1] Yerevan Physics Institute, 2 Alikhanyan br., 0036 Yerevan, Armenia
[2] Institut für Theoretische Physik, Universität Leipzig, 04009 Leipzig, Germany

E-mail: karakhan@yerphi.am

E-mail: roland.kirschner@itp.uni-leipzig.de

Abstract. We solve RLL-relation with fundamental orthogonal/symplectic R-matrix examining polynomial with respect to u expressions for L-operator. We use creation-annihilation operators to construct a realization of Yangian generators.

1. Introduction

The simplest and well known mechanism of appearance the projectivity of unitary representations is related to Heisenberg group. Upon passing to central extension in simplest quantization models the Planck constant appears as a measure of the central charge.

In infinite-dimensional case the projective representations appear as the groups of automprphysms of canonical (anti-)commutation relations, i.e. Metaplectic and Metagonal groups.

In other words, if the given group of automorphysms of some algebra (say, Clifford algebra) or group (say, Heisenberg group) maps the Fock representation (fermionic or bosonic) to itself, then in that representation space appears, generally speaking, projective representation of that group of automorphysm.

In finite-dimensional case the projective representation of the orthogonal group by automorphysms of Clifford algebra leads Cartan in 1913 to discovery of spinor group, which provides the twofold covering of the orthogonal group. Later A. Weil shown that S^1- covering of the symplectic group is reduced to the $\mathbb{Z}_2$ one [1]. The similarly S^1-extension of the orthogonal group is Metagonal group. It can be illustrated by the following scheme:

$$MO(2n,\mathbb{R}) \supset Spin(2n,\mathbb{R}) \longrightarrow SO(2n\mathbb{R}),$$

$$Mp(n,\mathbb{R}) \supset Sp_2(n,\mathbb{R}) \longrightarrow Sp(n\mathbb{R}),$$

here the Metagonal group $MO(2n,\mathbb{R})$ includes $Spin(2n,\mathbb{R})$ as S^1 covering, which is in turn is $\mathbb{Z}_2$ covering of $SO(2n\mathbb{R})$. Similarly the Metaplectic group contains *Spinsymplectic* $Sp_2(n,\mathbb{R})$ and symplectic $Sp(n\mathbb{R})$ groups [2], [3].

So one can conclude from these considerations that the algebra of fermion and boson oscillators is naturally related to orthogonal and symplectic symmetry.

2. Yang-Baxter relations

In this work we shall follow to notations of our previous work.

The fundamental Yang-Baxter equation:

$$\begin{gathered} R^{a_1a_2}_{b_1b_2}(u)R^{b_1a_3}_{c_1b_3}(u+v)R^{b_2b_3}_{c_2c_3}(v) = R^{a_2a_3}_{b_2b_3}(v)R^{a_1b_3}_{b_1c_3}(u+v)R^{b_1b_2}_{c_1c_2}(u) \Rightarrow \\ R_{12}(u)R_{13}(u+v)R_{23}(v) = R_{23}(v)R_{13}(u+v)R_{12}(u) \,. \end{gathered} \tag{1}$$

has following solution:

$$R^{a_1a_2}_{b_1b_2}(u) = u(u+\frac{n}{2}-\epsilon)I^{a_1a_2}_{b_1b_2} + (u+\frac{n}{2}-\epsilon)P^{a_1a_2}_{b_1b_2} - \epsilon\, u\, K^{a_1a_2}_{b_1b_2} \,, \tag{2}$$

where

$$I^{a_1a_2}_{b_1b_2} = \delta^{a_1}_{b_1}\delta^{a_2}_{b_2} \,, \quad P^{a_1a_2}_{b_1b_2} = \delta^{a_1}_{b_2}\delta^{a_2}_{b_1} \,, \quad K^{a_1a_2}_{b_1b_2} = \varepsilon^{a_1a_2}\,\varepsilon_{b_1b_2} \,, \tag{3}$$

and the choices $\epsilon = +1$ and $\epsilon = -1$ correspond to the $SO(n)$ and $Sp(n)$ cases respectively. We note that the R-matrix (2) is invariant under the adjoint action of any real form (related to the metric ε^{ab}) of the complex groups $SO(n,\mathbb{C})$ and $Sp(n,\mathbb{C})$. Here ε_{ab} is a non-degenerate invariant metric in V

$$\varepsilon_{ab} = \epsilon\,\varepsilon_{ba} \,, \quad \varepsilon_{ab}\varepsilon^{bd} = \delta^d_a \,, \tag{4}$$

which is symmetric $\epsilon = +1$ for $SO(n)$ case and skew-symmetric $\epsilon = -1$ for $Sp(n)$ case. We denote by ε^{bd} (with upper indices) the elements of the inverse matrix ε^{-1}. Namely the existence of the invariant tensors ε_{ab} leads to the above mentioned problems in $SO(n)$ and $Sp(n)$ cases, e.g., it causes a third term in the corresponding expressions of R-matrices and leads to the dependence on the spectral parameter of second power.

Let the index range be $a_1, a_2, \cdots = 1, \ldots, n$ for the $SO(n)$ case and $a_1, a_2, \cdots = -m, \ldots, -1, 1, \ldots, m$ – for the $Sp(n)$ $(n = 2m)$ case. For the choice $\varepsilon^{a_1a_2} = \delta^{a_1a_2}$ in the $SO(n)$ case we have

$$R^{a_1a_2}_{b_1b_2}(u) = u(u+\beta)\delta^{a_1}_{b_1}\delta^{a_2}_{b_2} + (u+\beta)\delta^{a_1}_{b_2}\delta^{a_2}_{b_1} - u\delta^{a_1a_2}\delta_{b_1b_2} \,, \quad \beta = (n/2-1) \,, \tag{5}$$

and for the choice $\varepsilon^{a_1a_2} = \epsilon_{a_2}\delta^{a_1,-a_2}$ (here $\epsilon_a = sign(a)$ and $\varepsilon^{ab} = -\varepsilon_{ab}$) in the $Sp(2m)$ case we have

$$R^{a_1a_2}_{b_1b_2}(u) = u(u+\beta)\delta^{a_1}_{b_1}\delta^{a_2}_{b_2} + (u+\beta)\delta^{a_1}_{b_2}\delta^{a_2}_{b_1} - u\epsilon_{a_2}\varepsilon_{b_2}\delta^{a_1,-a_2}\delta^{b_1,-b_2} \,, \quad \beta = (m+1) \,. \tag{6}$$

3. Linear resolution of $g\ell(n)$ Yangian

The simplest form fundamental R-matrix acquires in the case of $g\ell(n)$ symmetry:

$$R_{12}(u) = uI_{12} + P_{12}, \tag{7}$$

and choosing linear ansatz for L-operator: $L(u) = u + G$ one deduces that RLL-relation:

$$R_{12}(u)L_1(u+v)L_2(v) = L_2(v)L_1(u+v)R_{12}(u),$$

consists of contribution, proportional to u:

$$[G_1 + P_{12}, G_2] = 0 = [G_1, G_2 + P_{12}]. \tag{8}$$

Rewriting this relation in matrix indices:

$$[G^{a_1}{}_{c_1}, G^{a_1}{}_{c_1}] = \delta^{a_1}{}_{c_2}G^{a_2}{}_{c_1} - \delta^{a_2}{}_{c_1}G^{a_1}{}_{c_2}, \tag{9}$$

one immediately recognizes $g\ell(n)$-algebra.

So in the case of $g\ell(n)$ symmetry the linear ansatz for L-operator solves RLL-relations with fundamental R-matrix upon restriction that generators G realize $g\ell(n)$-algebra.

4. Truncated Yangians of *so* and *sp* types

Let $\mathcal{G}$ be the Lie algebra $so(n)$ or $sp(2m)$ $(2m = n)$. Then the corresponding Yangian $Y(\mathcal{G})$ is defined [10] as an associative algebra with the infinite number of generators $(L^{(k)})^a_b$ with $(L^{(0)})^a_b = I\delta^a_b$, where I is an unit element in $Y(\mathcal{G})$, and $(L^{(k)})^a_b$ for $k > 0$ satisfy the quadratic defining relations. The generators $(L^{(k)})^a_b \in Y(\mathcal{G})$ are considered as coefficients in the expansion of the L-operator

$$L^a_b(u) = \sum_{k=0}^{\infty} \frac{(L^{(k)})^a_b}{u^k} \; , \quad L^{(0)} = I \; , \tag{10}$$

which satisfies to the Yang-Baxter RLL-relations

$$\begin{gathered} R^{a_1 a_2}_{b_1 b_2}(u-v) L^{b_1}_{c_1}(u) L^{b_2}_{c_2}(v) = L^{a_2}_{b_2}(v) L^{a_1}_{b_1}(u) R^{b_1 b_2}_{c_1 c_2}(u-v) \;\Leftrightarrow \\ R_{12}(u-v)\, L_1(u)\, L_2(v) = L_2(v)\, L_1(u)\, R_{12}(u-v) \; . \end{gathered} \tag{11}$$

which play role of the Yangin $Y(\mathcal{G})$ defining relations.

Here $R^{a_1 a_2}_{b_1 b_2}(u-v)$ is the Yang-Baxter R-matrix (2).

The defining relations (11) are homogeneous in L and allow the redefinition $L(u) \to f(u)\, L(u + b_0)$ with any scalar function $f(u) = 1 + b_1/u + b_2/u^2 + \dots$, where b_i are parameters. The Yangian (2), (11) possesses the set of automorphisms

$$L(u) \to \frac{(u-a)^k}{u^k} L(u) \; , \qquad (k = 1, 2, \dots) \; ,$$

where a is a constant (in general a is a central element in $Y(\mathcal{G})$). At $k = 1$ the generators $L^{(j)}$ are transforming as

$$L^{(1)} \to L^{(1)} - aI_n \; , \quad L^{(2)} \to L^{(2)} - a\,L^{(1)} \; , \quad L^{(3)} \to L^{(3)} - aL^{(2)} \; , \quad \dots , \tag{12}$$

Taking $a = \frac{1}{n}\mathrm{Tr}(L^{(1)})$ one can fix $L^{(1)}$ to be traceless. Note that $\mathrm{Tr}(L^{(1)})$ is central element in $Y(\mathcal{G})$).

One has for the fundamental R-matrix (2):

$$\tfrac{1}{u^2v^2} R(u-v) = (\tfrac{1}{v} - \tfrac{1}{u})(\tfrac{1}{v} - \tfrac{1}{u} + \tfrac{\beta}{uv}) - (\tfrac{1}{uv^2} - \tfrac{1}{u^2v} + \tfrac{\beta}{u^2v^2})P - \epsilon(\tfrac{1}{uv^2} - \tfrac{1}{u^2v})K. \tag{13}$$

here $\beta = (\frac{n}{2} - \epsilon)$. Then the defining relations for the generators $(L^{(k)})^a_b$ of the Yangians $Y(\mathcal{G})$:

$$\begin{gathered} [L_1^{(k)},\, L_2^{(j-2)}] - 2[L_1^{(k-1)},\, L_2^{(j-1)}] + [L_1^{(k-2)},\, L_2^{(j)}] + \\ +\beta([L_1^{(k-1)},\, L_2^{(j-2)}] - [L_1^{(k-2)},\, L_2^{(j-1)}]) + \\ + P\Big(L_1^{(k-1)}\, L_2^{(j-2)} - L_1^{(k-2)}\, L_2^{(j-1)} + \beta L_1^{(k-2)}\, L_2^{(j-2)}\Big) - \\ - \Big(L_2^{(j-2)} L_1^{(k-1)} - L_2^{(j-1)} L_1^{(k-2)} + \beta L_2^{(j-2)}\, L_1^{(k-2)}\Big) P + \\ +\epsilon\Big(K\,(L_1^{(k-2)}\, L_2^{(j-1)} - L_1^{(k-1)}\, L_2^{(j-2)}) - (L_2^{(j-1)} L_1^{(k-2)} - L_2^{(j-2)} L_1^{(k-1)})\,K\Big) = 0 \; , \end{gathered} \tag{14}$$

where the operators K, P are given in (3), $\epsilon = +1$ for $\mathcal{G} = so(n)$ and $\epsilon = -1$ for $\mathcal{G} = sp(2m)$.

The defining relations are obviously completed for infinite set of generators, but it can be truncated at finite values as well, which will impose the additional restrictions on generators $(L^{(k)})^a_b$

At $k=1$ we obtain from (14) the set of relations

$$[L_1^{(1)},\, L_2^{(j-2)}] = -\left[(P_{12} - \epsilon K_{12}),\, L_2^{(j-2)}\right], \quad (\forall\, j), \tag{15}$$

which in particular lead to the statement that $\mathrm{Tr}(L^{(1)})$ is a central element in $Y(\mathcal{G})$: $[\mathrm{Tr}(L^{(1)}),\ (L^{(j)})^a{}_b] = 0\ (\forall j)$. For $j=3$ we deduce from (15) the defining relations for the Lie algebra generators $G^a{}_b \equiv -(L^{(1)})^a{}_b$:

$$[G_1,\, G_2] = [(P_{12} - \epsilon K_{12}),\, G_2]\ . \tag{16}$$

Permutation of indices $1 \leftrightarrow 2$ in this equation gives the consistency conditions and the same conditions are obtained from (14) directly.

$$K_{12}\,(G_1 + G_2) = (G_1 + G_2)\, K_{12}\ ,$$

Acting on this equation by K_{12} from the left (or by K_{12} from the right) we write it as

$$K_{12}\,(G_1 + G_2) = \frac{2}{n}\,\mathrm{Tr}(G)\, K_{12} = (G_1 + G_2)\, K_{12}\ , \tag{17}$$

where we have used

$$K_{12}^2 = \epsilon\, n\, K_{12}\ , \quad K_{12}\, N_1\, K_{12} = K_{12}\, N_2\, K_{12} = \epsilon\, \mathrm{Tr}(N)\, K_{12}\ , \tag{18}$$

Here N is any $n \times n$ matrix.

Then, according to the automorphism (12) we redefine the elements $G \to G - \frac{1}{n}\,\mathrm{Tr}(G)$ in such a way that for the new generators we have $\mathrm{Tr}(G) = 0$ This leads to the (anti)symmetry conditions for the generators:

$$\begin{aligned} K_{12}\,(G_1 + G_2) = 0 = (G_1 + G_2)\, K_{12} \quad &\Rightarrow \\ G^d{}_a\, \varepsilon_{db} + \varepsilon_{ad}\, G^d{}_b = 0\ . \end{aligned} \tag{19}$$

The equations (16) and (19) for $\epsilon = +1$ and $\epsilon = -1$ define the Lie algebra $\mathcal{G} = so(n)$ and $\mathcal{G} = sp(2m)$ $(2m = n)$, respectively. The defining relations (16) and (anti)symmetry condition (19) for the generators $G_{ab} = \varepsilon_{ad}\, G^d{}_b$ can be written in the familiar form

$$[G_{ab},\, G_{cd}] = \varepsilon_{cb} G_{ad} + \varepsilon_{db} G_{ca} + \varepsilon_{ca} G_{db} + \varepsilon_{da} G_{bc}\ , \qquad G_{ab} = -\epsilon\, G_{ba}\ . \tag{20}$$

This means (see [10]) that an enveloping algebra $\mathcal{U}(\mathcal{G})$ of the Lie algebra $\mathcal{G} = so(n), sp(2m)$ is always a subalgebra in the Yangian $Y(\mathcal{G})$.

5. L operators

The Yangian $Y(\mathcal{G})$ (14) with infinite number of generators can be truncated to some set with finite number of ones restricted to some special representation. We will suppose that the Yangian representation space is realized as a space of (matrix) polynomials of So (Sp)-generators G and start with consideration of some general properties of such monomials. Then, thinking generators G and spectral parameter as a dimensionfull variables (their ratio and L-operator are dimensionless) one will describe the Yangian as $\mathbb{Z}$ graded space with finite-dimensional components corresponding to irreducible representations.

1. Linear evaluation of $Y(\mathcal{G})$.

We put equal to zero all generators $L^{(k)} \in Y(\mathcal{G})$ with $k > 1$ and take the L-operator linear by spectral parameter:

$$L^a_b(u) = u\delta^a_b + G^a{}_b. \tag{21}$$

Then defining RLL-relation takes the form:

$$(u(u+\beta)I_{12} + (u+\beta)P_{12} - u\epsilon K_{12})(u+v+G_1)(v+G_2) =$$

$$= (v+G_2)(u+v+G_1)(u(u+\beta)I_{12} + (u+\beta)P_{12} - u\epsilon K_{12}).$$

This relation can be rewritten as follows:

$$(u+\beta)\Big([G_1, G_2] + (G_1 - G_2)P_{12} - \epsilon[K_{12}, G_2]\Big) - \epsilon v[K_{12}, G_1 + G_2] -$$

$$-\epsilon K_{12}(G_1 - \beta)G_2 + \epsilon G_2(G_1 - \beta)K_{12} = 0,$$

and has to take place identically by powers of u and v, implying three restrictions on generators G:

$$-v\mathfrak{C}^{(1,1)} = -\epsilon v[K_{12}, G_1 + G_2] = 0, \tag{22}$$

as a coefficient at v, which is takes place upon (22),

$$(u+\beta)\mathfrak{C}^{(1,2)} = (u+\beta)\Big([G_1, G_2] + (G_1 - G_2)P_{12} - \epsilon[K_{12}, G_2]\Big) = 0, \tag{23}$$

as a coefficient at u and expresses the SO (Sp)-algebra relations. The remaining terms are:

$$-\mathfrak{C}^{(1,3)} = -\epsilon\Big(K_{12}(G_1 - \beta)G_2 - G_2(G_1 - \beta)K_{12}\Big) = 0. \tag{24}$$

One can summarize constraints following from the Yang-Baxter RLL-relation for linear ansatz (21) as:

$$\mathfrak{C}^{(1,1)}[\epsilon K_{12}, G_1 + G_2] = 0, \tag{25}$$

$$\mathfrak{C}^{(1,2)} = [G_1 + P_{12} - \epsilon K_{12}, G_2] = 0, \tag{26}$$

$$\mathfrak{C}^{(1,3)} = [\epsilon K_{12}, G_1 G_2 + G_2 G_1] = 0. \tag{27}$$

Introducing graded oscillators $c_a = \varepsilon_{ab} c^b$:

$$c^a c^b + \epsilon c^b c^a = \varepsilon^{ab}, \qquad c_a c^b + \epsilon c^b c_a = \delta^b_a, \qquad c_a c_b + \epsilon c_b c_a = \varepsilon_{ba}, \tag{28}$$

in orthogonal case ($\epsilon = +1$) operators c_a will have Grassmann nature and representation is finite-dimensional, while in symplectic case one has the ordinary oscillators and an infinite-dimensional representation.

Note, that (28) is just "half" of set of oscillators. Indeed, the algebra of automorphysms, corresponding to Bogolyubov transformation of the set of Fermi (Bose) creation-annihilation operators is $O(2n)$ ($Sp(2n)$).

Then one can solve RLL-constraints by simple bilinear ansatz

$$G^{(1)a}{}_b = -c^a c_b + \frac{\epsilon}{2}\delta^a_b = \frac{\epsilon}{2}(c_b c^a - \epsilon c^a c_b) = \epsilon(c_b c^a - \frac{1}{2}\delta^a_b), \tag{29}$$

ensuring symmetry properties of G: expressions with lower and upper indices are $G^{(1)}{}_{ab} = -c_a c_b + \frac{\epsilon}{2}\varepsilon_{ab}$, $G^{(1)ab} = \epsilon(-c^a c^b + \frac{1}{2}\varepsilon^{ab})$. The overall coefficient is chosen according to (26).

IOP Conf. Series: Journal of Physics: Conf. Series **804** (2017) 012022 doi:10.1088/1742-6596/804/1/012022

Note that the metric is the unique invariant tensor and the single invariant combination:

$$\varepsilon^{ab}c_a c_b = \frac{1}{2}\varepsilon^{ab}(c_a c_b + \epsilon c_b c_a) = \frac{n}{2}, \tag{30}$$

can be constructed in the case of one set of oscillators.

2. Quadratic evaluation of $Y(\mathcal{G})$.

Now we put all generators $L^{(k)} \in Y(\mathcal{G})$ with $k > 2$ equal to zero. In this case the L-operator (10), after a multiplication by u^2, can be written in the form

$$L(u) = u(u-a) + u\,G' + H\,. \tag{31}$$

where we introduce

$$L^{(1)} = G' - a = G\,, \quad L^{(2)} = H\,, \tag{32}$$

where the constant a is chosen to be TrG', so the generator G is traceless.

We are going to consider the general quadratic ansatz

$$L(u) = u(u-a) + u\,G' + H = u^2 + uG + H\,, \tag{33}$$

and to solve the restrictions imposed by RLL Yang-Baxter relation.

The defining relation (11) has the form:

$$[u(u+\beta)I_{12} + (u+\beta)P_{12} - \epsilon u K_{12}]((u+v)^2 + (u+v)G_1 + H_1)(v^2 + vG_2 + H_2) - \tag{34}$$

$$(v^2 - vG_2 + H_2)((u+v)^2 - (u+v)G_1 + H_1)[u(u+\beta)I_{12} + (u+\beta)P_{12} - \epsilon u K_{12}] = 0.$$

This relation must take place at arbitrary values spectral parameters u and v, i.e. the coefficients at independent monomials $u^k v^r$ ($k + r \le 4$ must vanish. The l.h.s. of (34) can be represented as a sum of following eight combinations:

$$\mathfrak{C}^{(2,1)} = \epsilon[K_{12}, G_1 + G_2] = 0, \tag{35}$$

$$\mathfrak{C}^{(2,2)} = [G_1 + P_{12} - \epsilon K_{12}, G_2] = [G_1, G_2 - P_{12} - \epsilon K_{12}] = 0, \tag{36}$$

$$\mathfrak{C}^{(2,3)} = [K_{12}, H_1 - \frac{1}{2}(G_1^2 + G_2^2) + H_2] = 0, \tag{37}$$

$$\mathfrak{C}^{(2,4)} = [G_1 + P_{12} - \epsilon K_{12}, H_2] = 0, \tag{38}$$

$$\mathfrak{C}^{(2,5)} = [H_1, G_2 - P_{12} + \epsilon K_{12}] = 0, \tag{39}$$

$$\mathfrak{C}^{(2,6)} = [K_{12}, H_1 G_2 + H_2 G_1 + G_1 H_2 + G_2 H_1] = [K_{12}, \{H_1, G_2\} + \{H_2, G_1\}] = 0. \tag{40}$$

$$\mathfrak{C}^{(2,7)} = \Big([H_1, H_2] + (G_2 H_1 - H_2 G_1)P_{12} - \frac{\epsilon}{2}[K_{12}, \{G_1, H_2 - \frac{\epsilon}{2}G_2\}]\Big) = 0, \tag{41}$$

$$\mathfrak{C}^{(2,8)} = \epsilon\Big(K_{12}(H_1 - \beta G_1 + \beta^2)H_2 - H_2(H_1 - \beta G_1 + \beta^2)K_{12}\Big) = 0. \tag{42}$$

Let us now turn to oscillator realization of generators G and H.

Consider first the case of one set of oscillators. As we prove above the most general expression for G is given then by (29), for which constraints (35-36) take place. Then equation (35) allows to determine ε-symmetric traceless part of H, which is zero, because the only ε-symmetric tensor of second rank constructed from oscillators (28) is metric tensor. Then due to identity:

$$(G^{(1)})^2_{ab} + \beta G^{(2)}{}_{ab} = \frac{1}{4}(n\epsilon - 1)\varepsilon_{ab}, \tag{43}$$

one deduces:

$$H^{(1)}_{ab} = \alpha G^{(1)}{}_{ab} + \gamma \varepsilon_{ab},$$

where α and γ are arbitrary constants. In other words $H^{(1)}_{ab}$ is an arbitrary tensor, constructed from oscillators (28). Substituting $H^{(1)}$ and $G^{(1)}$ into remaining constraints one obtains identity at $\gamma = -\alpha^2$ and the quadratic ansatz for L-operator will take the form:

$$L(u) = u^2 + (u+\alpha)G^{(1)} - \alpha^2 = (u+\alpha)(u-\alpha+G),$$

which is just shifted linear ansatz up to the overall factor and one comes to:

Proposition *Operator dependence of RLL-relations on single set of oscillators leads to the linear solution.*

Let us now turn to the case of two sets of oscillators.

Oscillators have the natural grading (28). From the other hand in our previous work mentioned that a particular solution corresponding to quadratic ansatz is given by Jordan-Schwinger construction:

$$G_{ab} = x_b \partial_a - \epsilon x_a \partial_b, \tag{44}$$

where the variables x_a have an opposite grading (even for So- and odd for Sp-case):

$$\partial_a \partial_b - \epsilon \partial_b \partial_a = 0, \qquad x_a x_b - \epsilon x_b x_a = 0, \qquad \partial_a x_b - \epsilon x_b \partial_a = \varepsilon_{ab}. \tag{45}$$

Labeling oscillators by index $\alpha = 1, 2$:

$$d^1_a = x_a, \qquad d^2_a = \partial_a, \tag{46}$$

one can rewrite commutation relations (**??**) in more compact form:

$$d^\alpha_a d^\beta_b - \epsilon d^\beta_b d^\alpha_a = -\epsilon^{\alpha\beta} \varepsilon_{ab}, \tag{47}$$

where $\epsilon^{\alpha\beta}$ is two-dimensional Levi-Civita tensor ($\epsilon^{12} = 1 = -\epsilon^{21}$, $\epsilon^{11} = 0 = \epsilon^{22}$).

The bilinear ansatz (44) is compatible with cyclic identity:

$$\{G_{ab}, G_{cd}\} + \{G_{ca}, G_{bd}\} + \{G_{bc}, G_{cd}\} = 0, \tag{48}$$

upon substitution $G_{ab} = G^{(0)}_{ab}$ relation (48) becomes identity. The identity (48) is necessary to satisfy the constraints (35-42) following from the RLL-relation.

Then it is not hard to check that along with $G^{(0)}_{c_1 c_2} = \frac{1}{2}\epsilon_{\alpha\beta}(d^\alpha_{c_2} d^\beta_{c_1} - \epsilon d^\alpha_{c_1} d^\beta_{c_2})$ the relation:

$$H = G^2 + \beta G, \tag{49}$$

solves all constraints and provides the quadratic resolution through oscillators (47).

6. Oscillator realization

Let us consider the most general expression for G_{ab} depending on oscillators (28):

$$G_{ab} = \sum_{k=0}^{\infty} \mathrm{G}^{(k)}_{ab} = \sum_{k=0}^{\infty} \sum_{e_1,\ldots,e_k} A_{ab,e_1\ldots e_k} c^{e_1} \ldots c^{e_k}, \tag{50}$$

and try to solve constraints (25-27). In this expansion the operator G_{ab} and the coefficients $A_{ab,e_1\ldots e_k}$ is in one-to-one correspondence if the summation in (50) is ordered, say normally,

IOP Conf. Series: Journal of Physics: Conf. Series **804** (2017) 012022 doi:10.1088/1742-6596/804/1/012022

when all creation operators c_a with $a > 0$ stand from the left of annihilation ones c_a with $a < 0$. This restriction on summation rule can be lifted if we redefine coefficients A to be ε-antisymmetric:

$$A_{ab,e_1\ldots e_i\ldots e_j\ldots e_k} = -\epsilon A_{ab,e_1\ldots e_j\ldots e_i\ldots e_k}. \tag{51}$$

Indeed, any tensor $A_{\ldots e_i\ldots e_j\ldots}$ can be represented as the sum of ε-symmetric and ε-antisymmetric parts:

$$A_{\ldots e_i\ldots e_j\ldots} = \frac{1}{2}(A_{\ldots e_i\ldots e_j\ldots} + \epsilon A_{\ldots e_j\ldots e_i\ldots}) + \frac{1}{2}(A_{\ldots e_i\ldots e_j\ldots} - \epsilon A_{\ldots e_j\ldots e_i\ldots}) = A'_{\ldots e_i\ldots e_j\ldots} + A''_{\ldots e_i\ldots e_j\ldots}, \tag{52}$$

and upon contraction with oscillators the symmetric part becomes reducible

$$A'_{\ldots e_i\ldots e_j\ldots} c^{e_i} c^{e_j} = \frac{1}{2}(A_{\ldots e_i\ldots e_j\ldots} + \epsilon A_{\ldots e_j\ldots e_i\ldots}) c^{e_i} c^{e_j} = \frac{1}{2} A_{\ldots e_i\ldots e_j\ldots} \varepsilon^{e_i e_j}.$$

So without the loss of generality one can suppose coefficients $A_{ab,e_1\ldots e_i\ldots e_j\ldots e_k}$ ε-antisymmetric and traceless by indices e_i:

$$A_{ab,e_1\ldots e_i\ldots e_j\ldots e_k} \varepsilon^{e_i e_j} = 0. \tag{53}$$

In the first relation:

$$G_{c_1c_2} + \epsilon G_{c_2c_1} = G^b{}_b \varepsilon_{c_1c_2}, \tag{54}$$

generator G can be chosen traceless without loss of generality. Then substituting expansion (50) into (54) one obtains:

$$A_{c_1c_2,e_1\ldots e_k} + \epsilon A_{c_2c_1,e_1\ldots e_k} = 0, \quad \Rightarrow \quad \varepsilon^{ab} A_{ab,e_1\ldots e_k} = 0. \tag{55}$$

The first two indices a, b are different from the remaining ones $e_1 \ldots e_k$. Indeed, generator $G^a{}_b$ should rotate tensor quantities, which determine the possible form of corresponding commutator. In particular, the construction (29) can serve as generator, so we start with the simplest quadratic ansatz (29) and examine the most general expression quadratic by oscillators:

$$[c^{a_1} c_{c_1} - \frac{\epsilon}{2}\delta^{a_1}_{c_1}, (\mathrm{G}^{(k)})^{a_2}{}_{c_2}]. \tag{56}$$

The commutator with $A^a{}_b$ zeroth term of expansion (50) vanishes, because these are c-numbers, not operators. Similarly the first term is vanishes too, because we have no any chosen vector. For quadratic term of expansion one has:

$$[c^{a_1} c_{c_1} - \frac{\epsilon}{2}\delta^{a_1}_{c_1}, (\mathrm{G}^{(1)})^{a_2}{}_{c_2}] = [c^{a_1} c_{c_1} - \frac{\epsilon}{2}\delta^{a_1}_{c_1}, A^{a_2}{}_{c_2,bd} c^b c^d] = A^{a_2}{}_{c_2,c_1d} c^{a_1} c^d - A^{a_2}{}_{c_2,bd} \varepsilon^{ba_1} c_{c_1} c^d +$$

$$+ A^{a_2}{}_{c_2,bc_1} c^b c^{a_1} - A^{a_2}{}_{c_2,bd} \varepsilon^{da_1} c^b c_{c_1} = 2A^{a_2}{}_{c_2,ed}(\varepsilon^{ea_1}\varepsilon_{c_1b} - \delta^e_{c_1}\delta^{a_1}_b) c^b c^d =$$

$$= (\delta^{a_1}_{c_2} A^{a_2}{}_{c_1,bd} - \delta^{a_2}_{c_1} A^{a_1}{}_{c_2,bd} + \varepsilon^{a_2a_1} A_{c_1c_2,bd} - \varepsilon_{c_1c_2} A^{a_2a_1}{}_{bd}) c^b c^d.$$

The only chosen tensor we can operate is metric ε_{ab}, so the second term of expansion is given by:

$$\mathrm{G}^{(1)}_{ab} = A_{ab,e_1e_2} c^{e_1} c^{e_2} = -\frac{1}{2}(\varepsilon_{ae_1}\varepsilon_{be_2} - \epsilon\varepsilon_{be_1}\varepsilon_{ae_2}) c^{e_1} c^{e_2} = -c_a c_b - \frac{\epsilon}{2}\varepsilon_{ab}. \tag{57}$$

Following to these arguments one deduces that the higher term of expansion (50) vanish because commutator (56) contains too many terms and one deduces that the linear by oscillators term of the expansion is unappropriate

So the ansatz (29) is the most general in this case.

Suppose now that we have more than one set of oscillators, which are ε-commutative or ε-anticommutative each to other such that the bilinear combinations (57) are commutative:

$$\mathrm{G}_{i,ab}\mathrm{G}_{j,ab} = \mathrm{G}_{j,ab}\mathrm{G}_{i,ab}, \qquad i \neq j. \tag{58}$$

One can suppose the following commutation relations between oscillators:

$$c_i^a c_j^b + \epsilon c_j^b c_i^a = \delta_{ij}\varepsilon^{ab}, \qquad 1 \leq i,j \leq N. \tag{59}$$

One can construct the quadratic combinations from our oscillators:

$$\mathrm{C}_{ij}^{ab} = c_i^a c_j^b, \tag{60}$$

Consider first N diagonal ones C_{jj}^{ab}: the ε-antisymmetric by group indices combinations are:

$$\mathrm{G}_j^{ab} = \epsilon(c_j^a c_j^b - \frac{1}{2}\varepsilon^{ab}), \tag{61}$$

while the ε-symmetric ones are equal to metric tensor. The $N(N-1)/2$ non-diagonal combinations C_{ij}^{ab} $i<j$ can be separated by ε-symmetry:

$$\mathrm{S}_{ij}^{ab} = \frac{1}{2}(c_i^a c_j^b + \epsilon c_i^b c_j^a) = \epsilon \mathrm{S}_{ij}^{ba} = -\mathrm{S}_{ji}^{ab}, \tag{62}$$

$$\mathrm{A}_{ij}^{ab} = \frac{\epsilon}{2}(c_i^a c_j^b - \epsilon c_i^b c_j^a) = -\epsilon \mathrm{A}_{ij}^{ba} = \mathrm{A}_{ji}^{ab}, \tag{63}$$

here we used commutation rule (59). So we see that the remaining $N(N-1)/2$ combinations C_{ij}^{ab} with $i>j$ are expressed by ones with $i<j$. Using (62) one can construct $n(n-1)/2$ scalar combinations

$$s_{ij} = \frac{1}{2}\varepsilon_{ab}\mathrm{S}_{ij}^{ab} = \frac{1}{2}\varepsilon_{ab}c_i^a c_j^b, \tag{64}$$

contraction A_{ij}^{ab} with metric vanishes.

Definitions (62) and (63) can be extended to values $i=j$:

$$\mathrm{S}_{ii}^{ab} = \frac{1}{2}(c_i^a c_i^b + \epsilon c_i^b c_i^a) = \varepsilon^{ab}, \tag{65}$$

$$\mathrm{A}_{ii}^{ab} = \frac{\epsilon}{2}(c_i^a c_i^b - \epsilon c_i^b c_i^a) = \mathrm{G}_i^{ab}, \tag{66}$$

Then we again suppose the most general ansatz for generators G, depending on n sets of oscillators :

$$G^{ab} = \sum_{j=1}^{n} \gamma_j(s_{k\ell})\mathrm{G}_j^{ab} + \sum_{1\leq i<j\leq n} \alpha_{ij}(s_{k\ell})\mathrm{A}_{ij}^{ab} = \sum_{1\leq i\leq j\leq n} \alpha_{ij}(s_{k\ell})\mathrm{A}_{ij}^{ab}, \tag{67}$$

and calculate the following algebra between them:

$$[s_{ij}, \mathrm{G}_k^{ab}] = (\delta_{jk} - \delta_{ik})\mathrm{A}_{ij}^{ab}, \qquad \Rightarrow \qquad [s_{ij}, \sum_{k=1}^{n}\mathrm{G}_k^{ab}] = 0, \tag{68}$$

$$2[s_{ij}, s_{k\ell}] = \delta_{jk}s_{i\ell} - \delta_{ik}s_{j\ell} + \delta_{j\ell}s_{ki} - \delta_{i\ell}s_{kj}, \tag{69}$$

$$2[s_{ij}, \mathrm{A}_{k\ell}^{ab}] = \delta_{jk}\mathrm{A}_{i\ell}^{ab} - \delta_{ik}\mathrm{A}_{j\ell}^{ab} + \delta_{j\ell}\mathrm{A}_{ki}^{ab} - \delta_{i\ell}\mathrm{A}_{kj}^{ab}, \tag{70}$$

$$2[\mathrm{G}^{ab}_k, \mathrm{A}^{cd}_{ij}] = \epsilon(\delta_{ik} + \delta_{jk})\mathrm{A}^{bc,ad}_{ij} + (\delta_{ik} - \delta_{jk})\mathrm{S}^{bc,ad}_{ij}. \quad (71)$$

$$4[\mathrm{A}^{ab}_{ij}, \mathrm{A}^{cd}_{k\ell}] = \delta_{jk}S^{bc,ad}_{i\ell} + \delta_{ik}S^{bc,ad}_{j\ell} - \delta_{j\ell}S^{bc,ad}_{ki} - \delta_{i\ell}S^{bc,ad}_{kj} +$$
$$+\epsilon\Big(\delta_{jk}A^{bc,ad}_{i\ell} + \delta_{ik}A^{bc,ad}_{j\ell} + \delta_{j\ell}A^{bc,ad}_{ki} + \delta_{i\ell}A^{bc,ad}_{kj}\Big), \quad (72)$$

where

$$S^{bc,ad}_{ij} = \varepsilon^{bc}\mathrm{S}^{ad}_{ij} - \varepsilon^{ca}\mathrm{S}^{bd}_{ij} + \varepsilon^{ad}\mathrm{S}^{bc}_{ij} - \varepsilon^{db}\mathrm{S}^{ac}_{ij},$$

and

$$A^{bc,ad}_{ij} = \varepsilon^{bc}\mathrm{A}^{ad}_{ij} - \varepsilon^{ca}\mathrm{A}^{bd}_{ij} + \varepsilon^{ad}\mathrm{A}^{bc}_{ij} - \varepsilon^{db}\mathrm{A}^{ac}_{ij}.$$

7. Conclusion

The oscillator realization, presented in previous section being the most general way to describe an arbitrary operator dependence, allows to represent generators of the finite resolution of orthogonal or symplectic Yangian. Substituting it into the set of constraints (22-24), (35-42), etc. following from the RLL-relation for the particular truncation. This involving program will allow to construct the complete set of the particular representations compatible with finite resolutions of orthogonal and symplectic Yangians.

References

[1] Weil A. Sur certains groupes // Acta. Math. 1964. V. 111. P. 143-211.
[2] Konstant B. Symplectic spinors // Symposia math. 14, L., AP, 1974, P. 139-152.
[3] Vershik A. M. Metagonal and Metaplectic groups // Zap. Nauch. Sem. LOMI 1983 V. 123 P. 3-35.
[4] Faddeev L.D., Sklyanin E.K. and Takhtajan L.A., The Quantum Inverse Problem Method// Theor. Math. Phys. 1980 V. 40 P. 688 [Teor. Mat. Fiz. 40 (1979) 194]
[5] Zamolodchikov A.B., Zamolodchikov Al.B. Factorized S Matrices in Two-Dimensions as the Exact Solutions of Certain Relativistic Quantum Field Models,// Annals Phys. 1979 V. 120 P. 253
[6] Berg B., Karowski M., Weisz P., Kurak V., Factorized $U(n)$ Symmetric S Matrices in Two Dimensions,// Nucl. Phys. 1978 V. B 134 P. 125.
[7] Faddeev L.D., How Algebraic Bethe Ansatz works for integrable model, In: Quantum Symmetries/Symetries Quantiques, Proc.Les-Houches summer school, LXIV. Eds. A.Connes,K.Kawedzki, J.Zinn-Justin. North-Holland, 1998, P. 149-211, [hep-th/9605187]
[8] Kulish P.P., Sklyanin E.K., Quantum spectral transform method. Recent developments, Lect. Notes in Physics, 1982 V. 151 P. 61-119.
[9] Shankar R., Witten E. The S Matrix of the Kinks of the $(\psi\bar{\psi})^2$ Model,// Nucl. Phys. 1978 V. B 141 P. 349
[10] V.G. Drinfeld,Quantum Groups, in Proceedings of the Intern. Congress of Mathematics,Vol. 1 (Berkeley, 1986), p. 798.
[11] Zamolodchikov A B and Zamolodchikov A B 1979 Factorized s Matrices in Two-Dimensions as the Exact Solutions of Certain Relativistic Quantum Field Models *Annals Phys.* **120** 253
[12] Berg B Karowski M Weisz P and Kurak V 1978 Factorized U(n) Symmetric s Matrices in Two-Dimensions *Nucl. Phys. B* **134** 125
[13] Shankar R and Witten E 1978 The S Matrix of the Kinks of the $(\psi^-$bar $\psi)$**2 Model *Nucl. Phys. B* **141** 349 *[Erratum-ibid. B* **148** (1979) 538]
[14] Reshetikhin N Yu 1985 Integrable models of quantum one-dimensional models with $O(n)$ and $Sp(2k)$ symmetry *Theor. Math. Fiz.* **63** 347-366
[15] Karakhanyan D and Kirschner R 2010 Jordan-Schwinger representations and factorised Yang-Baxter operators *SIGMA* **6** 029 [arXiv:0910.5144 [hep-th]]
[16] Chicherin D Derkachov S and Isaev A P 2013 Conformal group: R-matrix and star-triangle relation arXiv:1206.4150 [math-ph] *and* The spinorial R-matrix *J. Phys. A* **46** 485201 arXiv:1303.4929 [math-ph]
[17] Isaev A P Karakhanyan D and Kirschner R *in preparation*

Orthogonal and symplectic Yangians

R Kirschner[1]

[1] Institut für Theoretische Physik, Universität Leipzig, 04009 Leipzig, Germany

E-mail: roland.kirschner@itp.uni-leipzig.de

Abstract. We consider Yang-Baxter relations with orthogonal or symplectic symmetry, in particular L matrices defining the related Yangian algebra. We study the conditions resulting from the truncation of the expansion of $L(u)$.

1. Introduction

In the case of general linear symmetry the Yang-Baxter (YB) relation of the RLL type, involving the fundamental R matrix and the L operator, is solved by the linear expression $L_{ab}(u) = u\delta_{ab} + M_{ab}$ with no restriction on the Lie algebra representation generated by M_{ab}. This is different in the cases of orthogonal or symplectic symmetries. Here the fundamental R matrix is quadratic in the spectral prameter u [1, 2, 4]. The restriction of the u expansion of $L(u)$ leads to constraints which cannot be fulfilled in all Lie algebra representations. The example of the spinor representation is well known [3]. This and more examples have been pointed out [4, 6]. We summarise some previous results [7, 8] and continue the study of the constraints arising from truncation at second order.

2. Fundamental YB matrices and L operators

We start from the fundamental Yang-Baxter matrix obeying the Yang-Baxter relation in the form

$$R^{a_1a_2}_{b_1b_2}(u)R^{b_1a_3}_{c_1b_3}(u+v)R^{b_2b_3}_{c_2c_3}(v) = R^{a_2a_3}_{b_2b_3}(v)R^{a_1b_3}_{b_1c_3}(u+v)R^{b_1b_2}_{c_1c_2}(u). \tag{1}$$

and symmetric with respect to the orthogonal or symeplectic Lie algebra. The fundamental representation of these Lie algebras is defined by matrices obeying

$$A^d{}_a\, \varepsilon_{db} + \varepsilon_{ad}\, A^d{}_b = 0\,, \tag{2}$$

where ε_{ab} is a non-degenerate invariant metric

$$\varepsilon_{ab} = \epsilon\, \varepsilon_{ba}\,, \quad \varepsilon_{ab}\varepsilon^{bd} = \delta^d_a\,, \tag{3}$$

which is symmetric $\epsilon = +1$ for $SO(n)$ case and skew-symmetric $\epsilon = -1$ for $Sp(n)$ case.

The well known fundamental R-matrix [1, 2, 3, 4] can be written in the unified form for an arbitrary metrics ε_{ab} (3) as follows (see, e.g., [5])

$$R^{a_1a_2}_{b_1b_2}(u) = u(u+\frac{n}{2}-\epsilon)I^{a_1a_2}_{b_1b_2} + (u+\frac{n}{2}-\epsilon)P^{a_1a_2}_{b_1b_2} - \epsilon\, u\, K^{a_1a_2}_{b_1b_2}\,, \tag{4}$$

IOP Conf. Series: Journal of Physics: Conf. Series **804** (2017) 012023 doi:10.1088/1742-6596/804/1/012023

where

$$I^{a_1 a_2}_{b_1 b_2} = \delta^{a_1}_{b_1}\delta^{a_2}_{b_2}\ , \quad P^{a_1 a_2}_{b_1 b_2} = \delta^{a_1}_{b_2}\delta^{a_2}_{b_1}\ , \quad K^{a_1 a_2}_{b_1 b_2} = \varepsilon^{a_1 a_2}\varepsilon_{b_1 b_2}\ , \tag{5}$$

We consider the RLL relation in the form

$$R^{a_1 a_2}_{b_1 b_2}(u) L^{b_1}_{c_1}(u+v) L^{b_2}_{c_2}(v) = L^{a_2}_{b_2}(v) L^{a_1}_{b_1}(u+v) R^{b_1 b_2}_{c_1 c_2}(u), \tag{6}$$

with the general form of the L operator expanded in the spectral parameter as

$$L^a_b(u) = \sum_{k=0}^{\infty} \frac{(L^{(k)})^a_b}{u^k}\ , \quad L^{(0)} = I. \tag{7}$$

The expansion coefficients $(L^{(k)})^a_b$ generate the Yangian algebra. The relation defining this algebra are encoded in the above RLL relation (6). We study the conditions on $(L^{(k)})^a_b$ arising in the case of truncation of this expansion at $k = 2$, i.e. if one imposes $(L^{(k)})^a_b = 0, k > 2$. In the study we change the notation to

$$L(u) = u^2 I + uM + N. \tag{8}$$

For this ansatz (6) implies

$$[K_{12}, M_1 + M_2] = 0, \tag{9}$$

$$[M_1, M_2] + [P_{12} - \varepsilon K_{12}, M_2] = 0, \tag{10}$$

$$[M_1, N_2] + [P_{12} - \varepsilon K_{12}, N_2] = 0, \tag{11}$$

$$K_{12}(-\beta M_2 + M_1 M_2 + N_1 + N_2) = (-\beta M_2 + M_2 M_1 + N_1 + N_2) K_{12}, \tag{12}$$

$$K_{12}(\beta(N_1 - N_2) + M_1 N_2 + N_1 M_2) = (\beta(N_1 - N_2) + N_2 M_1 + M_2 N_1) K_{12}, \tag{13}$$

$$[N_1, N_2] - (P_{12} - \varepsilon K_{12}) M_1 N_2 - N_2 M_1 (P_{12} - \varepsilon K_{12}) + \beta\varepsilon [K_{12}, N_2] = 0, \tag{14}$$

$$\beta[N_1, N_2] + \beta(P_{12} M_1 N_2 - N_2 M_1 P_{12}) - \varepsilon(K_{12} N_1 N_2 - N_2 N_1 K_{12}) = 0. \tag{15}$$

The relations are written in terms of (operator valued) matrices in the tensor product of two fundamental representation labeled by subscripts. Thus they can be written in components with upper indices a_1, a_2 and lower indices b_1, b_2, In particular we have

$$(M_1)^{a_1, a_2}_{b_1, b_2} = M^{a_1}_{b_1}\delta^{a_2}_{b_2}, \quad (M_2)^{a_1, a_2}_{b_1, b_2} = \delta^{a_1}_{b_1} M^{a_2}_{b_2},$$

$$(P_{12})^{a_1, a_2}_{b_1, b_2} = \delta^{a_1}_{b_2}\delta^{a_2}_{b_1}, \quad (K_{12})^{a_1, a_2}_{b_1, b_2} = \varepsilon^{a_1, a_2}\varepsilon_{b_1, b_2}.$$

Besides of the notations M_1, M_2 we use also

$$(M_+)^{a_1, a_2}_{b_1, b_2} = M^{a_1 a_2}\varepsilon_{b_1 b_2}, (M_-)^{a_1, a_2}_{b_1, b_2} = \varepsilon^{a_1, a_2} M_{b_1 b_2}. \tag{16}$$

We derive easily the following rules

$$KP = PK = \varepsilon K, K^2 = \varepsilon n K, \tag{17}$$

$$K A_1 B_2 = \varepsilon (A^t B)_-, A_1 B_2 K = \varepsilon (A B^t)_+,$$

$$K B_2 A_1 = (B^t A)^t_-, B_2 A_1 K = (B A^t)^t_+,$$

$$K B_2 = B_-, K A_1 = \varepsilon A^t_-, A_1 K = A_+, B_2 K = \varepsilon B^t_+,$$

$$A_+ B_- P = A_+ B^t_-, \ P A_+ B_- = A^t_+ B_-,$$

IOP Conf. Series: Journal of Physics: Conf. Series **804** (2017) 012023 doi:10.1088/1742-6596/804/1/012023

$$KA_1B_2K = KSp(AB^t) = \varepsilon KSp(A^tB).$$

The first condition (9) means that the algebra valued matrix M has no graded-symmetric contribution,

$$M - \varepsilon M^t = 0, \quad M = c_1 I + M'.$$

M' denotes the graded antisymmetric part of M. The second relation (10) represents the Lie algebra relations. It implies that the matrix elements of M' are the generators of the Lie algebra and c_1 represents a central extension commuting with the generators. The third relation (11) tells that N transforms by the adjoint action of the Lie algebra and c_1 commutes with N.

The remaining conditions express further constraints on M and N. The existence of further constraints on M implies in particular that the truncation is not allowed for arbitrary Lie algebra representations. Our aim is to identify admissable representations.

In the case of the Yangian based on the $s\ell$ type Lie algebras $Y(s\ell_n)$ no further constraints on M appear by truncation, i.e. $Y(s\ell_n)$ has the evaluation representation, where all higher terms in the expansion (7) with $k > 1$ vanish and the matrix elements $L^{(k)})^a_b M_{ab}$ are mapped into the Lie algebra generators. This is well known and can be checked from the conditions in the above formulation (9 - 15), because they reduce to the $s\ell_n$ case by the substitution $K_{12} \to 0$. Indeed, besides of the Lie algebra relations only the conditions (14, 15) remain which constrain N but not M.

3. The linear evaluation

The effect of truncation can be illustrated by going one further step of truncation, namely imposing $N = 0$ in (9- 15). Besides of the Lie algebra relations only the 4th condition (12) has a non-trivial remainder,

$$K_{12}(-\beta M_2 + M_1 M_2) = (-\beta M_2 + M_2 M_1) K_{12}.$$

It implies the relations for M^2,

$$M'^{\,2} + \beta M' = c_2 I, \tag{18}$$

where $nc_2 = tr(M'^{\,2})$. A representation obeying this constraint is formulated in terms of an undelying algebra generated by $c_a, a = 1, .., n$,

$$[c_a, c_b]_\epsilon \equiv c_a c_b + \epsilon\, c_b c_a = \varepsilon_{ba} = \epsilon\, \varepsilon_{ab}, \quad c_a c^a = \epsilon\, c^a c_a = \frac{1}{2}\varepsilon^{ab}. \tag{19}$$

The generators are

$$-M^a_b \to F^a_b = \frac{1}{2}(c^a c_b - \epsilon\, c_b c^a).$$

We find that the constraint (18) is fulfilled,

$$F^a{}_d F^d{}_b - \beta F^a{}_b = \frac{1}{4}(n\epsilon - 1)\,\delta^a_b. \tag{20}$$

In the orthogonal case ($\varepsilon = +1$) this is the spinorial representation and c_a are related to the Dirac Gamma matrices. The relation to the Jordan-Scwinger type representation will be discussed below.

4. The quadratic evaluation

Now we look for solutions of the conditions (9- 15) with $N \neq 0$. We write $M = c_1 I + M'$, $N = c_N I + N' + \tilde{N}$, where M', N' are graded anti-symmetric and $\tilde{N}$ is graded symmetric.

Actually, an important example of a representation resulting in a quadratic u dependence is just the fundamental one. Indeed the expression of the fundamental R matrix (4) can be rewritten in the form of the quadratic ansatz (8) by substituing $M = P - \epsilon K$ and perfoming a shift of u,

$$R(u - \frac{1}{2}\beta) = u^2 I + uM + \frac{1}{2}(M^2 + \beta M) + const\ I.$$

We have here $c_N = 0, N' = 0, M = M'$ and $N = \tilde{N} = \frac{1}{2}(M^2 + \beta M)$. This particular matrix of generators M obeys a cubic relation, since we have

$$(P - \epsilon K)^2 = 1 + \epsilon 2\beta\, K, \quad (P - \epsilon K)^3 = (P - \epsilon K) + (n\epsilon - 1)(n - 2\epsilon)K.$$

It is clear that an example of L quadratic in u is obtained by the product of the L operators of two linear ones. It is instructive to compare with the conditions in this case.

$$L(u) = L_1(u)L_2(u + \delta), \quad L_i(u) = Iu - G_i, \quad G_i^2 - \beta G_i = c_i^{(2)} I.$$

The product is performed in the fundamental representation and the subscript label two copies of representations allowing the first order evaluation, i.e. here the meaning of the subscripts differs from sect 2.

$$L(u) = I(u + \delta) - u(G_1 + G_2) - \delta G_1 + G_1 G_2 = Iu^2 + uM + N,$$

$$-M = G_1 + G_2 - \delta, \quad N = G_1 G_2 - \delta G_1.$$

We assume that G_i are graded-antisymmetric.

$$M = \delta + M', \quad -M' = G_1 + G_2.$$

We separate $G_1 G_2$ into the symmetric and anti-symmetric parts.

$$G_1 G_2 = \frac{1}{2}(G_1 G_2 + G_2 G_1) + \frac{1}{2}(G_1 G_2 - G_2 G_1),$$

$$N' = \frac{1}{2}(G_1 G_2 - G_2 G_1) - \delta G_1, \quad \tilde{N} = \frac{1}{2}(G_1 G_2 + G_2 G_1) = M'^{\,2} + \beta M'.$$

We have to check the consequences of the conditions (9-15) in particular the ones involving the graded anti-symmetric part N' following from (13),

$$2\beta N' = [c_1, c_N]_+ + [c_1, N']_- + [c_1, \tilde{N}]_+ - [c_N, M']_- - [M', N']_+ - [M', \tilde{N}]_-. \tag{21}$$

The second and the fourth term vanish. Further we have

$$[c_1, c_N]_+ + [c_1, \tilde{N}]_+ = -\delta(G_1 G_2 + G_2 G_1),$$

$$[M', \tilde{N}]_- = 0, \qquad [M', N']_+ = -2\beta N' - \delta(G_1 G_2 + G_1 G_2).$$

The consequence (21) is fulfilled.

More examples are obtained by fusion from the YB operators obeying an RLL type relation with the spinorial R matrix and acting in the tensor product of the spinorial and another representsation with generators G. The latter is constrained by the condition

$$[G_{[ae}, G_{f)b}]_+ = 0,$$

IOP Conf. Series: Journal of Physics: Conf. Series **804** (2017) 012023 doi:10.1088/1742-6596/804/1/012023

equivalent to

$$[G_{ae}, G_{fb}]_+ + [G_{ef}, G_{ab}]_+ + [G_{fa}, G_{eb}]_+ = 0. \tag{22}$$

Multiplication by G results in a cubic polynomial relation for the matrix of generators G [7],

$$G^3 + (\varepsilon - n)G^2 + (\varepsilon n - 2)G + \frac{1}{2} tr(G^2)(I_- \varepsilon G) = 0. \tag{23}$$

The Jordan Schwinger type representations discussed below (case $\eta\varepsilon = -1$) obey the condition (22) and provide more examples of quadratic Yangian evaluation with $N' = 0$. The JS type representations in the case $\eta\varepsilon = +1$ reduce to the spinorial representations allowing for the linear evaluation.

5. Jordan-Schwinger representations in metric formulation

Define n canonical pairs $x_a, \partial_a, a = 1, ..., n$, and the metric ε_{ab},

$$\varepsilon_{ab} = \varepsilon\varepsilon_{ba}, \varepsilon_{ac}\varepsilon^{cb} = \delta_a^{\ b} = \varepsilon\delta^b_{\ a} = \varepsilon\delta^a_{\ b},$$

$$[\partial_a, x^b]_\eta = \delta_a^{\ b}, \ [x^a, x^b]_\eta = 0, \ [\partial_a, \partial_b]_\eta = 0. \tag{24}$$

η distinguishes the normal bosonic case $\eta = -1$ from the fermionic/Grassmann case $\eta = +1$. ε distinguishes the parity of the metric.

We define the elementary canonical transformation preserving the Heisenberg commutation relation with metrics.

$$\mathcal{C}[\partial_a, x^b]_\eta \mathcal{C}^{-1} = \delta_a^{\ b}$$

by

$$\mathcal{C}_a \begin{pmatrix} x_a \\ \partial_a \end{pmatrix} \mathcal{C}_a^{-1} = \begin{pmatrix} \partial_a \\ \eta\varepsilon x_a \end{pmatrix}.$$

This implies in particular for the matrices of two versions of JS $g\ell_n$ generators

$$(L^+)_a^{\ b} = \partial_a x^b, \ \ (L^-)_a^{\ b} = \eta\varepsilon x_a \partial^b \tag{25}$$

the duality relation

$$L^+ = \mathcal{C}^{-1} L^- \mathcal{C}.$$

Further, the scalar product defined as

$$(\mathbf{x}^1 \cdot \mathbf{x}^2) = \sum \varepsilon_{ba} x^{1a} x^{2b} = x_{1a} x^{2a}$$

in terms of the dual coordinates, is invariant in the sense

$$[L_1^+ + L_2^-, (\mathbf{x}^1 \cdot \mathbf{x}^2)] = 0.$$

This distinguishes the linear combination of $g\ell_n$ generators defining the subalgebra so or sp,

$$M_{ab} = x_a\partial_b - \varepsilon x_b \partial_a, \ \ L^+ + L^- = I - M. \tag{26}$$

in two respects- invariance with respect to the canonical transformation and annihilation of the scalar product, i.e. upon operation on functions of the coordinates this combination is distinguished by the invariance of the scalar product $(\mathbf{x}^1 \cdot \mathbf{x}^2)$ with both coordinate vectors of the same space.

The generators in this metric form are related to the earlier used non-metric form [8] in the following way. In the sp case with the particular metric choice

$$\varepsilon_{ab} = \text{sign}(a)\delta_{a,-b}, \varepsilon^{ab} = -\text{sign}(a)\delta_{a,-b}, a, b = -m, ..., -1, +1, ... + m, n = 2m$$

the previous expression

$$\mathcal{M}_{ab} = x_a\partial_b - \text{sign}(a)\text{sign}(b)x_{-b}\partial_{-a}, \mathcal{M}_{ba} = -\mathcal{M}_{-a,-b}$$

is to be identified with the metric form

$$\mathcal{M}_{ab} \to M^a{}_b = x^a\partial_b - \varepsilon x_b\partial^a,$$

with

$$x_a = \varepsilon_{ac}x^c, \ \partial^a = \partial_c\varepsilon^{ca}, \ \varepsilon = -1.$$

The involved symmetry relation of $\mathcal{M}_{ab}$ is mapped into the simple ordinary symmetry relation obeyed by M_{ab}. In the case so and the metric $\varepsilon_{ab} = \delta_{ab}$ the correspondence is trivial.

We notice that in the case $\varepsilon\eta = 1$

$$\mathcal{C}(\partial_a \pm x_a)\mathcal{C}^{-1} = (\partial_a \pm x_a)$$

On the other hand with the definition

$$c_a = \frac{1}{2}(\partial_a + x_a), \ \ \bar{c}_a = \frac{1}{2}(\partial_a - x_a)$$

the canonical commutation relation imply

$$[c_a, c_b]_\eta = \frac{1}{4}(1+\varepsilon\eta)\varepsilon_{ab}, \ [\bar{c}_a, \bar{c}_b]_\eta = -\frac{1}{4}(1+\varepsilon\eta)\varepsilon_{ab}, \ [c_a\bar{c}_b]_\eta = -\frac{1}{4}(1-\varepsilon\eta)\varepsilon_{ab}.$$

Thus at $\varepsilon\eta = 1$ the Heisenberg algebra separates into independent subalgebras of Clifford form. The so/sp JS form of the Lie algebra also separates in this case. In general we have

$$M_{ab} = [c_a, c_b]_{-\varepsilon} - [\bar{c}_a, \bar{c}_b]_{-\varepsilon} + [\bar{c}_a, c_b]_{+\varepsilon} - [c_a, \bar{c}_b]_{+\varepsilon},$$

and in the case $\eta = \varepsilon$ the last two terms vanish. In this way the JS realisation of the so/sp algebra at $\varepsilon\eta = 1$ appears in two independent subalgebras,

$$M_{ab} = F_{ab} - \bar{F}_{ab}, F_{ab} = [c_a, c_b]_{-\eta}, [F_{ab}, \bar{F}_{cd}]_- = 0.$$

Above we have pointed out that the matrix with the elements F_{ab} obeys the restriction of the linear evaluation (20).

We calculate the squares of the JS generator matrices L^+, L^-.

$$L^{+\,2} = L^+((\mathbf{xp}) + 1), \quad L^{-\,2} = L^-(-(\mathbf{xp}) + 1 + \varepsilon\eta n)$$

We calculate also the products.

$$(L^+L^-)_{ab} = \varepsilon\eta\mathbf{x}^2\partial_a\partial_b + (1-\varepsilon\eta)L^-_{ab}, \ \ (L^-L^+)_{ab} = \varepsilon\eta\partial^2 x_a x_b + (1-\varepsilon\eta)L^+_{ab}.$$

We recall (26) that the sum $L^+ + L^-$ contains the generators of the so/sp subalgebra, $L^+ + L^- = I - M$, and consider its square.

$$(L^+ + L^-)^2 = [(x\partial) + 2 - \eta\varepsilon]L^+ + [\eta\varepsilon n + 2 - \eta\varepsilon - (x\partial)]L^- + \varepsilon\eta[\mathbf{x}^2\partial_a\partial_b + \partial^2 x_a x_b]. \quad (27)$$

IOP Conf. Series: Journal of Physics: Conf. Series **804** (2017) 012023 doi:10.1088/1742-6596/804/1/012023

The last term vanishes at $\eta\varepsilon = +1$. Now we proceed with the case $\eta\varepsilon = -1$. $L^+ + L^-$ commutes with $\mathbf{x}^2$ and ∂^2.

$$L^+_{ac}\partial^c\partial_b = [(x\partial)+1]\partial_a\partial_b = \partial_a\partial_b[(x\partial)-1],\ L^-_{ac}\partial^c\partial_b = \varepsilon\eta x_a\partial_b\partial^2,$$
$$L^+_{ac}x^c x_b = \partial_a x_b \mathbf{x}^2,\ L^-_{ac}x^c x_b = x_a x_b[\eta\varepsilon n - 1 - (x\partial)].$$

This is used in the calculation of the third power.

$$(L^+ + L^-)^3 = [(\mathbf{xp})+3][(\mathbf{xp})+1]L^+ + [(\mathbf{xp})+3][-\mathbf{p}^2 x_a x_b + 2L^+] +$$
$$[-n+3-(\mathbf{xp})][-n+1-(\mathbf{xp})]L^- + [-n+3-(\mathbf{xp})][-\mathbf{x}^2\partial_a\partial_b + 2L^-] -$$
$$-\mathbf{x}^2\partial_a\partial_b[(\mathbf{xp})-1] + \mathbf{x}^2 x_a\partial_b\mathbf{p}^2 - \mathbf{p}^2\partial_a x_b\mathbf{x}^2 + \mathbf{p}^2 x_a x_b[n+1+(\mathbf{xp})].$$

We have

$$\mathbf{x}^2 x_a\partial_b\mathbf{p}^2 = -\mathbf{x}^2\mathbf{p}^2 L^-_{ab} - 2\mathbf{x}^2\partial_a\partial_b,\quad -\mathbf{p}^2\partial_a\mathbf{x}^2 x_b = -\partial^2\mathbf{x}^2 L^+_{ab} - 2\mathbf{p}^2 x_a x_b$$

This results in

$$(L^+ + L^-)^3 = [((\mathbf{xp})+3)^2 - \mathbf{p}^2\mathbf{x}^2]L^+ + [(n+(\mathbf{xp})-3)^2 - \mathbf{x}^2\mathbf{p}^2]L^- + (n-4)[\mathbf{p}^2 x_a x_b + \mathbf{x}^2\partial_a\partial_b].$$

We use the result (27) for $(L^+ + L^-)^2$ to substitute the last term and obtain

$$(L^+ + L^-)^3 = -(n-1)(L^+ + L^-)^2 + A_+ L^+ + A_- L^-$$

The coefficients of L^+, L^- are equal

$$A_+ = A_- = (x\partial)^2 + (n-2)(x\partial) + n - 3 - \mathbf{x}^2\mathbf{p}^2,$$

where we have used $[\partial^2, \mathbf{x}^2] = 2n + 4(x\partial)$. We insert now the expression for the trace

$$Sp(L^+ + L^-)^2 = 2(\mathbf{xp})^2 + 2(n-2)(\mathbf{xp}) - 2\mathbf{x}^2\partial^2 - n$$

and obtain the polynomial relation in the matrix $(L^+ + L^-)$

$$(L^+ + L^-)^3 + (n-4)(L^+ + L^-)^2 - \frac{1}{2}[Sp(L^+ + L^-)^2 + n - 6)](L^+ + L^-) = 0 \qquad (28)$$

For comparison we write $L^+ + L^- = I - M$, (26), and obtain

$$M^3 - (n-1)M^2 - (\frac{1}{2}SpM^2 - n + 2)M + \frac{1}{2}SpM^2\ I. = 0 \qquad (29)$$

The resulting polynomial relation for the matrix of generators M compares with the relation for the JS generators G (23) with the substitutions

$$M \to \varepsilon G,\quad n \to \varepsilon n,\quad SpM^2 \to \varepsilon trG^2.$$

We use the relation for the graded anti-symmetric part of M^3 which follows from the Lie algebra relations,

$$M^3 - \varepsilon M^{3t} = 2M^3 - (n-1)(M^2 + \varepsilon M^{2t}) + ISp(M^2),$$

to rewrite the obtained condition as

$$M^3 - \varepsilon M^{3t} - Sp(M^2)M = 0. \qquad (30)$$

It shows that the polynomial condition constrains the graded anti-symmetric part of the third power. The graded symmetric part of M^3 can be written in terms of the power powers merely on the basis of the Lie algebra relations.

The calculation of $(L^+ + L^-)^3$ can be repeated in the case $\varepsilon\eta = +1$.

$$(L^+ + L^-)^3 = [(\mathbf{x}\partial)+1]^2 L^+ + [n - (\mathbf{x}\partial) + 1]^2 L^- =$$
$$-\frac{1}{2}[\varepsilon Sp(L^+ + L^-)^2 - n - 2](L^+ + L^-) + (2+n)(L^+ - L^-) + \frac{1}{2}n(n+2)[(L^+ + L^-) - (L^+ - L^-)].$$

In this case there is no analogous polynomial relation in terms of only $(L^+ + L^-)$.

6. Summary

We have discussed how the truncation of the expansion of $L(u)$ in inverse powers of u constrains the Lie algebra representations. We have formulated the condition of second order truncation and investigated their consequences.

The condition of first order evaluation can be formulated as the vanishing of a second order polynomial in the matrix of generators M. It expresses just the vanishing of the graded-symmetric part of M^2. In the second order evaluation this constraint is lifted, instead just this polynomial determines the graded-symmetric part of the second term N. The vanishing of a cubic polynomial in M is a consequence of the conditions in the case of the vanishing of the graded anti-symmetric part of N. It constrains the graded anti-symmetric part of M^3 to be proportional to M.

We have analysed the representations of Jordan-Schwinger type and shown how they result in examples of the first and second order evalations of the orthogonal and symplectic Yangian.

Acknowledgements

The author are grateful to A P Isaev and D Karakhanyan for discussions. The work is supported by JINR Dubna via a Heisenberg-Landau grant.

References

[1] Zamolodchikov A B and Zamolodchikov A B 1979 Factorized s Matrices in Two-Dimensions as the Exact Solutions of Certain Relativistic Quantum Field Models *Annals Phys.* **120** 253

[2] Berg B Karowski M Weisz P and Kurak V 1978 Factorized U(n) Symmetric s Matrices in Two-Dimensions *Nucl. Phys. B* **134** 125

[3] Shankar R and Witten E 1978 The S Matrix of the Kinks of the (ψ^-bar ψ)**2 Model *Nucl. Phys. B* **141** 349 *[Erratum-ibid. B* **148** (1979) 538]

[4] Reshetikhin N Yu 1985 Integrable models of quantum one-dimensional models with $O(n)$ and $Sp(2k)$ symmetry *Theor. Math. Fiz.* **63** 347-366

[5] Isaev A P 2004 Quantum groups and Yang-Baxter equations *preprint Bonn MPI* 2004-132 (http://webdoc.sub.gwdg.de/ebook/serien/e/mpi-mathematik/2004/132.pdf)

[6] Chicherin D Derkachov S and Isaev A P 2013 Conformal group: R-matrix and star-triangle relation *JHEP* **04** 020 [arXiv:1206.4150 [math-ph]] *and* The spinorial R-matrix *J. Phys. A* **46** 485201 arXiv:1303.4929 [math-ph]

[7] Isaev A P Karakhanyan D and Kirschner R 2016 Orthogonal and symplectic Yangians and Yang-Baxter R operators *Nucl. Phys.* **B904** 124147 arXiv:1511.06152

[8] Karakhanyan D and Kirschner R 2016 Yang-Baxter relations with orthogonal or symplectic symmetry *J. Phys. Conf. Ser.* **670** no 1 012029 doi:10.1088/1742-6596/670/1/012029

IOP Conf. Series: Journal of Physics: Conf. Series **804** (2017) 012024 doi:10.1088/1742-6596/804/1/012024

$N = 4$, $d = 3$ Born-Infeld theory in component approach

N Kozyrev and S Krivonos

Bogoliubov Laboratory for Theoretical Physics, JINR, 141980 Dubna, Russia

E-mail: nkozyrev,krivonos@theor.jinr.ru

Abstract. Using the formalism of nonlinear realizations we construct the component on-shell action of the $N = 4$, $d = 3$ Born-Infeld theory, which is the action of $N = 2$, $d = 3$ vector supermultiplet, fixed by invariance with respect to the additional spontaneously broken $N = 2$, $d = 3$ supersymmetry. Our construction shows that dealing with the systems with partial breaking of supersymmetry with vector fields in the multiplet, it is preferrable to use their formulation in terms of fermionic superfields only.

1. Introduction

The natural way to deal with the systems with spontaneous symmetry breaking is the formalism of nonlinear realizations as the appearing Goldstone fields can be identified with the parameters of coset space, which are associated with generators of broken symmetries [1], [2], [3]. Within this approach, it is possible to systematically find the transformation laws with respect to all the symmetries in the game, the differential Cartan forms covariant with respect to these symmetries, and, sometimes, the invariant actions as integrals of products of forms.

This approach has been applied to systems with partial breaking of supersymmetry, with Goldstone fermionic superfields being associated with half of supercharges. Then it appears to be possible to find, in addition, the covariant relations between Goldstone superfields (inverse Higgs effect - [4]), the irreducibility conditions on Goldstone multiplets and the covariant equations of motion by applying restrictions on some of the Cartan forms. While it is not possible to find the superfield action in this way (it is not an integral of Cartan form as the corresponding Lagrangian shifts under supersymmetry transformations), the construction of the component action is still possible. If one chooses the fermionic fields of the multiplet as the first components of the Goldstone fermionic superfields, they will be subjected to a simple transformation law with respect to the broken supersymmetry (just the shifts), while other fields will remain invariant. It allows one to conclude that these fields enter into action only through covariantized measure and derivatives (as Volkov and Akulov found [5]), as well as the Wess-Zumino terms. The component action, therefore, can be found as a simple generalization of the bosonic one, covariantized with respect to the broken supersymmetry. A few remaining coefficients can then be fixed by the unbroken supersymmetry invariance.

This approach allowed the construction of a few PBGS systems in three and four dimensions, with chiral and hypermultiplets as Goldstone superfields [6], [7], [8], with only scalar fields in the bosonic sector, as well as the $N = 2$, $d = 4$ Born-Infeld theory with the $N = 1$, $d = 4$ Goldstone vector multiplet [9]. All these constructions have their own difficulties. For

IOP Conf. Series: Journal of Physics: Conf. Series **804** (2017) 012024 doi:10.1088/1742-6596/804/1/012024

theories with hypermultiplets, it is required to solve the highly nonlinear relations between the multiplet components. For the $N = 2$, $d = 4$ Born-Infeld theory it is necessary to construct the irreducibility conditions for vector supermultiplet that are both off-shell and covariant with respect to the broken supersymmetry. Therefore, it would be desirable to understand how to construct the theories that contain both scalar and vector fields in the multiplet. One of the simplest examples of such systems is the $N = 4$, $d = 3$ Born-Infeld theory, Lagrangian for which can be found by either dimensional reduction from $d = 4$ or dualization of one scalar in the action of the membrane in $D = 5$ [6]. It would be desirable, therefore, to construct it from first principles.

2. The algebra, Cartan forms and derivatives

The starting point of the construction of the action of the $N = 4$, $d = 3$ Born-Infeld theory is the algebra of its symmetries, which is just the $N = 4$, $d = 3$ Poincaré superalgebra. Its bosonic generators are Lorentz rotations $M_{ab} = -(M_{ab})^\dagger$, translations $P_{ab} = (P_{ab})^\dagger$ and single central charge $Z = Z^\dagger$. Their commutation relations are

$$\begin{aligned}
\left[M_{ab}, M_{cd}\right] &= \epsilon_{ac}M_{bd} + \epsilon_{ad}M_{bc} + \epsilon_{bc}M_{ad} + \epsilon_{bd}M_{ac}, \\
\left[M_{ab}, P_{cd}\right] &= \epsilon_{ac}P_{bd} + \epsilon_{ad}P_{bc} + \epsilon_{bc}P_{ad} + \epsilon_{bd}P_{ac}.
\end{aligned} \tag{1}$$

Optionally, an automorphism generator K_{ab} could be added. With it, it would be just the bosonic algebra of the paper [10].

This bosonic algebra is accompanied by the supercharges Q_a, S_a, $\overline{Q}_a = (Q_a)^\dagger$, $\overline{S}_a = (S_a)^\dagger$. (Anti)commutation relations involving them read

$$\begin{aligned}
&\left\{Q_a, \overline{Q}_b\right\} = \left\{S_a, \overline{S}_b\right\} = 2P_{ab}, \quad \left\{Q_a, S_b\right\} = 2\epsilon_{ab}Z, \quad \left\{\overline{Q}_a, \overline{S}_b\right\} = 2\epsilon_{ab}Z, \\
&\left[M_{ab}, Q_c\right] = \epsilon_{ac}Q_b + \epsilon_{bc}Q_a, \quad \left[M_{ab}, S_c\right] = \epsilon_{ac}S_b + \epsilon_{bc}S_a.
\end{aligned} \tag{2}$$

The generator K_{ab} mixes Q_a with $\overline{S}_a$ and S_a with $\overline{Q}_a$.

The coset element can be chosen as

$$g = e^{\mathrm{i}x^{ab}P_{ab}} e^{\theta^a Q_a + \bar{\theta}^a \overline{Q}_a} e^{\mathrm{i}\mathbf{q}Z} e^{\psi^a S_a + \bar{\psi}^a \overline{S}_a}, \tag{3}$$

with M_{ab} and K_{ab} generators in the stability subgroup. The transformations in this coset space are induced by left multiplication $g_0 g = g' h$, where $h \sim e^{\gamma M + \sigma K}$. Most important transformation laws of the coordinates and fields are those with respect to the unbroken and broken supersymmetries:

$$\begin{aligned}
g_0 &= e^{\epsilon^a Q_a + \bar{\epsilon}^a \overline{Q}_a} \Rightarrow \delta_Q x^{ab} = \mathrm{i}\left(\epsilon^{(a}\bar{\theta}^{b)} + \bar{\epsilon}^{(a}\theta^{b)}\right),\ \delta_Q \theta^a = \epsilon^a, \\
g_0 &= e^{\varepsilon^a S_a + \bar{\varepsilon}^a \overline{S}_a} \Rightarrow \delta_S x^{ab} = \mathrm{i}\left(\varepsilon^{(a}\bar{\psi}^{b)} + \bar{\varepsilon}^{(a}\psi^{b)}\right),\ \delta_S \psi^a = \varepsilon^a,\ \delta_S \mathbf{q} = \mathrm{i}\left(\varepsilon_a \theta^a + \bar{\varepsilon}_a \bar{\theta}^a\right).
\end{aligned} \tag{4}$$

They ensure that unbroken supersymmetry is realized in the superspace in standard way. Also the imply that ψ_a is the Goldstone fermionic superfield, as expected.

The following differential forms are invariant with respect to the transformations in the coset:

$$\begin{aligned}
&g^{-1}dg = \mathrm{i}\triangle x^{ab} P_{ab} + \mathrm{i}\triangle \mathbf{q} Z + d\theta^a Q_a + d\bar{\theta}^a \overline{Q}_a + d\psi^a S_a + d\bar{\psi}^a \overline{S}_a, \\
&\triangle x^{ab} = dx^{ab} - \mathrm{i}\left(\theta^{(a} d\bar{\theta}^{b)} + \bar{\theta}^{(a} d\theta^{b)} + \psi^{(a} d\bar{\psi}^{b)} + \bar{\psi}^{(a} d\psi^{b)}\right),\ \triangle \mathbf{q} = d\mathbf{q} - 2\mathrm{i}\left(\psi_a d\theta^a + \bar{\psi}_a d\bar{\theta}^a\right).
\end{aligned} \tag{5}$$

$\triangle x^{ab}$, $d\theta^a$, $d\bar{\theta}^a$ can be used to define the useful covariant derivatives

$$\begin{aligned}
&dx^{ab}\,\partial_{ab} + d\theta^a \frac{\partial}{\partial\theta^a} + d\bar{\theta}^a \frac{\partial}{\partial\bar{\theta}^a} = \triangle x^{ab}\,\nabla_{ab} + d\theta^a\,\nabla_a + d\bar{\theta}^a\,\overline{\nabla}_a \Rightarrow \\
&\nabla_{ab} = \left(E^{-1}\right)_{ab}{}^{cd}\partial_{cd}, \quad E_{ab}{}^{cd} = \delta^c_{(a}\delta^d_{b)} - \mathrm{i}\left(\psi^{(c}\partial_{ab}\bar{\psi}^{d)} + \bar{\psi}^{(c}\partial_{ab}\psi^{d)}\right), \\
&\nabla_a = D_a - \mathrm{i}\left(\psi^c \nabla_a \bar{\psi}^d + \bar{\psi}^c \nabla_a \psi^d\right)\partial_{cd}, \quad \overline{\nabla}_a = \overline{D}_a - \mathrm{i}\left(\psi^c \overline{\nabla}_a \bar{\psi}^d + \bar{\psi}^c \overline{\nabla}_a \psi^d\right)\partial_{cd},
\end{aligned} \tag{6}$$

with

$$D_a = \frac{\partial}{\partial\theta^a} - \mathrm{i}\bar\theta^b\,\partial_{ab},\ \overline{D}_a = \frac{\partial}{\partial\bar\theta^a} - \mathrm{i}\theta^b\,\partial_{ab},\ \{D_a, \overline{D}_b\} = -2\mathrm{i}\partial_{ab}. \tag{7}$$

These derivatives form an algebra

$$\begin{aligned}
\{\nabla_a, \nabla_b\} &= -2\mathrm{i}\big(\nabla_a\boldsymbol{\psi}^c\,\nabla_b\bar{\boldsymbol{\psi}}^d + \nabla_b\boldsymbol{\psi}^c\,\nabla_a\bar{\boldsymbol{\psi}}^d\big)\nabla_{cd},\\
\{\nabla_a, \overline{\nabla}_b\} &= -2\mathrm{i}\nabla_{ab} - 2\mathrm{i}\big(\nabla_a\boldsymbol{\psi}^c\,\overline{\nabla}_b\bar{\boldsymbol{\psi}}^d + \overline{\nabla}_b\boldsymbol{\psi}^c\,\nabla_a\bar{\boldsymbol{\psi}}^d\big)\nabla_{cd},\\
[\nabla_{ab}, \nabla_c] &= -2\mathrm{i}\big(\nabla_{ab}\boldsymbol{\psi}^m\,\nabla_c\bar{\boldsymbol{\psi}}^n + \nabla_{ab}\bar{\boldsymbol{\psi}}^m\,\nabla_c\boldsymbol{\psi}^n\big)\nabla_{mn},\\
[\nabla_{ab}, \nabla_{cd}] &= 2\mathrm{i}\big(\nabla_{ab}\boldsymbol{\psi}^m\,\nabla_{cd}\bar{\boldsymbol{\psi}}^n - \nabla_{cd}\boldsymbol{\psi}^m\,\nabla_{ab}\bar{\boldsymbol{\psi}}^n\big)\nabla_{mn}.
\end{aligned} \tag{8}$$

Let us note that, as usual for such systems, one may define the covariant derivative $\mathcal{D}_{ab}$, which acts on the components, as the $\theta \to 0$ limit of ∇_{ab}:

$$\mathcal{D}_{ab} = \big(\mathcal{E}^{-1}\big)_{ab}{}^{cd}\partial_{cd},\quad \mathcal{E}_{(ab)}{}^{(cd)} = E_{ab}{}^{cd}|_{\theta\to 0} = \delta_a^{(c}\,\delta_b^{d)} - \mathrm{i}\big(\psi^{(c}\partial_{ab}\bar\psi^{d)} + \bar\psi^{(c}\partial_{ab}\psi^{d)}\big), \tag{9}$$

where $\psi_a = \boldsymbol{\psi}_a|_{\theta\to 0}$ is the first component of the fermionic Goldstone superfield.

3. Irreducibility conditions

As the Cartan forms (5) are invariant with respect to P_{ab}, Z, Q_a and S_a transformations, one may impose conditions, respected by all these symmetries, which express the $\boldsymbol{\psi}_a$ and $\bar{\boldsymbol{\psi}}_a$ in terms of the derivatives of the superfield $\mathbf{q}$:

$$\triangle\mathbf{q}|_{d\theta^a} = 0 \;\Rightarrow\; \nabla_a\mathbf{q} = -2\mathrm{i}\boldsymbol{\psi}_a,\quad \triangle\mathbf{q}|_{d\bar\theta^a} = 0 \;\Rightarrow\; \overline{\nabla}_a\mathbf{q} = -2\mathrm{i}\bar{\boldsymbol{\psi}}_a. \tag{10}$$

These constraints, however, do not imply any irreducibility conditions on the superfield $\mathbf{q}$. These conditions should be postulated independently.

The ordinary $N = 2$, $d = 3$ vector multiplet can be described by the following irreducibility conditions (see e.g.[11]):

$$D^a D_a\mathbf{q} = 0,\ \overline{D}^a\overline{D}_a\mathbf{q} = 0. \tag{11}$$

They leave in the superfield $\mathbf{q}$ the following independent components: a real scalar field $q = \mathbf{q}|_{\theta\to 0}$, two fermionic fields $\psi_a = \boldsymbol{\psi}_a|_{\theta\to 0}$, $\bar\psi_a = \bar{\boldsymbol{\psi}}_a|_{\theta\to 0}$, as well as the auxiliary field $A = \big[D^a, \overline{D}_a\big]\mathbf{q}|_{\theta\to 0}$ and a vector $V_{ab} = \mathrm{i}\big(\overline{D}_b\boldsymbol{\psi}_a + D_a\bar{\boldsymbol{\psi}}_b\big)|_{\theta\to 0}$. The real part of $D_a\bar{\boldsymbol{\psi}}_b$ reduces to a derivative of $\mathbf{q}$:

$$D_a\bar{\boldsymbol{\psi}}_b + \overline{D}_b\boldsymbol{\psi}_a = \frac{\mathrm{i}}{2}\{D_a, \overline{D}_b\}\mathbf{q} = \partial_{ab}\mathbf{q}. \tag{12}$$

The vector V_{ab} satisfies the Bianchi identity as a consequence of (11), and can be, therefore, identified with the electromagnetic field strength. This can be most easily shown by acting by D^2 on the constraint $\overline{D}^2\mathbf{q} = 0$ and commuting D^2 to the right:

$$0 = D^2\overline{D}^2\mathbf{q} \sim \partial_{ab}[D^a, \overline{D}^b]\mathbf{q} \;\Rightarrow\; \partial_{ab}V^{ab} = 0. \tag{13}$$

The most natural way to find the constraints covariant with respect to the broken supersymmetry is just to replace the D_a-derivatives by proper covariant ones

$$D^a D_a\mathbf{q} = 0 \;\to\; \nabla^a\nabla_a\mathbf{q} = 0,\quad \overline{D}^a\overline{D}_a\mathbf{q} = 0 \;\to\; \overline{\nabla}^a\overline{\nabla}_a\mathbf{q} = 0. \tag{14}$$

Note that, as ordinary derivatives D_a anticommute, the flat condition $D^a D_a\mathbf{q} = 0$ implies $D_a D_b\mathbf{q} = 0$. While the anticommutator of ∇_a-derivatives is nontrivial, one may note that $U_{ab} = \nabla_{(a}\boldsymbol{\psi}_{b)}$ satisfies the homogeneous equation as a result of the condition $\nabla^a\nabla_a\mathbf{q} = 0$:

$$-2\mathrm{i}U_{ab} = \frac{1}{2}\{\nabla_a, \nabla_b\}\mathbf{q} = -\mathrm{i}\big(U_{ac}\,\nabla_b\bar{\boldsymbol{\psi}}_d + U_{bc}\,\nabla_a\bar{\boldsymbol{\psi}}_d\big)\nabla^{cd}\mathbf{q}. \tag{15}$$

As the matrix of this equation is nondegenerate (it is just the unitary matrix in the lowest approximation, and $\nabla_b\bar{\psi}_d$ should not be singular in $\nabla_{ab}\mathbf{q}$), one may conclude that $U_{ab} = 0$ and present the constraints (14) as

$$\nabla_a\psi_b = 0,\ \overline{\nabla}_a\bar{\psi}_b = 0. \tag{16}$$

This is equivalent to putting to zero the $d\theta^a$, $d\bar{\theta}^a$ projections of the $d\psi_b$ $d\bar{\psi}_b$ Cartan forms (5), respectively. On these constraints, obviously, $\{\nabla_a, \nabla_b\} = 0$.

Relation (12) has an analog in the nonlinear theory

$$\nabla_a\bar{\psi}_b + \overline{\nabla}_b\psi_a = \frac{\mathrm{i}}{2}\{\nabla_a, \overline{\nabla}_b\}\mathbf{q}. \tag{17}$$

Then one may substitute here the explicit form of the anticommutator, (8) and the on-shell ansatz for $\nabla_a\bar{\psi}_b$,

$$\nabla_a\bar{\psi}_b|_{\theta\to 0} = \frac{1}{2}\big(X_{(ab)} + \mathrm{i}V_{(ab)}\big),\ \ \overline{\nabla}_b\psi_a|_{\theta\to 0} = \frac{1}{2}\big(X_{(ab)} - \mathrm{i}V_{(ab)}\big). \tag{18}$$

Let us note that according to this ansatz $\nabla_a\bar{\psi}^a = 0$, $\overline{\nabla}_a\psi^a = 0$, as the auxiliary field should not appear. Equation (17) in the limit $\theta \to 0$ implies

$$X_{ab} = \big(1 - \frac{1}{8}V_{cd}V^{cd} - \frac{1}{8}X_{cd}X^{cd}\big)\mathcal{D}_{ab}q + \frac{1}{4}\big(X^{cd}\mathcal{D}_{cd}q\big)X_{ab} + \big(V^{cd}\mathcal{D}_{cd}q\big)V_{ab}. \tag{19}$$

Equation (19) is actually a source of the significant problem: while its solution can be found analytically, it appears to be too complicated to be practically useful. More difficulties come as one tries to calculate the Bianchi identity simplifying the expression $\nabla^2\overline{\nabla}^2\mathbf{q} = 0$. In the bosonic limit, the result is

$$\begin{aligned}
&M_{ab}{}^{cd}\partial_{cd}V^{ab} - \frac{1}{2}M_{ab}{}^{cd}\partial_{mn}q\big(\partial_{cd}V^{am}\,X^{bn} - \partial_{cd}X^{bn}\,V^{am}\big) + \\
&+\frac{1}{8}V_a^c\,X^{ad}\,\partial_{cd}X^{km}\,X_k^n\,\partial_{mn}q + \frac{1}{8}V_a^c X^{ad}\partial_{cd}V^{km}V_k^n\partial_{mn}q = 0, \qquad (20)\\
&\text{where } M_{ab}{}^{cd} = \delta_a^{(c}\delta_b^{d)} + \frac{1}{4}X_a^{(c}X_b^{d)} + \frac{1}{4}V_a^{(c}V_b^{d)}.
\end{aligned}$$

Here also the auxiliary field was put to zero after calculation.

Identity (20), actually, should be brought to the canonical form $\partial_{ab}F^{ab} = 0$. Unfortunately, this cannot be done unless one substitutes X_{ab} as a function of $\partial_{ab}q$ and V_{ab}. In the third power in fields, it is just enough to take X_{ab} in the lowest approximation, $X_{ab} \approx \partial_{ab}q$. Then after a few integrations (20) can be shown to be equivalent to

$$\partial_{ab}\Big[V^{ab} + \frac{1}{8}V_{mn}V^{mn}\,V^{ab} + \frac{1}{4}\partial^{ac}q\,\partial^{bd}q\,V_{cd}\Big] = 0, \tag{21}$$

which assures that identity (20) is benign and the right irreducibility conditions are chosen. However, if one needs to go further and find the complete field strength, it is needed to use the solution of equation (19) with respect to X_{ab}. Note that solving (19) for $\partial_{ab}q$ to find identity (20) in terms of V_{ab}, X_{ab} would not do the job: in the third power in fields, it is required to take into account that $\partial_{cd}X_{ab} \approx \partial_{ab}X_{cd}$ to bring the identity to the desired form.

Let us mention that in the systems with higher supersymmetries, such as the $N = 4$, $d = 4$ Born-Infeld theory, the difficulties of solving the relations between the components and calculation of the Bianchi identities, are even more severe [12]. Therefore, another approach to such systems should be devised.

IOP Conf. Series: Journal of Physics: Conf. Series **804** (2017) 012024 doi:10.1088/1742-6596/804/1/012024

4. Alternative description

To avoid the need to solve the quadratic equation and to find Bianchi identity in terms of two, not three (V_{ab}, $\partial_{ab}q$, X_{ab}) vectors, one may try to find the formulation of the multiplet in terms of the fermionic fields only. Indeed, in such formulation, one deals only with X_{ab} and V_{ab} as bosonic components, and any sort of Bianchi identity will be found without $\partial_{ab}q$. Also, there is intrinsically no quadratic equations on components, instead, in this formulation some identity should appear that implies that X_{ab} is related to a derivative of a scalar. This identity should be some generalization of $\partial_{ab}X_{cd} - \partial_{cd}X_{ab} \approx 0$. Therefore, it may be used for simplification of Bianchi identity for the field strength, without need to solve this identity explicitly.

The irreducibility conditions (16) were obviously written in terms of fermionic superfields (16), but it would be impossible to derive the Bianchi identity from them without taking into account that $\boldsymbol{\psi}_a = \frac{\mathrm{i}}{2}\nabla_a \mathbf{q}$. Another condition on $\overline{\nabla}_a \boldsymbol{\psi}_b$ is, therefore, required. To find it, one may note that the relation in the linear limit (16) implies

$$D_a \bar{\boldsymbol{\psi}}_b + \overline{D}_b \boldsymbol{\psi}_a = \partial_{ab}\mathbf{q} \;\Rightarrow\; D_a \bar{\boldsymbol{\psi}}^a + \overline{D}^a \boldsymbol{\psi}_a = 0. \tag{22}$$

This relation works well in this limit: upon action of $D_a\overline{D}_b$ with the use of only $D_a\boldsymbol{\psi}_b = 0$, $\overline{D}_a\bar{\boldsymbol{\psi}}_b = 0$ it produces the condition

$$\partial_b^c D_{(a}\bar{\boldsymbol{\psi}}_{c)} + \partial_a^c \overline{D}_{(b}\boldsymbol{\psi}_{c)} = 0 \tag{23}$$

(antisymmetric parts of $\overline{D}_a\boldsymbol{\psi}_c$ are combined into a derivative of the original identity $\partial_{ab}(D_c\bar{\boldsymbol{\psi}}^c + \overline{D}^c\boldsymbol{\psi}_c) = 0$). Substituting here the real and imaginary parts of $D_a\bar{\boldsymbol{\psi}}_b$ just as in (18), one may immediately find

$$\partial_{bm}X_a^m + \partial_{am}X_b^m + \mathrm{i}\epsilon_{ab}\partial_{mn}V^{mn} = 0, \tag{24}$$

Therefore, V^{ab} satisfies the identity $\partial_{ab}V^{ab} = 0$ and appears to be the electromagnetic field strength in $d = 3$. For X_{ab}, the identity in vector notation is $\epsilon^{ABC}\partial_A X_B = 0$, and X_A is just a derivative of a scalar, as it should be [1].

Condition (22) looks very much like one of the $N = 1$, $d = 4$ vector supermultiplet irreducibility conditions [13] and can be obtained by dimensional reduction from it. Therefore, the full non-linear generalization of (22) should be sought with the use of the idea, which allowed us to derive the nonlinear irreducibility condition for the $N = 2$, $d = 4$ Born-Infeld theory [9].

The required irreducibility condition can be constructed as follows. At first, one should introduce the scalar automorphism generator of algebra (1), (2), that mixes Q_a and $\overline{S}_a$, and put it in the coset with the auxiliary field as a parameter. Then the irreducibility conditions can be found as projections of the resulting extended ω_S forms, with respect to new ω_Q forms. Indeed, the superalgebra (1), (2) can be extended with the automorphism generator with commutation relations

$$\left[U, Q_a\right] = \mathrm{i}\overline{S}_a,\; \left[U, S_a\right] = -\mathrm{i}\overline{Q}_a,\; \left[U, \overline{Q}_a\right] = \mathrm{i}S_a,\; \left[U, \overline{S}_a\right] = -\mathrm{i}Q_a. \tag{25}$$

Let us note that it satisfies the Jacobi identities involving K_{ab}, which had no analog in the $N = 2$, $d = 4$ Born-Infeld theory.

Introducing a new coset

$$g_U = e^{\mathrm{i}x^{ab}P_{ab}} e^{\theta^a Q_a + \bar{\theta}^a \overline{Q}_a} e^{\boldsymbol{\psi}^a S_a + \bar{\boldsymbol{\psi}}^a \overline{S}_a} e^{\mathrm{i}\varphi U}, \tag{26}$$

[1] The transfer to vector notation can be performed with the help of three real symmetric matrices $\left(\sigma^A\right)_{ab}$, $\left(\sigma^A\right)_{ab}\left(\sigma^B\right)^{bc} = \delta_a^c\eta^{AB} + \epsilon^{ABC}\left(\sigma_C\right)_a^{\;c}$, $\eta^{AB} = \mathrm{diag}(1,-1,-1)$.

one may calculate the Cartan forms and find that fermionic forms are modified to be

$$\begin{aligned}(\omega_Q)^a &= \cos\varphi\, d\theta^a + \sin\varphi\, d\bar{\psi}^a,\ (\bar{\omega}_Q)^a = \cos\varphi\, d\bar{\theta}^a + \sin\varphi\, d\psi^a,\\ (\omega_S)^a &= \cos\varphi\, d\psi^a - \sin\varphi\, d\bar{\theta}^a,\ (\bar{\omega}_S)^a = \cos\varphi\, d\bar{\psi}^a - \sin\varphi\, d\theta^a.\end{aligned} \tag{27}$$

This makes possible to introduce the covariant derivatives with respect to the ω_Q-forms

$$\begin{aligned}(\omega_Q)^a\mathbb{D}_a + (\bar{\omega}_Q)^a\overline{\mathbb{D}}_a + \triangle x^{ab}\mathbb{D}_{ab} &= \triangle x^{ab}\nabla_{ab} + d\theta^a\nabla_a + d\bar{\theta}^a\overline{\nabla}_a,\\ \mathbb{D}_a &= \frac{1}{\cos\varphi}\nabla_a - \tan\varphi\big(\mathbb{D}_a\bar{\psi}^b\,\nabla_b + \mathbb{D}_a\psi^b\,\overline{\nabla}_b\big),\\ \overline{\mathbb{D}}_a &= \frac{1}{\cos\varphi}\overline{\nabla}_a - \tan\varphi\big(\overline{\mathbb{D}}_a\bar{\psi}^b\,\nabla_b + \overline{\mathbb{D}}_a\psi^b\,\overline{\nabla}_b\big).\end{aligned} \tag{28}$$

The generalization of conditions (16), (22) can be found, if one puts to zero the ω_Q projection of ω_S, as well as trace of its $\bar{\omega}_Q$ projection. This results in the conditions

$$\mathbb{D}_a\psi_b = 0,\ \overline{\mathbb{D}}_a\bar{\psi}_b = 0,\ \overline{\mathbb{D}}_a\psi^a = -\mathbb{D}^a\bar{\psi}_a = 2\sin\varphi. \tag{29}$$

To use these conditions in actual calculations, one should formulate them in terms of more convenient ∇_a, $\overline{\nabla}_b$ derivatives and exclude the field φ. This calculation is analogous to the simplification of constraints of the $N=2$, $d=4$ Born-Infeld theory [9] and can be performed with the use of similar ideas. We, therefore, omit it and present only the final result:

$$\nabla_a\psi_b = 0,\ \overline{\nabla}_a\bar{\psi}_b = 0, \text{ and } (1 - 1/2\,\overline{\nabla}^m\psi^n\,\overline{\nabla}_m\psi_n)\nabla_a\bar{\psi}^a = (1 - 1/2\,\nabla^m\bar{\psi}^n\,\nabla_m\bar{\psi}_n)\overline{\nabla}_a\psi^a. \tag{30}$$

To derive the full nonlinear Bianchi identities for V_{ab}, X_{ab}, one needs to apply $\nabla_a\,\overline{\nabla}_b$ to the second of equations (30). Putting to zero the auxiliary field after differentiation and taking the bosonic limit, one may find the identity

$$\frac{\mathrm{i}}{2}(1\ \ 1/2\,\overline{\nabla}^m\psi^n\,\overline{\nabla}_m\psi_n)\{\nabla_c, \overline{\nabla}_b\}\nabla_a\bar{\psi}^c + \frac{\mathrm{i}}{2}(1 - 1/2\,\nabla^m\bar{\psi}^n\,\nabla_m\bar{\psi}_n)\{\nabla_a, \overline{\nabla}_c\}\overline{\nabla}^c\psi_b = 0. \tag{31}$$

Substituting here the on-shell expressions for the components of the fermionic superfields (18), one may find that the imaginary part, which begins with the derivative of V_{ab}, is automatically antisymmetric in (a, b), while the real part is automatically symmetric. It is more convenient to write them in the vector notation,

$$\begin{aligned}(\mathrm{BI}_{(1)}) &= \big(1 - \frac{1}{4}X^2 + \frac{1}{4}V^2\big)M_A{}^B\partial_B V^A + \frac{1}{2}(XV)M_A{}^B\partial_B X^A = 0, & (32)\\ (\mathrm{BI}_{(2)})^A &= \big(1 - \frac{1}{4}X^2 + \frac{1}{4}V^2\big)\Big(\epsilon^{ABC}M_B{}^D\partial_D X_C + \frac{1}{2}\epsilon^{MNP}\,\partial_M V^A\,X_N\,V_P\Big) - & (33)\\ &\quad -\frac{1}{2}(XV)\Big(\epsilon^{ABC}M_B{}^D\partial_D V_C - \frac{1}{2}\epsilon^{MNP}\,\partial_M X^A\,X_N\,V_P\Big) = 0,\\ M_A{}^B &= (1 - \frac{1}{4}V^2 - \frac{1}{4}X^2)\delta_A^B + \frac{1}{2}X_A X^B + \frac{1}{2}V_A V^B,\ X^2 = X_A X^A.\end{aligned}$$

As expected, in the lowest approximation they imply $\partial_A V^A \approx 0$ and $\epsilon^{ABC}\partial_B X_C \approx 0$, and X_A can be reduced to a function of a derivative of some scalar $\partial_A q$ and V_A.

It is worth noting that identity (33) is actually equivalent to quadratic equation (19). The simplest way to prove this is to solve equation (19) for $\partial_A q$ in the bosonic limit

$$\partial_A q = \frac{\big(1 - \frac{1}{4}X^2 + \frac{1}{4}V^2\big)X_A - \frac{1}{2}(XV)V_A}{1 - \frac{1}{16}\big(X^2 - V^2\big)^2 - \frac{1}{4}(XV)^2}. \tag{34}$$

For the left-hand side it, obviously, should be $\epsilon^{ABC}\partial_B\partial_C q = 0$. The same is true for the right-hand side if one takes into account both identities (32) and (33).

Let us now find the physical field strength in the bosonic limit. As it is easy to find, $\big(\mathrm{BI}_{(1)}\big)$ (32) cannot be brought to the standard form $\partial_A F^A = 0$ on its own, and it is necessary to study a combination of (32) and (33):

$$\mu \cdot \big(\mathrm{BI}_{(1)}\big) + \nu \cdot \epsilon^{ABC} X_B\, V_C \big(\mathrm{BI}_{(2)}\big)_A = \partial_A\big(f\, V^A + g\, X^A\big), \tag{35}$$

where μ, ν, f, g are the functions of $X^2 = X_A\, X^A$, $V^2 = V_A\, V^A$, $XV = X_A\, V^A$. Substituting the matrix $M_A{}^B$ explicitly, one may extract ten equations, these functions should obey, as coefficients of linearly independent combinations $\partial_A V^A$, $V^A\, X^B\, \partial_A V_B$, ... A careful analysis shows that three functions can be algebraically expressed in terms of one:

$$\begin{aligned} \mu &= 2\nu\Big(1+\frac{1}{4}X^2+\frac{1}{4}V^2\Big), \\ f &= 2\nu\Big(1-\frac{1}{4}X^2+\frac{1}{4}V^2\Big)\Big[1-\frac{1}{16}(X^2-V^2)^2-\frac{1}{4}(XV)^2\Big], \\ g &= \nu(XV)\Big[1-\frac{1}{16}(X^2-V^2)^2-\frac{1}{4}(XV)^2\Big]. \end{aligned} \tag{36}$$

Then the remaining differential equations imply

$$\nu = \frac{\text{const}}{\Big[1-\frac{1}{16}\big(X^2-V^2\big)^2-\frac{1}{4}(XV)^2\Big]^2}. \tag{37}$$

The physical field strength in the bosonic limit is, therefore, given by the equation

$$F^A = \frac{\Big(1-\frac{1}{4}X^2+\frac{1}{4}V^2\Big)V^A+\frac{1}{2}(XV)X^A}{1-\frac{1}{16}\big(X^2-V^2\big)^2-\frac{1}{4}(XV)^2}. \tag{38}$$

The last step in deriving the Bianchi identities is the restoration of the fermionic terms. Indeed, the Bianchi identities can be calculated with fermions taken into account. In the first approximation in bosons, the identity for X_A after a very long calculation can be found to be

$$\begin{gathered} \mathcal{D}_{ac}X^c_b + \mathcal{D}_{bc}X^c_a - \mathrm{i}\big(\mathcal{D}_{ak}\psi^m\, \mathcal{D}^k_b\bar\psi^n + \mathcal{D}_{bk}\psi^m\, \mathcal{D}^k_a\bar\psi^n\big)X_{mn} = 0 \Rightarrow \\ \epsilon^{ABC}\mathcal{D}_A X_B - 2\mathrm{i}\epsilon^{ABC}\mathcal{D}_A\psi^m\, \mathcal{D}_B\bar\psi^n\, \big(\sigma^M\big)_{mn} X_M \Rightarrow \epsilon^{MNP}\partial_M\big(\mathcal{E}_N{}^K\, X_K\big) = 0. \end{gathered} \tag{39}$$

Therefore, in this approximation $X_A \approx \mathcal{D}_A q$, and no new terms proportional to V_A and fermions arise. It can be generalized with the help of the bosonic limit (34) as

$$\mathcal{D}_A q = \frac{\Big(1-\frac{1}{4}X^2+\frac{1}{4}V^2\Big)X_A-\frac{1}{2}(XV)V_A}{1-\frac{1}{16}\big(X^2-V^2\big)^2-\frac{1}{4}(XV)^2}. \tag{40}$$

The same, obviously, follows from equation (19).

In the same approximation in bosons, the identity for the field strength is

$$\mathcal{D}_{ab}V^{ab} + 2\mathrm{i}V^{mn}\big(\mathcal{D}_{am}\,\psi_n\, \mathcal{D}^a_c\bar\psi^c - \mathcal{D}^a_c\psi^c\, \mathcal{D}_{am}\bar\psi_n\big) + 2X^{ab}\,\mathcal{D}^k_a\psi_m\, \mathcal{D}_{bk}\bar\psi^m = 0. \tag{41}$$

After some calculations it can be found equivalent to

$$\partial_A\big\{\det\mathcal{E}\,\big(\mathcal{E}^{-1}\big)_B{}^A\big[V^A + \epsilon^{BCD}\mathcal{D}_C q\,\big(\psi_m\,\mathcal{D}_D\bar\psi^m + \bar\psi^m\,\mathcal{D}_D\psi_m\big)\big]\big\} = 0, \tag{42}$$

with the nontrivial term containing $\mathcal{D}_A q$ and fermions (to prove this, one has to explicitly use $X_A \approx \mathcal{D}_A q$). Its generalization with the help of the bosonic limit (38) is not trivial. It can be argued that it should be generalized to

$$\partial_A\{\det\mathcal{E}(\mathcal{E}^{-1})_B{}^A[F^A + \epsilon^{BCD}\mathcal{D}_C q\,(\psi_m \mathcal{D}_D\bar{\psi}^m + \bar{\psi}^m \mathcal{D}_D\psi_m)]\}, \tag{43}$$

by just replacing $V_A \to F_A$, with the use of the idea of invariance with respect to broken supersymmetry. Indeed, a variation of the first part of (43) under the broken supersymmetry transformations

$$\delta_S\psi^a = \varepsilon^a,\ \delta_S x^A = \mathrm{i}(\varepsilon^a\bar{\psi}^b + \bar{\varepsilon}^a\psi^b)(\sigma^A)_{ab} \tag{44}$$

is proportional to itself

$$\delta_S\partial_A\{\det\mathcal{E}(\mathcal{E}^{-1})_B{}^A F^A\} = -\mathrm{i}(\varepsilon^a\partial_M\bar{\psi}^b + \bar{\varepsilon}^a\partial_M\psi^b)(\sigma^M)_{ab}\partial_A\{\det\mathcal{E}(\mathcal{E}^{-1})_B{}^A F^A\}, \tag{45}$$

For the same to happen with the second half of (43), $\mathcal{D}_A q$ should be kept during generalization, otherwise shifts of the explicitly contained fermions would ruin the symmetry:

$$\begin{aligned}
&\delta_S\partial_A\{\det\mathcal{E}(\mathcal{E}^{-1})_B{}^A\epsilon^{BCD}\mathcal{D}_C q\,(\psi_m \mathcal{D}_D\bar{\psi}^m + \bar{\psi}^m \mathcal{D}_D\psi_m)\} = \\
&= -\mathrm{i}(\varepsilon^a\partial_M\bar{\psi}^b + \bar{\varepsilon}^a\partial_M\psi^b)(\sigma^M)_{ab}\partial_A\{\det\mathcal{E}(\mathcal{E}^{-1})_B{}^A\epsilon^{BCD}\mathcal{D}_C q\,(\psi_m \mathcal{D}_D\bar{\psi}^m + \bar{\psi}^m \mathcal{D}_D\psi_m)\} + \\
&+\partial_A\{\det\mathcal{E}(\mathcal{E}^{-1})_B{}^A\epsilon^{BCD}\mathcal{D}_C q\,(\varepsilon_m \mathcal{D}_D\bar{\psi}^m + \bar{\varepsilon}^m \mathcal{D}_D\psi_m)\}.
\end{aligned} \tag{46}$$

If and only if $\mathcal{D}_A q$ is present, $\mathcal{E}_A{}^B$ can be removed from the last term, and then it would vanish due to the antisymmetric nature of ϵ^{ABC}.

5. The action and its unbroken supersymmetry invariance

The action of the $N = 4$, $d = 3$ Born-Infeld theory is dual to the action of supermembrane in $D = 5$. The dualization of one scalar field was performed in [6] with the result

$$\begin{aligned}
S &= -\int d^3x \det\mathcal{E}\Big[1 + \sqrt{1 - \mathcal{D}_A q\mathcal{D}^A q + F_A F^A - (F^B\mathcal{D}_B q)^2}\Big] = \\
&= -\int d^3x \det\mathcal{E}\Big[1 + \sqrt{\det\left(\eta_{AB} - \mathcal{D}_A q\,\mathcal{D}_B q + \epsilon_{ABC}F^C\right)}\Big].
\end{aligned} \tag{47}$$

(with some redefinition of the fields).

Let us prove that this action is actually invariant with respect to both supersymmetries. While invariance with respect to the broken supersymmetry in almost obvious, as the integration measure and the derivatives are properly covariantized, the unbroken supersymmetry deserves more detailed discussion.

With the use of expressions (38), (34), the action (47) can be written in terms of X^A, V^A or, most conveniently, $Y_A = X_A + \mathrm{i}V_A$:

$$S = -\int d^3x \det\mathcal{E}\frac{2 + \frac{1}{2}(V^2 - X^2)}{1 - \frac{1}{16}(V^2 - X^2)^2 - \frac{1}{4}(XV)^2} = -\int d^3x\, \det\mathcal{E}\frac{2 - \frac{1}{4}(Y^2 + \overline{Y}^2)}{1 - \frac{1}{16}Y^2\overline{Y}^2}. \tag{48}$$

Another advantage of using such variables is that they appear everywhere in the unbroken supersymmetry transformation laws. Indeed, these laws can be found using the formula

$$\begin{aligned}
&\delta_Q^\star f = -(\epsilon^a D_a + \bar{\epsilon}^a\overline{D}_a)\mathbf{f}|_{\theta\to 0} = -(\epsilon^a\nabla_a + \bar{\epsilon}^a\overline{\nabla}_a)\mathbf{f}|_{\theta\to 0} - H^A\partial_A f, \\
&H^A = -\frac{\mathrm{i}}{2}(\epsilon_k\psi^k Y^A + \bar{\epsilon}_k\bar{\psi}^k\overline{Y}^A) - \frac{\mathrm{i}}{2}\epsilon^{ABC}(\epsilon_k\psi^m Y_B + \bar{\epsilon}_k\bar{\psi}^m\overline{Y}_B)(\sigma_C)_m^k.
\end{aligned} \tag{49}$$

Then for all essential combinations of components the ϵ-part of these laws reads

$$\begin{aligned}
\delta_Q^\star \det\mathcal{E} &= -\partial_A\big(\det\mathcal{E}\, H^A\big) - \mathrm{i}\det\mathcal{E}\epsilon_k\, \mathcal{D}_A\psi^k\, Y^A - \mathrm{i}\det\mathcal{E}\epsilon^{ABC}\epsilon_k\, \mathcal{D}_A\psi^m\, Y_B\big(\sigma_C\big)_m^{\ k}, \\
\delta_Q^\star Y^2 &= -4\mathrm{i}\frac{1-1/4Y^2}{1-1/4\overline{Y}^2}\big[\epsilon_m\, \mathcal{D}_A\psi^m\, Y^A + \epsilon_m\, \epsilon^{ABC}\mathcal{D}_A\psi^n\, Y_B\,\big(\sigma_C\big)_n^{\ m}\big] - \\
&\quad -Y^2\frac{1-1/4Y^2}{1-1/4\overline{Y}^2}\big[\epsilon_m\, \mathcal{D}_A\psi^m\, \overline{Y}^A - \epsilon_m\, \epsilon^{ABC}\mathcal{D}_A\psi^n\, \overline{Y}_B\,\big(\sigma_C\big)_n^{\ m}\big] - H^C\partial_C Y^2, \\
\delta_Q^\star \overline{Y}^2 &= -4\mathrm{i}\epsilon_m\, \mathcal{D}_A\psi^m\, \overline{Y}^A + 4\mathrm{i}\epsilon_m\, \epsilon^{ABC}\, \mathcal{D}_A\psi^n\overline{Y}_B\,\big(\sigma_C\big)_n^{\ m} - \\
&\quad -\mathrm{i}\overline{Y}^2\cdot\epsilon_m\, \mathcal{D}_A\psi^m\, Y^A - \mathrm{i}\overline{Y}^2\cdot\epsilon_m\, \epsilon^{ABC}\mathcal{D}_A\psi^n\, Y_B\,\big(\sigma_C\big)_n^{\ m} - H^C\partial_C\overline{Y}^2.
\end{aligned} \tag{50}$$

Variation of Lagrangian in (48) under these transformations can be reduced just to

$$\delta_Q^\star\mathcal{L} = -2\det\mathcal{E}\,\epsilon_m\, \mathcal{D}_A\psi^m F^A + 2\mathrm{i}\det\mathcal{E}\,\epsilon_m\, \epsilon^{ABC}\mathcal{D}_A\psi^n\mathcal{D}_B q\big(\sigma_C\big)_n^{\ m}, \tag{51}$$

as due to (34) and (38)

$$\mathcal{D}_A q + \mathrm{i}F_A = Y_A\frac{1-\frac{1}{4}\overline{Y}^2}{1-\frac{1}{16}Y^2\overline{Y}^2}. \tag{52}$$

In the first approximation in fermions the second term in (51) reduces to the full derivative, while the first one, after integration by parts, produces the bosonic Bianchi identity $\partial_A F^A \approx 0$. A careful treatment of (51) with the help of (43) and explicit substitution of $\mathcal{E}_A{}^B$ show that fermionic terms also reduce to the full derivatives. The action (47) is, therefore, invariant.

6. Conclusion

In this paper, we constructed, from first principles, the component action of the $N=4$, $d=3$ Born-Infeld theory and found physical field strength of the corresponding vector supermultiplet. To perform this calculation, it was found to be necessary to carefully choose constraints, which define the multiplet, as most straightforward constraints result in unacceptably complicated relations between components of the multiplet, as well as the Bianchi identity for the field strength. To avoid these difficulties, the formulation of the multiplet, in which the fermionic superfield has fundamental role, was used. To ensure it has the proper component content, an additional irreducibility condition was provided. This condition, after applying covariant derivatives to it, produces the differential identities obeyed by the bosonic components of the multiplet. These identities finally allow identification of these components as the electromagnetic field strength and a derivative of a scalar.

Rather similar difficulties, solution of relation between the multiplet components and calculation of the Bianchi identity, can be found when considering the $N=4$, $d=4$ Born-Infeld theory. It is probable that they can be solved in a similar way by providing proper irreducibility conditions of the nonlinear $N=2$, $d=4$ vector multiplet.

Acknowledgments

The work of N.K. and S.K. was partially supported by the RSCF, grant 14-11-00598 and partially by the RFBR, grant 15-52-05022 Arm-a.

References

[1] Coleman S, Wess J and Zumino B 1969 Structure of phenomenological lagrangians. 1 *Phys. Rev.* **177** 2239, Callan C, Coleman S, Wess J and Zumino B 1969 Structure of phenomenological lagrangians. 2 *Phys. Rev.* **177** 2247

[2] Volkov D 1973 Phenomenological lagrangians *Sov. J. Part. Nucl.* **4** 3

[3] Ogievetsky V 1974 Nonlinear realizations of internal and spacetime symmetries *Acta Universitatis Wratislaviensis* **207** 117 (Proc. of the X'th Winter school of theoretical physics in Karpacz)

[4] Ivanov E and Ogievetsky V 1975 The inverse Higgs phenomenon in nonlinear realizations *Teor. Mat. Fiz.* **25** 164

[5] Volkov V and Akulov V 1972 Possible universal neutrino interaction *JETP Lett.* **16** 438, Volkov V and Akulov V 1973 Is the neutrino Goldstone particle? *Phys. Lett.* B **46** 109

[6] Bellucci S, Kozyrev N, Krivonos S and Yeranyan A 2014 Supermembrane in $D = 5$: component action *JHEP* **1405** 142 (*Preprint* 1312.0231[hep-th])

[7] Bellucci S, Kozyrev N, Krivonos S and Sutulin A 2015 Component on-shell actions of supersymmetric 3-branes: I. 3-brane in $D = 6$ *Class. Quant. Grav.* **32 no.3** 035025 (*Preprint* 1409.0641[hep-th])

[8] Bellucci S, Kozyrev N, Krivonos S and Sutulin A 2015 Component on-shell actions of supersymmetric 3-branes: II. 3-brane in $D = 8$ *Class. Quant. Grav.* **32 no.11** 115008 (*Preprint* 1411.7550[hep-th])

[9] Bellucci S, Kozyrev N, Krivonos S and Sutulin A 2015 Space-filling D3 brane within coset approach *JHEP* **1508** 094 (*Preprint* 1505.07386[hep-th])

[10] Ivanov E and Krivonos S 1999 $N = 1$ $D = 4$ supermembrane in coset approach *Phys. Lett.* B **453** 237-44 (*Preprint* hep-th/9901003)

[11] Zupnik B 2000 Partial spontaneous breakdown of 3-dimensional $N = 2$ supersymmetry *Theor. Mat. Phys.* **123** 463-77 (*Preprint* hep-th/9905108)

[12] Bellucci S, Ivanov E and Krivonos S 2001 $N = 2$ and $N = 4$ supersymmetric Born-Infeld theories from nonlinear realizations *Phys. Lett.* B **502** 279 (*Preprint* hep-th/0012236)

[13] Bagger J and Galperin A 1997 A new Goldstone multiplet for partially broken supersymmetry *Phys. Rev.* D **55** 1091 (*Preprint* hep-th/9608177)

Superintegrable systems on Poisson manifolds

A. Kurov, G. Sardanashvily

Department of Theoretical Physics, Moscow State University, 119991, Moscow, Russia

E-mail: kurov.aleksandr@physics.msu.ru, sardanashvi@phys.msu.ru

Abstract. The definition of superintegrable systems on a symplectic manifold implies a rather restrictive condition $2n = k + m$ where $2n$ is a dimension of a symplectic manifold, k is a dimension of a pointwise Lie algebra of a superintegrable system, and m is its corank. To solve this problem, we aim to consider partially superintegrable systems on Poisson manifolds where $k + m$ is the rank of a compatible Poisson structure. The according extensions of the Mishchenko–Fomenko theorem on generalized action-angle coordinates is formulated.

1. Introduction

The Liouville–Arnold theorem for completely integrable systems [1, 2], the Poincaré–Lyapounov–Nekhoroshev theorem for commutative partially integrable systems [3, 4] and the Mishchenko–Fomenko theorem (Theorem 2.4) for the superintegrable ones [5, 6, 7] on symplectic manifolds state the existence of action-angle coordinates around a compact invariant submanifold of an integrable system which is a torus T^m. These theorems have been extended to a general case of invariant submanifolds which need not be compact, but are diffeomorphic to a toroidal cylinder

$$\mathbb{R}^{m-r} \times T^r, \qquad T^r = \overset{r}{\times} S^1, \tag{1.1}$$

(Theorems 2.5, 2.6 and 2.3, respectively) [8, 9, 10, 11, 12, 13].

However, Definition 2.1 of a superintegrable system on a symplectic manifold is rather restrictive. Its item (iii) requires that the matrix function s_{ij} (2.3) must be of corank $m = 2n - k$ (2.4) where $2n$ is a dimension of a symplectic manifold and k is a number of independent generating functions of a superintegrable system. In particular, commutative partially integrable systems on a symplectic manifold fail to satisfy this condition (Remark 2.1).

Example 1.1 The Kepler problem on an $(n = 2)$-dimensional configuration space $\mathbb{R}^2 \setminus \{0\}$ possesses three integrals of motion: an orbital momentum M_{12} and two components of the Rung–Lenz vector A^a [11, 13]. They constitute two different superintegrable systems: (i) with a Lie algebra $so(3)$ on a domain U_- of a phase space $\mathbb{R}^4 \setminus \{0\}$ of negative energy, and (ii) with a Lie algebra $so(2,1)$ on a domain $U_+ \subset \mathbb{R}^4 \setminus \{0\}$ of positive energy. However, if $n > 2$, a number of integrals of motion (M_{ij}, A^i), $i = 1, \ldots, n$, of the Kepler system is more than $2n$, and they fail to form any superintegrable system. In this case, one however can consider a partially superintegrable system with the generating functions (M_{12}, A^1, A^2) because orbits of a motion of a Kepler system is well known to lie in a plane.

To avoid the restriction $m = 2n - k$ (2.4), we aim to consider partially superintegrable systems on Poisson manifolds (Definition 5.2). The following are their examples.

Example 1.2 Let $F = (F_1, \ldots, F_k)$ be a superintegrable system of corank $2n - k$ on a $2n$-dimensional connected symplectic manifold (Z, Ω). Let X be an r-dimensional manifold regarded as a Poisson manifold with a zero Poisson structure. A manifold product $Z \times Q$ can be endowed with a product Poisson structure w of rank $2n$. Then the pull-back of functions F_i onto $Z \times X$ exemplifies a partially superintegrable system of corank $m = 2n - k$ where $2n$ is a rank of a Poisson structure on a $(2n + r)$-dimensional Poisson manifold (Definition 5.2).

Example 1.3 Commutative partially integrable systems on Poisson manifolds (Definition 4.2) exemplify partially superintegrable systems.

A key point is that invariant submanifolds of a superintegrable system are integral manifolds of a certain commutative partially integrable system on a symplectic manifold (Remark 2.2). As a consequence, the proof of above mentioned generalized Mishchenko–Fomenko theorem (Theorem 2.3) for superintegrable systems is reduced to generalized Poincaré–Lyapounov–Nekhoroshev Theorem 2.6 for commutative partially integrable systems on a symplectic manifold (Section 2). Therefore, we start our investigation of partially superintegrable systems with commutative partially integrable systems on Poisson manifolds (Section 4).

Our goal is that, in a case of partially integrable systems on Poisson manifolds, the above mentioned restriction condition (2.4) comes to a form $k+m = r$ where r is the rank of a Poisson structure on a manifold Z, but not a dimension of Z (Lemma 5.1).

The extended Mishchenko–Fomenko theorem on action-angle coordinates in the case of symplectic superintegrable systems (Theorem 2.3) is generalized to partially superintegrable systems on Poisson manifolds (Theorem 5.2). Its proof also is reduced to Theorems 4.2 and 4.5 for commutative partially integrable systems on Poisson manifolds.

2. Superintegrable systems on symplectic manifolds

Throughout the work, all functions and maps are smooth, and manifolds are finite-dimensional smooth real and paracompact.

Definition 2.1 *Let (Z, Ω) be a $2n$-dimensional connected symplectic manifold, and let $(C^\infty(Z), \{,\})$ be a Poisson algebra of smooth real functions on Z. A subset*

$$F = (F_1, \ldots, F_k), \qquad n \leq k < 2n, \tag{2.1}$$

of a Poisson algebra $C^\infty(Z)$ is called a superintegrable system if the following conditions hold.

(i) All the functions F_i (called the generating functions of a superintegrable system) are independent, i.e., a k-form $\overset{k}{\wedge} dF_i$ nowhere vanishes on Z. It follows that a surjection

$$\widehat{F} : Z \to N = \underset{i}{\times} F_i(Z) \subset \mathbb{R}^k \tag{2.2}$$

is a submersion, i.e., a fibred manifold over a domain (contractible open subset) $N \subset \mathbb{R}^k$ endowed with the coordinates (x_i) such that $x_i \circ \widehat{F} = F_i$. Fibres of the fibred manifold $\widehat{F}$ (2.2) are called the invariant submanifolds of a superintegrable system.

(ii) There exist smooth real functions s_{ij} on N such that

$$\{F_i, F_j\} = s_{ij} \circ \widehat{F}, \qquad i, j = 1, \ldots, k. \tag{2.3}$$

(iii) The matrix function $\mathbf{s}$ with the entries s_{ij} (2.3) is of constant corank

$$m = 2n - k, \qquad 2n = \dim Z, \qquad k = \dim N, \tag{2.4}$$

at all points of N.

If $k > n$, the matrix $\mathbf{s}$ is necessarily non-zero. If $k = n$, then $\mathbf{s} = 0$, and we are in the case of completely integrable systems as follows.

Definition 2.2 *The subset* $(F_1, \ldots, F_n)$ *(2.1) of a Poisson algebra* $C^\infty(Z)$ *on a symplectic manifold* (Z, Ω) *is called the completely integrable system if* F_i *are independent functions in involution.*

Remark 2.1 A family $\{S_1, \ldots, S_m\}$ of $m \leq n$ independent smooth real functions in involution on a symplectic manifold (Z, Ω) is called the (commutative) partially integrable system. It should be emphasized that a commutative partially integrable system on a symplectic manifold fails to be a particular superintegrable system because the condition (2.4) in item (iii) of Definition 2.1 is not satisfied, unless $m = n$ and it is a completely integrable system.

The following two assertions clarify the structure of superintegrable systems [7, 10, 12, 13].

Lemma 2.1 *Given a symplectic manifold* (Z, Ω), *let* $\pi : Z \to N$ *be a fibred manifold such that, for any two functions* f, f' *constant on fibres of* π, *their Poisson bracket* $\{f, f'\}$ *also is well. Then* N *is provided with a unique coinduced Poisson structure* $\{,\}_N$ *such that* π *is a Poisson morphism.*

Since any function constant on fibres of π is a pull-back of some function on N, the superintegrable system (2.1) with $\pi = \widehat{F}$ satisfies the condition of Lemma 2.1 due to item (i) of Definition 2.1. Thus, a base N of the fibred manifold $\widehat{F}$ (2.2) is endowed with a coinduced Poisson structure of corank m. With respect to coordinates x_i in item (i) of Definition 2.1, its bivector field reads

$$w = s_{ij}(x_k)\partial^i \wedge \partial^j. \tag{2.5}$$

Lemma 2.2 *Given a fibred manifold* $\pi : Z \to N$ *with connected fibres in Lemma 2.1, the following conditions are equivalent [7, 16]:*

(i) the corank of the coinduced Poisson structure $\{,\}_N$ *on* N *equals* $m = \dim Z - \dim N$,

(ii) the fibres of π *are isotropic,*

(iii) the fibres of π *are maximal integral manifolds of the involutive distribution spanned by the Hamiltonian vector fields of the pull-back* π^*C *of Casimir functions* C *of a coinduced Poisson structure on* N.

It is readily observed that the fibred manifold $\widehat{F}$ (2.2) obeys condition (i) of Lemma 2.2 due to item (iii) of Definition 2.1, namely, $k - m = 2(k - n)$.

Remark 2.2 The pull-back π^*C of Casimir functions in item (iii) of Lemma 2.2 are in involution with all functions on Z which are constant on fibres of π. Let N admit a family of m independent Casimir functions (C_λ). Then their pull-back π^*C_λ constitute a commutative partially integrable system on a symplectic manifold (Z, Ω) (Remark 2.1). In this case it follows from item (iii) of Lemma 2.2 that invariant submanifolds of a superintegrable system are integral manifolds of this commutative partially integrable system. As a consequence, the proof of generalized Mishchenko–Fomenko Theorem 2.3 is reduced to the generalized Poincaré–Lyapounov–Nekhoroshev theorem (Theorem 2.6) for commutative partially integrable systems on a symplectic manifold.

Given a superintegrable system in accordance with Definition 2.1, the above mentioned extension of the Mishchenko – Fomenko theorem to non-compact invariant submanifolds states the following [10, 12, 11, 13].

Theorem 2.3 *Let the Hamiltonian vector fields* ϑ_i *of the generating functions* F_i *be complete, and let fibres of the fibred manifold* $\widehat{F}$ *(2.2) be connected and mutually diffeomorphic. Then the following hold.*

(I) The fibres of the fibred manifold $\widehat{F}$ *(2.2) are diffeomorphic to the toroidal cylinder (1.1) where* $m = 2n - k$.

(II) Given a fibre M of the fibration $\widehat{F}$ (2.2), there exists its open saturated neighborhood U_M (i.e., a fibre through a point of U_M belongs to U_M) which is a trivial principal bundle

$$U_M = N_M \times (\mathbb{R}^{m-r} \times T^r) \xrightarrow{\widehat{F}} N_M \tag{2.6}$$

with the structure additive group (1.1).

(III) A neighborhood U_M is provided with the bundle (generalized action-angle) coordinates $(I_\lambda, q^A, y^\lambda)$, $\lambda = 1, \dots, m$, $A = 1, \dots, 2(n-m)$, such that: (i) the generalized angle coordinates (y^λ) are coordinates on a toroidal cylinder, i.e., fibre coordinates on the fibre bundle (2.6); (ii) the (I_λ, q^A) are coordinates on its base N_M where the action coordinates (I_λ) are values of independent Casimir functions of the coinduced Poisson structure $\{,\}_N$ on N_M; and (iii) a symplectic form Ω on U_M reads

$$\Omega = dI_\lambda \wedge dy^\lambda + \Omega_{AB}(I_\beta, q^C) dq^A \wedge dq^B.$$

If the conditions of Theorem 2.3 are replaced with that fibres of the fibred manifold $\widehat{F}$ (2.2) are compact and connected, this theorem restarts the Mishchenko–Fomenko one as follows [5, 6, 7].

Theorem 2.4 *Let fibres of the fibred manifold $\widehat{F}$ (2.2) be compact and connected. Then they are diffeomorphic to a torus T^m, and statements (II) – (III) of Theorem 2.3 hold.*

Remark 2.3 In Theorem 2.4, the Hamiltonian vector fields v_λ are complete because fibres of the fibred manifold $\widehat{F}$ (2.2) are compact.

If F (2.1) is a completely integrable system, the coinduced Poisson structure on N equals zero, and the generating functions F_i are the pull-back of n independent functions on N. Then Theorems 2.4 and 2.3 come to the well-known Liouville–Arnold theorem [1, 2] and its extension (Theorem 2.5 below) to the case of non-compact invariant submanifolds [12, 13, 14], respectively.

Theorem 2.5 *Given a completely integrable system F in accordance with Definition 2.2, let the Hamiltonian vector fields ϑ_i of functions F_i be complete, and let fibres of the fibred manifold $\widehat{F}$ (2.2) be connected and mutually diffeomorphic. Then items (I) and (II) of Theorem 2.3 hold, and its item (III) is replaced with the following one.*

(III') The neighborhood U_M (2.6) where $m = n$ is provided with the bundle (generalized action-angle) coordinates (I_λ, y^λ), $\lambda = 1, \dots, n$, such that the angle coordinates (y^λ) are coordinates on a toroidal cylinder, and the symplectic form Ω on U_M reads

$$\Omega = dI_\lambda \wedge dy^\lambda. \tag{2.7}$$

In a general setting, one considers commutative partially integrable systems on a symplectic manifold (Remark 2.1). The Poincaré–Lyapounov–Nekhoroshev theorem [3, 4, 17] generalizes the Liouville–Arnold one to a commutative partially integrable system with compact invariant submanifolds. Forthcoming Theorem 2.6 is concerned with a generic commutative partially integrable system on a symplectic manifold [9, 11, 13].

Given a commutative partially integrable system $S = \{S_1, \dots, S_m\}$ of $m \leq n$ independent smooth real functions in involution on a symplectic manifold (Z, Ω), we have a fibred manifold

$$\widehat{S} : Z \to X = \overset{m}{\times} S_\lambda(Z) \subset \mathbb{R}^m \tag{2.8}$$

over a domain $X \subset \mathbb{R}^m$. We agree to call its fibres the invariant submanifolds of a commutative partially integrable system though it is not a superintegrable system (Remark 2.1), unless $m = n$ and it is a completely integrable system.

Hamiltonian vector fields v_λ of generating functions S_λ are mutually commutative and independent. They span an m-dimensional involutive distribution on Z whose maximal integral manifolds constitute an isotropic foliation $\mathcal{S}$ of Z. Its leaves are called the integral manifolds of a commutative partially integrable system. Because functions S_λ are constant on leaves of this foliation, each fibre of the fibred manifold $\widehat{S}$ (2.8) is foliated by the leaves of a foliation $\mathcal{S}$. If $m = n$, we are in the case of a completely integrable system, and its integral manifolds (i.e., leaves of $\mathcal{S}$) are connected components of its invariant submanifolds (i.e., fibres of the fibred manifold $\widehat{S}$ (2.8)).

Theorem 2.6 *Let a commutative partially integrable system $S = \{S_1, \ldots, S_m\}$ on a symplectic manifold (Z, Ω) satisfy the following conditions.*
(i) The Hamiltonian vector fields v_λ of S_λ are complete.
(ii) The foliation $\mathcal{S}$ is a fibred manifold $\pi_{\mathcal{S}} : Z \to N$ whose fibres are mutually diffeomorphic and, being integral manifolds, are connected.
Then the following hold [9, 12, 11, 13].
(I) The fibres of $\mathcal{S}$ are diffeomorphic to the toroidal cylinder (1.1).
(II) Given a fibre M of $\mathcal{S}$, there exists its open saturated neighborhood U_M such that the restriction of $\pi_{\mathcal{S}}$ to U_M is a trivial principal bundle with the structure additive group (1.1), and we have a composite fibre bundle

$$\widehat{S} : U_M \longrightarrow \pi_{\mathcal{S}}(U_M) \longrightarrow \widehat{S}(U_M) \subset \mathbb{R}^m. \tag{2.9}$$

(III) A neighborhood U_M is provided with the bundle (generalized action-angle) coordinates

$$(I_\lambda, q^A, y^\lambda) \to (I_\lambda, q^A) \to (I_\lambda), \quad \lambda = 1, \ldots, m, \quad A = 1, \ldots, 2(n-m), \tag{2.10}$$

such that: (i) the action coordinates (I_λ) on $\widehat{S}(U_M)$ are expressed into the values of generating functions S_λ; (ii) the angle coordinates (y^λ) are coordinates on the toroidal cylinder (1.1); and (iii) a symplectic form Ω on U_M reads

$$\Omega = dI_\lambda \wedge dy^\lambda + \Omega_{AB}(I_\beta, q^C) dq^A \wedge dq^B.$$

3. Lie algebra superintegrable systems

Following the original Mishchenko–Fomenko theorem [5, 6, 7], let us consider superintegrable systems whose generating functions $F = \{F_1, \ldots, F_k\}$ form a k-dimensional real Lie algebra $\mathfrak{g}$ of corank $m = 2n - k$ with the commutation relations

$$\{F_i, F_j\} = c_{ij}^h F_h, \qquad c_{ij}^h = \text{const}. \tag{3.1}$$

We agree to call them the Lie algebra superintegrable systems. In this case, the fibration $\widehat{F}$ (2.2) is a momentum mapping of Z onto a domain N of the Lie coalgebra $\mathfrak{g}^*$ (Section 6.2) which is provided with the coordinates x_i in item (i) of Definition 2.1 [12, 13, 18]. Accordingly, the coinduced Poisson structure $\{,\}_N$ on N coincides with the canonical Lie–Poisson structure on $\mathfrak{g}^*$, and it is given by a Poisson bivector field

$$w = \frac{1}{2} c_{ij}^h x_h \partial^i \wedge \partial^j.$$

Hamiltonian vector fields ϑ_i of generating functions F_i of the Lie algebra superintegrable system (3.1) make up a real Lie algebra $\mathfrak{g}$ with the commutation relations

$$[\vartheta_i, \vartheta_j] = c_{ij}^h \vartheta_h. \tag{3.2}$$

IOP Conf. Series: Journal of Physics: Conf. Series **804** (2017) 012025 doi:10.1088/1742-6596/804/1/012025

Following the conditions of Theorem 2.3, let us assume that Hamiltonian vector fields ϑ_i are complete. They define a Hamiltonian action on Z of a simply connected Lie group G whose Lie algebra is isomorphic to $\mathfrak{g}$. Since vector fields ϑ_i are independent, the action of G on Z is locally free, i.e., isotropy groups of points of U are discrete subgroups of G. Orbits of G coincide with k-dimensional maximal integral manifolds of a regular distribution on Z spanned by Hamiltonian vector fields ϑ_i [21]. They constitute a foliation $\mathcal{F}$ of Z. Then the fibration $\widehat{F}$ (2.2) sends its leaves $\mathcal{F}_z$ through points $z \in Z$ onto the orbits $\mathcal{G}_{\widehat{F}(z)}$ of the coadjoint action of G on $\mathfrak{g}^*$, which coincide with the canonical symplectic foliation $\mathcal{G}$ of $\mathfrak{g}^*$. Conversely, $\mathcal{F}_z = \widehat{F}^{-1}(\mathcal{G}_{\widehat{F}(z)})$ in accordance with item (iii) of Lemma 2.2.

It should be noted note that Casimir functions $C \in \mathcal{C}(\mathfrak{g}^*)$ of the Lie–Poisson structure on $\mathfrak{g}^*$ are exactly the coadjoint invariant functions on $\mathfrak{g}^*$. They are constant on orbits of the coadjoint action of G on $\mathfrak{g}^*$. Consequently, their pull-back $\widehat{F}^*C$ are constant on leaves of a foliation $\mathcal{F}$. Therefore, the real Lie algebra $\mathfrak{g}$ (3.2) is extended to a Lie algebra over a subring $\mathcal{C} = \widehat{F}^*\mathcal{C}(N) \subset C^\infty(Z)$ of the pull-back $\widehat{F}C$ of Casimir functions on $N \subset \mathfrak{g}^*$.

Now let us assume that a foliation $\mathcal{F}$ is a fibred manifold $\pi_\mathcal{F} : Z \to \pi_\mathcal{F}(Z)$ whose fibres are mutually diffeomorphic. This implies that a symplectic foliation $\mathcal{G}$ of $\widehat{F}(Z)$ also is a fibred manifold $\pi_\mathcal{G} : \widehat{F}(Z) \to \pi_{\mathcal{F}(Z)}$. Thus, we have a composite fibred manifold

$$\pi_\mathcal{F} = \pi_\mathcal{G} \circ \widehat{F} : Z \longrightarrow \widehat{F}(Z) \longrightarrow \pi_{\mathcal{F}(Z)}, \tag{3.3}$$

so that a fibration $\widehat{F}$ obeys the conditions of Theorem 2.3. It follows that, given a leaf V of $\mathcal{F}$, there exists its open saturated neighborhood U_V such that $\widehat{F}(U_V) \subset \mathfrak{g}^*$ is provided with some family of m independent Casimir functions $C = (C_1, \ldots, C_m)$ and, restricted to U_V, the composite fibred manifold $\pi_\mathcal{F}$ (3.3) becomes a composite bundle

$$\pi_\mathcal{F} = \widehat{C} \circ \widehat{F} : U_V \longrightarrow \widehat{F}(U_V) \longrightarrow \overset{m}{\times} C_\lambda(\widehat{F}(U_V)) = \overset{m}{\times} \widehat{F}^*C_\lambda(U_V) = \pi_\mathcal{F}(U_V)$$

in toroidal cylinders. In accordance with Remark 2.2, fibres of $\widehat{F}$ are integral manifolds of a commutative partially integrable system $S = (S_\lambda)$ on Z of the pull-back $S_\lambda = \widehat{F}^*C_\lambda$ of Casimir functions C_λ. Then the composite bundle (3.4) takes the form (2.9). Accordingly, it is endowed with the bundle (generalized action-angle) coordinates

$$(I_\lambda, q^A, y^\lambda) \to (I_\lambda, q^A) \to (I_\lambda), \quad \lambda = 1, \ldots, m, \quad A = 1, \ldots, k - m, \tag{3.4}$$

which are generalized action-angle coordinates (2.10) in Theorem 2.6 when S_λ are the pull-back $\widehat{F}^*C_\lambda$ of Casimir functions C_λ. Note that the latter in turn are the pull-back $C_\lambda = \pi_\mathcal{G}^*\Phi_\lambda$ of some functions Φ_λ on a base $\pi_\mathcal{F}(U_V)$ of the composite bundle (3.4).

4. Commutative partially integrable systems on Poisson manifolds

As was mentioned above invariant submanifolds of a superintegrable system are integral manifolds of a certain commutative partially integrable system on a symplectic manifold (Remark 2.2). Therefore, we start our analysis of partially superintegrable systems with commutative partially integrable systems on Poisson manifolds in Example 1.3 [9, 12, 13, 14, 15].

A key point is that a commutative partially integrable system admits different compatible Poisson structures (Theorem 4.2). Treating commutative partially integrable systems, we therefore are based on a wider notion of the commutative dynamical algebra [9].

Let we have m mutually commutative vector fields $\{\vartheta_\lambda\}$ on a connected smooth manifold Z which are functionally independent (i.e., $\overset{m}{\wedge} \vartheta_\lambda \neq 0$) everywhere on Z. We denote by $\mathcal{C} \subset C^\infty(Z)$

a $\mathbb{R}$-subring of smooth real functions f on Z whose derivations $\vartheta_\lambda \rfloor df$ along ϑ_i vanish for all ϑ_λ. Let $\mathcal{A}$ be an m-dimensional Lie $\mathcal{C}$-algebra generated by the vector fields $\{\vartheta_\lambda\}$.

Definition 4.1 *We agree to call $\mathcal{A}$ the commutative dynamical algebra.*

For instance, given a commutative partially integrable system S on a symplectic manifold (Remark 2.1), the Hamiltonian vector fields of its generating functions constitute a commutative dynamical algebra in accordance with Definition 4.1.

In a general setting, let us consider a commutative dynamical algebra on a Poisson manifold.

Definition 4.2 *Let (Z, w) be a (regular) Poisson manifold and $\mathcal{A}$ an m-dimensional commutative dynamical algebra on Z. A triple $(Z, \mathcal{A}, w)$ is said to be a commutative partially integrable system if the following hold.*

(a) Generators ϑ_λ are Hamiltonian vector fields of some independent functions $S_\lambda \in \mathcal{C}$ on Z.

(b) All elements of $\mathcal{C} \subset C^\infty(Z)$ are mutually in involution, i.e., their Poisson brackets $\{f, f'\}_w$, $f, f' \in \mathcal{C}$, equal zero.

It follows at once from this definition that the Poisson structure w is at least of rank $2m$, and that $\mathcal{C}$ is a commutative Poisson algebra. We call the functions S_λ in item (a) of Definition 4.2 the generating functions of a commutative partially integrable system, which is uniquely defined by a family $(S_1, \ldots, S_m)$ of these functions.

Definition 4.3 *We say that a Poisson structure in Definition 4.1 is compatible.*

One can show (Theorem 4.2) that a compatible Poisson structure is of rank $2m$.

Remark 4.1 If $2m = \dim Z$ in Definition 4.2, we have a completely integrable system on a symplectic manifold Z (Definition 2.2). However, a commutative partially integrable system on a symplectic manifold in Remark 2.1 fails to be well defined in accordance with Definition 4.2 because it does not satisfy item (b) of this Definition if $m < n$.

If $2m < \dim Z$, there exist different compatible Poisson structures on Z which bring a commutative dynamical algebra $\mathcal{A}$ into a commutative partially integrable system.

Forthcoming Theorem 4.1 shows that, under certain conditions, there exists a compatible Poisson structure on an open neighborhood of an invariant submanifold M of a commutative dynamical algebra. Theorems 4.2 – 4.3 describe all these Poisson structures around an invariant submanifold $M \subset Z$ of $\mathcal{A}$ [9]. Given a commutative partially integrable system $(w, \mathcal{A})$ in Theorem 4.2, the bivector field w (4.4) can be brought into the canonical form

$$w = \partial^\lambda \wedge \partial_\lambda \qquad (4.1)$$

with respect to generalized action-angle coordinates in Theorem 4.5. This theorem extends the above-mentioned Liouville–Arnold and Poincaré–Lyapounov–Nekhoroshev theorems to the case of a Poisson structure and a non-compact invariant submanifold [9, 12, 13].

Given a commutative dynamical algebra $\mathcal{A}$ on a manifold Z, let G be the group of local diffeomorphisms of Z generated by the flows of its elements. The orbits of G are maximal invariant submanifolds of $\mathcal{A}$ (we follow the terminology of [21]). Tangent spaces to these submanifolds form a (regular) distribution $\mathcal{V} \subset TZ$ whose maximal integral manifolds coincide with orbits of G. Being involutive, this distribution yields a foliation $\mathcal{S}$ of Z (Section 6.1).

Theorem 4.1 *Let $\mathcal{A}$ be a commutative dynamical algebra, M its invariant submanifold, and U a saturated open neighborhood of M. Let us suppose that:*

(i) vector fields ϑ_λ on U are complete,

(ii) a foliation $\mathcal{S}$ of U is a fibred manifold $\pi_\mathcal{S}$ with mutually diffeomorphic fibres.

Then the following hold [9, 12, 13].

(I) Leaves of $\mathcal{S}$ are diffeomorphic to the toroidal cylinder (1.1).

(II) There exists an open saturated neighborhood of M, say U again, which is a trivial principal bundle

$$U = N \times (\mathbb{R}^{m-r} \times T^r) \xrightarrow{\pi_S} N \tag{4.2}$$

with the structure additive group (1.1) over a domain $N \subset \mathbb{R}^{\dim Z - m}$.

(III) If $2m \leq \dim Z$, there exists a Poisson structure of rank $2m$ on U such that $\mathcal{A}$ is a commutative partially integrable system in accordance with Definition 4.2.

Remark 4.2 If fibres of a fibred manifold in item (ii) of Theorem 4.1 are assumed to be compact then this fibred manifold is a fibre bundle and vertical vector fields on it (e.g., in condition (i) of Theorem 4.1) are complete.

A Poisson structure in Theorem 4.1 is by no means unique as follows.

Theorem 4.2 *Given the toroidal domain U (4.2) provided with bundle coordinates (x^A, r^λ), it is readily observed that, if a Poisson bivector field on U satisfies Definition 4.2, it takes a form*

$$w = w_1 + w_2 = w^{A\lambda}(x^B)\partial_A \wedge \partial_\lambda + w^{\mu\nu}(x^B, r^\lambda)\partial_\mu \wedge \partial_\nu. \tag{4.3}$$

Conversely, given a Poisson bivector field w (4.3) of rank $2m$ on the toroidal domain U (4.2), there exists a toroidal domain $U' \subset U$ such that a commutative dynamical algebra $\mathcal{A}$ in Theorem 4.1 is a commutative partially integrable system on U'.

Remark 4.3 It is readily observed that any Poisson bivector field w (4.3) fulfils condition (b) in Definition 4.2, but condition (a) imposes a restriction on a toroidal domain U. A key point is that the characteristic foliation $\mathcal{F}$ of U yielded by the Poisson bivector fields w (4.3) is the pull-back of an m-dimensional foliation $\mathcal{F}_N$ of a base N, which is defined by the first summand w_1 (4.3) of w. With respect to the adapted coordinates (J_λ, z^A), $\lambda = 1, \ldots, m$, on the foliated manifold $(N, \mathcal{F}_N)$, the Poisson bivector field w reads

$$w = w^\mu_\nu(J_\lambda, z^A)\partial^\nu \wedge \partial_\mu + w^{\mu\nu}(J_\lambda, z^A, r^\lambda)\partial_\mu \wedge \partial_\nu. \tag{4.4}$$

Then condition (a) in Definition 4.2 is satisfied if $N' \subset N$ is a domain of a coordinate chart (J_λ, z^A) of the foliation $\mathcal{F}_N$. In this case, a commutative dynamical algebra $\mathcal{A}$ on a toroidal domain $U' = \pi^{-1}(N')$ is generated by the Hamiltonian vector fields

$$\vartheta_\lambda = -w\lfloor dJ_\lambda = w^\mu_\lambda \partial_\mu \tag{4.5}$$

of the m independent functions $S_\lambda = J_\lambda$.

Remark 4.4 Let us note that coefficients $w^{\mu\nu}$ in the expressions (4.3) and (4.4) are affine in coordinates r^λ due to the relation $[w, w]_{\mathrm{SN}} = 0$ and, consequently, they are constant on tori.

Now, let w and w' be two different Poisson structures (4.3) on the toroidal domain (4.2) which make a commutative dynamical algebra $\mathcal{A}$ into different commutative partially integrable systems $(w, \mathcal{A})$ and $(w', \mathcal{A})$.

Definition 4.4 *We agree to call a triple $(w, w', \mathcal{A})$ the bi-Poisson commutative partially integrable system if any Hamiltonian vector field $\vartheta \in \mathcal{A}$ with respect to w possesses the same Hamiltonian representation*

$$\vartheta = -w\lfloor df = -w'\lfloor df, \qquad f \in \mathcal{C}, \tag{4.6}$$

relative to w', and vice versa.

Definition 4.4 establishes *sui generis* an equivalence between the commutative partially integrable systems $(w, \mathcal{A})$ and $(w', \mathcal{A})$.

Theorem 4.3 *(I) The triple $(w, w', \mathcal{A})$ is a bi-Poisson partially integrable system in accordance with Definition 4.4 iff the Poisson bivector fields w and w' (4.3) differ in the second terms w_2 and w'_2. (II) These Poisson bivector fields admit a recursion operator.*

Given a smooth real manifold X, let w and w' be Poisson bivector fields of rank $2m$ on X, and let $w^\sharp$ and $w'^\sharp$ be the corresponding bundle homomorphisms. A tangent-valued one-form R on X yields bundle endomorphisms

$$R : TX \to TX, \qquad R^* : T^*X \to T^*X. \tag{4.7}$$

It is called the recursion operator if $w'^\sharp = R \circ w^\sharp = w^\sharp \circ R^*$.

Lemma 4.4 *A recursion operator between Poisson structures of the same rank exists iff their characteristic distributions coincide.*

For instance, the Poisson bivector field w (4.3) and the Poisson bivector field

$$w_0 = w^{A\lambda}\partial_A \wedge \partial_\lambda$$

admit a recursion operator $w_0^\sharp = R \circ w^\sharp$ whose entries are given by the equalities

$$R^A_B = \delta^A_B, \quad R^\mu_\nu = \delta^\mu_\nu, \quad R^A_\lambda = 0, \quad w^{\mu\lambda} = R^\lambda_B w^{B\mu}.$$

Given a commutative partially integrable system $(w, \mathcal{A})$ in Theorem 4.2, the bivector field w (4.4) can be brought into the canonical form (4.1) with respect to partial action-angle coordinates in forthcoming Theorem 4.5. This theorem extends the Liouville–Arnold theorem to the case of a Poisson structure and a non-compact invariant submanifold [9, 12, 13, 15].

Theorem 4.5 *Given a commutative partially integrable system $(w, \mathcal{A})$ on a Poisson manifold (U, w), there exists a toroidal domain $U' \subset U$ equipped with partial action-angle coordinates $(I_\lambda, z^A, r^\lambda)$ such that, restricted to U', a Poisson bivector field takes the canonical form (4.1) while a commutative dynamical algebra $\mathcal{A}$ is generated by Hamiltonian vector fields of the action coordinate functions $S_\lambda = I_\lambda$.*

5. Partially superintegrable systems on Poisson manifolds

Studying partially superintegrable systems, we bear in mind that in Example 1.2, but restrict our consideration to Lie algebra superintegrable systems whose generating functions constitute a real Lie algebra (Section 3).

Given a smooth manifold Z, let $\{\vartheta_i\}$ be k independent vector fields on Z (i.e., $\overset{k}{\wedge}\vartheta_i \neq 0$) which generate a real Lie algebra $\mathfrak{g}$ with commutation relations

$$[\vartheta_i, \vartheta_j] = c^h_{ij}\vartheta_h. \tag{5.1}$$

We denote by $\mathcal{C} \subset C^\infty(Z)$ a $\mathbb{R}$-subring of smooth real functions f on Z whose derivations $\vartheta_i \rfloor df$ vanish for all ϑ_i. Let $\mathcal{A}$ be a k-dimensional Lie $\mathcal{C}$-algebra generated by vector fields $\{\vartheta_i\}$.

Definition 5.1 *We agree to call $\mathcal{A}$ the dynamical algebra.*

In particular, this definition reproduces Definition 4.1 of a commutative dynamical algebra if vector fields $\{\vartheta_i\}$ mutually commute.

Definition 5.2 *Let (Z, w) be a Poisson manifold and $\mathcal{A}$ a k-dimensional dynamical algebra on Z. A triple $(Z, \mathcal{A}, w)$ is said to be a partially superintegrable system if the following hold.*

(a) Generators ϑ_i of $\mathcal{A}$ are Hamiltonian vector fields of some independent functions F_i on Z. These functions obey the commutation relations

$$\{F_i, F_j\}_w = c^h_{ij}F_h, \qquad \{F_i, f\}_w = 0, \qquad f \in \mathcal{C}. \tag{5.2}$$

IOP Conf. Series: Journal of Physics: Conf. Series **804** (2017) 012025 doi:10.1088/1742-6596/804/1/012025

(b) All elements of $\mathcal{C}\subset C^\infty(Z)$ are mutually in involution, i.e., their Poisson brackets $\{f,f'\}_w$ equal zero.

For instance, let $F=(F_i)$ be the Lie algebra superintegrable system (3.1) on a symplectic manifold (Z,Ω). Hamiltonian vector fields ϑ_i of its generating functions F_i obey the commutation relations (5.1) and yield a dynamical algebra. Then it is a partially superintegrable system in accordance with Definition 5.2 where a ring $\mathcal{C}$ consists of the pull-back of Casimir functions of the Lie–Poisson structure on a Lie coalgebra $\mathfrak{g}^*$.

A partially superintegrable systems on a product of manifolds in Examples 1.2 and commutative partially integrable systems in Example 1.3 also are well in accordance with Definition 5.2.

Certainly, a Poisson structure w in Definition 5.2 is not unique (see Theorem 4.2 for a case of commutative partially integrable systems). Following Definition 4.3, we agree to call it compatible. Generalizing Theorem 4.3, one can show that two Poisson structures are compatible only if they admit the recursion operator (4.7). In accordance with Lemma 4.4, their symplectic foliations coincide.

Lemma 5.1 *The rank of a compatible Poisson structure of a partially superintegrable system equals $k+m$ where m is a corank of the Lie algebra $\mathfrak{g}$ (5.1).*
Outline of proof: Let $\mathcal{W}$ be a symplectic foliation of Z and W its leaf. The generating vector field ϑ_i obviously are tangent to W. Restricted to W, they generate a Lie algebra superintegrable system so that $k+m$ is a dimension of a leaf W. □

Let $(Z,\mathcal{A},w)$ be a partially superintegrable system, and let its generating vector fields be complete. In accordance with the above-mentioned theorem [19, 20], they define a Hamiltonian action of a simply connected Lie group G whose Lie algebra is isomorphic to $\mathfrak{g}$ on Z. Since vector fields ϑ_i are independent, the action of G on Z is locally free, i.e., isotropy groups of points of U are discrete subgroups of G. Orbits of G coincide with k-dimensional maximal integral manifolds of a regular distribution $\mathcal{V}$ on Z spanned by vector fields ϑ_i [21]. They constitute a foliation $\mathcal{F}$ of Z. It is subordinate a symplectic foliation $\mathcal{W}$ whose leaves are foliated by leaves of $\mathcal{F}$.

Let both a foliation $\mathcal{F}$ and a foliation $\mathcal{W}$ be fibred manifolds $\pi_\mathcal{F}$ and $\pi_\mathcal{W}$ with mutually diffeomorphic fibres, respectively. Thus, we have a composite fibred manifold

$$\pi_\mathcal{W}: Z\longrightarrow \pi_\mathcal{F}(Z)\longrightarrow \pi_\mathcal{W}(Z). \tag{5.3}$$

Then one can show the following.

Theorem 5.2 *Let V be a leaf of $\mathcal{F}$. Then there exists an open saturated neighborhood U_V of V such that the composite fibred manifold $\pi_\mathcal{W}$ (5.3) becomes a composite bundle*

$$\pi_\mathcal{W}: U_V\longrightarrow \pi_\mathcal{F}(U_V)\longrightarrow \pi_\mathcal{W}(U_V) \tag{5.4}$$

which is endowed with bundle (generalized action-angle) coordinates

$$(I_\lambda,q^A,y^\lambda,w^a)\to(I_\lambda,q^A,x^a)\to(I_\lambda,x^a),\qquad \lambda=1,\ldots,m,\qquad A=1,\ldots,k-m,$$

where: (i) the angle coordinates y^i are coordinates on $\mathbb{R}^{m-r}\times T^r$; (ii) the (I_λ,x^a) are coordinates on $\pi_\mathcal{F}(U_V)$; the (x^a) are coordinates on $\pi_\mathcal{W}(U_V)$. With respect to these coordinates, the Poisson bivector field takes a form

$$w=\partial^\lambda\wedge\partial_\lambda+w^{AB}(q^C,x^a)\partial_A\wedge\partial_B.$$

Outline of proof: The proof of Theorem 5.2 is reduced to Theorems 4.2 and 4.5 for commutative partially integrable systems because the pull-back $\pi^*_\mathcal{F}C$ of functions on $\pi_\mathcal{F}(U_N)$ constitute a commutative partially integrable system on a Poisson manifold (Z,w) whose integral manifolds, diffeomorphic to toroidal cylinders, are invariant submanifolds of fibres of $\pi_\mathcal{F}$. □

References

[1] Arnold, V. (Ed.) (1990). *Dynamical Systems III, IV* (Springer, Berlin).
[2] Lazutkin, V. (1993). *KAM Theory and Semiclassical Approximations to Eigenfunctions* (Springer, Berlin).
[3] Nekhoroshev, N. (1994). The Poincaré – Lyapounov – Liuoville – Arnold theorem, *Funct. Anal. Appl.* **28**, 128.
[4] Gaeta, G. (2002). The Poincaré – Lyapounov – Nekhoroshev theorem, *Ann. Phys.* **297**, 157.
[5] Mishchenko, A. and Fomenko, A. (1978). Generalized Liouville method of integration of Hamiltonian systems, *Funct. Anal. Appl.* **12**, 113.
[6] Bolsinov, A. and Jovanović, B. (2003). Noncommutative integrability, moment map and geodesic flows, *Ann. Global Anal. Geom.* **23**, 305.
[7] Fassó, F. (2005). Superintegrable Hamiltonian systems: geometry and applications, *Acta Appl. Math.* **87**, 93.
[8] Vinogradov, A. and Kupershmidt, B. (1977). The structure of Hamiltonian mechanics, *Russian Math. Surveys* **32** (4), 177.
[9] Giachetta, G., Mangiarotti, L. and Sardanashvily, G. (2003). Bi-Hamiltonian partially integrable systems, *J. Math. Phys.* **44**, 1984; *arXiv*: math.DS/0211463.
[10] Fiorani, E. and Sardanashvily, G. (2006). Noncommutative integrability on noncompact invariant manifold, *J. Phys. A* **39**, 14035; *arXiv*: math.DS/0604104.
[11] Sardanashvily, G. (2009). Superintegrable Hamiltonian systems with noncompact invariant submanifolds. Kepler system, *Int. J. Geom. Methods Mod. Phys.* **6**, 1391; *arXiv*: 0911.0992.
[12] Giachetta, G., L.Mangiarotti, L. and Sardanashvily, G. (2010) *Geometric Formulation of Classical and Quantum Mechanics* (World Scientific, Singapore).
[13] Sardanashvily, G. (2015). Handbook on Integrable Hamiltonian Systems (URSS, M., 2015).
[14] Fiorani, E., Giachetta, G. and Sardanashvily, G. (2003). The Liouville – Arnold – Nekhoroshev theorem for noncompact invariant manifolds, *J. Phys. A* **36**, L101; *arXiv*: math.DS/0210346. .
[15] Kurov, A. (2016). Commutative partially integrable systems on Poisson manifolds, *Moscow Univ. Phys. Bull.* **2016**, N4.
[16] Libermann, P. and Marle, C-M. (1987). *Symplectic Geometry and Analytical Mechanics* (D.Reidel Publishing Company, Dordrecht).
[17] Gaeta, G. (2003). The Poincaré – Nekhoroshev map, *J. Nonlin. Math. Phys.* **10**, 51.
[18] Guillemin, V. and Sternberg, S. (1984). *Symplectic Techniques in Physics* (Cambr. Univ. Press, Cambridge).
[19] Palais, R. (1957). A global formulation of Lie theory of transformation groups, *Mem. Am. Math. Soc.* **22**, 1.
[20] Onishchik, A. (Ed.) (1993). *Lie Groups and Lie Algebras I. Foundations of Lie Theory, Lie Transformation Groups* (Springer, Berlin).
[21] Sussmann, H. (1973). Orbits of families of vector fields and integrability of distributions, *Trans. Amer. Math. Soc.* **180**, 171.
[22] A.Kurov, G.Sardanashvily, Partially superintegrable systems on Poisson manifolds, *arXiv*: 1606.03868.

IOP Conf. Series: Journal of Physics: Conf. Series **804** (2017) 012026 doi:10.1088/1742-6596/804/1/012026

Supercoset construction of Yang-Baxter deformed $AdS_5 \times S^5$ backgrounds

Hideki Kyono

Department of Physics, Kyoto University, Kitashirakawa Oiwake-cho, Kyoto 606-8502, Japan

E-mail: h_kyono@gauge.scphys.kyoto-u.ac.jp

Abstract. We consider Yang-Baxter deformations of the $AdS_5 \times S^5$ superstring theory. In previous works, the metric and B-field of some well-known string backgrounds concerned with the AdS/CFT correspondence have been obtained as deformations of $AdS_5 \times S^5$ based on r-matrices satisfying the homogeneous Yang-Baxter equation. Recently, the remaining fields including the Ramond-Ramond fields and the dilaton have been derived completely by performing the supercoset construction for abelian r-matrices. We also discuss the deformation with a non-abelian r-matrix and, in this case, the resulting background is not a solution of the type IIB supergravity. This article is based on the original paper [1].

1. Introduction

The Yang-Baxter sigma model [2] provides a good approach to study integrable deformations of two-dimensional non-linear sigma models. This deformed model was originally invented by Klimcik for principal chiral models. The deformation is characterized by an r-matrix satisfying the classical Yang-Baxter equation (CYBE). One can consider various deformations of the original model by changing the classical r-matrix. It ensures the existence of an associated Lax pair. In this sense, it is an integrable deformation.

Subsequently, this method has been generalized to symmetric (super)coset models, especially, to the type IIB superstring theory on the $AdS_5 \times S^5$ background for the case of r-matrices satisfying the modified classical Yang-Baxter equation (mCYBE) [3] or homogeneous CYBE [4, 5]. In fact, the Green-Schwarz (GS) type action can be constructed with a supercoset [6]. The supercoset has the $\mathbb{Z}_4$-grading structure which ensures the kinematical integrability of this system [7, 8]. Thus some integrable deformed string theories may be provided by Yang-Baxter deformations with some classical r-matrices.

So far, some Yang-Baxter deformed $AdS_5 \times S^5$ backgrounds have been studied. The deformed background with the Drinfel'd-Jimbo type r-matrix satisfying the mCYBE [9] was derived by Arutyunov, Borsato and Frolov, which is often called the η-deformed background [10]. They have obtained the deformed metric and Neveu-Schwarz-Neveu-Schwarz (NS-NS) two-form, and recently succeeded in deriving the Ramond-Ramond (R-R) fields and the dilaton by calculation including the fermionic components [11]. It was found that the resulting full background cannot satisfy the equations of motion of the type IIB supergravity, but it is a solution of the "generalized" type IIB supergravity [12]. On the "generalized" type IIB supergravity background the worldsheet theory does not preserve the Weyl invariance but only the scale invariance.

IOP Conf. Series: Journal of Physics: Conf. Series **804** (2017) 012026 doi:10.1088/1742-6596/804/1/012026

One can consider the deformations with r-matrices satisfying the homogeneous CYBE. The metrics and NS-NS two forms of some well-known string backgrounds concerned with the AdS/CFT correspondence [13] have been obtained as homogeneous Yang-Baxter deformations of the $AdS_5\times S^5$ background in the series of works [14–22]. Furthermore, we have succeeded in deriving the remaining fields including the R-R fields and the dilaton by performing the supercoset construction [1], for the cases including gravity duals of non-commutative gauge theories [23, 24], γ-deformations of S^5 [25, 26] and Schrödinger spacetimes [27]. These examples of deformed backgrounds satisfy the equations of motion of the usual type IIB supergravity.

As long as considering abelian r-matrices, the resulting backgrounds seem to be solutions of the usual type IIB supergravity. The backgrounds of our abelian examples can be also derived from TsT-transformations, that is, the deformations with abelian r-matrices may be equivalent to TsT-transformations. Recently, this correspondence has been proven generally in [28, 29]. On the other hand, the deformed backgrounds with non-abelian r-matrices seem not to be the usual supergravity solutions. We will give one of these examples which is not a supergravity solution. In [30] it was shown that this background satisfies the equations of motion of the "generalized" supergravity. More generally, the Yang-Baxter deformed backgrounds satisfy the "generalized" equations due to the kappa symmetry of the deformed action [31]. This fact is consistent with the result of the previous work [12]. In the end of this article, we mention briefly recent progress.

This article is organized as follows. Section 2 is a brief review of the Yang-Baxter deformed $AdS_5\times S^5$ superstring. Section 3 is devoted the discussion of the supercoset construction by following the procedure of [11]. Section 4 shows the concrete examples of classical r-matrices. Section 5 concludes the results and mentions briefly recent progress.

2. A brief review of the Yang-Baxter deformation

Here we introduce the Yang-Baxter deformed $AdS_5\times S^5$ superstring action. The discussion is based on the previous works for the mCYBE [3] or for the homogeneous CYBE [4, 5].

The Yang-Baxter deformed action of the $AdS_5\times S^5$ superstring is given by

$$S = -\frac{\sqrt{\lambda_c}}{4}\int_{-\infty}^{\infty} d\tau \int_0^{2\pi} d\sigma\, (\gamma^{ab} - \epsilon^{ab})\, \mathrm{STr}\Big[A_a\, d \circ \frac{1}{1-\eta R_g \circ d}(A_b)\Big]\,, \tag{1}$$

where the left-invariant one-form A_a is defined as

$$A_a(\tau,\sigma) \equiv -g^{-1}\partial_a g\,, \qquad g = g(\tau,\sigma) \in SU(2,2|4) \tag{2}$$

with the world-sheet index $a = (\tau,\sigma)$ and the deformation parameter η. When $\eta = 0$, this deformed action goes back to the original undeformed $AdS_5\times S^5$ GS action [6]. Here we have supposed the conformal gauge $\gamma^{ab} = \mathrm{diag}(-1,+1)$. Hence there is no coupling of the dilaton to the world-sheet scalar curvature. The anti-symmetric tensor ϵ^{ab} is normalized as $\epsilon^{\tau\sigma} = +1$ and the constant λ_c is the 't Hooft coupling.

A key ingredient in Yang-Baxter deformations is the operator R_g defined as

$$R_g(X) \equiv g^{-1}R(gXg^{-1})g\,, \qquad X \in \mathfrak{su}(2,2|4)\,, \tag{3}$$

where R is a linear map: $\mathfrak{su}(2,2|4) \to \mathfrak{su}(2,2|4)$ defined as follows[1]

$$R(X) = \mathrm{STr}_2[r(1\otimes X)] = \sum_i \left(a_i\,\mathrm{STr}[b_i X] - b_i\,\mathrm{STr}[a_i X]\right)\,. \tag{4}$$

[1] STr_2 is defined as $\mathrm{STr}_2[(W\otimes X)(Y\otimes Z)] = WY\,\mathrm{STr}[XZ]$.

Here r is a *skew-symmetric* classical r-matrix denoted as

$$r = \sum_i a_i \wedge b_i \equiv \sum_i (a_i \otimes b_i - b_i \otimes a_i) \qquad \text{with} \quad a_i,\ b_i \in \mathfrak{su}(2,2|4)\,. \tag{5}$$

These generators a_i and b_i characterize the directions of deformation. It is important for the integrability that this R-operator satisfies the following equation,

$$[R(X), R(Y)] - R([R(X), Y] + [X, R(Y)]) = c\,[X, Y] \qquad \text{for } {}^\forall X\,, Y \in \mathfrak{su}(2,2|4)\,. \tag{6}$$

When $c = 0 (\neq 0)$, it is called the homogeneous (modified) CYBE. In the following, we concentrate on r-matrices satisfying the homogeneous CYBE.

The projection operator d is defined as

$$d \ \equiv \ P_1 + 2P_2 - P_3\,, \tag{7}$$

where $P_\ell\,(\ell = 0, 1, 2, 3)$ are projections to the $\mathbb{Z}_4$-graded components of $\mathfrak{su}(2,2|4)$. In particular, $P_0(\mathfrak{su}(2,2|4))$ is a local symmetry of the classical action, $\mathfrak{so}(1,4) \oplus \mathfrak{so}(5)$. The numerical coefficients in the linear combination (7) are fixed by requiring the kappa symmetry [4, 6].

3. Supercoset construction

In this section, we introduce the detail of the supercoset construction of deformed models. The strategy is based on [11]. Our purpose here is to read off the type IIB supergravity fields from the deformed action (1) by comparing it with the canonical form of GS action [32]. Hence we will investigate the deformed action at the quadratic level of fermions.

The canonical form of the Lagrangian at order θ^2 [32] is given by[2],

$$\begin{aligned} \mathcal{L}^{(\theta^2)} \ &= \ -\frac{\sqrt{\lambda_c}}{2}\, i\bar{\Theta}_I(\gamma^{ab}\delta^{IJ} + \epsilon^{ab}\sigma_3^{IJ})\,\tilde{e}^m_a\Gamma_m \tilde{D}^{JK}_b \Theta_K\,, \\ \tilde{D}^{IJ}_a \ &\equiv \ \delta^{IJ}\left(\partial_a - \frac{1}{4}\tilde{\omega}^{mn}_a\Gamma_{mn}\right) + \frac{1}{8}\sigma_3^{IJ}\tilde{e}^m_a H_{mnp}\Gamma^{np} \\ &\quad -\frac{1}{8}e^\Phi\left[\epsilon^{IJ}\Gamma^p F_p + \frac{1}{3!}\sigma_1^{IJ}\Gamma^{pqr}F_{pqr} + \frac{1}{2\cdot 5!}\epsilon^{IJ}\Gamma^{pqrst}F_{pqrst}\right]\tilde{e}^m_a\Gamma_m\,. \end{aligned} \tag{8}$$

This Lagrangian contains the dilaton Φ, the three-form field strength $H_3 = dB_2$ (B_2 : NS-NS two-form), the one-form field strength $F_1 = d\chi$ (χ : axion or R-R scalar), the three-form field strength $F_3 = dC_2$ (C_2 : R-R two-form), and the five-form field strength $F_5 = dC_4$ (C_4 : R-R four-form). In the following steps, we rewrite the deformed Lagrangian (1) in order to make it easy to read off the type IIB supergravity fields by using the canonical form (8).

3.1. Left-invariant one-form

First of all, it is needed to determine the expression of the left-invariant one-form A. We parametrize the group element g as $g = g_b g_f$ where g_b includes only bosonic generators and g_f includes only fermionic ones. Then the bosonic left-invariant one-form $A_b \equiv g_b^{-1}\mathrm{d}g_b$ can be expressed as

$$A_b \ = \ e^m\mathbf{P}_m + \frac{1}{2}\omega^{mn}\,\mathbf{J}_{mn}\,, \tag{9}$$

[2] For the notation, please see the original paper [1].

where e^m is the vielbein for the $AdS_5 \times S^5$ spacetime and ω^{mn} is the spin connection. When the fermionic component g_f is parametrized as $g_f = \exp(\mathbf{Q}^I\theta_I)$, the full left-invariant one-form A can be written as follows,

$$\begin{aligned} A &= (e^m + \frac{i}{2}\bar{\theta}_I\gamma^m D^{IJ}\theta_J)\,\mathbf{P}_m - \mathbf{Q}^I\, D^{IJ}\theta_J + \frac{1}{2}\omega^{mn}\,\mathbf{J}_{mn} \\ &\quad -\frac{1}{4}\epsilon^{IJ}\bar{\theta}_I(\gamma^{\check{m}\check{n}}\,\mathbf{J}_{\check{m}\check{n}} - \gamma^{\hat{m}\hat{n}}\,\mathbf{J}_{\hat{m}\hat{n}})D^{JK}\theta_K\,. \end{aligned} \tag{10}$$

Here the covariant derivative for θ is defined as

$$D^{IJ}\theta_J = \delta^{IJ}\left(d\theta_J - \frac{1}{4}\omega^{mn}\gamma_{mn}\theta_J\right) + \frac{i}{2}\epsilon^{IJ}\,e^m\,\gamma_m\theta_J\,. \tag{11}$$

The last term represents the contribution of the R-R five-form field strength.

For later convenience, it is helpful to rearrange the above expansion of A with respect to the order of θ as follows,

$$A = A_{(0)} + A_{(1)} + A_{(2)}\,.$$

Here $A_{(p)}$ is the p-th order of θ and the explicit expressions of $A_{(p)}$ are given by

$$\begin{aligned} A_{(0)} &= e^m\,\mathbf{P}_m + \frac{1}{2}\omega^{mn}\,\mathbf{J}_{mn}\,, \\ A_{(1)} &= -\mathbf{Q}^I\,D^{IJ}\,\theta_J\,, \\ A_{(2)} &= \frac{i}{2}\bar{\theta}_I\gamma^m D^{IJ}\theta_J\,\mathbf{P}_m - \frac{1}{4}\epsilon^{IJ}\bar{\theta}_I(\gamma^{\check{m}\check{n}}\,\mathbf{J}_{\check{m}\check{n}} - \gamma^{\hat{m}\hat{n}}\,\mathbf{J}_{\hat{m}\hat{n}})D^{JK}\theta_K\,. \end{aligned} \tag{12}$$

3.2. Deformation operator

Furthermore, it is necessary to expand the deformation operator in terms of θ. θ is contained in the operator R_g through the group element g.

Let us define the deformation operator $\mathcal{O}$ and expand it in terms of θ as

$$\begin{aligned} \mathcal{O} &\equiv 1 - \eta R_g \circ d \\ &= \mathcal{O}_{(0)} + \mathcal{O}_{(1)} + \mathcal{O}_{(2)} + \mathcal{O}(\theta^3)\,. \end{aligned} \tag{13}$$

Similarly, for the inverse operator, $\mathcal{O}^{\rm inv}_{(i)}$ is defined as

$$\begin{aligned} \mathcal{O}^{\rm inv} &\equiv \frac{1}{1 - \eta R_g \circ d} \\ &= \mathcal{O}^{\rm inv}_{(0)} + \mathcal{O}^{\rm inv}_{(1)} + \mathcal{O}^{\rm inv}_{(2)} + \mathcal{O}(\theta^3)\,. \end{aligned} \tag{14}$$

Here, due to the relation $\mathcal{O}\circ\mathcal{O}^{\rm inv} = 1$, each of the components $\mathcal{O}^{\rm inv}_{(p)}$ $(p = 0, 1, 2)$ can be expressed as follows,

$$\begin{aligned} \mathcal{O}^{\rm inv}_{(0)} &= \frac{1}{1 - \eta R_{g_b} \circ d}\,, \\ \mathcal{O}^{\rm inv}_{(1)} &= -\mathcal{O}^{\rm inv}_{(0)} \circ \mathcal{O}_{(1)} \circ \mathcal{O}^{\rm inv}_{(0)}\,, \\ \mathcal{O}^{\rm inv}_{(2)} &= -\mathcal{O}^{\rm inv}_{(0)} \circ \mathcal{O}_{(2)} \circ \mathcal{O}^{\rm inv}_{(0)} - \mathcal{O}^{\rm inv}_{(1)} \circ \mathcal{O}_{(1)} \circ \mathcal{O}^{\rm inv}_{(0)}\,. \end{aligned} \tag{15}$$

IOP Conf. Series: Journal of Physics: Conf. Series **804** (2017) 012026 doi:10.1088/1742-6596/804/1/012026

In the following, we will concentrate only on the bosonic deformations, that is, the generators a_i and b_i are included in $\mathfrak{su}(2,2)\oplus\mathfrak{su}(4)$. Then the action of $R_{g_\mathrm{b}}\circ d$ for the basis of $\mathfrak{su}(2,2|4)$ can be evaluated as below,

$$
\begin{aligned}
R_{g_\mathrm{b}}\circ d(\mathbf{P}_m) &= 2\left(\lambda_m{}^n\mathbf{P}_n+\frac{1}{2}\lambda_m{}^{np}\mathbf{J}_{np}\right)\,,\\
R_{g_\mathrm{b}}\circ d(\mathbf{J}_{mn}) &= 0\,,\qquad R_{g_\mathrm{b}}\circ d(\mathbf{Q}^I)=0\,.
\end{aligned}
\tag{16}
$$

Here $\lambda_m{}^n$ and $\lambda_m{}^{np}$ are defined as below from the relation in (4),

$$
\begin{aligned}
\lambda_m{}^n &\equiv (a_i^{g_\mathrm{b}})^n\,(b_i^{g_\mathrm{b}})_m-(b_i^{g_\mathrm{b}})^n\,(a_i^{g_\mathrm{b}})_m\,,\\
\lambda_m{}^{np} &\equiv (a_i^{g_\mathrm{b}})^{np}\,(b_i^{g_\mathrm{b}})_m-(b_i^{g_\mathrm{b}})^{np}\,(a_i^{g_\mathrm{b}})_m\,,
\end{aligned}
\tag{17}
$$

where $(a_i^{g_\mathrm{b}})^m$, $(a_i^{g_\mathrm{b}})^{mn}$, $(b_i^{g_\mathrm{b}})^m$ and $(b_i^{g_\mathrm{b}})^{mn}$ are defined as

$$
\begin{aligned}
a_i^{g_\mathrm{b}} &\equiv g_\mathrm{b}^{-1}\,a_i\,g_\mathrm{b}=(a_i^{g_\mathrm{b}})^m\,\mathbf{P}_m+\frac{1}{2}(a_i^{g_\mathrm{b}})^{mn}\,\mathbf{J}_{mn}\,,\\
b_i^{g_\mathrm{b}} &\equiv g_\mathrm{b}^{-1}\,b_i\,g_\mathrm{b}=(a_i^{g_\mathrm{b}})^m\,\mathbf{P}_m+\frac{1}{2}(b_i^{g_\mathrm{b}})^{mn}\,\mathbf{J}_{mn}\,.
\end{aligned}
\tag{18}
$$

Now the action of $\mathcal{O}^{\mathrm{inv}}_{(0)}$, $\mathcal{O}^{\mathrm{inv}}_{(1)}$ and $\mathcal{O}^{\mathrm{inv}}_{(2)}$ can be examined as follows.

The action of $\mathcal{O}^{\mathrm{inv}}_{(0)}$ is given by

$$
\begin{aligned}
\mathcal{O}^{\mathrm{inv}}_{(0)}(\mathbf{P}_m) &\equiv k_m{}^n\,\mathbf{P}_n+\frac{1}{2}l_m{}^{np}\,\mathbf{J}_{np}\,,\\
\mathcal{O}^{\mathrm{inv}}_{(0)}(\mathbf{J}_{mn}) &= \mathbf{J}_{mn}\,,\qquad \mathcal{O}^{\mathrm{inv}}_{(0)}(\mathbf{Q}^I)=\mathbf{Q}^I\,,
\end{aligned}
\tag{19}
$$

where $k_m{}^n$ is determined from $\lambda_m{}^n$ by the following relation,

$$
k_m{}^n=(\delta-2\eta\lambda)^{-1}{}_m{}^n\,.
\tag{20}
$$

When $\eta=0$, $k_m{}^n$ is reduced to $\delta_m{}^n$. Here we have not displayed the explicit form of $l_m{}^{np}$ because it does not appear in the final expression due to the presence of the projection operators.

Then the action of $\mathcal{O}^{\mathrm{inv}}_{(1)}$ is written as

$$
\begin{aligned}
\mathcal{O}^{\mathrm{inv}}_{(1)}(\mathbf{P}_m) &= i\epsilon^{IJ}k_m{}^n\eta\lambda_n{}^p\,\mathbf{Q}^J\,\gamma_p\theta_I+\frac{1}{2}\delta^{IJ}k_m{}^n\eta\lambda_n{}^{pq}\,\mathbf{Q}^J\,\gamma_{pq}\theta_I\,,\\
\mathcal{O}^{\mathrm{inv}}_{(1)}(\mathbf{J}_{mn}) &= 0\,,\\
\mathcal{O}^{\mathrm{inv}}_{(1)}(\mathbf{Q}^I) &= i\sigma_3^{IJ}\,k_m{}^p\eta\lambda^{nm}\,\bar{\theta}_J\gamma_n\,\mathbf{P}_p+\frac{1}{2}\sigma_1^{IJ}\,k_m{}^q\eta\lambda^{m,np}\,\bar{\theta}_J\gamma_{np}\,\mathbf{P}_q+\text{terms with }\mathbf{J}\,.
\end{aligned}
\tag{21}
$$

Here the terms proportional to $\mathbf{J}_{mn}$ are not explicitly written down because they do not contribute to the final expression.

Finally, the action of $\mathcal{O}^{\mathrm{inv}}_{(2)}$ is evaluated as

$$
\begin{aligned}
\mathcal{O}^{\mathrm{inv}}_{(2)}(\mathbf{P}_m) &= \bar{\theta}_I\left[\delta^{IJ}(\mathcal{M}^{P\delta}_{(2)})_m{}^n+\epsilon^{IJ}(\mathcal{M}^{P\epsilon}_{(2)})_m{}^n+\sigma_1^{IJ}(\mathcal{M}^{P\sigma_1}_{(2)})_m{}^n+\sigma_3^{IJ}(\mathcal{M}^{P\sigma_3}_{(2)})_m{}^n\right]\theta_J\,\mathbf{P}_n\\
&\quad+\text{terms with }\mathbf{J}\,,\\
\mathcal{O}^{\mathrm{inv}}_{(2)}(\mathbf{J}_{mn}) &= 0\,,\qquad \mathcal{O}^{\mathrm{inv}}_{(2)}(\mathbf{Q}^I)=\text{irrelevant terms}\,,
\end{aligned}
\tag{22}
$$

where $\mathcal{M}^{P\delta}_{(2)}$, $\mathcal{M}^{P\epsilon}_{(2)}$, $\mathcal{M}^{P\sigma_1}_{(2)}$, and $\mathcal{M}^{P\sigma_3}_{(2)}$ are defined as

$$\begin{aligned}
(\mathcal{M}^{P\delta}_{(2)})_m{}^n &\equiv -\frac{i}{4}\Big[(k_r{}^n\gamma^r)(k_m{}^s\eta\lambda_s{}^{pq}\gamma_{pq}) - (k_r{}^n\eta\lambda^{r,\,pq}\gamma_{pq})(k_m{}^s\gamma_s)\Big]\,,\\
(\mathcal{M}^{P\epsilon}_{(2)})_m{}^n &\equiv -\frac{1}{2}\Big[(k_p{}^n\gamma^p)(k_m{}^s\eta\lambda_s{}^q\gamma_q) - (k_p{}^n\eta\lambda^{pq}\gamma_q)(k_m{}^s\gamma_s)\Big]\,,\\
(\mathcal{M}^{P\sigma_1}_{(2)})_m{}^n &\equiv (k_s{}^n\eta\lambda^{rs}\gamma_r)(k_m{}^q\eta\lambda_q{}^p\gamma_p) + \frac{1}{4}(k_s{}^n\eta\lambda^{s,rt}\gamma_{rt})(k_m{}^u\eta\lambda_u{}^{pq}\gamma_{pq})\,,\\
(\mathcal{M}^{P\sigma_3}_{(2)})_m{}^n &\equiv \frac{i}{2}\Big[(k_s{}^n\eta\lambda^{rs}\gamma_r)(k_m{}^t\eta\lambda_t{}^{pq}\gamma_{pq}) + (k_s{}^n\eta\lambda^{s,rt}\gamma_{rt})(k_m{}^q\eta\lambda_q{}^p\gamma_p)\Big]\,.
\end{aligned}\tag{23}$$

Here the terms proportional to $\mathbf{J}_{mn}$ have not been written down on the same reasoning. Furthermore, the explicit expression of $\mathcal{O}^{\rm inv}_{(2)}(\mathbf{Q}^I)$ is not necessary for our argument because it always leads to higher-order contributions with $\mathcal{O}(\theta^4)$ in the resulting Lagrangian.

3.3. The deformed Lagrangian at quadratic level of θ
Now the Lagrangian in (1) can be rewritten as

$$\mathcal{L} = -\frac{\sqrt{\lambda_c}}{4}\left(\gamma^{ab} - \epsilon^{ab}\right){\rm STr}\left[\tilde{d}(A_a)\mathcal{O}^{\rm inv}(A_b)\right]\,, \tag{24}$$

where $\tilde{d}$ is the transpose operator of d and it is defined as

$$\tilde{d} \equiv -P_1 + 2P_2 + P_3\,. \tag{25}$$

This Lagrangian can be expanded in terms of θ at quadratic level as

$$\mathcal{L} = \mathcal{L}_{(0)} + \mathcal{L}_{(2,0,0)} + \mathcal{L}_{(0,0,2)} + \mathcal{L}_{(1,1,0)} + \mathcal{L}_{(0,1,1)} + \mathcal{L}_{(0,2,0)} + \mathcal{L}_{(1,0,1)} + \mathcal{O}(\theta^4)\,. \tag{26}$$

Here, $\mathcal{L}_{(0)}$ does not include any θ. The second-order term $\mathcal{L}_{(l,m,n)}$ contains two θs. The set of subscripts (l,m,n) indicates the numbers of θ included in $\tilde{d}(A_a)$, $\mathcal{O}^{\rm inv}$ and A_b, respectively. For example, in the case of $\mathcal{L}_{(2,0,0)}$, the two θs are included in $\tilde{d}(A_a)$, and there is no θ in $\mathcal{O}^{\rm inv}$ and A_b. That is, $\mathcal{L}_{(2,0,0)}$ is given by

$$\mathcal{L}_{(2,0,0)} = -\frac{\sqrt{\lambda_c}}{4}\left(\gamma^{ab} - \epsilon^{ab}\right){\rm STr}\left[\tilde{d}((A_{(2)})_a)\mathcal{O}^{\rm inv}_{(0)}((A_{(0)})_b)\right]\,. \tag{27}$$

In the following, let us see each term of the expansion (26). The first one can be rewritten into the standard form as follows,

$$\begin{aligned}
\mathcal{L}_{(0)} &= -\frac{\sqrt{\lambda_c}}{4}(\gamma^{ab} - \epsilon^{ab})\,{\rm STr}\left[\tilde{d}((A_{(0)})_a)\mathcal{O}^{\rm inv}_{(0)}((A_{(0)})_b)\right]\\
&= -\frac{\sqrt{\lambda_c}}{2}(\gamma^{ab} - \epsilon^{ab})\,e^m_a e^n_b\,k_{nm}\\
&= -\frac{\sqrt{\lambda_c}}{2}\left[\gamma^{ab}e^m_\mu e^n_\nu\,k_{(mn)}\,\partial_a X^\mu\partial_b X^\nu - \epsilon^{ab}e^m_\mu e^n_\nu\,k_{[nm]}\,\partial_a X^\mu\partial_b X^\nu\right]\,.
\end{aligned}\tag{28}$$

Here we have used the relation $e^m_a = e^m_\mu\partial_a X^\mu$, and X^μ are the target-spacetime coordinates. From the last expression of (28), one can obtain the deformed spacetime metric $\widetilde{G}$ and NS-NS two-form B ,

$$\begin{aligned}
\widetilde{G}_{MN} &\equiv e^m_M e^n_N\,k_{(mn)} = \tilde{e}^m_M\tilde{e}_{mN}\,,\\
B_{MN} &\equiv e^m_M e^n_N\,k_{[nm]}\,.
\end{aligned}\tag{29}$$

Here, for our later convenience, we have introduced the vielbein $\tilde{e}^m_M$ for the deformed metric. Note that the index M is raised and lowered by $\widetilde{G}^{MN}$ and $\widetilde{G}_{MN}$, respectively.

Then let us evaluate the combination $\mathcal{L}_{(2,0,0)} + \mathcal{L}_{(0,0,2)}$. From the point of view of symmetry, this combination is convenient and can be evaluated as

$$\begin{aligned}
&\mathcal{L}_{(2,0,0)} + \mathcal{L}_{(0,0,2)} \\
=\;& -\frac{\sqrt{\lambda_c}}{4}(\gamma^{ab}-\epsilon^{ab})\,\mathrm{STr}\big[i\bar{\theta}_I\gamma^m D^{IJ}_a\theta_J\,\mathbf{P}_m\,e^n_b k_n{}^p\,\mathbf{P}_p + ie^m_a\mathbf{P}_m\,\bar{\theta}_I\gamma^n D^{IJ}_b\theta_J\,k_n{}^p\mathbf{P}_p\big] \\
=\;& -\frac{i\sqrt{\lambda_c}}{4}(\gamma^{ab}-\epsilon^{ab})\,\bar{\theta}_I\,(e^n_b k_{nm}\gamma^m D^{IJ}_a + e^m_a k_{nm}\gamma^n D^{IJ}_b)\,\theta_J\,. \qquad (30)
\end{aligned}$$

By the same reasoning, it is helpful to evaluate the combination $\mathcal{L}_{(1,1,0)} + \mathcal{L}_{(0,1,1)}$. The resulting expression is given by

$$\begin{aligned}
&\mathcal{L}_{(1,1,0)} + \mathcal{L}_{(0,1,1)} \qquad (31)\\
=\;& -\frac{\sqrt{\lambda_c}}{4}(\gamma^{ab}-\epsilon^{ab})\,\mathrm{STr}\big[\sigma^{IJ}_3\mathbf{Q}^J D^{IK}_a\theta_K\mathcal{O}^{\rm inv}_{(1)}(e^m_b\mathbf{P}_m) + 2e^m_a\mathbf{P}_m\mathcal{O}^{\rm inv}_{(1)}(-\mathbf{Q}^I D^{IJ}_b\theta_J)\big] \\
=\;& -\frac{\sqrt{\lambda_c}}{2}(\gamma^{ab}-\epsilon^{ab})\,\bar{\theta}_I\left[i\eta\lambda_n{}^p\gamma_p\sigma^{IJ}_3 - \frac{1}{2}\eta\lambda_n{}^{pq}\gamma_{pq}\sigma^{IJ}_1\right](e^m_b k_m{}^n D^{JK}_a + e^m_a k^n{}_m D^{JK}_b)\,\theta_K\,.
\end{aligned}$$

Finally, $\mathcal{L}_{(0,2,0)}$ and $\mathcal{L}_{(1,0,1)}$ are evaluated as, respectively,

$$\begin{aligned}
\mathcal{L}_{(0,2,0)} =\;& -\frac{\sqrt{\lambda_c}}{4}(\gamma^{ab}-\epsilon^{ab})\,\mathrm{STr}\left[2e^m_a\mathbf{P}_m\mathcal{O}^{\rm inv}_{(2)}(e^n_b\mathbf{P}_n)\right] \\
=\;& -\frac{\sqrt{\lambda_c}}{2}(\gamma^{ab}-\epsilon^{ab})\,e^m_a e^n_b\,\bar{\theta}_I\Big[\epsilon^{IJ}(\mathcal{M}^{P\epsilon}_{(2)})_{nm} + \delta^{IJ}(\mathcal{M}^{P\delta}_{(2)})_{nm} \\
&+\sigma^{IJ}_1(\mathcal{M}^{P\sigma_1}_{(2)})_{nm} + \sigma^{IJ}_3(\mathcal{M}^{P\sigma_3}_{(2)})_{nm}\Big]\theta_J\,, \qquad (32)\\
\mathcal{L}_{(1,0,1)} =\;& -\frac{\sqrt{\lambda_c}}{4}(\gamma^{ab}-\epsilon^{ab})\,\mathrm{STr}\left[\sigma^{IJ}_3\mathbf{Q}^J D^{IK}_a\theta_K(-\mathbf{Q}^L D^{LM}_b)\theta_M\right] \\
=\;& -\frac{i\sqrt{\lambda_c}}{2}\epsilon^{ab}\sigma^{IJ}_3\,\bar{\theta}_I e^m_a\gamma_m D^{JK}_b\theta_K\,. \qquad (33)
\end{aligned}$$

So far, we have derived the deformed Lagrangian at the quadratic level of θ. However, coordinate transformations of X^M and θ_I are still needed to recast the Lagrangian into the canonical form. In the following, we perform a shift of X^M and a rotation of θ_I.

3.4. Shift of X

Let us see the terms with $\gamma^{ab}\partial_b\theta$ in $\mathcal{L}$. The relevant parts are

$$(a)\quad \mathcal{L}^{\gamma}_{(2,0,0)} + \mathcal{L}^{\gamma}_{(0,0,2)} \qquad \text{and} \qquad (b)\quad \mathcal{L}^{\gamma}_{(1,1,0)} + \mathcal{L}^{\gamma}_{(0,1,1)}\,.$$

One can realize that the terms should appear with δ^{IJ} from the expression of the canonical form (8). There is no obstacle for (a), however (b) involves terms like

$$\frac{\sqrt{\lambda_c}}{2}\,\bar{\theta}_I\,\gamma^{ab}\,\sigma^{IJ}_1\,e^m_a\,k_{(mn)}\eta\,\lambda^{n,pq}\gamma_{pq}\partial_b\theta_J\,. \qquad (34)$$

Such terms proportional to σ^{IJ}_1 do not appear in the canonical form (8) and hence must be removed somehow. A possible resolution is to shift X as [11]

$$X^\mu \longrightarrow X^\mu + \bar{\theta}_I\delta X^{\mu IJ}\theta_J\,, \qquad \delta X^{\mu IJ} \equiv \frac{1}{4}\sigma^{IJ}_1 e^{n\mu}\,\eta\,\lambda_n{}^{pq}\gamma_{pq}\,. \qquad (35)$$

While this shift removes the problematic terms, it generates additional ones,

$$
\begin{aligned}
-\frac{\sqrt{\lambda_c}}{2}\, i\bar{\theta}_I\, \gamma^{ab}\delta^{IJ}\Big[&-\frac{i}{2}\,\sigma_1^{JK}\; e_a^m\, e_N^n\, k_{(mn)}\partial_b(e^{n\,N}\,\eta\,\lambda_n{}^{pq}\gamma_{pq}) \\
&-\frac{i}{4}\sigma_1^{JK}\,\partial_P\tilde{G}_{MN}\partial_a X^M\,\partial_b X^N\, e^{n\,P}\,\eta\,\lambda_n{}^{pq}\gamma_{pq}\Big]\theta_K\,. \qquad (36)
\end{aligned}
$$

Note here that these terms do not involve derivatives of θ.

At this stage, the quadratic Lagrangian including γ^{ab} is written down as

$$
\begin{aligned}
\mathcal{L}^{\gamma} \;=\; -i\frac{\sqrt{\lambda_c}}{2}\,\gamma^{ab}\delta^{IJ}\bar{\theta}_I\Bigg[& e_a^p\, k_{(pn)}\,(\eta^{nm}-(-1)^J 2\eta\lambda^{nm})\gamma_m D_b^{JK} \\
&+\frac{1}{2}\sigma_3^{JK}\; e_a^m k_{(mn)}\eta\lambda^{n,\,pq}\gamma_{pq}\; e_b^r\,\gamma_r-\frac{1}{4}\delta^{JK}\; e_a^m\; k_{nm}\,(\eta^{np}-(-1)^J\,2\eta\,\lambda^{np})\;\gamma_p\; e_b^q k_q{}^r\,\eta\,\lambda_r{}^{st}\,\gamma_{st} \\
&+\frac{1}{4}\delta^{JK}\; e_a^m k_{mn}\,\eta\,\lambda^{n,\,pq}\;\gamma_{pq}\, e_b^r\, k_{rs}\,(\eta^{st}-(-1)^J\,2\eta\,\lambda^{st})\;\gamma_t \\
&+\frac{i}{2}\epsilon^{JK} e_a^m\; k_{nm}\,(\eta^{np}-(-1)^J\,2\eta\,\lambda^{np})\;\gamma_p\; e_b^q\, k_q{}^r\,\eta\,\lambda_r{}^s\,\gamma_s \\
&-\frac{i}{4}\sigma_1^{JK}\; e_a^m k_{(mn)}\,\eta\,\lambda^{n,\,pq}\,\gamma_{pq}\;\omega_b^{rs}\;\gamma_{rs}-\frac{i}{2}\epsilon^{JK}\; e_a^m\; k_{nm}\,\eta\,\lambda^{np}\,\gamma_p\; e_b^q\, k_q{}^r\,\gamma_r \\
&-\frac{i}{4}\sigma_1^{JK}\; e_a^m\; k_{nm}\,\eta\,\lambda^{n,pq}\,\gamma_{pq}\; e_b^r\, k_r{}^s\,\eta\,\lambda_s{}^{tu}\,\gamma_{tu}-\frac{i}{2}\sigma_1^{JK}\; e_a^m\; e_M^n\; k_{(mn)}\;\partial_b\left(e^{p,\,M}\,\eta\,\lambda_p{}^{qr}\,\gamma_{qr}\right) \\
&-\frac{i}{4}\sigma_1^{JK}\partial_P\widetilde{G}_{MN}\partial_a\, X^M\,\partial_b\, X^N\, e^{m,\,P}\,\eta\,\lambda_m{}^{np}\,\gamma_{np}\Bigg]\theta_K\,. \qquad (37)
\end{aligned}
$$

The next step is to see the terms with $\epsilon^{ab}\partial_b\theta$ in $\mathcal{L}$. This part has the terms involving σ_1^{IJ} as well. The sift of X in (35) can eliminate the problematic terms simultaneously, while some additional terms including ϵ^{ab} are again generated. Then the quadratic Lagrangian including ϵ^{ab} is written down as

$$
\begin{aligned}
\mathcal{L}^{\epsilon} \;=\; -i\,\frac{\sqrt{\lambda_c}}{2}\,\epsilon^{ab}\,\bar{\theta}_I\Bigg[&\Big(\delta^{IJ}\; e_a^m\; k_{[mn]}\,\gamma^n+\sigma_3^{IJ}\left[e_a^m\; k_{[mn]}\,2\,\eta\,\lambda^{np}\,\gamma_p+e_a^m\,\gamma_m)\right]\Big)D_b^{JK} \\
&+i\,\sigma_1^{IJ}\; e_a^m\; k_{[mn]}\,\eta\,\lambda^{n\,,pq}\,\gamma_{pq}\left(-\frac{1}{4}\delta^{JK}\;\omega_b^{rs}\,\gamma_{rs}+\frac{i}{2}\epsilon^{JK} e_b^r\,\gamma_r\right) \\
&+\frac{1}{4}\,\delta^{IK} e_a^m\; k_{nm}\,(\eta^{np}-(-1)^I\,2\eta\,\lambda^{np})\;\gamma_p\; e_b^q k_q{}^r\,\eta\,\lambda_r{}^{st}\,\gamma_{st} \\
&-\frac{1}{4}\delta^{IK}\; e_a^m k_{mn}\,\eta\,\lambda^{n,\,pq}\,\gamma_{pq}\, e_b^r\, k_{rs}\,(\eta^{st}-(-1)^I\,2\eta\,\lambda^{st})\;\gamma_t \\
&-\frac{i}{2}\epsilon^{IK} e_a^m\; k_{nm}\,(\eta^{np}-(-1)^I\,2\eta\,\lambda^{np})\;\gamma_p\; e_b^q\, k_q{}^r\,\eta\,\lambda_r{}^s\,\gamma_s \\
&+\frac{i}{2}\epsilon^{IK}\; e_a^m\; k_{nm}\,\eta\,\lambda^{np}\,\gamma_p\; e_b^q\, k_q{}^r\,\gamma_r+\frac{i}{4}\sigma_1^{IK}\; e_a^m\; k_{nm}\,\eta\,\lambda^{n,pq}\,\gamma_{pq}\; e_b^r\, k_r{}^s\,\eta\,\lambda_s{}^{tu}\,\gamma_{tu} \\
&+\frac{i}{2}\sigma_1^{IK} B_{MN}\partial_a X^M\,\partial_b\left(e^{n\,N}\,\eta\,\lambda_n{}^{pq}\right)\gamma_{pq} \\
&+\frac{i}{4}\sigma_1^{IK}\partial_P B_{MN}\partial_a X^M\partial_b X^N e^{n\,P}\eta\lambda_n{}^{pq}\gamma_{pq}\Bigg]\theta_K\,. \qquad (38)
\end{aligned}
$$

For the next step, it is convenient to switch from the 16×16 gamma matrices γ to the 32×32 ones Γ, and hence we will work in the 32×32 notation in the following. The lift-up rule is summarized in [1], and it is straightforward to rewrite the Lagrangian.

IOP Conf. Series: Journal of Physics: Conf. Series **804** (2017) 012026 doi:10.1088/1742-6596/804/1/012026

3.5. Rotation of θ

After shifting X, the resulting derivative terms of θ involving γ^{ab} take the following form,

$$-\frac{\sqrt{\lambda_c}}{2}\, i\bar{\Theta}_I\, \gamma^{ab}\delta^{IJ}\, \tilde{e}_{(I)a}{}^{m}\, \Gamma_m \partial_b\, \Theta_J\,. \tag{39}$$

Here, the vielbeins[3] $\tilde{e}_{(I)a}{}^{m}$ $(I = 1,2)$ are defined as

$$\tilde{e}_{(I)a}{}^{m} \equiv e_a^p\, k_{(pn)}\left[\eta^{nm} - (-1)^I 2\eta\lambda^{nm}\right] \tag{40}$$

and depend on the index I. Hence we need to perform a Lorentz transformation for the spinor θ to remove the I dependence.

The first step is to determine the I-independent form of the vielbeins as a reference frame. Hereafter, it is fixed by taking $I = 1$ in (40) as

$$\tilde{e}_a^m = e_a^p\, k_{(pn)}\left[\eta^{nm} + 2\eta\lambda^{nm}\right]. \tag{41}$$

Then, by performing a Lorentz transformation for θ, this term can be rewritten as

$$\begin{aligned} &\bar{\Theta}_I\, \tilde{e}_{(I)a}{}^{m}\, \bar{U}_{(I)}\Gamma_m\, U_{(I)}\partial_b\Theta_I + (\text{the derivative term of } U) \\ =\ &\bar{\Theta}_I\, \tilde{e}_{(I)a}{}^{m}\, \Lambda_{(I)m}{}^{n}\Gamma_n\, \partial_b\Theta_I + (\text{the derivative term of } U)\,. \end{aligned} \tag{42}$$

Note that the Lorentz transformation performed here depends on the index I.

In order to realize the I-independent form (41), the transformation Λ should be taken as

$$\Lambda_{(I)m}{}^{n} = \left[\delta_m{}^p + (-1)^I 2\eta\lambda_m{}^p\right](\delta - 2\eta\lambda)^{-1}{}_p{}^n\,. \tag{43}$$

Then the spinor transformation $U_{(I)}$ and its inverse $\bar{U}_{(I)}$ have to be determined through the following relation,

$$\bar{U}_{(I)}\Gamma_m U_{(I)} = \Lambda_{(I)m}{}^{n}\Gamma_n\,. \tag{44}$$

After all this, we have obtained the canonical form of the Lagrangian[4]. In the canonical form of the Lagrangian (8), the dilaton and R-R field strength appear as the product of them. From this reason, we use the following formula to decide the dilaton,

$$e^{\Phi} = \frac{1}{\det_{10}(\delta_m{}^n + 2\,\eta\,\lambda_m{}^n)^{\frac{1}{2}}}\,, \tag{45}$$

where $\det_D$ means the determinant of a $D \times D$ matrix. Note that this is similar to the formula in [33] for λ-deformation. It works well for well-known examples, including the examples discussed in Sec. 4.

In the actual calculation, we used a concrete expression of a classical r-matrix and computation software like Mathematica or Maple to decide the R-R fields and the dilation. The resulting backgrounds for some examples are presented in Sec. 4[5].

[3] Note that $\tilde{e}_{(I)a}{}^{m}$ satisfy the relation

$$\tilde{e}_{(I)a}{}^{m}\, \tilde{e}_{(I)b\,m} = e_a^m e_b^n\, k_{(mn)} = \tilde{G}_{MN}\partial_a X^M \partial_b X^N \qquad (\text{for } I = 1,2)\,.$$

[4] To read off the R-R fields, we still have to take the trace of gamma matrices in the quadratic Lagrangian.

[5] The deformed R-R fields can also be determined by using the concrete form of the kappa symmetry of the deformed action as another method to determine. The results are consistent with those from the supercoset construction.

IOP Conf. Series: Journal of Physics: Conf. Series **804** (2017) 012026 doi:10.1088/1742-6596/804/1/012026

4. Examples

In this section, we show the results for some r-matrices satisfying the homogeneous CYBE. For the following argument, let us introduce the terms "abelian" and "non-abelian" classical r-matrices. Suppose that a classical r-matrix is given by $r = a \wedge b$. It is called "abelian" when a and b commute with each other. If not, it is "non-abelian".

4.1. Gravity duals of noncommutative gauge theories

Let us discuss the following classical r-matrix,

$$r = P_2 \wedge P_3 \,. \tag{46}$$

This is an abelian classical r-matrix and satisfies the homogeneous CYBE. The bosonic part of the deformation with this r-matrix has already been studied in [16]. Here P_μ are the translation operators of $\mathfrak{so}(2,2) \in \mathfrak{so}(2,2|4)$. By this r-matrix, the AdS_5 part is deformed.

Through the general argument in Sec. 3, the following deformed background can be obtained,

$$\begin{aligned}
ds^2 &= \frac{-(dx^0)^2 + (dx^1)^2}{z^2} + \frac{z^2\left[(dx^2)^2 + (dx^3)^2\right]}{z^4 + 4\eta^2} + \frac{dz^2}{z^2} + ds^2_{\mathrm{S}^5}\,, \\
B_2 &= \frac{2\eta}{z^4 + 4\eta^2}\, dx^2 \wedge dx^3\,, \\
F_3 &= \frac{8\,\eta}{z^5}\, dx^0 \wedge dx^1 \wedge dz\,, \\
F_5 &= 4\left(\mathrm{e}^{2\Phi}\,\omega_{\mathrm{AdS}_5} + \omega_{\mathrm{S}^5}\right)\,, \qquad \Phi = \frac{1}{2}\log\left(\frac{z^4}{z^4 + 4\eta^2}\right)\,.
\end{aligned} \tag{47}$$

This is nothing but the solution found in [23, 24] as a gravity dual of noncommutative gauge theories.

4.2. γ-deformations of S^5

We shall discuss three-parameter γ-deformations of S^5 with the following classical r-matrix,

$$r = \frac{1}{8}\left(\nu_3\, h_1 \wedge h_2 + \nu_1\, h_2 \wedge h_3 + \nu_2\, h_3 \wedge h_1\right)\,. \tag{48}$$

Here ν_i $(i = 1, 2, 3)$ are real constant parameters, and h_a $(a = 1, 2, 3)$ are the Cartan generators of $\mathfrak{su}(4)$. This is also an abelian classical r-matrix and satisfies the CYBE. The bosonic part has already been studied in [15]. The remaining task is to perform supercoset construction in order to determine the R-R sector and the dilaton.

Then, by following the general discussion, the full solution presented in [25, 26] can be reproduced as

$$ds^2 = ds^2_{\mathrm{AdS}_5} + \sum_{i=1}^{3}(d\rho_i^2 + G\,\rho_i^2 d\phi_i^2) + G\rho_1^2\rho_2^2\rho_3^2\left(\sum_{i=1}^{3}\hat{\gamma}_i\, d\phi_i\right)^2\,, \tag{49}$$

$$\begin{aligned}
B_2 &= G\left(\hat{\gamma}_3\,\rho_1^2\rho_2^2\, d\phi_1 \wedge d\phi_2 + \hat{\gamma}_1\,\rho_2^2\rho_3^2\, d\phi_2 \wedge d\phi_3 + \hat{\gamma}_2\,\rho_3^2\rho_1^2\, d\phi_3 \wedge d\phi_1\right)\,, \\
F_3 &= -4\sin^3\alpha\,\cos\alpha\,\sin\theta\,\cos\theta\left(\sum_{i=1}^{3}\hat{\gamma}_i\, d\phi_i\right) \wedge d\alpha \wedge d\theta \\
F_5 &= 4\left(\omega_{\mathrm{AdS}_5} + G\,\omega_{\mathrm{S}^5}\right)\,, \qquad \Phi = \frac{1}{2}\log G\,.
\end{aligned} \tag{50}$$

Here we have introduced a scalar function G and $\hat{\gamma}_i$ $(i = 1, 2, 3)$ defined as

$$G^{-1} \equiv 1 + \hat{\gamma}_3^2\,\rho_1^2\rho_2^2 + \hat{\gamma}_1^2\,\rho_2^2\rho_3^2 + \hat{\gamma}_2^2\,\rho_3^2\rho_1^2\,, \qquad \hat{\gamma}_i \equiv \eta\nu_i\,. \tag{51}$$

Three coordinates ρ_i satisfying the constraint $\sum_{i=1}^{3}\rho_i^2 = 1$ are parametrized by two angle variables α and θ through the relation,

$$\rho_1 \equiv \sin\alpha\cos\theta\,, \quad \rho_2 \equiv \sin\alpha\sin\theta\,, \quad \rho_3 \equiv \cos\alpha\,. \tag{52}$$

It should be remarked that the resulting background is non-supersymmetric other than for exceptional cases like $\nu_1 = \nu_2 = \nu_3$. But the supercoset construction still works well.

4.3. Schrödinger spacetimes

Let us consider Schrödinger spacetimes by employing the following classical r-matrix,

$$r = \frac{i}{4}\,P_- \wedge (h_1 + h_2 + h_3)\,. \tag{53}$$

Here $P_- \equiv (P_0 - P_3)/\sqrt{2}$ is a light-cone generator in $\mathfrak{su}(2,2)$, and h_1, h_2, h_3 are the Cartan generators in $\mathfrak{su}(4)$. This is also an abelian classical r-matrix and satisfies the CYBE. The bosonic part has already been studied in [18].

After all, the full solution [27] has been reproduced as

$$\begin{aligned} ds^2 &= \frac{-2dx^+dx^- + (dx^1)^2 + (dx^2)^2 + dz^2}{z^2} - \eta^2\frac{(dx^+)^2}{z^4} + ds^2_{\mathrm{S}^5}\,, \\ B_2 &= \frac{\eta}{z^2}\,dx^+ \wedge (d\chi + \omega)\,, \\ F_5 &= 4\,(\omega_{\mathrm{AdS}_5} + \omega_{\mathrm{S}^5})\,, \qquad \Phi = \text{const.}, \end{aligned} \tag{54}$$

and the other fields are zero. Here, the light-cone coordinates are defined as

$$x^\pm \equiv \frac{1}{\sqrt{2}}(x^0 \pm x^3)\,.$$

Note that the R-R sector has not been deformed and the dilaton remains constant, though the expression of the fermionic sector is very complicated in the middle of the computation.

4.4. A non-abelian classical r-matrix

So far, we have considered abelian classical r-matrices, for which it seems likely that the supercoset construction works well. Here we study a non-abelian classical r-matrix.

As for non-abelian classical r-matrices, there is no well-known example of the associated background. A nice candidate for non-abelian classical r-matrices [17] is given by

$$\begin{aligned} r &= \frac{1}{\sqrt{2}}E_{24} \wedge (c_1E_{22} - c_2E_{44}) \qquad \Big[(E_{ij})_{kl} \equiv \delta_{ik}\,\delta_{jl}\Big] \\ &= -\frac{1}{2}P_- \wedge \left[\frac{c_1 + c_2}{2}\,(D - L_{03}) + i\,\frac{c_1 - c_2}{2}\left(L_{12} - \frac{i}{2}\,\mathbf{1}_4\right)\right]\,. \end{aligned} \tag{55}$$

Note here that $\mathbf{1}_4$ is included in the expression and hence the image is extended from $\mathfrak{su}(2,2|4)$ to $\mathfrak{gl}(4|4)$. However, it can be ignored due to the presence of the projection operator in the classical action as pointed out in [19].

IOP Conf. Series: Journal of Physics: Conf. Series **804** (2017) 012026 doi:10.1088/1742-6596/804/1/012026

To ensure that the resulting metric and NS-NS two-form are real, it is necessary to impose the reality condition [17]

$$c_2 = c_1^* . \tag{56}$$

It is now convenient to introduce a_1 and a_2 as follows,

$$a_1 \equiv \frac{c_1 + c_2}{2} = \text{Re}(c_1)\,, \quad a_2 \equiv i\frac{c_1 - c_2}{2} = -\text{Im}(c_1)\,. \tag{57}$$

Note here that the classical r-matrix (55) is non-abelian in general. The case that c_1 is pure imaginary (i.e., $a_1 = 0$) is exceptional and it becomes abelian.

The bosonic part has already been studied well [14,17,19], and the R-R sector and the dilaton can be obtained by performing supercoset construction,

$$\begin{aligned}
ds^2 &= \frac{-2dx^+dx^- + d\rho^2 + \rho^2 d\phi^2 + dz^2}{z^2} - \eta^2\left[(a_1^2 + a_2^2)\frac{\rho^2}{z^6} + \frac{a_1^2}{z^4}\right](dx^+)^2 + ds^2_{\text{S}^5}\,,\\
B_2 &= \eta\left[\frac{a_1x^1 + a_2x^2}{z^4}dx^+\wedge dx^1 + \frac{a_1x^2 - a_2x^1}{z^4}dx^+\wedge dx^2 + a_1\frac{1}{z^3}dx^+\wedge dz\right],\\
F_3 &= 4\eta\left[\frac{a_2x^1 - a_1x^2}{z^5}dx^+\wedge dx^1\wedge dz + \frac{a_1x^1 + a_2x^2}{z^5}dx^+\wedge dx^2\wedge dz + \frac{a_1}{z^4}dx^+\wedge dx^1\wedge dx^2\right],\\
F_5 &= 4\left(\omega_{\text{AdS}_5} + \omega_{\text{S}^5}\right)\,, \qquad \Phi = \text{const.}\,,
\end{aligned} \tag{58}$$

and the other components are zero. Here the light-cone coordinates are defined in the same way as the previous example. Notice that the background (58) does not satisfy the equation of motion of B_2 because the Bianchi identity for F_3 is broken. Thus the classical r-matrix (55) does not lead to a solution of the usual type IIB supergravity.

It is worth noting that the pathology vanishes when $a_1 = 0$. This is an exceptional case in which the classical r-matrix becomes abelian and the background (58) is reduced to the Hubeny-Rangamani-Ross solution [34]. This correspondence was originally argued in [17] and elaborated in [19].

5. Conclusion and recent progress

In this article, we have discussed the supercoset construction in Yang-Baxter deformed $\text{AdS}_5\times\text{S}^5$ superstring theories based on the homogeneous CYBE. For abelian classical r-matrices, perfect agreements have been shown for well-known examples including gravity duals of non-commutative gauge theories, γ-deformations of S^5 and Schrödinger spacetimes. For non-abelian classical r-matrices, we have concentrated on a certain example. The resulting background does not satisfy the equation of motion of the NS-NS two-form. Thus, at this stage, it seems that there would be no problem for abelian classical r-matrices, while there are some potential problems in the non-abelian cases.

Recently, some essential progress has been made. Firstly, it is found that the deformed background (58) is a solution of the "generalized" supergravity [30,31] like as the η-deformation [12]. In general, the GS action on a type II supergravity background has the kappa-symmetry [35]. The converse, however, is not true. The presence of the kappa-symmetry ensures only that the background field is a solution of the "generalized" supergravity and, in general, it is not a usual solution [31]. Because the Yang-Baxter deformed action preserves the kappa-symmetry, the Yang-Baxter deformed backgrounds are solutions of the "generalized" supergravity [36].

In our result, the examples of abelian r-matrices lead supergravity solutions. It is known that these backgrounds can also be derived from TsT-transformations. In general, it have been shown that the Yang-Baxter deformation with an abelian r-matrix corresponds to a certain TsT-transformation [28].

Recently, Borsato and Wulff have discovered the condition for r-matrices to lead solutions of the usual supergravity and it is called the unimodularity condition [36]. Furthermore, they have proven that in the case of the S^5 deformations only abelian classical r-matrices lead the usual supergravity solutions. They have also conjectured that in the case of the deformations of the AdS_5 part, the deforamations are equivalent to sequences of TsT-transformations. Recently, Hoare and Tseytlin have conjectured the equivalence between the homogeneous Yang-Baxter deformations and non-abelian T-duals [29]. They have pointed out that a TsT-transformation can be interpreted as a special case of non-abelian duality and proven it for some examples.

By these works, a relation between the Yang-Baxter deformations with the homogeneous CYBE and (non-abelian) T-dual transformations has been revealed. It is interesting to search such a relation in the case of deformations based on the mCYBE. In [30], it is pointed out that some "generalized" supergravity backgrounds form the the YB deformation can be returned the original undeformed backgrounds by a formal T-duality. It seems also interesting to consider a string interpretation of the "generalized" supergravity background.

Acknowledgments
We are grateful to Andrzej Borowiec, Jerzy Lukierski, Takuya Matsumoto and Jun-ichi Sakamoto for the previous works. We are also very grateful Kentaroh Yoshida for the collaboration of the original work and for helpful comments on the draft. The work of H.K. is supported by the Japan Society for the Promotion of Science (JSPS).

References

[1] H. Kyono and K. Yoshida, "Supercoset construction of Yang-Baxter deformed $AdS_5 \times S^5$ backgrounds," 10.1093/ptep/ptw111 [arXiv:1605.02519 [hep-th]].
[2] C. Klimcik, "Yang-Baxter sigma models and dS/AdS T duality," JHEP **0212** (2002) 051 [hep-th/0210095]; "On integrability of the Yang-Baxter sigma-model," J. Math. Phys. **50** (2009) 043508 [arXiv:0802.3518 [hep-th]]; "Integrability of the bi-Yang-Baxter sigma model," Lett. Math. Phys. **104** (2014) 1095 [arXiv:1402.2105 [math-ph]].
[3] F. Delduc, M. Magro and B. Vicedo, "On classical q-deformations of integrable sigma-models," JHEP **1311** (2013) 192 [arXiv:1308.3581 [hep-th]]; "An integrable deformation of the $AdS_5 \times S^5$ superstring action," Phys. Rev. Lett. **112** (2014) 051601 [arXiv:1309.5850 [hep-th]].
[4] I. Kawaguchi, T. Matsumoto and K. Yoshida, "Jordanian deformations of the AdS_5xS^5 superstring," JHEP **1404**, 153 (2014) [arXiv:1401.4855 [hep-th]].
[5] T. Matsumoto and K. Yoshida, "Yang-Baxter sigma models based on the CYBE," Nucl. Phys. B **893** (2015) 287 [arXiv:1501.03665 [hep-th]].
[6] R. R. Metsaev and A. A. Tseytlin, "Type IIB superstring action in $AdS_5 \times S^5$ background," Nucl. Phys. B **533** (1998) 109 [hep-th/9805028].
[7] I. Bena, J. Polchinski and R. Roiban, "Hidden symmetries of the $AdS_5 \times S^5$ superstring," Phys. Rev. D **69** (2004) 046002 [hep-th/0305116].
[8] G. Arutyunov and S. Frolov, "Foundations of the $AdS_5 \times S^5$ Superstring. Part I," J. Phys. A **42** (2009) 254003 [arXiv:0901.4937 [hep-th]].
[9] V. G. Drinfel'd, "Hopf algebras and the quantum Yang-Baxter equation," Sov. Math. Dokl. **32** (1985) 254;"Quantum groups," J. Sov. Math. **41** (1988) 898 [Zap. Nauchn. Semin. **155**, 18 (1986)]. M. Jimbo, "A q difference analog of $U(g)$ and the Yang-Baxter equation," Lett. Math. Phys. **10** (1985) 63.
[10] G. Arutyunov, R. Borsato and S. Frolov, "S-matrix for strings on η-deformed $AdS_5 \times S^5$ s," JHEP **1404** (2014) 002 [arXiv:1312.3542 [hep-th]].
[11] G. Arutyunov, R. Borsato and S. Frolov, "Puzzles of eta-deformed $AdS_5 \times S^5$," arXiv:1507.04239 [hep-th].
[12] G. Arutyunov, S. Frolov, B. Hoare, R. Roiban and A. A. Tseytlin, "Scale invariance of the eta-deformed $AdS_5 \times S^5$ superstring, T-duality and modified type II equations," arXiv:1511.05795 [hep-th].

[13] J. M. Maldacena, "The large N limit of superconformal field theories and supergravity," Int. J. Theor. Phys. **38** (1999) 1113 [Adv. Theor. Math. Phys. **2** (1998) 231] [hep-th/9711200].
[14] I. Kawaguchi, T. Matsumoto and K. Yoshida, "A Jordanian deformation of AdS space in type IIB supergravity," JHEP **1406**, 146 (2014) [arXiv:1402.6147 [hep-th]].
[15] T. Matsumoto and K. Yoshida, "Lunin-Maldacena backgrounds from the classical Yang-Baxter equation - towards the gravity/CYBE correspondence," JHEP **1406** (2014) 135 [arXiv:1404.1838 [hep-th]].
[16] T. Matsumoto and K. Yoshida, "Integrability of classical strings dual for noncommutative gauge theories," JHEP **1406** (2014) 163 [arXiv:1404.3657 [hep-th]].
[17] T. Matsumoto and K. Yoshida, "Yang-Baxter deformations and string dualities," JHEP **1503** (2015) 137 [arXiv:1412.3658 [hep-th]].
[18] T. Matsumoto and K. Yoshida, "Schrödinger geometries arising from Yang-Baxter deformations," JHEP **1504** (2015) 180 [arXiv:1502.00740 [hep-th]].
[19] S. J. van Tongeren, "On classical Yang-Baxter based deformations of the $AdS_5 \times S^5$ superstring," JHEP **1506** (2015) 048 [arXiv:1504.05516 [hep-th]].
[20] S. J. van Tongeren, "Yang-Baxter deformations, AdS/CFT, and twist-noncommutative gauge theory," Nucl. Phys. B **904** (2016) 148 [arXiv:1506.01023 [hep-th]].
[21] T. Matsumoto and K. Yoshida, "Integrable deformations of the $AdS_5 \times S^5$ superstring and the classical Yang-Baxter equation -*Towards the gravity/CYBE correspondence*-," J. Phys. Conf. Ser. **563** (2014) 1, 012020 [arXiv:1410.0575 [hep-th]]; "Towards the gravity/CYBE correspondence – current status –," J. Phys. Conf. Ser. **670** (2016) no.1, 012033.
[22] T. Kameyama, H. Kyono, J. Sakamoto and K. Yoshida, "Lax pairs on Yang-Baxter deformed backgrounds," JHEP **1511** (2015) 043 [arXiv:1509.00173 [hep-th]].
[23] A. Hashimoto and N. Itzhaki, "Noncommutative Yang-Mills and the AdS / CFT correspondence," Phys. Lett. B **465** (1999) 142 [hep-th/9907166].
[24] J. M. Maldacena and J. G. Russo, "Large N limit of noncommutative gauge theories," JHEP **9909** (1999) 025 [hep-th/9908134].
[25] O. Lunin and J. M. Maldacena, "Deforming field theories with $U(1) \times U(1)$ global symmetry and their gravity duals," JHEP **0505** (2005) 033 [hep-th/0502086].
[26] S. Frolov, "Lax pair for strings in Lunin-Maldacena background," JHEP **0505** (2005) 069 [hep-th/0503201].
[27] C. P. Herzog, M. Rangamani and S. F. Ross, "Heating up Galilean holography," JHEP **0811** (2008) 080 [arXiv:0807.1099 [hep-th]]; J. Maldacena, D. Martelli and Y. Tachikawa, "Comments on string theory backgrounds with non-relativistic conformal symmetry," JHEP **0810** (2008) 072 [arXiv:0807.1100 [hep-th]]; A. Adams, K. Balasubramanian and J. McGreevy, "Hot Spacetimes for Cold Atoms," JHEP **0811** (2008) 059 [arXiv:0807.1111 [hep-th]].
[28] D. Osten and S. J. van Tongeren, "Abelian Yang-Baxter Deformations and TsT transformations," arXiv:1608.08504 [hep-th].
[29] B. Hoare and A. A. Tseytlin, "Homogeneous Yang-Baxter deformations as non-abelian duals of the AdS_5 sigma-model," arXiv:1609.02550 [hep-th].
[30] D. Orlando, S. Reffert, J. i. Sakamoto and K. Yoshida, "Generalized type IIB supergravity equations and non-Abelian classical r-matrices," arXiv:1607.00795 [hep-th].
[31] L. Wulff and A. A. Tseytlin, "Kappa-symmetry of superstring sigma model and generalized 10d supergravity equations," JHEP **1606**, 174 (2016) [arXiv:1605.04884 [hep-th]].
[32] M. Cvetic, H. Lu, C. N. Pope and K. S. Stelle, "T duality in the Green-Schwarz formalism, and the massless / massive IIA duality map," Nucl. Phys. B **573**, 149 (2000) [hep-th/9907202].
[33] T. J. Hollowood, J. L. Miramontes and D. M. Schmidtt, "An Integrable Deformation of the $AdS_5 \times S^5$ Superstring," J. Phys. A **47**, no. 49, 495402 (2014) [arXiv:1409.1538 [hep-th]].
[34] V. E. Hubeny, M. Rangamani and S. F. Ross, "Causal structures and holography," JHEP **0507** (2005) 037 [hep-th/0504034].
[35] M. T. Grisaru, P. S. Howe, L. Mezincescu, B. Nilsson and P. K. Townsend, "$\mathcal{N} = 2$ Super-strings in a Supergravity Background," Phys. Lett. B 162 (1985) 116.
[36] R. Borsato and L. Wulff, "Target space supergeometry of η and λ-deformed strings," arXiv:1608.03570 [hep-th].

IOP Conf. Series: Journal of Physics: Conf. Series **804** (2017) 012027 doi:10.1088/1742-6596/804/1/012027

$SL(2, C)$ Instanton Sheaves

Sheng-Hong Lai

Department of Electrophysics, National Chiao-Tung University, Hsinchu, Taiwan, R.O.C.

E-mail: xgcj.ep02g@nctu.edu.tw

Abstract. We constructed $SL(2, C)$ Yang-Mills instanton solutions which satisfy the complex ADHM equations and the monad construction by using the biquaternion method. And we also found that the $SL(2, C)$ instanton solutions can be used to explicitly construct instanton sheaves on CP^3 which similar to the case holomorphic vector bundles on CP^3 of $SU(2)$ ADHM construction. And the existence of these instanton sheaves is related to singularities on S^4 which do not exist in $SU(2)$ instanton case.

1. Introduction

In 1970s, searching classical exact solutions of Yang-Mills (SDYM) equation was an important issue. The instanton solutions are classical solutions of Euclidean $SU(2)$ (anti)self-dual Yang-Mills (SDYM). Yang-Mills instantons are not only important in quantum field theory for physicists but also influential on algebraic geometry for mathematicians. In nonperturbative quantum field theory, quantum instanton tunnelling has resolved the QCD $U(1)_A$ problem [1] and created the strong CP problem with associated QCD θ-vacua [2] structure. Mathematically, instanton can be used as a tool in algebraic geometry classify four-manifolds [3] is the key idea.

The first instanton solution [4] BPST 1-instanton was found in 1975. The $5k$ moduli parameters CFTW k-instanton solutions [5] were constructed later. And then JNR extended the number of moduli parameters of the k-instanton solutions from $5k$ to $5k + 4$ (5,13 for $k = 1$,2) [6] based on the $4D$ conformal symmetry group. Finally, mathematicians ADHM used method in algebraic geometry and worked out the complete $8k - 3$ moduli parameters instanton solutions for each k-th homotopy in 1978[7]. ADHM used the monad construction combining with the Penrose-Ward transform to construct the most general instanton solutions. One to one correspondence between anti-self-dual $SU(2)$-connections on S^4 and global holomorphic vector bundles of rank two on CP^3 is the key idea in ADHM construction. The explicit complete $SU(2)$ instanton solutions for $k \leq 3$ had been worked out in [8].

ADHM construction has been generalized to the cases of many other compact Lie groups[8, 9] SDYM theories. Recently, the present authors wrote a paper[10] and generalized the quaternion calculation in $SU(2)$ ADHM construction to the biquaternion calculation with biconjugation operation, and built a class of non-compact $SL(2, C)$ Yang-Mills instanton solutions with $16k-6$ parameters for each k-th homotopy class. The number of moduli parameters $16k-6$ is consistent with the conjecture made by Frenkel and Jardim in [11] and proved recently in this paper[12] from the mathematical sight. From these new $SL(2, C)$ instanton solutions, we can easily get $SL(2, C)$ (M, N) instanton solutions constructed in 1984 [13].

The discovery of instanton sheaf structure on the projective space in $SL(2, C)$ case is different from the holomorphic vector bundles on CP^3 in $SU(2)$ case. Comparing with the well know regular $SU(2)$ ADHM instanton solutions without any singularities on S^4 spacetime,there are singularities on S^4 for $SL(2, C)$ instanton solutions which can not be gauged away. Recalling that there is a fibration from CP^3 to S^4 with fibers being CP^1. A bundle E on CP^3 can descend down to a bundle over S^4 if and only if no fiber of the twistor fibration is a jumping line for E. This is precisely the case for the $SU(2)$ ADHM construction.

In $SL(2, C)$ instanton solutions, some twistor lines called jumping lines. After Penrose-Ward transform, we can expect $SL(2, C)$ instanton sheaf structure on CP^3. We will show the two examples of instanton sheaf in this paper. We discover some points on CP^3 for the 2-instanton case, the vector bundle description of $SL(2, C)$ 2-instanton on CP^3 breaks down, and we have to use the sheaves description or "sheaf instanton" on CP^3[11]

In section II and III of this paper, we will review the biquaternion ADHM construction of $SL(2, C)$ Yang-Mills instantons with $16k - 6$ moduli parameters [10]. In section IV, we will show how to express the biquaternion instanton solution constructed in section III to complex ADHM data $(B_{lm}, I_{m,} J_m)$ with $l, m = 1, 2$. which are solutions of the complex version of the ADHM equations [14],and we will also show the equivalence between biquaternion ADHM construction and complex ADHM equations. In section V, we will identify the relationship between α and β matrices in the monad construction and Kerβ/Imα the holomorphic vector bundles on CP^3.Then, we will extend the vector bundles case to the sheaves case which are the break down of vector bundle at some points of CP^3 and give two examples of sheaves.

2. Biquaternions

2.1. Quaternion

Now, we are going to review the construction of $SL(2, C)$ YM instantons [13, 10]. We will use the convention $\mu = 0, 1, 2, 3$ and $\epsilon_{0123} = 1$ for $4D$ Euclidean space. In contrast to the quaternion in the $Sp(1)$ $(= SU(2))$ ADHM construction, the authors of [10] used *biquaternion* to construct $SL(2, C)$ Yang-Mills instantons. A quaternion x can be written as

$$x = x_\mu e_\mu, \;\; x_\mu \in R, \;\; e_0 = 1, e_1 = i, e_2 = j, e_3 = k \tag{1}$$

where e_1, e_2 and e_3 anticommute and obey

$$e_i \cdot e_j = -e_j \cdot e_i = \epsilon_{ijk} e_k; \;\; i, j, k = 1, 2, 3, \tag{2}$$

$$e_1^2 = -1, e_2^2 = -1, e_3^2 = -1. \tag{3}$$

The conjugate quarternion is defined to be

$$x^\dagger = x_0 e_0 - x_1 e_1 - x_2 e_2 - x_3 e_3 \tag{4}$$

so that the norm square of a quarternion is

$$|x|^2 = x^\dagger x = x_0^2 + x_1^2 + x_2^2 + x_3^2. \tag{5}$$

Generally the unit quarternions can be expressed as Pauli matrices

$$e_0 \to \begin{pmatrix} 1 & 0 \\ 0 & 1 \end{pmatrix}, e_i \to -i\sigma_i \; ; \; i = 1, 2, 3. \tag{6}$$

2.2. Biquaternions

A biquaternion (or complex-quaternion) z can be written as

$$z = z_\mu e_\mu, \;\; z_\mu \in C, \tag{7}$$

which occasionally can be written as

$$z = x + yi \tag{8}$$

where x and y are quaternions and $i = \sqrt{-1}$, not to be confused with e_1 in Eq.(1). The biconjugation [16] of z is defined to be

$$z^{\circledast} = z_\mu e_\mu^\dagger = z_0 e_0 - z_1 e_1 - z_2 e_2 - z_3 e_3 = x^\dagger + y^\dagger i, \tag{9}$$

which was heavily used in the construction of $SL(2,C)$ instantons [10] in contrast to the complex conjugation

$$z^* = z_\mu^* e_\mu = z_0^* e_0 + z_1^* e_1 + z_2^* e_2 + z_3^* e_3 = x - yi. \tag{10}$$

The norm square of a biquarternion is defined to be

$$|z|_c^2 = z^{\circledast} z = (z_0)^2 + (z_1)^2 + (z_2)^2 + (z_3)^2, \tag{11}$$

which is a *complex* number in general so we use a subscript c in the norm.

3. Biquaternion ADHM

Now, we're going to review the biquaternion construction of $SL(2,C)$ instantons. The first step is to introduce the $(k+1) \times k$ biquarternion matrix $\Delta(x) = a + bx$

$$\Delta(x)_{ab} = a_{ab} + b_{ab}x, \; a_{ab} = a^\mu_{ab} e_\mu, b_{ab} = b^\mu_{ab} e_\mu, x = x^\mu e_\mu \tag{12}$$

where a^μ_{ab} and b^μ_{ab} are complex numbers, and a_{ab} and b_{ab} are biquarternions. x^μ is the position in 4D Euclidean space. The biconjugation of the $\Delta(x)$ matrix is defined to be

$$\Delta(x)^{\circledast}_{ab} = \Delta(x)^\mu_{ba} e_\mu^\dagger = \Delta(x)^0_{ba} e_0 - \Delta(x)^1_{ba} e_1 - \Delta(x)^2_{ba} e_2 - \Delta(x)^3_{ba} e_3. \tag{13}$$

The quadratic condition of $SL(2,C)$ instantons reads

$$\Delta(x)^{\circledast} \Delta(x) = f^{-1} = \text{symmetric, non-singular } k \times k \text{ matrix for } x \notin J, \tag{14}$$

from which we can deduce that $a^{\circledast}a, b^{\circledast}a, a^{\circledast}b$ and $b^{\circledast}b$ are all symmetric matrices. The choice of *biconjugation* operation was crucial for the construction of the $SL(2,C)$ instantons. On the other hand, for $x \in J$, $\det \Delta(x)^{\circledast}\Delta(x) = 0$. The set J is called singular locus or "jumping lines". There are no jumping lines for the case of $SU(2)$ instantons on S^4. In the $Sp(1)$ quaternion case, the symmetric condition on f^{-1} implies f^{-1} is real; while for the $SL(2,C)$ biquaternion case, it implies f^{-1} is *complex* which means $[\Delta(x)^{\circledast}\Delta(x)]^\mu_{ij} = 0$ for $\mu = 1,2,3$.

To construct the self-dual gauge field, we introduce a $(k+1) \times 1$ dimensional biquaternion vector $v(x)$ satisfying the following two conditions

$$v^{\circledast}(x)\Delta(x) = 0, \tag{15a}$$

$$v^{\circledast}(x)v(x) = 1 \tag{15b}$$

where $v(x)$ is fixed up to a $SL(2,C)$ gauge transformation

$$v(x) \longrightarrow v(x)g(x), \quad g(x) \in \; 1 \times 1 \text{ Biquaternion}. \tag{16}$$

Note that in general a $SL(2,C)$ matrix can be written in terms of a 1×1 biquaternion as

$$g = \frac{q_\mu e_\mu}{\sqrt{q^{\circledast} q}} = \frac{q_\mu e_\mu}{|q|_c}. \tag{17}$$

The next step is to define the gauge field

$$G_\mu(x) = v^{\circledast}(x)\partial_\mu v(x), \tag{18}$$

which is a 1×1 biquaternion. The $SL(2,C)$ gauge transformation of the gauge field is

$$\begin{aligned} G_\mu(x) -> G'(x) &= (g^{\circledast}(x)v^{\circledast}(x))\partial_\mu(v(x)g(x)) \\ &= g^{\circledast}(x)G_\mu(x)g(x) + g^{\circledast}(x)\partial_\mu g(x) \end{aligned} \tag{19}$$

where in the calculation Eq.(15b) has been used. Note that, unlike the case for $Sp(1)$, $G_\mu(x)$ needs not to be anti-Hermitian.

One can then define the $SL(2,C)$ field strength

$$F_{\mu\nu} = \partial_\mu G_\nu(x) + G_\mu(x)G_\nu(x) - [\mu \longleftrightarrow \nu], \tag{20}$$

and prove the self-duality of $F_{\mu\nu}$. To count the number of moduli parameters for the $SL(2,C)$ k-instantons, one can use transformations which preserve conditions Eq.(14), Eq.(15a) and Eq.(15b), and the definition of G_μ in Eq.(18) to bring a and b in Eq.(12) into the following simple canonical form

$$b = \begin{bmatrix} 0_{1\times k} \\ I_{k\times k} \end{bmatrix}, a = \begin{bmatrix} \lambda_{1\times k} \\ -y_{k\times k} \end{bmatrix} \tag{21}$$

where λ and y are biquaternion matrices with orders $1 \times k$ and $k \times k$ respectively, and y is symmetric

$$y = y^T. \tag{22}$$

Thus the constraints for the moduli parameters are

$$a^{\circledast}_{ci} a_{cj} = 0, i \neq j, \text{ and } y_{ij} = y_{ji}. \tag{23}$$

The total number of moduli parameters for k-instanton can be calculated through Eq.(23) to be

$$\#\text{ of moduli for } SL(2,C)\ k\text{-instantons} = 16k - 6, \tag{24}$$

which is twice of that of the case of $Sp(1)$. Roughly speaking, there are $8k$ parameters for instanton "biquaternion positions" and $8k$ parameters for instanton "sizes". Finally one has to subtract an overall $SL(2,C)$ gauge group degree of freedom 6.

We provide two explicit examples of $SL(2,C)$ instantons here. These will be used in section V for the discussion of instanton sheaves.

4. Complex ADHM Equations

In this section, we're going to show the equivalence between biquaternion ADHM construction complex ADHM equations.

IOP Conf. Series: Journal of Physics: Conf. Series **804** (2017) 012027 doi:10.1088/1742-6596/804/1/012027

4.1. From Biquaternion to ADHM Data

We begin with the canonical form (21) and using the explicit matrix representation(6) to express the biquaternions as Pauli matrices.

$$a = \begin{bmatrix} \lambda_1 & \lambda_2 & \cdots & \lambda_k \\ y_{11} & y_{12} & \cdots & y_{1k} \\ y_{21} & y_{22} & \cdots & y_{2k} \\ .. & \cdots & \cdots & \cdots \\ y_{k1} & y_{k2} & \cdots & y_{kk} \end{bmatrix} \tag{25}$$

$$a = \begin{bmatrix} \lambda_1^0 - i\lambda_1^3 & -(\lambda_1^2 + i\lambda_1^1) & \lambda_2^0 - i\lambda_2^3 & -(\lambda_2^2 + i\lambda_2^1) & \cdots & \lambda_k^0 - i\lambda_k^3 & -(\lambda_k^2 + i\lambda_k^1) \\ \lambda_1^2 - i\lambda_1^1 & \lambda_1^0 + i\lambda_1^3 & \lambda_2^2 - i\lambda_2^1 & \lambda_2^0 + i\lambda_2^3 & .. & \lambda_k^2 - i\lambda_k^1 & \lambda_k^0 + i\lambda_k^3 \\ y_{11}^0 - iy_{11}^3 & -(y_{11}^2 + iy_{11}^1) & y_{12}^0 - iy_{12}^3 & -(y_{12}^2 + iy_{12}^1) & \cdots & y_{1k}^0 - iy_{1k}^3 & -(y_{1k}^2 + iy_{1k}^1) \\ y_{11}^2 - iy_{11}^1 & y_{11}^0 + iy_{11}^3 & y_{12}^2 - iy_{12}^1 & y_{12}^0 + iy_{12}^3 & \cdots & y_{1k}^2 - iy_{1k}^1 & y_{1k}^0 + iy_{1k}^3 \\ y_{12}^0 - iy_{12}^3 & -(y_{12}^2 + iy_{12}^1) & y_{22}^0 - iy_{22}^3 & -(y_{22}^2 + iy_{22}^1) & \cdots & y_{2k}^0 - iy_{2k}^3 & -(y_{2k}^2 + iy_{2k}^1) \\ y_{12}^2 - iy_{12}^1 & y_{12}^0 + iy_{12}^3 & y_{22}^2 - iy_{22}^1 & y_{22}^0 + iy_{22}^3 & \cdots & y_{2k}^2 - iy_{2k}^1 & y_{2k}^2 - iy_{2k}^1 \\ .. & \cdots & \cdots & \cdots & \cdots & \cdots & \cdots \\ y_{1k}^0 - iy_{1k}^3 & -(y_{1k}^2 + iy_{1k}^1) & y_{2k}^0 - iy_{2k}^3 & -(y_{2k}^2 + iy_{2k}^1) & \cdots & y_{kk}^0 - iy_{2k}^3 & -(y_{kk}^2 + iy_{kk}^1) \\ y_{1k}^2 - iy_{1k}^1 & y_{1k}^0 + iy_{1k}^3 & y_{2k}^2 - iy_{2k}^1 & y_{2k}^0 + iy_{2k}^3 & \cdots & y_{kk}^2 - iy_{kk}^1 & y_{kk}^0 + iy_{kk}^3 \end{bmatrix}. \tag{26}$$

After rearrangement, we can reorganize this ADHM data into the block matrices.

$$a \to \begin{bmatrix} \lambda_1^0 - i\lambda_1^3 & \lambda_2^0 - i\lambda_2^3 & .. & \lambda_k^0 - i\lambda_k^3 & -(\lambda_1^2 + i\lambda_1^1) & -(\lambda_2^2 + i\lambda_2^1) & \cdots & -(\lambda_k^2 + i\lambda_k^1) \\ \lambda_1^2 - i\lambda_1^1 & \lambda_2^2 - i\lambda_2^1 & .. & \lambda_k^2 - i\lambda_k^1 & \lambda_1^0 + i\lambda_1^3 & \lambda_2^0 + i\lambda_2^3 & \cdots & \lambda_k^0 + i\lambda_k^3 \\ y_{11}^0 - iy_{11}^3 & y_{12}^0 - iy_{12}^3 & \cdots & y_{1k}^0 - iy_{1k}^3 & -(y_{11}^2 + iy_{11}^1) & -(y_{12}^2 + iy_{12}^1) & \cdots & -(y_{1k}^2 + iy_{1k}^1) \\ y_{12}^0 - iy_{12}^3 & y_{22}^0 - iy_{22}^3 & \cdots & y_{2k}^0 - iy_{2k}^3 & -(y_{12}^2 + iy_{12}^1) & -(y_{22}^2 + iy_{22}^1) & \cdots & -(y_{2k}^2 + iy_{2k}^1) \\ .. & \cdots & \cdots & \cdots & \cdots & \cdots & \cdots & \cdots \\ y_{1k}^0 - iy_{1k}^3 & y_{2k}^0 - iy_{2k}^3 & \cdots & y_{kk}^0 - iy_{kk}^3 & -(y_{1k}^2 + iy_{1k}^1) & -(y_{2k}^2 + iy_{2k}^1) & \cdots & -(y_{kk}^2 + iy_{kk}^1) \\ y_{11}^2 - iy_{11}^1 & y_{12}^2 - iy_{12}^1 & \cdots & y_{1k}^2 - iy_{1k}^1 & y_{11}^0 + iy_{11}^3 & y_{12}^0 + iy_{12}^3 & \cdots & y_{1k}^0 + iy_{1k}^3 \\ y_{12}^2 - iy_{12}^1 & y_{22}^2 - iy_{22}^1 & \cdots & y_{2k}^2 - iy_{2k}^1 & y_{12}^0 + iy_{12}^3 & y_{22}^0 + iy_{22}^3 & \cdots & y_{2k}^0 + iy_{2k}^3 \\ .. & \cdots & \cdots & \cdots & \cdots & \cdots & \cdots & \cdots \\ y_{1k}^2 - iy_{1k}^1 & y_{2k}^2 - iy_{2k}^1 & \cdots & y_{kk}^2 - iy_{kk}^1 & y_{1k}^0 + iy_{1k}^3 & y_{2k}^0 + iy_{2k}^3 & \cdots & y_{kk}^0 + iy_{kk}^3 \end{bmatrix} \tag{27}$$

$$= \begin{bmatrix} J_1 & J_2 \\ B_{11} & B_{21} \\ B_{12} & B_{22} \end{bmatrix} \tag{28}$$

here we have the *rearrangement rule* for an element z_{ij} in a

$$\begin{aligned} z_{2n-1,2m-1} &\to z_{n,m} \ , \\ z_{2n-1,2m} &\to z_{n,k+m} \ , \\ z_{2n,2m-1} &\to z_{k+n,m} \ , \\ z_{2n,2m} &\to z_{k+n,k+m}. \end{aligned} \tag{29}$$

and we also do the same process on $a^{\circledast}$

$$a^{\circledast} = \begin{bmatrix} \lambda_1^{\circledast} & y_{11}^{\circledast} & y_{12}^{\circledast} & .. & y_{1k}^{\circledast} \\ \lambda_2^{\circledast} & y_{12}^{\circledast} & y_{22}^{\circledast} & .. & y_{2k}^{\circledast} \\ .. & ... & ... & ... & ... \\ \lambda_k^{\circledast} & y_{1k}^{\circledast} & y_{2k}^{\circledast} & .. & y_{kk}^{\circledast} \end{bmatrix} \rightarrow \tag{30}$$

$$\tag{31}$$

$$= \begin{bmatrix} -I_2 & B_{22} & -B_{21} \\ I_1 & -B_{12} & B_{11} \end{bmatrix} \tag{32}$$

4.2. The Complex ADHM Equations

After the rearrangement and identification, we get symmetric condition as

$$a^{\circledast}a \rightarrow \begin{bmatrix} (a^{\circledast}a)_{11}^0 & 0 & ... & 0 & 0 & 0 & 0 & 0 \\ 0 & (a^{\circledast}a)_{22}^0 & ... & 0 & 0 & 0 & 0 & 0 \\ .. & ... & ... & ... & 0 & 0 & 0 & 0 \\ 0 & 0 & ... & (a^{\circledast}a)_{kk}^0 & 0 & 0 & 0 & 0 \\ 0 & 0 & 0 & 0 & (a^{\circledast}a)_{11}^0 & 0 & ... & 0 \\ 0 & 0 & 0 & 0 & 0 & (a^{\circledast}a)_{22}^0 & ... & 0 \\ 0 & 0 & 0 & 0 & ... & ... & ... & ... \\ 0 & 0 & 0 & 0 & 0 & 0 & ... & (a^{\circledast}a)_{kk}^0 \end{bmatrix} \tag{33}$$

$$= \begin{bmatrix} -I_2J_1 + B_{22}B_{11} - B_{21}B_{12} & -I_2J_2 + B_{22}B_{21} - B_{21}B_{22} \\ I_1J_1 + B_{11}B_{12} - B_{12}B_{11} & I_1J_2 + B_{11}B_{22} - B_{12}B_{21} \end{bmatrix}, \tag{34}$$

We can easily derive the complex ADHM equations

$$[B_{11}, B_{12}] + I_1J_1 = 0, \tag{35}$$

$$[B_{21}, B_{22}] + I_2J_2 = 0, \tag{36}$$

$$[B_{11}, B_{22}] + [B_{21}, B_{12}] + I_1J_2 + I_2J_1 = 0. \tag{37}$$

4.3. The Reality Conditions

We can impose the condition

$$\begin{aligned} I_1 &= J^{\dagger}, I_2 = -I, J_1 = I^{\dagger}, J_2 = J, \\ B_{11} &= B_2^{\dagger}, B_{12} = B_1^{\dagger}, B_{21} = -B_1, B_{22} = B_2 \end{aligned}$$

Then we can get the Real ADHM equation which are the constrain equations for $SU(2)$ case.

$$\begin{aligned} [B_1, B_2] + IJ &= 0, \\ \left[B_1, B_1^{\dagger}\right] + \left[B_2, B_2^{\dagger}\right] + II^{\dagger} - J^{\dagger}J &= 0. \end{aligned}$$

5. Monad Construction

In this section, we will talk about α and β matrices in the monad construction.

5.1. From Complex ADHM to Monad Construction

First, we define α and β as functions of homogeneous coordinates x, y, z, w on CP^3

$$\alpha = \begin{bmatrix} zB_{11} + wB_{21} + x \\ zB_{12} + wB_{22} + y \\ zJ_1 + wJ_2 \end{bmatrix}, \tag{38}$$

$$\beta = \begin{bmatrix} -zB_{12} - wB_{22} - y & zB_{11} + wB_{21} + x & zI_1 + wI_2 \end{bmatrix}. \tag{39}$$

Just like the real ADHM equations, the following equation is set up if and only if the ADHM data is satisfied in complex ADHM equations(37).

$$\beta\alpha = 0 \tag{40}$$

In the monad construction, Eq.(40) tells us that Im α is a subspace of Ker β which will make us to consider the quotient space Kerβ/Imα. The quotient space Kerβ/Imα is a function of each point at CP^3 contains two parts of linear transformations. If the transformation α is injective and the transformation β is surjective, then the dimension of Ker β/ Im α will be $k+2-k=2$ on all points of CP^3 ,so we can use holomorphic vector bundles to describe instantons. In $SU(2)$ case, we can always use holomorphic vector bundles to study instantons ,but for $SL(2,C)$ case, the situation is different. β and α may not be surjective and injective at some points of CP^3 for some ADHM data.The dimension of (Ker β/ Im α) may vary from point to point on CP^3, and we have to use sheaf description[11] instead of the holomorphic vector bundles in the non-compact $SL(2,C)$ instanton case.

5.2. Example of Sheaves

In this subsection, we will talk about the examples of $SL(2,C)$ Yang-Mills 2-instanton sheaves. In order to discuss explicitly, we define

$$l = \begin{vmatrix} \lambda_1^0 & \lambda_1^3 \\ \lambda_2^0 & \lambda_2^3 \end{vmatrix} - \begin{vmatrix} \lambda_1^1 & \lambda_1^2 \\ \lambda_2^1 & \lambda_2^2 \end{vmatrix}, \tag{41a}$$

$$n = \begin{vmatrix} \lambda_1^0 & \lambda_1^2 \\ \lambda_2^0 & \lambda_2^2 \end{vmatrix} - \begin{vmatrix} \lambda_1^0 & \lambda_1^1 \\ \lambda_2^0 & \lambda_2^1 \end{vmatrix}, \tag{41b}$$

$$m = \begin{vmatrix} \lambda_1^0 & \lambda_1^1 \\ \lambda_2^0 & \lambda_2^1 \end{vmatrix} - \begin{vmatrix} \lambda_1^2 & \lambda_1^3 \\ \lambda_2^2 & \lambda_2^3 \end{vmatrix}. \tag{41c}$$

And we know y_{12} can be express from other parameters λ_1 , λ_2 , y_{11} , and y_{22}

$$y_{12} = \frac{1}{2}\frac{(y_1 - y_2)}{|y_1 - y_2|_c^2}(\lambda_2^{\circledast}\lambda_1 - \lambda_1^{\circledast}\lambda_2). \tag{42}$$

5.2.1. Example One

For the first sample solution, we take the moduli parameters$\lambda_1^1 = \lambda_1^2 = \lambda_1^3 = \lambda_2^0 = \lambda_2^2 = \lambda_2^3 = 0$, then we get $l=0$, $n=0$ and $m = \lambda_1^0\lambda_2^1$. With these inputs, $w = \frac{1}{im}\left[0 \pm \sqrt{-m^2}\right] = \pm 1$ and the constraints from common eigenvector become

$$\begin{bmatrix} \pm\left(\frac{-i}{2d}\right)m \\ d - \sqrt{d^2 - \frac{m^2}{4d^2}} \end{bmatrix} \sim \begin{bmatrix} \left(\frac{-i}{2d}\right)m \\ \pm d - \sqrt{d^2 - \frac{m^2}{4d^2}} \end{bmatrix} \sim \begin{bmatrix} \mp i\lambda_2^1 \\ -\lambda_1^0 \end{bmatrix}. \tag{43}$$

If we choose $d^2 - \frac{m^2}{4d^2} = 0$, we have

$$m = 2d^2 = \lambda_1^0\lambda_2^1,\ \ \lambda_1^0 = -\lambda_2^1,\ x = y = 0. \tag{44}$$

IOP Conf. Series: Journal of Physics: Conf. Series **804** (2017) 012027 doi:10.1088/1742-6596/804/1/012027

Let's set $\lambda_1^0 = a, \lambda_2^1 = -a$ where a is a complex number and $a \neq 0$, then the corresponding solutions of moduli parameters are

$$\begin{bmatrix} \lambda_1 & \lambda_2 \\ y_{11} & y_{12} \\ y_{12} & y_{22} \end{bmatrix} \sim \begin{bmatrix} \lambda_1^0 - i\lambda_1^3 & -\left(\lambda_1^2 + i\lambda_1^1\right) & \lambda_2^0 - i\lambda_2^3 & -\left(\lambda_2^2 + i\lambda_2^1\right) \\ \lambda_1^2 - i\lambda_1^1 & \lambda_1^0 + i\lambda_1^3 & \lambda_2^2 - i\lambda_2^1 & \lambda_2^0 + i\lambda_2^3 \\ -d & 0 & \left(\frac{-i}{2d}\right) l & \left(\frac{-i}{2d}\right)(m - in) \\ 0 & -d & \left(\frac{-i}{2d}\right)(m + in) & -\left(\frac{-i}{2d}\right) l \\ \left(\frac{-i}{2d}\right) l & \left(\frac{-i}{2d}\right)(m - in) & d & 0 \\ \left(\frac{-i}{2d}\right)(m + in) & -\left(\frac{-i}{2d}\right) l & 0 & d \end{bmatrix}$$

$$= \begin{bmatrix} a & 0 & 0 & ia \\ 0 & a & ia & 0 \\ \frac{-i}{\sqrt{2}}a & 0 & 0 & \frac{a}{\sqrt{2}} \\ 0 & \frac{-i}{\sqrt{2}}a & \frac{a}{\sqrt{2}} & 0 \\ 0 & \frac{a}{\sqrt{2}} & \frac{i}{\sqrt{2}}a & 0 \\ \frac{a}{\sqrt{2}} & 0 & 0 & \frac{i}{\sqrt{2}}a \end{bmatrix}, a \neq 0. \tag{45}$$

Note that since $\lambda_2^1 \neq 0$, this set of ADHM data is outside of CFTW case. Thus, we have discovered that, for points $[x : y : z : w] = [0 : 0 : 1 : \pm 1]$ on CP^3 and the ADHM data given in Eq.(45), the vector bundle description of $SL(2, C)$ 2-instanton on CP^3 breaks down, and which is led to use sheaf description for these non-compact Yang-Mills instantons or "sheaf instantons" on CP^3[11]. Note in addition that for the case of $SU(2)$ 2-instanton, λ_1^0 , λ_2^1 and d are real numbers inconsistent with Eq.(44). This is consistent with the known vector bundle description of $SU(2)$ 2-instanton on CP^3.

5.2.2. Example Two For the second sample solution, we can take the moduli parameters as $\lambda_1^1 = \lambda_1^2 = \lambda_1^3 = \lambda_2^0 = \lambda_2^1 = \lambda_2^3 = 0$, then we get $l = 0$, $n = \lambda_1^0 \lambda_2^2$ and $m = 0$. With these inputs, $w = \pm i$ and the constraints from common eigenvector become

$$\begin{bmatrix} \pm\left(\frac{-i}{2d}\right) in \\ id - \sqrt{-d^2 + \frac{n^2}{4d^2}} \end{bmatrix} \sim \begin{bmatrix} \left(\frac{-i}{2d}\right) n \\ \pm d - \sqrt{d^2 - \frac{n^2}{4d^2}} \end{bmatrix} \sim \begin{bmatrix} \mp i\lambda_2^2 \\ -\lambda_1^0 \end{bmatrix}.$$

If we choose $d^2 - \frac{n^2}{4d^2} = 0$, we have

$$n = 2d^2 = \lambda_1^0 \lambda_2^2 \,, \ \ \lambda_1^0 = -\lambda_2^2, \ \ x = y = 0. \tag{46}$$

Let's set $\lambda_1^0 = a, \lambda_2^2 = -a$ where a is a complex number and $a \neq 0$, then the corresponding solutions of moduli parameters are

$$\begin{bmatrix} \lambda_1 & \lambda_2 \\ y_{11} & y_{12} \\ y_{12} & y_{22} \end{bmatrix} \sim \begin{bmatrix} \lambda_1^0 - i\lambda_1^3 & -\left(\lambda_1^2 + i\lambda_1^1\right) & \lambda_2^0 - i\lambda_2^3 & -\left(\lambda_2^2 + i\lambda_2^1\right) \\ \lambda_1^2 - i\lambda_1^1 & \lambda_1^0 + i\lambda_1^3 & \lambda_2^2 - i\lambda_2^1 & \lambda_2^0 + i\lambda_2^3 \\ -d & 0 & \left(\frac{-i}{2d}\right) l & \left(\frac{-i}{2d}\right)(m - in) \\ 0 & -d & \left(\frac{-i}{2d}\right)(m + in) & -\left(\frac{-i}{2d}\right) l \\ \left(\frac{-i}{2d}\right) l & \left(\frac{-i}{2d}\right)(m - in) & d & 0 \\ \left(\frac{-i}{2d}\right)(m + in) & -\left(\frac{-i}{2d}\right) l & 0 & d \end{bmatrix}$$

$$= \begin{bmatrix} a & 0 & 0 & a \\ 0 & a & -a & 0 \\ \frac{-i}{\sqrt{2}}a & 0 & 0 & \frac{-ia}{\sqrt{2}} \\ 0 & \frac{-i}{\sqrt{2}}a & \frac{ia}{\sqrt{2}} & 0 \\ 0 & \frac{-ia}{\sqrt{2}} & \frac{i}{\sqrt{2}}a & 0 \\ \frac{ia}{\sqrt{2}} & 0 & 0 & \frac{i}{\sqrt{2}}a \end{bmatrix}, a \neq 0. \tag{47}$$

Notice that $\lambda_2^2 \neq 0$, this ADHM data is again outside of CFTW case. Eq.(47) gives the second example of the $SL(2,C)$ 2-instanton sheaf solution on CP^3. Note again that for the case of $SU(2)$ 2-instanton, λ_1^0 , λ_2^2 and d are real numbers which are inconsistent with Eq.(44). We have discovered that, for points $[x:y:z:w] = [0:0:1:\pm i]$ on CP^3 and the ADHM data given in Eq.(47), the description of vector bundle of $SL(2,C)$ 2-instanton on CP^3 breaks down.

To the further discussion of the geometric picture we have shown above points $[x:y:z:w] = [0:0:1:\pm 1]$ or $[0:0:1:\pm i]$ (or setting $w=1$ instead of $z=1$ with similar formulas) the vector bundle description is valid. In general, any other biquaternion ADHM data a set of points can be found as above. So the vector bundle description is necessarily valid outside these finitely many points. (this does not mean, however, that the vector bundle description has to be broken at these finitely many points)

We can make a conclusion that a certain proper subset in our biquaternion ADHM data (including the examples above), the description of vector bundle will break down only at some finitely many points (which depend on the ADHM data). Nevertheless, for ADHM data out of this proper subset the vector bundle description remains valid on the whole CP^3. In mathematics[11], the breakdown of the vector bundle description is related to the third Chern number c_3 of the obtained sheaf, which could be nonzero in the sheaf case in contrast to the vector bundle case in which c_3 is necessarily zero because the bundle is of rank two (two dimensional).

Secondly, the one to one correspondence between ASD connections on the one side and certain holomorphic objects on the other side-twistor space in ADHM construction will help us to to understand ASD connections by using the knowledge and information on the twistor side which is mainly accomplished by Penrose-Ward transform. While this correspondence works well for the vector bundle case to the $SU(2)$ instantons, $SL(2,C)$ case the holomorphic objects are no longer vector bundles on the twistor space as we have discussed in this paper. For example, a vector bundle on CP^3 can descend down to S^4 if and only if its set of jumping lines does not include any fiber of the fibration map $CP^3 \to S^4$. It is worth working on examining the singularities of the Penrose-Ward transformed object on S^4. When the preceding jumping-line condition is not met, or when the holomorphic object on CP^3 is a sheaf instead of a vector bundle, the singularities may appear. The cases vary a lot and the nature of the problem appears really different from one case to another case. So, studying these singularities on S^4 is an important future work to do.

Finally, it is natural to ask if our biquaternion ADHM solutions give all solutions to complex ADHM equations. In our biquaternion construction, we can easily find the number $16k-6$ of parameters match the number expected by mathematicians[11]. An interested reader will easily find the solution for $k=1$ with $I_1 = [1\ 0]$, $I_2 = [0\ 1]$, $J_1 = J_2 = 0$ and all $B_{lm} = 0$ [11] seem to be out of the biquaternion ADHM data. Now if you take $I_1 = [t\ 0]$, $I_2 = [0\ t]$, $J_1 = -I_1^\dagger$, $J_2 = I_2^\dagger$ and all $B_{lm} = 0$, which is seen to be a biquaternion ADHM solution and is equivalent to $(B_{lm}, gI_m, J_m g^{-1})$ in general for any nonzero complex number g, then you will see, by letting $g = t^{-1}$ and setting $t \to 0$ the above biquaternion solution under equivalence indeed reproduces the above (non-biquaternion) solution.

6. Conclusion

In $SU(2)$ Yang-Mills instanton ADHM construction, an one to one correspondence between anti-self-dual $SU(2)$ connections on S^4 and global holomorphic vector bundles of rank two on CP^3 satisfying certain reality conditions. We are going to extend this correspondence to the case of non-compact $SL(2,C)$ Yang-Mills instanton. First, we use biquaternion ADHM construction to build $SL(2,C)$ Yang-Mills instanton solutions which can be shown the complex ADHM data satisfy the complex version of the ADHM equations and the monad construction.

The next step is to calculate whether the ADHM data makes α injective and β surjective. In the $SL(2,C)$ CFTW k-instanton solutions case with $10k$ moduli parameters, although the jumping lines exist on S^4, the corresponding ADHM data are locally free which means α is injective and β is surjective and the vector bundle description of $SL(2,C)$ CFTW k-instanton on CP^3 remains valid as in the case of $SU(2)$ instantons. We then continue to calculate the second case of complete known $SL(2,C)$ 2-instanton solutions with 26 moduli parameters. We find that, for some subset of the complex ADHM data of $SL(2,C)$ 2-instanton solutions on some points on CP^3, the vector bundle description of $SL(2,C)$ 2-instanton on CP^3 breaks down, and this is led to use sheaf description for these non-compact Yang-Mills instantons or "sheaf instantons" on CP^3.

Although we have found the instanton sheaves by using quaternion methods in the previous discussion, we have not worked out the explicit constructions of instanton sheaves yet. We hope the explicit forms of the $SL(2,C)$ Yang-Mills sheaf instanton solutions constructed in this paper will help us to dig more materials in both physical and mathematical fields.

7. acknowledgments

I collaborate with Jenchi Lee and I.H. Tsai on this work. I would like to thank the financial support of Ministry of Science and Technology in Taiwan.

References

[1] G. 't Hooft, "Computation of the quantum effects due to a four-dimensional pseudoparticle", Phys. Rev. D 14 (1976) 3432. G. 't Hooft, "Symmetry breaking through Bell-Jackiw anomalies", Phys. Rev. Lett. 37 (1976) 8.
[2] C. Callan Jr., R. Dashen, D. Gross, "The structure of the gauge theory vacuum", Phys. Lett. B 63 (1976) 334; "Toward a theory of the strong interactions", Phys. Rev. D 17 (1978) 2717. R. Jackiw, C. Rebbi, "Vacuum periodicity in a Yang-Mills quantum theory", Phys. Rev. Lett. 37 (1976) 172.
[3] S.K. Donaldson and P.B. Kronheimer, "The Geometry of Four Manifolds", Oxford University Press (1990).
[4] A. Belavin, A. Polyakov, A. Schwartz, Y. Tyupkin, "Pseudo-particle solutions of the Yang-Mills equations", Phys. Lett. B 59 (1975) 85.
[5] E.F. Corrigan, D.B. Fairlie, Phys. Lett. 67B (1977)69; G. 'tHooft, Phys. Rev. Lett., 37 (1976) 8; F. Wilczek, in "Quark Confinement and Field Theory", Ed. D.Stump and D. Weingarten, John Wiley and Sons, New York (1977).
[6] R. Jackiw, C. Rebbi, "Conformal properties of a Yang-Mills pseudoparticle", Phys. Rev. D 14 (1976) 517; R. Jackiw, C. Nohl and C. Rebbi, "Conformal properties of pseu-doparticle con gurations", Phys. Rev. D 15 (1977) 1642.
[7] M. Atiyah, V. Drinfeld, N. Hitchin, Yu. Manin, "Construction of instantons", Phys. Lett. A 65 (1978) 185.
[8] N. H. Christ, E. J. Weinberg and N. K. Stanton, "General Self-Dual Yang-Mills Solutions", Phys. Rev. D 18 (1978) 2013. V. Korepin and S. Shatashvili, "Rational parametrization of the three instanton solutions of the Yang-Mills equations", Math. USSR Izversiya 24 (1985) 307.
[9] R. Jackwi and C. Rebbi, Phys. Lett. 67B (1977) 189. C. W. Bernard, , N. H. Christ, A. H. Guth and E. J. Weinberg, Phys. Rev. D16 (1977) 2967.
[10] S. H. Lai, J. C. Lee and I. H. Tsai, "Biquaternions and ADHM Construction of Non-Compact SL(2,C) Yang-Mills Instantons", Annals Phys. 361 (2015) 14.
[11] I. Frenkel and M. Jardim, "Complex ADHM equations and sheaves on P^3", Journal of Algebra 319 (2008) 2913-2937. J. Madore, J.L. Richard and R. Stora, "An Introduction to the Twistor Programme", Phys. Rept. 49, No. 2 (1979) 113-130.
[12] M. Jardim and M. Verbitsky, "Trihyperkahler reduction and instanton bundles on CP^3", Compositio Math. 150 (2014) 1836.
[13] K. L. Chang and J. C. Lee, "On solutions of self-dual SL(2,C) gauge theory", Chinese Journal of Phys. Vol. 44, No.4 (1984) 59. J.C. Lee and K. L. Chang, "SL(2,C) Yang-Mills Instantons", Proc. Natl. Sci. Counc. ROC (A), Vol 9, No 4 (1985) 296.
[14] S. Donaldson, "Instantons and Geometric Invariant Theory", Comm. Math. Phys. 93 (1984) 453–460.
[15] Tai Tsun Wu and Chen Ning Yang, Phys. Rev. D12, 3843 (1975); Phys. Rev.D13, (1976) 3233.
[16] W. R. Hamilton, "Lectures on Quaternions", Macmillan & Co, Cornell University Library (1853).

KAPPA-DEFORMATIONS: HISTORICAL DEVELOPMENTS AND RECENT RESULTS

Jerzy Lukierski[1]

Institute for Theoretical Physics, University of Wrocław (Poland)

E-mail: jerzy.lukierski@ift.uni.wroc.pl

Abstract. I shall recall in historical perspective some results from nineties and show further how κ-deformed symmetries and κ-Minkowski space inspired DSR (Doubly of Deformed Special Relativity) approach proposed after 2000. As very recent development I shall show how to describe quantum-covariant κ-deformed phase spaces by passing from Hopf algebras to Hopf algebroids (arXiv:1507.02612) and I will briefly describe the κ-deformations of $AdS_5 \times S^5$ superstring target spaces (arXiv:1510.030.83).

1. Introduction

Transition from classical to quantum physics leads to the appearance of noncommutative algebraic structures. In standard quantum mechanics (QM) the canonical quantum phase space is described by Heisenberg algebra (HA) (i,j=1,2,3)

$$\text{HA in} \qquad [\hat{x}_i, \hat{x}_j] = [\hat{p}_i, \hat{p}_j] = 0 \tag{1a}$$

$$\text{standard QM} \qquad [\hat{x}_i, \hat{p}_j] = i\hbar\, \delta_{ij} \tag{1b}$$

The noncommutativity (1b) of positions $\hat{x}_i$ and momenta $\hat{p}_i$ can be used to derive the Heisenberg uncertainty relations which introduce the bounds on accuracy of simultaneous measurements of positions and momenta. Besides it follows from (1a) that in standard QM one can localize separately the positions or momenta with arbitrary accuracy, what is reflected in the use in QM of classical geometry, with commutative space and time.

The canonical nonrelativistic HA (1a–1b) is changed however if $(\hat{x}_i, \hat{p}_i)$ describe a dynamical system, e.g. point particles moving in a field-theoretic background. In particular, in the presence of electromagnetic (EM) fields the canonical momenta are becoming noncommutative (NC). Nonrelativistic EM background described by magnetic field $\vec{H} = \text{rot}\vec{A}$ ($H_i = \frac{1}{2}\varepsilon_{ijk}F_{jk}$ where $F_{jk} = \partial_j A_k - \partial_k A_j$), leads to the following modification of algebra (1a–1b) [1, 2]

$$\text{HA in the presence} \qquad [\hat{x}_i, \hat{x}_j] = 0 \qquad [\hat{p}_i, \hat{p}_j] = ie\, F_{ij} \tag{2a}$$

$$\text{of magnetic field H} \qquad [\hat{x}_i, \hat{p}_j] = i\hbar\, \delta_{ij} \tag{2b}$$

[1] Presented at ISQS24 Conference on Integrable Systems and Quantum Symmetries, Prague (Czech Republik), 14.06–18.06.2016, 3-rd POTOR (Polish Society on Relativity) Conference, Cracov (Poland), 25–29.09.2016 and 5-th Conference "New Trends in Field Theories", Varanasi (India), 06.11–10.11.2016.

In relativistic theories one uses the Poincaré algebra as describing the group of motions in Minkowski space-time. The presence of a constant relativistic EM background ($F_{\mu\nu} = \partial_\mu A_\nu - \partial_\nu A_\mu$; $\mu, \nu = 0, 1, 2, 3$) leads to the deformation of Poincaré algebra into Maxwell algebra [3]–[5], with noncommutative fourmomenta generators P_μ

$$\begin{array}{ccc} [P_\mu, P_\nu] = 0 & \xrightarrow{F_{\mu\nu} \neq 0} & [P_\mu, P_\nu] = iZ_{\mu\nu} \\ \text{(Poincaré algebra)} & & \text{(Maxwell algebra)} \end{array} \tag{3}$$

where new six Abelian generators $Z_{\mu\nu} = eF_{\mu\nu}$ describe the tensorial central charges.

Further let us consider quantum-mechanical system in the presence of gravitational field ($g_{\mu\nu}(x)$ or $e^a{}_\mu(x)$, where $g_{\mu\nu} = e_\mu{}^a \eta_{ab} e_\nu{}^b$; $\eta_{ab} = \mathrm{diag}(-1, 1, 1, 1)$). It has been argued [6, 7] that one gets in quantum regime for such a system the deformed canonical phase space with noncommutative positions sector, because the quantum fluctuations of gravitational field does not allow the localization of positions (space-time coordinates) with arbitrary accuracy. Such restriction of measurements has been physically interpreted in the framework of Einstein gravity as caused by the formation at Planckian distances ($\lambda_p \simeq 10^{-33}$cm) of mini-black holes, screening experimental visibility of subPlanckian distances ($\lambda < \lambda_p$)[2].

Summarizing, if we consider quantized dynamical systems interacting with gravity (QG) one should use new type of quantum phase spaces, with noncommuting coordinates, in relativistic quantum theories described by deformed noncommutative Minkowski spaces. We get

$$\begin{array}{ccc} [\hat{x}_\mu, \hat{x}_\nu] = 0 & \xrightarrow[\text{QG effects}]{\text{nonvanishing}} & [\hat{x}_\mu, \hat{x}_\nu] \neq 0 \\ \text{space-time without} & & \text{at Planckian} \\ \text{QG effects} & & \text{distances} \end{array} \tag{4}$$

One can point out that noncommutative (quantum) space-time coordinates represent new geometric paradigma in theoretical physics, providing the description of quantum dynamical systems with inclusion of quantum gravity effects.

Let me report briefly on the appearance of noncommutative geometry, quantum spaces and particularly quantum space-times in mathematical physics. First noncanonical noncommutative structures were introduced around 1980 as so-called quantum algebras describing the algebraic and geometric properties of quantum integrable system [9]–[11]. These algebro-geometric structures were introduced as one-parametric deformations of universal enveloping algebra for some low-dimensional complex Lie algebras, and supplemented with additional Hopf-algebraic structure, i.e. with coalgebras (coproducts), coinverses (antipodes) and counits[3]. The deformed quantum Hopf algebras were proposed as noncommutative and noncocommutative generalizations of the notions of classical Lie groups G and Lie algebras $\hat{g}$. First classification of deformed simple complex Lie algebras $\hat{g}$ (the Hopf deformations of enveloping Lie algebras $U(\hat{g})$) was given by Drinfeld [12] and Jimbo [15], who introduced standard quantum deformations, called as well Drinfeld-Jimbo (DJ) or q-deformation (q - complex deformation parameter). In eighties, in parallel way, there were also introduced the quantum groups as Hopf-algebraic deformations of matrix Lie groups G, described by functions on matrix groups with suitably introduced noncommutative matrix entries [16, 17, 13]. Subsequently, using Hopf-algebraic duality, one gets the description of quantum symmetries by pairs of dual Hopf algebras generalizing dual pairs of classical matrix Lie groups and classical Lie algebras, linked by exponential map (passage from Lie algebras to Lie group) or differentiation of Lie group elements at group unit (passage from Lie groups to Lie algebras).

[2] Such mechanism was firstly qualitatively predicted by Bronstein in 1936 [8]. He predicted that the geometry describing all interactions should incorporate three fundamental constants: $c, \hbar$ and Newton constant G; in place of G there was later introduced the Planck mass $m_{\rm pl} = (\frac{\hbar c}{G})^{1/2} \simeq 10^{-5}g$.

[3] For Hopf-algebraic description of deformed groups and Lie algebras see e.g. [12]–[14].

Quantum symmetries are described in their global (finite) and infinitesimal versions as follows:

$$\begin{array}{c}\text{infinitesimal}\\ \text{quantum symmetries}\\ \text{(quantum Lie}\\ \text{algebras)}\end{array} \quad \xleftrightarrow[\text{duality}]{\text{Hopf-algebraic}} \quad \begin{array}{c}\text{finite}\\ \text{quantum symmetries}\\ \text{(quantum matrix}\\ \text{groups)}\end{array} \tag{5}$$

We add that quantum matrix groups, with noncommutative group elements, were also introduced as the linear transformation groups describing the quantum symmetries of various noncommutative spaces (see e.g. [17]).

The basic problem at the end of eighties was the consistent quantum deformation of relativistic symmetries: $D = 4$ Lorentz and Poincaré algebras. The explicit q-deformation of Lorentz algebra was easily introduced as particular application of general DJ deformation framework [18]–[21]. However, to obtain quantum deformation of non-semisimple Poincaré algebra a new way of deriving Hopf-algebraic quantum deformations was needed[4].

The quantum deformations of important class of nonsemisimple classical Lie algebras, described by semidirect product of Abelian and simple Lie subalgebras, were obtained by the quantum modification of known Wigner-Inonu (WI) contraction procedure [22, 23]. Firstly such quantum WI contraction was applied to the derivation of quantum κ-deformation $U_\kappa(e(2))$ of D=2 Euclidean inhomogeneous algebra $e(2) = O(2) \ltimes T^2$ [24]

$$U_q(su(2)) \qquad \xrightarrow[R\to\infty]{q(R)} \qquad U_\kappa(e(2)) \tag{6}$$

where R denotes the WI contraction parameter, $U_q(su(2))$ describes DJ deformation of enveloping algebra $U(su(2))$ and $\xrightarrow[R\to\infty]{q(R)}$ denotes the quantum WI contraction limit $R \to \infty$ with $q(R) \to 1$, which requires the special choice of R-dependence of $q(R)$, provided by the following asymptotic expansion

$$q(R) = 1 + \frac{1}{\kappa R} + O\left(\frac{1}{(\kappa R)^2}\right) \quad \xrightarrow[R\to\infty]{} \quad 1 \tag{7}$$

The particular R-dependence given by (7) leads to finite but nonclassical (deformed) contraction limit, defining κ-deformed Hopf algebra $U_\kappa(e(2))$. Because for space-time symmetries R has the length dimension $[L]$ and $q(R)$ should be dimensionless, it follows that in (7) the parameter κ has the dimensionality $[L^{-1}]$, what (after putting $c = \hbar = 1$) implies the mass dimensionality of κ. In such a way there were introduced quantum-deformed D=2 inhomogeneous Euclidean symmetries with mass-like deformation parameter κ, which has been further identified with the Planck mass m_{pl}.

Our presentation reviews briefly three periods of research activity.

i) last decade of XXth century, when the description of κ-deformed quantum Hopf-algebraic symmetries was proposed and well established,

ii) first decade of present century, with the development of rather semi-phenomenological approach to the deformation of special relativity, called Doubly Special Relativity or later Deformed Special Relativity (DSR),

iii) recent years, when several new applications of κ-deformations were proposed.

The plan of our presentation is the following:

[4] Few years later it was shown that DJ q-deformation of D=4 Lorentz algebra can be as well extended to q-deformed D=4 Poincaré algebra, but in the framework of nonstandard braided Hopf algebras [21], with coalgebras defined with the use of nonstandard braided tensor product.

In Sect.2 I recall the introduction in 1991–92 of κ-deformed D=4 Poincaré algebra [25]–[27], obtained by quantum Wigner-Inonu contraction procedure (see (6–7)) applied to suitable Lie algebras of rank 2. Further it is presented briefly the problem of choice of algebraic basis for κ-deformed Poincaré-Hopf algebra, and we provide two important explicit choices: Majid-Ruegg bicrossproduct basis [28, 29] and classical Poincaré algebra basis [30]–[32].

In Sect.3 I describe the notion of duality of Hopf algebras and define κ-deformed Poincaré-Hopf group as dually determined by the κ-deformed Poincaré-Hopf algebra (see e.g. [33, 34]). It will be shown how by using κ-deformed Poincaré algebra in a selected basis one can obtain the finite κ-deformed quantum Poincaré group transformations.

In Sect.4 I consider briefly the postulates of DSR theories [35]–[37] and expose their link with earlier descriptions of κ-deformed Poincaré symmetries. One can show that basic two formulae of DSR theories providing the deformation of mass-shell condition and quantum deformation of finite Lorentz transformations were already present in earlier Hopf algebraic description of κ-deformed quantum symmetries (see e.g. [38]).

Further I shall recall subsequent development of DSR theories, with coalgebraic part of κ-deformed formalism represented as curved structures of classical momentum spaces [39]–[42]. I will comment as well on other deformations which introduce mass-like deformation parameter, in particular the oldest one, the Snyder deformation of space-time [43], which introduces mass-like deformation parameter (e.g. Planck mass) without violating the classical Lorentz symmetry.

In Sect.5 I shall describe briefly two recent applications of κ-deformations: the introduction of quantum-covariant κ-deformed phase space as an example of quantum space with Hopf algebroid structure [44, 45] and the insertion of κ-deformation into the Yang-Baxter (YB) sigma model defined on $\frac{SU(2,2;4)}{O(4,1)\times O(5)}$ coset space, which leads to the κ-deformation of D=10 GS superstring target space geometry [46].

Finally, in Sect. 6, an outlook is presented. The κ-deformation is an example of a global, nondynamical deformation of space-time algebra, which only approximates "physical" noncommutative structures in QG. More general deformations, local and of dynamical origin, should relate the parametrization of noncommutativity with dynamical degrees of freedom of quantum gravity + quantum matter systems. New formalism determining such relations will provide in future the final fundamental quantum theory, describing interacting QG with the quantized matter fields.

2. κ-deformed Poincaré-Hopf algebras

Drinfeld-Jimbo (DJ) q-deformation scheme [12] has been proposed only for complex simple Lie algebras. Because space-time groups of motions are real and nonsemisimple, in order to obtain e.g. quantum-deformed D=4 Poincaré algebra $\hat{p}_{3,1}$, which is a nonsemisimple cross product of real Lorentz and translation (Abelian) subalgebras ($\hat{p}_{3,1} = \hat{o}(3,1) \ni T_4$), one should apply the quantum WI contraction procedure. In four dimensions in place of (6) we perform the contraction[5]

$$U_q(\hat{o}(3,2)) \quad \xrightarrow[R\to\infty]{q(R)} \quad U_\kappa(\hat{p}_{3,1}) \tag{8}$$

where $q(R)$ has asymptotic expansion (7).

To perform explicitly the contraction (8) we should extend DJ deformation of Cartan-Chevaley (CC) basis of complexified $\hat{o}(3,2)$ algebra ($Sp(4;c) \simeq o(5;c)$), defined for 6 generators $h_i, e_{\pm i}$, ($i = 1,2$), to Cartan-Weyl (CW) Lie-algebraic basis (10 generators extending $\hat{o}(3,2)$ CC basis by four generators $e_{\pm 3}, e_{\pm 4}$ [47, 48]). We resolve in such a way the q-deformed Serre

[5] The contraction (8) provides undeformed $\hat{o}(3)$ rotation. If we apply quantum WI contraction to $U_q(o(4,1))$ we obtain κ-deformation of D=4 Poincaré algebra with undeformed $\hat{o}(2,1)$ subalgebra (see [47], table 1.)

relations formulated in DJ deformation scheme[6]. Further, because the D=4 physical space-time symmetry algebras ($\hat{p}_{3,1}, \hat{o}(3,2), \hat{o}(4,1), \hat{o}(4,2)$ etc.) are real, with Hermitean or antiHermitean generators, one should study all possible reality conditions, and consider corresponding real $\star$-Hopf algebras. There were used in literature mainly the following two types of real $\star$-Hopf algebras:

i) Standard $\star$-Hopf algebras, with involutive anticonjugation $(x,y)^\star = y^\star x^\star$, which maps tensor products in nonflipped way ($(x \otimes y)^\star = x^\star \otimes y^\star$) [27, 28]. Such real q-deformed $\star$-Hopf algebra $U_q(\hat{o}(3,2))$ with q real was used to obtain by quantum WI contraction the κ-deformed Poincaré algebra as standard real $\star$-Hopf algebra (for notion of standard $\star$-Hopf algebras see e.g. [49], [50]).

ii) nonstandard $\star$-Hopf algebra, with involutive conjugation $(xy)^\star = x^\star \cdot y^\star$ and flipped conjugation of tensor products $(x \otimes y)^\star = y^\star \otimes x^\star$, which is for example used in q-deformed models with q described by roots of unity [51, 52]. Such nonstandard reality conditions was also used in first derivations of κ-deformed Poincaré algebra [25, 26] obtained from q-deformed AdS algebra $\hat{o}(3,2)$ with $|q| = 1$.

The quantum WI contraction should be performed on all Hopf-algebraic operations, i.e.

$$\mathbb{H}_q = (A_q = U_q(\hat{o}(3,2)), m, \Delta_q, \epsilon, S_q) \xrightarrow[R\to\infty]{q(R)} \mathbb{H}_\kappa = (A_\kappa = U_\kappa(\hat{p}_{3,1}), m, \Delta_\kappa, \epsilon, S_\kappa) \tag{9}$$

where $m : A \otimes A \to A$ denotes multiplication, $\Delta : A \to A \otimes A$ the coproduct and $S : A \to A$ the antipode (coinverse).

The κ-deformation of Poincaré algebra obtained by (9) contains undeformed $\hat{o}(3) \supset\!\!\!\!+ \hat{T}_3$ subalgebra describing nonrelativistic D=3 Euclidean group of motions.

Using three-dimensional notation for D=4 Lorentz generators $M_{\mu\nu} = (M_i, N_i)$ (i=1,2,3) and fourmomenta $P_\mu = (P_i, P_0)$, there were obtained in [25]–[27] the following deformations

- modified commutators $[N_i, N_j]$ and $[P_i, N_j]$
- nonprimitive coproducts $\Delta(N_i)$, $\Delta(P_i)$
- modified antipodes (in undeformed case $S(P_\mu) = -P_\mu$, $S(M_{\mu\nu}) = -M_{\mu\nu}$)

Subsequently it was realized that the suitable change of linear Poincaré algebra generators

$$M_{\mu\nu} \to M'_{\mu\nu}(M_{\mu\nu}, P_\mu) \qquad\qquad P_\mu \to P'_\mu = P'_\mu(P_\mu) \tag{10}$$

can shift some deformations from algebra to coalgebra. The following two bases became very useful and subsequently used:

i) Majid-Ruegg basis [28]

In such a basis the κ-Poincaré-Hopf algebra is described as follows:

a) Algebra

- Lorentz algebra remains classical, i.e. described by $o(3,1)$ Lie algebra
- Space-time translations generators P_μ commute (i.e. also remain classical)
- The only κ-deformed Poincaré algebra commutator looks as follows:

[6] We point out that q-deformations of Cartan-Weyl bases for rank two Lie algebras and superalgebras has been firstly presented in explicite form in [48].

IOP Conf. Series: Journal of Physics: Conf. Series **804** (2017) 012028 doi:10.1088/1742-6596/804/1/012028

$$[N_i, P_j] = i\delta_{ij}\left[\frac{\kappa}{2}(1 - e^{-\frac{2P_0}{\kappa}}) + \frac{1}{2\kappa^2}\vec{P}^2\right] + \frac{1}{\kappa}P_iP_j \tag{11}$$

b) Coalgebra

The coproducts $\Delta(P_0)$ and $\Delta(M_i)$ remained primitive (classical), the ones which were deformed consistently with (11) are the following

$$\begin{aligned} \Delta P_i &= P_i \otimes 1 + e^{-\frac{P_0}{\kappa}} \otimes P_i \\ \Delta N_i &= N_i \otimes 1 + e^{-\frac{P_0}{\kappa}} \otimes N_i + \frac{1}{\kappa} \in_{ijk} P_j \otimes M_k \end{aligned} \tag{12}$$

c) Antipodes (coinverses)

$$\begin{aligned} S(M_i) &= -M_i & S(N_i) &= -N_i + \frac{3i}{\kappa}P_i \\ S(P_i) &= -e^{-\frac{P_0}{\kappa}}P_i & S(P_0) &= -P_0 \end{aligned} \tag{13}$$

d) Casimir operators

d1) The κ-deformed mass Casimir, defining κ-deformed mass-shell condition, takes the form

$$C_2 = \vec{P}^2\, e^{\frac{P_0}{\kappa}} - (2\kappa \sin\frac{P_0}{\kappa})^2 \tag{14}$$

d2) The κ-deformed spin-square Casimir looks as follows

$$C_4 = (\cosh\frac{P_0}{\kappa} - \frac{\vec{P}^2}{4\kappa^2})W_0^2 - \vec{W}_\kappa^2 \tag{15}$$

where the component W_0 is undeformed[7]

$$W_0 = \vec{P}\,\vec{M} \tag{16}$$

but three-vector $\vec{W}_\kappa$ is κ-dependent

$$\vec{W}_\kappa = \kappa\vec{M}\,\sinh\frac{P_0}{\kappa} + \vec{P}\times\vec{N} \tag{17}$$

ii) classical Poincaré basis [30]–[32]

In such a basis the whole Poincaré algebra is not deformed, and all deformations are incorporated into quite complicated coproducts (in particular the coproduct for energy generator $\Delta(P_0)$ becomes nonprimitive).

The κ-deformed Poincaré-Hopf algebras in Majid-Ruegg and classical Poincaré bases differ only by nonlinear change of the fourmomenta generators. In both bases the algebraic and coalgebraic sectors are described as κ-deformed bicrossproduct Hopf algebras. The choice of bicrossproduct basis appears to be convenient in the study of duality relations between κ-deformed Poincaré algebra and κ-deformed Poincaré group (see Sect. 3).

[7] The fourvector $W_\mu = (W_0, W_i)$ was introduced in undeformed case by Lubanski [53]. It is called Pauli-Lubanski fourvector.

It is known that leading $\frac{1}{\kappa}$ deformation term in the coproducts for quantum-deformed Lie algebras is provided for large class of bases by means of general formula ($\hat{g}$ describes Lie algebra generators)

$$\Delta(\hat{g}) = \Delta^0(\hat{g}) + \frac{1}{\kappa}\left[\Delta^{(0)}(\hat{g}), \hat{r}\right] + O(\frac{1}{\kappa^2}) \tag{18}$$

where $\hat{r} \in \hat{g} \otimes \hat{g}$ denotes the classical r-matrix which satisfies the classical Yang-Baxter equation (CYBE)

$$[[\hat{r}, \hat{r}]] \equiv [\hat{r}_{12}, \hat{r}_{13}] + [\hat{r}_{12}, \hat{r}_{23}] + [\hat{r}_{13}, \hat{r}_{23}] = \hat{\Omega}_3 \equiv \Omega_3^{ijk} I_i \wedge I_j \wedge I_k \tag{19}$$

In (19) the expression $[[\hat{r}, \hat{r}]]$ describes Schouten bracket (see e.g. [14]), and $\Omega_3 \in \hat{g} \otimes \hat{g} \otimes \hat{g}$ is a $\hat{g}$-invariant 3-form. One gets the following two classes of deformations:

a) satisfying standard CYBE with $\Omega_3 \equiv 0$.

In such a case it can be shown that the deformation (quantization) of Lie algebra $\hat{g}$ is determined by twist function $\hat{F} \in U(\hat{g}) \otimes U(\hat{g})$, determining the deformed coproducts as follows:

$$\Delta(\hat{g}) = \hat{F}(\hat{g}) \circ \Delta^{(0)}(\hat{g}) \circ \hat{F}(\hat{g}) \tag{20}$$

where $D^0(\hat{g}) = \hat{g} \otimes 1 + 1 \otimes \hat{g}$ and $A \circ B = A_{(1)}B_{(1)} \otimes A_{(2)}B_{(2)}$ where $A = A_{(1)} \otimes A_{(2)}$ etc.

b) satisfying modified CYBE with $\Omega_3 \neq 0$.

It appears that for D=4 Poincaré algebra we get unique choice of $\hat{g}$-invariant 3-form, which may occur in CYBE namely

$$\Omega_3(P_\mu, M_{\mu\nu}) = t\, P_\mu \wedge P_\nu \wedge M^{\mu\nu} \qquad t \text{ - complex parameter} \tag{21}$$

The κ-deformed Poincaré algebras obtained in [27, 28] are generated by the following classical r-matrix (see (18))

$$\hat{r} = N_i \wedge P_i \qquad (a \wedge b \equiv a \otimes b - b \otimes a) \tag{22}$$

In a middle of nineties (see e.g. [54]) there was introduced the generalization of κ- deformations depending on constant fourvector a_μ, generated by the following generalization of classical r-matrix (22)

$$\hat{r}(a_\mu) = \frac{1}{\kappa} P_\mu \wedge M^{\mu\nu} a_\nu \tag{23}$$

CYBE for the generalized r-matrix (23) takes the following form

$$[[\hat{r}(a_\mu), \hat{r}(a_\nu)]] = a_\mu a^\mu \cdot \Omega_3(P_\mu, M_{\mu\nu}) \tag{24}$$

For Lorentzian signature ($a_\mu a^\mu = \vec{a}^2 - a_0^2$) one obtains three different types of κ-deformations, which can not be related by any change of Poincaré algebra basis. We get namely that

1) If $a_\mu a^\mu = 1$ (one can choose $a_\mu = (1, 0, 0, 0)$) the generalized classical r-matrix (23) reduces to (22). We obtain in such a case standard or time-like κ-deformations (see e.g. (11)–(17)).

2) If $a_\mu a^\mu = -1$ (one can choose $a_\mu = (0, 1, 0, 0)$) one gets tachyonic κ-deformation [55].

3) If $a_\mu a^\mu = 0$ (one can choose light-like vector $a_\mu = (1, 1, 0, 0)$) one obtains the light-cone κ-deformation [56]. It follows from (24) that for light-cone κ-deformation the classical r-matrix (23) describes the solution of standard CYBE, with $\Omega_3 = 0$. Such quantization can be realized by twist factor, which has been explicitly calculated [57, 58].

IOP Conf. Series: Journal of Physics: Conf. Series **804** (2017) 012028 doi:10.1088/1742-6596/804/1/012028

3. Quantum κ-Poincaré group from Hopf-algebraic duality

If we consider classical symmetries in classical mechanics the finite symmetry transformations are described by elements g of classical matrix groups G ($g \in G$), and infinitesimal ones are calculated by using the matrix realizations of corresponding Lie algebra $\hat{g}$. The theory of quantum symmetries was analogously developed in two-fold way. Firstly, the classical matrix groups G after their consistent q-deformation were used to construct the Hopf algebras $\tilde{\mathbb{H}}$ of functions $F(\hat{G}_q)$, with noncommutative elements of quantum-deformed matrix group $\hat{G}_q$. In second way, Lie algebras $\hat{g}$ and their enveloping algebras $U(\hat{g})$ were deformed into the quantum enveloping algebras $U_q(\hat{g})$. After supplementing consistently the Hopf-algebraic operations (coproducts Δ, antipodes S and counits $\in$) one gets quantum Lie-Hopf algebra $\mathbb{H}$. In previous Section we presented $\mathbb{H}$ as κ-Poincaré-Hopf algebra describing deformed relativistic space-time symmetries; below we shall introduce quantum κ-Poincaré group $\tilde{\mathbb{H}}$ as its dual Hopf-algebra.

One says that the Hopf algebra $\mathbb{H} = (A = U_q(\hat{g}), m, \Delta, S, \in)$ is in duality with Hopf algebra $\tilde{\mathbb{H}} = (A^\star = F(\hat{G}_q), m^\star, \Delta^\star, S^\star, \in^\star)$ if there is a nondegenerate pairing $< \cdot, \cdot >$: $A \otimes A^\star \to \mathbb{C}$ such that ($a, b \in A, c, d \in A^\star$)

$$\langle ab, c\rangle = \langle a \otimes b, \Delta^\star(c)\rangle \equiv \langle a, c_{(1)}\rangle\langle b, c_{(2)}\rangle \tag{25a}$$

$$\langle \Delta(a), c \otimes d\rangle = \langle a, cd\rangle \equiv \langle a_{(1)}, c\rangle\langle a_{(2)}, d\rangle \tag{25b}$$

where $\Delta(a) = a_{(1)} \otimes a_{(2)}$ and $\Delta^\star(c) = c_{(1)} \otimes c_{(2)}$. Further we postulate that

$$\begin{gathered} \langle S(a), c\rangle = \langle a, S^\star(c)\rangle \\ \langle a, 1_{A^\star}\rangle = \in(a) \qquad \langle 1_A, c\rangle = \in^\star(c) \end{gathered} \tag{26}$$

The duality relations (25) link the multiplication (products) in $\mathbb{H}(\tilde{\mathbb{H}})$ and comultiplication (coproducts) in $\tilde{\mathbb{H}}(\mathbb{H})$. One obtains the following diagram

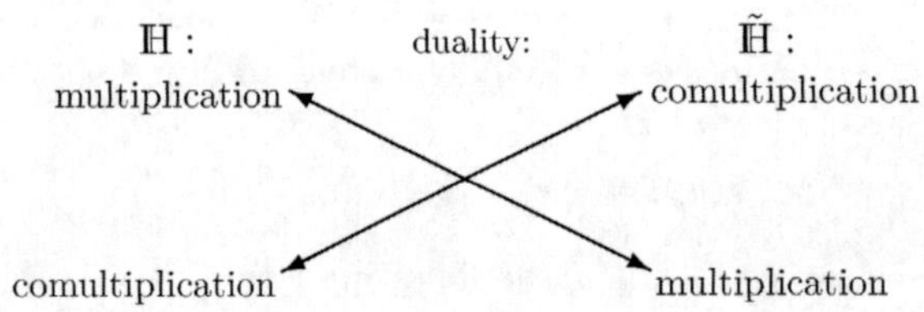

Two Hopf algebras $\mathbb{H}$ and $\tilde{\mathbb{H}}$ act one on another - we gets the formulae for right action $\triangleright$ and left action $\triangleleft$ (see e.g. [14])

$$\begin{gathered} a \triangleright c = c_{(1)}\langle a, c_{(2)}\rangle \\ a \triangleleft c = \langle a_{(1)}, c\rangle a_{(2)} \end{gathered} \tag{27}$$

The action of A on products of elements of $A^\star$ is given by the Hopf-algebraic formula

$$a \triangleright cd = (a_{(1)} \triangleright c)(a_{(2)} \triangleright d) \tag{28}$$

and similarly for the left action on the product in A. We say that $A^\star$ is a right $\mathbb{H}$ module, and A is a left $\tilde{\mathbb{H}}$-module.

Using the relations (25b) one can get from the coproducts of standard κ-deformed Poincaré-Hopf algebra the algebra of elements $\tilde{e}^B \in A^\star$ $(A = 1, \ldots, 10)$ spanning the orthonormal dual linear basis, satisfying the relation

$$\langle e_A, \tilde{e}^B \rangle = \delta_A^B \tag{29}$$

where $e_A = (P_\mu, M_{\mu\nu}) \in A,$, $\tilde{e}^B = (\hat{x}^\mu, \hat{\Lambda}^{\mu\nu}) \in \tilde{A}$. One gets explicitly [33, 28, 29]

$$[\hat{x}^\mu, \hat{x}^\nu] = -\frac{i}{\kappa}(\hat{x}^\mu \delta_0^\mu - \hat{x}^\nu \delta_0^\mu) \tag{30a}$$

$$\left[\hat{x}^\mu, \hat{\Lambda}_\rho^\nu\right] = \frac{i}{\kappa}(\Lambda_0^\nu - \delta_0^\nu)\Lambda^\mu{}_\rho + \eta^{\mu\nu}(\hat{\Lambda}_\rho^0 - \delta^0_\rho) \tag{30b}$$

$$\left[\hat{\Lambda}_\nu^\mu, \hat{\Lambda}_\tau^\rho\right] = 0 \tag{30c}$$

The relation (30a) describes the algebra of κ-deformed D=4 Minkowski space-time corresponding to the choice $a_\mu = (1, 0, 0, 0)$ of the constant fourvector introduced in formula (23). It should be also observed that due to the rhs of relation (30b) the formulae (30a–c) do not describe Lie algebra with linear commutators.

Similarly, using relations (25a) one obtains the following coproduct formulae in $\tilde{\mathbb{H}}$:

$$\begin{aligned} \Delta(\hat{x}_\mu) &= \hat{x}^\mu \otimes 1 + \hat{\Lambda}^\mu{}_\nu \otimes \hat{x}^\nu \\ \Delta(\hat{\Lambda}_\mu{}^\nu) &= \hat{\Lambda}_\mu{}^\rho \otimes \hat{\Lambda}_\rho{}^\nu \end{aligned} \tag{31}$$

The coproducts (31) describe the undeformed Poincaré group transformations. Indeed, if we put $\hat{b}^\mu = \hat{x}_\mu \otimes 1$, $\hat{\alpha}^\mu{}_\nu = \hat{\Lambda}_\mu{}^\nu \otimes 1$ and $\Delta(\hat{x}_\mu) = \hat{x}'_\mu$, $1 \otimes \hat{x}_\mu = \hat{x}_\mu$ one gets the standard transformation laws

$$\begin{aligned} \hat{x}'_\mu &= \hat{b}_\mu + \hat{\alpha}_\mu{}^\nu \hat{x}_\nu \\ \Lambda'_\mu{}^\nu &= \hat{\alpha}_\mu{}^\rho \hat{\Lambda}_\rho{}^\nu \end{aligned} \tag{32}$$

It should be observed that following [13, 16] the classical transformation laws (coproducts) remain valid after the quantum deformation (quantization) $G \to \hat{G}_q$ of any classical matrix group G. Denoting noncommutative matrix entries $\hat{G}^a{}_b \in \hat{G}_q$, we get general coproduct formula for quantum matrix groups

$$\Delta(\hat{G}^a{}_b) = \hat{G}^a{}_c \otimes \hat{G}^c{}_b \tag{33}$$

The pair of dual Hopf algebras can be calculated as well for generalized κ-deformations, described by the classical r-matrix (23). In particular one gets the following κ-deformed Minkowski space algebra depending on constant fourvector a_μ

generalized κ-deformed Minkowski space:

$$[\hat{x}^\mu, \hat{x}^\nu] = \frac{i}{\kappa}(\hat{x}^\mu a^\nu - \hat{x}^\nu a^\mu) \tag{34}$$

It appears that the noncommutativity in quantum space-time (34) is present in only one quantum coordinate $\hat{x}^\mu a_\mu$; if we introduce three linearly independent fourvectors $b_\mu^{(i)}$ $(i = 1, 2, 3)$ which are orthogonal to a^μ (i.e. $a^\mu b_\mu^{(i)} = 0$), the remaining three quantum coordinates $\hat{x}^{(i)} = b_\mu^{(i)} \hat{x}^\mu$ describe three-dimensional commutative manifold, i.e. $[\hat{x}^{(i)}, \hat{x}^{(j)}] = 0$ for any i, j.

The dual pair of Hopf algebras $\mathbb{H}, \tilde{\mathbb{H}}$ is useful in describing the algebraic quantum-deformed phase spaces. For that purpose one introduces cross-products of two dual Hopf algebras (see e.g. [14])

$$\mathcal{H} = \mathbb{H} \ltimes \tilde{\mathbb{H}} \qquad \text{(e.g. } U_\kappa(\hat{g}) \in \mathbb{H},\ F(\hat{G}_\kappa) \in \tilde{\mathbb{H}}) \tag{35}$$

which is called Heisenberg double algebra $\mathcal{H}$. The cross product multiplication rule for $a \in \mathbb{H}$ and $c \in \tilde{\mathbb{H}}$ is given by formula ($\mathcal{H} = \mathcal{A} \otimes \mathcal{A}^\star$)

$$a \cdot c \equiv (a \otimes 1) \cdot (1 \otimes c) = c_{(1)} \langle a_{(1)}, c_{(2)} \rangle a_{(2)} \tag{36}$$

what permits to calculate in $\mathcal{H}$ as well the cross commutators between $\mathbb{H}$ and $\tilde{\mathbb{H}}$. A simple application of Heisenberg double construction is provided by relativistic quantum-mechanical Heisenberg algebra. Such simple Heisenberg double is obtained from two dual Abelian Hopf algebras describing commuting space-time coordinates $\hat{x}_\mu \in \tilde{\mathbb{H}}_x$ and fourmomenta $\hat{p}_\mu \in \mathbb{H}_p$; the Heisenberg double $\mathcal{H}^{(0)}_{4,4} = \mathbb{H}_p \ltimes \tilde{\mathbb{H}}_x$ is characterized by the cross-product relations derived from (36) (we put $\hbar = 1$)

$$[\hat{x}_\mu, \hat{p}^\nu] = i\, \delta_\mu^{\ \nu} \tag{37}$$

i.e. we get relativistic quantum-mechanical phase space. Subsequently, various quantum-deformed phase spaces can be treated as consistent deformations of the classical Heisenberg double $\mathcal{H}^{(0)}_{4,4}$.

The Heisenberg doubles describe quantum spaces, however without Hopf-algebraic structure. In Sect. 5 we shall show that Heisenberg doubles $\mathcal{H}$ define $\mathbb{H}$-covariant quantum-deformed phase spaces, which belong to the algebraic category of quantum spaces with Hopf algebroid structure.

4. From κ-deformed Hopf-algebraic symmetries to DSR approach and curved momentum space

In search for an explanation of possible high energy effects of QG (see e.g. [59]) there was proposed around year 2000 the generalization of Einstein special relativity, with two observer-independent parameters - light velocity c and Planck length $L_p \simeq 10^{-33}$cm - named Doubly Special Relativity (DSR) [60]–[63][8].

In first formulation of DSR framework [60, 63] the basic formulae were coincident with the ones which follow from the formulation of κ-deformed Poincaré algebra in bicrossproduct basis, with supplementary identification of mass-like deformation parameter κ as the Planck mass m[9].

The basic two notions of DSR framework are the following

i) Deformed energy-momentum dispersion relation, which was called in [60] "key characteristic of DSR, both conceptually and phenomenologically". The deformed mass shell condition presented in [60] was earlier described by the formula for κ-deformed mass Casimir in bicrossproduct basis [28, 29]. Further it can be shown (see e.g. [64]) that various κ-deformed mass-shell conditions in DSR framework were directly related with κ-deformed Poincaré-Hopf algebra written in different algebra bases.

ii) Deformation of classical linear Lorentzian boost transformations of fourmomenta into the ones described by nonlinear formulae [65]. We add that DSR framework, by following the description of particle kinematics in scattering theory, is formulated in commutative fourmomentum space, with commuting space-time coordinates introduced by standard Fourier transform of the

[8] Later the notation "DSR" was used as well in more general sense, with the meaning "Deformed Special Relativity".
[9] For $\hbar = c = 1$ one gets $m_p = (L_p)^{-1}$.

momentum-dependent functions (see e.g. [66])[10]. It was shown however [68, 69] that the modified finite Lorentz transformations calculated in [65] can be also obtained by using the Hopf algebra structure of κ-deformed Poincaré algebra in bicrossproduct basis. The general Hopf-algebraic expression for finite κ-deformed boosts in arbitrary basis is given by the following formula (see [70], Sect. 2d)[11].

$$P_\mu(\alpha) = ad_{e^{i\alpha N_3}} P_\mu \equiv \sum_{k=0}^{\infty} \frac{i^k}{k!} \left(ad_{\alpha N_3}(ad_{\alpha N_3} \ldots (ad_{\alpha N_3} P_\mu) \ldots)\right) \tag{38}$$

where quantum adjoint action $ad_Y X$ defined as follows

$$ad_Y X = Y_{(1)} X S(Y_{(2)}) \qquad\qquad (\Delta(Y) = Y_{(1)} \otimes Y_{(2)}) \tag{39}$$

can be expressed in Majid-Ruegg bicrossproduct basis of $\mathcal{U}_\kappa(\hat{p}_{3,1})$ by known "classical" formula

$$P_\mu(\alpha) = e^{i\alpha N_3}\, P_\mu\, e^{-i\alpha N_3} \tag{40}$$

The differential equation following from formula (40) was used in [65] in order to calculate explicitly the nonlinear boost transformations in DSR theory.

Important question addressed as well by DSR approach is κ-deformed addition of fourmomenta describing modified conservation law. In Hopf-algebraic approach such addition is determined by the coproducts and from quantum algebra/quantum group duality follows the appearance of noncommutative quantum space-times. In DSR approach the choice of coproduct was rather ambiguous, because the symmetric and nonsymmetric coproducts, both allowed by deformed mass Casimirs, were used. In the case of symmetric addition law (see e.g. [63], where the postulates of DSR theory are discussed) the deformed relativistic symmetries of fourmomenta are described by the standard special relativity rules, however expressed by nonlinearly transformed classical fourmomenta generators (see [68], [71]–[73]). If we employ however the deformed Poincaré algebra as quantum group[12] the coproducts are necessarily nonsymmetric and dual space-time becomes noncommutative. It appears that for some large class of κ-Poincaré algebra bases the κ-deformed quantum space-time is described by κ-Minkowski space (see also (32a)).

In DSR formalism instead of full Hopf-algebraic description of κ-deformed relativistic symmetries one restricts usually the framework to the so-called DSR algebras [74]–[77] describing κ-deformation of classical semidirect product $R_{3,1} \rtimes \hat{p}_{3;1}$, where $R_{3,1}$ denotes Minkowski space and $\hat{p}_{3;1}$ the classical Poincaré algebra[13]. DSR algebra is spanned by generators $(\hat{X}_\mu, \hat{P}_\mu, \hat{M}_{\mu\nu})$, with $\hat{X}_\mu$ describing κ-deformed Minkowski space and $(\hat{P}_\mu, \hat{M}_{\mu\nu})$ given by quantum-deformed Poincaré algebra which can be endowed with Hopf algebra structure. In such algebraic (not Hopf-algebraic!) approach the deformations of DSR algebra are present in the commutators $[\hat{X}_\mu, \hat{P}_\nu]$ and $[\hat{X}_\mu, \hat{M}_{\rho,\tau}]$ (see (47)).

Because fourmomenta P_μ are Abelian, one can look also for the realizations of DSR algebra in terms of classical Heisenberg algebra generators $\hat{p}_\mu, \hat{x}_\mu$, satisfying the relations (37). If we use

[10] Analogous naive way of introducing space-time coordinates was used in early formalism of κ-deformed relativistic symmetries, in 1991–1993 [26, 27, 29, 67]. Firstly it was clearly stressed in [28] that the modules of κ-deformed Poincaré-Hopf algebra, e.g. describing κ-Minkowski spaces, should be necessarily noncommutative.

[11] We choose boosts along the third space axis.

[12] See Drinfeld' definition of quantum group in [12], described as noncocommutative Hopf algebras with nonsymmetric coproducts $(\Delta(x) = \Delta_{(1)}(x) \otimes \Delta_{(2)}(x) \neq \Delta_{(2)}(x) \otimes \Delta_{(1)}(x))$.

[13] Such algebras were also called Heisenberg-Poincaré algebras [78].

the Schrödinger realization of the algebra (37)

$$\hat{p}_\mu = p_\mu \qquad \hat{x}_\nu = x_\nu \equiv i\frac{\partial}{\partial p^\nu} \tag{41}$$

one can express the κ-Minkowski coordinates $\hat{X}_\mu$ by using of momentum-dependent tetrad $E_\mu^\alpha(p)$

$$\hat{X}_\mu = \hat{x}_\alpha\, E_\mu^\alpha(p) \tag{42}$$

with $E_\mu^\alpha(p)$ determined by the commutator $[\hat{X}_\mu, \hat{P}_\mu]$. Embedding of κ-deformed quantum symmetry algebras with Hopf algebra structure into enveloping classical Heisenberg algebra has been also used, however such procedure is questionable because such a realization does not preserve the Hopf-algebraic structure[14].

The non-Abelian fourmomenta coproducts, if applied to the description of two fourvectors p_μ, and dp_μ indicate the curved structure of fourmomentum space (see e.g. [80, 81]). In DSR approach instead of using coproducts and algebraic methods of Hopf algebra theory, the curved momentum space techniques are used as representing the non-Abelian fourmomenta addition law. It was realized that coproducts of Poincaré-Hopf algebras with associative composition law of fourmomenta describe curved Cartan-Riemann fourmomentum spaces, with nonvanishing torsion and vanishing curvature. The examples of such curved spaces are provided by group manifolds. Indeed, for κ-deformed Poincaré-Hopf algebra it was shown [82]–[84] that corresponding curved fourmomentum space is described by a four-dimensional Lie group $A\cdot N(3)$, entering into the Iwasawa decomposition of $SO(1,4)$ group manifold.

The use of nonlinear momentum space with nontrivial metric which leads to non-Abelian composition law leads to nonstandard description of dynamical systems. Such framework helped to introduce the notion of relative locality [85] providing space-time as derived concept which is described effectively by the interactions of particle probes in momentum space. It can be also added that recently the curved momentum space has been incorporated into dynamical curved phase space framework describing so-called meta-string theory (see e.g. [86, 87]).

5. Recent applications of κ-deformations - two examples

The applications of κ-deformed symmetries to physical models (e.g. deformed QFT) are somewhat limited because for κ-deformed Poincaré-Hopf algebra the universal R-matrix is not known. We shall provide two recent results in the framework of κ-deformations which use the κ-deformed Heisenberg double (see [88]) and classical κ-Poincaré r-matrices (see formulae (22),(23)).

5.1. κ-deformed phase space as Hopf bialgebroid

It has been recently observed [89]–[92], [44, 45] that quantum phase spaces can be supplemented with coalgebra structure in the framework of Hopf algebroids [93, 94]. It is known already a long-time that large class of quantum-deformed phase spaces, with coordinate sector described by Hopf-algebraic quantum group, can be identified as Heisenberg double [88]. Further, exploiting Liu theorem[15] that finite-dimensional Heisenberg double has the structure of a Hopf algebroid [94], there was shown [44, 45] that quantum κ-deformed phase space is the quantum space with Hopf algebroid structure.

[14] The inconsistency follows from the fact that Heisenberg algebra (37) is not endowed with bialgebra structure (see Sect. 5.1). An example of realization of Hopf algebra as embedding into enveloping Heisenberg algebra can be found e.g. in [79].

[15] This theorem is proved rigorously for finite-dimensional Hopf algebras, but formally it can be applied to infinite-dimensional case [45, 92].

While Hopf algebras are quantum analogues of groups, Hopf algebroids are quantum analogues of grupoids [95]. Hopf algebroids are bialgebroids supplemented with the antipode map and additional structures in algebra sector: base subalgebra B of total algebra A and two maps ($h \in A, b \in B$)

i) Source algebra map s: $b \cdot h = s(b)h \quad s(b) \in h$

ii) target antialgebra map t: $h \cdot b = t(b)h \quad t(b) \in h$

which commute, i.e.

$$[s(b), t(b')] = 0 \qquad b, b' \in B \tag{43}$$

The Hopf algebroid $\mathbb{H}$ is specified as follows:

$$\mathbb{H} = (A, m; B, s, t; \Delta, s, \in) \tag{44}$$

where B in the case of Hopf algebra reduces to the unity element $\mathbb{1}$ of algebra A, i.e. disappears.

The role of base algebra plays essential role in defining coalgebraic sector of bialgebroid: one defines coproducts Δ using new tensor product $A \underset{\mathrm{B}}{\otimes} A$ over generally noncommutative base ring B [96, 97]. The coproduct for bialgebroid can be also realized in terms of standard tensor products $A \otimes A$. If we introduce left ideal in algebra A defined in terms of source and target maps (s, t) as follows

$$I_L = s(b) \otimes 1 - 1 \otimes t(b) \tag{45}$$

the tensor product $A \underset{\mathrm{B}}{\otimes} A$ is defined by the following equivalence classes of standard tensor products $A \otimes A$ (see e.g. [97])

$$h \underset{\mathrm{B}}{\otimes} h' \simeq h \otimes h' \ \text{ iff } \ I_L \circ (h \otimes h') = s(b)h \otimes h' - h \otimes t(b)h' = 0 \tag{46}$$

for all $b \in B$. The ideal I_L generates in terms of standard tensor product the nonuniqueness, which is consistent with the homomorphism property of coalgebra reproducing the algebraic structure of A. In [44] such a freedom was called the coproduct gauge.

The simplest example of Hopf bialgebroid is a standard quantum phase space with canonical Heisenberg relations (see (37)), which is given as a Heisenberg double of a pair of Abelian and co-Abelian Hopf algebras $\tilde{\mathbb{H}}_x$ and $\mathbb{H}_p$, describing classical coordinates x_μ and classical momenta p^μ. The cross-product multiplication rule (36) applied to Heisenberg double $\mathbb{H}_p \ltimes \mathbb{H}_x$ leads to the relation (37), i.e. results in the quantization of Poisson structure on classical phase space. If we calculate the Heisenberg double of κ-deformed Poincaré-Hopf algebra $\mathbb{H}_\kappa$ and dual quantum κ-Poincaré group $\tilde{\mathbb{H}}_\kappa$, with generators $(P_\mu, M_{\mu\nu}, \hat{x}_\mu, \hat{\Lambda}_{\mu\nu})$, we obtain the following cross-commutators (see [88])

i) κ-deformed relations containing $\hat{x}_\mu$

$$\begin{aligned}
[P_\mu, \hat{x}_\rho] &= -i\,\eta_{\mu\rho} - \frac{i}{\kappa}(\eta_{\rho 0} P_\mu - \eta_{\mu 0} P_\rho) \\
[M_{\mu\nu}, \hat{x}_\rho] &= i(\eta_{\mu\rho}\hat{x}_\nu - \eta_{\nu\rho}\hat{x}_\mu) + \frac{i}{\kappa}(\eta_{0\mu} M_{\nu\rho} - \eta_{0\nu} M_{\mu\rho})
\end{aligned} \tag{47}$$

ii) nondeformed relations containing

$$\begin{aligned}
[P_\mu, \hat{\Lambda}_{\nu\rho}] &= -i(\eta_{\mu\nu} P_\rho - \eta_{\nu\rho} P_\nu) \\
[M_{\mu\nu}, \hat{\Lambda}_{\rho\tau}] &= i(\eta_{\mu\rho}\hat{\Lambda}_{\nu\tau} - \eta_{\nu\rho}\hat{\Lambda}_{\mu\tau}) - (\rho \leftrightarrow \tau)
\end{aligned} \tag{48}$$

The relations (47,48) supplemented with the algebraic relations of κ-deformed Poincaré algebra $\hat{\mathbb{H}}_\kappa$ (see (11)) and κ-deformed Poincaré group $\mathbb{H}_\kappa$ (see (30)) describe generalized κ-deformed quantum phase space $\mathcal{P}^{10;10}$. It can be shown that $\mathcal{P}^{10;10}$ is a Hopf algebra $\hat{\mathbb{H}}_\kappa$-module, or equivalently $\mathcal{P}^{10;10}$. is κ-covariant.

One can distinguish the following three subalgebras of $\mathcal{P}^{10;10}$.

i) κ-deformed DSR algebra $\mathcal{P}^{10;4}$, with base $(P_\mu, M_{\mu\nu}; \hat{x}_\nu)$, describing noncommutative κ-Minkowski space with acting covariantly the κ-deformed Poincaré-Hopf algebra $\mathbb{H}_\kappa$

ii) κ-deformed DSR group $\mathcal{P}^{4;10}$, with base $(\hat{P}_\mu; \hat{x}^\mu, \hat{\Lambda}^\mu{}_\nu)$, which is the quantum κ-Poincaré group $\mathbb{H}_\kappa$ with $\hat{p}_\mu$ acting in covariant way

iii) κ-deformed standard phase space $\mathcal{P}^{4;4}$, with base $(\hat{x}_\mu, \hat{p}_\nu)$, where $\hat{x}_\mu$ satisfy the algebra (32a) and commutative fourmomenta $\hat{p}_\mu$ have the following cross relations: $(\mu = 0, i; i = 1, 2, 3)$

$$\begin{aligned} [\hat{p}_i, \hat{x}_j] &= -i\,\delta_{ij} \qquad & [\hat{p}_i, \hat{x}_0] &= 0 \\ [\hat{p}_0, \hat{x}_0] &= -i \qquad & [\hat{p}_0, \hat{x}_i] &= \frac{i}{\kappa}\hat{p}_i \end{aligned} \tag{49}$$

We see that only last relation in (49) is κ-deformed, in comparison with standard relativistic Heisenberg algebra (37).

The generalized phase space $\mathcal{P}^{10;10}$ (with Lorentz spin sector $(M_{\mu\nu}, \hat{\Lambda}_\mu{}^\nu)$) as well as $\mathcal{P}^{4;4}$ are endowed with Hopf algebroid structure. The base algebra B is identified with quantum group algebra part ($\hat{x}_\mu$ for $\mathcal{P}^{4;4}$ and $(\hat{x}_\mu, \hat{\Lambda}_\mu{}^\nu)$ for $\mathcal{P}^{10;10}$).

The explicite Hopf algebroid structure for standard κ-deformed quantum phase space $\mathcal{P}^{4;4}$ has been obtained by purely algebraic considerations in [44, 45]. Following Lu theorem (see [94]) we have chosen the primary coproduct on base algebra as $\Delta(\hat{x}_\mu) = 1 \otimes \hat{x}_\mu$, with nonuniqueness described by coproduct gauge, e.g.

$$\Delta(\hat{x}^\mu) \xrightarrow[\text{gauge}]{\text{coproduct}} \Delta(\hat{x}^\mu) + \Lambda^\mu(\hat{x}, \hat{p}) = \hat{x}_\mu \otimes f_\mu{}^\nu(\hat{p}) \tag{50}$$

where $f_\mu{}^\nu(\hat{p})$ can be calculated by using for coproducts the algebraic structure (49). One gets

$$\begin{aligned} \Lambda_i(\hat{x}, \hat{p}) &= \hat{x}_i \otimes e^{\frac{p_0}{\kappa}} - 1 \otimes \hat{x}_i \\ \Lambda_0(\hat{x}, \hat{p}) &= \hat{x}_0 \otimes 1 + \frac{1}{\kappa}\hat{x}_i \otimes e^{\frac{p_0}{\kappa}}\,\hat{x}_i - 1 \otimes \hat{x}_0 \end{aligned} \tag{51}$$

The coproduct gauge (51) can be further generalized, in accordance with the homomorphism of coproducts with the algebraic relations in $\mathcal{P}^{4;4}$ (see [44]).

I would like to add that explicite Hopf algebroid structure of $\mathcal{P}^{10;10}$, with the description of coproduct gauges, is now under considerations.

5.2. κ-deformation of string target (super)spaces

The dynamics of (super) strings can be described by twodimensional σ-models, on suitable (super)group or (super)coset manifolds.

In particular the action describing D=10 Green-Schwarz superstring has been described as D=2 σ-model with D=10 N=2 super-Poincaré target manifold [98]. If one replaces in target superspace the D=10 Poincaré manifold by $AdS_5 \times S_5$, with S_5 describing an internal sector, one should consider D=2 σ-model on supercoset $\mathbb{K}$ [99].

$$AdS_5 \times S_5 \quad \mathrm{D}=10 \quad \text{superstring} \qquad \longleftrightarrow \qquad \begin{array}{c}\mathrm{D}{=}2\ \sigma\text{-model}\\ \text{supercoset } \mathbb{K} = \frac{PSU(2,2;4)}{O(4,1)\times O(5)}\end{array} \tag{52}$$

The σ-model action for strings with target space described by group manifold G or coset $\frac{G}{H}$ can be expressed by bilinars in D=2 Lie algebra - valued currents $\mathbb{J}_\alpha = J^a_\alpha I^a$ $(\alpha = 1, 2)$, where I^a describe the Lie algebra generators

$$S = -\frac{T}{2}\int d^2\xi \ (h^{\alpha\beta} - \in^{\alpha\beta})Tr[\mathbb{J}_\alpha P \mathbb{J}_\beta] \tag{53}$$

The trace is taken over matrix values of the generators I^a and P describes the projection operator $G \to \frac{G}{H}$. In the presence of quantum deformations described by classical r-matrix $\hat{r} = \hat{a}_i \wedge \hat{b}_i \in \hat{g} \otimes \hat{g}$ one can introduce the deformation of string action (53) by introducing Yang-Baxter YB σ-model [100, 101] with particular choice of deforming YB kernel $(1 - \xi R)^{-1}$

$$S_{v_B} = -\frac{T}{2}\int d^2\xi(h^{\alpha\beta} - \in^{\alpha\beta})Tr(\mathbb{J}_\alpha \frac{1}{1-\xi R} P \mathbb{J}_\beta) \tag{54}$$

The operator $\mathbb{R} = R(I_k) \cdot I_k$ is expressed by classical r-matrix as follows

$$R(I_k) = \sum \hat{a}_i \langle b_i, I_k\rangle = Str(\hat{r} \circ (1 \otimes I_k)) \tag{55}$$

and satisfies modified CYBE in Semenov-Tien-Shansky form [102, 103].

$$[R(I_j), R(I_k)] - R[R(I_j), I_k] + [I_j, R(I_k)] = c[I_j, I_k] \tag{56}$$

We add that if we consider superstrings, the generators $\mathbb{J}_\alpha$ describe both sets of bosonic (even) and fermionic (odd) currents, i.e. $\mathbb{J}_\alpha \to (J^B_\alpha, J^F_\alpha)$, and the trace Tr should be replaced in formulae (53–54) by supertrace Str. Further if we consider GS superstring with supercoset (52) as target space, one should introduce four projection operators, corresponding to Z_4-graded decomposition of $PSU(2, 2; 4)$ supergroup generators [104]–[106].

Because the conformal algebra $SU(2, 2)$ contains as subalgebras the commuting fourmomenta as well as D=4 Poincaré algebra, one can embedd D=4 Minkowski space-time into the coset $\mathbb{K}$ (see (52). One can ask further how the target superspace $\mathbb{K}$ will be modified and which NS-NS and R-R fluxes will be obtained if we insert in YB σ-model the classical r-matrices (22) or (23). Using suitable computer program such calculation have been performed ([107]; see also [108]). We calculated YB Minkowski space deformations for standard κ-deformation (see (23), with $a^2 > 0$), for tachyonic deformation ($a^2 < 0$) and for light-cone deformation ($a^2 = 0$). We obtained the following results (modulo Abelian T-duality transformations [109, 110]):

type of κ-def.	space-time geometry	remarks
$a^2 > 0$ (standard)	D=4 de-Sitter	κ^{-1} describes dS radius
$a^2 < 0$ (tachyonic)	D=4 AdS	κ^{-1} describes AdS radius
$a^2 = 0$ (light-cone)	$p - p$ wave background	x_+ - dependent [16] nonvanishing Ricci curvature; $R = 0$

[16] $x_+ = x_0 + x_3$ describes light-cone time.

It can be added that only in light-cone case one gets nonvanishing antisymmetric tensor field B (NS-NS 2-form), however in the form of a total derivative.

An interesting issue is the classification of integrable deformations of $AdS_5 \times S^5$ superstring. In [105] it has been shown that the generalized a_μ-dependent κ-deformations are integrable. For three types of κ-deformations we obtained the explicite formulae for Lax pairs, generating via Lax equations the infinite number of conserved currents. In particular it has been shown that zero curvature conditions for Lax pairs leads to the equations of motion for κ-deformed YB σ-model.

An important application of YB σ-models is the generalization of standard IIB SUGRA theory [111]. It appears that κ-invariance of D=10 IIB strings embrace as well some more extensions of IIB SUGRA solutions, which can be obtained from YB σ-models [112].

6. Final remarks

I would like to stress that there were presented only some aspects of the theoretical developments related with κ-deformations of relativistic space-time symmetries and we regret that the list of references is far from being complete. Let me mention some other important developments:

i) SUSY extension of κ-deformed framework.

It was shown already in 1993 [113, 114] that κ-deformation of N=1 Poincaré superalgebra can be calculated by considering quantum WI contraction of q-deformed $OSp(4|1)$ superalgebra.

ii) κ-deformed QM.

It has been realized that quantum-deformed particle model in NC phase space $(\hat{x}_\mu, \hat{p}_\mu)$ with classical momenta sector $(\hat{p}_\mu = p_\mu)$ is equivalent via noncanonical map $\hat{x}_\mu = \hat{F}_\mu(x,p)$ to the free particle model in standard quantum phase space (x_μ, p_μ) (for κ-deformed model see e.g. [115]). For standard time-like κ-deformations such noncanonical map is well-known from early nineties $(\hat{x}_i = x_i, \hat{x}_0 = x_0 - \frac{1}{\kappa}\vec{p}\vec{x})$.

iii) κ-deformed classical and quantum field theory.

If one introduces $\star$-product describing the algebra of NC functions on κ-Minkowski space in terms of classical fields [116] one can only obtain in such a way the κ-deformed classical theory. In quantum κ-deformed QFT the algebra of field oscillators is however κ-deformed as well [117]–[119] i.e. in construction of κ-deformed QFT one should introduce NCQFT with new notion of $\star$-product which takes into consideration as well the κ-deformation of oscillators algebra (see e.g. [120]).

iv) κ-deformed gravity action.

Seiberg-Witten map [121] applied to κ-deformed Einstein gravity, with local diffeomorphisms as local gauge group, produced higher order curvature corrections [122, 123]. The lowest nonvanishing order in $\frac{1}{\kappa}$ is second, i.e. one gets terms proportional to $\frac{1}{\kappa^2}$. If we put $\kappa = m_p$, these second order corrections are beyond observability limits in present experiments.

At present still the final choice of QG model is not established. Breaking through will arrive if we shall find experimentally some observable QG effects, what however still did not happened. At present we are therefore rather at early stage of the construction procedure of theoretically sound and phenomenologically confirmed quantum gravity model.

Acknowledgments
The author would like to thank prof. Cestimir Burdik for warm hospitality in Prague. The paper is supported by Polish NCN grant 2014/13/B/ST2 and EU Cost Action MP1405 QSPACE.

References

[1] H.R. Grümm, Acta Phys. Austr. **53**, 113(1981).
[2] R. Jackiw, Nucl. Phys. B (Proc. Suppl.) **108**, 30(2002); hep-th/0110057.
[3] H. Bacry, Ph. Combe, J.L. Richard, Nuovo Cim. **A67**, 267(1970).
[4] R. Schrader, Fortschr. Phys. **20**, 701(1972).
[5] J. Beckers, V. Hussin, J. Math. Phys. **24**, 1295(1983).
[6] L.J.Garay, Inst. J. Math. **A10**, 145(1995).
[7] S. Doplicher, K. Fredenhagen, J.E. Roberts, Comm. Math. Phys. **172**, 187(1995); hep-th/0608124.
[8] M.P. Bronstein, Zh.ETF (JETP), V**6** (1936) 195 (in Russian).
[9] L.D. Faddeev, Integrable models in 1+1-dimensional quantum field theory", Les Houches Lectures 1982, (ed. Elsevier Amsterdam, 1984).
[10] P.P. Kulish, E.K. Sklyanin, Lecture Notes in Physics, vol. **151**(1982), pp. 61–119.
[11] E.K. Sklyanin, Funkt. Anal. Prilozh. **16**, 27(1982); **17**, 34(1983).
[12] V.G. Drinfeld, Proc. of the Intern. Congress of Mathematicians, Berkeley, p. 786(1986).
[13] L.D. Faddeev, N. Reshetikhin, L. Takhtajan, Algebraic Analiz **1**, 178(1989) (in Russian); Leningrad Math. J. **1**, 193(1990).
[14] S. Majid, Int. J. Mod. Phys. **A5**, 1(1990); "Foundations of Quantum Group Theory", Cambridge Univ. Press, 1995.
[15] M. Jimbo, Lett. Math. Phys. **10**, 63(1985); **11**, 247(1986).
[16] S.L. Woronowicz, Comm. Math. Phys. **111**, 613(1987).
[17] Y.I. Manin, Quantum Groups and Noncommutative Geometry, Montreal Univ. preprint CRM1561 (1988).
[18] P. Podleś, S.L. Woronowicz, Comm. Math. Phys. **130**, 381(1990).
[19] U. Carow-Watamura, M. Schlieker, M. Scholl, S. Watamura, Int. J. Mod. Phys. **6**, 3081(1991).
[20] O. Ogievetsky, W.B. Schmidke, J. Wess, B. Zumino, Comm. Math. Phys. **150**, 495(1992).
[21] S. Majid, J. Math. Phys. **34**, 2045(1993).
[22] E. Inanu, E.P. Wigner Proc. Nat. Acad. Sci (USA) **36**, 510(1953).
[23] R. Gilmore, "Lie Groups, Lie Algebras and Some of Their Applications", John Willey and Sons (1974).
[24] E. Celeghini, R. Giachetti, E. Sorace, M. Tarlini, J. Math. Phys. **31**, 2548(1990).
[25] J. Lukierski, A. Nowicki, H. Ruegg, V.N. Tolstoy, Phys. Lett. **B264**, 331(1991).
[26] S. Giller, P. Kosiński, M. Majewski, P. Maślanka, J. Kunz, Phys. Lett. **B286**, 52(1992).
[27] J. Lukierski, A. Nowicki, H. Ruegg, Phys. Lett. **B293**, 344(1992).
[28] S. Majid, H. Ruegg, Phys. Lett. **B334**, 348(1994); hep-th/9405107.
[29] J. Lukierski, H. Ruegg, W.J. Zakrzewski, Ann. Phys. **243**, 90(1995); hep-th/9312153.
[30] H. Ruegg, V.N. Tolstoy, Lett. Math. Phys. **32**, 85(1994); hep-th/9406146.
[31] P. Kosiński, J. Lukierski, P. Maślanka, J. Sobczyk, Mod. Phys. Lett. **A10**, 2599(1995).
[32] A. Borowiec, A. Pachoł, J. Phys. **A43**; 045203(2010); arXiv:0903.5251 [hep-th].
[33] S. Zakrzewski, J. Phys. **A27**, 2075(1994).
[34] P. Kosiński, P. Maślanka, "The Duality between kappa-Poincare algebra and kappa-Poincaré group: hep-th/9411033, (unpublished).
[35] G. Amelino-Camelia, D. Benedetti, F. D'Andrea, A. Procaccini, Class. Quant. Grav. **20**, 5353(2003); hep-th/0201245.
[36] J. Kowalski-Glikman, Phys. Lett. **A286**, 391(2001).
[37] J. Magueijo, L. Smolin, Phys. Rev. Lett. **88**: 190403(2002).
[38] J. Lukierski, H. Ruegg, V.N. Tolstoy, "Quantum Kappa-Poincaré 1994", Proc. of XXXth Karpacz Winter School, Poland, Feb. 1994.
[39] S. Majid, Journ. of Algebra, **130**, 17(1990).
[40] G. Amelino-Camelia, S. Majid, Int. J. Mod. Phys. **15**, 4301(2000); hep-th/9907110.
[41] J. Kowalski-Glikman, S. Nowak, Class. Quant. Grav. **20**, 4799 (2003); hep-th/0304101.
[42] L. Freidel, T. Rempel, arXiv:1312.3674.
[43] H.S. Snyder, Phys. Rev. **71**, 38(1947).
[44] J. Lukierski, Z. Škoda, M. Woronowicz, Phys. Lett. **B750**, 401(2015); arXiv:1507.02612[hep-th].
[45] J. Lukierski, Z. Škoda, M. Woronowicz, Phys. Atom. Nucl. (English version of Yadernaya Fizika), vol. **80**, in press; arXiv:1601.01590[hep-th].
[46] A. Borowiec, H. Kyono, J. Lukierski, J. Sakamoto, K. Yoshida, **JHEP 04079** (2016); arXiv:1510.03083; arXiv: 1507.02612[hep-th].
[47] J. Lukierski, A. Nowicki, Phys. Lett. **271B**, 321(1991).
[48] S.M. Khoroshkin, V.N. Tolstoy, Comm. Math. Phys. **141**, 599(1991).
[49] S.L. Woronowicz, S. Zakrzewski, Composition Math. **90**, 211(1994).
[50] E. Twietmeyer, Lett. Math. Phys. **24**, 49(1992).

IOP Conf. Series: Journal of Physics: Conf. Series **804** (2017) 012028 doi:10.1088/1742-6596/804/1/012028

[51] Ph. Roche, D. Arnaudon, Lett. Math. Phys. **17**, 295(1989).
[52] C. De Concini, V.G. Kac, Progr. in Math. **92**, 471(1990).
[53] J.K. Lubański, Physica **9**, 310; **9**, 325(1942).
[54] S. Zakrzewski, Comm. Math. Phys. **178**, 285(1997).
[55] A. Ballesteros, F.J. Herranz, M.A. del'Olmo, M. Santander, J. Mod. Phys. **35**, 4928(1994).
[56] A. Ballesteros, F.J. Herranz, M.A. del'Olmo, M. Santander, Phys. Lett. **B351**, 137(1995).
[57] A. Mudrov, "Universal R-matrix for null-plane quantized Poincaré algebra", q-alg/9711001.
[58] A. Borowiec, J. Lukierski, V.N. Tostoy, Eur. Phys. J. **C44**, 139(2005); hep-th/0412131.
[59] G. Camelino-Camelia in "Towards Quantum Gravity", Proc. of XXXV-th Karpacz Winter School of Theor. Physics, February 1999, ed. Springer (2000), Lecture Notes in Physics No 514, p. 1.
[60] G. Camelino-Camelia, Phys. Lett. **B510**, 255(2001); hep-th/0012238.
[61] J. Kowalski-Glikman, Phys. Lett. **A286**, 391(2001); hep-th/0102098.
[62] J. Magueijo, L. Smolin, Phys. Rev. Lett. **88**: 190403(2002); hep-th/0112090.
[63] G. Camelino-Camelia, D. Benedetti, F.D'Andrea, Class. Quant. Grav. **20**, 5353(2003); hep-th/0201245.
[64] J. Kowalski-Glikman, S. Nowak, Phys. Lett. **B539**, 126(2002); hep-th/0203040.
[65] R. Bruno, G. Camelino-Camelia, J. Kowalski-Glikman, Phys. Lett. **B522**, 133(2001); hep-th/0107039.
[66] D. Kimberly, J. Magueijo, J. Medeiros, Phys. Rev. **D70**: 084007(2004); gr-qc/030367.
[67] J. Lukierski, H. Ruegg, W. Rühl, Phys. Lett. **B313**, 357(1993).
[68] J. Lukierski, A. Nowicki, Int. Journ. Mod. Phys. **A18**, 7(2003); hep-th/0203065.
[69] J. Lukierski, Proc. Intern. Workshop "Supersymmetries and Quantum Symmetries" (SQS'03), July 2003, publ. Dubna (JINR), p. 207(2004); hep-th/0402117.
[70] J. Lukierski, H. Ruegg, V.N. Tolstoy, in "Quantum Groups: Formalism and Applications", Proc. of XXX-th Karpacz Winter School, February 1994, ed. by Polish Scientific Publishers (1995), p. 359.
[71] S. Judes, M. Visser, Phys. Rev. **D68**, 045001(2003); gr-qc/0205067.
[72] D. Kimberly, J. Magueijo, J. Medeiros, Phys. Rev. **D70**, 084007(2004); gr-qc/0303067.
[73] R. Aloisio, A. Galante, A.F. Grillo, L. Luzio, F. Mendez, Phys. Lett. **B610**, 101(2005).
[74] J. Kowalski-Glikman, Phys. Lett. **B547**, 291(2002); hep-th/027279.
[75] S. Krešić–Jurić, S. Meljanac, M. Stojić, Eur. Phys. J. **C51**, 229(2007).
[76] A. Borowiec, A. Pachoł , SIGMA6, 086(2010); arXiv:1005.4429[math-ph].
[77] D. Kovačević , S. Meljanac, A. Pachoł, R. Štrajn, Phys. Lett. **B711**, 122(2012).
[78] J. Lukierski, P. Minnaert, A. Nowicki, "D=4 Quantum Poincaré-Heisenberg algeba" in "Symmetries in Science VI", ed. B. Gruber, Plenum Press, New York (1993), p.469–475.
[79] Zh. Kuznetsova, F. Toppan, Eur. Phys. J. **C73**, 2483(2013); arXiv: 1301.5501[hep-th].
[80] S. Majid, "Duality principle and braided geometry, Springer Lecture Notes in Phys. **147**, 125(1995).
[81] G. Amelino-Camelia, S. Majid, Int. J. Mod. Phys. **A15**, 4301(2000); hep-th/9907110.
[82] J. Kowalski-Glikman, S. Nowak, Class. Quant. Grav. **20**, 4799(2003); hep-th/0304101.
[83] G. Gubitosi, F. Mercati, Class. Quant. Grav. **30**; 145002(2013); arXiv:1106.5710[gr-qc].
[84] J. Kowalski-Glikman, Int. J. Mod. Phys. **A28**, 1330014(2013); arXiv:1303.0195[hep-th].
[85] G. Amelino-Camlia, L. Freidel, J. Kowalski-Glikman, L. Smolin, Phys. Rev. **D84**, 084010; arXiv: 1101.0931[hep-th].
[86] L. Freidel, R.G. Leight, D. Minic, "Quantum Gravity, Dynamical Phase Space and String Theory, arXiv:1405.3949[hep-th].
[87] L. Freidel, R.G. Leight, D. Minic,JHEP, **1506**:006(2015); arXiv:1502.-8005[hep-th].
[88] J. Lukierski, A. Nowicki, in Proc. XXI-th Int. Coll. on Group-Theor. Methods in Physics, ed. V.K. Dobrev, H.D. Doebner, Sofia, p.165 (1997); q-alg/9702003. [hep-th].
[89] S. Meljanac, A. Samsarov, R. Štrajn, JHEP **1208**, 127(2012); arXiv:1204.4324.
[90] T. Jurić, S. Meljanac, R. Štrajn, Phys. Lett. **A377**, 2472(2013); arXiv:1303.0994.
[91] T. Jurić, D. Kovačević, S. Meljanac, SIGMA **10**, 106(2014); arXiv:1402.0397.
[92] S. Meljanac, Z. Škoda, M. Stojić, "Lie algebra type noncommutative phase space as Hopf-algebroids"', Lett. Math. Phys., in press; arXiv: 1409.8188 [math.QA].
[93] G. Böhm, "Hopf algebroids", in Handbook of Algebra, Springer, 2008: arXiv: 0805.3806.
[94] J.-H. Lu, Intern. Journ. Math. **7**, 47(1996); q-alg/9505024.
[95] P. Xu, Comm. Math. Phys. **216**, 539(2001); q-alg/9905192.
[96] M. Takeuchi, J. Math. Soc. Japan **29**, 459(1977).
[97] T. Brzeziński, G. Militaru, J. Alg. **251**, 279(2002): math.QA/0012164.
[98] M. Henneaux, L. Mezinescu, Phys. Lett. **152B**, 340(1985).
[99] R.R. Metsaev, A.A. Tseytlin, Nucl. Phys. **B533**, 109(1998); hep-th/9805028.
[100] C. Klimčík, JHEP **12**, 051(2002); hep-th/0210095.
[101] F. Delduc, M. Magro, B. Vicedo, Phys. Rev. Lett. **112**: 051601(2014); arXiv: 1309.5850.

[102] M.A. Semenov-Tian-Shansky, Funct. Anal. Appl. **17**, 259(1983).
[103] M.A. Semenov-Tian-Shansky, nlin/0209057
[104] K. Zarembo, JHEP **1005**, 002(2010); arXiv:1003.0465 [hep-th].
[105] I. Bena, J. Polchinsky, R. Roiban, Phys. Rev. **D69**: 046002 (2004); hep-th/030511.
[106] G. Arutyunov, S. Frolov, J. Phys. **A42**: 254003(2009); arXiv: 0901.4937.
[107] A. Borowiec, H. Kyono, J. Lukierski, J. Sakamoto, K. Yoshida, JHEP**04**, 079(2016); arXiv: 1510.03083.
[108] A. Pachoł, S.J. van Tongeren, Phys. Rev. **93**: 026008(2016); arXiv: 1510.02389.
[109] T.H. Buscher, PLB **194**, 59(1987).
[110] B.Kulik, R. Roiban, JHEP **9**,007(2002); hep-th/0012010.
[111] G. Arutyunov, S. Frolov, B. Hoare, R. Roiban, A.A. Tseytlin, Nucl. Phys. **B903**, 262(2016); arXiv: 1511.05795 [hep-th].
[112] A.A. Tseytlin, L. Wulff, arXiv: 1605.04884 [hep-th].
[113] J. Lukierski, A. Nowicki, J. Sobczyk, J. Phys. **A26**, L1109(1993).
[114] P. Kosiński, J. Lukierski, P. Maślanka, J. Sobczyk, J. Phys. **A28**, 2255(1995); hep-th/9411115.
[115] F. Girelli, T. Konopka, J. Kowalski-Glikman, E.R. Livine, Phys. Rev. **D73**; 045009(2006); hep-th/0512107.
[116] P. Kosiński, J. Lukierski, P. Maślanka, Czech. J. Phys. **50**, 1283(2000).
[117] G. Fiore, J. Wess, Phys. Rev. **D75**; 105022(2007); hep-th/0701078.
[118] M. Daszkiewicz, J. Lukierski, M. Woronowicz, J. Phys. **A42**, 355201(2009); arXiv: 0807.1992[hep-th].
[119] M. Daszkiewicz, J. Lukierski, M. Woronowicz, Phys. Rev. **D77**; 105007(2008); arXiv: 0708.1561[hep-th].
[120] J. Lukierski, M. Woronowicz, Int. J. Mod. Phys. **A27**, 1250084(2012); arXiv: 1206.5656[hep-th].
[121] N. Seiberg, E. Witten, JHEP **9909** 032(1999); hep-th/9908142.
[122] P. Aschieri, M. Dimitrijevic, F. Meyer, J. Wess, Class. Quant. Grav. **23**, 1883(2006); hep-th/0510059.
[123] R. Banerjee, P. Mukherjee, S. Samanta, Phys. Rev. **D75**: 125020(2007); hep-th/0703128.

IOP Conf. Series: Journal of Physics: Conf. Series **804** (2017) 012029 doi:10.1088/1742-6596/804/1/012029

Anomalous transport and massive gravity theories

Eugenio Megías[1,2]

[1]Max-Planck-Institut für Physik (Werner-Heisenberg-Institut), Föhringer Ring 6, D-80805, Munich, Germany

[2]Departamento de Física Teórica, Universidad del País Vasco UPV/EHU, Apartado 644, 48080 Bilbao, Spain

E-mail: emegias@mppmu.mpg.de

Abstract. We review the Kubo formulae relevant to study anomalous transport properties of relativistic fluids. We apply this formalism to perform a computation of the transport coefficients in a holographic massive gravity model including vorticity and external electromagnetic fields. We find an interesting phase in which the electric DC conductivity is negligible, while the anomalous conductivities turn out to be nonvanishing.

1. Introduction

Quantum anomalies are one of the most intriguing properties of relativistic field theories [1]. In the past few years it has been extensively studied the role played by quantum anomalies in the hydrodynamics of relativistic fluids. The hydrodynamical systems should obey the conservation laws of the energy-momentum tensor and spin one currents. In presence of anomalies the currents are no longer conserved, and this has important consequences in the constitutive relations for hydrodynamics. Two relevant phenomena induced by anomalies appear at first order in the hydrodynamical expansion: the *chiral magnetic effect*, which is the responsible for the generation of an electric current parallel to a magnetic field [2], and the *chiral vortical effect*, in which the electric current is induced by a vortex [3]. It is believed that these phenomena can produce observable effects in heavy ion physics [4], as well as in condensed matter systems [5]. It has been proposed in the literature several methods to compute the transport coefficients from a microscopic theory, either dissipative or non-dissipative, including kinetic theory [6], Kubo formulae [7, 8], diagrammatic methods [9], fluid/gravity correspondence [10, 11, 12], and the partition function formalism [13, 14, 15].

In addition to anomalies, there are other sources of non-conservation of the currents. In this context it is worth mentioning *disorder* as a very interesting property of some condensed matter systems [16]. Physical systems affected by disorder include real-world crystals with impurities, and the disorder phenomena can be manifested in the form of metal-insulator transitions [17]. From a quantum field theory perspective, disorder has to do with explicit breaking of translation invariance, so that it leads to momentum dissipation and the conservation of the energy-momentum tensor is modified. The study of disorder is particularly interesting in holography, as the AdS/CFT correspondence is a powerful tool that might help to establish some universal properties of these systems. Several holographic massive gravity models in AdS spacetimes have been recently proposed as a new and interesting approach to tackle this problem [18, 19]. It has been studied in these models some dissipative transport coefficients, like the DC electric conductivity, which turns out to be decreasing when disorder increases.

In this work we will review the Kubo formalism to study anomalous transport in hydrodynamics. In a second step, this formalism will be applied to study anomalous transport phenomena in a massive gravity model. We will contrast the result for the DC electric conductivity, with the one obtained for anomalous conductivities.

2. Hydrodynamics of relativistic fluids

Hydrodynamics is a useful approach to study many phenomena for physical systems out of equilibrium. It can be applied when the mean free path of particles is much shorter than the characteristic size of the system. The basic ingredients are the constitutive relations, which are expressions of the energy-momentum tensor and the charge currents in terms of fluid quantities [20],

$$\begin{aligned} \langle T^{\mu\nu}\rangle &= (\varepsilon+P)u^\mu u^\nu + P g^{\mu\nu} + \langle T^{\mu\nu}\rangle_{\text{diss \& anom}} \,, \\ \langle J^\mu\rangle &= n u^\mu + \langle J^\mu\rangle_{\text{diss \& anom}} \,. \end{aligned}$$

Here ε is the energy density, P the pressure, n the charge density and u^μ the local fluid velocity. The equilibrium contributions are well known in the literature for many physical systems; however there are extra terms in the constitutive relations which lead to dissipative and anomalous effects. These extra contributions are in general difficult to compute, and the standard approach in hydrodynamics is to organize them in a derivative expansion, also called hydrodynamic expansion [11, 12, 21, 22]. In the Landau frame, defined as the one in which the energy flux vanishes at rest, the hydrodynamic expansion at first order in derivatives reads

$$\langle T^{\mu\nu}\rangle_{\text{diss \& anom}} = -\eta P^{\mu\alpha}P^{\nu\beta}\left(D_\alpha u_\beta + D_\beta u_\alpha - \frac{2}{3}g_{\alpha\beta}D^\lambda u_\lambda\right) - \zeta P^{\mu\nu}D^\alpha u_\alpha + \cdots \,, \quad (1)$$

$$\langle J^\mu\rangle_{\text{diss \& anom}} = -\sigma T P^{\mu\nu}D_\nu\left(\frac{\mu}{T}\right) + \sigma E^\mu + \sigma^{\mathcal{B}} B^\mu + \sigma^{\mathcal{V}}\omega^\mu + \cdots \,, \quad (2)$$

where $P^{\mu\nu} = g^{\mu\nu} + u^\mu u^\nu$ is the projector in the space orthogonal to the fluid velocity, $E^\mu = F^{\mu\nu}u_\nu$ and $B^\mu = \frac{1}{2}\epsilon^{\mu\nu\rho\lambda}u_\nu F_{\rho\lambda}$ are the electric and magnetic fields respectively, with the field strength of the gauge field defined as $F_{\mu\nu} = D_\mu A_\nu - D_\nu A_\mu$, and $\omega^\mu = \epsilon^{\mu\nu\rho\lambda}u_\nu D_\rho u_\lambda$ is the vorticity. The shear η and bulk ζ viscosities play a prominent role in heavy ion experiments, as they are dissipative coefficients that are responsible for most of the entropy production of the quark gluon plasma during the collision, see e.g. [23]. The electric conductivity σ is also dissipative, and its effects is to generate an electric current in the medium induced by an external electric field. Finally, the coefficients $\sigma^{\mathcal{B}}$ and $\sigma^{\mathcal{V}}$ are the so-called chiral magnetic [2] and chiral vortical [3] conductivities respectively, and they lead also to an electric current induced, either by an external magnetic field, or by a vortex in the fluid. As we will explain in Sec. 3, the coefficients $\sigma^{\mathcal{B}}$ and $\sigma^{\mathcal{V}}$ receive contributions from quantum anomalies, so that we will refer to them as anomalous conductivities.

In order to understand the (non)dissipative character of these transport coefficients, let us study their time reversal $\mathcal{T}$ properties. The second law of thermodynamics states that the entropy should increase with time, i.e.

$$\frac{\partial}{\partial t}s > 0 \,. \quad (3)$$

One can see from Eq. (3) that only $\mathcal{T}$-odd transport coefficients can contribute to the entropy production, while $\mathcal{T}$-even coefficients cannot. On the one hand the electric current $\vec{J}$ is $\mathcal{T}$-odd, while the electric field $\vec{E}$ is $\mathcal{T}$-even, and both the magnetic field $\vec{B}$ and vorticity $\vec{\omega}$ are $\mathcal{T}$-odd.

IOP Conf. Series: Journal of Physics: Conf. Series **804** (2017) 012029 doi:10.1088/1742-6596/804/1/012029

Then, from a quick inspection of Eq. (2) one concludes that the electric conductivity σ is $\mathcal{T}$-odd, while the chiral magnetic and vortical conductivities, $\sigma^{\mathcal{B}}$ and $\sigma^{\mathcal{V}}$, are $\mathcal{T}$-even. This explains the dissipative properties of σ, and the non-dissipative characters of $\sigma^{\mathcal{B}}$ and $\sigma^{\mathcal{V}}$. From a similar analysis of the parity $\mathcal{P}$ properties one finds that the chiral conductivities are related to $\mathcal{P}$-odd transport, see e.g. [21].

It is possible to observe non-dissipative effects also in the energy-momentum tensor under an appropriate choice of the frame (see Sec. 3.2), in particular the generation of an energy-flux induced by a magnetic field or by a vortex, i.e.

$$\langle T^{\mu\nu}\rangle_{\mathrm{anom}} = \sigma^{\mathcal{B}}_{\varepsilon}(B^\mu u^\nu + B^\nu u^\mu) + \sigma^{\mathcal{V}}_{\varepsilon}(\omega^\mu u^\nu + \omega^\nu u^\mu) . \tag{4}$$

We will refer to $\sigma^{\mathcal{B}}_{\varepsilon}$ ($\sigma^{\mathcal{V}}_{\varepsilon}$) as chiral magnetic (vortical) conductivity of energy current. In the rest of this work we will study the transport coefficients σ, $\sigma^{\mathcal{B}}$, $\sigma^{\mathcal{V}}$, $\sigma^{\mathcal{B}}_{\varepsilon}$ and $\sigma^{\mathcal{V}}_{\varepsilon}$. We will be especially focused on the chiral conductivities.

3. Kubo Formulae

On the basis of linear response theory, hydrodynamic transport coefficients can be extracted from the long-wavelength and low-frequency limits of some retarded Green functions, leading to the so called Kubo formulae. The Kubo formula for the chiral magnetic conductivity was derived in [24], while the one for the chiral vortical conductivity was studied in [7]. They read, respectively,

$$\sigma^{\mathcal{B}} = \lim_{p_c\to 0} \frac{i}{2p_c}\sum_{a,b}\epsilon_{abc}\langle J^a J^b\rangle|_{\omega=0} , \quad \sigma^{\mathcal{V}} = \lim_{p_c\to 0} \frac{i}{2p_c}\sum_{a,b}\epsilon_{abc}\langle J^a T^{0b}\rangle|_{\omega=0} . \tag{5}$$

Similar Kubo formulae have been derived for the transport coefficients related to the generation of energy flux, in particular $\sigma^{\mathcal{B}}_{\varepsilon} \sim \epsilon_{abc}\langle T^{0a}J^b\rangle$ and $\sigma^{\mathcal{V}}_{\varepsilon} \sim \epsilon_{abc}\langle T^{0a}T^{0b}\rangle$, see [7, 25, 8]. We will study in this section the result of the computation of Eq. (5) at weak and strong coupling.

3.1. Weak coupling results

Let us consider a theory of N free chiral fermions transforming under a global symmetry group G generated by matrices $(T_A)^f{}_g$. The chemical potential for the fermion Ψ^f is given by $\mu^f = \sum_A q^f_A \mu_A$, while the Cartan generator is $H_A = q^f_A \delta^f{}_g$ where q^f_A are the charges. It has been performed in the literature the 1-loop computation of the chiral magnetic [24] and the chiral vortical [26] conductivities by using the Kubo formulae of Eq. (5). The results read, respectively,

$$(\sigma^{\mathcal{B}})_{AB} = \frac{1}{4\pi^2} d_{ABC}\mu^C , \qquad (\sigma^{\mathcal{V}})_A = \frac{1}{8\pi^2}\sum_{B,C} d_{ABC}\,\mu^B\,\mu^C + \frac{T^2}{24} b_A , \tag{6}$$

where the coefficients

$$d_{ABC} = \frac{1}{2}\left[\mathrm{tr}(T_A\{T_B,T_C\})_R - \mathrm{tr}(T_A\{T_B,T_C\})_L\right] , \qquad b_A = \mathrm{tr}\,(T_A)_R - \mathrm{tr}\,(T_A)_L , \tag{7}$$

are related to the trace of the generators of the symmetry group. The subscripts R, L stand for the contributions of right-handed and left-handed fermions. One can easily identify d_{ABC} with the group theoretic factor related to the axial anomaly, which typically appears in the computation of the anomalous triangle diagram corresponding to three non-abelian gauge fields coupled to a chiral fermion. On the other hand, the coefficient b_A corresponds to the mixed gauge-gravitational anomaly [27], appearing in the anomalous triangle diagram with one non-abelian gauge field and two insertions of the energy-momentum tensor coupled to a chiral fermion. This means that the chiral magnetic and chiral vortical conductivities are induced by chiral anomalies, either the axial anomaly or the mixed gauge-gravitational anomaly.

Anomalies are also responsible for a non-vanishing value of the divergence of the currents, which in this case reads [28] [1]

$$D_\mu J^\mu_A = \epsilon^{\mu\nu\rho\lambda}\left(\frac{d_{ABC}}{32\pi^2}F^B_{\mu\nu}F^C_{\rho\lambda} + \frac{b_A}{768\pi^2}R^\alpha{}_{\beta\mu\nu}R^\beta{}_{\alpha\rho\lambda}\right). \tag{8}$$

A clear consequence of Eq. (8) is that anomalies modify the hydrodynamic equations, due to the non-conservation of the currents. So it is obvious that they should affect the hydrodynamic expansion with new contributions not appearing in absence of anomalies. We have identified above these new contributions at first order in derivatives as $\sigma^{\mathcal{B}}$ and $\sigma^{\mathcal{V}}$ in Eq. (2). [2]

3.2. Strong coupling results

The Kubo formulae Eq. (5) have been computed in [25, 22, 30] at strong coupling within a Einstein-Maxwell model in 5 dim. In order to mimic the anomalous effects, the model is supplemented with pure gauge and mixed gauge-gravitational Chern-Simons terms. The action reads

$$\begin{aligned} S &= \frac{1}{16\pi G}\int d^5x\sqrt{-g}\Bigg[R + 12 - \frac{1}{4}F_{MN}F^{MN} \\ &\quad + \epsilon^{MNPQR}A_M\left(\frac{\kappa}{3}F_{NP}F_{QR} + \lambda R^A{}_{BNP}R^B{}_{AQR}\right)\Bigg] + S_{GH} + S_{CSK}, \end{aligned} \tag{9}$$

where S_{GH} is the usual Gibbons-Hawking boundary term, and S_{CSK} is an extra boundary contribution needed to reproduce the mixed gauge-gravitational anomaly at a general hypersurface. A computation of the current with this model leads to

$$\tilde{J}^\mu = \frac{\partial S}{\partial A_\mu} = -\frac{\sqrt{-\gamma}}{16\pi G}\left[F^{r\mu} + \frac{4}{3}\kappa\epsilon^{\mu\nu\rho\lambda}A_\nu F_{\rho\lambda}\right]_\partial \equiv J^\mu + K^\mu. \tag{10}$$

Note that $\tilde{J}^\mu$ is not gauge covariant, and we refer to it as *consistent* current. The *covariant* version of the current, denoted by J^μ, is the usual one appearing in the constitutive relations. From Eq. (10), an on-shell computation of the divergence of the covariant current leads to the anomaly [3]

$$D_\mu J^\mu = -\frac{1}{16\pi G}\epsilon^{\mu\nu\rho\lambda}\left(\kappa\hat{F}_{\mu\nu}\hat{F}_{\rho\lambda} + \lambda\hat{R}^\alpha{}_{\beta\mu\nu}\hat{R}^\beta{}_{\alpha\rho\lambda}\right). \tag{11}$$

Finally, from a comparison with Eq. (8) one can identify the parameters κ and λ as the axial anomaly and mixed gauge-gravitational anomaly coefficients, respectively.

The holographic computation of the transport coefficients with the Kubo formulae follows from the study of retarded propagators by using the AdS/CFT dictionary [31, 32]. In general lines, we split the metric and gauge field into a background and a linear perturbation

$$g_{MN} = g^{(0)}_{MN} + \epsilon h_{MN}, \qquad A_M = A^{(0)}_M + \epsilon a_M. \tag{12}$$

[1] The gauge-gravitational anomaly is the statement that it is not possible to preserve at the same time the vanishing of the divergence of the energy-momentum tensor and of the chiral $U(1)$ currents.

[2] $\sigma^{\mathcal{B}}_\varepsilon$ and $\sigma^{\mathcal{V}}_\varepsilon$ in Eq. (4) are induced also by chiral anomalies. A computation of these conductivities at weak coupling in line with this section can be found in e.g. [8, 29].

[3] Quantities with hat $(\hat{F}, \hat{R}, \cdots)$ refer to their induced four dimensional objects at a cut-off surface.

Inserting these fields in the action and expanding up to second order in ϵ one can obtain the second order action, and from there the desired propagators [33]. The equations of motion of the fluctuations should obey appropriate boundary conditions, in particular: i) regularity at the horizon; and ii) vanishing at the boundary, as the fluctuations cannot modify the boundary values, which are given by the background. There is some freedom in one of the boundary conditions which is related to the choice of frame. This corresponds to the particular definition of the local fluid velocity.

We show in Table 1 the values of the anomalous conductivities computed with the holographic model of Eq. (9), corresponding to a system which realizes a single $U_A(1)$ symmetry. We display the results in three different frames. u^μ can be taken to be proportional to: i) the energy flux $\langle T^{0i}\rangle = (\varepsilon + P)u^i$ (Landau frame), and ii) the charge current $\langle J^i\rangle = nu^i$ (Eckart frame). In addition, the laboratory rest frame is the natural frame in the field theory computation, see Sec. 3.1, and it seems to be related to the entropy current $\langle J^i_S\rangle_{\text{anom}} = 0$, see e.g. [34].

Conductivities	Laboratory rest frame	Landau frame	Eckart frame
$(\sigma^{\mathcal{B}})_{\text{F}}$	$\sigma^{\mathcal{B}} = \frac{\mu}{4\pi^2}$	$\sigma^{\mathcal{B}} - \frac{n}{\varepsilon+P}\sigma^{\mathcal{B}}_\varepsilon$	0
$(\sigma^{\mathcal{V}})_{\text{F}}$	$\sigma^{\mathcal{V}} = \frac{\mu^2}{8\pi^2} + \frac{T^2}{24}$	$\sigma^{\mathcal{V}} - \frac{n}{\varepsilon+P}\sigma^{\mathcal{V}}_\varepsilon$	0
$(\sigma^{\mathcal{B}}_\varepsilon)_{\text{F}}$	$\sigma^{\mathcal{B}}_\varepsilon = \sigma^{\mathcal{V}}$	0	$\sigma^{\mathcal{B}}_\varepsilon - \frac{\varepsilon+P}{n}\sigma^{\mathcal{B}}$
$(\sigma^{\mathcal{V}}_\varepsilon)_{\text{F}}$	$\sigma^{\mathcal{V}}_\varepsilon = \frac{\mu^3}{12\pi^2} + \frac{\mu T^2}{12}$	0	$\sigma^{\mathcal{V}}_\varepsilon - \frac{\varepsilon+P}{n}\sigma^{\mathcal{V}}$

Table 1. Anomalous conductivities contributing to the constitutive relations, Eqs. (2) and (4), for a theory with a single chiral $U_A(1)$ symmetry. The result at weak and strong coupling agree.

Let us finally stress that the values of the anomalous conductivities at strong coupling coincide precisely with the ones obtained at weak coupling, and this a strong hint towards a non-renormalization theorem for the anomalous conductivities, see e.g. [26, 25, 30]. [4]

4. Transport properties in massive gravity

It has been recently studied in [18, 19] some minimal models for massive gravity in 4 dim. The momentum relaxation is described through the Stueckelberg mechanism with Goldstone modes corresponding to scalars X^I, which are related to spatial translations. In this section we will consider the model presented in [19]. We have explicitly checked that our conclusions are the same when considering the model of [18].

4.1. The model

In order to study anomalous transport in holographic massive gravity theories, one should consider the theory in odd dimensions, as only in these cases there is contribution from the chiral anomaly and we can introduce the corresponding Chern-Simons terms, see Eq. (9). We will consider the model of [19], and extend it to 5 dim. The action reads

$$S = \frac{1}{16\pi G}\int d^5x\sqrt{-g}\left[R + 12 - \frac{1}{2}\partial^M X^I \partial_M X^I - \frac{1}{4}F^2 - \frac{\mathcal{J}}{4}\partial_M X^I \partial_N X^I F^N{}_L F^{LM}\right], \quad (13)$$

with scalar fields $X^I = k\delta^I_i x^i$, where i denotes spatial directions. Momentum relaxation is implemented by giving the scalar fields a vacuum expectation value. The term $\sim \mathcal{J}$ is a higher derivative coupling between the charge and the scalar field. The parameter k controls the degree

[4] Note, however, that in presence of dynamical gluons some renormalization effects have been observed in $\sigma^{\mathcal{V}}$, see [35, 36].

of breaking of translational invariance, in particular when $k = 0$ one recovers the massless gravity theory. The equations of motion for the fields $(\delta g^{MN}, \delta A^M, \delta X^I)$ read

$$\begin{aligned} 0 &= G_{MN} - 6g_{MN} + \frac{1}{2}F_M{}^L F_{LN} - \frac{1}{8}g_{MN}F^2 - \frac{1}{2}\partial_M X^I \partial_N X^I + \frac{1}{4}g_{MN}(\partial_L X^I)(\partial^L X^I) \\ &\quad - \frac{\mathcal{J}}{4}\left(\tilde{X}.F.F + F.\tilde{X}.F + F.F\tilde{X}\right)_{MN} + \frac{\mathcal{J}}{8}g_{MN}\mathrm{tr}(\tilde{X}.F.F), \qquad (14) \\ 0 &= \nabla_N F^{NM} + \frac{\mathcal{J}}{2}\nabla_L(\tilde{X}.F)^{ML} - \frac{\mathcal{J}}{2}\nabla_L(\tilde{X}.F)^{LM}, \qquad (15) \\ 0 &= \Box X^I + \frac{\mathcal{J}}{2}\nabla_M(\partial_N X^I F^N{}_L F^{LM}), \qquad (16) \end{aligned}$$

where $\tilde{X}_{MN} \equiv \partial_M X^I \partial_N X^I$, and the products are $\tilde{X}.F.F = \partial_M X^I \partial_N X^I F^N{}_L F^{LM}$, etc. These equations can be solved with the black-brane ansatz:

$$ds^2 = -f dt^2 + \frac{dr^2}{f} + r^2 d\vec{x}^2, \qquad A = A_t\, dt. \qquad (17)$$

We find the following solution:

$$f = r^2 - \frac{M}{r^2} + \frac{\rho^2}{3r^4} + \frac{k^2}{2}, \qquad A_t = \mu - \frac{\rho}{r^2}, \qquad (18)$$

and the temperature and chemical potential can be related to the mass M and charge ρ of the black hole as

$$T = \frac{1}{2\pi r_h}\left(\frac{2M}{r_h^2} - \mu^2 - \frac{k^2}{2}\right), \qquad \mu r_h^2 = \rho, \qquad (19)$$

where r_h is the black hole horizon corresponding to the largest solution of $f = 0$.

4.2. Electric DC conductivity

As shown in Eq. (2), the electric conductivity measures the electric current J^μ induced by an electric field E^μ. There are several methods to compute the DC conductivity in holography, but one of the most straightforward is the one proposed in [37]. We consider the following ansatz for the fluctuations in the gauge field and the metric:

$$A_x = \epsilon(Et + a_x(r)), \qquad g_{tx} = r^2(1 + \epsilon h_{tx}(r)), \qquad g_{rx} = r^2(1 + \epsilon h_{rx}(r)), \qquad (20)$$

and work to first order in ϵ. The fluctuation equations can be written now as

$$\begin{aligned} 0 &= \tfrac{k^2}{2}\left(1 - \tfrac{2\mathcal{J}\rho^2}{r^6}\right)h_{rx} - \left(1 - \tfrac{\mathcal{J}k^2}{2r^2}\right)\tfrac{E\rho}{fr^3}, \qquad (21) \\ r^8 f(h''_{tx} + \frac{5}{r}h'_{tx}) &= k^2(r^6 + 2\mathcal{J}\rho^2) - \rho(2r^2 - k^2\mathcal{J}^2)(rfa'_x), \qquad (22) \\ 0 &= \left[\left(1 - \tfrac{\mathcal{J}k^2}{2r^2}\right)rfa'_x + \left(2 - \tfrac{\mathcal{J}k^2}{r^2}\right)\rho h_{tx}\right]'. \qquad (23) \end{aligned}$$

The DC conductivity can be obtained from a solution of these equations with appropriate boundary conditions at the horizon and the boundary, as explained in Sec. 3.2. At this point we will skip the technical details, that will be presented in [38]. The final result for the DC conductivity reads

$$\sigma_{DC} = \frac{J_x}{E} = r_h\left(1 - \frac{\mathcal{J}k^2}{2r_h^2}\right)\left[1 + \left(1 - \frac{\mathcal{J}k^2}{2r_h^2}\right)\frac{4\mu^2}{k^2 M(r_h)}\right], \qquad (24)$$

with $M(r_h) = 1 + \frac{2\mathcal{J}\rho^2}{r_h^6}$. This formula is the equivalent to the result presented in Eq. (3.15) of [19], but in 5 dim. We plot in Fig. 1 the value of the DC conductivity from Eq. (24) as a function of the parameter k (left) and the temperature T (right) for different values of the parameter $\mathcal{J}$. On the one hand, we observe that in the case $\mathcal{J} = 0$, the conductivity is bounded from below, so that there is no insulating behavior without $\mathcal{J}$. More important is the fact that when $\mathcal{J} \neq 0$, there is a particular value $k_* = k(\mathcal{J})$ at which the DC conductivity vanishes, so that the theory behaves as an insulator. This property is one of the main results of [18, 19] in 4 dim, and here we have reproduced it as well in 5 dim. Finally, we show in Fig. 2 the regime of parameters where the DC conductivity vanishes.

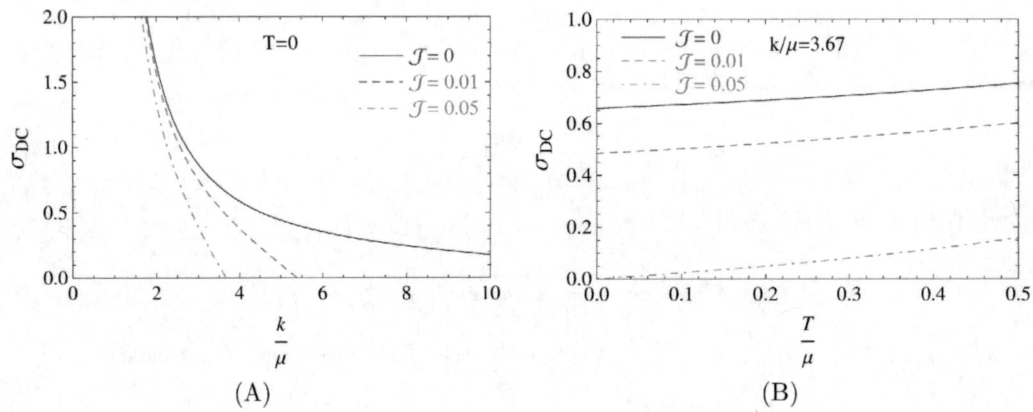

Figure 1. (A) DC conductivity at zero temperature as a function of k (normalized to the chemical potential μ). (B) DC conductivity (at fixed $k/\mu = 3.67$) as a function of temperature (normalized to μ). These results are obtained with Eq. (24) for different values of the parameter $\mathcal{J}$.

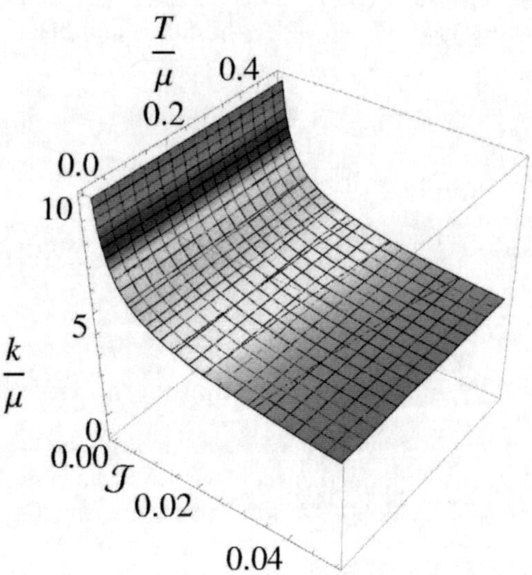

Figure 2. Region in the plane $(\mathcal{J}, T/\mu, k/\mu)$ where the DC conductivity of Eq. (24) vanishes.

4.3. Anomalous Transport

In Sec. 4.2 we have studied some dissipative properties of the 5 dim massive gravity model. In order to study non-dissipative transport coefficients, we should introduce anomalous effects in the theory. As we explained in Sec. 3, anomalies are mimicked in the gravity side through Chern-Simons terms. When extending the model of Eq. (13) with the Chern-Simons terms of Eq. (9), extra contributions appear in the equations of motion of the theory, in particular the term $-2\lambda\epsilon_{LPQR(M}\nabla_B\left(F^{PL}R^B{}_{N)}{}^{QR}\right)$ in the rhs of Eq. (14), and the terms $\epsilon^{MNPQR}(\kappa F_{NP}F_{QR}+\lambda R^A{}_{BNP}R^B{}_{AQR})$ in the rhs of Eq. (15). By using the following conventions

$$ds^2 = \frac{du^2}{4u^2 f(u)} + \frac{1}{u}\left(f(u)dt^2 + 2h_{ti}(u)e^{ipz}dtdx^i + d\vec{x}^2\right), \tag{25}$$

where $i = x, y$, the fluctuation equations are

$$0 = (a_i' f)' - \rho h_{ti}' - ip\epsilon_{ij}\left[4\kappa\rho a_j + \lambda\left(12u^2\rho^2 + \frac{24}{u}(1-f) + 5k^2\right)h_{tj}'\right], \tag{26}$$

$$\begin{aligned} 0 = {} & h_{ti}'' - \frac{1}{u}h_{ti}' - \frac{k^2}{4uf}h_{ti} - u\rho a_i' + ip\lambda\epsilon_{ij}\left[\left(24(1-f) - 12u^3\rho^2 - 5uk^2\right)a_j' + \right. \\ & \left. + \left(\frac{24}{u}(1-f) - 32u^2\rho^2 - 6k^2\right)a_j - 8\rho u(uh_{ti}')'\right]. \end{aligned} \tag{27}$$

Finally, from the procedure of Sec. 3 we get the result for the anomalous conductivities: [5]

$$\sigma^{\mathcal{B}} = \lim_{p\to 0}\frac{i}{p}\langle J^x J^y\rangle = \kappa 4\mu, \tag{28}$$

$$\sigma^{\mathcal{V}} = \lim_{p\to 0}\frac{i}{p}\langle J^x T^{0y}\rangle = \kappa 2\mu^2 + \lambda 16\pi^2 T^2, \tag{29}$$

$$\sigma_\varepsilon^{\mathcal{B}} = \lim_{p\to 0}\frac{i}{p}\langle T^{0x} J^y\rangle = \kappa 2\mu^2 + \lambda 16\pi^2 T^2 + \lambda 2k^2, \tag{30}$$

$$\sigma_\varepsilon^{\mathcal{V}} = \lim_{p\to 0}\frac{i}{p}\langle T^{0x} T^{0y}\rangle = \kappa\frac{4}{3}\mu^3 + \lambda 32\pi^2 T^2\mu. \tag{31}$$

Some remarks deserve to be mentioned at this point. On the one hand the chiral magnetic and vortical conductivities of charge currents, $\sigma^{\mathcal{B}}$ and $\sigma^{\mathcal{V}}$, and the chiral vortical conductivity of energy current, $\sigma_\varepsilon^{\mathcal{V}}$, are the same as in the massless gravity theory, see Table 1. This means that these anomalous conductivities are not affected by translational breaking effects, and this constitutes one of the most important results of this work. From this property, together with the result of Sec. 4.2, we conclude that there is a regime in the theory in which the DC conductivity vanishes, but the anomalous conductivities do not vanish. A consequence is that one can study the anomalous effects of these systems in this regime in a clean way. Let us finally mention that, contrary to the other coefficients, the chiral magnetic effect of energy current, $\sigma_\varepsilon^{\mathcal{B}}$, has some dependence on k. The interpretation of this result is currently in progress. [6]

[5] We show in these expressions the explicit dependence on the parameters κ and λ for an easier identification of the axial and gauge-gravitational anomaly contributions. One can compare with the results of Table 1 by considering $\kappa = 1/(16\pi^2)$ and $\lambda = 1/(384\pi^2)$ in Eqs. (28)-(31).

[6] Several hypotheses might be proposed for the dependence of $\sigma_\varepsilon^{\mathcal{B}}$ on the parameter k, including renormalization or gauge artefacts, as well as lack of unitarity of the theory. In addition, one cannot exclude the possibility that this might be due to a pathology of the model. In fact, it has been studied in the literature the thermodynamics of some massive gravity theories, and an intriguing result has been found related to the non-vanishing value of the entropy in the ground state ($T = 0$, $\mu = 0$), see [39].

5. Discussion and conclusions

In this work we have reviewed the role played by the chiral anomalies in the hydrodynamics of relativistic fluids. In particular, we have focused on the effects of external magnetic fields and vortices in the fluid, and characterized how the chiral anomalies contribute to the corresponding conductivities through the anomaly coefficients, Eq. (7). The anomalous conductivities turn out to be non-dissipative at first order in the hydrodynamic expansion.

Massive gravity models have been introduced in the literature as holographic duals of disorder in condensed matter systems. By using the Kubo formalism, we have computed the transport properties in a massive gravity model with higher derivative corrections. We found an interesting regime in which the electric DC conductivity vanishes, but the anomalous conductivities turn out to be nonvanishing. Moreover, the anomalous transport seem to be unaffected by translational breaking effects. These and other issues will be addressed in more detail in a forthcoming publication [38].

Acknowledgments

I would like to thank K. Landsteiner and O. Pujolàs for valuable discussions. I thank the Instituto de Física Teórica UAM/CSIC, Madrid, Spain, for their hospitality during the completion of the final stages of this work. Research supported by the European Union under a Marie Curie Intra-European fellowship (FP7-PEOPLE-2013-IEF) with project number PIEF-GA-2013-623006.

References

[1] Bertlmann R A Oxford, UK: Clarendon (1996) 566 p. (International series of monographs on physics: 91)
[2] Fukushima K, Kharzeev D E and Warringa H J 2008 *Phys. Rev.* **D78** 074033 (*Preprint* 0808.3382)
[3] Son D T and Surowka P 2009 *Phys. Rev. Lett.* **103** 191601 (*Preprint* 0906.5044)
[4] Keren-Zur B and Oz Y 2010 *JHEP* **06** 006 (*Preprint* 1002.0804)
[5] Basar G, Kharzeev D E and Yee H U 2014 *Phys. Rev.* **B89** 035142 (*Preprint* 1305.6338)
[6] Arnold P B, Moore G D and Yaffe L G 2000 *JHEP* **11** 001 (*Preprint* hep-ph/0010177)
[7] Amado I, Landsteiner K and Pena-Benitez F 2011 *JHEP* **1105** 081 (*Preprint* 1102.4577)
[8] Landsteiner K, Megias E and Pena-Benitez F 2013 *Lect. Notes Phys.* **871** 433–468 (*Preprint* 1207.5808)
[9] Manes J L and Valle M 2013 *JHEP* **01** 008 (*Preprint* 1211.0876)
[10] Bhattacharyya S, Hubeny V E, Minwalla S and Rangamani M 2008 *JHEP* **0802** 045 (*Preprint* 0712.2456)
[11] Erdmenger J, Haack M, Kaminski M and Yarom A 2009 *JHEP* **01** 055 (*Preprint* 0809.2488)
[12] Banerjee N *et al.* 2011 *JHEP* **01** 094 (*Preprint* 0809.2596)
[13] Banerjee N, Bhattacharya J, Bhattacharyya S, Jain S, Minwalla S and Sharma T 2012 *JHEP* **09** 046 (*Preprint* 1203.3544)
[14] Jensen K, Kaminski M, Kovtun P, Meyer R, Ritz A and Yarom A 2012 *Phys. Rev. Lett.* **109** 101601 (*Preprint* 1203.3556)
[15] Megias E and Valle M 2014 *JHEP* **11** 005 (*Preprint* 1408.0165)
[16] Lee P A and Ramakrishnan T V 1985 *Rev. Mod. Phys.* **57** 287–337
[17] Imada M, Fujimori A and Tokura Y 1998 *Rev. Mod. Phys.* **70** 1039–1263
[18] Baggioli M and Pujolas O 2016 (*Preprint* 1601.07897)
[19] Goutraux B, Kiritsis E and Li W J 2016 *JHEP* **04** 122 (*Preprint* 1602.01067)
[20] Kovtun P 2012 *J. Phys.* **A45** 473001 (*Preprint* 1205.5040)
[21] Kharzeev D E and Yee H U 2011 *Phys. Rev.* **D84** 045025 (*Preprint* 1105.6360)
[22] Megias E and Pena-Benitez F 2013 *JHEP* **05** 115 (*Preprint* 1304.5529)
[23] Bozek P 2010 *Phys. Rev.* **C81** 034909 (*Preprint* 0911.2397)
[24] Kharzeev D E and Warringa H J 2009 *Phys. Rev.* **D80** 034028 (*Preprint* 0907.5007)
[25] Landsteiner K, Megias E, Melgar L and Pena-Benitez F 2011 *JHEP* **09** 121 (*Preprint* 1107.0368)
[26] Landsteiner K, Megias E and Pena-Benitez F 2011 *Phys. Rev. Lett.* **107** 021601 (*Preprint* 1103.5006)
[27] Kumura T 1969 *Prog. Theor. Phys.* **42** 1191–1205
[28] Alvarez-Gaume L and Witten E 1984 *Nucl.Phys.* **B234** 269
[29] Landsteiner K, Megias E and Pena-Benitez F 2014 *Phys. Rev.* **D90** 065026 (*Preprint* 1312.1204)
[30] Landsteiner K, Megias E, Melgar L and Pena-Benitez F 2012 *J. Phys. Conf. Ser.* **343** 012073 (*Preprint* 1111.2823)
[31] Son D T and Starinets A O 2002 *JHEP* **09** 042 (*Preprint* hep-th/0205051)

IOP Conf. Series: Journal of Physics: Conf. Series **804** (2017) 012029 doi:10.1088/1742-6596/804/1/012029

[32] Herzog C P and Son D T 2003 *JHEP* **03** 046 (*Preprint* hep-th/0212072)
[33] Kaminski M, Landsteiner K, Mas J, Shock J P and Tarrio J 2010 *JHEP* **02** 021 (*Preprint* 0911.3610)
[34] Megias E, Landsteiner K and Pena-Benitez F 2013 *Acta Phys. Polon. Supp.* **6** 45–51
[35] Golkar S and Son D T 2015 *JHEP* **02** 169 (*Preprint* 1207.5806)
[36] Hou D F, Liu H and Ren H c 2012 *Phys. Rev.* **D86** 121703 (*Preprint* 1210.0969)
[37] Donos A and Gauntlett J P 2014 *JHEP* **11** 081 (*Preprint* 1406.4742)
[38] Megias E , *work in progress (2016).*
[39] Vegh D 2013 (*Preprint* 1301.0537)

IOP Conf. Series: Journal of Physics: Conf. Series **804** (2017) 012030 doi:10.1088/1742-6596/804/1/012030

Representations of the Bondi–Metzner–Sachs group in three space–time dimensions

Evangelos Melas

University of Athens, Department of Economics, Unit of Mathematics and informatics, Sofokleous 1, Athens 10559, Greece

E-mail: emelas@econ.uoa.gr

Abstract. The original Bondi–Metzner–Sachs group B is the common asymptotic symmetry group of all asymptotically flat Lorentzian 4–dim space–times. As such, B is the best candidate for the universal symmetry group of General Relativity (G.R.). In 1973, with this motivation, P. J. McCarthy classified all relativistic B–invariant systems in terms of strongly continuous irreducible unitary representations (IRS) of B. Here, we construct the IRS of $B(2,1)$, the analogue of B, in 3 space–time dimensions. The IRS are induced from 'little groups' which are compact. The finite 'little groups' are cyclic groups of even order. The inducing construction is exhaustive notwithstanding the fact that $B(2,1)$ is not locally compact in the employed Hilbert topology.

1. Introduction

The Bondi–Metzner–Sachs (BMS) group B is the common asymptotic group of all curved real Lorentzian space–times which are asymptotically flat in future null directions [1, 2], and is the best candidate for the universal symmetry group of G.R..

In 1939 Wigner laid the foundations of special relativistic quantum mechanics [3] and relativistic quantum field theory by constructing the Hilbert space IRS of the (universal cover) of the Poincare group P.

The universal property of B for G.R. makes it reasonable to attempt to lay a similarly firm foundation for quantum gravity by following through the analogue of Wigner's programme with B replacing P. Some years ago McCarthy constructed explicitly [4, 5, 6, 7, 8, 9, 10, 11] the IRS of B for exactly this purpose. This work was based on G.W.Mackey's pioneering work on group representations [3, 12, 13, 14, 15]; in particular McCarthy's work extended G.W.Mackey's work to the relevant infinite–dimensional case.

It is difficult to overemphasize the importance of Piard's results [16, 17] who soon afterwards proved that *all* the IRS of B, when this is equipped with the Hibert topology, are derivable by the inducing construction. This proves the exhaustivity of McCarthy's list of representations and renders his results even more important.

However, in quantum gravity, complexified or euclidean versions of G.R. are frequently considered and the question arises: Are there similar symmetry groups for these versions of the theory? McCarthy constructed [18], in abstract form, all possible analogues of B, both real and in any signature, or complex, with all possible notions of asymptotic flatness 'near infinity'. There are, in fact, forty–one such groups. These abstract constructions were given in a quantum setting; the paper was concerned with finding the IRS of these groups in Hilbert spaces (especially for the complexification $\mathcal{CB}$ of B itself). It has been argued [19, 20, 21, 22] that these Hilbert space representations are related to elementary particles and quantum gravity (via gravitational instantons).

Here, we follow this programme for 3−dim G.R. and construct in the Hilbert topology the IRS of $B(2,1)$, the analogue of B in three space−time dimensions. It is proved that all IRS of $B(2,1)$ are induced from IRS of *compact* 'little groups'. It follows that some IRS of $B(2,1)$ are controlled by IRS of the finite symmetry groups of regular polygons in ordinary euclidean 2−space. It is proved that *all* IRS of $B(2,1)$ are induced by the IRS of its little groups notwithstanding the fact that $B(2,1)$ is *not* locally compact in the employed Hilbert topology. The paper closes with the explicit construction of the IRS of $B(2,1)$.

We turn now to the study of to $B(2,1)$, the analogue of B in three space−time dimensions.

1.1. The group $B^{2,1}(N^+)$

Recall that the 2 + 1 Minkowski space is the vector space R^3 of row vectors with 3 real components, with the inner product defined as follows. Let $x, y \in R^3$ have components x^μ and y^μ respectively, where $\mu = 0, 1, 2$. Define the inner product $x.y$ between x and y by

$$x.y = x^0y^0 - x^1y^1 - x^2y^2. \tag{1}$$

Then the 2 + 1 Minkowski space, sometimes written $R^{2,1}$, is just R^3 with this inner product. The "2,1" refers to the one plus and two minus signs in the inner product. Let $SO(2,1)$ be the (connected component of the identity element of the) group of linear transformations preserving the inner product. Matrices $\Lambda \in SO(2,1)$ are taken as acting by matrix multiplication from the right, $x \longmapsto x\Lambda$, on row vectors $x \in R^{2,1}$.

The future null cone $N^+ \subset R^{2,1}$ is just the set of nonzero vectors with zero length and $x^0 > 0$:

$$N^+ = \left\{x \in R^{2,1} | x.x = 0, x^0 > 0\right\}. \tag{2}$$

Let R^*_+ denote the multiplicative group of all *positive* real numbers. Obviously, if $x \in N^+$, then $tx \in N^+$ for any $t \in R^*_+$. Let $F_1(N^+)$ denote the vector space (under pointwise addition) of all functions $f : N^+ \to R$ satisfying the homogeneity condition

$$f(tx) = tf(x) \tag{3}$$

for all $x \in N^+$ and $t \in R^*_+$. Define a representation T of $SO(2,1)$ on $F_1(N^+)$ by setting, for each $x \in N^+$ and $\Lambda \in SO(2,1)$,

$$(T(\Lambda)f)(x) = f(x\Lambda). \tag{4}$$

Now let $B^{2,1}(N^+)$ be the semi−direct product

$$B^{2,1}(N^+) = F_1(N^+) \rtimes_T SO(2,1). \tag{5}$$

That is to say, $B^{2,1}(N^+)$ is, as a set, just the product $F_1(N^+) \times SO(2,1)$, and the group multiplication law for pairs is

$$(f_1, \Lambda_1)(f_2, \Lambda_2) = (f_1 + T(\Lambda_1)f_2, \Lambda_1\Lambda_2). \tag{6}$$

IOP Conf. Series: Journal of Physics: Conf. Series **804** (2017) 012030 doi:10.1088/1742-6596/804/1/012030

1.2. The double cover $B^{2,1}(N^+)_c$

Let $\mathrm{SL}(2,R)$ be the group of all real 2×2 matrices with determinant one. $\mathrm{SL}(2,R)$ is sometimes denoted by G below. Let $\mathrm{M}_s(2,R)$ be the set of all 2×2 symmetric real matrices. We define a right action of G on $\mathrm{M}_s(2,R)$ by $\mathrm{M}_s(2,R)\times G\to \mathrm{M}_s(2,R)$ with

$$(m,g)\mapsto g^{\top}mg, \tag{7}$$

where the superscript $\top$ means transpose. Clearly any element $\mu\in \mathrm{M}_s(2,R)$ can be parameterized as follows:

$$\mu=\begin{bmatrix} x^{o}-x^{1} & x^{2}\\ x^{2} & x^{o}+x^{1}\end{bmatrix}$$

where $x^o, x^1, x^2\in R$. We now consider the map $b: R^3\to \mathrm{M}_s(2,R)$ defined by

$$b(x)=\begin{bmatrix} x^{o}-x^{1} & x^{2}\\ x^{2} & x^{o}+x^{1}\end{bmatrix}, \tag{8}$$

where the x^μ are the components of $x\in R^3$. This map is a linear bijection, so the right action of G on $\mathrm{M}_s(2,R)$ induces a linear right action of G on R^3. Since

$$\det(b(x))=x\cdot x \tag{9}$$

and the G action preserves determinants (indeed $\det g = 1$) in $\mathrm{M}_s(2,R)$, G acts as transformations from $SO(2,1)$. In fact, this construction gives an homomorphism

$$\gamma: G\to SO(2,1) \tag{10}$$

which is onto, and has kernel $Z_2=\{Id,-Id\}$ in G, Id denoting the identity element of G. Thus γ identifies G as the double cover of $SO(2,1)$

$$G=SO(2,1)_c. \tag{11}$$

Therefore, the double cover of the group $B^{2,1}(N^+)$, given in (5), has the form

$$B^{2,1}(N^+)_c=F_1(N^+)\rtimes_T \mathrm{SL}(2,R). \tag{12}$$

Strictly speaking, "T" should read "$T\gamma$", but the notation is simpler as above.

1.3. The group $B(2,1)$

So far, the supertranslation space $F_1(N^+)$ has been defined as a space of truly *arbitrary* homogeneous functions of degree one. This has been merely for clarity; for physical applications, it is necessary to give this space additional structure. For reasons discussed in detail in McCarthy [18], we now give a new realization of $B^{2,1}(N^+)_c$ where the supertranslation space is restricted to be the separable Hilbert space $L^2(P_1(R),\lambda,R)$ of real–valued functions defined on $P_1(R)\simeq S^1$; functions square integrable with respect to the standard normalized (Lesbegue) measure λ on $P_1(R)\simeq S^1$; $P_1(R)\equiv S^1/Z_2$ is the one–dimensional real projective space (the circle quotient the antipodal map).

In particular in [23] the following Theorem is proved:

Theorem 1.1 *The group $B^{2,1}(N^+)_c$ can be realised as*

$$B(2,1)=L^2(P_1(R),\lambda,R)\rtimes_T G \tag{13}$$

with semi–direct product specified by

$$(T(g)\alpha)(x)=\kappa_g(x)\alpha(xg), \tag{14}$$

where $G =$ SL(2,R), $g \in G$, $\alpha \in L^2(P_1(R), \lambda, R)$. Moreover, if

$$g = \begin{bmatrix} a & b \\ c & d \end{bmatrix}, \tag{15}$$

then the components x_1, x_2 of $\mathbf{x} \in R^2 - \{0\}$ transform linearly under g, so that the ratio $x = x_1/x_2,\ x_2 \neq 0,$ transforms fraction linearly under g. Writing xg for the transformed ratio,

$$xg = \frac{(\mathbf{x}g)_1}{(\mathbf{x}g)_2} = \frac{x_1 a + x_2 c}{x_1 b + x_2 d} = \frac{xa + c}{xb + d}. \tag{16}$$

The ratio $x = x_1/x_2,\ x_2 \neq 0$, is a local inhomogeneous coordinate of $P_1(R)$. We denote our final realization of our group by $B(2,1)$ to distinguish it from the previous realizations $B^{2,1}(N^+)$ and $B^{2,1}(N^+)_c$. In analogy to B, it is natural to choose a measure λ on $P_1(R)$ which is invariant under the maximal compact subgroup $SO(2)$ of G; we choose λ to be the standard normalized Lesbegue measure $\mathrm{d}\lambda = \frac{\mathrm{d}\theta}{2\pi}$.

The factor $\kappa_g(x)$ on the right hand side of (14) is defined by

$$\kappa_g(x) = \frac{(xb + d)^2 + (xa + c)^2}{1 + x^2}. \tag{17}$$

It is well known [24] that the topological dual of a Hilbert space can be identified with the Hilbert space itself, so that we have $L^{2'}(P_1(R), \lambda, R) \simeq L^2(P_1(R), \lambda, R)$. In fact, given a continuous linear functional $\phi \in L^{2'}(P_1(R), \lambda, R)$, we can write, for $\alpha \in L^2(P_1(R), \lambda, R)$

$$(\phi, \alpha) =< \phi, \alpha >, \tag{18}$$

where the function $\phi \in L^2(P_1(R), \lambda, R)$ on the right is uniquely determined by (and denoted by the same symbol as) the linear functional $\phi \in L^{2'}(P_1(R), \lambda, R)$ on the left. The representation theory of $B(2,1)$ is governed by the dual action T' of G on the topological dual $L^{2'}(P_1(R), \lambda, R)$ of $L^2(P_1(R), \lambda, R)$. The dual action T' is defined by:

$$< T'(g)\phi, \alpha >=< \phi, T(g^{-1})\alpha > \cdot \tag{19}$$

A short calculation gives

$$(T'(g)\phi)(x) = \kappa_g^{-2}(x)\phi(xg). \tag{20}$$

Now, this action T' of G on $L^{2'}(P_1(R), \lambda, R)$, given explicitly above is, like the action T of G on $L^2(P_1(R), \lambda, R)$, continuous. The 'little group' L_ϕ of any $\phi \in L^{2'}(P_1(R), \lambda, R)$ is the stabilizer

$$L_\phi = \{g \in G \mid T'(g)\phi = \phi\}. \tag{21}$$

By continuity, $L_\phi \subset G$ is a closed subgroup.

2. Representation theory

Let A and $\mathcal{G}$ be topological groups, and let T be a given homomorphism from $\mathcal{G}$ into the group of automorphisms Aut(A) of A. Suppose A is abelian and $\mathcal{H} = A \rtimes_T \mathcal{G}$ is the semi−direct product of A and $\mathcal{G}$, specified by the continuous action $T : \mathcal{G} \longrightarrow$ Aut(A). In the product topology of $A \times \mathcal{G}$, $\mathcal{H}$ then becomes a topological group. It is assumed that it becomes a separable locally compact topological group. In order to give the operators of the induced representations explicitly it is necessary ([3], [12], [13], [14], [15] and references therein) to give the following information:

(i) An irreducible unitary representation U of L_ϕ on a Hilbert space D for each L_ϕ.
(ii) A $\mathcal{G}$ –quasi–invariant measure μ on each orbit $\mathcal{G}\phi \approx \mathcal{G}/L_\phi$, where L_ϕ denotes the little group of the base point $\phi \in A^{'}$ of the orbit $\mathcal{G}\phi$; $A^{'}$ is the topological dual of A.

Let D_μ be the space of functions $\psi : \mathcal{G} \to D$ which satisfy the conditions

$$(a) \qquad \psi(gl) = U(l^{-1})\psi(g), \qquad g \in \mathcal{G},\ l \in L_\phi,$$

$$(b) \quad \int_{\mathcal{G}\phi} < \psi(q), \psi(q) > \mathrm{d}\mu(q) < \infty,$$

where the scalar product under the integral sign is that of D. Note, that the constraint (a) implies that $< \psi(gl), \psi(gl) >=< \psi(g), \psi(g) >$, and therefore the inner product $< \psi(g), \psi(g) >$, $g \in \mathcal{G}$, is constant along every element q of the coset space $\mathcal{G}/L_\phi \approx \mathcal{G}\phi$. This allows to assign a meaning to $< \psi(q), \psi(q) >$, where $q = gL_\phi$, by defining $< \psi(q), \psi(q) >:=< \psi(g), \psi(g) >$. Thus the integrand in (b) becomes meaningful due to the condition (a). A pre–Hilbert space structure can now be given to D_μ by defining the scalar product

$$< \psi_1, \psi_2 >= \int_{\mathcal{G}\phi} < \psi_1(q), \psi_2(q) > \mathrm{d}\mu(q), \tag{22}$$

where $\psi_1, \psi_2 \in D_\mu$. It is convenient to complete the space D_μ with respect to the norm defined by the scalar product (22). In the resulting Hilbert space, functions are identified whenever they differ, at most, on a set of μ–measure zero. Thus our Hilbert space is

$$D_\mu = L^2(\mathcal{G}\phi, \mu, D). \tag{23}$$

Define now an action of $\mathcal{H} = A \rtimes_T \mathcal{G}$ on D_μ by

$$(g_o\psi)(q) = \sqrt{\frac{\mathrm{d}\mu_{g_o}}{\mathrm{d}\mu}(q)}\psi(g_o^{-1}q), \tag{24}$$

$$\alpha\psi(q) = e^{i<g_o\phi,\alpha>}\psi(q), \tag{25}$$

where $g_o \in \mathcal{G}$, $q \in \mathcal{G}\phi$, and $\alpha \in A$. Eqs. (24) and (25) define the IRS of $B(2,1)$ induced from *each* $\phi \in A^{'}$ and *each* irreducible representation U of L_ϕ. The 'Jacobian' $\frac{\mathrm{d}\mu_{g_o}}{\mathrm{d}\mu}$ of the group transformation is known as the *Radon–Nikodym* derivative of μ_{g_o} with respect to μ and ensures that the resulting IRS of $B(2,1)$ are unitary.
The central results of induced representation theory ([3], [12], [13], [14], [15] and references therein) are the following:

(i) Given the topological restrictions on $\mathcal{H} = A \rtimes_T \mathcal{G}$ (separability and local compactness), any representation of $\mathcal{H}$, constructed by the method above, is irreducible if the representation U of L_ϕ on D is irreducible. Thus an irreducible representation of $\mathcal{H}$ is obtained for *each* $\phi \in A^{'}$ and *each* irreducible representation U of L_ϕ.
(ii) If $\mathcal{H} = A \rtimes_T \mathcal{G}$ is a regular semi–direct product (i.e., $A^{'}$ contains a Borel subset which meets each orbit in $A^{'}$ under $\mathcal{H}$ in just one point) then *all* of its irreducible representations can be obtained in this way.

3. Obstructions and resolutions

Two remarks are in order regarding the representations of $B(2,1)$ obtained by the above construction:

(i) As it is explained in [23] the subgroup $L^2(P_1(R), \lambda, R)$ of $B(2,1) = L^2(P_1(R), \lambda, R) \rtimes_T G$ is topologised as a (pre) Hilbert space by using a natural measure on $P_1(R)$ and by introducing a scalar product into $L^2(P_1(R), \lambda, R)$. If R^4 is endowed with the natural metric topology then the group $G = SL(2,R)$, considered as a subset of R^4, inherits the induced topology on G. In the product topology of $L^2(P_1(R), \lambda, R) \times G$, $B(2,1)$ is a non−locally compact group (the proof follows, without substantial change, Cantoni's proof [25], see also [4]). (In fact the subgroup $L^2(P_1(R), \lambda, R)$, and therefore the group $B(2,1)$ can be employed with many different topologies. The Hilbert type topology employed here appears to describe quantum mechanical systems in asymptotically flat space−times [10]). Since in the Hilbert type topology $B(2,1) = L^2(P_1(R), \lambda, R) \rtimes_T G$ is not locally compact the theorems dealing with the irreducibility of the representations obtained by the above construction no longer apply (see e.g. [13]). However, it can be proved that the induced representations obtained above *are* irreducible. The proof follows very closely the one given in [7] for the case of the original BMS group B.

(ii) Here it is assumed that $B(2,1)$ is equipped with the Hilbert topology. It is of outmost significance that it can be proved [23] that in this topology $B(2,1)$ is a regular semi−direct−product. The proof follows the corresponding proof [16, 17] for the group B. Regularity amounts to the fact [12] that $L^{2'}(P_1(R), \lambda, R)$ can have no equivalent classes of quasi−invariant measures μ such that the action of $\mathcal{G}$ is strictly ergodic with respect to μ. When such measures μ do exist it can be proved [12] that an irreducible representation of the group, with the semi−direct−product structure at hand, may be associated with each such measure μ, that is not equivalent to any of the IRS constructed by the Wigner−Mackey's inducing method. In a different topology it is not known if $B(2,1)$ is a regular or irregular semi−direct−product. Irregularity of $B(2,1)$ in a topology different from the Hilbert topology would imply that there are IRS of $B(2,1)$ that are not unitary equivalent to any of the IRS obtained by the inducing construction. Strictly ergodic actions are notoriously hard to deal with even in the locally compact case. Indeed, for locally compact non−regular semi−direct products, there is no known example for which all inequivalent irreducibles arising from strictly ergodic actions have been found. For the other 41 groups defined in [18] regularity has only been proved for B [16, 17] when B is equipped with the Hilbert topology. Similar remarks apply to all of them regarding IRS arising from strictly ergodic actions in a given topology.

4. Little groups and inducing construction

In [23] it is proved that when $B(2,1)$ is employed with the Hilbert topology *all* little groups of $B(2,1)$ are compact. In particular the following Theorem is proved:

Theorem 4.1 *The little groups L_ϕ for $B(2,1)$ are precisely the closed subgroups of $K = SO(2)$ which contain the element $-I$, I being the identity element of G. These are (A) K itself, and (B) the cyclic groups C_n of even order n.*

For a given little group L_ϕ the elements $\phi \in A'$ which are invariant under L_ϕ, i.e., the elements $\phi \in A'$ which satisfy

$$(T'(g)\phi)(x) = \phi(x) \tag{26}$$

form a subspace of $L^2(P_1(R), \lambda, R)$. We denote this subspace by $L^2(L_\phi)$. Then we have the following Theorem:

Theorem 4.2 *The Hilbert space $L^2(SO(2))$ of invariant vectors $\phi \in A'$ under $SO(2)$ is:*

$$L^2(SO(2)) = \{\phi \in L^2(P_1(R), \lambda, R) \mid \phi(x) = c,\ c \in R\}. \tag{27}$$

So $L^2(SO(2))$ is just the one–dimensional space of constant real–valued functions defined on $P_1(R)$. The Hilbert space $L^2(C_n)$ of invariant vectors $\phi \in A'$ under C_n is:

$$L^2(C_n) = L^2(E_n), \tag{28}$$

where $L^2(E_n)$ is the Hilbert space of square integrable real–valued functions defined on

$$E_n = \left\{\theta \in P_1(R) \mid 0 < \theta < \frac{4\pi}{n}\right\}. \tag{29}$$

Moreover in [23] it is shown that the Wigner–Mackey's inducing construction is exhaustive despite the fact that $B(2,1)$ is not locally compact in the employed Hilbert topology. This result is rather important because other group theoretical approaches to quantum gravity which invoke Wigner–Mackey's inducing construction (see e.g. [26]) are typically plagued by the non–exhaustiveness of the inducing construction which results precisely from the fact that the group in question is not locally compact in the prescribed topology. Exhaustiveness is not just a mathematical nicety: If the inducing construction is not exhaustive one cannot know if the most interesting information or part of it is coded in the irreducibles which cannot be found by the Wigner–Mackey's inducing procedure. These results, i.e. compactness of the little groups and exhaustiveness of the inducing construction, not only are significant for the group theoretical approach to quantum gravity advocated here, but also they have repercussions [23] for other approaches to quantum gravity.

5. Construction of the IRS of $B(2,1)$

To find explicitly the operators of the induced representations of $B(2,1)$, it suffices to provide the information cited in (i) and (ii) in section 2 for each of the orbit types. We note that all the little groups are abelian. All IRS of an abelian group are one–dimensional.

(i), (i$_\alpha$) $L_\phi = K = SO(2)$.
The IRS U of K are parameterized by an integer ν which for distinct representations takes the values $\nu = ..., -2, -1, 0, 1, 2, ...$. Denoting these representations by $U^{(\nu)}$, they are given by multiplication in one complex dimension $D \approx C$ by

$$U^{(\nu)}(R(\theta)) = e^{i\nu\theta}, \tag{30}$$

where $R(\theta) = \begin{pmatrix} cos\theta & -sin\theta \\ sin\theta & cos\theta \end{pmatrix} \in SO(2)$.

(i$_\beta$) $L_\phi = C_n$, n is even.
The IRS U of C_n are parameterized by an integer λ which for distinct representations takes the values $\lambda = 0, 1, 2, ..., n-1$. Denoting these representations by $U^{(\lambda)}$, they are given by multiplication in one complex dimension $D \approx C$ by

$$U^{(\lambda)}\left(R\left(\frac{2\pi}{n}j\right)\right) = e^{i\frac{2\pi}{n}\lambda j}, \tag{31}$$

where j parameterizes the elements of the group C_n.

(ii) A $\mathcal{G}$ −quasi−invariant measure μ on each orbit $\mathcal{G}\phi \approx \mathcal{G}/L_\phi$ is required, however, a $\mathcal{G}$ −invariant measure μ on each orbit $\mathcal{G}\phi \approx \mathcal{G}/L_\phi$ will be given in all cases. We note that when μ is $\mathcal{G}$ −invariant then $\frac{\mathrm{d}\mu_{g_o}}{\mathrm{d}\mu}(q) = 1$, and this is precisely what happens in the case of $B(2,1)$. Moreover, in the case of $B(2,1)$, $\mathcal{G} = G$ =SL(2,R), and the little groups L_ϕ are given in Theorem 4.1.

(ii$_\alpha$) In [23] it is proved that the construction of a unique (up to a constant factor) G−invariant measure on the orbit $01 \equiv G/L_\phi$, $L_\phi = K = SO(2)$, necessitates the construction of a G−invariant measure on G, and the construction of a K−invariant measure on K. A G−invariant measure on

$$G = \mathrm{SL}(2,R) = \left\{ \begin{pmatrix} a & b \\ c & d \end{pmatrix}, \quad a,b,c,d \in R, \quad ad - bc = 1 \right\},$$

is given by

$$\mathrm{d}g = \frac{\mathrm{d}a \wedge \mathrm{d}b \wedge \mathrm{d}c}{a}. \tag{32}$$

G−invariant measure means $\mathrm{d}g = \mathrm{d}(gg_o) = \mathrm{d}(g_o g)$, $g_o \in \mathrm{SL}(2,R)$. A K−invariant measure on K is given by the 1−form $\mathrm{d}\theta$ where θ is the usual angular coordinates which covers $P_1(R) \simeq S^1$.

(ii$_\beta$) The orbits $02 \equiv G/L_\phi$, $L_\phi = C_n$, where n is even, can be endowed with the G−invariant measure on G given in case 01. Indeed, for a given little group L_ϕ, the orbit $02 \equiv G/L_\phi$ is the space of orbits of the right action $T : G \times L_\phi \to G$ of the group L_ϕ on G given by

$$g \star c := g \cdot c, \tag{33}$$

where $g \in G$ and $c \in L_\phi$. Thus the action T denoted by $\star$ is identical to the group multiplication in G. Since the group L_ϕ is finite and since the action (33) is fixed point free the coset spaces G/L_ϕ inherit the measure on G.

This completes the necessary information in order to construct the induced representations of $B(2,1)$. As already stated, it is shown in [23] that when $B(2,1)$ is equipped with the Hilbert topology the inducing construction is exhaustive notwithstanding the fact that $B(2,1)$ is not locally compact. To conclude, in this paper *all* IRS of $B(2,1)$ have been constructed in the Hilbert topology.

[1] Bondi H, Van Der Berg M G J and Metzner A W K 1962 *Gravitational waves in general relativity VII. Waves from axi−symmetric isolated sources* Proc. R. Soc. Lond. A **269** 21−52
[2] Sachs R K 1962 *Gravitational waves in general relativity VIII. Waves in asymptotically flat space−time* Proc. R. Soc. Lond. A **270** 103−126
[3] Wigner E 1939 *On unitary represenations of the Inhomogeneous Lorentz Group* Annals of Mathematics **40** 149−204
[4] McCarthy P J 1972 *Representations of the Bondi−Metzner−Sachs group I. Determination of the representations* Proc. R. Soc. Lond. A **330** 517−535
[5] McCarthy P J 1972 *Structure of the Bondi−Metzner−Sachs group* J. Math. Phys. **13** 1837−1842
[6] McCarthy P J 1973 *Representations of the Bondi−Metzner−Sachs group II. Properties and classification of the of the representations* Proc. R. Soc. Lond. A **333** 317−336
[7] McCarthy P J and Crampin M 1973 *Representations of the Bondi−Metzner−Sachs group III. Poincare spin multiplicities and irreducibility* Proc. R. Soc. Lond. A **335** 301−311
[8] McCarthy P J 1975 *The Bondi−Metzner−Sachs group in the nuclear topology* Proc. R. Soc. Lond. A **343** 489−523.
[9] McCarthy P J 1978 *Lifting of projective representations of the Bondi-Metzner-Sachs group in the nuclear topology* Proc. R. Soc. Lond. A **358** 141−171
[10] McCarthy P J and Crampin M 1974 *Physical significance of the topology of the Bondi−Metzner−Sachs group* Phys. Rev. Lett. **33** 547−550

IOP Conf. Series: Journal of Physics: Conf. Series **804** (2017) 012030 doi:10.1088/1742-6596/804/1/012030

[11] McCarthy P J 1972 *Asymptotically Flat Space−Times and Elementary Particles* Phys. Rev. Lett. **29** 817−819

[12] Mackey G W 1968 *Induced representations of groups and quantum mechanics* (New york : Benjamin)

[13] Mackey G W 1955 *The theory of group representations* (Chicago : The University of Chicago Press)

[14] Simms D J 1968 *Lie groups and quantum mechanics* (Berlin : Springer)

[15] Isham C J 1984 *Relativity groups and topology,* ed B S DeWitt and R Stora (Amsterdam : North−Holland)

[16] Piard A 1977 *Unitary representations of semi−direct product groups with infinite dimensional abelian normal subgroup* Rep. Math. Phys. **11** 259−278

[17] Piard A 1977 *Representations of the Bondi−Metzner−Sachs group with the Hilbert topology* Rep. Math. Phys. **11** 279−283

[18] McCarthy P J 1992 *Real and Complex Asymptotic Symmetries in Quantum Gravity, Irreducible Representations, Polygons, Polyhedra, and the A,D, E Series* Phil. Trans. R. Soc. Lond. A **338** 271−299

[19] Melas E 2004 *The BMS group and generalized gravitational instantons* J. Math. Phys. **45** 996−1002

[20] Melas E 2006 *Construction of the irreducibles of* $B(2,2)$ J. Phys. A: Math. Gen. **39** 3341−3366

[21] Melas E 2013 *Representations of the ultrahyperbolic BMS group HB. II. Determination of the representations induced from infinite little groups* arXiv:1312.0532v1

[22] Melas E 2014 *Representations of the ultrahyperbolic BMS group HB. III. Determination of the representations induced from finite little groups* arXiv:1402.1428v1

[23] Melas E *The Bondi−Metzner−Sachs group* $B(2,1)$ *in the Hilbert topology* (unpublished)

[24] Rudin W 1991 *Functional Analysis* (New York : McGraw−Hill)

[25] Cantoni V 1967 *On the representations of the Bondi−Metzner−Sachs group* Ph.D. Thesis University of London

[26] Isham C J and Kakas A C 1984 *A Group Theoretical Approach to Canonical Quantization of Gravity II. Unitary Representations of the Canonical Group* Class. Quantum Grav. **1** 633−650

IOP Conf. Series: Journal of Physics: Conf. Series **804** (2017) 012031 doi:10.1088/1742-6596/804/1/012031

Quantum vorticity at positive temperature for spins systems with continuous symmetry

D Minenkov [1] **and M Rouleux** [2]

[1] Institute for Problems in Mechanics, RAS, Prosp. Vernadskogo 101-1, Moscow, 119526, Russia

[2] Aix Marseille University, University Toulon, CNRS, CPT, Marseille, France

E-mail: [1] minenkov.ds@gmail.com, [2] rouleux@univ-tln.fr

Abstract. We propose a definition of vorticity at inverse temperature β for Gibbs states in quantum XY or Heisenberg spin systems on the lattice by testing $\exp[-\beta H]$ on a complete set of observables ("one-point functions"). Imposing a compression of Pauli matrices at the boundary, which stands for the classical environment, we perform some numerical simulations on finite lattices in case of XY model, which exhibit usual vortex patterns.

Introduction.

Consider the quantum XY or Heisenberg spin model for $S = 1/2$ on the 2-D lattice $\mathbf{Z}^2$, with nearest neighbor interactions. Marmin-Wagner, and Hohenberg theorems tell us that Gibbs states, for all inverse temperature β, are invariant under simultaneous rotation of spins (absence of continuous symmetry breaking in two dimensions). In the classical case, we know a bit more : although there is a unique Gibbs state, with rotational symmetry, which rules out the existence of first order transitions, a particular form for phase transition exists, characterized by a change of behavior in the correlation functions. For the 2-D rotator, it has been described by Berezinskii [3], and Kosterlitz-Thouless [11] in term of topological excitations, called vortices [9]. For Heisenberg model, we observe higher order topological defects, called instantons [2].

In this Report, we make a first attempt to answer the natural question : *How can we observe vorticity in the quantum case?*

Let us first consider a system in finite volume $\Lambda \subset \mathbf{Z}^2$. The Hamiltonians we are interested in are of the form $H_\Lambda(\Phi) = -\sum_{X\subset\Lambda} \Phi(X)$, where Φ is an "interaction" between sites in Λ. For nearest neighbor interaction, the contributing X are pairs $\langle i, j\rangle$, and the Hamiltonian reads

$$H_\Lambda = -\frac{1}{2} \sum_{\langle i,j\rangle\subset\Lambda} (\sigma_i^x \otimes \sigma_j^x + \sigma_i^y \otimes \sigma_j^y + u\sigma_i^z \otimes \sigma_j^z) \tag{1}$$

where u is a coupling constant, $u > 0$ in the ferromagnetic case ($u = 1$ when isotropic), $u < 0$ in the anti-ferromagnetic case ($u = -1$ when isotropic), and $u = 0$ is the XY model. We could also add an external magnetic field $h \sum_{i\in\Lambda} \sigma_i^z$ to H_Λ, and in case $u = 0$, allow for anisotropy between x and y components.

Though vortices can merge spontenaously in infinite volume, there are external fields that would certainly enhance vorticity. General external fields are defined within the notion of a "state" [14], Sect.II.3.

Throughout we denote by Tr the ordinary trace, and by tr the normalized trace, as $\mathrm{tr}(A) = \frac{1}{d}\mathrm{Tr}(A)$, where d is the dimension. Recall from [14], Sect.II.1 the partial trace: if A is a linear operator on $\mathcal{K}_1 \otimes \mathcal{K}_2$, the partial trace $\mathrm{tr}_{\mathcal{K}_1}$ or simply tr_1 is an operator $\mathcal{L}(\mathcal{K}_1 \otimes \mathcal{K}_2) \to \mathcal{L}(\mathcal{K}_1)$ defined by the requirement :

$$\mathrm{tr}_{\mathcal{K}_1}\big(B(\mathrm{tr}_1(A))\big) = \mathrm{tr}_{\mathcal{K}_1 \otimes \mathcal{K}_2}((B \otimes 1)A), \quad B \in \mathcal{L}(\mathcal{K}_1) \tag{2}$$

A quantum state ρ assigns to each finite $X \subset \mathbf{Z}^2$ an operator ρ_X with $\mathrm{Tr}(\rho_X) = 1$ and $\mathrm{tr}_{\mathcal{H}_Y}(\rho_{X \cup Y}) = \rho_X$ on $\mathcal{H}_Y = \otimes_{i \in Y} \mathbf{C}_i^2$ for all disjoint $X, Y \subset \mathbf{Z}^2$. (Instead of $\mathrm{tr}_{\mathcal{H}_Y}(A)$, we use also the notation $\mathrm{tr}_Y(A)$.) Given a state and the finite interaction $\Phi(X)$, we define the Hamiltonian on all of $\mathbf{Z}^2$

$$H_\Lambda^\rho(\Phi) = -\sum_{X \cap \Lambda \neq \emptyset} \mathrm{Tr}_{X \setminus \Lambda}\big[(1 \otimes \rho_{X \setminus \Lambda})\Phi(X)\big] \tag{3}$$

that couples Λ with the external field ρ through its nearest neighbors at the boundary. For A a quasi-local observable on $\mathbf{Z}^2$, we define the expectation value

$$\langle A \rangle_{\beta,\Lambda}^\rho = \frac{\mathrm{Tr}_\Lambda\big(\exp[-\beta H_\Lambda^\rho(\Phi)]\, A\big)}{\mathrm{Tr}_\Lambda \exp[-\beta H_\Lambda^\rho(\Phi)]} \tag{4}$$

and $\langle A \rangle_{\beta,\Lambda}^\rho$ has a limit as $|\Lambda| \to \infty$. Such a state has been constructed in [1] for the XY chain.

From a practical point of vue however, it is suitable to produce explicit "approximate states" that will favour the existence of vortices in finite volume; we proceed in the following way. Let $\Lambda \subset \mathbf{Z^2}$ be the "small system", and $\partial\Lambda \subset \mathbf{Z^2}$ its "environment", both finite. On $\partial\Lambda$, we "compress" the spin operators, so that the measure of observable "direction of spin" is deterministic for $j \in \partial\Lambda$, and quantum for $j \in \Lambda$. The resulting Hamiltonian $H_{\Lambda\cup\partial\Lambda}(\Phi)$ accounts for the interaction with the approximate "external field" on $\partial\Lambda$, as in .

Now, in finite volume $\Lambda \cup \partial\Lambda$, the only (normalized) Gibbs state is given by

$$A \mapsto \omega_\beta(A) = \frac{\mathrm{tr}(\exp[-\beta H_{\Lambda\cup\partial\Lambda}]A)}{\mathrm{tr}(\exp[-\beta H_{\Lambda\cup\partial\Lambda}]} \tag{5}$$

and called the "canonical Gibbs state". We shall actually define vorticity at inverse temperature β by decomposing the linear form ω_β on a canonical (orthonormal) basis of observables.

1. Vorticity matrices

Gibbs state (1) for spin $S = 1/2$ systems, as a linear form on the $\mathbf{C}^*$-algebra of observables

$$\mathcal{O} = \otimes_{j \in \Lambda\cup\partial\Lambda} o_j, \quad o_j = \mathcal{M}_{2\times2}(\mathbf{C})$$

("quasi-local observables" if we were to consider the thermodynamical limit,) can be decomposed in a canonical basis. The simplest way is to restrict to "one-point functions", i.e. the set $\widetilde{\mathcal{O}} \subset \mathcal{O}$ of $2N \times 2N$, block-diagonal 2×2 matrices (Pauli matrices are spin representations of SU(2) of dimension $2S + 1 = 2$), supported on individual sites of $\Lambda \cup \partial\Lambda$, $N = |\Lambda \cup \partial\Lambda|$.

a) The XY model

IOP Conf. Series: Journal of Physics: Conf. Series **804** (2017) 012031 doi:10.1088/1742-6596/804/1/012031

We first consider the XY model, where we can restrict to real 2×2 matrices. The compression of Pauli matrices on $\partial\Lambda$ is given by the orthogonal projector

$$\Pi(\theta) = \begin{pmatrix} \cos^2\theta & \sin\theta\cos\theta \\ \sin\theta\cos\theta & \sin^2\theta \end{pmatrix} \tag{6}$$

where θ parametrizes a point on the unit circle, accounting for the prescribed direction of "vorticity" on $\partial\Lambda$. Let again $\widetilde{\mathcal{O}}_{\mathbf{R}} \subset \widetilde{\mathcal{O}}$ be a real sub-algebra $\widetilde{\mathcal{O}}$, of real dimension $4N$.

Example 1: $\widetilde{\mathcal{O}}_{\mathbf{R}}$ is the "canonical" algebra, generated by real matrices $(D^i)_{i\in\Lambda\cup\partial\Lambda}$, whose all non-diagonal 2×2 blocks vanish, and all diagonal 2×2 blocks vanish, except this supported on site i that takes values in $\{\delta_1,\delta_2,\delta_3,\delta_4\}$, where

$$\delta_1 = \begin{pmatrix} 1 & 0 \\ 0 & 0 \end{pmatrix}, \quad \delta_2 = \begin{pmatrix} 0 & 1 \\ 0 & 0 \end{pmatrix}, \quad \delta_3 = \begin{pmatrix} 0 & 0 \\ 1 & 0 \end{pmatrix}, \quad \delta_4 = \begin{pmatrix} 0 & 0 \\ 0 & 1 \end{pmatrix}$$

So the family of block-diagonal $2N \times 2N$ matrices with 2×2 entry δ_j, $1\leq j\leq 4$ at the i:th place, $1\leq i\leq N$

$$(D^i_j)_{i\in\Lambda\cup\partial\Lambda} = (0\oplus\cdots\oplus\delta_j\oplus\cdots\oplus 0) \tag{7}$$

gives an orthonormal basis (ONB) of 1-point functions $\widetilde{\mathcal{O}}_{\mathbf{R}}$.

Example 2: $\widetilde{\mathcal{O}}_{\mathbf{R}}$ is the (real) algebra generated by Pauli matrices $(\widetilde{D}^i)_{i\in\Lambda\cup\partial\Lambda}$ with diagonal block supported on site i that takes values in $\{\mathrm{Id}, i\sigma^x, i\sigma^y, i\sigma^z\}$.

We shall restrict to the canonical algebra, whose generators enjoy the nice property of being real matrices. Let also $o_{\mathbf{R}} \subset o$ be the algebra of 2×2 matrices with real coefficients, endowed with the scalar product $(A|B) = \mathrm{Tr}(B^*A)$, which is isometric with $\mathbf{R}^4$. By extension, the basis $\delta = \{\delta_1,\delta_2,\delta_3,\delta_4\}$ of $o_{\mathbf{R}}$ will be called an "elementary basis" of $\widetilde{\mathcal{O}}_{\mathbf{R}}$, since N copies of δ, attached to each site i, give a basis $(D^i_j)_{i\in\Lambda\cup\partial\Lambda,1\leq j\leq 4}$ of $\widetilde{\mathcal{O}}_{\mathbf{R}}$. We say the same thing of any other ONB $b = \{b_1,b_2,b_3,b_4\}$ of $o_{\mathbf{R}}$, and of the corresponding basis $(B^i_j)_{i\in\Lambda,1\leq j\leq 4}$ of $\widetilde{\mathcal{O}}_{\mathbf{R}}$, where B^i_j is defined as in (7), with b_j instead of δ_j. Actually, the order of the elements of b matters, so we prefer to think of b as an "array", namely

$$b = \begin{pmatrix} b_1 & b_2 \\ b_3 & b_4 \end{pmatrix} \tag{8}$$

where b_k is of the form

$$b_k = \begin{pmatrix} b^{1k} & b^{2k} \\ b^{3k} & b^{4k} \end{pmatrix} \in \mathcal{M}_{2\times 2}(\mathbf{R})$$

which we identify with the vector $b^k = {}^t(b^{1k}, b^{2k}, b^{3k}, b^{4k})$. We will not use the algebraic structure of $o_{\mathbf{R}}$. With the notations above (partial traces), we see easily that :

$$\mathrm{tr}_1(b) = \begin{pmatrix} \mathrm{tr}b_1 & \mathrm{tr}b_2 \\ \mathrm{tr}b_3 & \mathrm{tr}b_4 \end{pmatrix} \tag{9}$$

which justifies the interpretation of b as a matrix (operator). So tr_1 are the components, in some matrix representation, of the usual trace ("tracial state") on $o_{\mathbf{R}}$. For simplicity we set $T(b) = \mathrm{tr}_1(b)$ and call it the "matrix of traces". An important rôle will be played with symmetric basis.

Definition 1.1: *We call the ONB b symmetric iff the corresponding matrix b in* (9) *is Hermitian, i.e.* $b_1 = b_1^*$, $b_4 = b_4^*$, *and* $b_3 = b_2^*$. *We call it δ-symmetric if moreover b is real, and* $T(b)$

has a degenerate eigenvalue, that is, is a multiple of identity.

Most of the basis are not symmetric, but occasionally we can make them symmetric, by permuting or multiplying by -1 some elements. We can characterize δ-symmetric basis: namely, if b is δ-symmetric, then modulo such transformations, there exists discrete or one-parameter families $P_s \in O(2;\mathbf{R})$ such that

$$b = b_s = {}^t P_s \delta P_s \tag{10}$$

(where the product is understood as if b_j's were numbers).

So far we have constructed "one point functions", i.e. a basis of $\widetilde{\mathcal{O}}_{\mathbf{R}}$. In the sequel we content with Hamiltonians of type (1) which are of second order in the interactions; if we were to include the linear term $\sum_{i\in\Lambda}\sigma_i^z$ we would write it as $\sum_{\langle i,j\rangle} 1_i \otimes \sigma_j^z$. Embed $\widetilde{\mathcal{O}}_{\mathbf{R}}$ into $\widetilde{\mathcal{O}}_{\mathbf{R}} \otimes \widetilde{\mathcal{O}}_{\mathbf{R}}$ by the usual coproduct Δ, and set $\widetilde{x} = \Delta(x) = \frac{1}{2}(1 \otimes x + x \otimes 1) \in o_{\mathbf{R}} \otimes o_{\mathbf{R}}$, for $x \in o_{\mathbf{R}}$. So we have "lifted" $\widetilde{b} = \Delta(b)$ as a family of $\widetilde{\mathcal{O}}_{\mathbf{R}} \otimes \widetilde{\mathcal{O}}_{\mathbf{R}}$ by $(\widetilde{B}_j^i)_{i\in\Lambda, 1\leq j\leq 4}$, with $\widetilde{B}_j^i = \Delta(B_j^i)$. With the notations of (8) and (9) we have

$$\widetilde{B}^i = \begin{pmatrix} \widetilde{B}_1^i & \widetilde{B}_2^i \\ \widetilde{B}_3^i & \widetilde{B}_4^i \end{pmatrix} \in \mathcal{M}_{4N\times 4N}(\mathbf{R}), \quad \mathrm{tr}_1(\widetilde{B}^i) = \begin{pmatrix} \mathrm{tr}\widetilde{B}_1^i & \mathrm{tr}\widetilde{B}_2^i \\ \mathrm{tr}\widetilde{B}_3^i & \mathrm{tr}\widetilde{B}_4^i \end{pmatrix} \in \mathcal{M}_{2\times 2}(\mathbf{R}) \tag{11}$$

In the same way, we form $e^{-\beta H}\widetilde{B}_j^i$, so we can map to each site $i \in \Lambda \cup \partial\Lambda$ a 2×2 matrix :

$$\mathrm{tr}_1(e^{-\beta H}\widetilde{B}^i) = \mathrm{tr}(e^{-\beta H}) \begin{pmatrix} \omega_\beta(\widetilde{B}_1^i) & \omega_\beta(\widetilde{B}_2^i) \\ \omega_\beta(\widetilde{B}_3^i) & \omega_\beta(\widetilde{B}_4^i) \end{pmatrix}$$

Definition 1.2: *We call vorticity matrix at site i, relative to the basis b, at inverse temperature β, the matrix :*

$$\Omega_\beta^i(b) = \frac{\mathrm{tr}_1(e^{-\beta H}\widetilde{B}^i)}{\mathrm{tr}(e^{-\beta H})}$$

The traceless matrix

$$\widehat{\Omega}_\beta^i(b) = \Omega_\beta^i(b) - \mathrm{tr}\big(\Omega_\beta^i(b)\big)\mathrm{Id} \tag{12}$$

is called the reduced vorticity matrix at site i.

Example: $\Lambda = \{1,2\}$ is a lattice with 2 sites, $\partial\Lambda = \emptyset$, one has $\widehat{\Omega}_\beta^1(\delta) = \widehat{\Omega}_\beta^2(\delta) = 0$. This is observed also numerically for all Λ, with $\partial\Lambda = \emptyset$, although vortices could merge sponteanously in infinite volume. If b is a symmetric basis of $o_{\mathbf{R}}$, then $\Omega_\beta^i(b)$ and $\widehat{\Omega}_\beta^i(b)$ are Hermitean since H is self-adjoint (real symmetric if moreover H has real coefficients), and

$$\big(\widehat{\Omega}_\beta^i(b)\big)^2 = \det\widehat{\Omega}_\beta^i(b)\mathrm{Id} \tag{13}$$

Thus $\widehat{\Omega}_\beta^i(b)$ enjoys the nice property, to be diagonalizable with real (opposite) eigenvalues for all sites i, and all inverse temperature β. Viewing these as a field of matrices over the lattice, we can figure out the "vorticity" of the system, by simply looking at their principal directions. We will see that it also gives a measure of vorticity, i.e. numbers (integers) that should be independent of the choice of "elementary" basis b. Next we define vortices as the set of sites where the reduced vorticity matrix is singular.

Definition 1.3: *We say that $\xi \in \Lambda$ is a vortex at inverse temperature β relative to the δ-symmetric ONB b iff $\Omega_\beta^\xi(b)$ has a degenerate eigenvalue, i.e. $\widehat{\Omega}_\beta^\xi(b) = 0$. We call regular the* *other points.*

By construction, all sites are vortices when $\beta = 0$.

Now we turn to consistency of Definitions 1.2 and 1.3 relatively to the choice of b within δ-symmetric basis. That b is a δ-symmetric basis is a natural requirement for computing the degree of $\widehat{\Omega}^i(b)$, see Sect.2. With b written as in (8), and $P \in O(2;\mathbf{R})$, we set with the notations of (10) $a = {}^tPbP$. The same holds after taking the co-product Δ of each term, i.e. $\widetilde{a} = {}^tP\widetilde{b}P$. This defines conjugacy classes, which pass to the partial traces (9), i.e. $T(a) = {}^tPT(b)P$, and $T(\widetilde{a}) = {}^tPT(\widetilde{b})P$. Moreover, if $X \in \mathcal{L}(\mathbf{R}^2)$, we have

$$(1 \otimes X)b = \begin{pmatrix} Xb_1 & Xb_2 \\ Xb_3 & Xb_4 \end{pmatrix} = (1 \otimes X)Pa\,{}^tP \tag{14}$$

After lifting $\widetilde{a}$ and $\widetilde{b}$ to $\widetilde{\mathcal{O}}_{\mathbf{R}} \otimes \widetilde{\mathcal{O}}_{\mathbf{R}}$, (11) becomes

$$\widetilde{A}^i = {}^tP\begin{pmatrix} \widetilde{B}_1^i & \widetilde{B}_2^i \\ \widetilde{B}_3^i & \widetilde{B}_4^i \end{pmatrix} P \in \mathcal{M}_{4N\times 4N}(\mathbf{R}), \quad \mathrm{tr}_1(\widetilde{A}^i) = {}^tP\begin{pmatrix} \mathrm{tr}\widetilde{B}_1^i & \mathrm{tr}\widetilde{B}_2^i \\ \mathrm{tr}\widetilde{B}_3^i & \mathrm{tr}\widetilde{B}_4^i \end{pmatrix} P \in \mathcal{M}_{2\times 2}(\mathbf{R}) \tag{15}$$

and (14) also extends when taking $X \in \mathcal{L}(\mathbf{R}^{4N})$ and replacing a by $\widetilde{A}^i$, b by $\widetilde{B}^i$. Let now $X = e^{-\beta H}$, we obtain that conjugacy classes again pass to vorticity matrices, i.e.

$$\Omega_\beta^i(a) = {}^tP\Omega_\beta^i(b)P, \quad \widehat{\Omega}_\beta^i(a) = {}^tP\widehat{\Omega}_\beta^i(b)P \tag{16}$$

and (10) eventually gives :

Proposition 1.4: *Definitions 1.2 and 1.3 are consistent, i.e. vorticity matrices relative to all δ-symmetric ONB b are related by* (16) *for some $P_s \in O(2;\mathbf{R})$, and in particular ξ is a vortex relative to δ iff this is a vortex relatively to any δ-symmetric b.*

Moreover we have the numerical evidence that, among all δ-symmetric basis b, the canonical basis δ is most "faithful", in the sense that $\Omega_\beta^i(\delta)$ have on the boundary lattice $\partial\Lambda$ the same principal directions as the directions along which Pauli matrices are compressed (associated with the eigenprojector Π_j).

b) Heisenberg model

The algebra goes essentially along the same lines, except for the fact that the basis b cannot be real. Again, this relies on the observation that the spin representation of SU(2) is 2-D, so the "one-point functions" can be simply parametrized by 2×2 matrices. Instead of 4×4 array $b = \begin{pmatrix} \delta_1 & \delta_2 \\ \delta_3 & \delta_4 \end{pmatrix}$, we consider the 8×8 array $e = \begin{pmatrix} e_1 & e_2 \\ e_3 & e_4 \end{pmatrix} = \begin{pmatrix} e_1' & e_2' + ie_2'' \\ e_3' - ie_3'' & e_4' \end{pmatrix}$ with

$$e_1' = \begin{pmatrix} \delta_1 & 0 \\ 0 & 0 \end{pmatrix}, e_2' = \begin{pmatrix} 0 & \delta_2 \\ 0 & 0 \end{pmatrix}, \quad e_3' = (e_2')^* \tag{17}$$

$$e_4' = \begin{pmatrix} 0 & 0 \\ 0 & \delta_4 \end{pmatrix}, e_2'' = \begin{pmatrix} \delta_2 & 0 \\ 0 & 0 \end{pmatrix}, \quad e_3'' = (e_2'')^* \tag{18}$$

this choice being non-unique. Moreover the e_j's have the right dimension for quadratic interaction, so we don't need to take co-product as in the case of XY model. Compression of Pauli matrices on $\partial\Lambda$ can be obtained by the orthogonal projector

$$\Pi(\theta,\varphi)=\begin{pmatrix}\cos^2\frac{\theta}{2} & e^{-i\varphi}\sin\frac{\theta}{2}\cos\frac{\theta}{2}\\ e^{-i\varphi}\sin\frac{\theta}{2}\cos\frac{\theta}{2} & \sin^2\frac{\theta}{2}\end{pmatrix} \tag{19}$$

where (θ,φ) parametrizes a point on Bloch sphere, accounting for the prescribed direction of "vorticity" on $\partial\Lambda$.

For both XY and Heisenberg model, we observe that (reduced) vorticity matrices belong to $\mathrm{su}(2)=\{M\in\mathcal{M}_{2\times 2}(\mathbf{C}): M^*=M,\ \mathrm{tr}M=0\}$ which is the tangent Lie algebra of SU(2).

2. Topological degree and holonomy on SU(2)

The natural idea is to consider vorticity matrices as a map $\Lambda\cup\partial\Lambda\to\mathrm{su}(2)$ and "integrate" it, so to get some topologial invariant, such as the local degree. This assumes a thorough knowledge of discrete analysis on the lattice, with values in su(2). For advanced results on Differential Calculus on lattices in the scalar case, see [15]. The non-commutative discrete case has still to be set up, so we pass here to an idealistic continuous limit, where vorticity matrices would be defined as a smooth field on $\mathbf{R}^2$ (away from vortices), valued in the Lie algebra su(2). Our purpose is to integrate such fields vanishing at some points, and define a "non-commutative degree".

We consider Heisenberg model (Hermitean vorticity matrices), the XY model (real symmetric vorticity matrices) will be treated as a particular case. So let $M: D\subset\mathbf{R}^2\to\mathrm{su}(2), x\mapsto M(x)$ be a C^1 map, $M(x)^2=\lambda(x)\mathrm{Id}$, $\lambda(x)\geq 0$, and consider $\rho\in\Lambda^1(\mathbf{R}^2;\mathrm{su}(2))$ the 1-form

$$\rho(x)=\frac{1}{2}(M^{-1}(x)dM(x)-dM(x)M^{-1}(x)) \tag{20}$$

(anti-symmetrized Maurer-Cartan form). We have $M^{-1}(x)dM(x)+dM(x)M^{-1}(x))=\frac{d\lambda(x)}{\lambda(x)}$.

Let $M=\begin{pmatrix}a & b\\ c & -a\end{pmatrix}$, we compute

$$d\rho=-\lambda^{-2}(adb\wedge dc+bdc\wedge da+cda\wedge db)M \tag{21}$$

(so $d\rho=0$ if M is symmetric). On the other hand, computing the structure coefficients for the Lie algebra su(2), we find

$$d\rho+[\rho,\rho]=0 \tag{22}$$

Recall that if G is a Lie group, and $\mathcal{G}$ its Lie algebra, ω the canonical Maurer-Cartan form on G, invariant by left translations, we define Darboux differential of the map $f\in C^1(D;G)$ by $\pi_f=f^*\omega$. The fundamental existence theorem ("Poincaré lemma"), with a differential form $\rho\in\Lambda^1(D;\mathcal{G})$ verifying (22), assigns (locally) a map $f\in C^1(D;G)$, whose Darboux differential is precisely equal to ρ. Moreover this map is unique when prescribing its value at a point $x_0\in D$. Applying this result to (20), gives local primitives of ρ, whenever $\lambda(x)\neq 0$, called a "logarithm" of M, which belong to SU(2).

Remark: For these computations we can also use the isomorphism $\theta:(\mathbf{R}^3,\wedge)\to(\mathrm{su}(2),\frac{i}{2}[\cdot,\cdot])$, where $\wedge$ is the usual wedge-product on $\mathbf{R}^3$.

For the XY model on $\mathbf{R}^2$ ("idealized" lattice $\mathbf{Z}^2$) we can define the square of the "local degree" at an isolated singularity (vortex) $\xi\in\mathbf{R}^2$, by

$$s_\xi^2=\det\frac{1}{2\pi}\int_\gamma\rho(x) \tag{23}$$

IOP Conf. Series: Journal of Physics: Conf. Series **804** (2017) 012031 doi:10.1088/1742-6596/804/1/012031

where γ encircles ξ, and similarly, when M elliptic at infinity $|\lambda(x)| \geq C > 0$, $|x| \geq r_0$ the square of the "total degree"

$$s_\infty^2 = \det \frac{1}{2\pi} \int_{|x|=r} \rho(x), \quad r > r_0 \tag{24}$$

Since the fundamental group of the universal covering of SU(2) is $\mathbf{Z}$, we can conclude that $s_\xi, s_\infty \in \mathbf{Z}$.

Example: For the symmetric matrices

$$M_a(x) = \begin{pmatrix} a\cos n\theta & \sin n\theta \\ \sin n\theta & -a\cos n\theta \end{pmatrix}$$

we have $\deg_\infty(M) = n$. The 1-form ρ associated wih M_0 is simply $\begin{pmatrix} 0 & n \\ -n & 0 \end{pmatrix} d\theta$.

This makes sense also for Heisenberg model on Riemann's sphere $\mathbf{S}^2$, provided $\lambda(x) > 0$ everywhere, but we should speak of "instantons" rather than of "vortices", see [2], [7].

Example: For Hermitean matrices

$$M_a(x) = \begin{pmatrix} a\cos\theta & e^{id\varphi}\sin\theta \\ e^{-id\varphi}\sin\theta & -a\cos\theta \end{pmatrix} \tag{25}$$

where $x = \begin{pmatrix} \cos\theta & e^{i\varphi}\sin\theta \\ e^{-i\varphi}\sin\theta & -\cos\theta \end{pmatrix} \approx (\theta, \varphi) \in \mathbf{S}^2$, we have $\deg_\infty(M) = d$. The condition $M(0,\varphi) = -M(\pi,\varphi)$ reproduces the condition of [2] that all (classical) spins point upwards at infinity ($\theta = 0$) while they point downwards at 0 ($\theta = 0$), so that the equilibrium state at inverse temperature β is a spin wave, or "instanton", of degree d (the number of coverings of the sphere).

Other topological defects, such as "lines of vortices" occur in Heisenberg model on $\mathbf{R}^3$ ("idealized" lattice $\mathbf{Z}^3$).

Among possible extensions, we mention: (1) the orbital compass model, which has reflection posivity, but no rotation invariance [4]; (2) Hubbard model with continuous symmetry (hopping and spin interaction) [10], with an application to the dynamics of Cooper pairs in a supraconductor bulk, or the dynamics of electron/hole pairs in SNS junctions. Vorticity can also defined for maps $\mathbf{R}^{(2n+1)n} \to \mathrm{sp}(2n;\mathbf{R})$ (Hamiltonian matrices) and as well in general gauge sigma models [6].

3. Numerical tests for the XY model

Recall we have completed the lattice Λ with an environment $\partial\Lambda \subset \mathbf{Z}^2$ where Pauli matrices are compressed in directions $(\theta_j)_{j\in\partial\Lambda}$, i.e. we change σ by $\Pi_\theta\sigma\Pi_\theta$, where Π_θ as in (6). Thus

$$\sigma_i^x(\theta_i) = (\sin 2\theta_i)\Pi_{\theta_i}, \quad \sigma_i^y(\theta_i) = 0$$

Hamiltonian (1) with nearest neighbor interaction has too large a kernel, to be suitable for numerical simulations, even when modified by an external field. As in QFT we could try to remove the "artificial" part of $\ker H$ by reducing the Hilbert space $\mathcal{H} = \mathbf{C}^{4N}$ to a "physical space", but H is not positive in the form sense. So a first attempt to lift the degeneracy of the spectrum of the Hamiltonian, and enhance the effects of the external field on vorticity, is to change (1) to the anisotropic XY model. So for $n, k > 0$, we consider

$$H_{(n,k)}(\sigma|\partial\Lambda) = -\frac{1}{2(n+k)} \sum_{\langle i,j\rangle; i,j\in\Lambda} (n\,\sigma_i^x \otimes \sigma_j^x + k\,\sigma_i^y \otimes \sigma_j^y) \quad (26)$$
$$-\frac{1}{2(n+k)} \sum_{\langle i,j\rangle;(i,j)\in\Lambda\times\partial\Lambda} n\,(\sigma_i^x \otimes \sigma_j^x(\theta_j) + \sigma_j^x(\theta_j) \otimes \sigma_i^x) - \frac{1}{2(n+k)} \sum_{\langle i,j\rangle; i,j\in\partial\Lambda} n\,\sigma_i^x(\theta_i) \otimes \sigma_j^x(\theta_j)$$

so $H_{(n,k)}(\sigma|\partial\Lambda)$ is self-adjoint and real. For $k \neq 1$, we call $H_{(1,k)}(\sigma|\partial\Lambda)$ the *anisotropic XY model.* Only when $\partial\Lambda = \emptyset$, $H_{(1,k)}$ is unitarily equivalent to $H_{(k,1)}$. We consider rectangular lattices of minimal sizes to exclude important volume effects, with sufficiently large $\partial\Lambda$ to constrain the "quantum system" within Λ. We choose $\theta_j = d\omega_j + \phi$ where ω_j is the polar angle representing the vector $j \in \partial\Lambda$.

We study Gibbs state at inverse temperature β, with significant results provided β ranges in some interval, for which however, there is no evidence (even in an approximate sense) of a second order phase transition. Computing $\frac{1}{2\pi}\int_\gamma \rho$ as a discrete integral along a contour $\gamma \in \Lambda$, not too far from $\partial\Lambda$ (in practice, 2 or 3 layers), it turns out that the computed degree is close to this we would obtain in (25). The main flaw affecting the computations is due to the fact that eigenvalues of $\widehat{\Omega}^i_\beta(b)$ are decaying exponentially when approaching the center of Λ. This we partially compensate by considering the anisotropic model. To make our vorticity patterns more demonstrative we consider the case of high anisotropy with $k = 10$. In Table 1 below we give the results for degree, computed along a cycle $\gamma \subset \Lambda$ consisting of the rectangle of the first or the second neighbors to the boundary for different values of the anisotropy parameter $n = 1, k = 2; 10$. Inverse temperature is $\beta = 1$.

Table 1. Table of calculated degree for different values of anisotropy factor k for the first and the second neighbors to the boundary. Here the case of $|\Lambda \cup \partial\Lambda| = 23 \times 33$ with 2 boundary layers is considered, and $\beta = 1$.

Given degree	$k = 2$ 1st *neighbours*	$k = 10$ 1st *neighbours*	$k = 2$ 2nd *neighbours*	$k = 10$ 2nd *neighbours*
1	1.05	1.09	0.89	1.05
2	1.98	2.03	1.70	1.78
3	2.76	2.75	2.01	2.50

Let us finally discuss the antiferromagnetic model. It is known that on $\mathbf{Z}^2$, the unitary transformation U consisting in flipping the spins at sites i with i odd (i.e. indices $i = (i_1, i_2)$ such that $|i| = |i_1| + |i_2|$ is odd) intertwines the ferro with the antiferromagnetic models. More precisely, $-H = U^*HU$. The reason is that $\mathbf{Z}^2_e$ and $\mathbf{Z}^2_o$ (the even and odd lattices) are swapped into each other by symmetries on the lines $x = n + 1/2$ or $y = m + 1/2$ (called the "chessboard symmetry"). There follows that $\mathrm{tr}\exp[\beta H]A = \mathrm{tr}\exp[-\beta H]UAU^*$, and if $A = \widetilde{D}^i$ (the canonical basis), we can check $UAU^* = A$ so the matrices of vorticity (for the Hamiltonian with free boundary conditions) are the same. This equivalence holds also in the case of the torus, but not on $\Lambda \subset \mathbf{Z}^2$ with an odd number of sites. Of course, when $\partial\Lambda \neq \emptyset$, H and $-H$ are not so simply related; nevertheless, we may observe (numerically) that the relation $\Omega^i_\beta(\delta) = \Omega^i_{-\beta}(\delta)$ holds with a very good accuracy.

[1] Aschbacher W and Pillet C A 2003 *J. Stat. Phys.* **112** 1153
[2] Belavin A and Polyakov A 1975 *JETP Lett.* **22** **10** 245
[3] Berezinskii V.L 1975 Sov. Phys. *JETP* **32** 493
[4] Biskup M, Chayes L and Starr S 2007 *Commun. Math. Phys.* **269** **3** 611
[5] Bricmont J, Fontaine J and Landau J *Commun. Math. Phys.* **56** 281

IOP Conf. Series: Journal of Physics: Conf. Series **804** (2017) 012031 doi:10.1088/1742-6596/804/1/012031

[6] Cieliebak K and Salamon D 2002 *J. Symplectic Geom.* **1** **3** 543
[7] El-Bouanani H and Rouleux M 2007 *J. Stat. Physics.* **28** **3** 741
[8] Fröhlich J and Lieb E 1978 *Commun. Math. Phys.* **60** 233-267
[9] Fröhlich J and Spencer T 1981 *Commun. Math. Phys.* **81** 527
[10] Koma T and Tasaki H 1992 *Phys. Rev. Letters* **68** **21** 3248
[11] Kosterlitz J.M and Thouless D.J 1973 *J. Phys. C* **6**, 1181
[12] Malliavin P 1972 *Géométrie Différentielle Intrinsèque* (Hermann, Paris)
[13] Messager A, Miracle S and Pfister C 1978 *Commun. Math. Phys.* **58** 19
[14] Simon B 1993, *The Statistical Mechanics of Lattice Gas I* (Princeton Univ. Press)
[15] Smirnov S *Discrete complex analysis and probability*
http://www.unige.ch/ smirnov/slides/slides-hyderabad.pdf

Modular Spacetime and Metastring Theory

Laurent Freidel[a], **Robert G Leigh**[b] **and Djordje Minic**[c]

[a]Perimeter Institute for Theoretical Physics, Waterloo ON, N2L 2Y5, Canada
[b]Department of Physics, University of Illinois, Urbana IL 61801, U.S.A.
[c]Department of Physics, Virginia Tech, Blacksburg, VA 24061, U.S.A.

E-mail: `lfreidel@perimeterinstitute.ca, rgleigh@uiuc.edu, dminic@vt.edu`

Abstract. In this talk we review our recent work on metastring theory and its habitat, a new form of quantum spacetime, called modular spacetime. We emphasize that the geometry underlying modular spacetime, i.e. the background geometry of metastring theory is also the geometry underlying any quantum theory as formulated in terms of Aharonov's modular variables. Thus the metastring sheds light on the foundations of quantum theory, and it represents a new formulation of string theory and quantum gravity based on the principle of relative locality.

1. Introduction

Quantum gravity is arguably the theoretical conundrum *par excellence*. There are many approaches to this problem, and little in the way of broad consensus. For example, one might quantize metric fluctuations as one quantizes other quantum fields, but then one is faced with the problem of non-renormalizability, that some researchers try to avoid by looking for a UV fixed point (asymptotic safety) [1]. One might attempt to quantize geometry, as is done in the context of loop quantum gravity [1], where one emphasizes background independence and where space is fundamentally discrete, but where the challenge is in the continuum limit and in dynamics, as well as in the reconciliation with the general relativity principle outside the classical limit. One might attempt the extrapolation of quantum field theory, as offered by canonical string theory [2], where spacetime is probe-dependent, but where the quantum or discrete notion of spacetime is hard to grasp. One might attempt emergent models of quantum gravity based on causal dynamical triangulation [3], or causal sets [4], or condensed matter inspired approaches [5], or violation of Lorentz symmetry [6], or holography via the AdS/CFT correspondence [7]. One might attempt to formulate intrinsic non-commutative geometry of spacetime and matter [8], or look at auxiliary spaces, such as twistor spaces, which encode space-time in a non-local, but intrinsically geometric manner [9].

We take the lessons of all these approaches very seriously in our new formulation of string theory and quantum gravity that we call *metastring theory*, developed in [10, 11, 12, 13, 14]. In particular, we take into account the lessons of geometry (the view of general relativists interested in quantum gravity) and the lessons of the quantum (the view of quantum field theorists and string theorists interested in quantum gravity) very seriously and proceed with a fresh look at the problem. One message that is common to all these various approaches to quantum gravity is that of *non-locality*. There are non-local observables in background independent approaches, there are non-local probes such as strings or twistors, there is non-locality of holography or discrete space-time etc etc. Moreover, there is intrinsic non-locality in general relativity because of diffeomorphism symmetry, and there is quite different non-locality in quantum theory due to its linearity (leading to interference and entanglement).

In this talk we will see that the language of string theory is particularly appropriate for our unified approach, because it offers a view of both quantum theory of gravity and quantum theory of matter. According to canonical, textbook [2], string theory the underlying theory is quantum mechanically consistent (it is a two-dimensional worldsheet theory) and the worldsheet fields are interpreted as coordinatizing a target spacetime. This means, that the geometrical structures (such as the target space metric) are viewed as states of the string. In this approach to quantum gravity one uses a built-in interpretation: there is an emergent target space-time theory that at low energies reproduces local effective quantum field theories, and also includes (quantum) gravity. This is nice, but it cannot be the whole story. Our first insight towards a novel formulation of quantum gravity is the following observation. Any theory of gravity contains two parameters $\hbar$ and G_N which can be viewed as conversion parameters, based on two scales: the Planck constant is the product of fundamental length and fundamental energy scales, and the Newton constant is the ratio of the same. If we want to probe quantum gravity in all regimes, we have to take this into account. (Note that in string theory, the Newton constant involves both a fundamental scale of the string, α', as well as the string coupling. In that case, it is α' that is given as the ratio of the fundamental length and fundamental energy.) Also, usually G_N is treated as a coupling, and not a conversion parameter. However, in general relativity due to the equivalence principle, G_N can be understood as a universal inverse tension, and thus, as a conversion parameter between spacetime and energy-momentum space. This in turn leads to observer-dependent spacetimes and energy-momentum spaces and, in principle, to their unification in a new geometric structure associated with a relativistic phase space of individual quantum probes of spacetime physics. The running of a dimensionful coupling such as G_N is always fixed in terms of another dimensionful parameter. We choose Planck units as units of energy and time, where G_N does not run. Thus, we are extending the equivalence principle to quantum probes.

Given this insight, we also need to take into account that the concept of locality might be observer-dependent. We speak of *absolute locality* when spacetime is treated as an arena, independent of physical probes. Likewise, we speak of *relative locality* [15, 16] when spacetime depends on the nature of a probe (e.g. its energy). In metastring theory, relative locality is implemented by regarding spacetime as a subspace of the target space of the string. In other words, spacetime can be thought of as a choice of basis, that is, a choice of polarization of phase space. This means that *Born reciprocity* [17] can be built into the theory: a given space-time is only one choice of polarization, and energy-momentum space is on the same footing as spacetime, within a fully consistent quantum gravity. This ties nicely with the insight about G_N being a conversion parameter (which puts it on the same footing as $\hbar$ and c). Finally, we will see that there is a particular geometric structure that takes into account these insights, the one we call *Born geometry.*

In what follows, we present an implementation of these insights in string theory, in what we call a metastring formulation and then find, quite surprisingly, that the habitat of metastring theory is a new quantum space, called modular spacetime, which reveals novel geometric structures behind quantum theory, thus shedding light on the origin of quantum theory, string theory and quantum gravity.

2. Born Reciprocity, String theory and T-duality

Born reciprocity is the simple observation that the change of basis from the position to the energy basis ($P \to Q$ and $Q \to -P$) is a fundamental kinematical symmetry of quantum mechanics (QM). However, Born reciprocity is broken in general relativity (GR). In GR there

IOP Conf. Series: Journal of Physics: Conf. Series **804** (2017) 012032 doi:10.1088/1742-6596/804/1/012032

is a preferred polarization: the spacetime one. Thus Born reciprocity is broken by the choice of the principle of absolute locality, the statement that the same spacetime is shared by all probes independently of their energy state or history. From this viewpoint there exists a fundamental tension between GR and QM. On one hand GR admits curvature and the spacetime geometry can "flex" at will. On the other hand QM is algebraic and its underlying structure is rigid. From the geometrical point of view, the superposition principle of QM implies that its geometry is projective. That is, the QM geometry is almost linear and constantly curved. Unifying the two geometries (the dynamical GR geometry and the rigid QM geometry) requires to "gravitize quantum mechanics". In order to render QM more flexible, we need to relax its geometry. We will see that this is one of the messages of metastring theory.

We start by emphasizing the role of Born reciprocity, or duality (as a motivation for relative locality) in string theory. We begin our discussion [10] by examining the Polyakov action coupled to a flat metric h [2]

$$S_P(X) = \frac{1}{4\pi}\int_\Sigma h_{\mu\nu}(*\mathrm{d}X^\mu \wedge \mathrm{d}X^\nu), \tag{1}$$

where $*, \mathrm{d}$ denote the Hodge dual and exterior derivative on the worldsheet, respectively. We generally will refer to local coordinates on Σ as σ, τ, while it is traditional to interpret X^μ as local coordinates on a target space M, here with Minkowski metric $h_{\mu\nu}$. Since we are in Lorentzian signature, $*\mathrm{d}\tau = \mathrm{d}\sigma$ and $*^2 = 1$. Note that S_P has dimensions of length-squared if we take X^μ to have dimension of length, so appears in the path integral as e^{iS_P/λ^2}. λ is the string length which is related to the slope parameter by $\lambda^2 \equiv \alpha'\hbar$, where $\hbar$ is the Planck constant of the worldsheet quantum theory. With this definition S_P/λ^2 has the usual coefficient $\frac{1}{4\pi\alpha'}$ in units of $\hbar$. In order for the Polyakov action to be well-defined, one must demand that the integrand be single-valued on Σ. For example, on the cylinder $(\sigma,\tau) \in [0, 2\pi] \times \mathbb{R}$ it would be sufficient that $\mathrm{d}X^\mu(\sigma,\tau)$ is periodic[1] with respect to σ with period 2π. However, and this is a crucial point, this does not mean that $X^\mu(\sigma,\tau)$ has to be a periodic function, even if M is non-compact. Instead, it means that X^μ must be a *quasi-periodic* function which satisfies $X^\mu(\sigma + 2\pi, \tau) = X^\mu(\sigma,\tau) + \delta^\mu$. Here δ^μ is the quasi-period, or monodromy, of X^μ. If δ^μ is not zero, there is no *a priori* geometrical interpretation of a closed string propagating in a flat space-time – periodicity goes hand-in-hand with a space-time interpretation. Of course, such monodromies are encountered in textbook string theory [2] if the background is compact. However, we will see later in this talk that in the metastring formulation such monodromies are generic, even for non-compact backgrounds.

Now, let us consider the dimensionless momentum and position densities:

$$P \equiv \partial_\tau X/\lambda, \quad Q \equiv \partial_\sigma X/\lambda. \tag{2}$$

The dynamics of the canonical free string is characterized by the following Hamiltonian and differemorphism constraints:

$$H \equiv P^2 + Q^2 = 0, \quad D \equiv P \cdot Q = 0. \tag{3}$$

This is symmetric under the exchange $P \to Q, Q \to -P$, as noticed by Veneziano [18]. But this is just the statement of Born reciprocity: physics should be formulated in a way that is democratic from the position and momentum point of view [17]. This concept of Born reciprocity can be precisely related to one of the main features of string theory, to wit, that of T-duality [2]. In toroidal compactifications, the string does not distinguish radius R from α'/R: the spectrum of the string is simply, relabeled, with momentum and winding modes being exchanged. The

[1] The most general condition would be to ask that $\mathrm{d}X^\mu(\sigma+2\pi) = \Lambda^\mu_\nu \mathrm{d}X^\nu(\sigma)$ where Λ is a Lorentz transformation. For simplicity, in this talk we only consider the case where $\Lambda = 1$.

usual string solutions (the solution of the 2d wave equation) are chiral (i.e. left (L) and right (R)) combinations

$$X(\tau, \sigma) = X_L(\tau - \sigma) + X_R(\tau + \sigma). \tag{4}$$

In order to discuss T-duality we consider the other, dual, combination

$$\tilde{X}(\tau, \sigma) = X_L(\tau - \sigma) - X_R(\tau + \sigma). \tag{5}$$

Thus T-duality exchanges momentum density (P) with the position density (Q). As such, T-duality is a Born duality.

We can make this observation more precise. T-duality can be thought of as acting non-trivially on a choice of polarization. In fact, it can be thought of as a "Fourier transform" [10]. Usually in toroidal compactifications, we interpret short distance (radius) as long distance in a dual spacetime. However, we can also directly relate this to a Fourier transform. In particular, consider a string state

$$\Psi[x(\sigma)] = \int_{X|\partial\Sigma=x} [DXDg] e^{iS_P[X]/\lambda^2}. \tag{6}$$

Define a Fourier transform of this state by

$$\tilde{\Psi}[y(\sigma)] \equiv \int [Dx] e^{i/\hbar \int_{\partial\Sigma} x^\mu dy_\mu} \Psi[x(\sigma)]. \tag{7}$$

Extending $y(\sigma)$ to the worldsheet, we integrate out X to obtain a *dual* Polyakov path integral

$$\tilde{\Psi}[y(\sigma)] = \int_{Y|\partial\Sigma=y} [DYDg] e^{-iS_P[Y]/\epsilon^2}. \tag{8}$$

Note that $\alpha' p = \int_C *dX$ and $\delta = \int_C dX$ exchange roles, so that in the dual description $\delta = \alpha' \int_C *dY$ and $p = \int_C dY$.

Metastring theory developed in our papers [10, 11, 12, 13, 14], represents a reformulation of the Polyakov (bosonic) string theory, written in such a way that T-duality acts linearly on coordinates and that by so doing spacetime and energy-momentum space are put on equal footing in a unified phase space formulation. This re-interpretation of the string path integral is well defined as is the usual quantum string theory. The principal difference, however, is that in the case of metastring theory, the boundary conditions are relaxed, allowing for general monodromies $X^\mu(\sigma + 2\pi, \tau) = X^\mu(\sigma, \tau) + \delta^\mu$.

3. Metastring Theory: classical and quantum

At the classical level, metastring theory [10, 11, 12, 13, 14] can be thought of as a formulation in which the target space is doubled in such a way that T-duality acts linearly on the coordinates. This doubling means that momentum and winding modes appear on an equal footing. We refer to the target space as a phase space since the metastring action requires the presence of a background symplectic form ω. The metastring formulation also requires the presence of geometrical structures that generalize to phase space the spacetime metric and the B-field. In fact, in the metastring we have not one but *two* notions of a metric. The first metric η is a neutral metric that defines a bi-Lagrangian structure and allows to define the classical spacetime as a Lagrangian sub-manifold[2] — more precisely, the classical space-time is defined as a null subspace for η which is also Lagrangian for ω. The second metric H is a metric of signature

[2] We remind the reader that in symplectic geometry, a Lagrangian subspace is a half-dimensional submanifold of phase space upon which the symplectic form pulls back to zero. In simple terms, a Lagrangian submanifold might be the subspace coordinatized by the q's within the phase space coordinatized by q's and p's.

$(2, 2(D-1))$ that encodes the geometry along the classical spacetime (of dimension D) as well as the transverse energy-momentum space geometry. In this formulation, T-duality exchanges the Lagrangian sub-manifold with its image under $J = \eta^{-1}H$. Classical metastring theory is defined by Tseytlin's action [19, 20, 21]

$$\hat{S} = \frac{1}{4\pi}\int_{\Sigma} d^2\sigma\Big(\partial_\tau \mathbb{X}^A(\eta_{AB} + \omega_{AB})\partial_\sigma \mathbb{X}^B - \partial_\sigma \mathbb{X}^A H_{AB}\partial_\sigma \mathbb{X}^B\Big), \tag{9}$$

where $\mathbb{X}^A$ are dimensionless coordinates on phase space and the fields η, H, ω are all dynamical (i.e., in general dependent on $\mathbb{X}$) phase space fields. In the context of a flat metastring we have constant η_{AB}, H_{AB} and ω_{AB}

$$\eta_{AB} \equiv \begin{pmatrix} 0 & \delta \\ \delta^T & 0 \end{pmatrix}, \quad H_{AB} \equiv \begin{pmatrix} h & 0 \\ 0 & h^{-1} \end{pmatrix}, \quad \omega_{AB} = \begin{pmatrix} 0 & \delta \\ -\delta^T & 0 \end{pmatrix}, \tag{10}$$

where δ^μ_ν is the d-dimensional identity matrix and $h_{\mu\nu}$ is the d-dimensional Lorentzian metric, T denoting transpose.

In this formulation it is convenient, as suggested by the double field formalism [22, 23], to introduce dimensionless coordinates $\mathbb{X}^A \equiv (X^\mu/\lambda, P_\mu/\varepsilon)^T$ on phase space, where λ and ϵ represent the fundamental spacetime and energy-momentum scales. As already stated, $\hbar = \lambda\epsilon$ and $\alpha' = \frac{\lambda}{\epsilon}$. Given a pair (H, η) it is natural to consider the operator $J \equiv \eta^{-1}H$. The consistency of string theory requires J to be a chiral structure, that is, a real structure ($J^2 = 1$) compatible with η, implying that J is an $O(D, D)$ transformation (realizing generalized T-duality in target space). These three structures, the symplectic $Sp(2D)$ ω, the $O(D, D)$ η and the $SO(2, 2(D-1))$ H, define the new concept of Born geometry [10, 11, 12, 13, 14] which unifies the complex geometry of quantum theory with the metrical geometry of general relativity and the symplectic geometry of canonical Hamiltonian dynamics [24, 25, 26, 27, 28]. Note that in the phase space formulation the local phase space coordinates $\mathbb{X}$ are *quasiperiodic*

$$\mathbb{X}^A(\sigma + 2\pi) = \mathbb{X}^A(\sigma) + \Delta^A, \tag{11}$$

where Δ^A is the corresponding quasiperiod (which either vanishes for the canonical Polyakov string or is given by the winding number in the usual treatment of T-duality on compact spaces [2].)

The worldsheet Σ is a topological surface of genus g which possesses n boundaries. The first important point is that, in order to formulate the Tseytlin action we need to distinguish between σ and τ. This means that (the Lorentzian worldsheet) Σ is equipped with what we call a *causal structure*: That is, we assume that there exists a time function $\tau : \Sigma \to \mathbb{R}$ such that τ is a Morse function and such that $\partial\Sigma = \partial\Sigma_- \cup \partial\Sigma_+$ where $\partial\Sigma_\pm = \tau^{-1}(\pm\infty)$ [12]. By demanding that Tseytlin's action is well defined on Σ we have to impose that $\partial_\tau\mathbb{X}$ and $\partial_\sigma\mathbb{X}$ are single valued on Σ, i.e. they are periodic. But as we already emphasized, this *does not mean* that $\mathbb{X}$ is a periodic function. This is a crucial point on which our generalization rests. The proper mathematical implementation of this idea is that instead of parametrizing the action by a sets of coordinates $\mathbb{X}^A$ on $\mathcal{P}$, we need to parametrize it by a *closed one form* $\delta^A = \delta^A_\sigma \mathrm{d}\sigma + \delta^A_\tau \mathrm{d}\tau$ valued into the tangent space $T\mathcal{P}$, $\mathrm{d}\delta^A = 0$. Such a form possesses *monodromies*. For each cycle γ we have

$$\Delta^A_\gamma = \int_\gamma \delta^A. \tag{12}$$

Since δ^A is closed the monodromy depends only on the homology of C. In particular, this means that if the string goes through an interaction point and splits, then the monodromy after the

IOP Conf. Series: Journal of Physics: Conf. Series **804** (2017) 012032 doi:10.1088/1742-6596/804/1/012032

splitting is the sum of the monodromies before the split, $\Delta_{\gamma_{12}} = \Delta_{\gamma_1} + \Delta_{\gamma_2}$. Therefore, the set of monodromies should form a lattice Γ. That is, if $\Delta, \Delta' \in \Gamma$ then $\Delta + \Delta' \in \Gamma$. This implies that the Tseytlin action on a generic surface depends on monodromies $\Delta \in \Gamma$ and that the flat Tseytlin action on a generic surface should be written as

$$S_\Delta \equiv \frac{1}{2\pi} \int_\Sigma \left((\eta_{AB} + \omega_{AB}) \delta_\tau^A \delta_\sigma^B - H_{AB} \delta_\sigma^A \delta_\sigma^B \right), \tag{13}$$

where δ^A is a closed form with fixed monodromy $\gamma \in H_1(\Sigma) \to \Delta_\gamma^A$.

Now we can state the quantum definition of the metastring: The quantum metastring amplitude associated with Σ is a functional that depends on n_+ initial configurations $\boldsymbol{x}_i$ and n_+ out configurations $\boldsymbol{x}_o$, where $\boldsymbol{x}(\sigma) = \int_0^\sigma \delta_\sigma^A \mathrm{d}\sigma$ for each boundary circle. It is defined as

$$A_\Sigma(\boldsymbol{x}_i, \boldsymbol{x}_o) \equiv \sum_{\Delta \in \Gamma} \int_C D\tau D\sigma \int_{H^1(\Sigma, \Delta)} e^{iS_\Delta(\delta)} D\delta. \tag{14}$$

Here the sum is over all the monodromies in Γ and the first integral is over the moduli space of causal structure, which is the Lorentzian analog of the moduli space of complex structure by the GWKN theorem [29, 30]. The last integration is over all closed one forms with prescribed holonomies. This prescription is the metastring generalization of Polyakov's prescription for the fundamental string amplitude. There are two interesting things to notice about this amplitude. The first one is that $A_\Sigma(\boldsymbol{x}_i, \boldsymbol{x}_o)$ is necessarily a function periodic with respect to translations along the lattice Γ

$$A_\Sigma(\boldsymbol{x}_i + \Delta, \boldsymbol{x}_o + \Delta') = A_\Sigma(\boldsymbol{x}_i, \boldsymbol{x}_o), \qquad \forall \Delta, \Delta' \in \Gamma. \tag{15}$$

The second one is that the form of the lattice Γ is restricted by the demand of worldsheet diffeomorphism invariance. The requirement that the coupling of the 2-dimensional causal structure to the metastring is invariant under worldsheet diffeomorphisms leads to *two* constraints, the Hamiltonian and diffeomorphism constraints. These can be expressed simply at the classical level as

$$H = \delta_\sigma^A H_{AB} \delta_\sigma^B = 0, \qquad D = \delta_\sigma^A \eta_{AB} \delta_\sigma^B = 0. \tag{16}$$

At the quantum level they imply that

$$\frac{1}{2} \Delta^A \eta_{AB} \Delta^B = N - \tilde{N}, \qquad \frac{1}{2} \Delta^A H_{AB} \Delta^B = N + \tilde{N} - 2, \tag{17}$$

where $N, \tilde{N} \in \mathbb{N}$ are positive integers. Imposing this conditions forces the lattice Γ to be a self-dual even lattice. Note that the constant metric η defines a non-degenerate pairing $X \cdot Y = \mathbb{X}^A \eta_{AB} \mathbb{Y}^B$. Then the first condition of (17) implies that if $\Delta, \Delta' \in \Gamma$ and if we define $\Delta_A \equiv \eta_{AB} \Delta^B$ and $\Delta \cdot \Delta' = \Delta^A \Delta'_A$, then $\Delta \cdot \Delta' = \frac{1}{2}(\Delta + \Delta') \cdot (\Delta + \Delta') - \frac{1}{2} \Delta \cdot \Delta - \frac{1}{2} \Delta' \cdot \Delta' \in \mathbb{Z}$. This means that Δ_A belongs to the dual lattice Γ^*. Note that the second constraints implies that the map $\Delta \mapsto J(\Delta)$ is an endomorphism of Γ to itself [31]. In fact we can show that the condition of modular invariance implies that $\Gamma = \Gamma^*$ [12].

Here is a summary of the crucial features of the metastring:

1) The worldsheet formulation of the metastring is chiral. Thus, even though the fields are doubled the central charges (left and right) are $c_L = c_R = D$ and we still have $D = 26$ for criticality.

2) The metastring is not manifestly invariant under the worldsheet Lorentz transformations.

3) The metastring contains monodromies $\mathbb{X}^A(\sigma + 2\pi) = \mathbb{X}^A(\sigma) + \Delta^A$.

IOP Conf. Series: Journal of Physics: Conf. Series **804** (2017) 012032 doi:10.1088/1742-6596/804/1/012032

4) The usual Polyakov string can be obtained by integrating out the dual $\tilde{X}$, for constant η and H backgrounds, and by supposing that the monodromies are in the kernel of $(\eta - \omega)$.

5) T-duality is implemented in target space by the action of the chiral J operator ($J \equiv \eta^{-1}H$, $J^2 = 1$): $\mathbb{X} \to J(\mathbb{X})$.

6) The target space of the metastring is not spacetime, but, to first order, a chiral phase space $\mathcal{P}$ equipped by the symplectic structure ω, and the bilagrangian structure, and in particular, the polarization metric η which relates to the symplectic connection of the Fedosov deformation quantization [32] and thus leads to the star product of deformation quantization, and finally, the quantum H metric which relates to the complex structure in the context of geometric quantization [33], leading to the concept of Hilbert spaces. This classical Born geometry implements the ideas of Born duality in string theory.

7) The classical equations of motion of the metastring reads

$$\partial_\tau \mathbb{X}^A - (J\partial_\sigma \mathbb{X})^A = 0, \tag{18}$$

also implies the relation between momenta and monodromies $2\pi P = J(\Delta)$. There is *soldering* between worldsheet null coordinates $\sigma^\pm \equiv \sigma \pm \tau$ and the chiral target space structure $\partial_\pm \mathbb{X}^A - (P_\pm \mathbb{X})^A = 0$, where the chiral projector is defined as $2P_\pm = (1 \pm J)$. This allows us to liberate the left geometry from the right geometry (which is reminiscent of twistor theory).

8) The careful analysis of the metastring action [12] shows that its symplectic form is

$$\Omega = \frac{1}{4\pi} \int \delta\mathbb{X}^A \eta_{AB} \nabla_\sigma \delta\mathbb{X}^B, \tag{19}$$

where ∇ is the generalized Fedosov connection found in the Fedosov deformation quantization approach [32].

9) The operator product expansion of the metastring vertex operators

$$V_k = \epsilon_k e^{i\mathbb{K}\mathbb{X}}, \tag{20}$$

leads to the restriction of $\mathbb{K}$ on a double Lorentzian integral lattice Γ, that by modular invariance, must be self-dual. These exist in $D = 2mod(8)$, and are unique. Criticality gives a very unique lattice

$$\Gamma = \Pi_{1,25} \times \Pi_{1,25}. \tag{21}$$

This fact, in turn, leads to the large symmetry structure found by Borcherds in the study of the monstrous moonshine [34, 35][3].

10) As already noted, the metastring is chiral. This requires the introduction of a preferred worldsheet time coordinate which is fundamentally Lorentzian [12]. How can this be consistent with modular invariance? The answer is given by employing the Giddings-Wolpert-Krichever-Novikov (GWKN) construction [29, 30]: given a Riemann surface, provided a choice of local coordinates around punctures is labeled by one scalar, there exists a unique Abelian differentional e with imaginary periods. The real part of this Abelian differential is the modular invariant time $\tau = Re(e)$. The zeros of e represent interaction points where the worldsheet Lorentzian cones double. Cutting the Riemann surface along the real trajectory of e we obtain a string decomposition of the surface. The Nakamura graphs [31] encode this decomposition and give a very effective cell decomposition of moduli space. Thus Nakamura graphs are the natural Feynman diagrams for closed strings [37].

[3] For a string theory related discussion, see [36].

3.1. More General Backgrounds and Born Geometries

Although in this talk we will mainly address the flat metastring, it is instructive to briefly consider more general structures. In fact, it is quite straightforward to include a (constant) background in the Polyakov action

$$S_P(X) = \frac{1}{4\pi}\int (G_{\mu\nu}(X){*}\mathrm{d}X^\mu \wedge \mathrm{d}X^\nu + B_{\mu\nu}(X)\mathrm{d}X^\mu \wedge \mathrm{d}X^\nu). \tag{22}$$

The simplest way to incorporate this curved background is to apply to the flat Tseytlin action an $O(d,d)$ transformation of the form

$$O^T = \begin{pmatrix} 1 & B \\ 0 & 1 \end{pmatrix}\begin{pmatrix} e^T & 0 \\ 0 & e^{-1} \end{pmatrix}. \tag{23}$$

This acts trivially on the constant η and has the effect of mapping

$$H \mapsto H = O^T H O \equiv \begin{pmatrix} [G - BG^{-1}B] & [BG^{-1}] \\ -[G^{-1}B] & [G^{-1}] \end{pmatrix}, \quad \omega \mapsto \omega = O^T \omega O \equiv \begin{pmatrix} -2B & \delta \\ -\delta & 0 \end{pmatrix}, \tag{24}$$

resulting in the general metastring action (9). Alternatively, this can be derived from the first order formulation of the Polyakov action

$$\hat{S} = \frac{1}{2\pi}\int \left(\frac{1}{\lambda\varepsilon}\boldsymbol{P}_\mu \wedge \mathrm{d}X^\mu + \frac{1}{2\varepsilon^2}(\hat{G}^{\mu\nu} * \boldsymbol{P}_\mu \wedge \boldsymbol{P}_\nu + \hat{B}^{\mu\nu}\boldsymbol{P}_\mu \wedge \boldsymbol{P}_\nu)\right), \tag{25}$$

where we have defined

$$[\hat{G} + \hat{B}] \equiv [(G+B)^{-1}], \tag{26}$$

although one must be careful to preserve boundary terms in order to obtain the correct modification of ω.

Thus, the usual string theory in curved backgrounds corresponds to making ω and the H dynamical (but not the η). Let us discuss further generalizations. Given the existence of ω and H, there is a natural way to understand this geometrical structure (ω, H) from the point of view of quantum mechanics. If one takes the point of view of geometric quantization [33], the construction of the Hilbert space associated with a phase space $(\mathcal{P}, \omega)$ requires the introduction of a complex structure I compatible with ω. Such a complex structure defines the notion of coherent states as holomorphic functionals and equips the phase space with a quantum-metric via the relation $HI = \omega$ [10]. This structure is, in effect, what Born suggested to be part of quantum gravity in the 1930's [17]. In the string case, H is related to ω via a complex structure.

However, the Born proposal is not enough. As we have pointed out in [10], in metastring theory we must take η to be dynamical as well. As we have seen above, it is η that governs the splitting of phase space into space and momentum space. In particular, one can think of spacetime as a Lagrangian submanifold, that is a manifold of maximal dimension on which the symplectic structure vanishes. Analogously, momentum space is just another Lagrangian submanifold $\tilde{L}$ in this description, which is transverse to the spacetime Lagrangian submanifold. Thus we end up with a bilagrangian structure on $\mathcal{P}$. That is a decomposition of $\mathcal{P}$ into two transverse Lagrangian manifolds: $T\mathcal{P} = T\mathcal{L} \oplus T\tilde{\mathcal{L}}$ and $TL \cap T\tilde{L} = \{0\}$. What is remarkable is the fact that a bilagrangian structure is uniquely characterized by a polarization metric η. This metric is characterized by the fact that $L = \ker(\eta + \omega)$ and $\tilde{L} = \ker(\eta - \omega)$. In other words, the geometrical notion of η is to provide a bilagrangian decomposition of phase space. The neutral metric η that seems like a purely stringy metric is in fact a very natural object from the point of view of phase space, in that it labels its decomposition into space and momentum. In order to prove this, let us introduce a structure K which is $+1$ on the vectors tangent to the spacetime Lagrangian L and -1 on the momentum space Lagrangian $\tilde{L}$. This is a real structure which satisfies $K^2 = 1$. Since L and $\tilde{L}$ are Lagrangians K also satisfies an anti-compatibility condition with ω: $K^T \omega K = -\omega$. These two properties in turn show that $\eta = \omega K$ is a neutral metric. We have already emphasized the importance of the endomorphism $J = \eta^{-1} H$, which relates the two metrics. Its properties enforce the chirality of the σ-model. We thus suppose that the geometry of $\mathcal{P}$ should be constrained by the property $J^2 = 1$.

It is relative locality that suggests that both η and H be dynamical. In particular, in canonical quantum theory H is a purely kinematical structure and η, which describes the choice of polarization, can be modified by unitary dynamics. Conversely, in the context of gravitational dynamics, η is a purely kinematical structure (because spacetime provides the preferred basis or polarization), while H, through its spacetime part, can be made dynamical. According to Born, when we introduce gravity into quantum theory we have to make H into a truly dynamical quantity. When we introduce quantum theory into gravity, we have to make the neutral metric η dynamical, and thus in the context of quantum gravity, both H and η have to be dynamical.

The neutral metric η is, together with the generalized phase space metric H, indispensable for the definition of space-time as a maximally null subspace of η with the spacetime metric given by the restriction of the H metric to this η-null subspace [10]. The structure (ω, η, K) can also be described in terms of the two real structures J, K and the map I. We can check that the relation between these maps is given by

$$JK = I. \tag{27}$$

If, in addition, we assume that B vanishes we have that I is a complex structure and that $JK = -KJ$. Phase space geometries that have I, J, K satisfying these conditions were referred to as Born geometries in [10]. These possess para-quaternionic structure (because $I^2 = -1$, but $J^2 = 1 = K^2$ and they anticommute). Born geometry represents a natural unification of quantum and spacetime and phase space geometries, and it implies a new view on the kinematical and dynamical structure of quantum gravity.

Note that the above is a traditional sigma-model discussion, with some apparent relation to generalized geometry. However, in metastring theory we speak in terms of phase space $\mathcal{P}$ instead of a double spacetime, as is the case in generalized geometry and double field theory. In fact, now we think that the proper interpretation of the phase space $\mathcal{P}$ equipped with a triplet of structures, the symplectic structure ω, the polarization metric η and the generalized metric H, can be found in the *geometry of modular quantization* [12, 13, 14]. Thus we turn to the subject of geometry of quantum theory.

4. Quantum Geometry: Phase Space, Heisenberg Group and Modular Variables

In this section we re-examine the geometry of quantization, in order to find a very revealing and unsuspected connection with metastring theory. We focus on the Heisenberg (or Weyl-Heisenberg) group, which is generated, on the level of the corresponding algebra, by familiar position $\hat{q}^a$ and momentum $\hat{p}_b$ operators:

$$[\hat{q}^a, \hat{p}_b] = i\hbar\delta^a_b. \tag{28}$$

It will be convenient to introduce a length scale λ and a momentum scale ϵ, with $\lambda\epsilon = \hbar$. Then, let us introduce the following suggestive notation

$$\hat{x}^a \equiv \hat{q}^a/\lambda, \quad \hat{\tilde{x}}_a \equiv \hat{p}_a/\epsilon, \quad [\hat{x}^a, \hat{\tilde{x}}_b] = i\delta^a_b. \tag{29}$$

Even more compactly let us, once again, suggestively write

$$\mathbb{X}^A \equiv (x^a, \tilde{x}_a)^T, \quad [\hat{\mathbb{X}}^a, \hat{\mathbb{X}}^b] = i\omega^{AB}, \tag{30}$$

with $\frac{1}{2}\omega_{AB} dX^A dX^B = \frac{1}{\hbar} dp_a \wedge dq^a$, where $\omega_{AB} = -\omega_{BA}$ is the canonical symplectic form on phase space $\mathcal{P}$. The Heisenberg group $H_{\mathcal{P}}$ is generated by Weyl operators [38]

$$W_{\mathbb{K}} \equiv e^{2pi i\omega(\mathbb{K},\mathbb{X})}. \tag{31}$$

These form a central extension of the translation algebra

$$W_{\mathbb{K}} W_{\mathbb{K}'} = e^{2pi i\omega(\mathbb{K},\mathbb{K}')} W_{\mathbb{K}+\mathbb{K}'}. \tag{32}$$

The projection $\pi : H_{\mathcal{P}} \to \mathcal{P}$ (where $\pi : W_{\mathbb{K}} \to \mathbb{K}$) defines a line bundle over $\mathcal{P}$ (in principle a covariant phase space of quantum probes). In this formulation, states are sections of degree one

$$W_{\mathbb{K}'} \Phi(\mathbb{K}) = e^{2\pi i\omega(\mathbb{K},\mathbb{K}')} \Phi(\mathbb{K} + \mathbb{K}'). \tag{33}$$

In this language, geometric quantization means to take a Lagrangian $L \in \mathcal{P}$, so that states descent to square integrable functions on L.

A Lagrangian submanifold L is a maximally isotropic subspace L with $\omega|_L = 0$, and thus $\{\partial/\partial q^a\} \in T\mathcal{P}$ defines a Lagrangian submanifold, or "space". (Indeed, $\omega(\partial/\partial q^a, \partial/\partial q^b) = 0$.) This can be understood as a classical characterization of space (and in the covariant context, of spacetime), as a "slice" of phase space. How about a purely quantum characterization of space?

Note that for space-like separations the operators of a local quantum field theory commute. Thus in order to understand the meaning of quantum spacetime (quantum Lagrangian), we need to look at a maximally commuting subalgebra of the Heisenberg algebra and the representation that diagonalizes it. Thus, borrowing from notions of non-commutative algebra and non-commutative geometry [39] (such as the theorem of Gelfand-Naimark [40]), we can say that a Lagrangian submanifold is a maximally commutative subgroup of the Heisenberg group. If we accept this notion of a Lagrangian, then the quantum regime is very different from the classical regime. In particular the vanishing Poisson bracket $\{f(q), g(p)\}$ requires either f or g to be constant. However, the vanishing commutator $[f(\hat{q}), g(\hat{p})] = 0$ requires only that the functions be commensurately periodic

$$e^{i\alpha\hat{p}} e^{i\beta\hat{q}} = e^{i\hbar\alpha\beta} e^{i\beta\hat{q}} e^{i\alpha\hat{p}}, \quad \alpha\beta = 2\pi/\hbar. \tag{34}$$

What is interesting here is that similar considerations led Aharonov to introduce *modular variables* to describe purely quantum phenomena, such as interference [41].

4.1. Modular variables

Modular variables are described in great detail in the very insightful book by Aharonov and Rohrlich [41], where one can find detailed bibliography on this subject[4]. The fundamental question posed there was as follows: how does one capture interference effects (due to the fundamental linearity of quantum theory) in terms of Heisenberg operators? For example, what are the quantum observables that can measure the relative phase responsible for interference in a double-slit experiment? No polynomial functions of the operators $\hat{q}$ and $\hat{p}$ can detect such phases, but operators that translate in space, such as $e^{iR\hat{p}/\hbar}$, do. Thus the modular variables denoted $[\hat{q}]$ and $[\hat{p}]$, which are defined modulo a length scale R (the slit spacing being a natural choice), play a central role, where

$$[p]_{h/R} = p \mod (h/R)\,, \qquad [q]_R = q \mod (2\pi R)\,, \tag{35}$$

and $h = 2\pi\hbar$. The shift operator $e^{iR\hat{p}/\hbar} = e^{iR[\hat{p}]/\hbar}$ shifts the position of a particle state (say an electron in the double-slit experiment) by a distance R and is a function of the modular momenta.

[4] See also [42] and [43].

IOP Conf. Series: Journal of Physics: Conf. Series **804** (2017) 012032 doi:10.1088/1742-6596/804/1/012032

These modular variables satisfy non-local operator equations of motion. For example, given the Hamiltonian, $\hat{H} = \hat{p}^2/2m + V(\hat{q})$, the Heisenberg equation of motion for the shift operator is,

$$e^{-iR\hat{p}/\hbar} \frac{d}{dt} e^{iR\hat{p}/\hbar} = -\frac{iR}{\hbar} \left(\frac{V(\hat{q}+R) - V(\hat{q})}{R} \right). \tag{36}$$

Modular variables are fundamentally non-local in a non-classical sense, since we see here that their evolution depends on the value of the potential at distinct locations. Remarkably, thanks to the uncertainty principle, this dynamical non-locality does not lead to a violation of causality [41]. One of the characteristic features of these variables is that they do not have classical analogues; indeed, the limit $\hbar \to 0$ of $[p]_{h/R}$ is ill-defined.

Note that modular variables are contextual in general[5]. In other words, they carry specific experimental information, such as the length R between the two-slits. However, in the context of quantum gravity such scales are automatically built in, and the contextuality is in principle removed. Also, the fundamental dynamical equations for modular variables are non-local in quantum gravity because of the presence of the fundamental length.

When exponentiated (i.e. when understood as particular Weyl operators), the modular variables naturally commute. In other words, given $[x^a, \tilde{x}_b] = \frac{i}{2\pi}\delta^a_b$, the following commutator of modular operators vanishes [14]

$$[e^{2\pi i x}, e^{2\pi i \tilde{x}}] = 0. \tag{37}$$

Thus a quantum algebra of modular variables possesses more commutative directions than a classical Poisson algebra, because the Poisson bracket of modular variables does not vanish, $\{e^{2\pi i x}, e^{2\pi i \tilde{x}}\} \neq 0$.

Here we make a historical note [44]: The above non-local equations of motion were essentially written by Max Born, in the very first paper which used the phrase "Quantum Mechanics" in its title, in 1924, one year before the Heisenberg breakthrough paper. Actually, Heisenberg crucially used Born's prescription of replacing classical equations by the corresponding difference equations, in order to derive what we now call the canonical commutation relations (properly written by Born and Jordan) from the Bohr-Sommerfeld quantization conditions.

5. Modular Space and Geometry of Quantum Theory

Returning to the subject of quantum Lagrangians, note that the quantum Lagrangian is analogous to a Brillouin cell in Condensed Matter physics. The volume and shape of the cell are given by λ and ϵ (i.e. $\hbar$ and G_N (α')) The uncertainty principle is implemented in a subtle way: we can specify a point in modular cell, but if so, we can't say *which* cell we are in.

This means that there is a more general notion of quantization, beyond that of geometric quantization. Instead of selecting a classical polarization L (the arguments of the wave function, or the arguments of a local quantum field) we can choose a *modular polarization.* In terms of the Heisenberg group all that is happening is that in order to have a commutative algebra, we need only

$$\omega(\mathbb{K}, \mathbb{K}') \in 2Z, \quad W_{\mathbb{K}} W_{\mathbb{K}'} = e^{2pi i\omega(\mathbb{K},\mathbb{K}')} W_{\mathbb{K}+\mathbb{K}'} = W_{\mathbb{K}'} W_{\mathbb{K}'}. \tag{38}$$

This defines a lattice Λ in phase space $\mathcal{P}$. Finally, we specify a "lift" of the lattice from the phase space $\mathcal{P}$ to the Heisenberg group $H_{\mathcal{P}}$.

Maximally commuting subgroups $\hat{\Lambda}$ of the Heisenberg group correspond to lattices that are integral and self-dual with respect to ω [45, 46]. Given W_λ where $\lambda \in \Lambda$ there is a lift to $\hat{\Lambda}$ which defines "modular polarization"

$$U_\lambda = \alpha(\lambda) W_\lambda, \tag{39}$$

[5] Aharonov and collaborators have pushed the logic associated with modular variables to argue for a new kind of weak measurements of such non-local variables that capture the superposition principle of quantum theory. Similarly, Aharonov and collaborators argue for a time symmetric formulation of quantum theory [41].

where $\alpha(\lambda)$ satisfies the co-cycle condition

$$\alpha(\lambda)\alpha(\mu)e^{piiω(\lambda,\mu)} = \alpha(\lambda+\mu), \quad \lambda, \alpha \in \Lambda. \tag{40}$$

One can parametrize a solution to the co-cycle condition by introducing a symmetric bilinear from η and setting

$$\alpha_\eta(\lambda) \equiv e^{i\frac{pi}{2}\eta(\lambda,\lambda)}. \tag{41}$$

Finally, when we choose a classical Lagrangian L, there is a special state that we associate with the vacuum: it is translation invariant (which in our context can be interpreted as "empty space"). In modular quantization, there is no such translation invariant state (because of the lattice structure). The best we can do is to choose a state that minimizes an "energy", which requires the introduction of another symmetric bilinear form, that we call, again suggestively, H. This means, first, that we are looking for operators such that

$$[\hat{\mathbb{P}}_A, \Phi] = \frac{i}{2\pi}\partial_A\Phi, \qquad \Phi(\hat{\mathbb{X}}+\lambda) = \Phi(\hat{\mathbb{X}}), \tag{42}$$

where the modular observables $\Phi(\hat{\mathbb{X}}+\lambda) = \Phi(\mathbb{X})$ are generated by the lattice observables U_λ with $\lambda \in \Lambda$. Translation invariance would be the condition $\hat{\mathbb{P}}|0\rangle = 0$. Since this is not possible the next natural choice, is to minimize the translational energy. Therefore we pick a positive definite metric H_{AB} on $\mathcal{P}$, and we define

$$\hat{E}_H \equiv H^{AB}\hat{\mathbb{P}}_A\hat{\mathbb{P}}_B, \tag{43}$$

and demand that $|0\rangle_H$ be the ground state of $\hat{E}_H$. This is indeed the most natural choice and it shows that we cannot fully disentangle the kinematics (i.e., the definition of translation generators) from the dynamics. In the Schrödinger case, since the translation generators commute, the vacuum state $\hat{E}|0\rangle = 0$ is also the translation invariant state and it carries no memory of the metric H needed to define the energy. In our context, due to the non-commutativity of translations, the operators $\hat{E}_H$ and $\hat{E}_{H'}$ do not commute. As a result the vacuum state depends on H, in other words $|0\rangle_H \neq |0\rangle_{H'}$, and it also possesses a non-vanishing zero point energy.

Thus, modular quantization involves the introduction of three quadratic forms (ω, η, H), i.e. Born geometry. Therefore, quite surprisingly, Born geometry underlies the concept of Aharonov's modular variables.

5.1. Born Geometry and Quantum Theory

This triple of structures has appeared in the context of metastring theory. In that context a choice of polarization is a choice of a spacetime within $\mathcal{P}$ but the most general choice is a *modular polarization* that we have discussed above. From this new point of view Born geometry (ω, η, H) arises as a parametrization of such quantizations, which results in a notion of quantum spacetime, that we call *modular spacetime*. Finally, large spacetimes of canonical string theory and general relativity result as a "many-body" phenomenon, through a process of tensoring of unit modular cells, that we refer to as "extensification" [14].

In particular, the symplectic structure ω

$$ds^2_\omega = \frac{1}{2}\omega_{AB}d\mathbb{X}^A d\mathbb{X}^B = \frac{1}{\hbar}dp_a \wedge dq^a, \tag{44}$$

is encoded in the canonical Heisenberg commutator between q^a and p_a. The generalized, quantum, metric H comes from the Born rule in quantum theory

$$ds^2_H = H_{AB}d\mathbb{X}^A d\mathbb{X}^B = \frac{1}{\hbar}(\frac{dq_a dq^a}{G_N} + G_N dp_a dp^a). \tag{45}$$

For weak gravity, this metric reduces to the spacetime metric (where spacetime can be viewed as a slice of phase space). Due to gravity's extreme weakness, we only see spacetime metric at low energies. (The ratio ϵ/λ defines a tension; if this is identified with c^3/G_N, it is enormous, $\sim 10^{32} kg/sec$.) Therefore, in this formulation the usual dynamical spacetime metric is the low energy leftover of the quantum metric. Finally, the polarization (or locality metric) η encodes the distinction between spacetime-like and energy-momentum-like aspects of phase space (and in this sense it defines an analog of the "causal" structure in phase space)

$$ds_\eta^2 = \eta_{AB} d\mathbb{X}^A d\mathbb{X}^B = \frac{2}{\hbar} dp_a dq^a. \tag{46}$$

This new metric captures the essence of relative locality - when η is constant we have absolute locality. Curving η also means "gravitizing the quantum". In general all three elements of Born geometry, ω, η and H are dynamical and curved in metastring theory, as we will discuss in section 7.

Before doing so, let us end by a few comments regarding the Stone-von Neumann theorem [47, 48, 49] which asserts that all representations of the Heisenberg group are unitarily equivalent. Normally, we think of this as a choice of basis in phase space (a choice of polarization or classical Lagrangian), and all such choices are related by Fourier transform. Similarly, one can pass from a classical polarization (such as the Schrödinger representation) to a modular polarization via the Zak transform [50]. Note that, there is a connection on the line bundle over phase space that has unit flux through a modular cell. (This is very similar to integer quantum Hall effect systems.) A modular wave function is quasi-periodic

$$\Psi(x+a,\tilde{x}) = e^{2i\pi a\tilde{x}}\Psi(x,\tilde{x}), \quad \Psi(x,\tilde{x}+\tilde{a}) = \Psi(x,\tilde{x}). \tag{47}$$

The quasi-periods correspond to the tails of an Aharonov-Bohm [51] potential attached to a unit flux. In particular, vacuum states must have at least one zero in a cell, which leads to theta functions (the Zak transforms of Gaussians). Note that from the point of modular polarization, the familiar Schrödinger polarization is just a singular limit.

6. Modular Spacetime and Metastring Theory

Now we come to our main point. Is there any real evidence that we should regard the above discussion of modular polarization as connected to quantum gravity in some way? In fact, the triple of data $(\mathcal{P};\omega,\eta,\mathcal{H})$ is in one-to-one correspondence with the geometric data underlying metastring, which indeed has the string length built in (λ). As well, the vertex operator algebra of the metastring gives copies of the Heisenberg group (in the zero mode sector, with decorations by oscillator modes) [12].

When gravity is turned on, we should regard modular quantization as a gravitization of the quantum, which makes the whole of Born geometry dynamical. According to the picture presented above, the modular cell is a quantum unit of spacetime - the spacetime qubit. More generally, we can extensify the modular cell, by tensoring many units together, resulting in the large flux. The procedure of extensification can lead to large spacetimes, and in principle, to arbitrary spacetime backgrounds (including known and unknown compactifications).

To summarize: Usually, the notion of space determines what set of commuting measurement can be performed. We have reversed this logic and defined modular space as the maximal set of commutative operations ($[e^{2\pi ix}, e^{2\pi i\tilde{x}}] = 0$). In this sense modular spaces are fundamentally quantum. Also, modular spaces turn out to have, by construction, built in length and energy scales. Note that in contrast to a one dimensional Euclidean space, which is non-compact and simply connected, one dimensional modular space is actually two dimensional, it is compact and not simply-connected (it carries Aharonov-Bohm flux [51]). One can view the unit phase space cells equipped with the unit Aharonov-Bohm flux lines, as the spacetime qubits, the basic building blocks of quantum spacetime. In principle, we now have superpositions and entanglements of quantum spacetimes.

The surprising fact is that modular spaces are realized in string theory, in its metastring formulation, because the target space of a closed metastring is a modular space. (That is, it is not a doubled space, and it is not a compactified space. Also, modularity is the target space realization of T-duality.) In the canonical Polyakov string, the target space is interpreted as spacetime. In the metastring, the target space is a modular space that comes equipped with the triple structure of Born geometry $(\mathcal{P};\omega,\eta,\mathcal{H})$. From this point of view, the usual spacetime backgrounds are just Lagrangian submanifolds. Note that if we suppose this to be a modular Lagrangian, it is associated with a lattice which turns out to be unique. This comes from a careful study of the operator algebra of the string, closely associated with the analysis of the Heisenberg group [12]. In particular, mutual locality of physical states and modular invariance, require that lattice be Lorentzian even and self-dual. This is the unique Lorentzian lattice associated with the famous mathematical work of Borcherds [34, 35] $\Lambda = \Pi_{1,25}\times\Pi_{1,25}$. Note that this is the same as "Narain lattice of fully compactified space-time" [36]. However, the traditional interpretation of this lattice, is nonsensical because there is no apparent causal interpretation - i.e. the time is compact as well as space [36]. Our point is that in the modular spacetime interpretation, such problems are lifted. Moreover, the modular quantization, reconciles the apparent lack of Lorentz symmetry of this lattice structure, taken as the basic building block of quantum spacetime, as we will discuss below.

The metastring possesses a huge symmetry group generated by all dimension $(1,0)+(0,1)$ operators, that is, the products of Borcherds algebras. (In principle, all the usual string backgrounds, a presumably many more, are extensifications of these algebras.) The simple roots of the Borcherds algebra in the lattice L are either tachyonic or in L. The tachyonic roots generate a compact subgroup. Also, T-duality is represented by a rotation by an angle π. The lattice L is given as

$$L \equiv \{K \in I_{1,25} | K^2 = 2, K\cdot\rho = -1\} \tag{48}$$

and is isomorphic to the Leech lattice $\rho \equiv (0,1,2,...,24|70)$. The null roots generate the Heisenberg group needed for the modular construction. The Borcherds symmetry is effectively broken in the usual Polyakov string by choosing various subsectors. How does one recover the usual string? The monodromies of the usual string are $\Delta^A = (0,\tilde{\delta}_\mu)$. The momenta of the usual string are $K^A \equiv J(\Delta)/2\pi = (\tilde{\delta}_\mu/2\pi, 0)$. For the metastring we have $K_A = (k_\mu, \tilde{k}^\mu)$. The usual string spectrum is recovered if we truncate the metastring spectrum to operators such that $|k| >> |\tilde{k}|$. The usual locality (absolute locality) limit is a limit of large quantum numbers, i.e. it is a semi-classical limit.

6.1. Modular space, Lorentz symmetry and Relative locality

So how does our construction resolve the fundamental conundrum of any quantum theory of gravity: how to make continuous symmetries (such as the Lorentz symmetry) consistent with the existence of a fundamental scale that is characteristic of quantum (modular) spaces [14] ?

A useful analogy for the reconciliation of discreteness of quantum spacetime and continuity of Lorentz symmetry is provided by angular momentum in quantum mechanics. (In some sense, the spacetime qubits are the direct analog of spin.) Classically, the angular momentum describes a point on a sphere, and its canonical actions are given by rotations which are symmetries of the sphere. Quantum mechanically, representations are discrete; the eigenvalue of the angular momentum is evenly spaced and this can be interpreted as a discretization of the sphere, in which it is replaced by a discrete set of circles equally spaced along, say, the z-axis. The lattice apparently destroys the rotational symmetry of the classical description. However, we know that is not true: quantum theory restores the rotational symmetry by arranging for a superposition of spin states which in turn parametrize a sphere's-worth of states. The spin states are merely a basis for the entire state space, and the basis is not invariant under rotations.

How does this analogy work in the context of modular spacetime? In the Schrödinger representation, rotations act within the set of states $|x\rangle$ (using a fully covariant notation), giving back a linear combination of these states. In the modular polarization, clearly, we should regard the Lorentz rotations as acting on the basis, resulting in a new choice of modular cell, i.e., a new 'quantization axis'. Such a transformation corresponds to a canonical transformation.

As shown in [14], in the context of spatial rotations but fully applicable to Lorentz rotations, once we introduce a fundamental scale, the natural polarization that respects the presence of this scale introduces a bilagrangian lattice [45] $\Lambda = \ell \oplus \tilde{\ell}$ embedded in phase space $\mathcal{P}$, and equipped with a neutral metric η. Moreover, the Hilbert space is given by a space of sections of a line bundle L_Λ over the torus $T_\Lambda = \mathcal{P}/\Lambda$ and that the embedding $T_\Lambda \hookrightarrow L_\Lambda$ is characterized by a lift, $\alpha_\eta \in \mathrm{U}(1)$. Lorentz rotations act on this fully covariant phase space as symplectic transformations

$$\mathbb{X} = (x, \tilde{x}) \mapsto O \cdot \mathbb{X} = (Ox, O^T \tilde{x}), \qquad O \in \mathrm{O}(1, d-1). \tag{49}$$

These transformations not only preserve ω, but they also preserve the polarization metric η and the quantum metric H. Remarkably, we can show that Lorentz rotations are in fact the *only* transformations that preserve the given Born geometry (ω, η, H) of the modular space. Let us consider the basic triplet of structures associated with Born geometry, $(\mathcal{P}, \omega, \eta, H)$, where ω is a symplectic structure, η a polarisation metric and H a quantum metric. In Darboux coordinates, we have

$$\omega^0 = \mathrm{d}x^a \wedge \mathrm{d}\tilde{x}_a, \quad \eta^0 = \mathrm{d}x^a \mathrm{d}\tilde{x}_a, \quad H^0 = \mathrm{d}x^a h_{ab} \mathrm{d}x^b + \mathrm{d}\tilde{x}_a \tilde{h}^{ab} \mathrm{d}\tilde{x}_b, \tag{50}$$

where $h_{ab} = \mathrm{diag}(-1, 1, \cdots, 1)$ is a flat Lorentzian metric and $\tilde{h}^{ab}$ its inverse. We now consider a linear coordinate transformation

$$\mathbb{U} := \begin{pmatrix} \alpha^a{}_b & \gamma^{ab} \\ \beta_{ab} & \delta_a{}^b \end{pmatrix}, \qquad \mathbb{X} \mapsto \mathbb{U}\mathbb{X} = \begin{pmatrix} \alpha x + \gamma \tilde{x} \\ \beta x + \delta \tilde{x} \end{pmatrix}. \tag{51}$$

Under such a transformation the geometrical structure is modified to $(\mathcal{P}, \omega, \eta, H)$, where

$$\omega = [\alpha^T \beta]_{[ab]} \mathrm{d}x^a \wedge \mathrm{d}x^b + [\gamma^T \delta]^{[ab]} \mathrm{d}\tilde{x}_a \wedge \mathrm{d}\tilde{x}_b + [\alpha^T \delta - \beta^T \gamma]_a{}^b \mathrm{d}x^a \wedge \mathrm{d}\tilde{x}_b, \tag{52}$$

as well as

$$\eta = [\alpha^T \beta]_{(ab)} \mathrm{d}x^a \mathrm{d}x^b + [\gamma^T \delta]^{(ab)} \mathrm{d}\tilde{x}_a \mathrm{d}\tilde{x}_b + [\alpha^T \delta + \beta^T \gamma]_a{}^b \mathrm{d}x^a \mathrm{d}\tilde{x}_b, \tag{53}$$

and finally,

$$H = [\alpha^T h \alpha + \beta^T \tilde{h} \beta]_{(ab)} \mathrm{d}x^a \mathrm{d}x^b + [\delta^T \tilde{h} \delta + \gamma^T h \gamma]^{(ab)} \mathrm{d}\tilde{x}_a \mathrm{d}\tilde{x}_b + 2[\alpha^T h \gamma + \beta^T \tilde{h} \delta]_a{}^b \mathrm{d}x^a \mathrm{d}\tilde{x}_b. \tag{54}$$

The subset of transformations that preserves ω, η and H satisfy

$$\beta = \gamma = 0, \qquad \alpha^T = \delta^{-1}, \qquad \alpha^T h \alpha = h. \tag{55}$$

So we conclude that the group of transformations that preserves the background Born geometry is the Lorentz group! In other words, we have that the Lorentz group lies at the intersection of the symplectic, neutral and doubly orthogonal groups,

$$\mathrm{O}(1, d-1) = \mathrm{Sp}(2d) \cap \mathrm{O}(d, d) \cap \mathrm{O}(2, 2(d-1)). \tag{56}$$

Here Sp($2d$) is the symplectic group preserving ω, O(d,d) is the neutral group preserving η and O($2,2(d-1)$) the orthogonal (conformal) group preserving H. It is also interesting to understand the pairwise intersections. For instance, the group of transformations that preserves the symplectic structure and the polarization metric η is given by

$$\mathrm{GL}(d) = \mathrm{Sp}(2d) \cap \mathrm{O}(d,d). \tag{57}$$

Here GL(d) is the general linear group that represents the sets of frames e_i^a determining the vacuum metric

$$H(x,\tilde{x}) = x^a q_{ab} x^b + \tilde{x}_a q^{ab} \tilde{x}_b, \tag{58}$$

with $q^{ab} = e_i^a \delta^{ij} e_j^a$ and q_{ab} its inverse, while the space of vacuum metrics associated with a given Hilbert space is encoded in GL(d)/O(d). The other pairwise intersections are

$$\mathrm{U}(d) = \mathrm{Sp}(2d) \cap \mathrm{O}(2d), \quad \mathrm{O}(d) \times \mathrm{O}(d) = \mathrm{O}(d,d) \cap \mathrm{O}(2d). \tag{59}$$

The space U(d)/O(d) represents the choice of inequivalent lifts η that preserve the symplectic structure and the vacuum state, while the quotient of the double orthogonal group by O(d) represent the choice of inequivalent symplectic structures, hence Weyl groups, that possess the same Hilbert space with the same vacuum.

Given this new view of the Lorentz symmetry, as defined by the intersection of the triplet of geometric structures of Born geometry, how does modular quantization avoid breaking the Lorentz group into a discrete subgroup? The answer is offered by relative locality. Relative locality implies that this breaking cannot take place: a spacetime is a choice of polarization, and thus each observer makes his/her own choice. Each observer sees the discreteness, but Lorentz symmetry acts to change the polarization, restoring Lorentz invariance. (This is precisely what happens in the above spin analogy: choose a quantization axis, get discrete spin states, which viewed as a discretization of the sphere, break the rotation group, but rotations act to rotate the quantization axis, which is a choice of polarization, and because the spin operators do not commute the states of one axis are unitarily related to those of the other.) Thus, for Lorentz symmetry to be unbroken, we need, non-commutative torus in phase space and relative locality, that is, spacetime not being an arena shared by all observers, the arena being only a "many-body" effect in this interpretation/approximation.

7. Metastring theory and the two scale Renormalizaton Group (RG)

Finally, we explain how metastring theory makes the Born geometry dynamical by extending the usual argument in canonical string theory, according to which the Einstein equations are derived from the requirement of conformal invariance of the worldsheet sigma model [2].

One of the central features of the metastring is the following phase space commutator:

$$\{X^A(\sigma), X^B(\sigma')\} = \eta^{AB}\theta(\sigma-\sigma') + \omega^{AB} \tag{60}$$

where θ is the staircase distribution and ω^{AB} term, that is indicative of the phase space interpretation, captures the non-commutativity of the zero modes. This can be derived from the Tseytlin action [12] (and will be elaborated on in our upcoming publications). Thus at the quantum level the metastring target variables are non-commutative. In particular, both the zero mode and oscillator sectors are non-commutative. This fact is important because it is known that in the context of non-commutative field theory one needs to generalize the concept of the renormalization group (RG) to a two scale RG in order to address the question of the existence of the continuum limit [52]. Thus, in the context of the metasting in which the doubled coordinates do not commute, one expects a similar (however in this case, covariant) two scale RG.

First, we recall the fundamental idea of Friedan [53, 54]: Demanding invariance under scale transformation of the probe is equivalent to Einstein's equations. Let us look at a relativistic particle (with the action $W_p = e^{i\frac{Length}{2\lambda}}$), coupled to a non trivial metric g [55]. In order to define the propagator let us specify a discretization of the path where the proper time steps are taken to be $\Delta T = \epsilon$, so that $T = N\delta$. We have that

$$G(y,x) = \int dw \lim_{N\to\infty, N\delta=w} \int \prod_{i=1}^{N} dx_i \prod_{i=0}^{N} G_\delta(x_{i+1}, x_i), \tag{61}$$

where the particle Green's function is

$$G_\delta(x_{i+1}, x_i) = (\sqrt{\frac{m}{2\pi\hbar\delta}})^D \exp[-\frac{m}{\hbar\delta} d_\delta^2(x_{i+1}, x_i)]. \tag{62}$$

If we coarse-grain by a factor of two, we effectively get a change of scale $\delta \to 2\delta$. Under this change the effective metric changes by

$$g_\delta \to g_{2\delta} = g_\delta - \frac{\hbar\delta}{6m} R, \tag{63}$$

where R denotes the Ricci scalar. Therefore we see that if the vacuum Einstein equations are satisfied, then we are at a fixed point of the RG flow. Or, we may reverse the logic and say that if we demand that the physics of the probe is independent of its finer details, then Einstein's equations are satisfied [53]. Friedan's idea is of particular relevance for string theory, because the string propagates in the background that is a condensation of the string's own excitations. In the Polyakov string theory case the quantum weight of a string is $W_s = e^{i\frac{Area(\Sigma)}{2\lambda^2}}$, where λ is the string length. Note that in order to define the quantum string we only need a length scale, whereas to define quantum gravity we need two: $\hbar$ and G_N. If we discretize the string along a square lattice with the condition $N^2\delta^2 = A$, the important point is that the scale δ drops out. If we perform a decimation we find that the metric is renormalized by

$$g_{ab} \to g_{ab} - \frac{\lambda^2}{12} R_{ab} + ... \tag{64}$$

Thus, the vacuum Einstein equation follows from demanding the independence of the physics under the probe's small distance structure (or discretization) [53].

This procedure has to be generalized in the context of metastring theory. The admissible backgrounds are now Lorentz invariant CFT's that correspond to metastring sigma models invariant under Weyl (W) and Lorentz (L) transformations. If we write the metric in terms of frame fields $ds^2 = e_+ e_-$, we have

$$W : (e_+, e_-) \to (e^\sigma e_+, e^\sigma e_-), \quad L : (e_+, e_-) \to (e^\theta e_+, e^{-\theta} e_-). \tag{65}$$

In the metastring sigma model the first symmetry is kinematical, while the second one is implemented as the end point symmetry of the RG flow. Indeed, it is well known that even if the generic theories are not conformally invariant, the end point of the RG flow admits an enhanced, conformal symmetry. The general formulation of metastring theory necessarily needs to involve non-Lorentz covariant sigma-models and thus we should find an RG flow whose endpoint admits an enhanced symmetry in two ways: it is enhanced under Weyl transformations and it is enhanced under Lorentz transformations. This means that we need to double the RG flow into the following transformations of the metastring partition function Z

$$\delta_{\sigma(x)} Z = 0 \to (\partial^2 + \tilde{\partial}^2)\delta\eta = 0, \tag{66}$$

as well as

$$\delta_{\theta(x)} Z = 0 \to (\partial \cdot \tilde{\partial})\delta\eta + 2(J\delta\eta) = 0. \tag{67}$$

These flow equations determine the fixed points under the double RG. In the context of the metastring this should be the proper formulation of closed string field theory. Fixed points which are Lorentz invariant but not free (i.e. have non-trivial Lorentz dimensions) correspond to backgrounds where η is curved, in the same way that fixed points which are conformally invariant but not free, correspond to backgrounds where H is curved. Now why do we expect a double flow in relativistic theories? In a relativistic theory if we want to put a cutoff we need generically two scales

$$E < \Lambda, \quad |\vec{p}| < \Lambda'. \tag{68}$$

In 2d this means that we have one scale for the left momenta and one for the right momenta, and we do not need to assume that they are equal. The conformal invariance should then be doubled to a fixed point symmetry that dynamically implements both the conformal (Weyl) and Lorentz symmetries.

Following the usual RG derivation of Einstein's equations in the sigma model approach to string theory, this new formulation of quantum gravity via metastring theory gives new equations for η_{AB} and H_{AB}, for both dynamical space-time and dynamical momentum space, which, in turn, consistently define a dynamical phase space. At the linearized level, these equations can be also obtained from conformal perturbation techniques[6], again, following the usual string theory technique. The resulting equations for H and η read as follows:

$$H^{AB}\partial_A\partial_B H_{CD} = 0, \quad H^{AB}\partial_A\partial_B \eta_{CD} = 0, \quad \eta^{AB}\partial_A\partial_B H_{CD} = 0. \tag{69}$$

The first two equations show that both metrics H and η are indeed dynamical, and they reduce to the linearized Einstein equations in the spacetime limit. At the non-linear level, unless we force η to be flat, we have a coupling between the phase space metric H and the polarization/localization metric η. Finally, note that, in general, even the symplectic form ω becomes dynamical, which is not taken into account in this calculation.

8. Outlook

In this presentation we have given a reformulation of string theory and quantum gravity that generalizes the concept of locality (and causality) and incorporates T-duality. It also provides a specific quantum version of spacetime, the modular spacetime. This formulation possesses deep links with the geometry of quantization and exhibits a large symmetry algebra of Borcherds. The metastring integrates many of either new or recently developed ideas: generalized geometry, relative locality, Nakamura strips, modular variables and double RG flow. The formulation is essentially non-commutative (even though consistent with Lorentz symmetry) and non-local (but consistent with causality), and it sheds new light on the foundations and origins of quantum theory, including quantum field theory. Using the double RG analysis of the metastring we have obtained a generalization of linearized Einstein equation for modular spacetime with implications for fully non-linear Born geometry. Also, in this formulation, there should exist a new effective modular quantum field theory that is consistent with the two scale RG.

In this talk we have concentrated our attention on a new concept of quantum spacetime, called modular spacetime that also appears as a habitat for metastring theory. Note that this concept stems from a quantization of spacetime, and not from quantization of gravitational field/metric. Even the flat space is quantized according to our approach to quantum gravity. This allows for superposition and entanglement of spacetimes. Also, this formulation provides for an explicit construction of spacetime quanta or qubits.

[6] Such computations were performed in detail by Barak Shoshany, a student at Perimeter Institute.

We conclude with a few remarks. A very basic question is the role of causality in this formalism. For example, in the context of ordinary quantum mechanics, Aharonov has argued that the presence of modular time $([t],[E])$ does not lead to a violation of causality [41]. We think it is plausible that the same is true in our more general context; at the very least, the usual field-theoretic notion of causality emerges with extensification. Nevertheless, it is intriguing to contemplate the purely stringy/quantum effects that modular spacetime offers. In ordinary quantum mechanics, modular variables were introduced in order to describe purely quantum phenomena in a simple way, diffraction gratings and Aharonov-Bohm effects being paradigmatic examples. One might say facetiously that modular spacetime is the diffraction grating of quantum gravity. Are there such effects that might be observable? Do they become important in extreme situations, such as in black hole physics or cosmology? For example, non-local effects have been expected for long time in the context of the vacuum (dark) energy problem [56], and more recently in the context of black hole physics [57]. This is the natural arena for the application of the metastring physics. Similarly, quantum effects associated with "fluctuation" of spacetime are being sought for in the context of gravitational interferometry (LIGO, Fermilab's Holometer). Such experiments provide an observational platform needed for detecting modular spacetime. Similarly, gravitization of the quantum should be constrained in the experimental tests of quantum theory in gravitational environments [58]. Finally, we think that there are implications of the metastring for the problems of dark matter and the origin of the standard model of particle physics or the physics of the early universe. We believe that modular space-time may illuminate some of the long-standing puzzles in the foundations of quantum theory, such as the measurement problem [59], as well as the outstanding questions of string theory concerning the short distance [60], high energy and high temperature limits [61] and target space properties of strings [62].

Acknowledgements: We thank Giovanni-Amelino Camelia, Philip Argyres, Nathan Berkovits, David Berman, Martin Bojowald, Cestmir Burdik, Thomas Curtright, Sumit Das, Jan de Boer, Bianca Dittrich, Davide Gaiotto, Murat Gunaydin, Vishnu Jejjala, Jerzy Kowalski-Glikman, Jerzy Lukierski, Kostas Sfetsos, Lee Smolin, Alfred Shapere, Thomas Thiemann, Erik Verlinde and other organizers of various conferences at Perimeter Institute, CERN, Erlangen, Penn State, La Sapienza University in Rome, the Simons Center for Geometry and Physics, the Born meetings in Wroclaw, the winter meetings in Miami, the Amsterdam string meetings, the Great Lakes meetings, the SPOCK meetings on string theory and particle physics, the Mandelstam Institute for Theoretical Physics at the University of Witwatersrand, Johannesburg, South Africa, the Sao Paulo ICTP, and last but not least, the conference on Integrable Systems and Quantum Symmetries in Prague, for inviting us to present the results of this research. RGL and DM thank Perimeter Institute for hospitality. LF, RGL and DM thank the Banff Center for providing an inspiring environment for work. RGL is supported in part by the U.S. Department of Energy contract DE-FG02-13ER42001 and DM by the U.S. Department of Energy under contract DE-FG02-13ER41917. Research at Perimeter Institute for Theoretical Physics is supported in part by the Government of Canada through NSERC and by the Province of Ontario through MRI.

References

[1] Ashtekar, A, Reuter M and Rovelli C 2014 From General Relativity to Quantum Gravity *Preprint* arXiv:1408.4336 [gr-qc]
[2] Polchinski, J 1998 *String theory. Vol. 1: An introduction to the bosonic string* (Cambridge: Cambridge University Press)
[3] Ambjorn J, Gorlich A, Jurkiewicz J and Loll R 2013 *Int. J. Mod. Phys.* D **22** 1330019
[4] Sorkin R D 2003 Causal sets: Discrete gravity, (*Preprint* gr-qc/0309009)
[5] Volovik, G E 2006 *The Universe in a helium droplet* (Oxford: Oxford Science Publications)
[6] Horava, P 2009 *Phys. Rev.* D **79** 084008

[7] Natsuume, M 2015 *AdS/CFT Duality User Guide* (Berlin: Springer)
[8] Chamseddine, A H and Connes, A 1997 *Commun. Math. Phys.* **186** 731
[9] Penrose, R and Rindler, W 1986 *Spinors And Space-time. Vol. 2: Spinor And Twistor Methods In Space-time Geometry* (Cambridge: Cambridge University Press)
[10] Freidel, L, Leigh R G and Minic, D 2014 *Phys. Lett.* B **730** 302
[11] Freidel, L, Leigh R G and Minic, D 2014 *Int. J. Mod. Phys.* D **23** 1442006
[12] Freidel, L, Leigh R G and Minic, D 2015 *JHEP* **1506** 006
[13] Freidel, L, Leigh R G and Minic, D 2015 *Int. J. Mod. Phys.* D **24** 1544028
[14] Freidel, L, Leigh R G and Minic, D 2015 Quantum Spaces are Modular (*Preprint* arXiv:1606.01829 [hep-th])
[15] Amelino-Camelia, G, Freidel, L, Kowalski-Glikman, J and Smolin, L 2011 *Phys. Rev.* D **84** 084010
[16] Amelino-Camelia, G, Freidel, L, Kowalski-Glikman, J and Smolin, L 2011 *Gen. Rel. Grav.* **43** 2547
[17] Born, M 1949 *Rev. Mod. Phys.* **21** 463
[18] Veneziano, G 1986 *Europhys. Lett.* **2** 199
[19] Tseytlin, A A 1991 *Nucl. Phys.* B **350** 395
[20] Tseytlin, A A 1990 *Phys. Lett.* B **242** 163
[21] Floreanini, R and Jackiw, R 1987 *Phys. Rev. Lett.* **59** 1873
[22] Siegel, W 1993 it Phys. Rev. D **47** 5453
[23] Hull, C and Zwiebach, B 2009 *JHEP* **0909** 099
[24] Gibbons, G W 1992 *J. Geom. Phys.* **8** 147
[25] Ashtekar, A and Schilling, T A 1997 Geometrical formulation of quantum mechanics *Preprint* gr-qc/9706069
[26] Minic, D and Tze, H C 2003 *Phys. Rev.* D **68** 061501
[27] Minic, D and Tze, H C 2004 *Phys. Lett.* B **581** 111
[28] Jejjala, V, Kavic, M and Minic, D 2007 *Int. J. Mod. Phys.* A **22** 3317
[29] Giddings, S B and Wolpert, S A 1987 *Commun. Math. Phys.* **109** 177
[30] Krichever, I M and Novikov, S P 1987 *Funct.Anal.Appl.* **21** 294
[31] Nakamura, S 2000 *Tokyo J. Math.* **23** 87
[32] Fedosov, B 1996 *Deformation quantization and index theory* (Berlin: Akademie-Verlag)
[33] Woodhouse, N 1991 *Geometric Quantization* (Oxford: Clarendon Press)
[34] Borcherds, R E 1986 *Proc. Nat. Acad. Sci.* **83** 3068
[35] Borcherds, R E 1992 *Inventiones Mathematicae* **109** 405
[36] Moore, G W 1993 Finite in all directions *Preprint* [hep-th/9305139]
[37] Freidel, L, Garner, D and Ramgoolam, S 2015 *Phys. Rev.* D **91** 126001
[38] Weyl, H 1931 *The theory of groups and quantum mechanics* (New York: Dover)
[39] Connes, A 1994 *Noncommutative Geometry* (Cambridge: Academic Press)
[40] Gelfand, I M and Naimark, MA 1943 *Mat.Sbornik* **12** 197
[41] Aharonov, Y and Rohrlich, D 2005 *Quantum Paradoxes: Quantum Theory for the Perplexed* (New York: Wiley)
[42] Schwinger, J 2000 *Quantum kinematics and dynamics* (Boulder: Westview Press)
[43] 't Hooft, 2014 The Cellular Automaton Interpretation of Quantum Mechanics. A View on the Quantum Nature of our Universe, Compulsory or Impossible? *Preprint* arXiv:1405.1548 [quant-ph]
[44] van der Waerden, B L 1968 *Sources of Quantum Mechanics* (New York: Dover)
[45] Mackey, G 1958 *Acta Math.* **99** 265
[46] Mackey, G 1952 *Annals Math.* **55** 101
[47] Stone, M 1930 *Proc.Nat.Acad.Sci.* **16** 172
[48] von Neuman, J 1931 *Math.Ann.* **104** 570
[49] Mackey, G 1949 *Proc.Nat.Acad.Sci.* **35** 537
[50] Zak, J 1967 *Phys.Rev.Lett.* **19** 1385
[51] Aharonov, Y and Bohm, D 1959 *Phys. Rev.* **115** 485
[52] Grosse, H and Wulkenhaar, R 2005 *Commun. Math. Phys.* **256** 305
[53] Friedan, D H 1985 *Annals Phys.* **163** 318
[54] Callan, C G, Martinec, E J, Perry, M. J and Friedan, D 1985 *Nucl. Phys.* B **262** 593
[55] Konishi, K, Paffuti, G and Provero, P 1990 *Phys. Lett.* B **234** 276
[56] Weinberg, S 1989 *Rev. Mod. Phys.* **61** 1
[57] A. Almheiri, D. Marolf, J. Polchinski and J. Sully, JHEP **1302**, 062 (2013)
[58] Chen, Y 2013 *J. Phys.* B **46** 104001
[59] Penrose, R 2014 *Found. Phys.* **44** 557
[60] Gross, D J and Mende, P F 1988 *Nucl. Phys.* B **303** 407
[61] Atick, J J and Witten, E 1988 *Nucl. Phys.* B **310** 291
[62] Klebanov, I K and Susskind, L 1988 *Nucl. Phys.* B **309** 175

ISQS IOP Publishing
IOP Conf. Series: Journal of Physics: Conf. Series **804** (2017) 012033 doi:10.1088/1742-6596/804/1/012033

Nonlinear Schrödinger Equation, 2D $\mathcal{N} = (2,2)^*$ Topological Yang-Mills-Higgs Theory and Their Gravity Dual

Jun Nian

Institut des Hautes Études Scientifiques, Le Bois-Marie, 35 route de Chartres, 91440 Bures-sur-Yvette, France

E-mail: nian@ihes.fr

Abstract. The duality between the N-particle sector of quantum nonlinear Schrödinger equation and the 2D $\mathcal{N} = (2,2)^*$ $U(N)$ topological Yang-Mills-Higgs theory was found by Gerasimov and Shatashvili some time ago. At the large N and large 't Hooft coupling limit, the gravity dual of the 2D $\mathcal{N} = (2,2)^*$ $U(N)$ topological Yang-Mills-Higgs theory can be constructed. Consequently, as a first example, one can formulate a triangle relation between integrable model, gauge theory and gravity. We present the results of the gravity dual in this paper, and make some checks at classical level between the gravity dual and the nonlinear Schrödinger equation. This paper is based on the talk given by the author at the 24th International Conference on Integrable Systems and Quantum Symmetries, and more details can be found in Ref. [1].

1. Introduction

Many interesting and profound relations between integrable models and gauge theories have been revealed in recent years [2, 3, 4]. The integrable models are defined in (1+1)D, and they can be nonlinear partial differential equations or lattice spin models. The corresponding 2D gauge theories have $\mathcal{N} = (2,2)^*$ supersymmetry.

Among these relations, the simplest example is the one between the nonlinear Schrödinger equation and the 2D $\mathcal{N} = (2,2)^*$ $U(N)$ topological Yang-Mills-Higgs theory found by Gerasimov and Shatashvili [5, 6]. From the wave function of the 2D $\mathcal{N} = (2,2)^*$ $U(N)$ topological Yang-Mills-Higgs theory one can reproduce the wave function of the quantum nonlinear Schrödinger equation in the N-particle sector.

The 2D $\mathcal{N} = (2,2)^*$ $U(N)$ Yang-Mills-Higgs theory was constructed in Ref. [7]. It can be viewed as the dimensional reduction of 4D topologically twisted $\mathcal{N} = 2$ $U(N)$ super Yang-Mills theory with a deformation term, which breaks 8 supercharges into 4 supercharges. Alternatively, it can also be viewed as the 2D $\mathcal{N} = (2,2)^*$ $U(N)$ super Yang-Mills theory with some supersymmetry exact terms, which do not modify the theory at quantum level. Hence, the 2D $\mathcal{N} = (2,2)^*$ $U(N)$ topological Yang-Mills-Higgs theory is equivalent to the 2D $\mathcal{N} = (2,2)^*$ $U(N)$ super Yang-Mills theory, and we can study the latter one instead.

Based on the principle of gauge/gravity correspondence, we can construct the gravity dual of the 2D $\mathcal{N} = (2,2)^*$ $U(N)$ super Yang-Mills theory in the large N and large 't Hooft coupling limit. Together with the relation between the nonlinear Schrödinger equation and the 2D

IOP Conf. Series: Journal of Physics: Conf. Series **804** (2017) 012033 doi:10.1088/1742-6596/804/1/012033

$\mathcal{N} = (2,2)^*$ $U(N)$ topological Yang-Mills-Higgs theory, now we have a triangle relation among three theories (see Fig. 1).

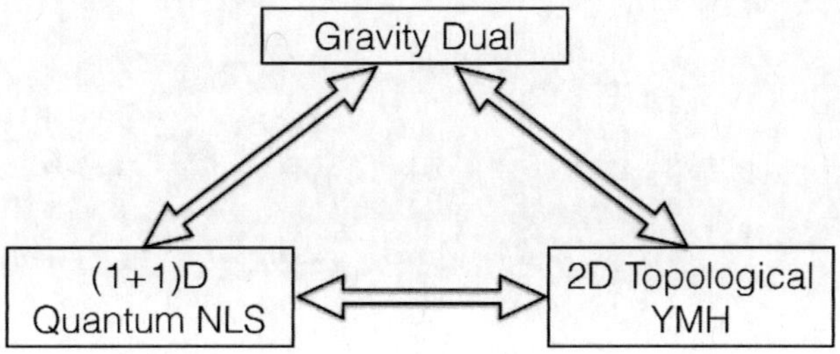

Figure 1. The relation between different theories

By setting up this triangle relation, in principle we can study the integrability of the nonlinear Schrödinger equation on the gravity side, and at the same time study some properties of the gravity on the integrable model side. For the moment, we can first take the classical limit in both the gravity dual and the nonlinear Schrödinger equation, and we can immediately find the correspondence between N D-branes in the supergravity and N solitons in the classical nonlinear Schrödinger equation. The correspondence at quantum level will be investigated in the future.

This paper is based on the talk given by the author at the 24th International Conference on Integrable Systems and Quantum Symmetries, and more details can be found in Ref. [1]. The paper is organized as follows. In Section 2 the 2D $\mathcal{N} = (2,2)^*$ $U(N)$ topological Yang-Mills-Higgs theory, the nonlinear Schrödinger equation and the relation between them will be reviewed. In Section 3, we discuss the construction of the gravity dual of the 2D $\mathcal{N} = (2,2)^*$ $U(N)$ super Yang-Mills theory, and the triangle relation demonstrated in Fig. 1 will be set up. Finally, in Section 4 the prospect for the future research will be discussed.

2. Gerasimov-Shatashvili Duality

In this section, we briefly review the 2D $\mathcal{N} = (2,2)^*$ $U(N)$ topological Yang-Mills-Higgs theory, the nonlinear Schrödinger equation and the relation between them.

2.1. 2D Topological Yang-Mills-Higgs Theory

The 2D $\mathcal{N} = (2,2)^*$ $U(N)$ topological Yang-Mills-Higgs theory was first constructed in Ref. [7]. It is given by the path integral

$$Z_{YMH}(\Sigma_h) = \frac{1}{\mathrm{Vol}(\mathcal{G}_{\Sigma_h})} \int D\varphi_0\, D\varphi_\pm\, DA\, D\Phi\, D\psi_A\, D\psi_\Phi\, D\chi_\pm\, e^S \,, \tag{1}$$

where

$$S = S_0 + S_1 \tag{2}$$

with

$$S_0 = \frac{1}{2\pi}\int_{\Sigma_h} d^2z \left[\mathrm{Tr}\left(i\varphi_0(F(A) - \Phi\wedge\Phi) - c\,\Phi\wedge *\Phi\right) + \varphi_+\nabla_A^{(1,0)}\Phi^{(0,1)} + \varphi_-\nabla_A^{(0,1)}\Phi^{(1,0)}\right], \tag{3}$$

$$S_1 = \frac{1}{2\pi}\int_{\Sigma_h} d^2z\, \mathrm{Tr}\left[\frac{1}{2}\psi_A\wedge\psi_A + \frac{1}{2}\psi_\Phi\wedge\psi_\Phi + \chi_+\left[\psi_A^{1,0},\, \Phi^{(0,1)}\right] + \chi_-\left[\psi_A^{(0,1)},\, \Phi^{(1,0)}\right] + \chi_+\nabla_A^{(1,0)}\psi_\Phi^{(0,1)} + \chi_-\nabla_A^{(0,1)}\psi_\Phi^{(1,0)}\right). \tag{4}$$

In the absence of the deformation term $c\,\mathrm{Tr}(\Phi\wedge *\Phi)$ the theory preserves $\mathcal{N} = (4,4)$ supersymmetry. For generic values of $c \neq 0$ the theory preserves $\mathcal{N} = (2,2)$ supersymmetry, and the $\mathcal{N} = (2,2)$ supersymmetry transformations are given by

$$QA = i\psi_A\,, \quad Q\psi_A = -D\varphi_0\,, \quad Q\varphi_0 = 0\,, \tag{5}$$

$$Q\Phi = i\psi_\Phi\,, \tag{6}$$

$$Q\psi_\Phi^{(1,0)} = [\Phi^{(1,0)},\, \varphi_0] + c\Phi^{(1,0)}\,, \quad Q\psi_\Phi^{(0,1)} = [\Phi^{(0,1)},\, \varphi_0] + c\Phi^{(0,1)}\,, \tag{7}$$

$$Q\chi_\pm = i\varphi_\pm\,, \quad Q\varphi_\pm = [\chi_\pm,\, \varphi_0] \pm c\chi_\pm\,. \tag{8}$$

This theory can also be understood as the dimensional reduction of the 4D topologically twisted $\mathcal{N} = 2$ $U(N)$ super Yang-Mills theory with a deformation term. The details will be discussed in Ref. [1].

From the supersymmetric transformations (5) $\sim$ (8), one can show that the 2D $\mathcal{N} = (2,2)^*$ Yang-Mills-Higgs theory can be written as the 2D $\mathcal{N} = (2,2)^*$ super Yang-Mills theory with a supersymmetry exact deformation as follows:

$$S_{YMH} = S_{YM} + \left[Q,\, \int_{\Sigma_h} d^2z\, \mathrm{Tr}\left(\frac{1}{2}\Phi\wedge\psi_\Phi + \chi_+\nabla_A^{(1,0)}\Phi^{(0,1)} + \chi_-\nabla_A^{(0,1)}\Phi^{(1,0)}\right)\right]. \tag{9}$$

In Ref. [5] the supersymmetry exact deformation term was changed without modifying the theory at quantum level. Based on this fact, in the following, when we construct the gravity dual of the 2D $\mathcal{N} = (2,2)^*$ Yang-Mills-Higgs theory, we will consider the gravity dual of the 2D $\mathcal{N} = (2,2)^*$ super Yang-Mills theory instead.

2.2. Nonlinear Schrödinger Equation

The (1+1)D nonlinear Schrödinger equation is

$$i\partial_t\phi = -\frac{1}{2}\partial_x^2\phi + 2c(\phi^*\phi)\phi\,. \tag{10}$$

The Hamiltonian of the theory is given by

$$\mathcal{H} = \int dx\, \left[\frac{1}{2}\frac{\partial\phi^*}{\partial x}\frac{\partial\phi}{\partial x} + c\,(\phi^*\phi)^2\right], \tag{11}$$

where the field ϕ has the Poisson structure

$$\{\phi^*(x),\, \phi(x')\} = \delta(x - x')\,. \tag{12}$$

In (1+1)D, this theory is integrable both at the classical level and at the quantum level.

For the (1+1)D quantum nonlinear Schrödinger equation, if we consider the N-particle sector in the domain $x_1 \leq x_2 \leq \cdots \leq x_N$, the N-particle wave function satisfies the equation

$$\left(-\frac{1}{2}\sum_{i=1}^{N}\frac{\partial^2}{\partial x_i^2}\right)\Phi_\lambda(x) = 2\pi^2\left(\sum_{i=1}^{N}\lambda_i^2\right)\Phi_\lambda(x)\,, \tag{13}$$

where λ_i denotes the momentum of the i-th particle. The normalized wave function is given by

$$\Phi_\lambda(x) = \sum_{\omega\in W}(-1)^{l(w)}\prod_{i<j}\left(\frac{\lambda_{\omega(i)}-\lambda_{\omega(j)}+ic\,\mathrm{sgn}(x_i-x_j)}{\lambda_{\omega(i)}-\lambda_{\omega(j)}-ic\,\mathrm{sgn}(x_i-x_j)}\right)^{\frac{1}{2}}\exp\left(2\pi i\sum_i\lambda_{\omega(k)}\,x_k\right), \tag{14}$$

where λ_i obey the Bethe Ansatz equation:

$$e^{2\pi i\lambda_j}\prod_{k\neq j}\frac{\lambda_k-\lambda_j-ic}{\lambda_k-\lambda_j+ic} = 1\,,\quad j=1,\cdots,N. \tag{15}$$

2.3. The Duality

Using the technique of cohomological localization, one can compute the partition function of the 2D $\mathcal{N}=(2,2)^*$ $U(N)$ topological Yang-Mills-Higgs theory exactly:

$$Z_{YMH}(\Sigma_h) = e^{(1-h)\,a(c)}\sum_{\lambda\in\mathcal{R}_N}D_\lambda^{2-2h}\,e^{-\sum_{k=1}^{\infty}t_k\,p_k(\lambda)}\,, \tag{16}$$

where $p_k(\lambda)$ is defined as

$$\frac{1}{(2\pi)^k}\,\mathrm{Tr}\,\varphi^k\,\Psi_\lambda(x_1,\cdots,x_N) = p_k(\lambda)\,\Psi_\lambda(x_1,\cdots,x_N)\,, \tag{17}$$

and the factor D_λ is given by

$$D_\lambda = \mu(\lambda)^{-1/2}\prod_{i<j}(\lambda_i-\lambda_j)\left(c^2+(\lambda_i-\lambda_j)^2\right)^{1/2}\,, \tag{18}$$

while $\mathcal{R}_N$ denotes the set of λ_i's satisfying the same Bethe Ansatz equation:

$$e^{2\pi i\lambda_j}\prod_{k\neq j}\frac{\lambda_k-\lambda_j+ic}{\lambda_k-\lambda_j-ic} = 1\,,\quad k=1,\cdots,N. \tag{19}$$

From this analysis, we see the equivalence between the wave function of the 2D $\mathcal{N}=(2,2)^*$ $U(N)$ topological Yang-Mills-Higgs theory and the wave function of the (1+1)D quantum nonlinear Schrödinger equation in the N-particle sector. Hence, the duality between these two theories at quantum level is implied.

3. Gravity Dual

Now we turn to the construction of the gravity dual of the 2D $\mathcal{N}=(2,2)^*$ $U(N)$ super Yang-Mills theory. We follow the same way how the gravity dual of 4D $\mathcal{N}=2^*$ super Yang-Mills theory was constructed [8, 9]. In that case, one can start with the 5D $\mathcal{N}=8$ gauged supergravity, which is a consistent truncation of the 10D type IIB supergravity on $AdS_5\times S^5$. After turning on additional scalar fields and choosing an appropriate scalar potential, one can break the supersymmetry from

$\mathcal{N}=4$ to $\mathcal{N}=2$, which corresponds to a mass deformation on the field theory side. Finally, the solution in the 5D $\mathcal{N}=8$ gauged supergravity can be uplifted to 10D.

In our case, the gravity dual of the 2D $\mathcal{N}=(4,4)$ $U(N)$ super Yang-Mills theory was constructed in Ref. [10] using the brane construction, and it was shown that by changing variables the solutions of the gravity dual can also be obtained from the 5D $\mathcal{N}=2$ gauged supergravity, which is a further consistent truncation of the 5D $\mathcal{N}=8$ gauged supergravity. Therefore, following the same logic of Ref. [8, 9], we can turn on an additional scalar field and choose an appropriate scalar potential to break the supersymmetry from $\mathcal{N}=(4,4)$ to $\mathcal{N}=(2,2)$ in the 5D $\mathcal{N}=2$ gauged supergravity. After that, the 5D gravity solution can also be uplifted to 10D and rewritten into an expression, from which the brane construction is more transparent.

3.1. *Gravity Dual of 2D $\mathcal{N}=(4,4)$ Super Yang-Mills Theory*

As explained in the beginning of this section, to construct the gravity of the 2D $\mathcal{N}=(2,2)^*$ $U(N)$ super Yang-Mills theory, we start with the known gravity dual of the 2D $\mathcal{N}=(4,4)$ $U(N)$ super Yang-Mills theory, which was constructed in Ref. [10]. Let us briefly review the construction in this subsection.

To realize the $\mathcal{N}=(4,4)$ supersymmetry, one considers a D3-brane wrapped on the 2-cycle of a CY 2-fold, which can be seen from the following table:

	$\mathbb{R}^{1,1}$		S^2		N_2		$\mathbb{R}^4$			
D3	−	−	○	○	·	·	·	·	·	·

Locally, the CY 2-fold is $S^2 \times N_2$. From the brane construction, one can propose an Ansatz of the metric in 10 type IIB supergravity:

$$ds^2 = H^{-\frac{1}{2}}\left[dx_{1,1}^2 + \frac{z}{m^2}\left(d\theta^2 + \sin^2\theta\,(d\phi)^2\right)\right] + H^{\frac{1}{2}}\left[\frac{1}{z}d\sigma^2 + \frac{\sigma^2}{z}\left(d\psi + \cos\theta\, d\phi\right)^2 + d\rho^2 + \rho^2 d\Omega_3^2\right], \tag{20}$$

where

$$0 \leq \theta \leq \pi\,, \quad 0 \leq \phi,\, \psi < 2\pi\,, \quad 0 \leq \rho,\, \sigma < \infty\,. \tag{21}$$

In addition, the RR 5-form in the 10D type IIB supergravity is given by

$$F_5 = \mathcal{F}_5 + *\mathcal{F}_5\,, \tag{22}$$

where $\mathcal{F}_5 = d\mathcal{C}_4$ with

$$\mathcal{C}_4 = g(\rho,\sigma)\,\omega_3 \wedge (d\psi + \cos\theta\, d\phi) \tag{23}$$

and ω_3 is the volume element of the 3-sphere, i.e.,

$$\omega_3 = \sin^2\alpha_1\,\sin\alpha_2\, d\alpha_1 \wedge d\alpha_2 \wedge \alpha_3\,. \tag{24}$$

The constant m in the metric (20) is fixed by the quantization condition of the RR 5-form F_5:

$$\frac{1}{2\kappa_{10}^2}\int_{\mathcal{M}_5} F_5 = N_c T_3 \tag{25}$$

with

$$2\kappa_{10}^2 = (2\pi)^7 g_s^2 (\alpha')^4\,, \quad T_3 = \frac{1}{(2\pi)^3 g_s (\alpha')^2}\,. \tag{26}$$

IOP Conf. Series: Journal of Physics: Conf. Series **804** (2017) 012033 doi:10.1088/1742-6596/804/1/012033

After some analyses, one finds that the constant m is fixed by

$$\frac{1}{m^2} = \sqrt{4\pi g_s N_c \alpha'}\,, \tag{27}$$

where g_s and α' are the string coupling constant and the Regge slope respectively.

From the metric (20) and the flux (22), one can write down the BPS equations and try to solve them. It turns out that the BPS equations can be solved by using the results from the 5D $\mathcal{N}=2$ gauged supergravity discussed in Ref. [11]. This is due to the fact that the metric (20) can also be constructed from the 5D $\mathcal{N}=2$ gauged supergravity.

The bosonic part of the 5D $\mathcal{N}=2$ gauged supergravity is given by

$$\mathcal{L} = R - \frac{1}{2}(\partial_\mu\phi_1)^2 - \frac{1}{2}(\partial_\mu\phi_2)^2 + 4\sum_{i=1}^{3} e^{\alpha_i} - \frac{1}{4}\sum_{i=1}^{3} e^{2\alpha_i} F^i_{\mu\nu}F^{i,\mu\nu} + \frac{1}{4}\epsilon^{\mu\nu\alpha\beta\rho}F^1_{\mu\nu}F^2_{\alpha\beta}A^3_\rho\,, \tag{28}$$

where

$$\alpha_1 = \frac{\phi_1}{\sqrt{6}} + \frac{\phi_2}{\sqrt{2}}\,, \quad \alpha_2 = \frac{\phi_1}{\sqrt{6}} - \frac{\phi_2}{\sqrt{2}}\,, \quad \alpha_3 = -\frac{2}{\sqrt{6}}\phi_1\,. \tag{29}$$

For the 5D gauged supergravity compactified on a surface H^2, there is the following condition:

$$a_1 + a_2 + a_3 = 1\,. \tag{30}$$

It was explained in Ref. [11] that one can obtain the solution for the compactification on S^2 by replacing $\theta \to i\theta$.

By choosing $a_I = (0,\,0,\,1)$, it was constructed in Ref. [10] the gravity solution with $\mathcal{N}=(4,4)$ supersymmetry in the 5D $\mathcal{N}=2$ gauged supergravity:

$$ds_5^2 = e^{2f(r)}\left(dx_{1,1}^2 + dr^2\right) + \frac{e^{2g(r)}}{m^2}\left[d\theta^2 + \sin^2\theta\,(d\phi)^2\right]\,. \tag{31}$$

The gauge fields are chosen to be

$$A^1 = 0\,, \quad A^2 = 0\,, \quad A^3 = \frac{1}{m}\cos\theta\, d\phi\,. \tag{32}$$

This solution can be uplifted to 10D in the following way:

$$ds_{10}^2 = \sqrt{\Delta}ds_5^2 + \frac{3}{m^2\sqrt{\Delta}}\sum_{i=1}^{3} X_i\left[d\mu_i^2 + \mu_i^2\left(d\phi^i + mA^i\right)^2\right]\,, \tag{33}$$

where

$$\Delta = \sum_{i=1}^{3} X^i\mu_i^2\,, \tag{34}$$

and

$$\sum_{i=1}^{3}\mu_i^2 = 1\,. \tag{35}$$

One can parametrize μ_i's as follows:

$$\mu_1 = \cos\widetilde{\theta}\sin\widetilde{\psi}\,, \quad \mu_2 = \cos\widetilde{\theta}\cos\widetilde{\psi}\,, \quad \mu_3 = \sin\widetilde{\theta}\,, \tag{36}$$

where $0 \le \widetilde{\theta} \le \pi/2$ and $0 \le \widetilde{\psi} \le 2\pi$. The quantities X_i and X^i are defined by

$$X_i = \frac{1}{3}\left(e^{\varphi},\, e^{\varphi},\, e^{-2\varphi}\right)\,, \quad X^i = \left(e^{-\varphi},\, e^{-\varphi},\, e^{2\varphi}\right)\,, \tag{37}$$

It was shown in Ref. [10] that indeed the metric (33) can be rewritten into the expression of the metric (20) discussed before by changing variables.

3.2. Gravity Dual of 2D $\mathcal{N} = (2,2)^*$ Super Yang-Mills Theory
Let us consider the gravity dual of the 2D $\mathcal{N} = (2,2)^*$ $U(N)$ super Yang-Mills theory. As mentioned in the beginning of this section, we can start with the the gravity dual of the 2D $\mathcal{N} = (4,4)$ $U(N)$ super Yang-Mills theory discussed in the previous section, and break the supersymmetry from $\mathcal{N} = (4,4)$ to $\mathcal{N} = (2,2)$ in the framework of the 5D $\mathcal{N} = 2$ gauged supergravity [11]. To break supercharges in the 5D $\mathcal{N} = 2$ gauged supergravity, one can simply choose the parameters a_I to be

$$a_I = (\widetilde{c},\, 0,\, 1 - \widetilde{c}) \tag{38}$$

with $0 \leq \widetilde{c} < 1$. When $\widetilde{c} = 0$, it returns to the case analyzed in the previous subsection, which preserves $\mathcal{N} = (4,4)$ supersymmetry. When $0 < \widetilde{c} < 1$, the gravity solution preserves $\mathcal{N} = (2,2)$ supersymmetry. The gauge fields now become

$$A^1 = \frac{\widetilde{c}}{m}\cos\theta\, d\phi\,, \quad A^2 = 0\,, \quad A^3 = \frac{1-\widetilde{c}}{m}\cos\theta\, d\phi\,. \tag{39}$$

Similar to Ref. [11], one can also solve the BPS equations near the boundary $r = 0$ (see Ref. [1] for more details). The asymptotic solutions are

$$g(r) = -\log(r) + \frac{7}{36}r^2 + \cdots\,, \tag{40}$$

$$f(r) = -\log(r) - \frac{1}{18}r^2 + \cdots\,, \tag{41}$$

$$\phi_1(r) = -\frac{1-3a_3}{\sqrt{6}}r^2\log(r) + \cdots\,, \tag{42}$$

$$\phi_2(r) = -\frac{a_1 - a_2}{\sqrt{2}}r^2\log(r) + \cdots\,, \tag{43}$$

which suggest that the scalar ϕ_2 corresponds to the dimension-2 operator $c\,\mathrm{Tr}(\Phi\wedge *\Phi)$ in the field theory. In the absense of this mass deformation, the field theory has $\mathcal{N} = (4,4)$ supersymmetry, and the scalar ϕ_2 is turned off on the gravity side. When the mass deformation is turned on, the scalar ϕ_2 should also be turned on correspondingly. Hence, at least for small values of c and $\widetilde{c}$, there should be

$$c \propto \widetilde{c}\,. \tag{44}$$

Based on our previous analyses, the 10D metric preserving $\mathcal{N} = (2,2)^*$ supersymmetry can be given in the following way:

$$\begin{aligned} ds_{10}^2 = {} & \sqrt{\Delta}\left[e^{2f}(dx_{1,1}^2 + dr^2) + \frac{e^{2g}}{m^2}\left(d\theta^2 + \sin^2\theta\,(d\phi)^2\right)\right] \\ & + \frac{1}{m^2\sqrt{\Delta}}\Bigg[e^{\varphi_1+\varphi_2}d\mu_1^2 + e^{\varphi_1-\varphi_2}d\mu_2^2 + e^{-\varphi_2}d\mu_3^2 + e^{\varphi_1+\varphi_2}\cos^2\widetilde{\theta}\sin^2\widetilde{\psi}\left(d\phi^1 + \widetilde{c}\cos\theta\, d\phi\right)^2 \\ & \qquad + e^{\varphi_1-\varphi_2}\cos^2\widetilde{\theta}\cos^2\widetilde{\psi}(d\phi^2)^2 + e^{-2\varphi_1}\sin^2\widetilde{\theta}\left(d\phi^3 + (1-\widetilde{c})\cos\theta\, d\phi\right)^2\Bigg]\,, \end{aligned} \tag{45}$$

where $\varphi_1 = \phi_1/\sqrt{6}$, $\varphi_2 = \phi_2/\sqrt{2}$, and Δ has the same definition as Eq. (34).

In principle, this 10D metric (45) can be rewritten into the form similar to the one given by Eq. (20), from which the brane construction is more transparent. However, for generic values of $\widetilde{c}$, the explicit form of the metric can be very complicated. To simplify our discussions, let us consider a special case $\widetilde{c} = 1/2$. For this case, the metric (45) can be rewritten into a form similar to Eq. (20) by changing variables (for more details see Ref. [1]):

$$\begin{aligned} ds^2 = & H^{-\frac{1}{2}} \left[dx_{1,1}^2 + \frac{z}{m^2} \left(d\theta^2 + \sin^2\theta \, (d\phi)^2 \right) \right] \\ & + H^{\frac{1}{2}} \left[\frac{1}{\sqrt{z}} d\sigma^2 + \frac{\sigma^2}{\sqrt{z}} \left(d\widetilde{\psi}^2 + \sin^2\widetilde{\psi} \left(d\phi^3 + \frac{1}{2}\cos\theta \, d\phi \right)^2 + \cos^2\widetilde{\psi} \left(d\phi^1 + \frac{1}{2}\cos\theta \, d\phi \right)^2 \right) \right. \\ & \left. + d\rho^2 + \rho^2 d\psi^2 \right] . \end{aligned} \tag{46}$$

From this metric one can read off that the brane construction for this case ($\widetilde{c} = 1/2$) becomes

	$\mathbb{R}^{1,1}$		S^2		N_4				$\mathbb{R}^2$	
D3	−	−	○	○	·	·	·	·	·	·

Hence, now the solutions describes a D3-brane wrapped on the 2-cycle of a CY 3-fold, which is locally $S^2 \times N_4$. This configuration realizes $\mathcal{N} = (2, 2)$ supersymmetry. By turning on the mass deformation on the field theory side, we see a change of topology on the gravity side, i.e., from a CY 2-fold at $\widetilde{c} = 0$ becomes a CY 3-fold at $\widetilde{c} \neq 0$, and consequently the supersymmetry is broken from $\mathcal{N} = (4, 4)$ to $\mathcal{N} = (2, 2)$.

To have the complete gravity solution in 10D type IIB supergravity, we still need to specify the RR 5-form F_5. It is given by

$$F_5 = \mathcal{F}_5 + *\mathcal{F}_5 \, , \tag{47}$$

where

$$\mathcal{F}_5 = \sum_{i=1}^{3} \left[2X^i (X^i \mu_i^2 - \Delta) \epsilon_5 + \frac{1}{2(X^i)^2} d(\mu_i^2) \left((d\phi_i + A^i) \wedge *_5 F^i + X^i *_5 dX^i \right) \right] , \tag{48}$$

and ϵ_5 and $*_5$ are the volume form of ds_5^2 and the Hodge dual in ds_5 respectively, while $F^i = dA^i$ are the field strengths of the gauge fields given by Eq. (39). Similar to the $\mathcal{N} = (4, 4)$ case, the quantization condition of F_5 will fix the constant m as in Eq. (27).

There is a subtle point that we would like to emphasize. In the original construction in Ref. [11], the gravity solution has a 2-form field strength, which should satisfy a quantization condition when integrated on the Riemann surface. This condition imposes the constraint on the parameters a_I's that they should have rational instead of real values. However, in our discussion above, we only have a flux quantization condition on the 5-form F_5, which fixes the constant m, hence, the a_I's can be real in our solutions.

3.3. The Triangle Relation

After finding the gravity dual, we have set up a triangle relation among the N-particle sector of quantum nonlinear Schrödinger equation, the 2D $\mathcal{N} = (2, 2)^*$ $U(N)$ topological Yang-Mills-Higgs theory and their gravity dual. This relation should hold at quantum level in the limit $N \to \infty$, $g_{YM}^2 \to 0$ and $g_{YM}^2 N \to \infty$. As a first check, let us take the classical limit $\hbar \to 0$.

It was known in the literature that when $N \to \infty$ the N-particle sector of the quantum nonlinear Schrödinger equation with an attractive potential becomes the N-soliton sector of the classical nonlinear Schrödinger equation. Therefore, in the classical limit we can identify the N-solitons in the nonlinear Schrödinger equation with the N D3-branes from the brane construction on the gravity side, while N becomes the rank of the gauge group on the gauge theory side.

4. Discussions

In this paper, we illustrate the way of setting up a triangle relation between three theories by finding the gravity dual of the 2D $\mathcal{N} = (2,2)^* \ U(N)$ topological Yang-Mills-Higgs theory. More details of this approach can be found in Ref. [1]. We believe this result opens up a new direction of research, and a lot of interesting questions should be addressed and clarified. Hopefully, it can provide some deep insights into the connections of integrable models, gauge theories and gravities.

An immediate generalization is to add matters in the fundamental representation of the gauge group, i.e., add flavors in the gravity dual. This will allow us to find the gravity dual of the more general duality found by Nekrasov and Shatashvili, and consequently to study the more general triangle relation. Another related question is to study the integrability on the gravity side both at the classical level and at the quantum level. We would like to investigate these perspectives in the future research.

A little unexpected relation comes from the boson/vortex duality discussed in Refs. [12, 13] and recently revisited in Refs. [14, 15]. Using this duality, one can show that in (3+1)D nonlinear Schrödinger equatoin can be mapped into an effective string theory. Correspondingly, different solutions to the (3+1)D nonlinear Schrödinger equatoin (vortex lines, vortex rings, dark solitons) can be mapped into some configurations in the effective string theory (open string, closed strings, D-branes). To relate this duality to the (1+1)D nonlinear Schrödinger equatoin, which is an integrable model, one can either perform a dimensional reduction or apply the duality map directly in (1+1)D. In this way a lot of interesting features emerge, and many apparently different theories are related in a larger duality web. We would like to present this work elsewhere.

Acknowledgements

We would like to thank Daniel Areán, André Coimbra, Heng-Yu Chen, Ilmar Gahramanov, Albrecht Klemm, Vladimir Korepin, Peter Koroteev, Hamid Noori, Vasily Pestun, Nuno Romão, Vatche Sahakian, Olof Ohlsson Sax, Linus Wulff and Takehiko Yasuda for many useful discussions.

References

[1] Nian J and Pestun V *to appear soon*
[2] Nekrasov N A and Shatashvili S L 2009 *Nucl. Phys. Proc. Suppl.* **192-193** 91–112 (*Preprint* 0901.4744)
[3] Nekrasov N A and Shatashvili S L 2009 *Prog. Theor. Phys. Suppl.* **177** 105–119 (*Preprint* 0901.4748)
[4] Nekrasov N A and Shatashvili S L 2009 *Proceedings, 16th International Congress on Mathematical Physics (ICMP09): Prague, Czech Republic, August 3-8, 2009* pp 265–289 (*Preprint* 0908.4052)
[5] Gerasimov A A and Shatashvili S L 2008 *Commun. Math. Phys.* **277** 323–367 (*Preprint* hep-th/0609024)
[6] Gerasimov A A and Shatashvili S L 2007 *Proceedings of Symposia in Pure Mathematics, May 25-29 2007, University of Augsburg, Germany* (*Preprint* 0711.1472)
[7] Moore G W, Nekrasov N and Shatashvili S 2000 *Commun. Math. Phys.* **209** 97–121 (*Preprint* hep-th/9712241)
[8] Freedman D Z, Gubser S S, Pilch K and Warner N P 1999 *Adv. Theor. Math. Phys.* **3** 363–417 (*Preprint* hep-th/9904017)
[9] Pilch K and Warner N P 2001 *Nucl. Phys.* **B594** 209–228 (*Preprint* hep-th/0004063)
[10] Arean D, Merlatti P, Nunez C and Ramallo A V 2008 *JHEP* **12** 054 (*Preprint* 0810.1053)
[11] Maldacena J M and Nunez C 2001 *Int. J. Mod. Phys.* **A16** 822–855 [182(2000)] (*Preprint* hep-th/0007018)
[12] Zee A 1994 *Nucl.Phys.* **B421** 111–124
[13] Gubser S S, Nayar R and Parikh S 2015 *Nucl. Phys.* **B892** 156–180 (*Preprint* 1408.2246)
[14] A Muñoz Mateo, Yu X and Nian J 2016 (*Preprint* 1606.02776)
[15] Giaccari S and Nian J 2016 (*Preprint* 1608.07262)

IOP Conf. Series: Journal of Physics: Conf. Series **804** (2017) 012034 doi:10.1088/1742-6596/804/1/012034

Minimal conformal supergravity invariants in six dimensions

J Novak

Max-Planck-Institut für Gravitationsphysik, Albert-Einstein-Institut,
Am Mühlenberg 1, D-14476 Golm, Germany

E-mail: joseph.novak@aei.mpg.de

Abstract. The off-shell superspace formulation for six-dimensional conformal supergravity given in arXiv:1606.02921 is presented here. We describe how the formulation may be used to describe two invariants for $\mathcal{N} = (1, 0)$ conformal supergravity in six dimensions. One of the invariants contains a C^3 term at the component level, while the other contains a $C\Box C$ term. It is demonstrated that any invariant for minimal conformal supergravity must be a linear combination of these two invariants.

1. Introduction

Conformal gravity invariants play an important role in conformal field theories, where on an arbitrary background anomalies appear in the Ward identities which follow from the conformal symmetry. These anomalies express the violation of the conservation and tracelessness of the stress-energy tensor in certain correlation functions. It is well known [1] that there are two main types of conformal anomalies in even dimensions: type A and type B. There is always exactly one type A anomaly but an increasing number of type B anomalies by increasing spacetime dimension D. The type B anomalies are associated with conformal gravity invariants, which are locally diffeomorphism and Weyl invariant functions of the metric and its derivatives. In four dimensions (4D) there is only one conformal gravity invariant (or type B anomaly), while in six dimensions (6D) there are three.

In 6D and in the presence of supersymmetry conformal anomalies acquire further significance. Supersymmetry places additional restrictions on the general structure of correlation functions. Remarkably, 6D proves to be the highest spacetime dimension in which superconformal field theories exist [2]. Furthermore, the only known non-trivial unitary conformal field theories in 6D are supersymmetric and arise in string theory, and realize either $\mathcal{N} = (2, 0)$ and $\mathcal{N} = (1, 0)$ superconformal symmetry. This places 6D superconformal field theories at a special point of study in the space of conformal field theories.

An important step in the study of the general anomaly structure of 6D superconformal field theories is the construction of the conformal supergravity invariants, which are supersymmetric extensions of the conformal gravity invariants. For the minimally supersymmetric version, $\mathcal{N} = (1, 0)$, superconformal tensor calculus was developed over thirty years ago in [3] and later applied to describe certain 6D $\mathcal{N} = (1, 0)$ supergravity invariants, such as a supersymmetric Riemann curvature squared term [4]. However, it has not provided insight into the construction of further higher derivative invariants, in particular the minimal conformal supergravity invariants, which are important ingredients in the description of general supergravity-matter systems.

Apart from superconformal tensor calculus, there is another approach to $\mathcal{N}$-extended conformal supergravity for $D \leq 6$ based on using a curved $\mathcal{N}$-extended superspace $\mathcal{M}^{D|\delta}$, where δ is the number of fermionic dimensions. Within the superspace setting, one can choose to gauge only part of the superconformal algebra and use the structure group $\mathrm{SO(D-1,1)} \times G_R$, where G_R is some R-symmetry group. The superspace geometry is then constrained to describe conformal supergravity and thus permits additional transformations known as super-Weyl transformations, which are generated by a real unconstrained superfield parameter.[1] However, for applications involving the construction of super-Weyl invariant objects, it is often more useful to make use of another superspace formulation, known as conformal superspace [9, 10, 11, 12], which is based on gauging the entire superconformal group and is more general since the conventional formulation may be obtained by partially fixing the gauge freedom. Conformal superspace is technically easier to reduce to components and has already proved useful in the construction of higher derivative invariants, e.g. the 5D supersymmetric R^2 invariant in the standard Weyl multiplet [12]. For these reasons it is useful to have a 6D $\mathcal{N} = (1,0)$ conformal superspace formulation for the construction of 6D minimal conformal supergravity invariants.

Before moving on to presenting the 6D conformal superspace formulation, we will first illustrate a conceptual difference in the 6D $\mathcal{N} = (1,0)$ case from its 4D $\mathcal{N} = 2$ counterpart in the construction of the conformal supergravity invariants. The invariant for 4D $\mathcal{N} = 2$ conformal supergravity is a supersymmetric C^2 term and is remarkably simple.[2] It is given by a chiral integral of the form

$$I_{C^2} = \int \mathrm{d}^4x\, \mathrm{d}^{2\mathcal{N}}\theta\, \mathcal{E}\, \mathcal{L}_{\mathrm{c}} + \mathrm{c.c.} = \int \mathrm{d}^4x\, \mathrm{d}^{2\mathcal{N}}\theta\, \mathcal{E}\, W^{\alpha\beta} W_{\alpha\beta} + \mathrm{c.c.}\,, \tag{1.1}$$

where $\mathcal{E}$ is the chiral measure and $W^{\alpha\beta}$ is the super-Weyl tensor. However, the 6D $\mathcal{N} = (1,0)$ case is conceptually different as there is no covariant analogue to the chiral action. Furthermore, although it is natural to consider a full superspace integral of the form $\int \mathrm{d}^6x \mathrm{d}^8\theta E \mathcal{L}$, where $\mathcal{L}$ is a real primary superfield of dimension 2, an appropriate superspace Lagrangian $\mathcal{L}$ cannot be constructed from the 6D $\mathcal{N} = (1,0)$ super-Weyl tensor $W^{\alpha\beta}$. This makes addressing the 6D case non-trivial as one needs to find an appropriate action principle analogous to the chiral action.

In this paper we present the results of [13] in which the 6D minimal conformal supergravity invariants were constructed. The invariants were constructed by achieving the following: (i) developing the 6D $\mathcal{N} = (1,0)$ conformal superspace; (ii) deriving action principles capable of supporting the Weyl invariants; and (iii) applying the action principles to describe the Weyl invariants. We will describe each of these points in the main body of this paper.

This paper is organized as follows. In section 2 we give a brief review of 6D conformal gravity highlighting important points of relevance for section 3. In section 3 we present the conformal superspace formulation of conformal supergravity and demonstrate that there should only be two invariants for minimal conformal supergravity. Section 4 is devoted to a review on the superform approach to the construction of locally supersymmetric invariants and how it can be used to construct the invariants for minimal conformal supergravity. Finally, conclusions are presented in section 5.

2. Conformal gravity in six dimensions

Before diving into superspace it is useful to briefly review 6D conformal gravity and the conformal gravity invariants emphasising some important points for the subsequent presentation.

[1] This is known as the conventional superspace approach. It was developed for the 4D $\mathcal{N} = 1$ and $\mathcal{N} = 2$ cases in [5, 6] and presented for the 5D $\mathcal{N} = 1$ case in [7] which was extended to the 6D $\mathcal{N} = (1,0)$ case in [8].

[2] Here C schematically represents the Weyl tensor C_{abcd}.

Conformal gravity in 6D may be viewed as being based on gauging the entire conformal group SO(6,2), which contains the generators $X_{\underline{A}} = \{P_a, M_{ab}, \mathbb{D}, K_a\}$ corresponding to translation, Lorentz, dilatation and special conformal transformations, respectively.[3] To gauge the conformal group one begins with a 6D manifold parametrised by local coordinates x^m, $m = 0, 1, \cdots 5$. Then one introduces the inverse vielbein $e_a = e_a{}^m \partial_m$ associated with P_a, and the Lorentz connection $\omega_a{}^{bc}$, the dilatation connection b_a and the special conformal connection f_{ab} associated with each of their respective generators. The transformations laws for the vielbein and the connections are chosen such that

$$\nabla_a = e_a - \frac{1}{2}\omega_a{}^{bc} M_{bc} - b_a \mathbb{D} - f_{ab} K^b \tag{2.1}$$

transform covariantly. Their transformation laws under the gravity gauge group may be summarized by[4]

$$\delta_{\mathcal{K}} \nabla_a = [\mathcal{K}, \nabla_a] \ , \quad \mathcal{K} := \xi^a \nabla_a + \Lambda^{\underline{a}} X_{\underline{a}} := \xi^a \nabla_a + \frac{1}{2}\Lambda(M)^{ab} M_{ab} + \sigma \mathbb{D} + \Lambda(K)_a K^a \ . \tag{2.2}$$

It is worth mentioning that the transformation laws of the connections are such that the generators $X_{\underline{a}}$ act on the covariant derivatives in the same way as they do on P_a, except with P_a replaced by ∇_a.

In order to describe conformal gravity it is necessary to impose some constraints on the geometry. These constraints fix the entire covariant derivative algebra in terms of the Weyl tensor as follows

$$[\nabla_a, \nabla_b] \equiv -\mathcal{F}_{ab} = -\frac{1}{2} C_{ab}{}^{cd} M_{cd} - \frac{1}{6} \nabla^d C_{abcd} K^c \ , \tag{2.3}$$

where C_{abcd} is the Weyl tensor which is primary,[5] $K_f C_{abcd} = 0$, and satisfies the symmetry properties, $C_{abcd} = C_{[ab][cd]}$ and $C_{[abc]d} = 0$. The Bianchi identity $[\nabla_{[a}, \mathcal{F}_{bc]}] = 0$ requires the following differential constraint on the Weyl tensor

$$\nabla_{[a} C_{bc]}{}^{de} = -\frac{2}{3} \nabla_f C_{[ab}{}^{f[d} \delta^{e]}_{c]} \ . \tag{2.4}$$

It is important to note that the covariant derivative algebra is expressed entirely in terms of the Weyl tensor, which provides a very simple way of seeing that a vanishing Weyl tensor implies a conformally flat geometry.

All conformal gravity invariants may be written down as integrals of the form

$$I = \int \mathrm{d}^6 x \, e \, L \ , \quad K_a L = 0 \ , \quad \mathbb{D} L = 6L \ , \tag{2.5}$$

where the Lagrangian L is a dimension 6 primary field constructed out of the Weyl tensor. There are exactly three possible Lagrangians that one may construct:

$$L^{(1)}_{C^3} = C_{abcd} C^{aefd} C_e{}^{bc}{}_f \ , \tag{2.6a}$$

$$L^{(2)}_{C^3} = C_{abcd} C^{cdef} C_{ef}{}^{ab} \ , \tag{2.6b}$$

$$L_{C\Box C} = C^{abcd} \Box C_{abcd} + \frac{1}{2} (\nabla_e C_{abcd}) \nabla^e C^{abcd} + \frac{8}{9} (\nabla^d C_{abcd}) \nabla_e C^{abce} \ , \tag{2.6c}$$

where $\Box := \nabla^a \nabla_a$.

[3] We refer the reader to [13] for the conformal algebra and conventions.
[4] A tensor field U transforms as $\delta_{\mathcal{K}} U = \mathcal{K} U$.
[5] A superfield U is primary if $K_a U = 0$.

We highlight that once we have gauged the entire conformal algebra the curvature and torsion tensors are constrained such that they are expressed entirely terms of the Weyl tensor and its covariant derivatives. Furthermore, the conformal gravity invariants are all described by primary Lagrangians. In the remaining sections we seek to find an analogous description for the supersymmetric case.

3. $\mathcal{N} = (1,0)$ conformal superspace

In lower dimensions [9, 10, 11, 12] conformal superspace possesses the following key properties: (i) the entire superconformal algebra is gauged in superspace; (ii) the torsion and curvature tensors are built out of covariant derivatives of a single primary superfield; and (iii) the covariant derivative algebra resembles that of supersymmetric Yang-Mills theory. In this section we present the 6D $\mathcal{N} = (1,0)$ conformal superspace developed in [13], which satisfies the three aforementioned properties. For the sake of brevity we refer the reader to [13] for our conventions and the superconformal algebra, which now contains, in addition to the generators in the bosonic case, the fermionic generators Q^i_α and S^α_i, denoting the Q-supersymmetry and S-supersymmetry generators, as well as J^{ij} the SU(2) R-symmetry generator.

3.1. Gauging the superconformal algebra

To gauge the superconformal algebra one takes a $\mathcal{N} = (1,0)$ curved superspace $\mathcal{M}^{6|8}$ parametrised by coordinates $z^M = (x^m, \theta^\mu_{\mathtt{i}})$, where $m = 0,1,2,3,4,5$, $\mu = 1,2,3,4$ and $\mathtt{i} = \underline{1}, \underline{2}$. One now associates with each generator $X_{\underline{a}} = (M_{ab}, J_{ij}, \mathbb{D}, S^\gamma_k, K^c)$ a connection one-form $\omega^{\underline{a}} = (\Omega^{ab}, \Phi^{ij}, B, \mathfrak{F}^k_\gamma, \mathfrak{F}_c) = \mathrm{d}z^M \omega_M{}^{\underline{a}}$ and with P_A the vielbein $E^A = (E^\alpha_i, E^a)$. They may be used to construct the covariant derivatives

$$\nabla_A = E_A{}^M \partial_M - \frac{1}{2}\Omega_A{}^{ab} M_{ab} - \Phi_A{}^{ij} J_{ij} - B_A \mathbb{D} - \mathfrak{F}_A{}^j_\beta S^\beta_j - \mathfrak{F}_{Ab} K^b \ . \tag{3.1}$$

It is important to note that the generators appearing in the structure group act on the covariant derivatives in the same way as they do on P_A except with P_A replaced by ∇_A.

The supergravity gauge transformations of the covariant derivatives may be summarised as

$$\delta_\mathcal{K} \nabla_A = [\mathcal{K}, \nabla_A] \ , \quad \mathcal{K} := \xi^A \nabla_A + \frac{1}{2}\Lambda^{bc} M_{bc} + \Lambda^{ij} N_{ij} + \sigma \mathbb{D} + \Lambda^i_\alpha S^\alpha_i + \Lambda_a K^a \ , \tag{3.2}$$

where the gauge parameters associated with $\mathcal{K}$ satisfy natural reality properties. It should be mentioned that the supergravity transformations act on a tensor superfield U as $\delta_\mathcal{K} U = \mathcal{K} U$. The superfield U is said to be *primary* and of dimension Δ if $K_A U = 0$ and $\mathbb{D} U = \Delta U$.

The torsion and curvatures appear in the the (anti-)commutation relations

$$\begin{aligned}[\nabla_A, \nabla_B\} \equiv -\mathcal{F}_{AB} = -T_{AB}{}^C \nabla_C - \frac{1}{2} R(M)_{AB}{}^{cd} M_{cd} - R(N)_{AB}{}^{kl} J_{kl} \\ - R(\mathbb{D})_{AB} \mathbb{D} - R(S)_{AB}{}^k_\gamma S^\gamma_k - R(K)_{ABc} K^c \ ,\end{aligned} \tag{3.3}$$

where the torsion and curvatures are constrained to satisfy the Bianchi identities

$$[\nabla_{[A}, \mathcal{F}_{BC\}}\} = 0 \ . \tag{3.4}$$

3.2. Conformal supergravity

It is clear that the geometric setup of the previous subsection contains too many fields to describe conformal supergravity. In order to describe conformal supergravity it is necessary to impose further constraints on the covariant derivative algebra. As in lower dimensions, suitable constraints prove to resemble those of the Yang-Mills multiplet. Here we explicitly present the consequences in the 6D $\mathcal{N} = (1,0)$ case.

The covariant derivative algebra is constrained to resemble that of Yang-Mills theory:

$$\begin{aligned}
\{\nabla_\alpha^i, \nabla_\beta^j\} &= -2\mathrm{i}\varepsilon^{ij}(\gamma^c)_{\alpha\beta}\nabla_c\ , \\
[\nabla_a, \nabla_\beta^j] &= -\mathcal{F}_{a\beta}{}^j = (\gamma_a)_{\alpha\beta}\mathcal{W}^{\beta i}\ , \\
[\nabla_a, \nabla_b] &= -\mathcal{F}_{ab} = -\frac{\mathrm{i}}{8}(\gamma_{ab})_\alpha{}^\beta\{\nabla_\beta^k, \mathcal{W}_k^\alpha\}\ ,
\end{aligned} \tag{3.5}$$

where the operator $\mathcal{W}^{\alpha i}$ is constrained by the Bianchi identity

$$[\nabla_{[A}, \mathcal{F}_{BC\}}\} = 0 \quad \Longrightarrow \quad \{\nabla_\alpha^{(i}, \mathcal{W}^{\beta j)}\} = \frac{1}{4}\delta_\alpha^\beta\{\nabla_\gamma^{(i}, \mathcal{W}^{\gamma j)}\}\ . \tag{3.6}$$

The operator $\mathcal{W}^{\alpha i}$ is further constrained to be built out of the super-Weyl tensor $W^{\alpha\beta}$ [8] as follows

$$\mathcal{W}^{\alpha i} = W^{\alpha\beta}\nabla_\beta^i + \frac{1}{2}\mathcal{W}(W)^{\alpha i ab}M_{ab} + \mathcal{W}(W)^{\alpha i}\mathbb{D} + \mathcal{W}(W)^{\alpha i}{}_B K^B\ , \tag{3.7}$$

where super-Weyl tensor $W^{\alpha\beta}$ is a symmetric primary superfield of dimension 1,

$$W^{\alpha\beta} = W^{\beta\alpha}\ , \quad K^A W^{\beta\gamma} = 0\ , \quad \mathbb{D}W^{\alpha\beta} = W^{\alpha\beta}\ . \tag{3.8}$$

The Bianchi identity (3.6) is now equivalent to the following differential constraints on the super-Weyl tensor:

$$\nabla_\alpha^{(i}\nabla_\beta^{j)}W^{\gamma\delta} = -\delta_{[\alpha}^{(\gamma}\nabla_{\beta]}^{(i}\nabla_\rho^{j)}W^{\delta)\rho}\ , \tag{3.9a}$$

$$\nabla_\alpha^k\nabla_{\gamma k}W^{\beta\gamma} - \frac{1}{4}\delta_\alpha^\beta\nabla_\gamma^k\nabla_{\delta k}W^{\gamma\delta} = 8\mathrm{i}\nabla_{\alpha\gamma}W^{\gamma\beta}\ . \tag{3.9b}$$

In order to give the operator $\mathcal{W}^{\alpha i}$ more explicitly it is useful to introduce the dimension 3/2 superfields

$$X_\gamma^{k\alpha\beta} = -\frac{\mathrm{i}}{4}\nabla_\gamma^k W^{\alpha\beta} - \delta_\gamma^{(\alpha}X^{\beta)k}\ , \quad X^{\alpha i} := -\frac{\mathrm{i}}{10}\nabla_\beta^i W^{\alpha\beta} \tag{3.10}$$

together with the dimension 2 descendant superfields:

$$Y_\alpha{}^{\beta ij} := -\frac{5}{2}\Big(\nabla_\alpha^{(i}X^{\beta j)} - \frac{1}{4}\delta_\alpha^\beta\nabla_\gamma^{(i}X^{\gamma j)}\Big) = -\frac{5}{2}\nabla_\alpha^{(i}X^{\beta j)}\ , \tag{3.11a}$$

$$Y := \frac{1}{4}\nabla_\gamma^k X_k^\gamma\ , \tag{3.11b}$$

$$Y_{\alpha\beta}{}^{\gamma\delta} := \nabla_{(\alpha}^k X_{\beta)k}{}^{\gamma\delta} - \frac{1}{6}\delta_\beta^{(\gamma}\nabla_\rho^k X_{\alpha k}{}^{\delta)\rho} - \frac{1}{6}\delta_\alpha^{(\gamma}\nabla_\rho^k X_{\beta k}{}^{\delta)\rho}\ . \tag{3.11c}$$

The action of the spinor covariant derivatives, as well as with the S-supersymmetry generators, on the defined superfields may be found in [13]. In terms of these fields one finds

$$\begin{aligned}
\mathcal{W}^{\alpha i} = {} & W^{\alpha\beta}\nabla_\beta^i + X_\gamma^{i\alpha\beta}M_\beta{}^\gamma - \frac{1}{8}X^{\beta i}M_\beta{}^\alpha - \frac{5}{4}X_j^\alpha J^{ij} + \frac{5}{16}X^{\alpha i}\mathbb{D} \\
& + \frac{1}{16}Y_\beta{}^{\alpha ij}S_j^\beta + \frac{\mathrm{i}}{4}\nabla_{\beta\gamma}W^{\gamma\alpha}S^{\beta i} - \frac{5}{64}YS^{\alpha i} \\
& - \frac{1}{12}(\gamma^{ab})_\beta{}^\gamma(\nabla_b X_\gamma^{i\beta\alpha} - \frac{3}{4}\delta_\gamma^\alpha\nabla_b X^{\beta i})K_a\ .
\end{aligned} \tag{3.12}$$

IOP Conf. Series: Journal of Physics: Conf. Series **804** (2017) 012034 doi:10.1088/1742-6596/804/1/012034

The remaining covariant derivative algebra is constrained in terms of the super-Weyl tensor, its descendents and their vector covariant derivatives [13].

It is important to stress that the entire covariant derivative algebra is expressed directly in terms of the super-Weyl tensor and its descendents. In particular, it is trivial to see that when $W^{\alpha\beta}$ vanishes the supergeometry is superconformally flat. Furthermore, the standard Weyl multiplet of $6D$ $\mathcal{N} = (1,0)$ conformal supergravity is encoded in the superspace geometry. The component fields can be readily identified as certain $\theta = 0$ parts of the superspace gauge one-forms and descendants of $W^{\alpha\beta}$.

3.3. An immediate application: The number of independent invariants

Now that we have presented the conformal superspace formulation, it is possible to determine what the maximum number of invariants for conformal supergravity is. The maximum number of invariants was determined in [13] by making use of an argument based on analysing possible supercurrents. It is illustrative to reproduce the argument below. Before moving on to the constructive proof it is worth first recalling the definition of the supercurrent in 6D $\mathcal{N} = (1,0)$ supersymmetry.

The supercurrent is a supermultiplet described by a scalar superfield $\mathcal{J}$ containing, amongst its component content, the energy momentum tensor and the supersymmetry current(s), together with additional components, e.g. R-symmetry current. In the curved case it is described by a scalar primary superfield $\mathcal{J}$ satisfying the differential constraint

$$\nabla^{(i}_{[\alpha}\nabla^{j}_{\beta}\nabla^{k)}_{\gamma]}\mathcal{J} = 0 \ . \tag{3.13}$$

Since this constraint must be annihilated by the S-supersymmetry generator to be consistent with the superconformal symmetry, $\mathcal{J}$ must be of dimension 4.

Each of the invariants for conformal supergravity have an associated non-trivial supercurrent. Thus a bound on the number of possible invariants for conformal supergravity is given by the maximum number of possible supercurrents (up to irrelevant normalizations) one can construct out of the Weyl tensor and its covariant derivatives. To determine the number of supercurrents possible, one simply writes down the most general possible Ansatz which is

$$\begin{aligned} \mathcal{J} \ = \ & c_1\nabla^a\nabla_aY + c_2Y^2 + \mathrm{i}\,c_3\,X^{\alpha i}\nabla_{\alpha\beta}X^{\beta}_i + \mathrm{i}\,c_4\,X^{i\,\beta\gamma}_{\alpha}\nabla_{\gamma\delta}X_{\beta i}{}^{\alpha\delta} + c_5\,Y_\alpha{}^{\beta ij}Y_\beta{}^\alpha{}_{ij} \\ & + c_6\,Y_{\alpha\beta}{}^{\gamma\delta}Y_{\gamma\delta}{}^{\alpha\beta} + c_7\,W^{\alpha\gamma}\nabla_{\alpha\beta}\nabla_{\gamma\delta}W^{\delta\beta} + c_8\nabla_{\beta\alpha}W^{\alpha\gamma}\nabla_{\gamma\delta}W^{\delta\beta} \\ & + c_9\varepsilon_{\alpha_1\cdots\alpha_4}\varepsilon_{\beta_1\cdots\beta_4}W^{\alpha_1\beta_1}\dots W^{\alpha_4\beta_4} \ , \end{aligned} \tag{3.14}$$

where c_n are arbitrary real coeffcients. Requiring that $\mathcal{J}$ be primary and satisfy the differential constraint yields

$$\begin{aligned} & c_3 = -\frac{8}{3}c_2 - 5c_1 \ , \qquad c_4 = -\frac{32}{15}c_2 - 16c_1 \ , \quad c_5 = \frac{2}{15}c_2 + \frac{6}{5}c_1 \ , \\ & c_6 = \frac{2}{45}c_2 + \frac{1}{3}c_1 \ , \qquad c_7 = -\frac{2}{15}c_2 - \frac{1}{5}c_1 \ , \quad c_8 = \frac{1}{2}c_7 = -\frac{1}{15}c_2 - \frac{1}{10}c_1 \ , \\ & c_9 = 0 \ , \end{aligned} \tag{3.15}$$

which tells us that we have a two parameter family of solutions and at most two invariants for conformal supergravity.

4. Superform action principles and the invariants

Now that we have demonstrated that there can only be at most two invariants for conformal supergravity, we would like to describe how they were constructed in [13] in this section. It turns out that there are exactly two invariants, one of which contains a certain combination of C^3 terms at the component level, while the other contains a $C\Box C$ term at the component level. Their construction relies heavily on the use of the superform approach to constructing supersymmetric invariants [14, 15, 16], which we now turn to briefly reviewing in the context of 6D $\mathcal{N}=(1,0)$ conformal supergravity.

4.1. The superform approach to constructing supersymmetric invariants

The superform approach to constructing supersymmetric invariants is a general approach based on the idea that a closed super D-form automatically leads to a supersymmetric invariant. In six dimensions, one introduces a six-form $J=\frac{1}{6!}\mathrm{d}z^{M_6}\wedge\cdots\wedge\mathrm{d}z^{M_1}J_{M_1\cdots M_6}$ in 6D $\mathcal{N}=(1,0)$ superspace satisfying the closure condition,

$$\mathrm{d}J=\frac{1}{6!}\mathrm{d}z^{M_6}\wedge\cdots\wedge\mathrm{d}z^{M_0}\partial_{M_0}J_{M_1\cdots M_6}=0\ . \tag{4.1}$$

Such a superform immediately leads to the action principle

$$S=\int_{\mathcal{M}^6}i^*J=\int\mathrm{d}^6x\,e\,{}^*J|_{\theta=0}\ ,\qquad {}^*J:=\frac{1}{6!}\varepsilon^{mnpqrs}J_{mnpqrs}\ , \tag{4.2}$$

where $i:\mathcal{M}^6\to\mathcal{M}^{6|8}$ is the inclusion map and i^* is its pullback, the effect of which is to project $\theta^\mu_i=\mathrm{d}\theta^\mu_i=0$. Invariance under supersymmetry follows since under a superdiffeomorphism with $\xi=\xi^AE_A=\xi^M\partial_M$, the superform J transforms as

$$\delta_\xi J=\mathcal{L}_\xi J\equiv i_\xi\mathrm{d}J+\mathrm{d}i_\xi J=\mathrm{d}i_\xi J\ , \tag{4.3}$$

which corresponds to a total derivative. It is important to note that the component action may be obtained by expressing the action in terms of the tangent frame

$$\begin{aligned}S&=\int\mathrm{d}^6x\,\frac{1}{6!}\varepsilon^{m_1\cdots m_6}E_{m_6}{}^{A_6}\cdots E_{m_1}{}^{A_1}J_{A_1\cdots A_6}|_{\theta=0}\ ,\\&=\frac{1}{6!}\int\mathrm{d}^6x\,\mathrm{e}\,\varepsilon^{a_1\cdots a_6}\Big[J_{a_1\cdots a_6}+3\psi_{a_1}{}^\alpha_i J^i_{\alpha a_2\cdots a_6}+\frac{15}{4}\psi_{a_2}{}^\beta_j\psi_{a_1}{}^\alpha_i J^{ij}_{\alpha\beta a_3\cdots a_6}\\&\quad+\frac{5}{2}\psi_{a_3}{}^\gamma_k\psi_{a_2}{}^\beta_j\psi_{a_1}{}^\alpha_i J^{ijk}_{\alpha\beta\gamma a_4a_5a_6}+\mathcal{O}(\psi^4)\Big]|_{\theta=0}\ .\end{aligned} \tag{4.4}$$

In order to have a sensible action it must be invariant under all other gauge transformations, which form the subgroup $\mathcal{H}$. Therefore, we require J to also transform by (at most) an exact form under these transformations, $\delta_{\mathcal{H}}J=\mathrm{d}\Theta(\Lambda^{\underline{a}})$, where Θ is some five-form. For conformal supergravity this requires invariance under the standard superconformal transformations. Under the superconformal transformations, it is worth distinguishing two cases: (i) $\Theta=0$; and (ii) $\Theta\neq0$. The first case corresponds to when the closed six-form is itself invariant, which is of relevance to the construction of the C^3 invariant, while the second case is of relevance to the $C\Box C$ invariant. We now move on to describing the first case and the construction of the C^3 invariant.

4.2. A primary superform and the C^3 invariant

A special case is given by when the superform J is itself closed. This implies that the components $J_{A_1\cdots A_6}$ of $J=\frac{1}{6!}E^{A_6}\cdots E^{A_1}J_{A_1\cdots A_6}$ transform covariantly. In particular, with respect to the special conformal transformations we have the conditions

$$S^\beta_j J_{a_1\cdots a_n}{}^{i_1}_{\alpha_1}\cdots{}^{i_{6-n}}_{\alpha_{6-n}}=-\mathrm{i}\,n\,(\tilde\gamma_{[a_1})^{\beta\gamma}J_{\gamma j a_2\cdots a_n]}{}^{i_1}_{\alpha_1}\cdots{}^{i_{6-n}}_{\alpha_{6-n}}\ ,\qquad K^bJ_{A_1\cdots A_6}=0\ , \tag{4.5}$$

IOP Conf. Series: Journal of Physics: Conf. Series **804** (2017) 012034 doi:10.1088/1742-6596/804/1/012034

which allows us to express the closure condition (4.1) in terms of the tangent frame as follows

$$\nabla_{[A_1} J_{A_2 \cdots A_7\}} + 3T_{[A_1 A_2}{}^B J_{|B| A_3 \cdots A_7\}} = 0 . \tag{4.6}$$

We define *primary* superforms as those superforms satisfying the condition (4.5).

Under certain weak assumptions one can show that a closed primary six-form in 6D $\mathcal{N} = (1,0)$ superspace is unique [13], and is given in [17] in the context of SU(2) superspace. In conformal superspace, the six-form corresponds to choosing the lower dimension components of J to vanish and the dimension 9/2 component as follows:

$$J_{abc}{}_{\alpha}^{i}{}_{\beta}^{j}{}_{\gamma}^{k} = 3(\gamma_{abc})_{(\alpha\beta} A_{\gamma)}{}^{ijk} , \tag{4.7}$$

where, due to eq. (4.5), $A_\alpha{}^{ijk}$ is primary $K^B A_\alpha{}^{ijk} = 0$. The closure condition (4.6) can be shown to hold identically once one imposes the following differential condition on $A_\alpha{}^{ijk}$:

$$\nabla_{(\alpha}^{(i} A_{\beta)}{}^{jkl)} = 0 . \tag{4.8}$$

Closure (4.6) also fixes the remaining components of the superform, e.g. the top component of the superform is

$$J_{abcdef} = -\frac{\mathrm{i}}{2^4 4!} \varepsilon^{\alpha\beta\gamma\delta} \varepsilon_{abcdef} \nabla_{\alpha i} \nabla_{\beta j} \nabla_{\gamma k} A_\delta{}^{ijk} . \tag{4.9}$$

The remaining components of the superform are given in [13].

The superform constructed in terms of $A_\alpha{}^{ijk}$ can be used to describe supersymmetric invariants in an analogous way as the chiral action principle in the 4D $\mathcal{N} = 1$ and $\mathcal{N} = 2$ cases. In particular, all that one must do is search for a composite expression for $A_\alpha{}^{ijk}$ such that it is built out of the superfields of the theory in consideration. This very approach can be used to describe the supersymmetric C^3 invariant. By allowing $A_\alpha{}^{ijk}$ to be built out of the super-Weyl tensor and its descendents yields the following unique solution [13]

$$\begin{aligned} A_\alpha{}^{ijk} &= 5\mathrm{i}\varepsilon_{\alpha\beta\gamma\delta} X^{\beta(i} X^{\gamma j} X^{\delta k)} - 8\mathrm{i}\varepsilon_{\alpha\beta\gamma\delta} X^{\beta(i} X_{\alpha'}^{j}{}^{\gamma\beta'} X_{\beta'}^{k)\delta\alpha'} \\ &\quad + \frac{64\mathrm{i}}{3} \varepsilon_{\alpha\beta\gamma\delta} X_{\alpha'}^{(i}{}^{\beta\beta'} X_{\beta'}^{j}{}^{\gamma\gamma'} X_{\gamma'}^{k)\delta\alpha'} \\ &\quad + 4\varepsilon_{\alpha\beta\gamma\delta} Y_\rho{}^{\beta(ij} X_\eta^{k)\rho\gamma} W^{\eta\delta} - 3\varepsilon_{\alpha\beta\gamma\delta} Y_\rho{}^{\beta(ij} X^{\gamma k)} W^{\rho\delta} . \end{aligned} \tag{4.10}$$

The solution (4.10) automatically describes a supersymmetric invariant. By plugging the result (4.10) into eq. (4.9) and keeping in mind the component reduction formula (4.4), one can show that the component action contains a C^3 term proportional to

$$-\frac{1}{8} \varepsilon^{abcdef} \varepsilon_{a'b'c'd'e'f'} C_{ab}{}^{a'b'} C_{cd}{}^{c'd'} C_{ef}{}^{e'f'} = 4L_{C^3}^{(2)} - 8L_{C^3}^{(1)} . \tag{4.11}$$

4.3. A non-primary superform and the $C\Box C$ invariant

We have presented a primary superform that may be used as an action principle in constructing invariants. However, it becomes apparent that the primary superfield $A_\alpha{}^{ijk}$ cannot be used to describe a supersymmetric $C\Box C$ invariant due to its uniqueness. In particular, one can construct another object of the right index structure as $A_\alpha{}^{ijk}$ that satisfies the right differential constraint, but it can be shown to not be primary. This suggests that we should generalise our approach to look for non-primary closed superforms.

The obstruction we encounter in using the previous action principle for the construction of another conformal supergravity invariant is the special conformal transformations. To remedy

IOP Conf. Series: Journal of Physics: Conf. Series **804** (2017) 012034 doi:10.1088/1742-6596/804/1/012034

this we introduce the special conformal connections $\mathfrak{F}_A$ into our ansatz for the closed superform as follows:

$$J = J_0 + \mathfrak{F}_\alpha^i \wedge J_{S_i^\alpha} + \mathfrak{F}_a \wedge J_K{}^a \ , \tag{4.12}$$

where J_0 is a six-form, and $J_{S_i^\alpha}$ and $J_K{}^a$ are some 5-forms. The purpose of the five-forms $J_{S_i^\alpha}$ and $J_K{}^a$ is to capture the "anomalous terms" arising from the transformations of J under S-supersymmetry and special conformal transformations such that

$$\delta_S J = -\mathrm{d}(\Lambda_{S_\alpha^i} J_{S_i^\alpha}) \ , \qquad \delta_K J = -\mathrm{d}(\Lambda_{K a} J_K{}^a) \ . \tag{4.13}$$

Given our ansatz (4.12), it is still necessary to know how to constrain the components of the superform. A clue comes from considering the full superspace integral

$$S = \int \mathrm{d}^6 x \, \mathrm{d}^8 \theta \, E \, \mathcal{L} \ , \tag{4.14}$$

where $\mathcal{L}$ is a primary superfield of dimension 2. If we replace $\mathcal{L}$ by a tensor superfield Φ, which satisfies

$$B^{\alpha\beta ij} = \frac{\mathrm{i}}{2} \varepsilon^{\alpha\beta\gamma\delta} \nabla_\gamma^{(i} \nabla_\delta^{j)} \Phi = 0 \ , \tag{4.15}$$

the full superspace integral (4.14) vanishes since Φ has a prepotential formulation [18]. This suggests that there should be an action principle based on a primary superfield of the form $B^{\alpha\beta ij} \leftrightarrow B_a{}^{ij}$, which provides a way of describing the full superspace integral as a closed superform.

A solution based on the ansatz (4.12) was found in [13] such that all components of J_0, $J_{S_i^\alpha}$ and $J_K{}^a$ are constructed in terms of a primary superfield of dimension 3 satisfying the constraint

$$\nabla_\alpha^{(i} B^{\beta\gamma jk)} = -\frac{2}{3} \delta_\alpha^{[\beta} \nabla_\delta^{(i} B^{\gamma]\delta jk)} \ . \tag{4.16}$$

The superfield $B^{\alpha\beta ij}$ appears in the component of the J_K with lowest mass dimension.

$$J_K{}_{bcd}{}_{\alpha\beta}^{i\,j\,a} = -64\mathrm{i} (\gamma_{bcd})_{\alpha\beta} \, B^{a\,ij} \ , \tag{4.17}$$

while all other higher dimension components are expressed in terms of spinor derivatives of $B^{\alpha\beta ij}$, e.g.

$$J_0{}_{a_1 a_2 a_3 a_4 a_5 a_6} = -\varepsilon_{a_1 a_2 a_3 a_4 a_5 a_6} \Big(-\frac{1}{5} \nabla^4_{ijkl} C^{ijkl} + \cdots \Big) \ , \tag{4.18}$$

where $C^{ijkl} := -\frac{\mathrm{i}}{12} \nabla_\alpha^{(i} \nabla_\beta^j B^{\alpha\beta kl)}$ and the ellipses represents terms that directly involve the Weyl tensor. We refer the reader to [13] for the remaining components and details.

The supeform based on $B_{\alpha\beta}{}^{ij}$ provides an action principle analogous to the one in the previous subsection, in the sense that an invariant action may be constructed by allowing $B^{\alpha\beta ij}$ to be composed of the superfields involved in the theory of interest. For example, if one requires $B^{\alpha\beta ij}$ to be expressed in terms of the super-Weyl tensor and its covariant derivatives, one finds the following unique solution:

$$B^{\alpha\beta\, ij} = W^{\gamma[\alpha} Y_\gamma{}^{\beta]ij} + 8\mathrm{i}\, X_\gamma{}^{\delta[\alpha(i} X_\delta{}^{\beta]\gamma j)} - \frac{5\mathrm{i}}{2} X^{[\alpha(i} X^{\beta]j)} \ . \tag{4.19}$$

One can check that by plugging this result into the superform (4.18), one finds (upon integration by parts) a $C_{abcd} \Box C^{abcd}$ term at the component level.

IOP Conf. Series: Journal of Physics: Conf. Series **804** (2017) 012034 doi:10.1088/1742-6596/804/1/012034

5. Conclusion

In conclusion, we presented the 6D $\mathcal{N} = (1, 0)$ conformal superspace constructed in [13] and demonstrated it is well adapted to describing the invariants for conformal supergravity. The formulation was used to give a simple proof that there can only be a maximum of two invariants for conformal supergravity. We described a primary superform that can be used as an action principle and showed how it can be used to describe the supersymmetric C^3 invariant. We also showed that a non-primary superform action principle was required to construct the supersymmetric $C\Box C$ invariant. Due to the supercurrent analysis presented here, all invariants for minimal conformal supergravity must be a linear combination of the two invariants. The complete component actions for the invariants are given by their component projections and the bosonic sectors will be explicitly given in a forthcoming publication.

Acknowledgments

It is a pleasure to thank the Galileo Galilei Institute for Theoretical Physics for hospitality and the INFN for partial support during the completion of this work. This work is supported by GIF – the German-Israeli Foundation for Scientific Research and Development.

References

[1] Deser S and Schwimmer A 1993 *Phys. Lett.* B **309** 279
[2] Nahm W 1978 *Nucl. Phys.* B **135** 149
[3] Bergshoeff E, Sezgin E and Van Proeyen A 1986 *Nucl. Phys.* B **264** 653
[4] Bergshoeff E, Salam A and Sezgin E 1986 *Phys. Lett.* B **173** 73
[5] Howe P S 1981 *Phys. Lett.* B **100** 389
[6] Howe P S 1982 *Nucl. Phys.* B **199** 309
[7] Kuzenko S M and Tartaglino-Mazzucchelli G 2008 *JHEP* **0804** 032
[8] Linch III W D and Tartaglino-Mazzucchelli G 2012 *JHEP* **1208** 075
[9] Butter D 2010 *Annals Phys.* **325** 1026
[10] Butter D 2011 *JHEP* **1110** 030
[11] Butter D, Kuzenko S M, Novak J and Tartaglino-Mazzucchelli G 2013 *JHEP* **1309** 072
[12] Butter D, Kuzenko S M, Novak J and Tartaglino-Mazzucchelli G 2015 *JHEP* **1502**, 111
[13] Butter D, Kuzenko S M, Novak J and Theisen S 2016 Invariants for minimal conformal supergravity in six dimensions *Preprint* arXiv:1606.02921
[14] Hasler M F 1998 *Eur. Phys. J.* C **1** 729
[15] Gates Jr S J 1999 *Supersymmetries and Quantum Symmetries* ed J Wess and E A Ivanov (Berlin: Springer) p 46
[16] Gates Jr S J 1999 *Nucl. Phys.* B **541** 615
[17] Arias C, Linch III W D and Ridgway A K 2016 JHEP **1605** 016
[18] Sokatchev E 1988 *Class. Quant. Grav.* **5** 1459

Hirota's method for a spin model with self-consistent potential

G. Nugmanova, Zh. Sagidullayeva, R. Myrzakulov

L.N. Gumilyov Eurasian National University,
Satpayev Str., 2, Astana, Kazakhstan

E-mail: nugmanovagn@gmail.com

Abstract. In this work an integrable generalization of the Landau-Lifshitz equation with self-consistent vector potential is studied. It was established that self-consistence of spin vector and potential happen by the relation between solutions the potential and linear system, the compatibility condition of which corresponds to the equation considered by us. By generalizing Hirota method, it's exact solutions, defining self-consistent motion of the potential and soliton, are constructed.

1. Introduction

Solutions of integrable nonlinear differential equations which is a solitary wave and have elastic properties of interaction with the same another solution have a variety of applications in many areas of the natural sciences. Analytical studies of processes of interaction of the solitary waves is one of the main tasks of the theory of solitons. The development of nonlinear theory of magnetism, in turn, has put the problem of constructing integrable generalization of the Landau-Lifshitz equation with self-consistent potential.

One of such generalizations with self-consistent scalar potential was proposed in [1]. Various algebraic-geometric aspects of such models were studied in [2-5]. The generalized Landau-Lifshitz equation with self-consistent vector potential was obtained in [6], and also their connection with the movement of curves and surfaces are established.

2. Generalized Landau-Lifshitz equation

The Generalized Landau-Lifshitz equation (GLL) equation with self-consistent potential reads as

$$iS_t + \frac{1}{2}[S, S_{xx}] + \frac{1}{a}[S, W] = 0, \tag{2.1}$$

$$iW_x + a[S, W] = 0, \tag{2.2}$$

where $a = const$, $S = \Sigma_{j=1}^{3} S_j(x, y, t)\sigma_j$ is a matrix analogue of the spin vector, W - potential with the matrix form $W = \Sigma_{j=1}^{3} W_j(x, y, t)\sigma_j$, and

$$\sigma_1 = \begin{pmatrix} 0 & 1 \\ 1 & 0 \end{pmatrix}, \quad \sigma_2 = \begin{pmatrix} 0 & -i \\ i & 0 \end{pmatrix}, \quad \sigma_3 = \begin{pmatrix} 1 & 0 \\ 0 & -1 \end{pmatrix}$$

are Pauli matrixes.

The GLL equation with self-consistent potential is integrable by the IST. Its Lax representation can be written in the form

$$\Phi_x = U\Phi, \tag{2.3}$$
$$\Phi_t = V\Phi, \tag{2.4}$$

where the matrix operators U and V have the form

$$U = -i\lambda S, \tag{2.5}$$
$$V = \lambda^2 V_2 + \lambda V_1 + \left(\frac{i}{\lambda + a} - \frac{i}{a}\right) W. \tag{2.6}$$

Here

$$V_2 = -2iS, \quad V_1 = SS_x. \tag{2.7}$$

3. Bilinearization

In this section we construct a bilinear form of equations (2.1) - (2.2). Now for our convenience we write systems (2.1) - (2.2) in the component form

$$iS_t^+ + S^+ S_{3xx} - S_3 S_{xx}^+ + \frac{2}{a}(S^+ W_3 - S_3 W^+) = 0, \tag{3.1}$$
$$iS_t^- - S_{3xx} S^- + S_3 S_{xx}^- - \frac{2}{a}(S^- W_3 - S_3 W^-) = 0, \tag{3.2}$$
$$2iS_{3t} + S_{xx}^+ S^- - S^+ S_{xx}^- + \frac{2}{a}(S^- W^+ - S^+ W^-) = 0, \tag{3.3}$$
$$iW_x^+ - 2a(S_3 W^+ - S^+ W_3) = 0, \tag{3.4}$$
$$iW_x^- - 2a(S^- W_3 - S_3 W^-) = 0, \tag{3.5}$$
$$iW_{3x} - a(S^+ W^- - S^- W^+) = 0. \tag{3.6}$$

Self-consistent potential W is given as

$$W_3 = |\varphi_1|^2 - |\varphi_2|^2, \quad W^+ = 2\varphi_1^* \varphi_2, \quad W^- = 2\varphi_1 \varphi_2^*, \tag{3.7}$$

where φ_1 and φ_2 are solutions of the matrix systems (2.3)-(2.4). With considering (3.7) we rewrite equations (3.1)-(3.6) as

$$iS_t^+ + S^+ S_{3xx} - S_3 S_{xx}^+ + \frac{2}{a}\left(S^+(|\varphi_1|^2 - |\varphi_2|^2) - 2S_3 \varphi_1^* \varphi_2\right) = 0, \tag{3.8}$$
$$iS_t^- - S_{3xx} S^- + S_3 S_{xx}^- - \frac{2}{a}(S^-(|\varphi_1|^2 - |\varphi_2|^2) - 2S_3 \varphi_1 \varphi_2^*) = 0, \tag{3.9}$$
$$2iS_{3t} + S_{xx}^+ S^- - S^+ S_{xx}^- + \frac{4}{a}(S^- \varphi_1^* \varphi_2 - S^+ \varphi_1 \varphi_2^*) = 0, \tag{3.10}$$
$$\varphi_{1x} - ia(S_3 \varphi_1 + S^- \varphi_2) = 0, \tag{3.11}$$
$$\varphi_{2x} - ia(S^+ \varphi_1 - S_3 \varphi_2) = 0. \tag{3.12}$$

Let us consider the following stereographic transformation:

$$S^+ = \frac{2\omega}{1 + |\omega|^2}, \quad S^- = \frac{2\omega^*}{1 + |\omega|^2}, \quad S_3 = \frac{1 - |\omega|^2}{1 + |\omega|^2}. \tag{3.13}$$

Hence we obtain

$$\omega = \frac{S^+}{1+S_3}. \tag{3.14}$$

So we have

$$\begin{aligned} i\omega_t - \omega_{xx} + \frac{2\omega^*\omega_x^2}{1+|\omega|^2} + \frac{2}{a}(\omega\varphi_2^* + \varphi_1^*)(\omega\varphi_1 - \varphi_2) &= 0, & (3.15)\\ \varphi_{1x} - \frac{ia}{1+|\omega|^2}\left((1-|\omega|^2)\varphi_1 + 2\omega^*\varphi_2\right) &= 0, & (3.16)\\ \varphi_{2x} - \frac{ia}{1+|\omega|^2}\left(2\omega\varphi_1 - (1-|\omega|^2)\varphi_2\right) &= 0. & (3.17) \end{aligned}$$

Functions φ_1 and φ_2 can be given by

$$\varphi_1 = \frac{p}{f}e^{iax}, \quad \varphi_2 = \frac{q}{f}e^{iax}, \tag{3.18}$$

and $\omega = \frac{g}{f}$, where f is a real function, g, p, q are complex functions. In view of this further transforming equations (3.15) - (3.17) we obtain bilinear form for equations (3.8) - (3.12)

$$\begin{aligned} [iD_t + D_x^2](g\cdot f) - \frac{2}{a}p^*q &= 0, & (3.19)\\ D_x^2(f\cdot f) - \frac{2}{a}q^*q &= 0, & (3.20)\\ D_x(p\cdot f) - 2iag^*q &= 0, & (3.21)\\ D_x(q\cdot f) + 2iaqf &= 0, & (3.22) \end{aligned}$$

where Hirota's operators defined by

$$D_x^l D_t^n f(x,t)\cdot g(x,t) = (\partial_x - \partial_{x'})^l(\partial_t - \partial_{t'})^n f(x,t)\cdot g(x',t')|_{x=x',t=t'}. \tag{3.23}$$

4. Solutions

Using above obtained the bilinear forms, now we can construct soliton solutions for spin systems. For this, we expand the functions g, f, p and q in formal series on an arbitrary constant ε:

$$\begin{aligned} g(x,t) &= \varepsilon g_1(x,t) + \varepsilon^3 g_3(x,t) + \cdots, & (4.1)\\ f(x,t) &= 1 + \varepsilon^2 f_2(x,t) + \varepsilon^4 f_4(x,t) + \cdots, & (4.2)\\ p(x,t) &= 1 + \varepsilon^2 p_2(x,t) + \varepsilon^4 p_4(x,t) + \cdots, & (4.3)\\ q(x,t) &= \varepsilon q_1(x,t) + \varepsilon^3 q_3(x,t) + \cdots. & (4.4) \end{aligned}$$

In order to obtain one-soliton solution for equations (3.19) - (3.22), we choose $g = \varepsilon g_1$, $p = 1 + \varepsilon^2 p_2$, $q = \varepsilon q_1$, $f = 1 + \varepsilon^2 f_2$.

Substituting expressions into bilinear forms (3.19) - (3.22) we have

$$ig_{1t} + g_{1xx} - \frac{2}{a}q_1 = 0, \tag{4.5}$$

$$ig_{1t}f_2 - ig_1f_{2t} + g_{1xx}f_2 - 2g_{1x}f_{2x} + g_1f_{2xx} - \frac{2}{a}p_2^*q_1 = 0, \tag{4.6}$$

$$f_{2xx} - \frac{1}{a}q_1^*q_1 = 0, \tag{4.7}$$

$$f_2f_{2xx} - f_{2x}f_{2x} = 0, \tag{4.8}$$

$$p_{2x} - f_{2x} - 2iag_1^*q_1 = 0, \tag{4.9}$$

$$p_{2x}f_2 - p_2f_{2x} = 0, \tag{4.10}$$

$$q_{1x} + 2iaq_1 = 0, \tag{4.11}$$

$$q_{1x}f_2 - q_1f_{2x} + 2iaq_1f_2 = 0. \tag{4.12}$$

If we choose function g_1 as $g_1 = e^{i\theta}$, with $\theta = k_1x + l_1t + m_1$, where k_1,l_1 and m_1 are complex constants. Then we can construct solutions for f, q_1, p_2 in forms:

$$q_1 = -\frac{a}{2}(l_1 + k_1^2)e^{i\theta}, \tag{4.13}$$

$$f_2 = -\frac{a(l_1 + k_1^2)(l_1^* + k_1^{*2})}{(k_1 - k_1^*)^2}e^{i(\theta-\theta^*)}, \tag{4.14}$$

$$p_2 = -\frac{a(l_1 + k_1^2)}{(k_1 - k_1^*)}\left(\frac{a(l_1^* + k_1^{*2})}{4(k_1 - k_1^*)} + a\right)e^{i(\theta-\theta^*)}. \tag{4.15}$$

Equation (3.13) based on components of matrix S can be given as

$$S^+ = \frac{2fg}{f^2 + |g|^2}, \quad S^- = \frac{2fg^*}{f^2 + |g|^2}, \quad S_3 = \frac{f^2 - |g|^2}{f^2 + |g|^2}. \tag{4.16}$$

Finally, we can find solutions for equations (2.1)-(2.2)

$$S^+ = \frac{2\left(1 - \frac{a(k_1^2+l_1)(k_1^{*2}+l_1^*)e^{i(\theta-\theta^*)}}{4(k_1-k_1^*)^2}\right)e^{i\theta}}{\left(1 - \frac{a(k_1^2+l_1)(k_1^{*2}+l_1^*)e^{i(\theta-\theta^*)}}{4(k_1-k_1^*)^2}\right)^2 + e^{i(\theta-\theta^*)}}, \tag{4.17}$$

$$S_3 = \frac{\left(1 - \frac{a(k_1^2+l_1)(k_1^{*2}+l_1^*)e^{i(\theta-\theta^*)}}{4(k_1-k_1^*)^2}\right)^2 - e^{i(\theta-\theta^*)}}{\left(1 - \frac{a(k_1^2+l_1)(k_1^{*2}+l_1^*)e^{i(\theta-\theta^*)}}{4(k_1-k_1^*)^2}\right)^2 + e^{i(\theta-\theta^*)}}. \tag{4.18}$$

From equation (3.7) for the potential we get

$$W^+ = -\frac{a\left(1 - \frac{a(l_1+k_1^2)}{(k_1-k_1^*)}\left(\frac{a(l_1^*+k_1^{*2})}{4(k_1-k_1^*)} + a\right)e^{i(\theta-\theta^*)}\right)(l_1 + k_1^2)e^{i\theta}}{\left(1 - \frac{a(k_1^2+l_1)(k_1^{*2}+l_1^*)e^{i(\theta-\theta^*)}}{4(k_1-k_1^*)^2}\right)^2}, \tag{4.19}$$

$$
\begin{aligned}
W_3 &= \frac{\left(1-\frac{a(l_1+k_1^2)}{(k_1-k_1^*)}\left(\frac{a(l_1^*+k_1^{*2})}{4(k_1-k_1^*)}+a\right)e^{i(\theta-\theta^*)}\right)}{\left(1-\frac{a(k_1^2+l_1)(k_1^{*2}+l_1^*)e^{i(\theta-\theta^*)}}{4(k_1-k_1^*)^2}\right)^2} * \\
&* \frac{\left(1-\frac{a(l_1^*+k_1^{*2})}{(k_1^*-k_1)}\left(\frac{a(l_1+k_1^2)}{4(k_1^*-k_1)}+a\right)e^{i(\theta-\theta^*)}\right)}{\left(1-\frac{a(k_1^2+l_1)(k_1^{*2}+l_1^*)e^{i(\theta-\theta^*)}}{4(k_1-k_1^*)^2}\right)^2} + \\
&+ \frac{(a^2(l_1+k_1^2)(l_1^*+k_1^{*2})e^{i(\theta-\theta^*)})}{4\left(1-\frac{a(k_1^2+l_1)(k_1^{*2}+l_1^*)e^{i(\theta-\theta^*)}}{4(k_1-k_1^*)^2}\right)^2}. \qquad (4.20)
\end{aligned}
$$

Graphics for obtained solutions (4.18)-(4.20), constant parameters: $k_1 = 25 - i$, $l_1 = 10 - i$,

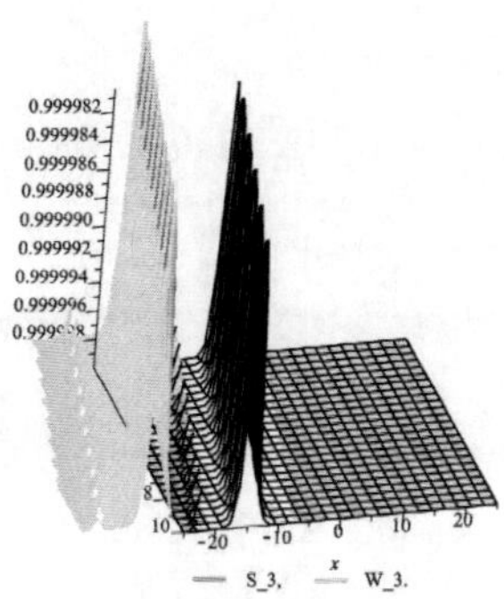

Figure 1. $a = 1$

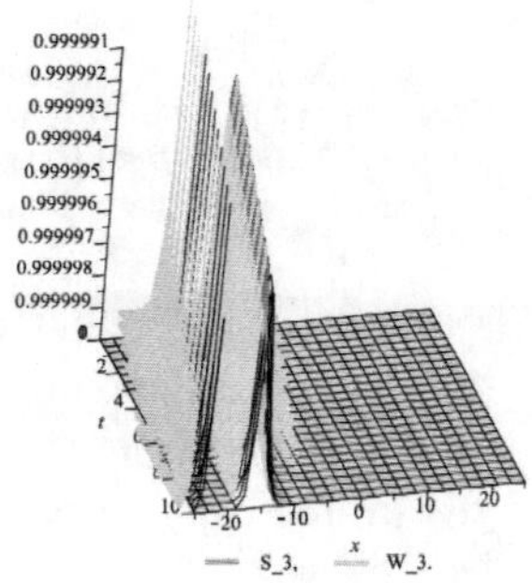

Figure 2. $a = 1$

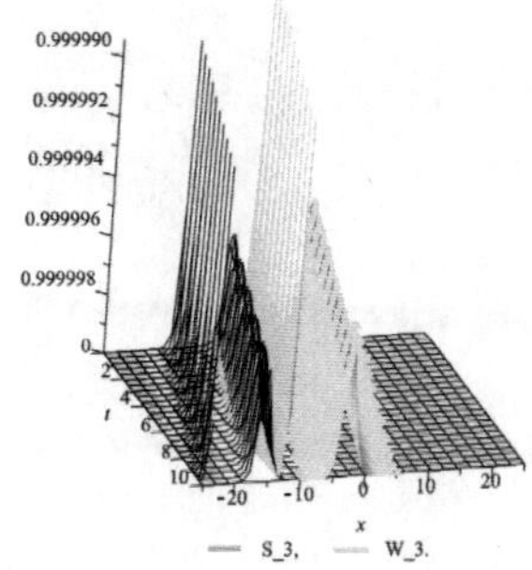

Figure 3. $a = 1$

$m_1 = 1$. We can clearly see self consistent behavior of spin matrix S and potential W, as shown on figures (1)- (3) in the time interval from 0 to 10.

5. Conclusions

The present work is a continuation of our previous researches in the area of integrable spin systems, specifically, generalized Landau-Lifshitz equation with self-consistent scalar potential. The object of the current research was the generalization of the Landau-Lifshitz equation with self-consistent source, the role of which is played by three-component vector with variable length. The components of this vector are the functions of two variables. For this reason the generalized Landau-Lifshitz equation with the self-consistent source considered by us is not limited by conclusions and laws formulated for spin model with the scalar potential. The main result of this work is finding the relation defining self-consistent motion of the vector and soliton wave.

6. References

[1] Myrzakulov R, Vijayalakshmi S, Nugmanova G and Lakshmanan M 1999 *Phys. Lett. A* **233** 391
[2] Myrzakulov R, Nugmanova G and Danlybaeva A 1999 *Theoretical and Mathematical Physics* **118** 441
[3] Myrzakulov R, Nugmanova G, Syzdykova R 1998 *J. Phys. A: Math. & Theor.* **31** 147
[4] Zhang Zh-H, Deng M, Zhao W-Zh, Ke W 2006 arXiv:nlin/0603069v1 [nlin.SI]
[5] Chen Ch, Zhou Zi-X 2009 *Chin. Phys. Lett.* **26** 080504
[6] Myrzakulov R, Mamyrbekova G, Nugmanova G, Lakshmanan M 2015 *Symmetry* **7** 1352

IOP Conf. Series: Journal of Physics: Conf. Series **804** (2017) 012036 doi:10.1088/1742-6596/804/1/012036

Generalized Jack and Macdonald polynomials arising from AGT conjecture

Y Ohkubo

Graduate School of Mathematics, Nagoya University, Nagoya 464-8602, Japan

E-mail: m12010t@math.nagoya-u.ac.jp

Abstract. We investigate the existence and the orthogonality of the generalized Jack symmetric functions which play an important role in the AGT relations. We show their orthogonality by deforming them to the generalized Macdonald symmetric functions.

1. Introduction

In [1], Aldey, Gaiotto and Tachikawa discovered surprising relations (called AGT relations) between 2D CFTs and 4D gauge theories. Since then, a number of studies on AGT relations have been carried out by mathematicians and physicists. One of them, a q-analogue of these relations is given by [2, 3], in which connections between the deformed Virasoro/W algebra and 5D gauge theories are proposed. Moreover in [4], it is conjectured with the help of the Ding-Iohara-Miki algebra [5, 6] that the matrix elements of the vertex operators with respect to the generalized Macdonald symmetric functions reproduce the Nekrasov factors of the 5D instanton partition function. In [7], an M-theoretic derivation of it is given. These generalized Macdonald functions are the eigenfunctions of a certain operator in the Ding-Iohara-Miki algebra. At last, the refined topological vertex of Awata-Kanno or Iqbal-Kozcaz-Vafa is deriven from another representation of Ding-Iohara-Miki algebra [8]. This result show that the correlation functions of the Ding-Iohara-Miki algebra coincide with the 5D instanton partition functions.

On the other hand, the 4D $SU(2)$ AGT relation as Hubbard-Stratanovich (HS) duality is proved in [9] with the help of the generalized Jack symmetric functions. (The formula of their selberg averages is still conjecture.) In their previous paper [10], the integrand of the Dotsenko-Fateev (DF) representation of the 4-point conformal block is expanded by the ordinary Jack symmetric polynomials and compared with the Nekrasov partition function. They show that, in the case $\beta = 1$, every term of the DF-integral coincides with the Nekrasov formula parametrized by pairs of Young diagrams. However at $\beta \neq 1$ it is not working. At this, a generalization of the Jack symmetric polynomials is introduced in [9] such that each term of DF-integral expanded by the generalized Jack polynomials corresponds to each term of Nekrasov partition function. However, the existence of the generalized Jack symmetric functions is unproven. Also, it is a little hard to give a mathematical, general proof of the orthogonality [11, Section 2] of these functions, because they have degenerate eigenvalues.

The generalized Macdonald and Jack symmetric functions were introduced independently and their relation had not been known. However, we found that in the limit $q \to 1$, the generalized Macdonald symmetric functions are reduced to the generalized Jack symmetric functions naturally. This leads the orthogonality and the existence of the generalized Jack symmetric functions. Since the eigenvalues of the generalized Macdonald symmetric functions are non-degenerate, we can prove their orthogonality without any other commutative operators, i.e., higher rank hamiltonians. Moreover we can describe 5D AGT relations as HS duality by using the generalized Macdonald symmetric functions [12] in the same way of [9].

This letter is organized as follows. In section 2, we give a short summary of the generalized Macdonald symmetric functions. In section 3, we calculate the limit $q \to 1$ of the operator whose eigenfunctions are the generalized Macdonald functions and prove the existence and the orthogonality of the generalized Jack symmetric functions. In section 4 we present several examples.

2. Generalized Macdonald symmetric function

The generalized Macdonald symmetric functions are described by N kinds of independent variables $\{x_n^{(i)} \mid n \in \mathbb{N}, i = 1, \ldots, N\}$. Hence we use N kinds of power sum symmetric functions $p_n^{(i)} = \sum_{k \geq 1} \left(x_k^{(i)}\right)^n$. Using this notation, set

$$\eta^{(i)}(z) := \exp\left(\sum_{n=1}^{\infty} \frac{1-t^{-n}}{n} z^n p_n^{(i)}\right) \exp\left(-\sum_{n=1}^{\infty}(1-q^n) z^{-n} \frac{\partial}{\partial p_n^{(i)}}\right), \tag{1}$$

$$\xi^{(i)}(z) := \exp\left(-\sum_{n=1}^{\infty} \frac{1-t^{-n}}{n} (t/q)^{\frac{n}{2}} z^n p_n^{(i)}\right) \exp\left(\sum_{n=1}^{\infty}(1-q^n)(t/q)^{\frac{n}{2}} z^{-n} \frac{\partial}{\partial p_n^{(i)}}\right), \tag{2}$$

$$\varphi_+^{(i)}(z) := \exp\left(-\sum_{n=1}^{\infty}(1-q^n)(1-t^n q^{-n})(t/q)^{-\frac{n}{4}} z^{-n} \frac{\partial}{\partial p_n^{(i)}}\right), \tag{3}$$

$$\varphi_-^{(i)}(z) := \exp\left(\sum_{n=1}^{\infty} \frac{1-t^{-n}}{n}(1-t^n q^{-n})(t/q)^{-\frac{n}{4}} z^n p_n^{(i)}\right), \tag{4}$$

where q and t are independent parameters. Then these operators represent the Ding-Iohara-Miki algebra [13]. Moreover for complex parameters u_i $(i = 1, 2, \ldots, N)$, we define

$$X(z) := \sum_{i=1}^{N} u_i \widetilde{\Lambda}_i, \tag{5}$$

$$\widetilde{\Lambda}_i := \varphi_-^{(1)}((t/q)^{\frac{1}{4}} z)\, \varphi_-^{(2)}((t/q)^{\frac{3}{4}} z) \cdots \varphi_-^{(i)}((t/q)^{\frac{2i-3}{4}} z)\, \eta^{(i)}((t/q)^{\frac{i-1}{2}} z). \tag{6}$$

The operator $X(z)$ is obtained by so-called the level N representation of Ding-Iohara-Miki algebra. The deformed W_N algebra can be generated by its representation [14]. In the $N = 1$ case, the coefficient in front of z^0 of $X(z) = \eta(z)$ is Macdonald's difference operator and its eigenfunctions are the Macdonald symmetric functions. The generalized Macdonald functions are the eigenfunctions of the operator

$$X_0 := \oint \frac{dz}{2\pi\sqrt{-1}z} X(z) \tag{7}$$

in the general N case.

The generalized Macdonald symmetric functions are parametrized by N-tuples of partitions $\vec{\lambda} = (\lambda^{(1)}, \ldots, \lambda^{(N)})$, where a partition $\lambda^{(i)} = (\lambda_1^{(i)}, \lambda_2^{(i)}, \ldots)$ is a sequence of non-negative integers such that $\lambda_1^{(i)} \geq \lambda_2^{(i)} \geq \cdots \geq 0$. For each partition λ and an N-tuple of partitions

$\vec{\lambda} = (\lambda^{(1)}, \ldots, \lambda^{(N)})$, we use the symbols : $|\lambda| := \sum_{k\geq 1} \lambda_k$, $\ell(\lambda) := \#\{k \mid \lambda_k \neq 0\}$ and $|\vec{\lambda}| := \sum_{i=1}^N |\lambda^{(i)}|$. We write $\vec{\lambda} \geq^{\mathrm{L}} \vec{\mu}$ (resp. $\vec{\lambda} \geq^{\mathrm{R}} \vec{\mu}$) if and only if $|\vec{\lambda}| = |\vec{\mu}|$ and

$$|\lambda^{(N)}| + \cdots + |\lambda^{(j+1)}| + \sum_{k=1}^{i} \lambda_k^{(j)} \geq |\mu^{(N)}| + \cdots + |\mu^{(j+1)}| + \sum_{k=1}^{i} \mu_k^{(j)} \tag{8}$$

$$\left(\text{resp.} \quad |\lambda^{(1)}| + \cdots + |\lambda^{(j-1)}| + \sum_{k=1}^{i} \lambda_k^{(j)} \geq |\mu^{(1)}| + \cdots + |\mu^{(j-1)}| + \sum_{k=1}^{i} \mu_k^{(j)}\right) \tag{9}$$

for all $i \geq 1$ and $1 \leq j \leq N$. Then "$\geq^{\mathrm{L}}$" and "$\geq^{\mathrm{R}}$" are generalized dominace partial orderings on the N-tuples of partitions. By using these partial orderings we can triangulate the representation matrix of X_0.

Fact 2.1 ([4]). Let $m_{\vec{\lambda}}$ be the product of monomial symmetric functions $\prod_{i=1}^N m_{\lambda^{(i)}}^{(i)}$ ($m_{\lambda^{(i)}}^{(i)}$ is an usual monomial symmetric function of variables $\{x_n^{(i)} \mid n\}$). Then

$$X_0 m_{\vec{\lambda}} = \sum_{\vec{\mu} \leq^{\mathrm{L}} \vec{\lambda}} c_{\vec{\lambda}\vec{\mu}} m_{\vec{\mu}}, \quad c_{\vec{\lambda}\vec{\mu}} \in \mathbb{Q}(q^{\frac{1}{2}}, t^{\frac{1}{2}}, u_1, \ldots, u_N) \tag{10}$$

and the eigenvalues are

$$c_{\vec{\lambda}\vec{\lambda}} = \sum_{i=1}^{N} u_i \, \epsilon_{\lambda^{(i)}}, \quad \epsilon_\lambda = 1 + (t-1) \sum_{k=1}^{\ell(\lambda)} (q^{\lambda_k} - 1) t^{-k}. \tag{11}$$

Thus we have the following existence theorem.

Fact 2.2 ([4]). For each N-tuple of partitions $\vec{\lambda}$, there exists a unique symmetric function $P_{\vec{\lambda}}$ satisfying the following two conditions:

$$P_{\vec{\lambda}} = m_{\vec{\lambda}} + \sum_{\vec{\mu} <^{\mathrm{L}} \vec{\lambda}} d_{\vec{\lambda}\vec{\mu}} m_{\vec{\mu}}, \qquad d_{\vec{\lambda}\vec{\mu}} \in \mathbb{Q}(q^{\frac{1}{2}}, t^{\frac{1}{2}}, u_1, \ldots, u_N); \tag{12}$$

$$X_0 P_{\vec{\lambda}} = e_{\vec{\lambda}} P_{\vec{\lambda}}, \qquad e_{\vec{\lambda}} \in \mathbb{Q}(q^{\frac{1}{2}}, t^{\frac{1}{2}}, u_1, \ldots, u_N). \tag{13}$$

In the case $N = 1$, the symmetric functions $P_{\vec{\lambda}}$ are the usual Macdonald symmetric functions. Hence we call these symmetric functions $P_{\vec{\lambda}}$ the generalized Macdonald symmetric functions.

The partial ordering "$\geq^{\mathrm{R}}$" gives the similar existence theorem of the dual symmetric functions $P_{\vec{\lambda}}^*$ for the adjoint operator X_0^* of X_0 with respect to Macdonald's scalar product $\langle -, - \rangle_{q,t}$. The scalar product $\langle -, - \rangle_{q,t}$ is defined by

$$\langle p_{\vec{\lambda}}, p_{\vec{\mu}} \rangle_{q,t} = \delta_{\vec{\lambda},\vec{\mu}} \prod_{i=1}^{N} z_{\lambda^{(i)}} \prod_{k=1}^{\ell(\lambda^{(i)})} \frac{1 - q^{\lambda_k^{(i)}}}{1 - t^{\lambda_k^{(i)}}}, \qquad z_{\lambda^{(i)}} := \prod_{k \geq 1} k^{m_k} \, m_k!, \tag{14}$$

where m_k is the number of entries in $\lambda^{(i)}$ equal to k and $p_{\vec{\lambda}} := \prod_{i=1}^N p_{\lambda^{(i)}}^{(i)} := \prod_{i=1}^N \prod_{k \geq 1} p_{\lambda_k^{(i)}}^{(i)}$. When q and t are generic, the eigenvalues $e_{\vec{\lambda}}$ of the generalized Macdonald symmetric functions are non-degenerate, that is

$$\vec{\lambda} \neq \vec{\mu} \quad \Rightarrow \quad e_{\vec{\lambda}} \neq e_{\vec{\mu}}. \tag{15}$$

Therefore we have the following orthogonality of them.

Fact 2.3 ([4]). If $\vec{\lambda} \neq \vec{\mu}$ then

$$\left\langle P_{\vec{\lambda}}^*, P_{\vec{\mu}} \right\rangle_{q,t} = 0. \tag{16}$$

3. Limit to β deformation

In this section, we set $u_i = q^{u'_i}$ $(i = 1, \ldots, N)$, $t = q^\beta$, $q = e^\hbar$ and take the limit $\hbar \to 0$ with β fixed in order to consider the specialization to β-deformation. Since $(1-t^{-n})(1-t^n q^{-n})/n = \mathcal{O}(\hbar^2)$ and $(1-t^{-n})/n$, $(1-q^n) = \mathcal{O}(\hbar)$,

$$\begin{aligned}\oint \frac{dz}{2\pi\sqrt{-1}z}\widetilde{\Lambda}_i(z) =& 1 + \sum_{n=1}^{\infty}\left(-\frac{(1-t^{-n})(1-q^n)}{n}p_n^{(i)}\frac{\partial}{\partial p_n^{(i)}}\right) \\ &+ \sum_{k=1}^{i-1}\sum_{n=1}^{\infty}\left(-\frac{(1-t^{-n})(1-t^n q^{-n})(1-q^n)}{n}(t/q)^{-\frac{(i-k-1)n}{2}}p_n^{(k)}\frac{\partial}{\partial p_n^{(i)}}\right) \\ &+ \frac{1}{2}\sum_{n,m}\left(-\frac{(1-t^{-n})(1-t^{-m})(1-q^{n+m})}{nm}p_n^{(i)}p_m^{(i)}\frac{\partial}{\partial p_{n+m}^{(i)}}\right) \\ &+ \frac{1}{2}\sum_{n,m}\left(\frac{(1-t^{-n-m})(1-q^n)(1-q^m)}{(n+m)}p_{n+m}^{(i)}\frac{\partial}{\partial p_n^{(i)}}\frac{\partial}{\partial p_m^{(i)}}\right) + \mathcal{O}(\hbar^4).\end{aligned} \tag{17}$$

Hence the $\hbar$ expansion is

$$\begin{aligned}\oint \frac{dz}{2\pi\sqrt{-1}z}\widetilde{\Lambda}_i(z) =& 1 + \hbar^2\left\{\beta\sum_{n=1}^{\infty} n\, p_n^{(i)}\frac{\partial}{\partial p_n^{(i)}}\right\} + \hbar^3\left\{\beta(1-\beta)\sum_{k=1}^{i-1}\sum_{i=1}^{\infty} n^2 p_n^{(k)}\frac{\partial}{\partial p_n^{(i)}}\right. \\ &+ \frac{\beta^2}{2}\sum_{n,m}(n+m)p_n^{(i)}p_m^{(i)}\frac{\partial}{\partial p_{n+m}^{(i)}} + \frac{\beta}{2}\sum_{n,m} n\, m\, p_{n+m}^{(i)}\frac{\partial^2}{\partial p_n^{(i)}\partial p_m^{(i)}} \\ &\left.+ \frac{\beta(1-\beta)}{2}\sum_{n=1}^{\infty} n^2 p_n^{(i)}\frac{\partial}{\partial p_n^{(i)}}\right\} + \mathcal{O}(\hbar^4).\end{aligned} \tag{18}$$

Thus we get

$$\begin{aligned}u_i\oint \frac{dz}{2\pi\sqrt{-1}z}\widetilde{\Lambda}_i(z) =& 1 + u'_i\,\hbar + \hbar^2\left\{\beta\sum_{n=1}^{\infty} n\, p_n^{(i)}\frac{\partial}{\partial p_n^{(i)}} + \frac{1}{2}u_i'^2\right\} \\ &+ \hbar^3\left\{\beta\,\mathcal{H}_\beta^{(i)} + \beta\sum_{k=1}^{i-1}\mathcal{H}_\beta^{(i,k)} + \frac{u_i'^3}{6}\right\} + \mathcal{O}(\hbar^4),\end{aligned} \tag{19}$$

where

$$\mathcal{H}_\beta^{(i)} := \frac{1}{2}\sum_{n,m}\left(\beta(n+m)p_n^{(i)}p_m^{(i)}\frac{\partial}{\partial p_{n+m}^{(i)}} + n\, m\, p_{n+m}^{(i)}\frac{\partial^2}{\partial p_n^{(i)}\partial p_m^{(i)}}\right) + \sum_{n=1}^{\infty}\left(u'_i + \frac{1-\beta}{2}n\right)np_n^{(i)}\frac{\partial}{\partial p_n^{(i)}}, \tag{20}$$

$$\mathcal{H}_\beta^{(i,k)} := (1-\beta)\sum_{n=1}^{\infty} n^2 p_n^{(k)}\frac{\partial}{\partial p_n^{(i)}}. \tag{21}$$

For $k = 0, 1, 2, \ldots$, we define operators H_k by

$$X_0 =: \sum_{k=0}^{\infty}\hbar^k H_k. \tag{22}$$

IOP Conf. Series: Journal of Physics: Conf. Series **804** (2017) 012036 doi:10.1088/1742-6596/804/1/012036

All homogeneous symmetric functions of the same degree belong to the same eigenspace of H_0, H_1 and H_2. Even without the operators H_0, H_1 and H_2, the eigenfunctions of X_0 do not change. In addition, we have

$$\lim_{\hbar\to 0}\left(\frac{X_0-(H_0+\hbar H_1+\hbar^2 H_2)}{(t-1)(q-1)^2}\right)=\mathcal{H}_\beta+\frac{1}{6\beta}\sum_{i=1}^{N}u_i'^3, \tag{23}$$

$$\mathcal{H}_\beta:=\sum_{i=1}^{N}\mathcal{H}_\beta^{(i)}+\sum_{i>j}\mathcal{H}_\beta^{(i,j)}. \tag{24}$$

Consequently the limit $q\to 1$ of the generalized Macdonald functions are eigenfunctions of the differential operator $\mathcal{H}_\beta$. $\mathcal{H}_\beta$ plus the momentum $(\beta-1)\sum_{i=1}^{N}\sum_{n=1}^{\infty}np_n^{(i)}\frac{\partial}{\partial p_n^{(i)}}$ corresponds to the differential operator of [9, 11], the eigenfunctions of which are called generalized Jack symmetric functions.

As the Fact 2.1, we can triangulate $\mathcal{H}_\beta$ similarly. Moreover if $\vec{\lambda}\geq^{\mathrm{L}}\vec{\mu}$ and β is generic, then $e'_{\vec{\lambda}}\neq e'_{\vec{\mu}}$. ($e'_{\vec{\lambda}}$, $e'_{\vec{\mu}}$ are eigenvalues of $\mathcal{H}_\beta$.) Therefore we get the existence theorem of the generalized Jack symmetric functions.

Proposition 3.1. There exists a unique symmetric function $J_{\vec{\lambda}}$ satisfying the following two conditions:

$$J_{\vec{\lambda}}=m_{\vec{\lambda}}+\sum_{\vec{\mu}<^{\mathrm{L}}\vec{\lambda}}d'_{\vec{\lambda}\vec{\mu}}m_{\vec{\mu}},\qquad d'_{\vec{\lambda}\vec{\mu}}\in\mathbb{Q}(\beta,u'_1,\ldots,u'_N); \tag{25}$$

$$\mathcal{H}_\beta J_{\vec{\lambda}}=e'_{\vec{\lambda}}J_{\vec{\lambda}},\qquad e'_{\vec{\lambda}}\in\mathbb{Q}(\beta,u'_1,\ldots,u'_N). \tag{26}$$

From the above argument and the uniqueness in this proposition we get the following important result.

Proposition 3.2. The limit of the generalized Macdonald symmetric functions $P_{\vec{\lambda}}$ to β-deformation coincide with the generalized Jack symmetric functions $J_{\vec{\lambda}}$. That is

$$P_{\vec{\lambda}}\quad\underset{u_i=q^{u'_i},\,t=q^\beta,\,q=e^\hbar}{\underset{\hbar\to 0,}{\longrightarrow}}\quad J_{\vec{\lambda}}. \tag{27}$$

Remark 3.3. For the dual functions $P^*_{\vec{\lambda}}$ and $J^*_{\vec{\lambda}}$, the similar proposition holds.

By Fact 2.3, Proposition 3.2 and the fact that the scalar product $\langle -,-\rangle_{q,t}$ reduces to the scalar product $\langle -,-\rangle_\beta$ which is defined by

$$\langle p_{\vec{\lambda}},p_{\vec{\mu}}\rangle_\beta=\delta_{\vec{\lambda},\vec{\mu}}\prod_{i=1}^{N}z_{\lambda^{(i)}}\beta^{-\ell(\lambda^{(i)})}, \tag{28}$$

we obtain the orthogonality of the generalized Jack symmetric functions.

Proposition 3.4. If $\vec{\lambda}\neq\vec{\mu}$, then

$$\left\langle J^*_{\vec{\lambda}},J_{\vec{\mu}}\right\rangle_\beta=0. \tag{29}$$

By this proposition, we can prove the chauchy formula of generalized Jack symmetric functions in the usual way. For example, in the $N=2$ case, we have

$$\sum_{\vec{\lambda}}\frac{J_{\vec{\lambda}}(x^{(1)},x^{(4)})\,J^*_{\vec{\lambda}}(x^{(2)},x^{(3)})}{v_{\vec{\lambda}}}=\exp\left(\beta\sum_{n\geq 1}\frac{1}{n}p_n^{(1)}p_n^{(2)}\right)\exp\left(\beta\sum_{n\geq 1}\frac{1}{n}p_n^{(3)}p_n^{(4)}\right), \tag{30}$$

where $v_{\vec{\lambda}}:=\langle J^*_{\vec{\lambda}},J_{\vec{\lambda}}\rangle_\beta$. This is the essential formula used in the scenario of proof of the AGT conjecture [9].

4. Example

We give examples of Proposition 3.2 in the case $N = 2$. The generalized Macdonald symmetric functions of level 1 and 2 have the forms:

$$\begin{pmatrix} P_{(0),(1)} \\ P_{(1),(0)} \end{pmatrix} = M^1_{q,t} \begin{pmatrix} m_{(0),(1)} \\ m_{(1),(0)} \end{pmatrix}, \qquad M^1_{q,t} := \begin{pmatrix} 1 & (t/q)^{\frac{1}{2}} \frac{(t-q)u_2}{t(u_1-u_2)} \\ 0 & 1 \end{pmatrix}, \tag{31}$$

$$\begin{pmatrix} P_{(0),(2)} \\ P_{(0),(1,1)} \\ P_{(1),(1)} \\ P_{(2),(0)} \\ P_{(1,1),(0)} \end{pmatrix} = M^2_{q,t} \begin{pmatrix} m_{(0),(2)} \\ m_{(0),(1,1)} \\ m_{(1),(1)} \\ m_{(2),(0)} \\ m_{(1,1),(0)} \end{pmatrix}, \qquad M^2_{q,t} := \tag{32}$$

$$\begin{pmatrix} 1 & \frac{(1+q)(t-1)}{qt-1} & \frac{(t/q)^{-\frac{1}{2}}(1+q)(q-t)(t-1)u_2}{(1-qt)(u_1-qu_2)} & \frac{(q-t)((1-q^2)tu_1-q(t^2-q(1+q)t+q)u_2)u_2}{qt(qt-1)(u_1-u_2)(u_1-qu_2)} & \frac{(1+q)(q-t)(t-1)((q-1)tu_1+q(q-t)u_2)u_2}{qt(qt-1)(u_1-u_2)(u_1-qu_2)} \\ 0 & 1 & (t/q)^{\frac{1}{2}}\frac{(t-q)u_2}{t(tu_1-u_2)} & \frac{(q-t)u_2}{q(tu_1-u_2)} & \frac{(q-t)(qu_2-t((t-1)u_1+u_2))u_2}{qt(u_1-u_2)(tu_1-u_2)} \\ 0 & 0 & 1 & (t/q)^{\frac{1}{2}}\frac{(t-q)u_2}{t(qu_1-u_2)} & (t/q)^{\frac{1}{2}}\frac{(q-t)((1+q+(q-1)t)u_1-2tu_2)}{t(qu_1-u_2)(-u_1+tu_2)} \\ 0 & 0 & 0 & 1 & \frac{(1+q)(t-1)}{qt-1} \\ 0 & 0 & 0 & 0 & 1 \end{pmatrix}$$

Also the generalized Jack symmetric functions have the forms:

$$\begin{pmatrix} J_{(0),(1)} \\ J_{(1),(0)} \end{pmatrix} = M^1_{\beta} \begin{pmatrix} m_{(0),(1)} \\ m_{(1),(0)} \end{pmatrix}, \qquad M^1_{\beta} := \begin{pmatrix} 1 & \frac{1-\beta}{-u'_1+u'_2} \\ 0 & 1 \end{pmatrix}, \tag{33}$$

$$\begin{pmatrix} J_{(0),(2)} \\ J_{(0),(1,1)} \\ J_{(1),(1)} \\ J_{(2),(0)} \\ J_{(1,1),(0)} \end{pmatrix} = M^2_{\beta} \begin{pmatrix} m_{(0),(2)} \\ m_{(0),(1,1)} \\ m_{(1),(1)} \\ m_{(2),(0)} \\ m_{(1,1),(0)} \end{pmatrix}, \tag{34}$$

$$M^2_{\beta} := \begin{pmatrix} 1 & \frac{2\beta}{1+\beta} & \frac{2\beta(1-\beta)}{(1+\beta)(1-u'_1+u'_2)} & \frac{(1-\beta)(2+\beta-\beta^2-2u'_1+2u'_2)}{(1+\beta)(u'_1-u'_2)(-1+u'_1-u'_2)} & \frac{2\beta(2-3\beta+\beta^2)}{(1+\beta)(u'_1-u'_2)(-1+u'_1-u'_2)} \\ 0 & 1 & \frac{1-\beta}{-\beta-u'_1+u'_2} & \frac{1-\beta}{\beta+u'_1-u'_2} & \frac{-1+3\beta-2\beta^2}{(u'_1-u'_2)(-\beta-u'_1+u'_2)} \\ 0 & 0 & 1 & \frac{1-\beta}{-1-u'_1+u'_2} & \frac{2(1-\beta)(-1+\beta-u'_1+u'_2)}{(-1-u'_1+u'_2)(\beta-u'_1+u'_2)} \\ 0 & 0 & 0 & 1 & \frac{2\beta}{1+\beta} \\ 0 & 0 & 0 & 0 & 1 \end{pmatrix}. \tag{35}$$

If we take the limit $q \to 1$ of $M^i_{q,t}$, then M^i_{β} appears.

Acknowledgements

The author would like to show his deepest gratitude to his supervisor H. Awata for his enormous help and also thanks H. Kanno and A. Morozov very much for their valuable comments. This work is supported in part by Grant-in-Aid for JSPS Fellow 26-10187.

References

[1] Alday L F, Gaiotto D and Tachikawa Y 2010 *Lett. Math. Phys.* **91** 167–197
[2] Awata H and Yamada Y 2010 *JHEP* **01** 125
[3] Awata H and Yamada Y 2010 *Prog. Theor. Phys.* **124** 227–262
[4] Awata H, Feigin B, Hoshino A, Kanai M, Shiraishi J and Yanagida S 2011 *RIMS kokyuroku* **1765** 12–32 (*Preprint* 1106.4088)
[5] Ding J and Iohara K 1997 *Lett. Math. Phys.* **41** 181–193
[6] Miki K 2007 *J. Math. Phys.* **48** 123520, 35

[7] Tan M C 2013 *JHEP* **12** 031
[8] Awata H, Feigin B and Shiraishi J 2012 *JHEP* **03** 041
[9] Morozov A and Smirnov A 2014 *Lett. Math. Phys.* **104** 585–612
[10] Mironov A, Morozov A and Shakirov S 2011 *JHEP* **02** 067
[11] Mironov S, Morozov A and Zenkevich Y 2014 *JETP Lett.* **99** 109–113
[12] Zenkevich Y 2015 *JHEP* **05** 131
[13] Feigin B, Hashizume K, Hoshino A, Shiraishi J and Yanagida S 2009 *J. Math. Phys.* **50** 095215, 42
[14] Feigin B, Hoshino A, Shibahara J, Shiraishi J and Yanagida S 2010 *RIMS kokyuroku* **1689** 133–152 (*Preprint* 1002.2485)

Integrability in out-of-equilibrium systems

Eric RAGOUCY
Laboratoire de Physique Théorique LAPTh, CNRS and USBM.
9 chemin de Bellevue, BP 110, F-74941 Annecy-le-Vieux Cedex, France.

Abstract. This short note presents a summary of the articles arXiv:1408.5357, arXiv:1412.5939, arXiv:1603.06796, arXiv:1606.01018, arXiv:1606.08148 that were done in collaboration with N. CRAMPE, M. EVANS, C. FINN, K. MALLICK and M. VANICAT. It presents an approach to the matrix ansatz of out-of-equilibrium systems with boundaries, within the framework of integrable systems.

1. General context

Statistical thermodynamics relies on the existence of a thermodynamical equilibrium stationary state, where the probability of a configuration is given by the Boltzmann distribution:

$$P_{eq}(\mathcal{C}) \sim e^{-\frac{E(\mathcal{C})}{k_B T}}. \tag{1}$$

The existence of this distribution allows one to compute many physical data and makes contact with the 'usual' thermodynamical laws. However, the existence of the thermodynamical equilibrium stationary state is quite restrictive, since it implies that there is no particle or energy flow. Then, simple situations such as a metal stick in between two reservoirs with different temperatures cannot be studied within the framework of thermodynamical equilibrium. Nowadays, to generalize this framework, people are considering non-equilibrium stationary states, where the state under consideration does not envolve in time but allows particle or energy currents. Unfortunately, the configuration probability $P_{stat}(\mathcal{C})$ of such non-equilibrium stationary state is not known in general. However for some models, the **matrix product ansatz** allows exact computations to get $P_{stat}(\mathcal{C})$. The goal of our work is to understand this matrix ansatz (when it exists) within the integrable system framework and to generalize it to other integrable models, where the matrix ansatz was not known up to now.

To present this 'integrable approach to matrix ansatz', we will first work with one of the simplest (though paradigmatic) model, the TASEP model with open boundaries (see definition below). We will introduce our general approach for this simple model (section 2), and then present it in its full generality (section 3). The general approach applies obviously to more models, but with more generators than the ones used in the TASEP model: we illustrate it on another (reaction-diffusion) model, that we solve using our technics (section 4). However, although rather general, our technics still lacks from general principles to fix the form of the matrix ansatz: we show it on a third example, based on multi-species TASEP (section 5). Finally, we conclude in section 6. An appendix present to some technical results.

We would like to stress that this short note does not intend to be exhaustive, and many works on the subject should be cited. We refer the interested reader to our original articles, or to reviews such as [1, 2, 3, 4, 5] for more references.

2. TASEP and Matrix Ansatz

2.1. The TASEP model

The acronym TASEP stands Totally Asymmetric Simple Exclusion Process, where 'totally asymmetric' indicates that particles are flowing in one direction only, 'simple' means that the jump can be only one step forwards, and 'exclusion' that two particles cannot be simultaneously at the same site. We thus have a one-dimenional lattice on which 'particles' are flowing in one direction only. We will consider the case of open boundaries, meaning that the lattice is connected to two infinite dimensional reservoirs, from which the particle enters the lattice (on the left) and leave it (on the right). The different possible rates of transition are show in figure 1. A rate $M(\mathcal{C},\mathcal{C}')$ corresponds to the probablity $M(\mathcal{C},\mathcal{C}')\,dt$ that the system jumps from the configuration $\mathcal{C}'$ to the configuration $\mathcal{C}$ during a small interval of time dt.

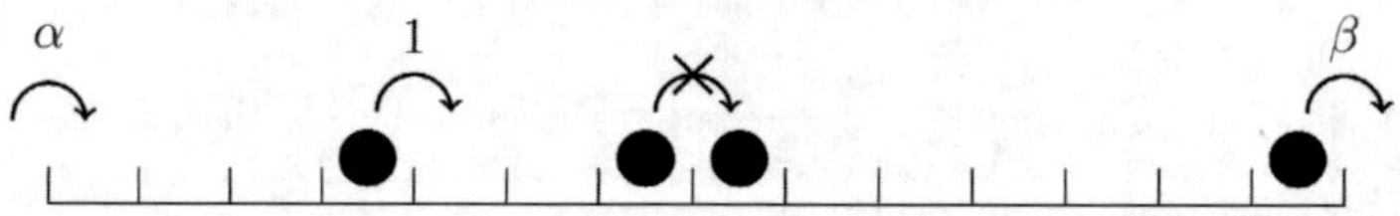

Figure 1. The different transition rates in the TASEP model

The TASEP is an out-of-equilibrium system, meaning that there exists a stationnary state, although there is a particle current (from the left to the right on figure 1). As we shall see there is a matrix ansatz that describes this stationnary state. Moreover the model is integrable, which makes it interesting for our study.

2.2. Markov matrix and master equation

To construct the stationary state, we need to formalize the presentation of the model. We denote a configuration of the system by $\mathcal{C} = (\tau_1, \tau_2, \ldots, \tau_L)$, with $\tau_j = 0$ if the site j is empty and $\tau_j = 1$ if a particle stands at site j. As an example, we have:

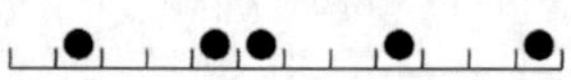 $\longrightarrow \quad (0,1,0,0,1,1,0,0,1,0,0,1).$

With this notation, the rates presented in figure 1 can be recasted as

- In the bulk, particles can jump to the right at the rate: $M(\{....,0,1,....\};\{....,1,0,...\}) = 1$
- On the left boundary, particles enter at the rate: $M(\{1,....\};\{0,...\}) = \alpha$
- On the right boundary, particles leave at the rate: $M(\{....,0\};\{....,1\}) = \beta$

where dots stand for arbitrary values of the τ_j's.

We denote by $P_t(\mathcal{C})$ the probability to be in configuration $\mathcal{C}$ at time t. The time evolution of this probability is governed by the different rates that define the model. In fact, performing a balance among the possible transitions between different configurations, we get

$$P_{t+dt}(\mathcal{C}) = \sum_{\mathcal{C}'\neq\mathcal{C}} M(\mathcal{C},\mathcal{C}')dt\, P_t(\mathcal{C}') + \Big(1 - \sum_{\mathcal{C}'\neq\mathcal{C}} M(\mathcal{C}',\mathcal{C})dt\Big) P_t(\mathcal{C}). \tag{2}$$

In words, it just states that the probalility to be in configuration $\mathcal{C}$ at time $t+dt$ is the sum of two terms: (i) the probability to jump in this configuration starting from another one between t and $t+dt$, (ii) the probability to stay in this configuration between t and $t+dt$.

For $dt \to 0$ this equation can be recasted as

$$\frac{dP_t(\mathcal{C})}{dt} = \sum_{\mathcal{C}'\neq\mathcal{C}} M(\mathcal{C},\mathcal{C}')P_t(\mathcal{C}') - \sum_{\mathcal{C}'\neq\mathcal{C}} M(\mathcal{C}',\mathcal{C})P_t(\mathcal{C}). \tag{3}$$

IOP Conf. Series: Journal of Physics: Conf. Series **804** (2017) 012037 doi:10.1088/1742-6596/804/1/012037

Equation (3) is called the Master equation.

Since at each site, we have 2 possibilities (0 or 1, i.e. empty or occupied), there is a $\mathbb{C}^2$ "local" space to describe the different probabilities. Then, for L sites we get a $(\mathbb{C}^2)^{\otimes L}$ total space, so that we can gather all the probabilities in a vector

$$|P_t\rangle = \begin{pmatrix} P_t(\,(0,\dots,0,0,0)\,) \\ P_t(\,(0,\dots,0,0,1)\,) \\ P_t(\,(0,\dots,0,1,0)\,) \\ P_t(\,(0,\dots,0,1,1)\,) \\ \vdots \\ P_t(\,(1,\dots,1,1,1)\,) \end{pmatrix} \in \quad (\mathbb{C}^2)^{\otimes L} \tag{4}$$

and the Master equation takes a vectorial form

$$\frac{d|P_t\rangle}{dt} = M\ |P_t\rangle \in (\mathbb{C}^2)^{\otimes L}, \tag{5}$$

where the Markov matrix M can be written in terms of local jump operators

$$M = B_1 + \sum_{\ell=1}^{L-1} m_{\ell,\ell+1} + \overline{B}_L \in (End(\mathbb{C}^2))^{\otimes L}. \tag{6}$$

In (6) we have used the standard auxiliary space notation for tensor products of $\mathbb{C}^2$ spaces, i.e. the indices indicate on which copies of $\mathbb{C}^2$ in the total space $(\mathbb{C}^2)^{\otimes L}$, the matrices act non-trivially:

$$m_{\ell,\ell+1} = \underbrace{\mathbb{I}\otimes\cdots\otimes\mathbb{I}}_{\ell-1}\otimes \underbrace{m}_{\ell,\ell+1} \otimes \underbrace{\mathbb{I}\otimes\cdots\otimes\mathbb{I}}_{L-1-\ell}\ ; \tag{7}$$

$$B_1 = \underbrace{B}_{1}\otimes\underbrace{\mathbb{I}\otimes\cdots\otimes\mathbb{I}}_{L-1} \quad ; \quad \overline{B}_L = \underbrace{\mathbb{I}\otimes\cdots\otimes\mathbb{I}}_{L-1}\otimes\underbrace{\overline{B}}_{L} \tag{8}$$

where $\mathbb{I}$ is the 2×2 identity matrix.

The local matrices m, B and $\overline{B}$ corresponding to the rates described above take the following form:

$$\underbrace{B = \begin{pmatrix} -\alpha & 0 \\ \alpha & 0 \end{pmatrix}}_{\in End(\mathbb{C}^2)}\ ;\ \underbrace{m = \begin{pmatrix} 0 & 0 & 0 & 0 \\ 0 & 0 & 1 & 0 \\ 0 & 0 & -1 & 0 \\ 0 & 0 & 0 & 0 \end{pmatrix}}_{\in End(\mathbb{C}^2\otimes\mathbb{C}^2)}\ ;\ \underbrace{\overline{B} = \begin{pmatrix} 0 & \beta \\ 0 & -\beta \end{pmatrix}}_{\in End(\mathbb{C}^2)}. \tag{9}$$

We remind that our goal is to compute the probability distribution corresponding to the (time independent) stationary state, $\mathcal{S}(\mathcal{C})$, which in the vectorial form amounts to solve the zero-eigenvalue problem for the Markov matrix (6):

$$M|\mathcal{S}\rangle = 0.$$

2.3. Matrix ansatz

The matrix ansatz was introduced in [6], and then developped by many authors for different models. In this formalism, to get the probability of a given configuration $\mathcal{C}$ in the stationary state, one first associates to each empty site an operator E, and to each occupied site an operator D, forming an algebra (see below). Then, one can uniquely associate to the configuration $\mathcal{C}$ a 'word' $\mathsf{w}(\mathcal{C})$ of length L in E and D. One also considers two vectors $\langle\langle W|$ and $|V\rangle\rangle$ belonging to two additional[1] spaces that form two representations of the (E, D) algebra. Let us stress that the additional spaces have nothing to do with the space $(\mathbb{C}^2)^{\otimes L}$ introduced before. In particular, the vectors $\langle\langle W|$ and $|V\rangle\rangle$ are scalar with respect to the $(\mathbb{C}^2)^{\otimes L}$ space, and commute with the matrices m, B and $\overline{B}$.

Now, the matrix ansatz states that the probability to be in the configuration $\mathcal{C}$ in the stationary state is given by

$$\mathcal{S}(\mathcal{C}) = \frac{\langle\langle W|\mathsf{w}(\mathcal{C})|V\rangle\rangle}{Z_L} \quad \text{with} \quad Z_L = \langle\langle W|(D+E)^L|V\rangle\rangle, \tag{10}$$

provided E and D obey the following algebraic relations (for the TASEP model)

$$DE = D + E \tag{11}$$

and the vectors $\langle\langle W|$ and $|V\rangle\rangle$ are such that

$$D|V\rangle\rangle = \frac{1}{\beta}|V\rangle\rangle \quad ; \quad \langle\langle W|E = \frac{1}{\alpha}\langle\langle W|. \tag{12}$$

Let us stress that not only there is a proof that the matrix ansatz provides the right weights for the probabilities (see below where the proof is reminded), but also it gives an explicit and recursive way to explicitly compute them. We illustrate it on an example.

Example If one consider a model with $L = 5$ sites and a configuration $\mathcal{C} = (0, 1, 0, 1, 1)$, one has $\mathsf{w}(0, 1, 0, 1, 1) = EDEDD$, so that

$\mathcal{S}($  $)$

$$= \frac{\langle\langle W|EDEDD|V\rangle\rangle}{Z_5} \quad \text{with} \quad Z_5 = \langle\langle W|(D+E)^5|V\rangle\rangle.$$

Using first relation (11) on the product DE and then relations (12) for E on the left and for DD on the right one easily computes that

$$\langle\langle W|EDEDD|V\rangle\rangle = \langle\langle W|E(D+E)DD|V\rangle\rangle = \frac{1}{\alpha\,\beta^2}\langle\langle W|(D+E)|V\rangle\rangle = \frac{\alpha+\beta}{\alpha^2\,\beta^3}\langle\langle W|V\rangle\rangle.$$

In the same way, one gets

$$\begin{aligned} Z_5 \;=\; & \Big\{\beta^5 + \alpha\beta^4(1+4\beta) + \alpha^2\beta^3(1+4\beta+9\beta^2) + \alpha^3\beta^2(1+4\beta+9\beta^2+14\beta^3) + \\ & +\alpha^4\beta(1+4\beta+9\beta^2+14\beta^3+14\beta^4) + \alpha^5(1+4\beta+9\beta^2+14\beta^3+14\beta^4)\Big\}\frac{\langle\langle W|V\rangle\rangle}{\alpha^5\beta^5} \end{aligned}$$

Then, up to the proof that there exist vectors $\langle\langle W|$ and $|V\rangle\rangle$ such that $\langle\langle W|V\rangle\rangle \neq 0$, one gets an explicit expression for $\mathcal{S}($ 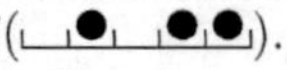 $)$. The construction of such vectors is also known for the TASEP, see [6] for more details.

[1] These spaces are usually called auxiliary spaces by the statistical physicists, but they are different from the auxiliary spaces used in integrable systems, so that we choose to call them additional.

IOP Conf. Series: Journal of Physics: Conf. Series **804** (2017) 012037 doi:10.1088/1742-6596/804/1/012037

2.4. The matrix ansatz gives the right weights of the stationary state

We reproduce this proof, not in its original version, but using an integrable system formalism that will be adapted to our generalisation. The vector $|\mathcal{S}\rangle$ can be written in the compact form

$$|\mathcal{S}\rangle = \frac{1}{Z_L}\langle\langle W|\, A_1\, A_2 \cdots\, A_L\, |V\rangle\rangle \in (\mathbb{C}^2)^{\otimes L} \quad \text{with} \quad A = \begin{pmatrix} E \\ D \end{pmatrix} \tag{13}$$

where we have again used the auxiliary space notation for tensor products of $\mathbb{C}^2$ spaces, meaning that

$$A_\ell = \underbrace{\mathbb{I} \otimes \cdots \otimes \mathbb{I}}_{\ell-1} \otimes A \otimes \underbrace{\mathbb{I} \otimes \cdots \otimes \mathbb{I}}_{L-\ell}, \quad \forall \ell \quad \text{so that} \quad A_1\, A_2 \cdots\, A_L = \underbrace{A \otimes A \otimes \cdots \otimes A}_{L}.$$

Using the form of the local markov matrix m, see (9), one can show that the bulk relation $DE = D + E$ is equivalent to

$$m\; A \otimes A = \bar{A} \otimes A - A \otimes \bar{A} \quad \text{with} \quad \bar{A} = \begin{pmatrix} 1 \\ -1 \end{pmatrix}. \tag{14}$$

Then

$$\begin{aligned}
\Big(m_{12} + m_{23} + \cdots + m_{L-1,L}\Big)\, A_1\, A_2 \cdots\, A_L \;&=\; \Big(\bar{A}_1\, A_2 - A_1\, \bar{A}_2\Big) A_3 \cdots\, A_L \\
&\quad + A_1 \Big(\bar{A}_2\, A_3 - A_2\, \bar{A}_3\Big) A_4 \cdots\, A_L \\
&\quad\; \vdots \\
&\quad + A_1 \cdots\, A_{L-2} \Big(\bar{A}_{L-1}\, A_L - A_{L-1}\, \bar{A}_L\Big)
\end{aligned}$$

where the first line corresponds to the action of m_{12}, the second line to the action of m_{23}, ..., up to the action of $m_{L-1,1}$. One sees that we get a telescoping sum, so that finally only the first and the last terms remain:

$$\Big(m_{12} + m_{23} + \cdots + m_{L-1,L}\Big)\, A_1\, A_2 \cdots\, A_L \;=\; \bar{A}_1\, A_2 \cdots\, A_L - A_1 \cdots\, A_{L-1}\, \bar{A}_L$$

This equality holds in the algebra, without the use of the vectors $\langle\langle W|$ and $|V\rangle\rangle$. Obviously it is still valid when applied to these vectors.

One can also show that the left (resp. right) boundary condition $\langle\langle W|E = \frac{1}{\alpha}\langle\langle W|$ (resp. $D|V\rangle\rangle = \frac{1}{\beta}|V\rangle\rangle$) are equivalent to

$$\langle\langle W|B\, A = -\langle\langle W|\bar{A} \quad \text{and} \quad \bar{B}\, A|V\rangle\rangle = \bar{A}|V\rangle\rangle \tag{15}$$

with again $A = \begin{pmatrix} E \\ D \end{pmatrix}$ and $\bar{A} = \begin{pmatrix} 1 \\ -1 \end{pmatrix}$. The relations (15) are recasted as

$$\begin{aligned}
&\langle\langle W|B_1\, A_1\, A_2 \cdots\, A_L = -\langle\langle W|\bar{A}_1\, A_2 \cdots\, A_L \\
&\bar{B}_L\, A_1\, A_2 \cdots\, A_L|V\rangle\rangle = A_1\, A_2 \cdots\, A_{L-1}\, \bar{A}_L|V\rangle\rangle.
\end{aligned}$$

Putting together these algebraic relations we get

$$M\langle\langle W|\, A_1\, A_2 \cdots\, A_L\, |V\rangle\rangle \;=\; \left(\sum_{\ell=1}^{L-1} m_{\ell,\ell+1} + B_1 + \overline{B}_L\right) \langle\langle W|\, A_1\, A_2 \cdots\, A_L\, |V\rangle\rangle = 0.$$

This shows that the steady state $|\mathcal{S}\rangle$ is proportional to $\langle\langle W|\, A_1\, A_2 \cdots\, A_L\, |V\rangle\rangle$. One fixes the normalisation by demanding that all the probabilities add up to 1, so that

$$|\mathcal{S}\rangle \;=\; \frac{1}{Z_L}\langle\langle W|\, A_1\, A_2 \cdots\, A_L\, |V\rangle\rangle = \frac{1}{Z_L}\langle\langle W|\, A \otimes A \otimes \cdots \otimes A\, |V\rangle\rangle$$

where we remind that $Z_L = \langle\langle W|(D + E)^L|V\rangle\rangle$.

IOP Conf. Series: Journal of Physics: Conf. Series **804** (2017) 012037 doi:10.1088/1742-6596/804/1/012037

2.5. Integrability of the TASEP model

As already mentioned, the model is integrable, and we briefly summarize the basic ingredients that are related to its integrability.

2.5.1. In the bulk: the R-matrix

$$R(x) = \begin{pmatrix} 1 & 0 & 0 & 0 \\ 0 & 0 & x & 0 \\ 0 & 1 & 1-x & 0 \\ 0 & 0 & 0 & 1 \end{pmatrix} \in End(\mathbb{C}^2 \otimes \mathbb{C}^2)$$

where $x \in \mathbb{C}$ is the spectral parameter. It is related to the local matrix m in the usual way

$$P\,\frac{d}{dx}R(x)\Big|_{x=1} = -m \quad \text{with} \quad P = \begin{pmatrix} 1 & 0 & 0 & 0 \\ 0 & 0 & 1 & 0 \\ 0 & 1 & 0 & 0 \\ 0 & 0 & 0 & 1 \end{pmatrix}.$$

Integrability in the bulk is ensured by the Yang-Baxter equation

$$R_{12}\Big(\frac{x_1}{x_2}\Big)\,R_{13}\Big(\frac{x_1}{x_3}\Big)\,R_{23}\Big(\frac{x_2}{x_3}\Big) = R_{23}\Big(\frac{x_2}{x_3}\Big)\,R_{13}\Big(\frac{x_1}{x_3}\Big)\,R_{12}\Big(\frac{x_1}{x_2}\Big)$$

where we used again the auxiliary space notation, e.g. $R_{12}(x) = R(x)\otimes\mathbb{I}$ and $R_{23}(x) = \mathbb{I}\otimes R(x)$. The R-matrix is unitary and regular:

$$R_{12}(x)R_{21}(1/x) = 1 \quad \text{and} \quad R(1) = P.$$

2.5.2. On the boundaries: K-matrices

$$K(x) = \begin{pmatrix} \dfrac{(-x\alpha+\alpha-1)x}{x\alpha-x-\alpha} & 0 \\ \dfrac{\alpha(x^2-1)}{x\alpha-x-\alpha} & 1 \end{pmatrix}, \quad \bar{K}(x) = \begin{pmatrix} 1 & -\dfrac{(x^2-1)\beta}{-x^2\beta+x\beta-x} \\ 0 & \dfrac{x\beta-x-\beta}{-x^2\beta+x\beta-x} \end{pmatrix} \tag{16}$$

They are connected to the local matrices B and $\overline{B}$:

$$\frac{d}{dx}K(x)\Big|_{x=1} = -2B \quad \text{and} \quad \frac{d}{dx}\bar{K}(x)\Big|_{x=1} = 2\bar{B}.$$

They satisfy the reflection equation

$$R_{12}\left(\frac{x_1}{x_2}\right)K_1(x_1)R_{21}(x_1x_2)K_2(x_2) = K_2(x_2)R_{12}(x_1x_2)K_1(x_1)R_{21}\left(\frac{x_1}{x_2}\right) \tag{17}$$

and are unitarity and regular

$$K(x)K(1/x) = 1 \quad \text{and} \quad K(1) = 1.$$

2.5.3. Integrability: We define the usual double row transfer matrix [7]

$$t(x) = tr_0\Big(\widetilde{K}_0(x)\; R_{0L}(x)\cdots R_{01}(x)\; K_0(x)\; R_{10}(x)\cdots R_{L0}(x)\Big), \tag{18}$$

where $\widetilde{K}$ is linked to $\bar{K}$ in the following way

$$\bar{K}_1(x) = tr_0\Big(\widetilde{K}_0(\frac{1}{x})R_{01}(\frac{1}{x^2})P_{01}\Big). \tag{19}$$

It generates the Markov matrix

$$-\frac{1}{2}\frac{dt(x)}{dx}\bigg|_{x=1} = -\frac{1}{2}K_1'(1) - \sum_{k=1}^{L-1} P_{k,k+1}R'_{k,k+1}(1) + \frac{1}{2}\bar{K}_L'(1) = M \tag{20}$$

and integrability of the model is ensured by the commutativity of the transfer matrix $[t(x), t(x')] = 0$. This last relation is proven in the usual way [7].

Integrability ensures that we can construct exactly the eigenvectors of the transfer matrix (and thus of the Markov matrix) using for instance the Bethe ansatz. The Bethe vectors are eigenvector of the transfer matrix $t(x)\,\mathbb{B}(\bar{u}) = \lambda(x,\bar{u})\,\mathbb{B}(\bar{u})$ when the set of parameters $\bar{u}$ obeys the so-called Bethe equations. However the stationary state is not easily obtained in this way since one should solve the Bethe equations to identify it. Our goal is to recover the matrix ansatz solution from R and K matrices.

2.6. Matrix ansatz relations from R and K matrices

Let us define the vector

$$A(x) = \begin{pmatrix} E-1+x \\ D-1+\frac{1}{x} \end{pmatrix} \quad\Rightarrow\quad A(1) = \begin{pmatrix} E \\ D \end{pmatrix} \equiv A,\; A'(1) = \begin{pmatrix} 1 \\ -1 \end{pmatrix} \equiv \bar{A}. \tag{21}$$

We impose on the vector $A(x)$ the following well-known relations.

Zamolodchikov-Faddeev (ZF) relation [8]:

$$R_{12}\Big(\frac{x}{y}\Big)\,A_1(x)\,A_2(y) = A_2(y)\,A_1(x). \tag{22}$$

Ghoshal-Zamolodchikov (GZ) relations [9]:

$$\langle\!\langle W|A(x) = \langle\!\langle W|K(x)A\Big(\frac{1}{x}\Big) \quad\text{and}\quad A(x)|V\rangle\!\rangle = \bar{K}(x)A\Big(\frac{1}{x}\Big)\,|V\rangle\!\rangle. \tag{23}$$

It turns out that these relations, although they depend on spectral parameters simplify drastically (due to the explicit expressions of $R(x)$, $K(x)$, $\overline{K}(x)$ and $A(x)$), and are equivalent to the matrix ansatz relations

$$DE = D+E\,,\quad \langle\!\langle W|E = \frac{1}{\alpha}\langle\!\langle W|\,,\quad D|V\rangle\!\rangle = \frac{1}{\beta}|V\rangle\!\rangle. \tag{24}$$

Moreover, taking the derivative with respect to x and setting $x = y = 1$ in (22) and (23), one obtains exactly the relations (14) and (15) that are needed for the matrix ansatz. This suggest the following general "integrable approach" to the matrix ansatz.

3. Integrable approach to matrix ansatz

We start with an integrable model described by a unitary R-matrix $R_{12}(x)$ of size $r^2 \times r^2$, which obeys YBE, and the corresponding unitary boundary matrices $K(x)$ and $\bar{K}(x)$, of size $r \times r$, obeying the reflection equation. Integrability is ensured by the double-row transfer matrix (18).

3.1. Construction of the matrix ansatz

We introduce the vector $A(x) = \begin{pmatrix} X_1(x) \\ X_2(x) \\ \vdots \\ X_r(x) \end{pmatrix}$ where the elements $X_1(x), \ldots, X_r(x)$ belong to some algebra, defined by the following relations

ZF relations (in the bulk):

$$R_{12}\left(\frac{x}{y}\right) A_1(x)A_2(y) = A_2(y)A_1(x) \tag{25}$$

The associativity of this algebra ensured by the Yang-Baxter equation. Another consistency relation for this algebra is ensured by the unitarity of the R-matrix.

The ZF relation implies (through derivation w.r.t. x and setting $x = y = 1$)

$$m\ A_1(1)\,A_2(1) = A_1'(1)\,A_2(1) - A_1(1)\,A_2'(1) \quad \text{with} \quad m = -PR_{12}'(1). \tag{26}$$

N.B.: Nothing in this construction implies that $\bar{A} = A'(1)$ is necessarily scalar, despite it was scalar for the TASEP model. We illustrate it on an example in section 4.

GZ relation (on the boundaries):

$$\langle\!\langle W|A(x) = \langle\!\langle W|K(x)A\left(\frac{1}{x}\right) \quad \text{and} \quad A(x)|V\rangle\!\rangle = \bar{K}(x)A\left(\frac{1}{x}\right)|V\rangle\!\rangle \tag{27}$$

Again the consistency of these relations is ensured by the reflection equation and the unitarity of the K matrices.

They imply (through derivation w.r.t. x and setting $x = 1$)

$$\langle\!\langle W|\,BA(1) = -\langle\!\langle W|A'(1) \quad \text{with} \quad B = -\frac{1}{2}K'(1), \tag{28}$$

$$\bar{B}A(1)|V\rangle\!\rangle = A'(1)|V\rangle\!\rangle \quad \text{with} \quad \bar{B} = \frac{1}{2}\bar{K}'(1). \tag{29}$$

Then the vector

$$|\mathcal{S}\rangle = \frac{1}{Z_L}\,\langle\!\langle W|A_1(1)A_2(1)\cdots A_L(1)|V\rangle\!\rangle, \tag{30}$$

is the stationary state of the process (we have again a telescopic sum):

$$M|\mathcal{S}\rangle = 0 \quad \text{for} \quad M = B_1 + \sum_{\ell=1}^{L-1} m_{\ell,\ell+1} + \overline{B}_L. \tag{31}$$

3.2. Inclusion of inhomogeneities

This approach can be further generalized. We define the vector

$$|\mathcal{S}(\theta_1,\ldots,\theta_L)\rangle = \langle\!\langle W|A_1(\theta_1)\cdots A_L(\theta_L)|V\rangle\!\rangle. \tag{32}$$

Starting from $|\mathcal{S}(\theta_1,\ldots,\theta_L)\rangle$ one chooses a vector $A_i(\theta_i)$ for some i. We move it to the right using the ZF algebra, bounce it on $|V\rangle\!\rangle$ thanks to GZ relation, then move it to the left (again with the ZF algebra) and bounce it on $\langle\!\langle W|$ (with the GZ relation) before moving it back to its original place using the ZF algebra. Gathering all the R and K matrices one gets through this procedure, one shows that[2]

$$|\mathcal{S}(\theta_1,\dots,\theta_L)\rangle\!\rangle = t(\theta_i|\bar\theta)\ |\mathcal{S}(\theta_1,\dots,\theta_L)\rangle\!\rangle,$$
$$|\mathcal{S}(\theta_1,\dots,\theta_L)\rangle\!\rangle = t(1/\theta_i|\bar\theta)\ |\mathcal{S}(\theta_1,\dots,\theta_L)\rangle\!\rangle,$$

with the inhomogeneous transfer matrix

$$t(x|\bar\theta) = tr_0\Big(\ \widetilde{K}_0(x)\ R_{0L}(\frac{x}{\theta_L})\cdots R_{01}(\frac{x}{\theta_1})\ K_0(x)\ R_{10}(x\theta_1)\cdots R_{L0}(x\theta_L)\ \Big). \tag{33}$$

If one assumes moreover a crossing symmetry for the R and K matrices, one gets a crossing relation on the transfer matrix:

$$t(x|\bar\theta) = \Big(\lambda(x|\bar\theta) - 1\Big)\ t(1/xq|\bar\theta) \tag{34}$$

where $\lambda(x|\bar\theta)$ is a function that depends on the model. Then, one can perform an interpolation in x to obtain

$$t(x|\bar\theta)\ |\mathcal{S}(\theta_1,\dots,\theta_L)\rangle\!\rangle = \lambda(x|\bar\theta)|\mathcal{S}(\theta_1,\dots,\theta_L)\rangle. \tag{35}$$

This shows that the 'generalized steady state' is an eigenvector of the transfer matrix $t(x|\bar\theta)$. Remark that, contrarily to the Bethe ansatz, there is no Bethe parameters (and no Bethe eqs). However, we get only one state.

4. Example: a reaction-diffusion model (DiSSEP)

We illustrate our method on a new model that we constructed from R and K matrices [10].

4.1. Description of the model, R and K matrices

We start by introducing the following matrices.

$$R(x) = \begin{pmatrix} \frac{\kappa(x+1)}{\kappa(x+1)+x-1} & 0 & 0 & \frac{x-1}{\kappa(x+1)+x-1} \\ 0 & \frac{\kappa(x-1)}{\kappa(x-1)+x+1} & \frac{x+1}{\kappa(x-1)+x+1} & 0 \\ 0 & \frac{x+1}{\kappa(x-1)+x+1} & \frac{\kappa(x-1)}{\kappa(x-1)+x+1} & 0 \\ \frac{x-1}{\kappa(x+1)+x-1} & 0 & 0 & \frac{\kappa(x+1)}{\kappa(x+1)+x-1} \end{pmatrix}. \tag{36}$$

This R-matrix satisfies the Yang-Baxter equation. It is unitary and regular.

$$K(x) = \begin{pmatrix} \frac{(x^2+1)((x^2-1)(\gamma-\alpha)+4x\kappa)}{2x((x^2-1)(\alpha+\gamma)+2\kappa(x^2+1))} & \frac{(x^2-1)((x^2+1)(\gamma-\alpha)+2x(\alpha+\gamma))}{2x((x^2-1)(\alpha+\gamma)+2\kappa(x^2+1))} \\ -\frac{(x^2-1)((x^2+1)(\gamma-\alpha)-2x(\alpha+\gamma))}{2x((x^2-1)(\alpha+\gamma)+2\kappa(x^2+1))} & -\frac{(x^2+1)((x^2-1)(\gamma-\alpha)-4x\kappa)}{2x((x^2-1)(\alpha+\gamma)+2\kappa(x^2+1))} \end{pmatrix}, \tag{37}$$

$$\bar{K}(x) = \begin{pmatrix} \frac{(x^2+1)((x^2-1)(\delta-\beta)+4x\kappa)}{2x(-(x^2-1)(\delta+\beta)+2\kappa(x^2+1))} & \frac{(x^2-1)((x^2+1)(\delta-\beta)-2x(\delta+\beta))}{2x(-(x^2-1)(\delta+\beta)+2\kappa(x^2+1))} \\ -\frac{(x^2-1)((x^2+1)(\delta-\beta)+2x(\delta+\beta))}{2x(-(x^2-1)(\delta+\beta)+2\kappa(x^2+1))} & -\frac{(x^2+1)((x^2-1)(\delta-\beta)-4x\kappa)}{2x(-(x^2-1)(\delta+\beta)+2\kappa(x^2+1))} \end{pmatrix}. \tag{38}$$

These K-matrices satisfy the reflection equation. They are unitary and regular.

From these matrices, we get a Markov matrix

$$M = B_1 + \sum_{\ell=1}^{L-1} m_{\ell,\ell+1} + \overline{B}_L$$

[2] The second relation is obtained in the same way, but moving to the left first.

where the local jump operators read

$$B = \kappa\, K'(1) = \begin{pmatrix} -\alpha & \gamma \\ \alpha & -\gamma \end{pmatrix}, \quad \overline{B} = -\kappa \bar{K}'(1) = \begin{pmatrix} -\delta & \beta \\ \delta & -\beta \end{pmatrix} \tag{39}$$

$$m = 2\kappa\, PR'(1) = \begin{pmatrix} -1 & 0 & 0 & 1 \\ 0 & -\kappa^2 & \kappa^2 & 0 \\ 0 & \kappa^2 & -\kappa^2 & 0 \\ 1 & 0 & 0 & -1 \end{pmatrix}. \tag{40}$$

The corresponding rates of transition are summarized in figure 2. One sees that the model is of SSEP type (i.e. a model where the particles move to the right or to the left with the same rate) but where pairs of particles can condensate or evaporate from the bulk, hence the name DiSSEP, "Di" standing for dissipative.

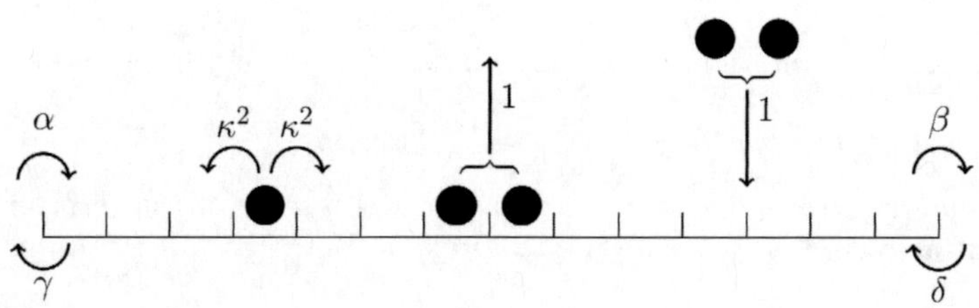

Figure 2. The transition rates in the DiSSEP model

4.2. Construction of the algebra needed for the matrix ansatz
We introduce the vector $A(x)$

$$A(x) = \begin{pmatrix} G_1 x + G_2 + \frac{G_3}{x} \\ -G_1 x + G_2 - \frac{G_3}{x} \end{pmatrix} \tag{41}$$

and impose the ZF and GZ relations. We obtain

$$R_{12}\left(\frac{x_1}{x_2}\right) A_1(x_1)\, A_2(x_2) = A_2(x_2)\, A_1(x_1) \;\Leftrightarrow\; \begin{cases} \phi\; G_1 G_2 = G_2 G_1 \\ G_1 G_3 = G_3 G_1 \\ \phi\; G_2 G_3 = G_3 G_2 \end{cases} \quad \text{with} \quad \phi = \frac{\kappa - 1}{\kappa + 1}. \tag{42}$$

$$\langle\!\langle W|A(x) \;=\; \langle\!\langle W|K(x) A\left(\frac{1}{x}\right) \;\Leftrightarrow\; \langle\!\langle W|(G_1 - c\,G_2 - a\,G_3) = 0 \tag{43}$$

$$A(x)|V\rangle\!\rangle \;=\; \bar{K}(x) A\left(\frac{1}{x}\right)|V\rangle\!\rangle \;\Leftrightarrow\; (G_3 - b\,G_1 - d\,G_2)|V\rangle\!\rangle = 0 \tag{44}$$

with $a = \dfrac{2\kappa - \alpha - \gamma}{2\kappa + \alpha + \gamma}$, $c = \dfrac{\gamma - \alpha}{2\kappa + \alpha + \gamma}$, $b = \dfrac{2\kappa - \delta - \beta}{2\kappa + \delta + \beta}$ and $d = \dfrac{\beta - \delta}{2\kappa + \delta + \beta}$.

IOP Conf. Series: Journal of Physics: Conf. Series **804** (2017) 012037 doi:10.1088/1742-6596/804/1/012037

4.3. Matrix ansatz:

$$A(1) = \begin{pmatrix} G_2 + (G_1 + G_3) \\ G_2 - (G_1 + G_3) \end{pmatrix} \quad \text{and} \quad A'(1) = (G_1 - G_3)\begin{pmatrix} 1 \\ -1 \end{pmatrix} \tag{45}$$

From $A(1) \equiv A$, we get the expressions $D = G_2 - G_1 - G_3$ and $E = G_2 + G_1 + G_3$. Remark that $A'(1) \equiv \bar{A}$ is not a scalar anymore, since it involves the generator $H = G_1 - G_3$. Nevertheless one can still apply the matrix ansatz to get the steady state. Note that the calculations are easier in the G basis, rather than in the E, D, H basis. For instance the normalization factor reads

$$Z_L = \langle\langle W|(D+E)^L|V\rangle\rangle = 2^L \langle\langle W|G_2^L|V\rangle\rangle. \tag{46}$$

In fact we are able to compute any word in G_1, G_2, G_3, for more details see [11].

4.4. Calculation of physical observables

One can compute the densities at one site, defined as ($i = 1, 2, ..., L$):

$$\langle n_i \rangle \;=\; \frac{\langle\langle W|(D+E)^{i-1}D(D+E)^{L-i}|V\rangle\rangle}{\langle\langle W|(D+E)^L|V\rangle\rangle} = \frac{1}{2}\frac{\langle\langle W|G_2^{i-1}(G_2 - G_1 - G_3)G_2^{L-i}|V\rangle\rangle}{\langle\langle W|G_2^L|V\rangle\rangle}.$$

Using solely the relation (42), (43) and (44), we get

$$\langle n_i \rangle = \frac{1}{2} - \frac{c\phi^{i-1} + ad\phi^{L+i-2} + d\phi^{L-i} + bc\phi^{2L-i-1}}{2(1 - ab\phi^{2L-2})}. \tag{47}$$

Remark that, due to the condensation/evaporation process, the density depends on the site position i. In fact it shows a Friedel like oscillations on the boundaries, see [10]. Note that to obtain (47), we have simplified by the factor $Z_L = \langle\langle W|G_2^L|V\rangle\rangle$. For consistency, we constructed a representation where this factor is non-zero, see appendix.

We can also compute the value of the currents. There are two types of currents: the diffusion current in the bulk, and the evaporation current. The diffusion current from the site i to $i+1$ reads

$$\langle J^{lat}_{i\to i+1} \rangle = \frac{\kappa^2}{\kappa+1} \; \frac{d\phi^{L-i-1} + bc\phi^{2L-i-2} - c\phi^{i-1} - ad\phi^{L+i-2}}{1 - ab\phi^{2L-2}}. \tag{48}$$

The evaporation current at sites $(i, i+1)$ has the form

$$\langle J^{eva}_{i,i+1} \rangle = -\frac{\kappa}{\kappa+1} \; \frac{c\phi^{i-1} + ad\phi^{L+i-2} + d\phi^{L-i-1} + bc\phi^{2L-i-2}}{1 - ab\phi^{2L-2}}. \tag{49}$$

We performed several other calculations, such as the variance of the lattice current, the thermodynamical limit, and a comparison with the Macroscopic Fluctuation Theory (MFT, [12]). For more details about the model presented in this section, see [11].

5. 2-species TASEP with boundaries

The 2-species TASEP is not a new model, it was known for a long time. It is depicted in figure 3. However, for generic values of the boundary parameters α_j and β_j the model is not integrable, and moreover the matrix ansatz was known only for "simple" boundaries, where one species is trapped, or when it is absent in the steady state of the model, see for instance [13, 14]. We classified the integrable boundaries of the model, discovering new (integrable) boundaries where both species can jump in and out of the lattice. Moreover, using the "integrable approach" we performed a matrix ansatz for these integrable boundaries.

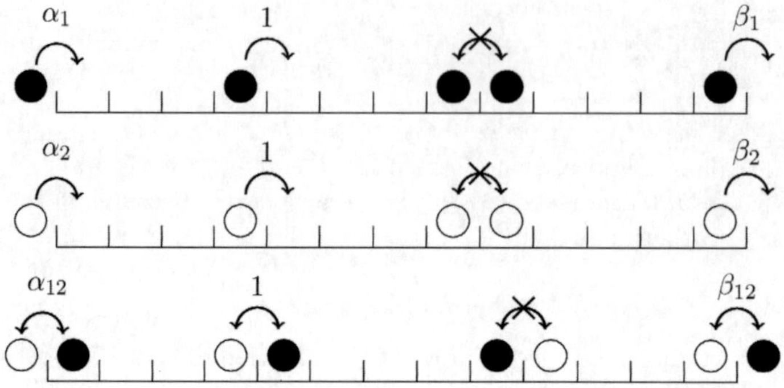

Figure 3. The transition rates in the 2-species TASEP

5.1. Integrable boundaries

To keep this review note short, we do not show the integrable K-matrices, but rather give the integrable transition rates. They are gathered in (50) and (51), where 0 stands for holes, and 1 (resp. 2) label the black (resp. white) particles of figure 3. The interested reader can find in [15] the explicit expressions for the K-matrices.

L_1	L_2	L_3	L_4
$0 \xrightarrow{\mu} 1$	$0 \xrightarrow{1-\alpha} 1$ $0 \xrightarrow{\alpha} 2$ $1 \xrightarrow{\alpha} 2$	$0 \xrightarrow{\alpha} 2$ $1 \xrightarrow{\alpha} 2$	$0 \xrightarrow{\alpha} 2$

Possible integrable rates (left boundary) (50)

R_1	R_2	R_3	R_4
$2 \xrightarrow{\nu} 1$	$1 \xrightarrow{\beta} 0$ $2 \xrightarrow{\beta} 0$ $2 \xrightarrow{1-\beta} 1$	$1 \xrightarrow{\beta} 0$ $2 \xrightarrow{\beta} 0$	$2 \xrightarrow{\beta} 0$

Possible integrable rates (right boundary) (51)

5.2. Matrix ansatz

Obviously, the matrix ansatz depends on which boundary conditions one chooses, and some (but not all) correspond to already studied cases. We take as an example the case of boundaries $L_2 - R_3$, that was not studied up to now. We introduce the following vector

$$A(x) = \begin{pmatrix} x^2 + G_9 x + G_8 + \dfrac{G_7}{x} \\ G_6 x + G_5 + \dfrac{G_4}{x} \\ G_3 x + G_2 + \dfrac{G_1}{x} + \dfrac{1}{x^2} \end{pmatrix} \quad \Rightarrow \quad A(1): \begin{array}{l} X_0 = 1 + G_9 + G_8 + G_7 \\ X_1 = G_6 + G_5 + G_4 \\ X_2 = G_3 + G_2 + G_1 + 1 \end{array} \tag{52}$$

Note the expansion in the spectral parameter x which is different from previous examples. In fact, there is up-to-now no deductive way to know the expansion that one should take for $A(x)$.

We did it by try-and-errors and brute force calculation on 'words' of 1, 2 and 3 letters: if the expansion is too restrictive (for instance stop at x instead of x^2 in the present case), then the values of the words on the vectors $\langle\langle W|$ and $|V\rangle\rangle$ will simply vanish, and we are led to take more terms in the expansion. The present expansion leads to an algebra with 9 generators which is much more involved than the usual (one-species) TASEP. However, everything needed for the matrix ansatz is still encoded by the ZF and GZ relations. Again, to keep this article short, we do not reproduce all the exchange relations for the 9 generators, nor the boundary relations. The interested reader can refer to the original article [15].

5.3. Physical quantities

We were able to compute the following quantities, that for simplicity we present in the particular case $\alpha = \frac{1}{2}$ and $\beta = 1$.

The partition function:

$$Z_L = \langle\langle W|C^L|V\rangle\rangle = (2L+1)\mathfrak{c}_L\mathfrak{c}_{L+1}\langle\langle W|V\rangle\rangle, \tag{53}$$

where $C = X_0 + X_1 + X_2$ and $\mathfrak{c}_L = \frac{1}{L+1}\begin{pmatrix} 2L \\ L \end{pmatrix}$ is the Catalan number.

The average density of black or white particles at site k:

$$\begin{aligned} n_1^{(k)} &= \frac{1}{Z_L}\langle\langle W|C^{k-1}\,X_1\,C^{L-k}|V\rangle\rangle = \frac{1}{\mathfrak{c}_{L+1}}\sum_{i=k}^{L}\frac{L-i+1}{L+2}\mathfrak{c}_i\mathfrak{c}_{L-i}, \\ n_2^{(k)} &= \frac{1}{Z_L}\langle\langle W|C^{k-1}\,X_2\,C^{L-k}|V\rangle\rangle = \frac{1}{\mathfrak{c}_{L+1}}\sum_{i=k}^{L}\frac{i+1}{L+2}\mathfrak{c}_i\mathfrak{c}_{L-i}\,. \end{aligned}$$

The current j_i of the particles of type $i = 1, 2$:

$$\begin{aligned} j_1 &= \frac{1}{Z_L}\langle\langle W|C^{k-1}(X_1X_0 - X_2X_1)C^{L-k-1}|V\rangle\rangle = \frac{1}{2(2L+1)}, \\ j_2 &= \frac{1}{Z_L}\langle\langle W|C^{k-1}X_2(X_0 + X_1)C^{L-k-1}|V\rangle\rangle = \frac{L+1}{2(2L+1)}. \end{aligned}$$

We also computed a representation for the matrix ansatz algebra that ensures that it is not trivial (for all values of α and β), i.e. which is such that $\langle\langle W|V\rangle\rangle \neq 0$, see [16].

6. Conclusion and perspectives

We presented a comprehensive approach of the matrix ansatz in the integrable systems framework. Two key ingredients are needed: the ZF algebra (in the bulk) and the GZ relations (on the boundaries). In principle it allows us to construct a matrix ansatz for *any* integrable reaction-diffusion process (if we know the R-matrix). However, in practice the matrix ansatz algebra can be very complicated and the computation of observables can be very hard.

Several points remain to be clarified. First of all, we wish to find a prescription for the expansion in the vector $A(x)$. As already mentioned, the truncation in the expansion of $A(x)$ is done case by case, on a try-and-error basis. A general (deductive) prescription would be a major step in this 'integrable matrix ansatz'. Secondly, the construction of the representation that ensures the validity of the matrix anstz is also done case by case: a general principle to construct it would be of great help. Apart from these two points, we wish also to make a better use of the spectral parameter (with the ZF and GZ relations) to extract property of the stationary state and find a more efficient way to compute observables. Note also that there are connections with orthogonal polynomials: either in computing physical data using the fact that they are orthogonal polynomials, or to compute explicitly orthogonal polynomials starting from these physical data, see e.g. [17, 18, 19]. Surely, a generalisation of the matrix ansatz that would allow to get other (excited) states above the steady states is desirable. For the moment, the only way to get them is to use the Bethe ansatz. Finally, we would like to apply our technique to solve more complicated models: for instance we are presently working on N-species ASEP with boundaries, see [20] for a presentation of integrable boundaries.

Appendix A. Representation for the algebra of the reaction-diffusion model

We want to construct the two representations associated to the vectors $\langle\langle W|$ and $|V\rangle\rangle$, for the algebra introduced in section 4:

$$\phi\, G_1G_2 = G_2G_1\ ;\quad G_1G_3 = G_3G_1\ ;\quad \phi\, G_2G_3 = G_3G_2. \tag{A.1}$$

We first introduce a realisation of this algebra, given by

$$G_1 = e\otimes 1,\quad G_2 = A(\phi)\otimes A(\phi),\quad G_3 = 1\otimes d, \tag{A.2}$$

where

$$e = \sum_{n=0}^{+\infty} |n+1\rangle\rangle\langle\langle n| \quad ;\quad A(\phi) = \sum_{n=0}^{+\infty} \phi^n |n\rangle\rangle\langle\langle n| \quad ;\quad d = \sum_{n=0}^{+\infty} |n\rangle\rangle\langle\langle n+1|. \tag{A.3}$$

We have used $\{|n\rangle\rangle \mid n\geq 0\}$ as an infinite basis of the additional space, and e (resp. d) are the lowering (resp. raising) operators. They obey

$$de = 1,\quad ed = 1 - A(0),\quad A(\phi)e = \phi\, e\, A(\phi) \quad\text{and}\quad d\, A(\phi) = \phi\, A(\phi)d. \tag{A.4}$$

These relations are sufficient to show that (A.2) forms a representation of the algebra (A.1).

Now, within this realization, we seek for vectors $\langle\langle W|$ and $|V\rangle\rangle$ such that

$$\langle\langle W|(G_1 - cG_2 - aG_3) = 0 \quad\text{and}\quad (G_3 - bG_1 - dG_2)|V\rangle\rangle = 0. \tag{A.5}$$

A solution is given by the infinite series

$$|V\rangle\rangle \;=\; \sum_{n,m=0}^{+\infty} v_{n,m}\, |n\rangle\rangle\otimes|m\rangle\rangle \quad\text{with}\quad v_{n,m} = d^{m-n}b^n \frac{\phi^{\frac{(m-n)(m-n-1)}{2}}}{(1-\phi^2)\cdots(1-\phi^{2n})}, \tag{A.6}$$

$$\langle\langle W| \;=\; \sum_{n,m=0}^{+\infty} w_{n,m}\, \langle\langle n|\otimes\langle\langle m| \quad\text{with}\quad w_{n,m} = c^{n-m}a^m \frac{\phi^{\frac{(n-m)(n-m-1)}{2}}}{(1-\phi^2)\cdots(1-\phi^{2m})}. \tag{A.7}$$

Then, one can compute that

$$Z_L \;=\; \langle\langle W|G_2^L|V\rangle\rangle = \sum_{n,m=0}^{+\infty} \frac{(\phi^L\, cb/d)^n\ (\phi^L\, da/c)^m\ \phi^{(m-n)^2}}{(1-\phi^2)\cdots(1-\phi^{2n})\times(1-\phi^2)\cdots(1-\phi^{2m})}.$$

The series is convergent when

$$|\phi|<1\,,\quad \left|\frac{b\,c\,\phi^L}{(1-\phi^2)d}\right|<1 \quad\text{and}\quad \left|\frac{a\,d\,\phi^L}{(1-\phi^2)c}\right|<1.$$

Thanks to the symmetry $\kappa\to-\kappa$, we can choose $|\phi|<1$. Then, when L is large enough, the conditions are always fulfilled (when $c, d, \kappa\neq 0$).

Bibliography

[1] B. Derrida, *Non-equilibrium steady states: fluctuations and large deviations of the density and of the current*, J. Stat. Mech. **0707** (2007) P07023.

[2] R.A. Blythe and M.R. Evans, *Nonequilibrium Steady States of Matrix Product Form: A Solver's Guide*, J. Phys. **A40** (2007) R333-R441 and `arXiv:0706.1678`.

[3] T. Kriecherbauer and J. Krug, *A pedestrian's view on interacting particle systems, KPZ universality and random matrices*, J. Phys. A: Math. Theor. **43**, 2010 403001.

[4] T. Chou, K. Mallick and R. K. P. Zia, *Non-equilibrium statistical mechanics: From a paradigmatic model to biological transport*, Rep. Prog. Phys. **74**, (2011) 116601.

[5] E. Bertin, *Theoretical approaches to the statistical physics of interacting non-conservative units*, `arXiv:1608.08507`.

[6] B. Derrida, M. Evans, V. Hakim and V. Pasquier, *Exact solution of a 1d asymmetric exclusion model using a matrix formulation*, J. Phys. **A26** (1993) 1493.

[7] E.K. Sklyanin, *Boundary conditions for integrable quantum systems,* J. Phys. **A21** (1988) 2375.

[8] A.B. Zamolodchikov and A.B. Zamolodchikov, *Factorized S-matrices in two dimensions as the exact solutions of certain relativistic quantum field theory models,* Ann. Phys. (N.Y.) **120** (1979) 253;
L.D. Faddeev, *Quantum completely integrable models in field theory,* Sov. Sci. Rev. **C1** (1980) 107.

[9] S. Ghoshal and A.B. Zamolodchikov, *Boundary S-matrix and boundary state in two-dimensional integrable quantum field theory,* Int. J. Mod. Phys. **A9** (1994) 3841 and `arXiv:hep-th/9306002`.

[10] N. Crampe, E. Ragoucy, M. Vanicat, *Integrable approach to simple exclusion processes with boundaries. Review and progress,* `arXiv:1408.5357` and J. Stat. Mech. **1411** (2014) P11032

[11] N. Crampe, E. Ragoucy, V. Rittenberg, M. Vanicat, *An integrable dissipative exclusion process: correlation functions and physical properties,* `arXiv:1603.06796` and Phys. Rev. **E94** (2016) 032102

[12] L. Bertini, A. De Sole, D. Gabrielli, G. Jona-Lasinio and C. Landim, *Macroscopic Fluctuation Theory for stationary non-equilibrium states,* J. Stat. Phys. **107**, (2002) 635.

[13] M. Uchiyama, *Two-Species Asymmetric Simple Exclusion Process with Open Boundaries,* Chaos, Solitons & Fractals **35**, (2008) 398 and `arXiv:cond-mat/0703660`.

[14] C. Arita, *Exact analysis of two-species asymmetric exclusion process with open boundary conditions,* J. Phys. Soc. Jpn. **75**, (2006) 065003.

[15] N. Crampe, K. Mallick, E. Ragoucy, M. Vanicat, *Open two-species exclusion processes with integrable boundaries,* `arXiv:1412.5939` and J. Phys. **A48** (2015) 175002

[16] N. Crampe, M. Evans, K. Mallick, E. Ragoucy, M. Vanicat, *Matrix product solution to a 2-species TASEP with open integrable boundaries,* `arXiv:1606.08148`

[17] M. Uchiyama, T. Sasamoto and M. Wadati, *Asymmetric simple exclusion process with open boundaries and Ashkey-Wilson polynomials,* J. Phys. A: Math. Gen. **37**, (2004) 4985.

[18] N. Crampe, K. Mallick, E. Ragoucy, M. Vanicat, *Inhomogeneous discrete-time exclusion processes,* `arXiv:1506.04874` and J. Phys. **A48** (2015) 484002

[19] L. Cantini, J. de Gier, M. Wheeler, *Matrix product formula for Macdonald polynomials,* J. Phys. **A48** (2015) 384001 and `arXiv:1505.00287`

[20] N. Crampe, C. Finn, E. Ragoucy, M. Vanicat, *Integrable boundary conditions for multi-species ASEP,* `arXiv:1606.01018`, J. Phys. **A** (2016) to appear

IOP Conf. Series: Journal of Physics: Conf. Series **804** (2017) 012038 doi:10.1088/1742-6596/804/1/012038

On nonlinear Fourier transform: towards the nonlinear superposition

Pavle Saksida[1,2]
[1]Faculty of Mathematics and Physics, University of Ljubljana
[2]Institute of Mathematics, Physics and Mechanics, Ljubljana
E-mail: pavle.saksida@fmf.uni-lj.si

Abstract. In the paper we consider the nonlinear Fourier transform associated to the AKNS-ZS systems. In particular, we discuss the construction of the nonlinear Fourier modes of this transform by means of a perturbation scheme. The linearization of the AKNS-ZS nonlinear Fourier transform is the usual, linear Fourier transform and the linearization of a nonlinear Fourier mode of frequency d is the linear Fourier mode of the same frequency. We show that the first non-trivial term in the perturbation expression of any nonlinear Fourier mode is given by the dilogarithm function.

1. Introduction

A well-known and important class of integrable partial differential equations are the AKNS-ZS systems. The theory of these systems grew from the poineering work of Zakharov and Shabat, [19] and Ablowitz, Kaup Newell and Segur, [1], [2]. The salient members of this class are the nonlinear Schrödinger equation and the sine-Gordon equation in the light-cone coordinates. Being integrable, the AKNS-ZS equations are equivalent to the zero-curvature equations of the form

$$(L_q)_t - (A_q)_x + [L_q, A_q] = 0,$$

where $L_q(x,t;z)$ and $A_q(x,t;z)$ are two suitable $\mathfrak{sl}(2,\mathbb{C})$-valued functions of time t, of one spatial variable x and of the spectral parameter z. The L-matrices of all the zero-curvature conditions, corresponding to the AKNS-ZS equations are of the form

$$L_q(x,t;z) = \begin{pmatrix} \frac{iz}{2} & q(x,t) \\ -\overline{q(x,t)} & -\frac{iz}{2} \end{pmatrix}. \tag{1}$$

A function $q(x,t)$ solves the AKNS-ZS equation at hand if and only if L_q, together with the suitably chosen A_q, satisfies the zero-curvature condition.

Let $l^2_{\mathbb{Z}}$ and $L^2[0,2\pi]$ denote the Hilbert spaces of the square-integrable bi-infinite sequences and of the square integrable complex-valued functions on $[0,2\pi]$, respectively. Our object of interest in this note will be the nonlinear operator

$$\mathcal{H} : L^2[0,2\pi] \times l^2_{\mathbb{Z}} \longrightarrow l^2_{\mathbb{Z}} \times l^2_{\mathbb{Z}}$$

given as follows. For $q(x) \in L^2[0,2\pi]$, let $\Phi_q(x;z)$ be the solution of the linear initial problem

$$(\Phi_q)_x = L_q(x;z) \cdot \Phi_q; \quad \Phi_q(0;z) = I,$$

IOP Conf. Series: Journal of Physics: Conf. Series **804** (2017) 012038 doi:10.1088/1742-6596/804/1/012038

where I denotes the identity 2×2 matrix. Let us denote

$$\mathrm{Hol}[q](z) = \Phi_q(x = 2\pi; z) = \begin{pmatrix} \mathcal{D}(z) & \mathcal{F}(z) \\ \mathcal{E}(z) & \mathcal{G}(z) \end{pmatrix} \in SL(2, \mathbb{C}).$$

Let now $\sigma = \{\sigma_n\}_{n\in\mathbb{Z}}$ be a sequence in $l^2_{\mathbb{Z}}$. Then $\mathcal{H}[q, \sigma]$ is given by

$$\mathcal{H}[q, \sigma](n) = \big(\mathcal{D}(z_n),\ \mathcal{F}(z_n)\big) = \big(\mathcal{D}(n + \sigma_n),\ \mathcal{F}(n + \sigma_n)\big). \tag{2}$$

It can be shown that the image of $\mathcal{H}$ is indeed in $l^2_{\mathbb{Z}} \times l^2_{\mathbb{Z}}$.

We shall call the operator $\mathcal{H}$ the nonlinear Fourier transform. The name is motivated by the fact that the Fourier method for solving the linear partial differential equations can be viewed as the linearization of the inverse scattering method for solving the nonlinear integrable equations. Schematically, the inverse scattering method can be represented by the following diagram:

$$q(x, 0) \overset{\mathrm{Hol}}{\longmapsto} \mathrm{Hol}[q(x, 0)](z) \overset{time\ evolution}{\longmapsto} \mathrm{Hol}[q(x, t)](z) \overset{\mathrm{Hol}^{-1}}{\longmapsto} q(x, t).$$

The Fourier method for solving linear partial differential equations has the same structure:

$$q(x, 0) \overset{F}{\longmapsto} F[q(x, 0)](z) \overset{time\ evolution}{\longmapsto} F[q(x, t)](z) \overset{F^{-1}}{\longmapsto} q(x, t).$$

Here F stands for the usual (linear) Fourier transform. It is not difficult to see that the entire holonomy $\mathrm{Hol}[q](z)$ can be reproduced from its first row. Therefore, the operators $\mathcal{H}$ and Hol are essentially equivalent, and it makes sense to call $\mathcal{H}$ the nonlinear Fourier transform. A similar operator is called nonlinear Fourier transform by Tao and Thiele in their text [17], and also by Fokas (see, e.g. [7], [6], [8]). Various analytical properties of the operators similar to $\mathcal{H}$ were studied by many authors. Important results are presented in the papers [9], [10], [4], [3], to name but a few.

The Fourier transform F can be thought of as a decomposition of the initial condition $q(x, 0)$ into the elementary Fourier modes. Calculating the time evolution of an elementary mode is an easy task. To obtain the solution $q(x, t)$, one has to perform the inverse Fourier transform. The inverse scattering method relies upon the same strategy. However, the operators $\mathcal{H}$ and $\mathcal{H}^{-1}$ are nonlinear, as opposed to F and F^{-1}. In particular, finding $\mathcal{H}^{-1}$ is a difficult inverse problem. In the case of the differential equations with periodic boundary conditions, the calculation of the time evolution of $\mathcal{H}[q(x, 0)](z)$ is also a difficult problem, not explicitly solvable for an arbitrary choice of the initial value $q(x, 0)$. The known solutions of the integrable nonlinear equations with periodic boundary conditions are constructed by means of the algebro-geometric tools, in particular by the use of spectral curves, their Jacobian tori and the associated theta functions. But this approach yields only the so-called *finite degrees of freedom solutions.* The linear analog of this class are the solutions, expressible with finite Fourier series. A more important limitation of the algebro-geometric approach is the fact that the solutions it provides are not solutions to given initial problems. Of course, one can determine the initial condition, corresponding to a given solution, but only *after* this solution was found. The initial condition does not figure in any way in the construction of such a solution.

One way to remedy this unfavorable situation might be to follow the steps of the Fourier method for the linear case more closely. For this purpose, one has to construct the nonlinear Fourier modes and then find the nonlinear superposition rules by means of which simpler solutions of a given nonlinear integrable equation yield more complex solutions. In this paper we shall only consider the first of the above two tasks.

The simplest sequence of the complex valued Fourier modes in the linear Fourier analysis $\{u_d(x)\}_{d\in\mathbb{Z}}$ can be defined by the equation

$$F[u_d](n) = a_d\ \delta_{n,d},$$

where $\delta_{n,d}$ denotes the Kronecker delta and a_d is a complex constant. The analogous definition of the modes $\{q_d(x)\}_{\mathbb{Z}}$ in the nonlinear Fourier analysis is then given by

$$\mathcal{H}[q_d, \sigma](n) = \mathrm{Hol}[q_d](z_n) = I + A_d\ \delta_{n,d},$$

where A_d is a matrix in $SL(2,\mathbb{C})$. Here $\sigma = \{\sigma_n\}_{n\in\mathbb{Z}}$ is a suitably chosen square-integrable sequence and $z_n = n + \sigma_n$, as in (2). The reason for the introduction of σ will be explained in section 2 below. Let us note already now that σ depends on A_d.

We shall see that the Fourier transform F is the linearization of $\mathcal{H}$ at the origin of $L^2[0,2\pi]\times l^2_{\mathbb{Z}}$. Therefore it makes sense to try to construct the nonlinear Fourier modes by means of a perturbation scheme. The main result that we describe in this note is the following.

Theorem 1 *Up to the third order in the perturbational parameter s, the nonlinear Fourier mode of frequency d is given by the formula*

$$q_d(x) = e^{idx}\Big[sA_1 + s^2A_2 + s^3\Big(A_3 + A_4\Big(\mathrm{Li}_2(e^{ix}) + \mathrm{Li}_2(e^{-ix})\Big)\Big)\Big] + \mathcal{O}(s^4), \tag{3}$$

where A_1, A_2, A_3 and A_4 are suitable complex constants.

The first non-trivial term in the perturbation expansion is of order 3 and it is given by the salient dilogarithm function

$$\mathrm{Li}_2(x) = \sum_{n=1}^{\infty}\frac{x^n}{n^2} = -\int_0^x \log(1-x)\,\frac{dx}{x}. \tag{4}$$

In the last decade or two, this classical special function, together with its relatives polylogarithms and multiple polylogarithms, gained much attention in mathematical and physical literature. These functions figure prominently in many important and diverse topics in number theory, geometry, topology, and mathematical physics. Appearances of dilogarithms include K-theory, volumes of hyperbolic three-manifolds and a variety of number theory topics. Probably the most well-known recent occurrences in mathematical physics were provided by W. Nahm and coworkers, [11], [12] who studied certain aspects of conformal field theories, and by L. D. Faddeev and R. M. Kashaev, [5] who introduced the quantum version of the dilogarithm. Their work was followed by a vast number of important contributions by many authors. A beautiful account on some of these topics is the paper [18] by Don Zagier.

Formula (3) above provides another addition to the wealth of interesting occurrences of the dilogarithm. The higher order terms in the perturbation expansion of $q_d(x)$ are expressible by polylogarithms and multiple polylogarithms. In this paper, we shall confine ourselves to the third-order formula (3), for the sake of brevity.

2. Linear and nonlinear Fourier transforms

The essential part of the definition (2) is the linear initial problem

$$(\Phi_q)_x = L_q(x;z)\cdot\Phi_q;\quad \Phi_q(0;z) = I, \tag{5}$$

for $L_q(x;z)$ given by (1). The relation to the linear Fourier transform can be exhibited more clearly if we change the gauge of the above equation. Let the gauge transformation be given by the matrix

$$G(x:z) = \begin{pmatrix} e^{-izx} & 0 \\ 0 & e^{izx} \end{pmatrix}.$$

The matrix $\Phi_q^G(x;z) = G(x;z)\cdot\Phi_q(x;z)$ solves the problem

$$(\Phi_q^G)_x = L_q^G(x;z)\cdot\Phi_q^G; \quad \Phi_q^G(0;z) = I \tag{6}$$

for

$$L_q^G(x;z) = G_x G^{-1}(x;z) + (G\cdot L_q\cdot G^{-1})(x;z) = \begin{pmatrix} 0 & e^{-izx}q(x) \\ -e^{izx}\overline{q(x)} & 0 \end{pmatrix}$$

if and only if Φ_q solves (5). Let us write the solution Φ_q^G of (6) in the form of Dyson's expansion and evaluate it at $x = 2\pi$, and at the integer values $n \in \mathbb{Z}$ of the spectral parameter. We get

$$\mathrm{Hol}^G[q](n) = I + \sum_{k=1}^{\infty}\int_{\triangle_k} L_q^G(x_1)\cdot L_q^G(x_2)\cdots L_q^G(x_k)\, d\vec{x},$$

where the integration domains $\triangle_k$ are the ordered simplices

$$\triangle_k = \{(x_1, x_2, \ldots, x_k);\ 2\pi \geq x_1 \geq x_2 \ldots \geq x_k \geq 0\}.$$

For every $n \in \mathbb{Z}$, the matrix $\mathrm{Hol}^G[q](n)$ is an element of $SU(2)$. Let us denote the first row of $\mathrm{Hol}^G[q](n)$ by $(\mathcal{D}_{\mathbb{Z}}^G[q](n), \mathcal{F}_{\mathbb{Z}}^G[q](n))$. We have

$$\mathcal{D}_{\mathbb{Z}}^G[q](n) = 1 + \sum_{k=1}^{\infty}\int_{\triangle_{2k}} e^{-in(x_1-x_2+\ldots-x_{2k})} q(x_1)\overline{q(x_2)}\cdots\overline{q(x_{2k})}\, d\vec{x} \tag{7}$$

$$\mathcal{F}_{\mathbb{Z}}^G[q](n) = \sum_{k=1}^{\infty}\int_{\triangle_{2k-1}} e^{-in(x_1-x_2+\ldots+x_{2k-1})} q(x_1)\overline{q(x_2)}\cdots q(x_{2k-1})\, d\vec{x}. \tag{8}$$

It is proved in [15] that $\mathcal{D}_{\mathbb{Z}}$ and $\mathcal{F}_{\mathbb{Z}}$ are bounded operators from $L^2[0,2\pi]$ to $l^2_{\mathbb{Z}}$. Deriving these operators at the origin in the direction v gives

$$D_0\mathcal{D}_{\mathbb{Z}}^G[v](n) = 0, \quad D_0\mathcal{F}_{\mathbb{Z}}^G[v](n) = \int_0^{2\pi} e^{-inx}q(x)\, dx = F[q](n).$$

So the classical Fourier transform is the linearization at the origin of the bounded operator $\mathcal{F}_{\mathbb{Z}}^G$. It is therefore tempting to adopt the operator $\mathcal{F}_{\mathbb{Z}}^G$ as the nonlinear Fourier transform. Indeed, this operator has many useful properties. In [15] it is proved that it is not only bounded but also analytic in the vicinity of the origin. In addition to that, it is locally invertible and the inverse $(\mathcal{F}_{\mathbb{Z}}^G)^{-1}[\{\tau_n\}]$ can be iteratively computed to any desired accuracy for *any* sequence $\{b_n\}$ close enough to the origin of $l^2_{\mathbb{Z}}$.

An important drawback of the operator $\mathcal{F}_{\mathbb{Z}}^G$ is the fact that there is no obvious way of constructing the nonlinear Fourier modes by using $\mathcal{F}_{\mathbb{Z}}^G$ alone. The equations

$$\mathcal{D}_{\mathbb{Z}}[q](n) = 1 + a_d\,\delta_{n,d}, \quad \mathcal{F}_{\mathbb{Z}}[q](n) = b_d\,\delta_{n,d}, \quad n \in \mathbb{Z} \tag{9}$$

cannot be solved for q. This will become clear below. From

$$\Phi_q^G(x;z) = G(x;z)\cdot\Phi_q(x;z)$$

it follows that

$$\mathcal{D}_{\mathbb{Z}}[q](n) = (-1)^n\mathcal{D}_{\mathbb{Z}}^G[q](n), \quad \mathcal{F}_{\mathbb{Z}}[q](n) = (-1)^n\mathcal{F}_{\mathbb{Z}}^G[q](n),$$

thus the operators $\mathcal{F}_{\mathbb{Z}}^G$ and $\mathcal{F}_{\mathbb{Z}}$ suffer from the same problem.

3. Nonlinear Fourier transform and nonlinear Fourier modes

As we mentioned already in the introduction, the defining property of a nonlinear Fourier mode q_d of frequency d should be

$$\mathrm{Hol}[q_d](n) = I + A_d\ \delta_{n,d}, \quad \text{or equivalently} \quad \mathrm{Hol}^G[q_d](n) = (-1)^n(I + A_d\ \delta_{n,d}). \tag{10}$$

The reason for this is the obvious stipulation that a Fourier mode must remain a Fourier mode under the time evolution, prescribed by the suitable integrable partial differential equation. The time evolution of the holonomy $\mathrm{Hol}[q](t;z)$ associated to any initial condition $q(x)$ and for any value z of the spectral parameter is always given by the adjoint action of a suitable time-dependent matrix $M(t;z)$. Explicitly, we have

$$\mathrm{Hol}[q](t;z) = M(t;z)\cdot \mathrm{Hol}[q](0;z)\cdot M(t;z)^{-1}.$$

Confining the spectral parameter to the integers, we would expect of a nonlinear Fourier mode $q_d(x)$ to satisfy the condition

$$\mathrm{Hol}[q_d](t;n) = M(t;n)\cdot \mathrm{Hol}[q_d](0;n)\cdot M[t;n]^{-1} = I + \widetilde{A}_d(t)\ \delta_{n,d}$$

for some matrix-valued and time-dependent function $\widetilde{A}_d(t)$. This condition would certainly be satisfied, if (10) held for q_d. However, (10) is not satisfied by any function, because, as we noted above, the system of equations (9) is not solvable. If $q = \mathcal{F}_{\mathbb{Z}}^{-1}(a_d\ \delta_{n,d})$, then $\mathcal{D}_{\mathbb{Z}}[q](n)$ is different from 1 for many n, not only for $n = d$. Thus $\mathcal{F}_{\mathbb{Z}}$ is not a good candidate for the nonlinear Fourier transform if we want to work with the nonlinear Fourier modes modes.

The problem can be remedied by allowing the values of the spectral parameter to vary. So, instead of evaluating at n, we evaluate at $z_n = n + \sigma_n$ for some complex correction term σ_n. It is not difficult to see that

$$\lim_{n\to\infty} \mathrm{Hol}[q](n) = I,$$

for every function $q \in L^2[0,2\pi]$. Therefore, it makes sense to stipulate that the sequences $\{\sigma_n\}_{n\in\mathbb{Z}}$ of corrections are square integrable. This brings us to the definition (2)

$$\mathcal{H}[q,\sigma](n) = \Big(\mathcal{D}[q](z_n), \mathcal{F}[q](z_n)\Big) = \Big(\mathcal{D}[q](n+\sigma_n), \mathcal{F}[q](n+\sigma_n)\Big)$$

of the nonlinear Fourier transform.

Without much trouble the following proposition can be proved.

Proposition 1 *For any function $q \in L^2[0,2\pi]$ and for any value of the spectral parameter, the holonomy $\mathrm{Hol}[q](z)$ is determined uniquely by its first row $(\mathcal{D}[q](z), \mathcal{F}[q](z))$, provided the holonomy is close enough to the identity matrix.*

The essential result concerning $\mathcal{H}$ is the following theorem.

Theorem 2 *The nonlinear operator*

$$\mathcal{H} : L^2[0,2\pi] \times l^2_{\mathbb{Z}} \longrightarrow l^2_{\mathbb{Z}} \times l^2_{\mathbb{Z}}$$

is locally bounded and invertible in the vicinity of the origin. Moreover, $\mathcal{H}$ and its local inverse are real analytic operators.

The proof is given in [16]. Important ingredients of this proof are the boundedness and analyticity of the operator $\mathcal{F}_{\mathbb{Z}}$. This is proved in [15]. Strictly speaking, we shall not need this result to prove theorem 1, therefore we omit the proof here. It is important to note, however, that in the above theorem the space $L^2[0,2\pi]$ cannot be replaced by $L^2(\mathbb{R})$. Muscalu,

IOP Conf. Series: Journal of Physics: Conf. Series **804** (2017) 012038 doi:10.1088/1742-6596/804/1/012038

Tao and Thiele have shown in [10] that the operator $\mathcal{F}$, defined on $L^2(\mathbb{R})$, is not even three times differentiable, let alone analytic. However, it is not too difficult to see that replacing $L^2(\mathbb{R})$ by the Schwartz class resolves this issue.

The above theorem and proposition show that the nonlinear modes can be constructed by means of solving the equations of the form

$$\mathcal{H}[q,\sigma](n) = (1 + a_d\ \delta_{n,d},\ b_d\ \delta_{n,d}), \tag{11}$$

for q and σ. We are really only interested in q but we cannot get it without also getting σ.

We shall need the following specific description of $\mathcal{H}$, also important in the proof of the above theorem.

Lemma 1 *Let $q \in L^2[0,2\pi]$ and $\{\sigma_n\}_{n\in\mathbb{Z}} \in l^2_{\mathbb{Z}}$. Then*

$$\mathcal{D}[q,\sigma](n) = (-1)^n e^{i\pi\sigma_n} + \sum_{k=-\infty}^{\infty} P_{n,k}[\sigma]\cdot \mathcal{D}_{\mathbb{Z}}[q](k), \quad \mathcal{F}[q,\sigma](n) = \sum_{k=-\infty}^{\infty} P_{n,k}[\sigma]\cdot \mathcal{F}_{\mathbb{Z}}[q](k).$$

Here $P_{n,k}[\sigma]$ is an infinite matrix, given by

$$P_{n,k}[\sigma] = \frac{1}{2\pi}\int_{-\pi}^{\pi} e^{-i(n-k)x} e^{-i\sigma_n x}\ dx.$$

Proof: We shall sketch the proof only for $\mathcal{D}$. The proof for $\mathcal{F}$ is the same. From the gauge relation between $\mathcal{D}^G$ and $\mathcal{D}$, and from Dyson's expansion we get

$$\mathcal{D}[q](z) = e^{i\pi z}\Big(1 + \sum_{k=1}^{\infty}\int_{\triangle_{2k}} e^{-iz(x_1-x_2+\ldots-x_{2k})}\ q(x_1)\overline{q(x_2)}\cdots\overline{q(x_{2k})}\ d\vec{x}\Big).$$

In every term of the above sum we can introduce the new variable $u = x_1 - x_2 + \ldots - x_{2k}$ and express the integrals in the form

$$\int_{\triangle_{2k}} e^{-iz(x_1-x_2+\ldots-x_{2k})}\ q(x_1)\overline{q(x_2)}\cdots\overline{q(x_{2k})}\ d\vec{x} = \int_0^{2\pi} e^{-izu}\ T_{2k}(u)\ du,$$

where

$$T_{2k}(u) = \int_{D(u)} q(x_1)\overline{q(x_2)}\cdots\overline{q(x_{2k})}\ d\vec{x}$$

and

$$D(u) = \{(x_1, x_2, \ldots x_{2k}) \in \triangle_{2k}; x_1 - x_2 + \ldots - x_{2k} = u\}.$$

We can therefore write

$$\mathcal{D}[q](z) = e^{i\pi z}\Big(1 + \int_0^{2\pi} e^{-izu}\ \widetilde{T}(u)\ du\Big) = e^{i\pi z} + \int_{-\pi}^{\pi} e^{-izv}\ \widetilde{T}(v+\pi)\ dv. \tag{12}$$

Above we denoted $\widetilde{T}(u) = \sum_{k=1}^{\infty} T_{2k}(u)$ and introduced the new integration variable $v = u - \pi$.

For every n, we can expand the function $e^{-iz_n x}$, $z_n = n + \sigma_n$ into the Fourier series:

$$e^{-iz_n x} = \sum_{k=-\infty}^{\infty} P_{n,k}[\sigma]\ e^{-ik}, \quad P_{n,k}[\sigma] = \frac{1}{2\pi}\int_{-\pi}^{\pi} e^{-iz_n x} e^{ikx}\ dx = \frac{1}{2\pi}\int_{-\pi}^{\pi} e^{-i(n-k)x} e^{-i\sigma_n x}\ dx.$$

Inserting this expansion into (12) and a short calculation give

$$\mathcal{D}[q,\sigma](n) = \mathcal{D}[q](z_n) = (-1)^n e^{i\pi\sigma_n} + \sum_{k=-\infty}^{\infty} P_{n,k}[\sigma] \cdot \mathcal{D}_{\mathbb{Z}}[q](k)$$

as claimed. Above, we used

$$\int_{-\pi}^{\pi} e^{-ikv}\, \widetilde{T}(v+\pi)\, dv = (-1)^k \int_0^{2\pi} e^{-iku}\, \widetilde{T}(u) = \mathcal{D}_{\mathbb{Z}}[q](k).$$

□

4. One-parameter families of Fourier modes

Fourier modes appear in one-parameter families. In the case of the linear Fourier transform, this parameter is related to the energy (or the amplitude) of the mode. Concretely, if we set the values of the spectral parameter to be the integers, the complex valued linear Fourier modes are

$$u_d(x) = a_d\, e^{idx}; \quad d \in \mathbb{Z}, \quad a_d \in \mathbb{C}.$$

In cases of linear integrable partial differential equations, the decomposition of solutions into the linear Fourier modes actually provides the action-angle variables of the system (see [13] and [14] for the case of the Klein-Gordon equation). Roughly speaking, the modulus of a_d is an action and the argument of a_d is the associated angle variable. In the nonlinear systems the situation is more involved. Clearly, the dependence of nonlinear Fourier modes of the action and angle variables will not be linear. Still, the linear case suggests that it should make sense to construct the one-parameter families of the nonlinear Fourier modes and that the parameter can be thought of as the energy (or some function of the energy) of the mode. Our strategy will therefore be the following. Let $s \mapsto (a_d(s), b_d(s))$ be a real analytic curve in $\mathbb{C}^2$, such that $(a_d(0), b_d(0)) = (0,0)$. We are searching for the curve $s \mapsto el_d(s) = (q(x;s),\ \{\sigma_n(s)\}) = \mathcal{H}^{-1}(a_d(s), b_d(s))$ in $L^2[0,2\pi] \times l^2_{\mathbb{Z}}$. By theorem 2, a unique such curve exists for every $(a_d(s), b_d(s))$ lying close enough to the identity. In addition, the curve $s \mapsto el_d(s)$ is analytic. It can therefore be expressed in the form of a convergent power series

$$\mathcal{H}^{-1}\Big(a_d(s), b_d(s)\Big) = \Big(q(x;s), \{\sigma_n(s)\}_{n\in\mathbb{Z}}\Big) = \Big(\sum_{m=1}^{\infty} \frac{s^m}{m!}\, q^{(m)}(x),\ \sum_{m=1}^{\infty} \frac{s^m}{m!}\, \{\sigma_n^{(m)}\}_{n\in\mathbb{Z}}\Big). \tag{13}$$

Indeed, formula (7) shows that also the curves $s \mapsto q(x;s)$ and $s \mapsto \{\sigma_n(s)\}$ begin at the origin, so $q^{(0)}(x) = 0$ and $\{\sigma_n^{(0)}\} = 0$.

Theorem 3 *Let $s \mapsto (a_d(s), b_d(s))$ in $l^2_{\mathbb{Z}} \times l^2_{\mathbb{Z}}$ be an analytic curve as above. Then the terms $q^{(m)}(x)$ and $\sigma_n^{(m)}$ given by (13) can be calculated iteratively from the equations*

$$\frac{d^m}{ds^m}\Big|_{s=0}\, \mathcal{H}\Big(q(x;s), \{\sigma_n(s)\}_{n\in\mathbb{Z}}\Big) = (a_d^{(m)}, b_d^{(m)}), \quad m = 1, 2, \ldots\,. \tag{14}$$

Above $a_d^{(m)}$, $b_d^{(m)}$ denote the terms in the Taylor expansion

$$\Big(a_d(s), b_d(s)\Big) = \Big(\sum_{m=1}^{\infty} \frac{s^m}{m!}\, a_d^{(m)},\ \sum_{m=1}^{\infty} \frac{s^m}{m!}\, b_d^{(m)}\Big).$$

The above theorem is proved in [16]. The general formulae for $q^{(m)}$ and $\sigma_n^{(m)}$ are quite involved and we shall not present them here.

We shall now prove the main result of this paper.

Proof of theorem 1: In order to prove our theorem 1, it is enough to solve explicitly equations (14) for $m = 1, 2, 3$. Solving these equations illustrates nicely the plausibility of theorem 3, and also shows that the formulae for the general $q^{(m)}$, $\sigma_n^{(m)}$ demand much more space and work.

Lemma 1 tells us that for every $m \in \mathbb{N}$ the equation (14) is equivalent to the system

$$(-1)^n \frac{d^m}{ds^m}|_{s=0}\, e^{i\pi\sigma_n(s)} + \sum_{k=-\infty}^{\infty}\sum_{l=0}^{m}\binom{m}{l} P_{n,k}^{(l)}[\sigma]\cdot \mathcal{D}_{\mathbb{Z}}^{(m-l)}[q](k) = a_d^{(m)}\,\delta_{n,d} \tag{15}$$

$$\sum_{k=-\infty}^{\infty}\sum_{l=0}^{m}\binom{m}{l} P_{n,k}^{(l)}[\sigma]\cdot \mathcal{F}_{\mathbb{Z}}^{(m-l)}[q](k) = b_d^{(m)}\,\delta_{n,d}, \tag{16}$$

where

$$P_{n,k}^{(l)}[\sigma] = \frac{d^l}{ds^l}|_{s=0}P_{n,k}[\sigma(s)],\quad \mathcal{D}_{\mathbb{Z}}^{(l)}[q] = \frac{d^l}{ds^l}|_{s=0}\mathcal{D}_{\mathbb{Z}}[q(x;s)],\quad \mathcal{F}_{\mathbb{Z}}^{(l)}[q] = \frac{d^l}{ds^l}|_{s=0}\mathcal{F}_{\mathbb{Z}}[q(x;s)].$$

From our assumption that the analytic curve $s \mapsto (a_q(s),\ b_d(s))$ starts at the origin, it follows that

$$\mathcal{D}_{\mathbb{Z}}^{(0)}[\sigma] = \mathcal{D}_{\mathbb{Z}}^{(1)}[\sigma] = 0,\quad \mathcal{F}_{\mathbb{Z}}^{(0)}[q] = 0.$$

Let us denote $\widetilde{\sigma}_n^{(l)} = i\pi\sigma_n^{(l)}$. We shall also use the following expressions which are easily computable from (7) and (8):

$$\mathcal{F}_{\mathbb{Z}}^{(1)}[q](n) = (-1)^n F[q^{(1)}](n),\quad \mathcal{F}_{\mathbb{Z}}^{(2)}[q](n) = (-1)^n F[q^{(2)}](n)$$

and

$$\mathcal{F}_{\mathbb{Z}}^{(3)}[q](n) = (-1)^n F[q^{(3)}](n) + (-1)^{(n+1)} \int_{\triangle_3} e^{-in(x_1-x_2+x_3)} q^{(1)}(x_1)\overline{q^{(1)}(x_2)}q^{(1)}(x_3)\, d\vec{x}.$$

As before, F denotes the linear Fourier transform. On the diagonal part of the holonomy we shall need only the expression

$$\mathcal{D}_{\mathbb{Z}}^{(2)}[n] = (-1)^n \int_{\triangle_2} e^{-in(x_1-x_2)} q^{(1)}(x_1)\overline{q^{(1)}(x_2)}\, d\vec{x}.$$

Below, we shall solve the first five of the equations (15) and (16), two at a time.

s^1:

$$\widetilde{\sigma}_n^{(1)} = a_d^{(1)}\,(-1)^n\delta_{n,d}$$
$$F[q^{(1)}](n) = b_d^{(1)}\,(-1)^n\delta_{n,d}.$$

The solution of the above is

$$\sigma_n^{(1)} = \frac{(-1)^{d+1}}{\pi}\,a_d^{(1)}\,\delta_{n,d} = \widetilde{a}_d^{(1)}\,\delta_{n,d}$$
$$q^{(1)}(x) = (-1)^d\, b_d^{(1)}\, e^{idx} = \widetilde{b}_d^{(1)}\, e^{idx}.$$

s^2:

$$\begin{aligned}\widetilde{\sigma}_n^{(2)} + (\widetilde{\sigma}_n^{(1)})^2 + (-1)^n \mathcal{D}_{\mathbb{Z}}^{(2)}(n) &= \widetilde{a}_d^{(2)}\ \delta_{n,d}\\ F[q^{(2)}](n) + 2(P^{(1)}[\sigma]\cdot F[q^{(1)}])(n) &= \widetilde{b}_d^{(2)}\ \delta_{n,d}.\end{aligned}$$

The solution is

$$\begin{aligned}\sigma_n^{(2)} &= \frac{\alpha_2}{n-d}(1-\delta_{n,d}) + \beta_2\ \delta_{n,d}\\ q^{(2)}(x) &= \gamma_2\ e^{idx},\end{aligned}$$

where α_2, β_2 and γ_2 are suitably chosen complex constants.

s^3:

$$\mathcal{F}_{\mathbb{Z}}[q^{(3)}, q^{(1)}] + 3P^{(1)}[\sigma^{(1)}]\cdot F[q^{(2)}] + 3P^{(2)}[\sigma^{(1)}, \sigma^{(2)}]\cdot F[q^{(1)}] = b_d^{(3)}. \tag{17}$$

Calculation shows that the sequence $P^{(1)}[\sigma^{(1)}]F[q^{(2)}]$ is equal to zero. For $P^{(2)}$ we get

$$P_{n,k}^{(2)}[\sigma] = -i\sigma_n^{(2)} m_{n,k}^{(1)} + (-i\sigma_n^{(1)})^2 m_{n,k}^{(2)},$$

where

$$m_{n,k}^{(1)} = \frac{1}{2\pi}\int_{-\pi}^{\pi} x\ e^{i(n-k)x}\ dx = \frac{i(-1)^{n+k}}{k-n}, \qquad m_{n,k}^{(2)} = \frac{1}{2\pi}\int_{-\pi}^{\pi} x^2\ e^{i(n-k)x}\ dx = \frac{2(-1)^{n+k}}{(n-k)^2}.$$

Because $F[q^{(1)}](n)$ is different from zero only for $n = d$, we are interested only in d-th row of the matrix $P^{(2)}$. Collecting everything and inserting into (17) gives

$$F[q^{(3)}](n) = \frac{\alpha_3}{(n-d)^2}(1-\delta_{n,d}) + \beta_3\ \delta_{n,d},$$

for a suitable pair of constants α_3 and β_3. By applying the inverse (linear) Fourier transform F^{-1}, we finally get

$$q^{(3)}(x) = e^{idx}\left(\beta_3 + \alpha_3\ (\mathrm{Li}_2(e^{ix}) + \mathrm{Li}_2(e^{-ix}))\right).$$

Collecting the $q^{(1)}$, $q^{(2)}$ and $q^{(3)}$ together yields the one parameter family of the elementary modes $q_d(x;s)$ of frequency d up to the third power of the perturbation parameter

$$q_d(x;s) = e^{idx}\left[sA_1 + s^2A_2 + s^3\left(A_3 + A_4\ (\mathrm{Li}_2(e^{ix}) + \mathrm{Li}_2(e^{-ix}))\right)\right] + \mathcal{O}(s^4).$$

□

The above proves our theorem 1.

We conclude with a brief remark on the choice of the parameter s and on the role of the constants A_i.

Remark 1 *The form of the expression (3) does not change if we replace the parameter s by another $t = f(s)$, where f is an analytic function, and $f(0) = 0$. Reparametrization affects only the choice of the constants $A_1, \ldots, A_4$. The freedom that we have in choosing the constants A_i shows that there exists a wealth of different nonlinear Fourier modes. This makes sense, since the L-matrix (1) appears in the zero-curvature conditions for many different integrable equations. However, the first nontrivial power in the perturbation expansion is the third power and the first nontrivial term is given by the dilogarithm* in all the cases.

References

[1] Ablowitz M J, Kaup D J, Newell A C and Segur H 1973 Nonlinear-evolution equations of physical significance *Phys. Rev. Lett.* **31** 125 - 27

[2] Ablowitz M J, Kaup D J, Newell A C and Segur H 1974 The inverse scattering transform - Fourier analysis for nonlinear problems *Stud. Appl. Math.* **53** 249-315

[3] Christ M and Kiselev A 1998 Absolutely continuous spectrum for one-dimensional Schrödinger operators with slowly decaying potentials: some optimal results *J. Amer. Math. Soc.* **11**, 771 - 97

[4] Christ M and Kiselev A 2001 Maximal functions associated to filtrations *J. Funct. Anal.* **179**, 409 - 25

[5] Faddeev L D, Kashaev R M, 1994 Quantum dilogarithm *Modern Phys. Lett. A* **9**, 427-434

[6] Fokas A S 2002 Integrable nonlinear evolution equations on the half-line *Commun. Math. Phys.* **230** 1 - 39

[7] Fokas A S and Gelfand I M 1994 Integrability of linear and lonlinear evolution equations and the associated nonlinear Fourier transform *Lett. Math. Phys.* **32** 189 - 210

[8] Fokas A S and Sung L Y 2005 Generalized Fourier transforms, their nonlinearization and the imaging of the Brain *Notices Amer. Math. Soc.*, **52** 1178 - 92

[9] Muscalu C, Tao T and Thiele C 2003 A Carleson type theorem for a Cantor group model of the scattering transform *Nonlinearity*, **16**, 219 - 46

[10] Muscalu C, Tao T and Thiele C 2003 A counterexample to a multilinear end-point question by Christ and Kiselev *Math. Res. Lett.* **10**, 237 - 46

[11] Nahm W, 1994 Conformal field theory, dilogarithms and three dimensional manifolds *Interface between physics and mathematics, Proceedings, Conference in Hangzhou, China September 1993*, ed. Nahm W, Shen, J-M (Singapore: World Scientific), 154-165

[12] Nahm W, Recknagel A and Terhoeven M, 1993 Dilogarithm identities in conformal field theory *Mod. Phys. Lett. A* **8**, 1835-1847

[13] Saksida P 2011 On zero-curvature condition and Fourier analysis *J. Phys. A: Math. Gen.* **44** 85203-85222

[14] Saksida P 2012 Lax pairs and Fourier analysis: The case of sine-Gordon and Klein-Gordon equations *J. Phys.: Conf.* Ser. 343 012109, (10pp)

[15] Saksida P, 2013 On the nonlinear Fourier transform associated with periodic AKNS-ZS system and its inverse *J. Phys. A: Math. Theor.* **46** 465204 (22pp)

[16] Saksida P, 2016 Elementary modes of the nonlinear Fourier transform *In preparation* (25pp)

[17] Tao T, Thiele C 2012 Nonlinear Fourier analysis, . arXiv:1201.5129, 1 - 94

[18] Zagier D, 2007 The dilogarithm function *Frontiers in Number Theory, Physics, and Geometry II - On Conformal Theories, Discrete Groups and Renormalization* ed P. Cartier *et al* (Berlin: Springer), 3-65

[19] Zakharov V E, Shabat A B, 1972 Exact theory of two-dimensional self-focusing and one-dimensional self-modulation of waves in nonlinear media *Soviet Physics JETP* **34** 62-69

Bi-Hamiltonian structure of the general heavenly equation

M B Sheftel[1], A A Malykh[2] and D Yazıcı[3]

[1] Department of Physics, 34342 Bebek, Boğaziçi University, Istanbul, Turkey
[2] Department of Numerical Modeling, Russian State Hydrometeorlogical University, Malookhtinsky prospect 98, 195196 St. Petersburg, Russia
[3] Department of Physics, Yıldız Technical University, 34220 Esenler, Istanbul, Turkey

E-mail: [1]mikhail.sheftel@boun.edu.tr [2]andrei-malykh@mail.ru [3]yazici@yildiz.edu.tr

Abstract. We discover two additional Lax pairs and three nonlocal recursion operators for symmetries of the general heavenly equation introduced by Doubrov and Ferapontov. Converting the equation to a two-component form, we obtain Lagrangian and Hamiltonian structures of the two-component general heavenly system. We discover that in the two-component form we have only a single nonlocal recursion operator. Composing the recursion operator with the first Hamiltonian operator we obtain second Hamiltonian operator. Thus, the general heavenly equation in the two-component form is a bi-Hamiltonian system completely integrable in the sense of Magri.

1. Introduction

In paper [1], Doubrov and Ferapontov introduced the general heavenly equation (GHE)

$$\alpha u_{12}u_{34} + \beta u_{13}u_{24} + \gamma u_{14}u_{23} = 0, \qquad \alpha + \beta + \gamma = 0 \tag{1.1}$$

where α, β and γ are arbitrary constants satisfying one linear relation given above. Here $u = u(z_1, z_2, z_3, z_4)$ is a holomorphic function of four complex variables. They also obtained the Lax pair for equation (1.1)

$$\begin{aligned} X_1 &= u_{34}\partial_1 - u_{13}\partial_4 + \gamma\lambda(u_{34}\partial_1 - u_{14}\partial_3), \\ X_2 &= u_{23}\partial_4 - u_{34}\partial_2 + \beta\lambda(u_{34}\partial_2 - u_{24}\partial_3), \end{aligned} \tag{1.2}$$

where ∂_i means $\partial/\partial z_i$, while $u_{ij} = \partial^2 u/\partial z_i \partial z_j$.

There are only few examples of multi-dimensional integrable systems. The so-called heavenly equations make up an important class of such integrable systems since they are obtained by a reduction of the Einstein equations with Euclidean (and neutral) signature for (anti-)self-dual (ASD) gravity which includes the theory of gravitational instantons. All of these equations are of Monge-Ampère type, so that the only nonlinear terms are Hessian 2×2 determinants. General heavenly equation is an important example of such equations. An explicit description of ASD Ricci-flat vacuum metric governed by GHE, null tetrad and basis 1-forms for this metric were obtained in our paper [2]. Recently, Bogdanov [3] showed important relations between GHE and heavenly equations of Plebański and developed $\bar{\partial}$-dressing scheme for GHE in the context of the inverse scattering method. This stresses the importance of a further study of the Doubrov-Ferapontov's general heavenly equation.

In this paper, we obtain recursion operators for symmetries of GHE (1.1) and discover its bi-Hamiltonian structure in a two-component form, where the single second-order PDE (1.1) is presented as an evolutionary system of two first-order PDEs with two unknowns. Therefore, by the theorem of Magri [4] this general heavenly (GH) system is completely integrable.

While completing this paper, we became aware of the preprint [5] by A. Sergyeyev where he discovers a recursion operator for the one-component version of GHE, which coincides with the first one of our recursion operators, as an example of application of his general method for constructing recursion operators for dispersionless integrable systems.

If φ is a symmetry characteristic for (1.1), it satisfies the symmetry condition

$$\hat{A}\varphi = \alpha(u_{34}\varphi_{12} + u_{12}\varphi_{34}) + \beta(u_{24}\varphi_{13} + u_{13}\varphi_{24}) + \gamma(u_{23}\varphi_{14} + u_{14}\varphi_{23}) = 0. \quad (1.3)$$

A recursion operator maps any symmetry again into a symmetry and, as a consequence, it commutes with the operator $\hat{A}$ on solutions. For all other heavenly equations in the classification [1] of Doubrov and Ferapontov the symmetry condition has a two-dimensional divergence form which allows us to introduce partner symmetries [6–10], a powerful tool for finding recursion operators and noninvariant solutions [11, 12] which are necessary for the construction of the famous gravitational instanton $K3$. It is easy to check that the symmetry condition (1.3) for the general heavenly equation cannot be presented in a two-dimensional divergence form but it can be presented in a three-dimensional divergence form.

Therefore, the method of partner symmetries does not work any more for GHE, so that we have to use here a different approach in order to find a recursion operator which could be regarded as a generalization of the method of partner symmetries. This approach is based on splitting each of the two Lax operators with respect to the spectral parameter λ in two operators and multiplying the first operator by the inverse of the second operator in each pair, obtained by the splitting in λ. This method was presented in [5] using somewhat more geometric language. The idea of obtaining recursion operators from Lax pairs was used in 2003 in our paper on partner symmetries of the complex Monge-Ampère equation [7].

For a single-component equation (1.1), we discover three Lax pairs, related by discrete symmetries of both GHE and its symmetry condition, and three recursion operators corresponding to them. However, a two component form of the equation (1.1) is not invariant under these permutations of indices, so two other recursion operators are related to two other 2-component systems. We do not consider them here because they are obtained from our two-component system just by the permutations of indices.

Another important property of heavenly equations is that they can be presented in a two-component evolutionary form as bi-Hamiltonian systems [13–16]. We show here that GHE also possesses this property. In a two-component form we construct a Lagrangian for this system and discover its symplectic and Hamiltonian structure. Composing the recursion operator in a 2×2 matrix form with the Hamiltonian operator J_0 we generate the second Hamiltonian operator $J_1 = RJ_0$. We have checked directly that the Jacobi identities for the linear combination $aJ_0 + bJ_1$ with arbitrary constant coefficients a and b (obviously skew-symmetric) are satisfied which proves the Hamiltonian property of J_1 and the compatibility of the two Hamiltonian operators J_0 and J_1. The proof of the Jacobi identities will be presented in a future publication. We find the corresponding Hamiltonian density H_0 such that the original general heavenly flow is generated by the action of J_1 on variational derivatives of the Hamiltonian functional $\mathcal{H}_0$, so that the GH system turns out to be a bi-Hamiltonian system.

In section 2, we obtain two more Lax pairs for GHE in addition to the Doubrov-Ferapontov Lax pair. We show how to use these three Lax pairs for constructing three nonlocal recursion operators for GHE.

In section 3, we present the general heavenly equation in a two-component evolutionary form and obtain a Lagrangian for this GH system.

In section 4, we discover symplectic and Hamiltonian structure of the GH system.

In section 5, we derive a nonlocal recursion operator for the two component GH system.

In section 6, acting by the recursion operator on the first Hamiltonian operator J_0 we obtain second nonlocal Hamiltonian operator J_1. We find the corresponding Hamiltonian density which generates the GH system with the aid of the second Hamiltonian operator. Thus, we obtain a bi-Hamiltonian representation for the general heavenly equation in a two-component form.

2. Recursion operators

We introduce the following first-order differential operators from which the Lax operators (1.2) are constructed

$$L_{14(3)} = u_{34}D_1 - u_{13}D_4, \quad L_{13(4)} = u_{34}D_1 - u_{14}D_3,$$
$$L_{24(3)} = u_{34}D_2 - u_{23}D_4, \quad L_{23(4)} = u_{34}D_2 - u_{24}D_3 \tag{2.1}$$

where D_i denotes total derivative with respect to z_i. The general definition is $L_{ij(k)} = u_{jk}D_i - u_{ik}D_j$, so that $L_{ji(k)} = -L_{ij(k)}$. Lax operators take the form

$$X_1 = L_{14(3)} + \lambda\gamma L_{13(4)}, \qquad X_2 = -L_{24(3)} + \lambda\beta L_{23(4)}. \tag{2.2}$$

The symmetry condition (1.3), where we use $\alpha = -(\beta+\gamma)$, becomes

$$\hat{A}\varphi = \{\beta(L_{14(2)}D_3 - L_{14(3)}D_2) + \gamma(L_{24(1)}D_3 - L_{24(3)}D_1)\}\varphi = 0$$
$$\equiv \{\beta(D_3L_{14(2)} - D_2L_{14(3)}) + \gamma(D_3L_{24(1)} - D_1L_{24(3)})\}\varphi = 0 \tag{2.3}$$
$$\text{where} \quad L_{14(2)} = u_{24}D_1 - u_{12}D_4, \quad L_{24(1)} = u_{14}D_2 - u_{12}D_4. \tag{2.4}$$

To arrive at two different Lax pairs, we apply two permutations of indices which leave invariant both the equation (1.1) and its symmetry condition (1.3) but which do change the Lax pair and recursion operators. The permutation $1 \leftrightarrow 3$, $2 \leftrightarrow 4$ yields the second Lax pair

$$X_1^{(2)} = L_{23(1)} + \lambda\gamma L_{13(2)}, \qquad X_2^{(2)} = L_{24(1)} - \lambda\beta L_{14(2)} \tag{2.5}$$
$$\text{where} \quad L_{13(2)} = u_{23}D_1 - u_{12}D_3, \qquad L_{23(1)} = u_{13}D_2 - u_{12}D_3 \tag{2.6}$$

and operators $L_{14(2)}$, $L_{24(1)}$ are defined in (2.4). In (2.5) the overall minus in $X_1^{(2)}$ is skipped.

The permutation $1 \leftrightarrow 4$, $2 \leftrightarrow 3$ yields the third Lax pair

$$X_1^{(3)} = L_{41(2)} + \lambda\gamma L_{42(1)}, \qquad X_2^{(3)} = -L_{31(2)} + \lambda\beta L_{32(1)} \tag{2.7}$$
$$\text{where} \quad L_{41(2)} = u_{12}D_4 - u_{24}D_1, \qquad L_{42(1)} = u_{12}D_4 - u_{14}D_2,$$
$$L_{31(2)} = u_{12}D_3 - u_{23}D_1, \qquad L_{32(1)} = u_{12}D_3 - u_{13}D_2. \tag{2.8}$$

We note that the symmetry condition (2.3) has a three-dimensional divergence form and therefore our definition of partner symmetries, which requires a two-dimensional divergence form of the symmetry condition [6], does not work here, so that we need here a different approach to obtain a recursion operator which is applied to the proof of the following theorem. It is based on our observation that recursion operators for partner symmetries of the heavenly equations, which we studied before, are composed from the operators obtained by splitting the Lax operators with respect to spectral parameter λ and multiplying the first operator by the inverse of the second operator in each pair, that was obtained by splitting in λ [7–9, 14, 15].

IOP Conf. Series: Journal of Physics: Conf. Series **804** (2017) 012039 doi:10.1088/1742-6596/804/1/012039

Theorem 2.1 *The general heavenly equation admits the following three different Lax pairs (2.9a), (2.10a), (2.11a) and three respective recursion operators defined by the relations (2.9b), (2.10b), (2.11b)*

$$X_1^{(1)} = L_{14(3)} + \lambda\gamma L_{13(4)}, \quad X_2^{(1)} = -L_{24(3)} + \lambda\beta L_{23(4)} \tag{2.9a}$$

$$L_{14(3)}\varphi = \gamma L_{13(4)}\psi, \quad L_{24(3)}\varphi = -\beta L_{23(4)}\psi. \tag{2.9b}$$

$$X_1^{(2)} = L_{23(1)} + \lambda\gamma L_{13(2)}, \quad X_2^{(2)} = L_{24(1)} - \lambda\beta L_{14(2)} \tag{2.10a}$$

$$L_{23(1)}\varphi = \gamma L_{13(2)}\psi, \quad L_{24(1)}\varphi = -\beta L_{14(2)}\psi. \tag{2.10b}$$

$$X_1^{(3)} = L_{41(2)} + \lambda\gamma L_{42(1)}, \quad X_2^{(3)} = -L_{31(2)} + \lambda\beta L_{32(1)} \tag{2.11a}$$

$$L_{41(2)}\varphi = \gamma L_{42(1)}\psi, \quad L_{31(2)}\varphi = -\beta L_{32(1)}\psi. \tag{2.11b}$$

Proof

The Lax pair (2.9a) is known from the paper [1] while two other Lax pairs are obtained from (2.9a) by permutations of indices which do not change the equation (1.1).

To prove that the relations (2.9b) indeed represent a recursion operator, we analyze their integrability conditions. We have two sets of integrability conditions for (2.9b)

$$[L_{24(3)}, L_{14(3)}]\varphi = (\gamma L_{24(3)}L_{13(4)} + \beta L_{14(3)}L_{23(4)})\psi \tag{2.12}$$

$$\beta\gamma[L_{23(4)}, L_{13(4)}]\psi = (\beta L_{23(4)}L_{14(3)} + \gamma L_{13(4)}L_{24(3)})\varphi. \tag{2.13}$$

In condition (2.12), the commutator in the left-hand-side expands as

$$[L_{24(3)}, L_{14(3)}]\varphi = \frac{1}{u_{34}}\{(u_{34}u_{234} + u_{23}u_{344})L_{14(3)} + (u_{34}u_{134} - u_{13}u_{344})L_{24(3)}\}\varphi$$

which, with the use of equations (2.9b) converts to derivatives of ψ. In condition (2.13), the commutator in the left-hand-side expands as

$$[L_{23(4)}, L_{13(4)}]\psi = \frac{1}{u_{34}}\{(u_{34}u_{234} - u_{24}u_{344})L_{13(4)} - (u_{34}u_{134} - u_{14}u_{334})L_{23(4)}\}\psi$$

which again with the use of (2.9b) converts to derivatives of φ. Thus, the condition (2.12) becomes the equation for ψ only, which can be straightforwardly checked to coincide with $\hat{A}\psi = 0$, while the condition (2.13) is the equation $\hat{A}\varphi = 0$ for φ only. Therefore, the integrability conditions of equations (2.9b) are symmetry conditions for ψ and φ, which means that both ψ and φ are symmetry characteristics for GHE (1.1). Hence, the relations (2.9b) are recursion relations between symmetries ψ and φ and, consequently, the relations (2.9b) determine a recursion operator. Since two other relations (2.10b) and (2.11b) are obtained from the recursion relations (2.9b) by permutations of indices, which leave invariant the equation (1.1) and its symmetry condition, these relations also determine recursion operators. □

We note that the recursion relations (2.9b) look like an auto-Bäcklund transformation between the symmetry conditions $\hat{A}\varphi = 0$ and $\hat{A}\psi = 0$. Later, we have found that the same observation was made earlier in [5] in a more general context. We have checked that our recursion relations (2.9b) coincide with the ones obtained somewhat earlier by A. Sergyeyev in the version 1 of [5]. The approach we used here can be regarded as a generalization of the concept of partner symmetries.

IOP Conf. Series: Journal of Physics: Conf. Series **804** (2017) 012039 doi:10.1088/1742-6596/804/1/012039

3. Two-component evolutionary form of the general heavenly equation

In order to discover Hamiltonian structure of GHE we need to convert it to a two-component evolutionary form. For this purpose, we transform z_1, z_2 into the "time" and "space" variables t and x as $t = z_1 + z_2, \quad x = z_1 - z_2, \quad y = z_3, \quad z = z_4$. GHE (1.1) becomes

$$\alpha(u_{tt} - u_{xx})u_{yz} + \beta(u_{ty} + u_{xy})(u_{tz} - u_{xz}) + \gamma(u_{ty} - u_{xy})(u_{tz} + u_{xz}) = 0. \tag{3.1}$$

To convert (3.1) into an evolutionary system, we define the second component as $v = u_t$ with the following final result for (3.1)

$$\begin{cases} u_t = v \\ v_t = \dfrac{1}{u_{yz}}\left[u_{xx}u_{yz} - u_{xy}u_{xz} + v_y v_z + b(v_y u_{xz} - v_z u_{xy})\right] \end{cases} \tag{3.2}$$

where $b = \dfrac{\beta - \gamma}{\alpha}$ is the single parameter constructed from the coefficients of GHE (1.1). It is easy to check that (3.2) are Euler-Lagrange equations with the Lagrangian density

$$L = \left(vu_t - \frac{v^2}{2}\right)u_{yz} - \frac{1}{2}u_y u_z u_{xx} + \frac{b}{3}u_t(u_z u_{xy} - u_y u_{xz}). \tag{3.3}$$

4. First Hamiltonian structure

The momenta are defined as

$$\Pi_u = \frac{\partial L}{\partial u_t} = vu_{yz} + b(u_z u_{xy} - u_y u_{xz}), \quad \Pi_v = \frac{\partial L}{\partial v_t} = 0. \tag{4.1}$$

Hence, the Lagrangian (3.3) is degenerate since the momenta cannot be inverted for the velocities. Therefore, following Dirac [17] we impose (4.1) as constraints

$$\Phi_u = \Pi_u - vu_{yz} - b(u_z u_{xy} - u_y u_{xz}), \quad \Phi_v = \Pi_v \tag{4.2}$$

and calculate the Poisson bracket of the constraints
$K_{ij} = [\Phi_i(x,y,z), \Phi_j(x',y',z')]$, where $i,j = 1,2,\ u_1 = u,\ u_2 = v,\ \Phi_1 = \Phi_u,\ \Phi_2 = \Phi_v$ using

$$[\Pi_i(x,y,z), u_k(x',y',z')] = \delta_i^k \delta(x - x')\delta(y - y')\delta(z - z').$$

The result in a matrix form reads

$$K = \begin{pmatrix} b(u_{xz}D_y - u_{xy}D_z) + v_z D_y + v_y D_z + v_{yz}, & -u_{yz} \\ u_{yz} & 0 \end{pmatrix} \tag{4.3}$$

which is obviously a skew symmetric matrix. The symplectic structure is defined in terms of (4.3) by the differential two-form $\omega = \frac{1}{2}du_i \wedge K_{ij}du_j$ with the final result

$$\omega = \frac{b}{2}(u_{xz}du \wedge du_y - u_{xy}du \wedge du_z) + \frac{1}{2}(v_z du \wedge du_y + v_y du \wedge du_z) - u_{yz}du \wedge dv. \tag{4.4}$$

It is easy to check that the two-form (4.4) is closed, $d\omega = 0$, up to a total divergence and hence it determines a symplectic structure. The inverse to the symplectic operator is the Hamiltonian operator J_0 because the closed condition for ω is equivalent to satisfaction of Jacobi identities for J_0 [18]. Thus, we obtain the first Hamiltonian operator in the form

$$J_0 = K^{-1} = \frac{1}{\sqrt{\det K_{ij}}}\begin{pmatrix} 0 & u_{yz} \\ -u_{yz} & K_{11} \end{pmatrix}\frac{1}{\sqrt{\det K_{ij}}} \tag{4.5}$$

where $\det K_{ij} = u_{yz}^2$. In a final form, the first Hamiltonian operator reads

$$J_0 = \begin{pmatrix} 0 & \dfrac{1}{u_{yz}} \\ -\dfrac{1}{u_{yz}} & J_0^{22} \end{pmatrix} \tag{4.6}$$

where

$$J_0^{22} = \frac{1}{u_{yz}}\left[b(u_{xz}D_y - u_{xy}D_z) + v_zD_y + v_yD_z + v_{yz}\right]\frac{1}{u_{yz}}, \qquad b = \frac{\beta-\gamma}{\alpha}. \tag{4.7}$$

We note that J_0 is skew symmetric. The Hamiltonian density for the system (3.2) corresponding to the Lagrangian (3.3) is determined by

$$H_1 = \Pi_u u_t + \Pi_v v_t - L = \frac{1}{2}\left(v^2u_{yz} + u_yu_zu_{xx}\right) \iff H_1 = \frac{1}{2}\left(v^2 + u_x^2\right)u_{yz} \tag{4.8}$$

the two expressions being equivalent because their difference is a total divergence. The system (3.2) in the first Hamiltonian form becomes

$$\begin{pmatrix} u_t \\ v_t \end{pmatrix} = J_0\begin{pmatrix} \delta_u H_1 \\ \delta_v H_1 \end{pmatrix} = \begin{pmatrix} 0 & \dfrac{1}{u_{yz}} \\ -\dfrac{1}{u_{yz}} & J_0^{22} \end{pmatrix}\begin{pmatrix} \delta_u H_1 \\ \delta_v H_1 \end{pmatrix} \tag{4.9}$$

with J_0^{22} defined in (4.7). Here δ_u and δ_v are Euler-Lagrange operators with respect to u and v, respectively, closely related to the variational derivatives of the functional $\mathcal{H}_1 = \int_{-\infty}^{+\infty} H_1\, dxdydz$ with respect to u and v [18]. Here we change the notation $\mathbf{E}_\alpha$ of [18] to δ_{u^α}.

5. Recursion operator for the two-component form of the general heavenly equation

Recursion operators in a two-component form are required for constructing new Hamiltonian operators for the two-component system (3.2).

Recursion relations (2.9b) become

$$\begin{aligned} u_{yz}(\varphi_t + \varphi_x) - (v_y + u_{xy})\varphi_z &= \gamma\{u_{yz}(\psi_t + \psi_x) - (v_z + u_{xz})\psi_y\} \\ u_{yz}(\varphi_t - \varphi_x) - (v_y - u_{xy})\varphi_z &= -\beta\{u_{yz}(\psi_t - \psi_x) - (v_z - u_{xz})\psi_y\}. \end{aligned} \tag{5.1}$$

Now we change the notation: ψ is now reserved for the second component of the symmetry characteristic, so that Lie equations read

$$\begin{pmatrix} u \\ v \end{pmatrix}_\tau = \begin{pmatrix} \varphi \\ \psi \end{pmatrix} \tag{5.2}$$

where $v = u_t$ implies $\psi = \varphi_t$ and

$$\begin{pmatrix} \tilde\varphi \\ \tilde\psi \end{pmatrix} = R\begin{pmatrix} \varphi \\ \psi \end{pmatrix} \tag{5.3}$$

where $\varphi_t = \psi$ and $\tilde\varphi_t = \tilde\psi$. Recursion relations (5.1) take the form

$$\begin{aligned} u_{yz}(\psi + \varphi_x) - (v_y + u_{xy})\varphi_z &= \gamma\{u_{yz}(\tilde\psi + \tilde\varphi_x) - (v_z + u_{xz})\tilde\varphi_y\} \\ (v_y - u_{xy})\varphi_z - u_{yz}(\psi - \varphi_x) &= \beta\{u_{yz}(\tilde\psi - \tilde\varphi_x) - (v_z - u_{xz})\tilde\varphi_y\} \end{aligned}$$

which in a two-component matrix form becomes

$$\begin{pmatrix} u_{yz}D_x - (v_y + u_{xy})D_z & u_{yz} \\ (v_y - u_{xy})D_z + u_{yz}D_x & -u_{yz} \end{pmatrix} \begin{pmatrix} \varphi \\ \psi \end{pmatrix} = \begin{pmatrix} \gamma\{u_{yz}D_x - (v_z + u_{xz})D_y\} & \gamma u_{yz} \\ \beta\{-u_{yz}D_x + (u_{xz} - v_z)D_y\} & \beta u_{yz} \end{pmatrix} \begin{pmatrix} \tilde{\varphi} \\ \tilde{\psi} \end{pmatrix} \tag{5.4}$$

To obtain the recursion operator R explicitly, we need to invert the matrix differential operator on the right-hand side of the relation (5.4). Note that the inverse $\begin{pmatrix} e & f \\ g & h \end{pmatrix}$ for the matrix $\begin{pmatrix} a & b \\ c & d \end{pmatrix}$ with non-commuting entries a, b, c, d, defined by the equation $\begin{pmatrix} e & f \\ g & h \end{pmatrix}\begin{pmatrix} a & b \\ c & d \end{pmatrix} = \begin{pmatrix} 1 & 0 \\ 0 & 1 \end{pmatrix}$, is determined by the formulas

$$e = (a - bd^{-1}c)^{-1},\ f = (c - db^{-1}a)^{-1},\ g = (b - ac^{-1}d)^{-1},\ h = (d - ca^{-1}b)^{-1}. \tag{5.5}$$

The result for R is convenient to express in terms of the two differential operators

$$w = u_{yz}D_x - u_{xz}D_y, \qquad \zeta = u_{yz}D_x - u_{xy}D_z. \tag{5.6}$$

Using the corresponding entries of the matrix on the right-hand side of (5.4) for a, b, c, d in the equations (5.5), we obtain e, f, g, h

$$e = \frac{1}{2\gamma}w^{-1},\quad f = -\frac{1}{2\beta}w^{-1},\quad g = \frac{1}{2\gamma u_{yz}}(1 + v_zD_yw^{-1})\quad h = \frac{1}{2\beta u_{yz}}(1 - v_zD_yw^{-1}).$$

The recursion operator is explicitly defined by the relation

$$R = \frac{1}{\alpha}\begin{pmatrix} \beta & -\gamma \\ \frac{\beta}{u_{yz}}(w + v_zD_y) & \frac{\gamma}{u_{yz}}(w - v_zD_y) \end{pmatrix} w^{-1} \begin{pmatrix} \zeta - v_yD_z & u_{yz} \\ \zeta + v_yD_z & -u_{yz} \end{pmatrix}$$

which we have multiplied by the overall constant factor $(2\beta\gamma)/\alpha$. After performing multiplication in (5.7) we obtain the matrix elements of R explicitly

$$\begin{aligned} R_{11} &= w^{-1}(b\zeta + v_yD_z), \qquad R_{12} = -w^{-1}u_{yz} \\ R_{21} &= \frac{1}{u_{yz}}\left(v_zD_yw^{-1}v_yD_z - \zeta\right) + \frac{b}{u_{yz}}\left(v_zD_yw^{-1}\zeta - v_yD_z\right) \\ R_{22} &= b - \frac{v_z}{u_{yz}}D_yw^{-1}u_{yz}, \qquad b = (\beta - \gamma)/\alpha. \end{aligned} \tag{5.7}$$

We do not consider the second and third recursion operators R^2 and R^3 defined in (2.10b) and (2.11b) in the two-component form because they refer to two different two-component systems.

The definition of the recursion operator in the two-component form (5.7) contains arbitrariness related to the definition of the inverse operator w^{-1} which involves addition of an arbitrary element of the kernel of w defined by $w(f) = (u_{yz}D_x - u_{xz}D_y)f = 0$. The general solution for $w(f) = 0$ reads $\ker w = \{f(z, u_z)\}$ with an arbitrary smooth f. Thus, w^{-1} is defined up to the addition of an arbitrary function $f(z, u_z)$ which plays the role of an arbitrary integration constant. This arbitrariness is eliminated by the condition that not only relation $ww^{-1} = I$ is satisfied, but also $w^{-1}w = I$ has to be satisfied.

Let us analyze the definition of the inverse operator w^{-1} in more detail. It determines a solution $f = w^{-1}g$ to the first-order ODE: $wf = u_{zy}f_x - u_{zx}f_y = g$ for any given g. Introduce parameter s along the curves tangent to the vector field w, $x = x(s), y = y(s)$ on the plane $z = const$

$$wf = \frac{df(x(s), y(s), z)}{ds} = \frac{dx}{ds} f_x + \frac{dy}{ds} f_y = u_{zy} f_x - u_{zx} f_y = g$$

which implies $\frac{dx}{ds} = u_{zy}$, $\frac{dy}{ds} = -u_{zx}$ and so

$$ds = \frac{dx}{u_{zy}} = -\frac{dy}{u_{zx}}. \tag{5.8}$$

The second of these equations is the characteristic ODE with the integrals

$$z = const, \quad u_z(x, y, z) = const \tag{5.9}$$

while the first one determines the parameter s along the characteristics

$$s = \int \frac{dx}{u_{zy}} = -\int \frac{dy}{u_{zx}} = \frac{1}{2} \int \left(\frac{dx}{u_{zy}} - \frac{dy}{u_{zx}} \right) \tag{5.10}$$

where the integrals are taken under conditions (5.9) and the equations (5.8) have been used. Under the constraints (5.9), we have $wf = \frac{df}{ds} \equiv D_s f = g$, so that $f = w^{-1} g = D_s^{-1} g = \int ds\, g$.

We define w^{-1} as the integral operator $\int ds$ with ds defined by (5.8)

$$w^{-1} = \left\{ \int_a^x \frac{d\xi}{u_{zy}(\xi, y(\xi, c, z), z)} \right\}_{c=u_z(x,y,z)} \quad \text{or} \quad w^{-1} = -\left\{ \int_b^y \frac{d\eta}{u_{zx}(x(\eta, c, z), \eta, z)} \right\}_{c=u_z(x,y,z)} \tag{5.11}$$

where all the integrals are taken at the conditions (5.9) so that $y = y(x, c, z)$ and $x = x(y, c, z)$ are determined by the equation $u_z(x, y, z) = c = const$ and z is a parameter of the integrations. Here a and b are arbitrarily fixed points such that the functions $f(x, y, z)$ are subject to the boundary condition $f|_{x=a} = 0$ or $f|_{y=b} = 0$, respectively. Then for these classes of functions one can check that $w^{-1} w f(x, y, z) = f$ and also that $w w^{-1} f(x, y, z) = f$, so that $w^{-1} w = I$ and $w w^{-1} = I$.

Indeed, using characteristic equations (5.8) and definition (5.6) of w for the first integral in (5.11) we have

$$\begin{aligned}
w^{-1} w f(x, y, z) &= \left\{ \int_a^x \frac{d\xi}{u_{zy}} u_{zy} \frac{\partial}{\partial \xi} f(\xi, y, z) - \int_a^x \frac{d\xi}{u_{zy}} u_{zx} \frac{\partial}{\partial y} f(\xi, y, z) \right\}_{c=u_z} \\
&= \int_{\substack{a \\ \{y=y(\xi, c, z)\}}}^x d\xi \left\{ \partial_\xi f(\xi, y, z) + \partial_\xi y\, \partial_y f(\xi, y, z) \right\}_{c=u_z} \\
&= \left\{ \int_a^x d\xi D_\xi f(\xi, y(\xi, c, z), z) \right\}_{c=u_z} = [f(x, y(x, c, z), z)|_{c=u_z}]|_a^x \\
&= f(x, y, z) - f(a, y, z) = f(x, y, z)
\end{aligned}$$

for all f satisfying $f|_{x=a} = 0$, so that $w^{-1} w = I$. Next, define

$$F(x, u_z(x, y, z), z) = \left\{ \int_a^x d\xi \frac{f(\xi, y(\xi, c, z), z)}{u_{zy}(\xi, y(\xi, c, z), z)} \right\}_{c=u_z(x,y,z)}$$

IOP Conf. Series: Journal of Physics: Conf. Series **804** (2017) 012039 doi:10.1088/1742-6596/804/1/012039

where we should use the equations of the characteristics (5.8) for the integral. Then

$$ww^{-1}f(x,y,z) = (u_{zy}D_x - u_{zx}D_y)\left\{\int_a^x d\xi \frac{f(\xi, y(\xi,c,z),z)}{u_{zy}(\xi, y(\xi,c,z),z)}\right\}_{c=u_z(x,y,z)}$$
$$= (u_{zy}D_x - u_{zx}D_y)F(x, u_z(x,y,z),z) = u_{zy}F_x + u_{zy}F_{u_z}u_{zx} - u_{zx}F_{u_z}u_{zy}$$
$$= u_{zy}F_x = u_{zy}\frac{f(x,y,z)}{u_{zy}} = f(x,y,z)$$

so that $ww^{-1} = I$. The check of the second definition of w^{-1} in (5.11) could be made in a similar way. With a more symmetric definition

$$w^{-1} = \frac{1}{2}\left\{\int_{-a}^x \frac{d\xi}{u_{zy}(\xi, y(\xi,c,z),z)} - \int_x^a \frac{d\xi}{u_{zy}(\xi, y(\xi,c,z),z)}\right\}_{c=u_z(x,y,z)}$$

$w^{-1}w = I$ implies the boundary condition $f(-a,y,z) = -f(a,y,z)$ for an arbitrarily fixed a.

6. Second Hamiltonian structure

The second Hamiltonian operator J_1 is obtained by composing the recursion operator (5.7) with the first Hamiltonian operator $J_1 = RJ_0$ or explicitly

$$\begin{pmatrix} J_1^{11} & J_1^{12} \\ J_1^{21} & J_1^{22} \end{pmatrix} = \begin{pmatrix} R_{11} & R_{12} \\ R_{21} & R_{22} \end{pmatrix}\begin{pmatrix} 0 & \frac{1}{u_{yz}} \\ -\frac{1}{u_{yz}} & J_0^{22} \end{pmatrix} \tag{6.1}$$

where we have used the formula (4.6) and J_0^{22} is defined in (4.7). Utilizing also the expressions (5.7), we obtain the matrix elements of J_1 in the form

$$J_1^{11} = w^{-1}, \quad J_1^{12} = \frac{b}{u_{yz}} - w^{-1}D_y\frac{v_z}{u_{yz}} \quad J_1^{21} = -\frac{b}{u_{yz}} + \frac{v_z}{u_{yz}}D_y w^{-1}$$
$$J_1^{22} = -\frac{1}{u_{yz}}\zeta\frac{1}{u_{yz}} + \frac{b^2}{u_{yz}}(\zeta - w)\frac{1}{u_{yz}} + \frac{b}{u_{yz}}(D_y v_z + v_z D_y)\frac{1}{u_{yz}}$$
$$-\frac{v_z}{u_{yz}}D_y w^{-1}D_y\frac{v_z}{u_{yz}}, \qquad \text{where} \quad b = \frac{(\beta-\gamma)}{\alpha} \tag{6.2}$$

which is explicitly skew-symmetric. We have also proved the Hamiltonian property of the operator J_1 and the compatibility of the two Hamiltonian structures J_0 and J_1 by checking directly the Jacobi identities for the operator $J_1 + aJ_0$ with an arbitrary constant a. This proof will be presented in a future publication.

The Hamiltonian density corresponding to the general heavenly system (3.2) with respect to the second Hamiltonian operator J_1 is

$$H_0 = \frac{1}{2(b^2-1)}\left[2u_x v + b(v^2 + u_x^2)\right]u_{yz}. \tag{6.3}$$

Here we eliminate the case of the degenerate GHE with $\beta \cdot \gamma = 0$, so that $b^2 - 1 \neq 0$. Thus, the general heavenly flow takes the bi-Hamiltonian form

$$\begin{pmatrix} u_t \\ v_t \end{pmatrix} = J_0\begin{pmatrix} \delta_u H_1 \\ \delta_v H_1 \end{pmatrix} = J_1\begin{pmatrix} \delta_u H_0 \\ \delta_v H_0 \end{pmatrix} \tag{6.4}$$

and hence presents a multidimensional completely integrable system in the sense of Magri [4].

Conclusion

We have constructed two additional Lax pairs and three nonlocal recursion operators for the general heavenly equation (GHE) obtained by the splitting of the Lax pairs with respect to the spectral parameter. Converting GHE to a two-component evolutionary form, we have discovered Lagrangian, symplectic and Hamiltonian structures of this GH system. We have converted the first recursion operator to a matrix 2×2 form appropriate for our two-component evolutionary system, while the other two operators refer to different two-component systems obtained from the first one by permutations, so that we end up with a single nonlocal recursion operator for the first GH system. Composing the recursion operator R with the first Hamiltonian operator J_0 we have obtained the second Hamiltonian operator $J_1 = RJ_0$ and found the corresponding Hamiltonian density generating the two-component flow of GHE. We have checked the Jacobi identities for J_1 and compatibility of the two Hamiltonian structures J_0 and J_1. The proof will be presented in a future publication. Thus, we have shown that the GHE two-component flow is a bi-Hamiltonian system completely integrable in the sense of Magri.

Acknowledgments

The research of M. B. Sheftel is partly supported by the research grant from Boğaziçi University Scientific Research Fund (BAP), research project No. 11643.

References

[1] Doubrov B and Ferapontov E V 2010 *J. Geom. Phys.* **60** 1604
[2] Malykh A A and Sheftel M B 2011 *J. Phys. A: Math. Theor.* **44** 155201
[3] Bogdanov L V 2015 Doubrov-Ferapontov general heavenly equation and the hyper-Kähler hierarchy *Preprint* nlin.SI/1412.7376
[4] Magri F 1978 *J. Math. Phys.* **19** 1156
[5] Sergyeyev A 2016 Recursion operators for multidimensional integrable systems *Preprint* math/1501.01955
[6] Sheftel M B and Malykh A A 2009 *J. Phys. A: Math. Theor.* **42** 395202
[7] Malykh A A, Nutku Y and Sheftel M B 2003 *J. Phys. A: Math. Gen.* **36** 10023
[8] Malykh A A, Nutku Y and Sheftel M B 2004 *J. Phys. A: Math. Gen.* **37** 7527
[9] Malykh A A, Nutku Y and Sheftel M B 2007 *J. Phys. A: Math. Theor.* **40** 9371
[10] Sheftel M B and Yazıcı D 2014 *J. Geom. Phys.* **85** 252
[11] Malykh A A and Sheftel M B 2011 *SIGMA* **7** 043
[12] Sheftel M B and Malykh A A 2013 *SIGMA* **9** 075
[13] Neyzi F, Nutku Y and Sheftel M B 2005 *J. Phys. A: Math. Gen.* **38** 8473
[14] Nutku Y, Sheftel M B, Kalayci J and Yazıcı D 2008 *J. Phys. A: Math. Theor.* **41** 395206
[15] Sheftel M B and Yazıcı D 2010 *J. Nonlin. Math. Phys.* **17** No. 4, 453
[16] Yazıcı D 2016 *J. Geom. Phys.* **101** 11
[17] Dirac P A M 1964 *Lectures on Quantum Mechanics* (Belfer Graduate School of Science Monographs, series 2), (New York: Yeshiva University Press)
[18] Olver P J 1986 *Applications of Lie Groups to Differential Equations* (New York: Springer-Verlag)

Relativistic wave equations of arbitrary spin in quantum mechanics and field theory, example spin s=2

Volodimir Simulik

Principal research associate, Institute of Electron Physics, National Academy of Sciences of Ukraine, 21 Universitetska Str., Uzhgorod 88000, Ukraine

E-mail: vsimulik@gmail.com

Abstract. Relativistic wave equations for the particles of arbitrary spin suggested by Bhabha, Bargmann–Wigner, Rarita–Schwinger (for spin s = 3/2) and other authors are under consideration. The comparison with the equations introduced recently by the author in Ukr. J. Phys., Vol. 60, N 10. 985–1006 (2015) and J. Phys.; Conf. Ser. Vol. 670, 012047(1–16) (2016) is given. The advantages of the new equations are considered briefly. The three level consideration (relativistic canonical quantum mechanics, canonical Foldy–Wouthuysen type field theory, locally covariant field theory) is presented. The operator link between the relativistic canonical quantum mechanics and locally covariant field theory of arbitrary spin is found. This link is given by the extended Foldy–Wouthuysen transwormation between the 2(2s+1)-component local field theory and the corresponding relativistic canonical quantum mechanics. The important partial example of spin s=2 case is considered in details. The 10-component Dirac-like wave equation for the spin s=(2,2) particle-antiparticle doublet is suggested. The link with 10-component Maxwell-like equation for this doublet is considered brifly. The hypothesis on the spin s=5/2 particle is discussed.

1. Introduction

The investigation of a particle equations for an arbitrary spin is the interesting problem of contemporary theoretical physics. The big zoo of known elementary particles (and corresponding fields) prefers a systematic description, which can be given by a suitable field equation for an arbitrary spin. Therefore, the particle equations for an arbitrary spin are the subject for careful investigations from the start, given by Paul Dirac [1] in 1936, until today.

Of course, the first-order partial differential equations are preferable. The canonical nonlocal pseudo-differential representations of such equations are under consideration as well. Thus, the second order equations (like the Klein–Gordon–Fock equation) are not the subject of the given below investigation.

Different approaches to the description of the field theory of an arbitrary spin can be found in [1–16]. Our contribution is presented in [15, 16]. Here and in [15, 16] only the approach started in [4] is the basis for further application. Other results given in [1–3, 5–14] are not used here. Below we continue the investigations of [16].

Note only some general deficiencies of the known equations for arbitrary spin [1–3, 5–14]. The consideration of the partial cases, when the substitution of the fixed value of spin is fulfilled, is not successful in all approaches. For example, for the spin $s > 1$ existing equations have the redundant components and should be complemented by some additional conditions. Indeed, the known equations [17, 18] for the spin s=3/2 (and their confirmation in [19]) should be essentially complemented by the additional conditions. The main difficulty in the models of an arbitrary spin is the interaction between the fields of higher-spin. Even the quantization of higher-spin fields generated the questions. These and other deficiencies of the known equations for higher-spin are considered, e. g., in [20–31] (a brief review of deficiencies see in [24]).

Even this brief analysis makes us sure in the prospects of the investigations started in [32, 33] and continued in [15, 16]. The successful description of the arbitrary spin field models is not the solved problem today. There is a place for new approaches among the known equations for an arbitrary spin.

Equations suggested in [32, 33] and [15, 16] are free of the mentioned above deficiencies. The equations from [32, 33] have been derived directly from the well-defined equations of relativistic canonical quantum mechanics. Therefore, the partial differential equations for arbitrary spin in [15, 16, 32, 33] are without redundant components. It is the advantage of equations from [15, 16] in comparison with equations considered in [17–31]. Indeed, the Rarita–Schwinger equation for spin s=3/2 has 16 components, our equation for spin s=(3/2,3/2) particle-antiparticle doublet has 8 components. The Bargman–Wigner equation in partial case s=3/2 has 12 components. Bhabha itself [34] have analyzed the partial case s=3/2 for his equation [2]. He have found [34] that in this case his equation [2] coincides with the Rarita–Schwinger equation, i. e. has 16 components. Therefore, in [32, 33] and [15, 16] the new equation for arbitrary spin has been suggested. The partial case of spin s=(3/2,3/2) particle-antiparticle doublet has been considered in [15, 16]. Below the partial case of spin s=(2,2) particle-antiparticle doublet is under consideration.

The start of our consideration is taken from [4], where the main foundations of the relativistic canonical quantum mechanics (RCQM) are formulated. In the articles [32, 33] and [15, 16] the results of [4] are generalized and extended. The operator link between the results of [35] and [4] (between the canonical Foldy–Wouthuysen type field theory and the relativistic canonical quantum mechanics) is suggested. Note that the cases s=3/2 and s=2 are not presented in [4], especially in explicit demonstrative forms.

The main idea and main new feature of our approach to the field theory of arbitrary spin is the start from the arbitrary spin relativistic canonical quantum mechanics. Therefore on every level ((i) relativistic canonical quantum mechanics, (ii) the canonical Foldy–Wouthuysen type field theory, (iii) locally covariant field theory) we are dealing with the equations without redundant components.

The concepts, definitions and notations here are the same as in [16]. For example, in the Minkowski space-time

$$\mathrm{M}(1,3) = \{x \equiv (x^\mu) = (x^0 = t,\, \overrightarrow{x} \equiv (x^j))\}; \quad \mu = \overline{0,3},\, j = 1,2,3, \tag{1}$$

x^μ denotes the Cartesian (covariant) coordinates of the points of the physical space-time in the arbitrary-fixed inertial reference frame. We use the system of units with $\hbar = c = 1$. The metric tensor is given by

$$g^{\mu\nu} = g_{\mu\nu} = g^\mu_\nu,\, (g^\mu_\nu) = \mathrm{diag}\,(1,-1,-1,-1)\,; \quad x_\mu = g_{\mu\nu}x^\mu, \tag{2}$$

and summation over the twice repeated indices is implied.

Minimal necessary information on the square-root operator $\widehat{\omega} \equiv \sqrt{\widehat{\overrightarrow{p}}^2 + m^2} = \sqrt{-\Delta + m^2} \geq m > 0$, where $\widehat{\overrightarrow{p}} \equiv (\widehat{p}^j) = -i\nabla$, $\nabla \equiv (\partial_\ell)$, has been considered in [16], see the section 2 in [16].

2. Briefly on our equations for an arbitrary spin

This section represents main information on the three level arbitrary spin model given in [16, 33].

2.1. Level of relativistic canonical quantum mechanics

The **arbitrary spin Schrödinger–Foldy equation of motion** is given by

$$(i\partial_0 - \widehat{\omega})f(x) = 0, \tag{3}$$

where the wave function has the form

$$f \equiv \text{column}(f^1, f^2, ..., f^{\text{N}}), \quad \text{N} = 2s+1, \quad f \in \text{H}^{3,\text{N}}. \tag{4}$$

Note that here the operator $\widehat{\omega} \equiv \sqrt{-\Delta + m^2}$ is the relativistic analog of the energy operator (Hamiltonian) of nonrelativistic quantum mechanics, $\text{H}^{3,\text{N}}$ is the Hilbert space of N-component functions. The step from the particle singlet of arbitrary spin to the corresponding particle-antiparticle doublet is evident.

Thus, for the arbitrary spin particle-antiparticle doublet the system of two N-component equations $(i\partial_0 - \widehat{\omega})f(x) = 0$ and $(i\partial_0 - \widehat{\omega})f(x) = 0$ is used. Therefore, the corresponding Schrödinger–Foldy equation is given by (3), where the 2N-component wave function is the direct sum of the particle and antiparticle wave functions, respectively. Due to the historical tradition of the physicists the antiparticle wave function is put in the down part of the 2N-column.

The **general solution** of the Schrödinger–Foldy equation of motion (3) (in the case of particle-antiparticle arbitrary spin doublet) has the form

$$f(x) = \frac{1}{(2\pi)^{\frac{3}{2}}} \int d^3k e^{-ikx} a^{2\text{N}}\left(\overrightarrow{k}\right) \text{d}_{2\text{N}}, \quad kx \equiv \omega t - \overrightarrow{k}\overrightarrow{x}, \quad \omega \equiv \sqrt{\overrightarrow{k}^2 + m^2}, \tag{5}$$

where the orts of the 2N-dimensional Cartesian basis are given by

$$\text{d}_1 = \begin{vmatrix} 1 \\ 0 \\ 0 \\ \cdot \\ \cdot \\ \cdot \\ 0 \end{vmatrix}, \ \text{d}_2 = \begin{vmatrix} 0 \\ 1 \\ 0 \\ \cdot \\ \cdot \\ \cdot \\ 0 \end{vmatrix}, \ ...., \ \text{d}_{2\text{N}} = \begin{vmatrix} 0 \\ 0 \\ 0 \\ \cdot \\ \cdot \\ \cdot \\ 1 \end{vmatrix}. \tag{6}$$

All necessary information on the relativistic canonical quantum mechanics of an arbitrary spin can be found in section 4 of [16].

2.2. Level of the canonical Foldy–Wouthuysen type field theory of an arbitrary spin

The operator, which transform the relativistic canonical quantum mechanics of the arbitrary spin particle-antiparticle multiplet into the corresponding canonical particle-antiparticle field, is given by

$$v_{2\text{N}} = \begin{vmatrix} \text{I}_\text{N} & 0 \\ 0 & \hat{C}\text{I}_\text{N} \end{vmatrix}, \quad v_{2\text{N}}^{-1} = v_{2\text{N}}^{\dagger} = v_{2\text{N}}, \quad v_{2\text{N}}v_{2\text{N}} = \text{I}_{2\text{N}}, \quad \text{N} = 2s+1, \tag{7}$$

where $\widehat{C}\text{I}_\text{N}$ is the $\text{N} \times \text{N}$ operator of complex conjugation. Indeed, the operator (7) translates any operator from canonical field FW representation into the relativistic canonical quantum mechanics representation and vice versa:

$$v_{2\text{N}}\hat{q}_{\text{cf}}^{\text{anti-Herm}}v_{2\text{N}} = \hat{q}_{\text{qm}}^{\text{anti-Herm}}, \quad v_{2\text{N}}\hat{q}_{\text{qm}}^{\text{anti-Herm}}v_{2\text{N}} = \hat{q}_{\text{cf}}^{\text{anti-Herm}}. \tag{8}$$

IOP Conf. Series: Journal of Physics: Conf. Series **804** (2017) 012040 doi:10.1088/1742-6596/804/1/012040

Here $\hat{q}_{\rm qm}^{\rm anti-Herm}$ is an arbitrary operator from the relativistic canonical quantum mechanics of the 2N-component particle-antiparticle doublet in the anti-Hermitian form, e. g., the operator $(\partial_0 + i\widehat{\omega})$ of equation of motion (3), the operator of spin taken in anti-Hermitian form, etc., $\hat{q}_{\rm cf}^{\rm anti-Herm}$ is an arbitrary operator from the canonical field theory of the 2N-component particle-antiparticle doublet in the anti-Hermitian form. Thus, the only warning is that operators here must be taken in anti-Hermitian form, see section 9 in [32] for the details and see [36, 37] for the mathematical correctness of anti-Hermitian operators application.

Note briefly that on this basis we are able to find the general transformation, which gives the direct relationship between the relativistic canonical quantum mechanics and the Dirac model, see, e. g., the formulae (41)–(45) in [38] and section 6 of [38] in general. In such approach the Dirac Hamiltonian has an anti-Hermitian form. Thus we have suggested extended non-Hermitian Foldy–Wouthuysen transformation and not only the Foldy–Wouthuysen transformation for non-Hermitian Hamiltonians as in [39]. For an arbitrary spin such extended non-Hermitian Foldy–Wouthuysen transformation has been suggested in [32, 33] as well.

Further, the operator (7) translates

$$\phi = v_{\rm 2N} f, \quad f = v_{\rm 2N}\phi, \tag{9}$$

the general solution (5) of the Schrödinger–Foldy equation (3) into the **general solution**

$$\phi(x) = \frac{1}{(2\pi)^{\frac{3}{2}}} \int d^3k \left[e^{-ikx} a^{\rm N}(\overrightarrow{k}) {\rm d}_{\rm N} + e^{ikx} a^{*\check{\rm N}}(\overrightarrow{k}) {\rm d}_{\check{\rm N}} \right], \tag{10}$$

$\mathrm{N} = 1, 2, ..., \mathrm{N}, \quad \check{\mathrm{N}} = \mathrm{N}+1, \mathrm{N}+2, ..., 2\mathrm{N}$, of the **Foldy–Wouthuysen type equation for an arbitrary spin**

$$(i\partial_0 - \Gamma^0_{\rm 2N}\widehat{\omega})\phi(x) = 0, \quad \Gamma^0_{\rm 2N} \equiv \sigma^3_{\rm 2N} = \begin{vmatrix} {\rm I_N} & 0 \\ 0 & -{\rm I_N} \end{vmatrix}, \tag{11}$$

$\widehat{\omega} = \sqrt{-\Delta + m^2}, \quad \mathrm{N} = 2s + 1$, and *vice versa*. Note that 2N-dimensional Cartesian basis is not changed under the transformation (9). Thus, the transformation (7), (8) translates the matrices $\Gamma^0_{\rm 2N}$ and

$$\Gamma^j_{\rm 2N} = \begin{vmatrix} 0 & \Sigma^j_{\rm N} \\ -\Sigma^j_{\rm N} & 0 \end{vmatrix}, \quad j = 1, 2, 3, \tag{12}$$

into the relativistic canonical quantum mechanics representation and *vice versa*

$$\bar{\Gamma}^{\bar{\mu}}_{\rm 2N} = v_{\rm 2N}\Gamma^{\bar{\mu}}_{\rm 2N}v_{\rm 2N}, \quad \Gamma^{\bar{\mu}}_{\rm 2N} = v_{\rm 2N}\bar{\Gamma}^{\bar{\mu}}_{\rm 2N}v_{\rm 2N}. \tag{13}$$

In (12) $\Sigma^j_{\rm N}$ are the N × N Pauli matrices. Matrices $\Gamma^\mu_{\rm 2N}$ (11), (12) together with matrix $\Gamma^4_{\rm 2N} \equiv \Gamma^0_{\rm 2N}\Gamma^1_{\rm 2N}\Gamma^2_{\rm 2N}\Gamma^3_{\rm 2N}$ satisfy the anticommutation relations of the Clifford–Dirac algebra.

All necessary information on the canonical Foldy–Wouthuysen type field theory of an arbitrary spin can be found in subsection 5.1 of [16].

2.3. Level of the locally covariant arbitrary spin particle-antiparticle field

The operator, which transform the canonical (Foldy–Wouthuysen type) model of the arbitrary spin particle-antiparticle field into the corresponding locally covariant particle-antiparticle field, is the generalized FW operator and is given by

$$V^{\mp} = \frac{\mp \overrightarrow{\Gamma}_{\rm 2N} \cdot \overrightarrow{p} + \widehat{\omega} + m}{\sqrt{2\widehat{\omega}(\widehat{\omega} + m)}}, \quad V^- = (V^+)^\dagger, \quad V^-V^+ = V^+V^- = {\rm I_{2N}}, \quad {\rm N} = 2s+1, \tag{14}$$

where Γ^j_{2N} are known from (12) and Σ^j_N are the N × N Pauli matrices.

Note that in formulas (14) and in all formulas before the end of the subsection the values of N are only even. Therefore, the canonical field equation (11) describes the larger number of multiplets then the generalized Dirac equation (15) given below.

The formulas (15), (16) below are found from the corresponding formulas (11) and (10) of canonical field model on the basis of the operator (14).

Thus, for the general form of arbitrary spin locally covariant particle-antiparticle field **the arbitrary spin Dirac-like equation of motion** follows from the equation (11) after the transformation (14) and is given by

$$\left[i\partial_0 - \Gamma^0_{2N}(\overrightarrow{\Gamma}_{2N} \cdot \overrightarrow{p} + m)\right]\psi(x) = 0. \tag{15}$$

The **general solution** has the form

$$\psi(x) = V^- \phi(x) = \frac{1}{(2\pi)^{\frac{3}{2}}}\int d^3k \left[e^{-ikx} a^{\mathrm{N}}(\overrightarrow{k})\mathrm{v}^-_{\mathrm{N}}(\overrightarrow{k}) + e^{ikx} a^{*\check{\mathrm{N}}}(\overrightarrow{k})\mathrm{v}^+_{\check{\mathrm{N}}}(\overrightarrow{k})\right], \tag{16}$$

where amplitudes and notation $\check{\mathrm{N}}$ are the same as in (10); $\left\{\mathrm{v}^-_{\mathrm{N}}(\overrightarrow{k}),\ \mathrm{v}^+_{\check{\mathrm{N}}}(\overrightarrow{k})\right\}$ are 2N-component Dirac basis spinors with properties of orthonormalisation and completeness similar to 4-component Dirac spinors from [40].

The corresponding equations for **the spin s=(3/2,3/2) particle-antiparticle doublet** are the partial cases of the equations (3), (11) and (15). This partial case has been considered in [15, 16]. Below we presented another partial case, i. e. the explicit form of the equations for **the spin s=(2,2) particle-antiparticle doublet**.

3. A brief scheme of the relativistic canonical quantum mechanics of the 10-component spin s=(2,2) particle-antiparticle bosonic doublet

The Schrödinger–Foldy equation is given by

$$(i\partial_0 - \widehat{\omega})f(x) = 0, \quad f \equiv \mathrm{column}(f^1, f^2, f^3, ..., f^{10}). \tag{17}$$

The general solution of the equation (17) has the form

$$f(x) = \left|\begin{array}{c} f_{\mathrm{part}} \\ f_{\mathrm{antipart}} \end{array}\right| = \frac{1}{(2\pi)^{\frac{3}{2}}}\int d^3k e^{-ikx} g^{\mathrm{A}}(\overrightarrow{k})\mathrm{d}_{\mathrm{A}}. \tag{18}$$

The orts of the 10-component Cartesian basis are given by

$$\mathrm{d}_1 = \mathrm{column}(1,0,0,...,0), \quad \mathrm{d}_2 = \mathrm{column}(0,1,0,...,0), \ ...., \quad \mathrm{d}_{10} = \mathrm{column}(0,0,0,...,1). \tag{19}$$

The space of the states is the rigged Hilbert space $\mathrm{S}^{3,10} \subset \mathrm{H}^{3,10} \subset \mathrm{S}^{3,10*}$. The SU(2)-spin generators have the form

$$s^1_{10} = s_{23} = \frac{1}{2}\left|\begin{array}{cccccccccc}
0 & 2 & 0 & 0 & 0 & 0 & 0 & 0 & 0 & 0 \\
2 & 0 & \sqrt{6} & 0 & 0 & 0 & 0 & 0 & 0 & 0 \\
0 & \sqrt{6} & 0 & \sqrt{6} & 0 & 0 & 0 & 0 & 0 & 0 \\
0 & 0 & \sqrt{6} & 0 & 2 & 0 & 0 & 0 & 0 & 0 \\
0 & 0 & 0 & 2 & 0 & 0 & 0 & 0 & 0 & 0 \\
0 & 0 & 0 & 0 & 0 & 0 & -2 & 0 & 0 & 0 \\
0 & 0 & 0 & 0 & 0 & -2 & 0 & -\sqrt{6} & 0 & 0 \\
0 & 0 & 0 & 0 & 0 & 0 & -\sqrt{6} & 0 & -\sqrt{6} & 0 \\
0 & 0 & 0 & 0 & 0 & 0 & 0 & -\sqrt{6} & 0 & -2 \\
0 & 0 & 0 & 0 & 0 & 0 & 0 & 0 & -2 & 0
\end{array}\right|,$$

$$s_{10}^2 = s_{31} = \frac{i}{2}\begin{vmatrix} 0 & -2 & 0 & 0 & 0 & 0 & 0 & 0 & 0 & 0 \\ 2 & 0 & -\sqrt{6} & 0 & 0 & 0 & 0 & 0 & 0 & 0 \\ 0 & \sqrt{6} & 0 & -\sqrt{6} & 0 & 0 & 0 & 0 & 0 & 0 \\ 0 & 0 & \sqrt{6} & 0 & -2 & 0 & 0 & 0 & 0 & 0 \\ 0 & 0 & 0 & 2 & 0 & 0 & 0 & 0 & 0 & 0 \\ 0 & 0 & 0 & 0 & 0 & 0 & -2 & 0 & 0 & 0 \\ 0 & 0 & 0 & 0 & 0 & 2 & 0 & -\sqrt{6} & 0 & 0 \\ 0 & 0 & 0 & 0 & 0 & 0 & \sqrt{6} & 0 & -\sqrt{6} & 0 \\ 0 & 0 & 0 & 0 & 0 & 0 & 0 & \sqrt{6} & 0 & -2 \\ 0 & 0 & 0 & 0 & 0 & 0 & 0 & 0 & 2 & 0 \end{vmatrix}, \quad (20)$$

$$s_{10}^3 = s_{12} = \begin{vmatrix} 2 & 0 & 0 & 0 & 0 & 0 & 0 & 0 & 0 & 0 \\ 0 & 1 & 0 & 0 & 0 & 0 & 0 & 0 & 0 & 0 \\ 0 & 0 & 0 & 0 & 0 & 0 & 0 & 0 & 0 & 0 \\ 0 & 0 & 0 & -1 & 0 & 0 & 0 & 0 & 0 & 0 \\ 0 & 0 & 0 & 0 & -2 & 0 & 0 & 0 & 0 & 0 \\ 0 & 0 & 0 & 0 & 0 & -2 & 0 & 0 & 0 & 0 \\ 0 & 0 & 0 & 0 & 0 & 0 & -1 & 0 & 0 & 0 \\ 0 & 0 & 0 & 0 & 0 & 0 & 0 & 0 & 0 & 0 \\ 0 & 0 & 0 & 0 & 0 & 0 & 0 & 0 & 1 & 0 \\ 0 & 0 & 0 & 0 & 0 & 0 & 0 & 0 & 0 & 2 \end{vmatrix},$$

and satisfy the commutation relations $\left[s^j, s^\ell\right] = i\varepsilon^{j\ell n}s^n$ of SU(2) algebra generators. The Casimir operator of the SU(2)-algebra representation is given by $\overrightarrow{s}^2 = 6\mathrm{I}_{10} = 2\,(2+1)\,\mathrm{I}_{10}$. The stationary complete set of operators is as follows: $\overrightarrow{p} \equiv (\widehat{p}^j) = -i\partial_j, \quad s_{10}^3 \in (20)$. The equations on the momentum operator eigenvalues have the form

$$\overrightarrow{p} e^{-ikx}\mathrm{d}_\mathrm{A} = \overrightarrow{k} e^{-ikx}\mathrm{d}_\mathrm{A}, \quad \mathrm{A} = \overline{1, 10}, \quad (21)$$

and the equations on the spin projection operator s_{10}^3 (20) eigenvalues are given by

$$s_{10}^3\mathrm{d}_1 = 2\mathrm{d}_1,\ s_{10}^3\mathrm{d}_2 = \mathrm{d}_2,\ s_{10}^3\mathrm{d}_3 = 0,\ s_{10}^3\mathrm{d}_4 = -\mathrm{d}_4,\ s_{10}^3\mathrm{d}_5 = -2\mathrm{d}_5, \quad (22)$$

$$s_{10}^3\mathrm{d}_6 = -2\mathrm{d}_6,\ s_{10}^3\mathrm{d}_7 = -\mathrm{d}_7,\ s_{10}^3\mathrm{d}_8 = 0,\ s_{10}^3\mathrm{d}_9 = \mathrm{d}_9,\ s_{10}^3\mathrm{d}_{10} = 2\mathrm{d}_{10}.$$

The interpretation of the amplitudes in the general solution (18) follows from the equations (21), (22). The functions $g^1(\overrightarrow{k}), g^2(\overrightarrow{k}), g^3(\overrightarrow{k}), g^4(\overrightarrow{k}), g^5(\overrightarrow{k})$ in the solution (18) are the momentum-spin quantum-mechanical amplitudes of the particle (boson) with the momentum $\overrightarrow{p}$ and spin projection eigenvalues (+2, +1, 0, -1, -2), respectively, and the functions $g^6(\overrightarrow{k}), g^7(\overrightarrow{k}), g^8(\overrightarrow{k}), g^9(\overrightarrow{k}), g^{10}(\overrightarrow{k})$ are the momentum-spin quantum-mechanical amplitudes of the antiparticle with the momentum $\overrightarrow{p}$ and spin projection eigenvalues (-2, -1, 0, +1, +2), respectively. The generators of the reducible unitary bosonic representation of the Poincaré group $\mathcal{P}$ are given by

$$\widehat{p}_0 = \widehat{\omega} \equiv \sqrt{-\Delta + m^2}, \quad \widehat{p}_\ell = i\partial_\ell, \quad \widehat{j}_{\ell n} = x_\ell\widehat{p}_n - x_n\widehat{p}_\ell + s_{ln} \equiv \widehat{m}_{\ell n} + s_{\ell n}, \quad (23)$$

$$\widehat{j}_{0\ell} = -\widehat{j}_{\ell 0} = t\widehat{p}_\ell - \frac{1}{2}\{x_\ell, \widehat{\omega}\} - \left(\frac{s_{\ell n}\widehat{p}_n}{\widehat{\omega} + m} \equiv \breve{s}_\ell\right), \quad (24)$$

whereas the spin s=(2,2) SU(2) generators $\overrightarrow{s} = (s^{\ell n})$ have the form (20). The Schrödinger–Foldy equation (17) (and the set {f} of its solutions (18)) is invariant with respect to the reducible unitary fermionic representation of the Poincaré group $\mathcal{P}$, whose Hermitian 10×10 matrix-differential nonlocal generators are given in (23), (24). The corresponding Casimir operators

$$p^2 = \widehat{p}^\mu\widehat{p}_\mu = m^2\mathrm{I}_{10}, \quad W = w^\mu w_\mu = m^2\overrightarrow{s}_{10}^2 = 2\,(2+1)\,m^2\mathrm{I}_{10}. \quad (25)$$

4. Canonical Foldy–Wouthuysen type field model for the 10-component spin s=(2,2) particle-antiparticle doublet

The canonical Foldy–Wouthuysen type field equation and its solution are given by

$$(i\partial_0 - \Gamma^0_{10}\widehat{\omega})\phi(x) = 0, \quad \Gamma^0_{10} = \begin{vmatrix} \mathrm{I}_5 & 0 \\ 0 & -\mathrm{I}_5 \end{vmatrix}, \quad \widehat{\omega} \equiv \sqrt{-\Delta + m^2}, \tag{26}$$

$$\phi(x) = \frac{1}{(2\pi)^{\frac{3}{2}}} \int d^3k \left[e^{-ikx} g^{\mathrm{A}}(\overrightarrow{k})\mathrm{d}_{\mathrm{A}} + e^{ikx} g^{*\mathrm{B}}(\overrightarrow{k})\mathrm{d}_{\mathrm{B}} \right], \tag{27}$$

$\mathrm{A} = 1,2,3,4,5,\quad \mathrm{B} = 6,7,8,9,10$, the Cartesian orts are known from (19). The link with the relativistic canonical quantum mechanics is given by the operator

$$v_{10} = \begin{vmatrix} \mathrm{I}_5 & 0 \\ 0 & C\mathrm{I}_5 \end{vmatrix}, \quad v_{10}^{-1} = v_{10}^{\dagger} = v_{10},\ v_{10}v_{10} = \mathrm{I}_{10}. \tag{28}$$

The spin operators of the canonical field theory found from the quantum-mechanical SU(2) spin (20) on the basis of the transformation (28) satisfy the commutation relations $\left[s^j, s^\ell\right] = i\varepsilon^{j\ell n}s^n$ of SU(2) algebra generators and have the following form

$$\overrightarrow{s}_{10} = \begin{vmatrix} \overrightarrow{s}_5 & 0 \\ 0 & \overrightarrow{s}_5 \end{vmatrix}, \tag{29}$$

where the spin s=2 5×5 SU(2) generators $\overrightarrow{s}_5$ are given by the up 5×5 diagonal blocks in (20). The corresponding Casimir operator is given by $\overrightarrow{s}^{\,2}_{10} = 6\mathrm{I}_{10} = 2(2+1)\mathrm{I}_{10}$, where I_{10} is the 10×10- unit matrix. The stationary complete set of operators is given by the operators $\overrightarrow{p}$, $s^3_{10} = s_z$ of the momentum and spin projection. The equations on the eigenvectors and eigenvalues of the operators $\overrightarrow{p}$ and $s^3_{10} = s_z$ have the form

$$\begin{gathered} \overrightarrow{p}e^{-ikx}\mathrm{d}_{\mathrm{A}} = \overrightarrow{k}e^{-ikx}\mathrm{d}_{\mathrm{A}}, \quad \mathrm{A} = \overline{1,5}, \quad \overrightarrow{p}e^{ikx}\mathrm{d}_{\mathrm{B}} = -\overrightarrow{k}e^{ikx}\mathrm{d}_{\mathrm{B}}, \quad \mathrm{B} = \overline{6,10}, \\ s^3_{10}\mathrm{d}_1 = 2\mathrm{d}_1,\ s^3_{10}\mathrm{d}_2 = \mathrm{d}_2,\ s^3_{10}\mathrm{d}_3 = 0,\ s^3_{10}\mathrm{d}_4 = -\mathrm{d}_4,\ s^3_{10}\mathrm{d}_5 = -2\mathrm{d}_5, \\ s^3_{10}\mathrm{d}_6 = 2\mathrm{d}_6,\ s^3_{10}\mathrm{d}_7 = \mathrm{d}_7,\ s^3_{10}\mathrm{d}_8 = 0,\ s^3_{10}\mathrm{d}_9 = -\mathrm{d}_9,\ s^3_{10}\mathrm{d}_{10} = -2\mathrm{d}_{10}, \end{gathered} \tag{30}$$

and determine the interpretation of the amplitudes in the general solution (27). Note that the direct quantum-mechanical interpretation of the amplitudes $g^{\mathrm{A}}(\overrightarrow{k})$, $g^{\mathrm{B}}(\overrightarrow{k})$ in solution (27) should be taken from the quantum-mechanical equations (21), (22) and is given in section 3. The canonical field equation (26) and the set $\{\phi\}$ of its solutions (27) are invariant with respect to the reducible unitary bosonic representation of the Poincaré group $\mathcal{P}$, whose Hermitian 10×10 matrix-differential nonlocal generators are given by

$$\widehat{p}^0 = \Gamma^0_{10}\widehat{\omega} \equiv \Gamma^0_{10}\sqrt{-\Delta + m^2}, \quad \widehat{p}^\ell = -i\partial_\ell, \quad \widehat{j}^{\ell n} = x^\ell\widehat{p}^n - x^n\widehat{p}^\ell + s^{\ell n}_{10} \equiv \widehat{m}^{\ell n} + s^{\ell n}_{10}, \tag{31}$$

$$\widehat{j}^{0\ell} = -\widehat{j}^{\ell 0} = x^0\widehat{p}^\ell - \frac{1}{2}\Gamma^0_{10}\left\{x^\ell, \widehat{\omega}\right\} + \Gamma^0_{10}\frac{(\overrightarrow{s}_{10} \times \overrightarrow{p})^\ell}{\widehat{\omega} + m}, \tag{32}$$

where the spin s=(2,2) SU(2) generators $\overrightarrow{s}_{10} = (s^{\ell n}_{10})$ have the form (29). It is easy to prove by the direct verification that generators (31), (32) commute with the operator $(i\partial_0 - \Gamma^0_6\widehat{\omega})$ of the canonical field equation (26) and satisfy the commutation relations of the Lie algebra of the Poincaré group $\mathcal{P}$. The corresponding Casimir operators are given by

$$p^2 = \widehat{p}^\mu\widehat{p}_\mu = m^2\mathrm{I}_{10}, \quad W = w^\mu w_\mu = m^2\overrightarrow{s}^{\,2}_{10} = 2\,(2+1)\,m^2\mathrm{I}_{10}, \tag{33}$$

where I_{10} is the 10×10 unit matrix.

5. Locally covariant field equations for spin s=(2,2) particle-antiparticle doublet

The only small difficulty is as follows. The covariant field theory equation for the spin s=(2,2) particle-antiparticle doublet can not be presented in terms of Γ matrices. Indeed, the 10×10 Clifford–Dirac Γ matrices are not exist as an mathematical objects. Therefore, one need to look for the locally covariant field equations for spin s=(2,2) particle-antiparticle doublet not in the terms of Γ matrices.

Consider two ways of the derivation of the covariant field equations for spin s=(2,2) particle-antiparticle doublet. The first way is based on the application of the corresponding Foldy–Wouthuysen transformation directly to the canonical field equation (26). Note that there is a nonzero chance to find the appropriate transformation among a big number of so-called Foldy–Wouthuysen transformations for an arbitrary spin [41–47]. Unfortunately for our approach, many authors considered for the spin s=1 case the 4-vector potential (see, e. g., [47]) and not 3-component wave function $\overrightarrow{\psi} = \overrightarrow{E} - i\overrightarrow{E}$ as it is in [15, 16] and [32, 33]. Therefore, the objects for spin s=2 case are different as well. Indeed, our approach for arbitrary spin equations differs from other known approaches.

Hence, below another way of the derivation of the covariant field equations for spin s=(2,2) particle-antiparticle doublet is realized. It is easy to construct the formalism of the relativistic canonical quantum mechanics of the spin s=(2,0,2,0) particle-antiparticle multiplet. The corresponding procedure is similar to the above given in section 3. Further, the canonical Foldy–Wouthuysen type model of the spin s=(2,0,2,0) particle-antiparticle field can be formulated. It is similar to the given in section 4 consideration. Moreover, both models are presented already in sections 16 and 26 of [32].'After that the 12×12 operator transformation (14) should be applied for the canonical Foldy–Wouthuysen type equation of the spin s=(2,0,2,0) particle-antiparticle multiplet.

Found by the above considered way (or found as the partial 12×12 case of arbitrary spin equation (15)) covariant field equation for the spin s=(2,0,2,0) particle-antiparticle multiplet is given by

$$\left[i\partial_0 - \Gamma^0_{12}(\overrightarrow{\Gamma}_{12} \cdot \overrightarrow{p} + m)\right]\psi(x) = 0, \quad \Gamma^0_{12} = \begin{vmatrix} I_6 & 0 \\ 0 & -I_6 \end{vmatrix}, \quad \Gamma^j_{12} = \begin{vmatrix} 0 & \Sigma^j_6 \\ -\Sigma^j_6 & 0 \end{vmatrix}, \tag{34}$$

where Σ^j_6 are the 6×6 Pauli matrices $\Sigma^1_6 = \begin{vmatrix} 0 & I_3 \\ I_3 & 0 \end{vmatrix}$, $\Sigma^2_6 = \begin{vmatrix} 0 & -iI_3 \\ iI_3 & 0 \end{vmatrix}$, $\Sigma^3_6 = \begin{vmatrix} I_3 & 0 \\ 0 & -I_3 \end{vmatrix}$.
The general solution of the equation (34) is given by the formula (327) in [32].

In order to find the covariant field equations for spin s=(2,2) particle-antiparticle doublet one can put equal to zero the scalar components in (34). The place of the scalar components is evident from the corresponding quantum-mechanical consideration (section 16 of [32]).

The similar way was demonstrated in [32] (section 23) to find the covariant Maxwell-like field equations for the spin s=(1,1) particle-antiparticle doublet from the known Dirac-like equation (252) in [32] for the spin s=(1,0,1,0) particle-antiparticle multiplet. Similarly to equation (275) in [32] the covariant field equation (35) (or (36)) for the spin s=(2,2) particle-antiparticle doublet can be presented in terms of **curl**, **div** and **grad** operators as well.

The locally covariant field equations for 10 component spin s=(2,2) particle-antiparticle doublet found by this way has the form

$$i\partial_0\psi^1 - p^3\psi^6 - m\psi^1 = 0,$$

$$i\partial_0\psi^2 - p^3\psi^7 + ip^2\psi^{10} - p^1\psi^{10} - m\psi^2 = 0,$$
$$i\partial_0\psi^3 - p^3\psi^8 + ip^2\psi^9 - p^1\psi^9 - m\psi^3 = 0,$$
$$i\partial_0\psi^4 + p^3\psi^9 - ip^2\psi^8 - p^1\psi^8 - m\psi^4 = 0,$$

$$
\begin{gathered}
i\partial_0\psi^5 + p^3\psi^{10} - ip^2\psi^7 - p^1\psi^7 - m\psi^5 = 0,\\
i\partial_0\psi^6 - p^3\psi^1 + m\psi^6 = 0,\\
i\partial_0\psi^7 - p^3\psi^2 + ip^2\psi^5 - p^1\psi^5 + m\psi^7 = 0,\\
i\partial_0\psi^8 - p^3\psi^3 + ip^2\psi^4 - p^1\psi^4 + m\psi^8 = 0,\\
i\partial_0\psi^9 + p^3\psi^4 - ip^2\psi^3 - p^1\psi^3 + m\psi^9 = 0,\\
i\partial_0\psi^{10} + p^3\psi^5 - ip^2\psi^2 - p^1\psi^2 + m\psi^{10} = 0,\\
-ip^2\psi^1 - p^1\psi^1 = 0, \quad -ip^2\psi^6 - p^1\psi^6 = 0,
\end{gathered} \tag{35}
$$

or the Dirac-like form

$$
\left[i\partial_0 - \Gamma_8^0(\overrightarrow{\Gamma}_8 \cdot \overrightarrow{p} + m)\right]\psi(x) = 0, \quad (i\partial_0 - \sigma^1 p^3 - \sigma^3 m)\chi = 0, \quad (p^1 + ip^2)\chi = 0, \tag{36}
$$

where $\psi = \text{column}(\psi^1, \psi^2, ..., \psi^8)$, $\chi = \begin{vmatrix} \psi^9 \\ \psi^{10} \end{vmatrix}$, $\Gamma_8^0 = \begin{vmatrix} \mathrm{I}_4 & 0 \\ 0 & -\mathrm{I}_4 \end{vmatrix}$, $\Gamma_8^j = \begin{vmatrix} 0 & \Sigma_4^j \\ -\Sigma_4^j & 0 \end{vmatrix}$, and Σ_4^j are the 4×4 Pauli matrices $\Sigma_4^1 = \begin{vmatrix} 0 & \mathrm{I}_2 \\ \mathrm{I}_2 & 0 \end{vmatrix}$, $\Sigma_4^2 = \begin{vmatrix} 0 & -i\mathrm{I}_2 \\ i\mathrm{I}_2 & 0 \end{vmatrix}$, $\Sigma_4^3 = \begin{vmatrix} \mathrm{I}_2 & 0 \\ 0 & -\mathrm{I}_2 \end{vmatrix}$. The form similar to (275) of [32] is found by the method similar to one presented in section 23 of [32].

6. Brief conclusions

It is demonstrated that our approach for the construction of the arbitrary spin field equations is different in comparison with other known approaches. The explicit comparison of our equation for the particles with arbitrary spin with the Bhabha and Bargman–Wigner equations is fulfilled.

The Dirac-like locally covariant field equations for 10 component spin s=(2,2) particle-antiparticle doublet is introduced. The found equation does not contain the redundant components. Moreover, the equation (35) (or (36)) for the 10 component spin s=(2,2) particle-antiparticle doublet can be interpreted as the equation for the classical gravitational field. Furthermore, this equation can be written in the Maxwell-like form. Massless case can be achieved very easy by putting to zero the mass term. Nevertheless, the massive case may have some special meaning as well. Additional forces (seven components) to the Newton force (three components) can be useful for the explanation of the tiny gravitational effects both in solar system and in the universe in general.

Important result is in further approbation of the effective method of deriving of arbitrary spin field equations. The three level table (relativistic canonical quantum mechanics, canonical Foldy–Wouthuysen type field, manifestly covariant field) of arbitrary spin S=(2s+1.2s+1) particle-antiparticle doublets is constructed. Thus, the equation for any spin can be written down very easy. For example, the three level model (relativistic canonical quantum mechanics, canonical field, covariant field), e. g., of the spin s=(5/2,5/2) particle-antiparticle doublet is formulated similarly to the given in [15, 16] three level spin s=(3/2,3/2) model. In the approach under consideration the particle with spin s=5/2 is the same theoretical reality as the particle with spin s=1/2, or s=3/2 (the particle with spin s=2 is the same theoretical reality as the particle with spin s=1). Hence, we can suggest the hypothesis that the particle with spin s=5/2 can exists in reality. Note that the spin s=5/2 case is considered in the article [14] as well.'The time of life of such new particle is smaller then the time of life of spin s=3/2 hyperons.

References

[1] Dirac P A M 1936 *Proc. R. Soc. Lond.* A **155** 447
[2] Bhabha H J 1945 *Rev. Mod. Phys.* **17** 200
[3] Bargman V and Wigner E P 1948 *Proc. Nat. Acad. Sci. USA.* **34** 211
[4] Foldy L L 1956 *Phys. Rev.* **102** 568
[5] Weaver D L, Hammer C L and Good R H Jr. 1964 *Phys. Rev.* **135** B241
[6] Pursey D L 1964 *Nucl. Phys.* **53** 174
[7] Pursey D L 1965 *Ann. Phys. (N.Y.)* **32** 157
[8] Mathews P M 1966 *Phys. Rev.* **143** 978
[9] Nelson T J and Good R H Jr. 1968 *Rev. Mod. Phys.* **40** 508
[10] Guertin R F 1974 *Ann. Phys. (N.Y.)* **88** 504
[11] Kraicik R A and Nieto M M 1977 *Phys. Rev.* D **15** 433
[12] Loide R-K, Ots I and Saar R 1997 *J. Phys.* A **30** 4005
[13] Gitman D M and Shelepin A L 2001 *Int. J. Theor. Phys.* **40** 603
[14] Shi-Zhong H et al 2003 *Eur. Phys. J.* C **26** 609
[15] Simulik V M 2015 *Ukr. J. Phys.* **60** 985
[16] Simulik V M 2016 *J. Phys. Conf. Ser.* **670** 012047
[17] Fierz M and Pauli W 1939 *Proc. R. Soc. Lond.* A **173** 211
[18] Rarita W and Schwinger J 1941 *Phys. Rev.* **60** 61
[19] Davydov A S 1943 *J. Exper. Theor. Phys.* **13** 313 (in Russian)
[20] Johnson K and Sudarshan E C G 1961 *Ann. Phys. (N.Y.)* **13** 126
[21] Velo G and Zwanziger D 1969 *Phys. Rev.* **186** 1337
[22] Velo G and Zwanziger D 1969 *Phys. Rev.* **188** 2218
[23] Hagen C R 1971 *Phys. Rev.* D **4** 2204
[24] Capri A Z and Kobes R L 1980 *Phys. Rev.* D **22** 1967
[25] Darkhosh T 1985 *Phys. Rev.* D **32** 3251
[26] Cox W 1989 *J. Phys.* A **22** 1599
[27] Pilling T 2005 *Int. J. Mod. Phys.* A **20** 2715
[28] Napsuciale M, Kirchbach M and Rodriguez S 2006 *Eur. Phys. J.* A **29** 289
[29] Kaloshin A E and Lomov V P 2011 *Particles and Nuclei, Letters* **8** 517
[30] Red'kov V M, Kisel V V and Ovsiyuk E M 2011 Particle with spin 2 and anomalous magnetic moment in external electromagnetic and gravitational fields *arXiv:* 1109.1715v1 [math-ph]
[31] Red'kov V M 2011 Particle with spin S=3/2 in Riemannian space-time *arXiv:* 1109.3871v1 [math-ph]
[32] Simulik V M 2014 Covariant local field theory equations following from the relativistic canonical quantum mechanics of arbitrary spin *arXiv:* 1409.2766v2 [quant-ph, hep-th]
[33] Simulik V M 2015 General form of the covariant field equations of arbitrary spin and the relativistic canonical quantum mechanics *arXiv:* 1509.04630 [quant-ph, hep-th]
[34] Bhabha H J 1951 *Proc. Indian Acad. Sci.* A **34** 335
[35] Foldy L L and Wouthuysen S A 1950 *Phys. Rev.* **78** 29
[36] Elliott J and Dawber P 1979 *Symmetry in physics, vol 1* (London: Macmillian Press)
[37] Wybourne B 1974 *Classical groups for Physicists* (New York: John Wiley and sons)
[38] Simulik V M and Krivsky I Y 2013 *Ukr. J. Phys.* **58** 1192
[39] Alexandre J and Bender C M 2015 *J. Phys.* A **48** 185403
[40] Bogoliubov N N and Shirkov D V 1980 *Introduction to the theory of quantized fields* (New York: John Wiley and Sons)
[41] Garrido L M and Oliver L 1967 *Nuov. Cim.* A **LII** 588
[42] Ramanujam G A 1974 *Nuov. Cim.* A **20** 27
[43] Nagpal A K and Kaushal R S 1974 *Lett. Nuov. Cim.* **9** 391
[44] Tekumalla A R, Santhanam T S 1974 *Lett. Nuov. Cim.* **10** 737
[45] Jayaraman J 1975 *J. Phys.* A **8** L1
[46] Leon J, Quiros M and Mittelbrunn J R 1979 *J. Math. Phys.* **20** 1068
[47] Silenko A J 2016 *Phys. Rev.* A **94** 032104

Ordinary matter, dark matter, and dark energy on normal Zeeman space-times

Zoltán Imre Szabó

Graduate Center and Lehman College of the City University of New York, 365 5th Ave, New York, NY 10016, USA and Alfréd Rényi Institute of Mathematics, 13-15 Reáltanoda utca, 1053 Budapest, Hungary

E-mail: `zoltan.szabo@lehman.cuny.edu`

Abstract. Zeeman space-times are new, relativistic, and operator based Hamiltonian models representing multi-particle systems. They are established on Lorentzian pseudo Riemannian manifolds whose Laplacian immediately appears in the form of original quantum physical wave operators. In classical quantum theory they emerge, differently, from the Hamilton formalism and the correspondence principle. Nonetheless, this new model does not just reiterate the well known conceptions but holds the key to solving open problems of quantum theory. Most remarkably, it represents the dark matter, dark energy, and ordinary matter by the same ratios how they show up in experiments. Another remarkable agreement with reality is that the ordinary matter appears to be non-expanding and is described in consent with observations. The theory also explains gravitation, moreover, the Hamilton operators of all energy and matter formations, together with their physical properties, are solely derived from the Laplacian of the Zeeman space-time. By this reason, it is called Monistic Wave Laplacian which symbolizes an all-comprehensive unification of all matter and energy formations. This paper only outlines the normal case where the particles do not have proper spin but just angular momentum. The complete anomalous theory is detailed in [Sz2, Sz3, Sz4, Sz5, Sz6, Sz7].

PACS numbers: 12.10.-g

Keywords: Operator based Hamiltonian models, dark matter, dark energy, ordinary matter, all-comprehensive unification.

1. Introduction

Dirac adamantly believed that the great challenges of particle physics should have been attacked by quantum wave operators which were explorable by the Hamilton formalism and the correspondence principle. Nonetheless, almost all modern theories such as the Standard Model or String Theory have been established by Lagrangian methods, where, as opposed to the waves, the particles are conceived as tiny billiard balls moving according to Yang-Mills equations established on suitable $SU(n)$ models by the principle of least action.

The Lagrangian method has become the major tool ever since Schwinger recognized the importance of gauge invariance in solving the infinity puzzle in QED and proposed to apply this principle to describe the weak interactions, as well. Dirac, however, most impatiently urged to return back to the complete Hamiltonian formalism and describe the quantum world in terms of natural Hamilton operators

IOP Conf. Series: Journal of Physics: Conf. Series **804** (2017) 012041 doi:10.1088/1742-6596/804/1/012041

and the associated matter waves. He himself implemented this tool in a circuitous way, which started out with a relativistic Lagrange function and the desired Hamilton operator was found after associating Hamilton functions to the Lagrangian considered in the first place [D]. By doing so, he wanted to see quantum theory as a new discipline but which is emerging in strong relations with the classical theories. Since this idea was not particularly helpful in solving the infinity puzzle, it has largely been ignored, since the mid 1940's.

The Zeeman manifold theory is a new operator based Hamiltonian model of a completely unified quantum theory which breaks away not just from the traditional Lagrangian but Dirac's circuitous Hamilton formalism as well. Namely, the desired Hamilton and the corresponding wave operators appear in the very first place as Laplace operators on particular, so called Zeeman manifolds and the corresponding Zeeman space-times. The main Lagrangian objects are introduced only in the second place which also include a natural compound scalar boson associated with the Hamilton and wave operators introduced in the first place. The wave operators are Laplacians on particular space-time models consistent with Einstein's general relativity. In order to distinguish this from the Hamilton formalism, it is called a "direct Einstein-Hamiltonian method" which could also be called "reversed Hamiltonian method" or operator based Hamiltonian model.

A clincher for this model passes the physical reality test is that all Hamilton operators established by Dirac's Hamilton formalism - which mostly arise from electromagnetic field theory - appear on Zeeman manifolds in the same classical form and one obtains relevant Hamilton operators even for weak, strong, and gravitational interactions, which have been explored, sofar, in SM and String Theory, just by Lagrangian methods. Although there are differences between them, the Zeeman manifold and space-times theories do not contradict SM but rather describes the electromagnetic, weak, and strong interactions in consent with this classical theory which has constantly been tested by experiments.

The metric on Zeeman manifolds - where the Hamilton operators are defined as Laplace operators - is still positive definite (Riemannian) and the relativistic Zeeman space-time having indefinite metric of Lorentz signature is introduced by appropriate extensions of Zeeman manifolds into the time direction. Then the indefinite (hyperbolic) Laplace operator appears as a natural relativistic quantum wave operator there. The quantum phenomena are described in terms of eigenwaves of this wave Laplacian. But it must be emphasized that the Zeeman space-time is only a stage; the true carrier of quantum physical contents are the waves and the probabilistic densities defined there.

The theory also implies a new cosmological model, expanding at an exponentially accelerating rate. It breaks away from the Big Bang theory in the sense that it describes the evolution of the universe as a dynamical "Eternal Whizz", started out infinitely long time ago and lasting forever. However, this model does not contradicts the Big Bang scenario. Namely, it is the Zeeman space-time, the stage of the quantum events, which has been existing forever. Even this stage is dynamical, going through several natural periods in which the quantum phenomena are described by the type of eigenfunctions characteristic for the period. There is also described a time-point, separating two periods, which could very well be the same as the time of the Big Bang. The reality of this model will be demonstrated by pointing out that the ordinary matter, dark matter, and dark energy appear according to the same ratios how they show up in Nature.

The prototype Zeeman manifolds are introduced on 2-step nilpotent Lie groups whose static and solvable extensions are the prototype Zeeman space-times. The metric on a two-step nilpotent Lie group is still Riemannian – namely, the positive definite left-invariant metric naturally present there – and the relativistic metric of Lorentz signature appears on the static and solvable extensions, respectively. The latter gives rise to the accelerating model where the exponentially accelerating expansion can be observed relative to the Euclidean metrics induced on the horospheres, defined as level sets by fixed time-values, such that the length of the horocycles' segments changes according to an exponentially accelerating expansion.

The main idea for realizing the unification is that the Hamilton and wave operators of all fundamental forces, as well as all matter-energy formations, arise from this single Laplace operator. More precisely, the wave functions associated with distinct matter-formations form distinct invariant subspaces and the corresponding Hamilton operators emerge as restrictions of the same Laplacian onto these invariant subspaces. The Zeeman Hamilton operator of charged point-particles orbiting in constant magnetic field emerges on torus bundles, defined by factoring the center of two step nilpotent Lie algebra by a lattice considered in the Z-space. The strong, weak, and gravitational Hamilton operators are associated with extended particles, whose wave functions are defined not on torus but Z-ball-bundles, where the balls are considered in the center.

Since this Laplacian gives rise to all Hamiltonians of particle theory, it is called Monistic Hamiltonian. The Laplacian on the Zeeman space-time is called Monistic Wave Operator. They work together with the Monistic Scalar Boson, comprising all bosons needed to carry out the actions of the Monistic Operator. The constituent bosons are the carriers of the corresponding fundamental forces. Their appearance as components of the Monistic Boson symbolizes the unification of fundamental forces.

This paper is focusing on the Hamilton and wave operators defined for the normal setting, where the particles can only have angular momentum. This is in contrast with those having proper spin where the corresponding operators act not on functions but spinors. The anomalous operators together with the Monistic Boson are outlined in [Sz2]. The full mathematical foundation of the theory is evolved in [Sz3]-[Sz7].

The paper consists of two major sections. The first one describes the Hamilton operators on Zeeman manifolds, including the explanation as to how does Schrödinger's classical Hamilton operator emerge from the Laplacian acting on functions periodic regarding a Z-lattice. These are the waves of point particles. By contrast, the waves of extended particles are defined on Z-ball bundles. The Monistic Hamilton Operator restricted onto this invariant function space defines the Hamiltonians by which the strong, weak, and gravitational interactions can be described. The second section is devoted for describing the expanding Zeeman space-time. The most important problem to be solved is to find appropriate decomposition of the Laplacian which respectively correspond to the wave operators of dark matter dark energy and ordinary matter. The investigations are carried out in relation to a new spectral mass assignment procedure which is analogous to the Higgs mechanism. One of its distinguishing feature is that it is not based on symmetry breaking but is purely established by the Monistic Wave Operator. Nonetheless, it relates to symmetry breaking.

IOP Conf. Series: Journal of Physics: Conf. Series **804** (2017) 012041 doi:10.1088/1742-6596/804/1/012041

2. Hamilton operators on Zeeman manifolds.

Zeeman space-times are established by relativistic extensions of Zeeman manifolds into the time direction. The latters are still endowed with positive definite Riemannian metrics whose Laplace-Beltrami operators serve as Monistic Hamilton Operators from which the Hamilton operators of the represented particle systems are derived. Because of the positiveness of the energy levels, the Hamilton operators must be elliptic, this is why the metric on a Zeeman manifold should be positive definite. The hyperbolic wave operator, which also includes the Monistic Hamilton Operator, will appear as Laplace operator on the Lorentzian Zeeman space-times, obtained by extension.

2.1. Mathematical construction of Zeeman manifolds.

In order to introduce the theory in a relatively easy way, this paper focuses mostly on prototype Zeeman manifolds, which, however, do not oversimplify the presentation. Quite to the contrary, these particular examples form a rather large subclass of the general category exhibiting all important features there. They are defined on two step nilpotent Lie groups, particularly on Heisenberg type groups endowed with natural left-invariant metrics. The prototype space-time indicated above will be introduced by the static resp. solvable (expanding) extensions of two step nilpotent Lie groups. The rudiments about H-type groups are as follows [K].

The Lie algebra of these groups are defined on the Cartesian product $\mathcal{X} \times \mathcal{Z} = \mathbb{R}^k \times \mathbb{R}^l$ of Euclidean spaces, where the X-space $\mathcal{X}$ is of even dimensional, by a one-to-one linear map $J : \mathcal{Z} \to \mathrm{SkewEnd}(\mathcal{X})$ associating with any Z-vector $Z \in \mathcal{Z}$ a skew endomorphisms, J_Z, acting on the X-space. Then, for any two vectors $X, Y \in \mathcal{X}$, the Lie bracket $[X, Y] \in \mathcal{Z}$ is defined by $\langle [X, Y], Z\rangle = \langle J_Z(X), Y\rangle$, for any $Z \in \mathcal{Z}$. This, together with $[\mathcal{X} \oplus \mathcal{Z}, \mathcal{Z}] = 0$, completely determines a two step nilpotent Lie algebra. The particular H-type Lie algebras are introduced by the Clifford condition $J_Z^2 = -|Z|^2\mathrm{Id}$ which identifies them with Clifford modules whose well known classification also gives rise to classifying all possible H-type algebras.

The Lie group, N, determined by this Lie algebra is a Hadamard manifold, also defined on $\mathbb{R}^k \times \mathbb{R}^l$, where the group product is given by:

$$(X, Z)(X', Z') = (X + \frac{1}{2}X', Z + Z' + \frac{1}{2}[X, X']). \tag{1}$$

The Lie algebra is identified with the tangent space at 0, which carries the natural Euclidean inner product $\langle ., .\rangle = g_0(., .)$ whose left-invariant extension is the Riemannian metric $g_p(., .)$ considered on the metric Heisenberg type Lie group.

Above, the left-invariant metric is defined by the left-invariant extension of the natural inner product given on the tangent space at the origin, nonetheless, it can also be described in terms of the unique connection, whose horizontal subspaces are spanned by the left-invariant extensions of the X-vectors in the Lie algebra. The connection form over an X-vector is defined in terms of an orthonormal basis $\{e_\alpha\}$ of the Z-space by $\omega_X(Y) = -\frac{1}{2}\sum_\alpha \langle J_\alpha(X), Y\rangle e_\alpha$. Then the invariant metric on the total space can also be described such that the metric of the X-space is uplifted to the horizontal subspaces, the horizontal and vertical subspaces are perpendicular, moreover, on the vertical subspaces, it agrees with the Euclidean metric of the Z-space. Since the connection is of non-zero curvature, this metric is of non-Euclidean. According to these, a H-type group appears on the total space of a trivial vector bundle $\mathcal{X} \times \mathcal{Z}$, where $\mathcal{X}$ is the base and $\mathcal{Z}$ is the fiber, which is endowed with a Yang-Mills type metric. Yet, it is not a particular Yang-Mills model. One of the distinguishing features is that the main physical object – the Monistic Hamilton Operator – appears on the total space, as opposed to the Yang-Mills models where the gauge invariant objects are defined on the base.

The theory extends to certain trivial vector bundles defined over certain Kähler manifolds. There is only a limited collection of manifolds where the Laplace operator appears as a realistic physical Hamilton operator. It turns out that the HyperKähler-Zeeman manifolds, which also include the Calabi-Yau-Zeeman manifolds, are the only Riemannian manifolds where the Laplace-Beltrami operator meets the criteria of physical reality. Later arguments show that the weak interactions can only be modelled by such manifolds on which the same Riemannian metric is Kähler regarding not just one but more independent complex structures. Such are exactly the hyperKähler manifolds where these complex structures define higher dimensional Z-spaces and hereby giving rise to appropriate HyperKähler-Zeeman manifolds exhibiting all aspects needed for a full-fledged physical model. An other point showing that the prototype models are not generic is that they describe just magnetic aspects but the electric ones are beyond their scope. The detailed general theory, which reflects all possible aspects of physics, is described in [Sz6]. Except for the remarks in Section 2.4, this paper does not pursue this topic further.

2.2. The Monistic Hamilton Operator: $\mathcal{MH}$.

The unification is established such that the Hamilton operators of distinct matter formations are derived from the very same Laplace operator existing on Zeeman manifolds. By this reason it is called Monistic Hamiltonian and it is denoted by $\mathcal{MH} = -\Delta$. Its negative is the Beltrami-Laplace operator Δ, which appears on prototype manifolds in the form:

$$\Delta = \Delta_X + (1 + \frac{1}{4}|X|^2)\Delta_Z + \sum_{\alpha=1}^{l} \partial_\alpha D_\alpha \bullet\,. \tag{2}$$

Symbol $D_\alpha\bullet$ denotes directional derivative on the X-space by the Hopf vector fields, defined at X by $J_\alpha(X)$. The physically most important part of Δ is the last one that involves $D_\alpha\bullet$. It is a compound angular momentum operator defining orbiting spin for all particles appearing on Zeeman manifolds.

Formula (2) can be established by the identity $\Delta = \sum_i \mathbf{X}_i^2 + \sum_\alpha \mathbf{Z}_\alpha^2$, valid for left-invariant metrics defined on a Lie group. Here $\mathbf{X}_i$ and $\mathbf{Z}_\alpha$ denote orthonormal left-invariant vector fields, respectively. The negative sign before Δ insures the positiveness of the eigenvalues of $\mathcal{MH}$.

This is the Laplacian which directly emerges on particular Riemann manifolds, without using any kind of Hamilton formalism. Since this operator is defined on the total space, it can actually not be established by the traditional Hamilton formalism. In Yang-Mills and the related Hamilton formalism, the fundamental objects - such as gauge field - are defined on the base and must be gauge invariant, meaning that they do not change if they are defined regarding a new connection form $\tilde{\omega}$, obtained by adding a closed form to the original connection form ω.

This change is called gauge transformation of ω. Gauge transformation of the metric means defining the metric regarding the new connection. That is, by uplifting the metric from the base to the new horizontal subspaces, which stand perpendicular to the vertical subspaces, furthermore, one keeps the original inner product on the vertical subspaces. It is rather remarkable that operator (2) is gauge invariant in the following sense: Metric $\tilde{g}$, defined from g by gauge transformation, is isometric to g, thus Laplacians Δ and $\tilde{\Delta}$ are isometrically equivalent.[Sz6]

2.3. Particle interpretations; Quarks vs. Splinters.

The model represents particles in two different ways. Primarily, it depicts them as "whole particles" but which naturally decompose into fractions - called splinters - which actually are the correspondents of quarks on Zeeman manifolds. The objects are preferably pictured by waves, defined on the total (X, Z)-space. Nevertheless, the wave-packets, visualizing the particles as tiny billiard balls, also are legitimate concepts, which can be included by the duality principle. On Zeeman manifolds, the laws of the quantum world are explored by the action of the Hamilton operator on the whole and splinter waves. This is opposed to the Lagrangian theories where the laws are explored by Lagrange equations describing the motions of the tiny billiard balls according to the principle of least action. The technical description of waves is provided in Sections 2.5 and 2.6.1. For better understanding, these particle interpretations are described in this section by less technical terms.

The whole particle waves are defined by functions which are invariant under the action of the compound angular momentum operator $\sum_{\alpha=1}^{l} \partial_\alpha D_\alpha \bullet$. A given whole particle system is defined by a system $\mathbf{Q} = \{Q_1, \dots, Q_{k/2}\}$ of orthonormal X-vectors. For a unit Z-vector $Z_u = Z/|Z|$, the J_{Z_u} is a complex structure in terms of which the $\mathbf{Q}$ becomes a complex vector system. The required functions are introduced in terms of holomorphic and antiholomorphic polynomials defined for $\mathbf{Q}$. They portray the physical situation in $\mathcal{X} \times \mathcal{Z}$ so that the X-space, whose real dimension is $k = \dim_{\mathbb{R}}(\mathcal{X})$, appears as the exterior space for $\kappa = k/2$ number of "whole" particles which live in complex planes defined by the complex structures J_{Z_u} and system $\mathbf{Q} = \{Q_1, \dots, Q_{k/2}\}$. Whereas, the Z-space is the common inner space shared by all of these particles. The particular form of the whole particle waves also allows to associate these particles with positive and negative charges as well.

The matter-particles appear in two forms, called point and extended particles, respectively. The waves of point particles are defined on torus bundles obtained by factoring the center by a lattice $\{Z_\gamma\}$ of the Z-space. Whereas, the waves of extended particles live on Z-ball bundles, defined by smooth fields of balls, $B_{\delta_Z(|X|)}(X)$, of radius $\delta_Z(|X|)$, considered - over each X-vector - about the origin in the Z-space.

The waves of point particles will be described in terms of the discrete Z-Fourier transform, defined for the lattice $\{Z_\gamma\}$ in the Z-space. They form an invariant subspace regarding the action of the Monistic Hamiltonian, where it appears as the Landau-Zeeman operator of charged particles orbiting in constant magnetic fields. The waves in this function space are "whole", consistent with the above description, which clearly exhibit the indivisibility of point particles.

The extended matter waves, however, will be introduced in terms of twisted Z-Fourier transforms, where an appropriate L^2-version of the Z-Fourier transforms is applied over each X-vector in the Z-space. They are appropriate adoptions of de Broglie's matter waves to the Zeeman setting. A particular feature of these waves is that they must satisfy prescribed boundary conditions, but which can not be enforced by "whole functions". The extended whole particles, also called Rutherford particles, form a special class including the neutrons, protons, and their combinations. In order to describe all extended particle waves satisfying the boundary conditions, the whole waves must be decomposed into fractions, called splinter waves, which provide a profound insight into the inner structure of extended particles. They actually are waves of quarks, but, to avoid confusion, the particles arising from them are called splinter-particles, or splinters. A particularly important property is that all splinter waves are needed to determine a complete invariant subspace for the Monistic Hamilton operator, where its actions on the waves reproduce the electromagnetic, weak, strong, and gravitational interactions.

Summing up, the prototype Zeeman manifold models represent multi particle systems not by the traditional Cartesian product but by a certain modification where only the exterior world appears as the Cartesian product of 2-dimensional exterior worlds associated with individual whole particles. Their interior world, however, is not added to the X-space by the Cartesian product of the individual interior worlds but by a single Z-space which serves as a common interior world for all of the particles present on the manifold. It should strongly be emphasized that the Cartesian product strictly refers to whole-particles. This also means that the proper splinter-particles are added to the system not by Cartesian product but they live together with their whole parent particles and sibling splinters on the same Zeeman manifold. An other new feature is that not just the X- but the Z-space as well is considered to be space-like. Furthermore, the basic objects such as the Hamilton operators are considered on the total space and not just on the X-space. This is in sharp contrast with the Yang-Mills theory, where the basic objects appear on the base, and also contrasts the traditional view of Heisenberg groups where the Z-space (center) is the time axis.

2.4. Schwinger-type gauge invariance on Zeeman manifolds.

In QED, the importance of gauge invariance was recognized by Schwinger, in the late 40's. The presence of this property in a theory means that the results do not depend on the particular accountancy scheme used for their establishments. Schwinger exploited this principle in QED for maintaining the invariance of electric charge. His theory also implied the existence of a massless agent - the photon - that is carrying the information about the invariant charge from one electron to the other. He also suggested to work out these ideas for the much more complicated weak and strong forces and foresaw the possibility for the electroweak unification. His ideas inspired the Yang-Mills models of general gauge theories, as well as the Glashow-Salam-Weinberg realization of the electroweak unification. In Yang-Mills theory, the gauge bosons - carrying the informations about invariants - are of zero masses. This implication caused very serious problems in building up the weak interaction theory, where the gauge bosons must carry charges, thus their rest masses must be of non-zero. The difficulties have been resolved by theorizing the Higgs boson, providing mass to each particle of non-zero rest mass.

By an earlier remark, the Monistic Hamiltonian is gauge invariant, thus there is no fundamental contradiction with the Yang-Mills models. But there is an other important question to be answered, namely, how does Schwinger-type gauge invariance manifest itself, in other words, how is the information about invariant quantities distributed on Zeeman manifolds or Zeeman space-times? The answer is as follows.

As it is pointed out later, the charge can be introduced on Zeeman manifolds by the complex structures J_{Z_u}, defined for the unit Z-vectors Z_u, and the holomorphic and antiholomorphic linear functions appearing in whole particle waves. These complex structures actually give rise to a natural tool for delivering all invariant quantities

IOP Conf. Series: Journal of Physics: Conf. Series **804** (2017) 012041 doi:10.1088/1742-6596/804/1/012041

to each point of the X-space. They define the very same identification of Z-spaces over two distinct X-vectors than what is defined by adding the Z-space to the X-space by Cartesian product. Thus the distribution of invariant quantities by the complex structures or by the Z-spaces are equivalent procedures. Schwinger's gauge-invariance can completely be established on Zeeman manifolds by the Z-space, which, speaking topologically, is added to the X-space by Cartesian product. In other words, the common interior space gives rise to a natural tool delivering the very same information right into the insides of particles, over each point X of the exterior space. This information-distribution does not contradicts relativity, due to that that the matter waves described in terms of these invariants can not travel with speed greater than c. It neither implies that the bosons emerging in the theory move with speed c, therefore, their resting masses must a priori be zero. Contrary to this, there is a natural Hamiltonian mass-assignment process on the Zeeman space-time which attributes masses to charge carrying bosons. This process is analogous to the Higgs mechanism.

Schwinger's gauge invariance is furnished on general Zeeman manifolds, very similarly, by considering such Riemannian metrics on the X-space which are Kähler regarding not just one but number of complex structures. They determine a trivial Z-space over the X-space in the same way as on prototype manifolds. In the compact case, exactly the Calabi-Yau manifolds are those which allow such complex structures defining multidimensional trivial Z-spaces needed in the theory. The most generic versions are introduced in [Sz6], whose discussion is beyond the scope of this paper. Anyhow, the HyperKähler Zeeman manifolds, which include the Calabi-Yau-Zeeman manifolds, are the only general Zeeman manifolds which allow a completely unified relativistic quantum theory, similarly to that yielded by prototype manifolds.

The theory is explained in this paper on prototype manifolds only. In the next sections, the Hamilton operators of distinct particles will be established from the very same Monistic Laplacian (2). The main idea in carrying out this scheme is that the Hamilton waves of distinct particle systems define distinct invariant subspaces and the specific Hamilton operators are obtained by restricting the Monistic Hamiltonian onto the invariant subspace associated with the given particle system. One of the strongest indications for these investigations are heading into the right direction is that the Hamilton operator derived for a point particle system from the Monistic Hamilton Operator is equal to the Zeeman operator that has been established in classical quantum theory quite differently, by the Hamilton formalism and the correspondence principle.

2.5. Hamilton operators of point particles.

The Hamilton waves of a point particle system associated with a torus bundle $\mathcal{X}\times(\Gamma_Z\backslash\mathcal{Z})$ are defined by the discrete Z-Fourier transform: $\Psi = \sum_\gamma \psi_\gamma(X)e^{2\pi\mathbf{i}\langle Z_\gamma,Z\rangle}$, where $\Gamma_Z = \{Z_\gamma\}$ is a lattice in the Z-space defining the Z-torus bundle over the X-space. This formula gives representation for all Γ_Z-periodic functions. For any fixed lattice point Z_γ, functions of the form $\psi(X)e^{2\pi\mathbf{i}\langle Z_\gamma,Z\rangle}$ span a sub-space denoted by W_γ. They define the Fourier-Weierstrass decomposition, $\sum_\gamma W_\gamma$, of Γ_Z-periodic functions.

The Laplacian acts on W_γ according to the formula: $\Delta(\psi(X)e^{2\pi\mathbf{i}\langle Z_\gamma,Z\rangle}) = \diamondsuit_\gamma(\psi(X))e^{2\pi\mathbf{i}\langle Z_\gamma,Z\rangle}$, where:

$$\diamondsuit_\gamma = \Delta_X + 2\pi\mathbf{i}D_\gamma \bullet -4\pi^2|Z_\gamma|^2(1+\frac{1}{4}|X|^2), \tag{3}$$

IOP Conf. Series: Journal of Physics: Conf. Series **804** (2017) 012041 doi:10.1088/1742-6596/804/1/012041

It shows that the Laplacian leaves each of the subspaces W_γ invariant. Operator $\diamondsuit_\gamma$ is known in quantum physics as the Landau-Zeeman operator of charged particles orbiting in constant magnetic fields. The second term is the angular momentum operator associated with the magnetic dipole momentum.

To see this matching, recall that the 2D-Landau-Zeeman operator of a particle of charge C, orbiting in the (x, y)-plane in constant magnetic field directed toward the z-axis, is of the form [B, LL, P]:

$$-\frac{\hbar^2}{2m}\Delta_{(x,y)} - \mathbf{i}\frac{\hbar CB}{2mc}D_z \bullet + \frac{C^2B^2}{8mc^2}(x^2+y^2), \tag{4}$$

where $D_z\bullet = x\partial_y - y\partial_x$. Thus, by choosing $\pi|Z_\gamma| = \mu = CB/2\hbar c$ and multiplying the whole operator with $-\hbar^2/2m$, the $\lhd_\gamma$ (also denoted by $\lhd_\mu$) is transformed to the Landau-Zeeman operator.

In quantum theory this operator is established by Maxwell equations, Hamilton formalism, and correspondence principle. Whereas on Zeeman manifolds, its action is equal to that of the Monistic Hamilton Operator on Γ_Z-periodic functions - the Hamilton waves of point particles - which constitute an invariant subspace regarding the Monistic Hamiltonian. This coincidence with realistic quantum Hamilton operators clearly reveals that the mathematical Zeeman manifolds should be considered as profound quantum physical models. It immediately raises the question as to which quantum physical contents are revealed by the action of the Monistic Hamilton on the waves of extended particles. The answer is explored in the following section where it turns out that that action corresponds to the electroweak and strong interactions and also exhibits gravitation, while the action on the waves of point particle systems relates Δ to the electromagnetic phenomena.

By closing this section, let it be pointed out again that all Hamilton and wave operators established sofar act on functions thus the waves of particles can not carry mutual spin but just angular momentum. The anomalous theory is established in [Sz2]-[Sz6] under most general circumstances where Coulomb and other generic electric potential functions also appear in the Monistic Hamilton Operator. It is an important point that these potential functions not just simply added to the magnetic operator, as it is usually done in the literature today, but they appear as inherent parts in the Monistic Laplacian.

2.6. Extended particles on Zeeman manifolds.

The theory of extended particles are worked out on Z-ball bundles, defined by smooth fields, $B_{\delta(|X|)}(X)$, of Z-balls over the X-space where $\delta(|X|)$ denotes the radius of the ball over X. Since the Z-balls represent the common inside of extended particles, their size gives rise to a physical invariant. Thus, in physics, the $\delta(|X|)$ should be constant characteristic for the particle system. Nonetheless, the following considerations can also be carried out for varying radius-functions only depending on $|X|$.

2.6.1. Hamilton waves of extended particles. The Hamilton operators of extended particle systems are obtained by restricting the very same Monistic Hamilton Operator to an other invariant subspace, consisting of appropriate functions defined on the Z-ball bundles, which emerge then as the Hamilton waves of extended particle systems. Like for point particles, they also are represented by a sort of Z-Fourier transform operating on L^2_Z functions over each X-vector in the Z-space, however, this version is far from being just a simple generalization, passing from the discrete to the L^2_Z-case. Complications arise from the boundary conditions these waves must comply with which inquire the implementation of new operations – such as the below described projections $\Pi^{\mathbf{s}}_{K_u}(\psi(K_u, X)$ – which do not show up in discrete Z-Fourier transforms. To answer all these requirements, the Hamilton waves, $\Psi(X, Z)$, of extended particle systems are introduced by the so called twisted Z-Fourier transforms in the form:

$$\int_{\mathbb{R}^l} e^{\mathbf{i}\langle K,Z\rangle}\phi_{\mathbf{s}}(|X|,|K|)\Pi^{\mathbf{s}}_{K_u}\psi(K_u, X)dK = \tag{5}$$

$$= \Phi_{\mathbf{s}}(|X|,|Z|)\Pi^{\mathbf{s}}_{Z_u}\psi(Z_u, X), \quad \text{where} \tag{6}$$

$$\psi(K_u, X) = \varphi(K_u)\Pi^{(n)}_X(\prod z_i^{p_i}\overline{z}_i^{q_i})(K_u, X), \quad n = \sum_{i=1}^{k/2}(p_i + q_i). \tag{7}$$

The position of indices $\mathbf{s}$ in terms like $\phi_{\mathbf{s}}\Pi^{\mathbf{s}}_{K_u}$ indicates summation regarding $\mathbf{s}$. The other constituents are explained as follows.

The formula is a Fourier transform written up in the Z-space in terms of polynomials $\prod z_i^{p_i}\overline{z}_i^{q_i}$ defined by the complex structures J_{K_u}, associated with unit vectors $K_u = K/|K|$, and a fixed system $\mathbf{Q} = \{Q_1, \dots, Q_{k/2}\}$ of orthonormal X-vectors representing $k/2$ number of whole particles. The holomorphic linear functions $z_i(K_u, X) = \langle Q_i + \mathbf{i}J_{K_u}(Q_i), X\rangle$ are defined regarding these vectors. The $\Pi^{(n)}_X$ projects the nth-order homogeneous polynomial of the X-variable standing in the argument to homogeneous harmonic polynomials of the same order. It can be expressed in the form $\sum_p a_p|X|^{2p}\Delta^p_X$, with recursively defined coefficients a_p [Sz4]. Function $\varphi(K_u)$ is the restriction of a homogeneous polynomial $\varphi(K)$, defined on the Z-space, onto the unit sphere of the Z-space.

Waves defined by the simpler formula:

$$\int_{\mathbb{R}^l} e^{\mathbf{i}\langle K,Z\rangle}\phi(|X|,|K|)\psi(K_u, X)dK \tag{8}$$

are the whole extended waves, which are also called Rutherford waves. They are eigenfunctions of the compound angular momentum operator, spanning a function space invariant under the action of the complete Monistic Hamilton Operator. The restriction defines now the Rutherford Hamilton Operator, which is very similar to that of Landau-Zeeman operator of point particles. But there arises a serious problem about these waves, namely, the boundary conditions on Z-ball bundles can not be enforced by them. This is the primary reason as to why the functions yielding the various boundary conditions should be sought by more sophisticated integral formulas like (5).

Such general waves are defined in terms of projections $\Pi^{\mathbf{s}}_{K_u}(\psi(K_u, X))$ in (5) which project ψ to spherical harmonics defined on the unit K-spheres, over each X-vector. Their explicit formulae, expressed in terms of the polynomials of the powered Laplacian $\Delta^r_{K_u}$, need numerous explanations. Namely, such formulae can primarily be established for projecting restrictions, $p(K_u)$, of homogeneous polynomials, $p(K)$, to spherical harmonics, and they depend both on the degrees of the function being projected and the function resulted by the projection. This is why index $\mathbf{s}$ must be compound, indicating both of these two degrees.

But this argument gives rise to an other complication, namely polynomials $z_i^{p_i}\overline{z}_i^{q_i}(K_u, X)$ are not restrictions of homogeneous polynomials onto the K_u-spheres but it is the sum of components having distinct degrees of homogeneity. The parts of

IOP Conf. Series: Journal of Physics: Conf. Series **804** (2017) 012041 doi:10.1088/1742-6596/804/1/012041

the same degree can be found by decomposing $z_i^{p_i}\overline{z}_i^{q_i}(K_u, X)$, where $z_i(K_u, X) = \langle Q_i + \mathbf{i}J_{K_u}(Q_i), X\rangle$, according to the powering and the distributivity laws into homogeneous components. Then $\Pi^{\mathbf{s}}_{K_u}$ acts on that component whose degree is indicated in $\mathbf{s}$. Projections $\Pi^{\mathbf{s}}_K$ acting on homogeneous polynomials can similarly be established as polynomials of the powered Laplacians Δ^r_K. Double radial function $\Phi_{\mathbf{s}}(|X|, |Z|)$ is established from $\phi_{\mathbf{s}}(|X|, |K|)$ by the Hankel transform, which describes the action of the Fourier integral standing on the left side of (5) in terms of radial functions [Sz2, Sz4, Sz6].

General waves (5) emerge from the Rutherford waves (8) by decomposing ψ by the projections $\Pi^{\mathbf{s}}_{K_u}$. Spherical harmonics $\Pi^{\mathbf{s}}_{K_u}(\psi(K_u, X))$ are called splinters of 0-generation, which are not "whole functions", like $\psi(K_u, X)$, but which describe the waves also in the insides of extended particles. Name "0-generation" indicates that the ψ is still a whole function. Higher generation functions $\psi^{(g)}$ are resulted below by successive application of the Monistic Hamiltonian on the above wave functions. The functions of higher order generations can not be considered whole anymore.

Projected functions $\Pi^{\mathbf{s}}_{K_u}(\psi^{(g)}(K_u, X))$ correspond to the Hamilton waves of quarks, the building blocks of the nucleus. For their deeper understanding, consider the homogeneous harmonic polynomials $\mathbf{P}^{\mathbf{s}}(K, X) = \Pi^{\mathbf{s}}_K(\psi^{(g)}(K, X))$ regarding the K-variable over each point X in the whole K-space, where the gradient vector field $\mathbf{B}^{\mathbf{s}}(K, X) = \mathrm{grad}_K(\Pi^{\mathbf{s}}_K(\psi^{(g)}(K, X))$ also is harmonic and corresponds to a closed 1-form in that Euclidean space. Thus we have: $\mathrm{div}_K(\mathbf{B}^{\mathbf{s}}) = 0$; $\mathrm{rot}_K(\mathbf{P}^{\mathbf{s}}) := *_K(d_K\mathbf{P}^{\mathbf{s}}) = 0$, that is, vector field $\mathbf{B}^{\mathbf{s}}$ satisfies the Maxwell equations of a pure magnetic field with which vanishing electric field, $\mathbf{E}^{\mathbf{s}} = 0$, is associated.

The whole system, $\{\mathbf{B}^{\mathbf{s}}\}$, defined for all compound index $\mathbf{s}$, is called magnetic labyrinth in the inside of extended particles. This version corresponds to the "little magnet" imagined by the physicists in the inside of point particles in order to visualize the spin-concept. Function $\mathbf{P}^{\mathbf{s}}$ represent the potential for the magnetic field $\mathbf{B}^{\mathbf{s}}$. The splinters - the wave functions of quarks - are the restrictions of individual magnetic potential fields onto the unit sphere of the K-space. Due to the Hankel transform, these objects appear on the Z-space in the very same form, where just K is substituted by Z.

Double radial functions $\Phi_{\mathbf{s}}(|X|, |Z|)$, defined from $\phi_{\mathbf{s}}(|X|, |K|)$ by the Hankel transform, are potential functions of the strong force keeping the quarks inside of particles. This errand is carried out by smooth functions defined inside of extended particles, whereas they vanish on the outside. If the degree of $\mathbf{P}^{\mathbf{s}}$ is s, then at the origin they behave asymptotically as $|Z|^s$. Thus the strong force is very weak inside - allowing the quarks moving almost freely there - whereas it gets infinitely strong at the boundary - not allowing to move the quarks to the outside. This paradoxical feature was recognized by Bjorken, in the 60's, and confirmed thereafter by numerous experiments.

2.6.2. Hamilton operators of extended particles. The very same Laplacian $\Delta = -\mathcal{M}\mathcal{H}$ acting on extended matter waves (5) appears as the sum of operators:

$$(\Delta_X + (1 + \frac{1}{4}|X|^2)\Delta_Z)\Psi + (\mathbf{q} - \mathbf{p})\sqrt{-\Delta_Z}(\Phi_{\mathbf{v}} + \bigodot|^{\mathbf{s}}_{\mathbf{v}}\Phi_{\mathbf{s}})\Pi^{(\mathbf{v})}_{Z_u}\psi, \quad (9)$$

$$\text{and} \qquad \sqrt{-\Delta_Z}\,\bigcirc|^{\mathbf{s}}_{\mathbf{w}}(\Phi_{\mathbf{s}})\,\Pi^{(\mathbf{w})}_{Z_u}(\mathbf{W}^+ + \mathbf{W}^-)\psi, \qquad (10)$$

called Stabler and Decay Hamiltonian, respectively.

IOP Conf. Series: Journal of Physics: Conf. Series **804** (2017) 012041 doi:10.1088/1742-6596/804/1/012041

The proof of these formulas explores the action of Δ on wave functions (5). The most complicated computational details arise from the action of the compound angular momentum operator, which appears behind the integral sign as $\mathbf{i}|K|D_{K_u}\bullet$. Term $\sqrt{-\Delta_Z}$ origins from multiplication with $|K|$. The other terms are computed from the relation:

$$[D_{K_u}\bullet, \Pi^{(\mathbf{s})}_{Z_u}] = -\bigodot|^{\mathbf{s}}_{\mathbf{v}}\Pi^{(\mathbf{v})}_{K_u} D_{K_u}\bullet + \bigcirc|^{\mathbf{s}}_{\mathbf{w}}\Pi^{(\mathbf{w})}_{K_u}\sum_{\beta=1}^{l-1}\partial_\beta D_\beta\bullet, \tag{11}$$

where $\{e_\beta(K_u)|\beta = 1, \ldots, l-1\}$ is an orthonormal basis in the tangent space of the unit sphere defined by unit vectors K_u.

The Stabler is associated with the term containing $D_{K_u}\bullet$, which results $(\mathbf{q}-\mathbf{p})$ in (9). Whereas, operator $(\mathbf{W}^+ + \mathbf{W}^-)$ in the Decay Hamiltonian is originated from $\sum_\beta \partial_\beta D_\beta\bullet$, which is involved to the second term standing on the right side of (11). It is a highly transmuting operator, indicating that it really is a decay-operator. It decreases the degrees of spherical harmonics and transmutes a holomorphic linear function to an antiholomorphic one and vice versa. Since these functions can be associated with positive and negative charges, the $D_\beta\bullet$ is considered to be acting as $\mathbf{W}^+$ on antiholomorphic functions and as $\mathbf{W}^-$ on holomorphic functions. When, due to the actions of these operators, the waves go through quantum leaps, then the change of energy levels is balanced by corresponding absorption or emission of bosons relating to $W^\pm$, introduced in the weak interaction theory.

Since the $\psi(K_u, X)$ is an eigenfunction of $-\mathbf{i}D_{K_u}\bullet$ with eigenvalue $\mathbf{q}-\mathbf{p}$, the Stabler does not transmutes ψ but it rather acts like a stabilizer, explaining its name. When the waves go through quantum leaps, due to the actions of this operator, then the change of energy levels is balanced by absorption resp. emission of Z-relating bosons, the other main objects in weak interaction theory. Explicit computations reveal that the spectrum is parity violating and the Stabler prefers to build up heavy stable protons accompanied with much lighter electrons. The Stabler actually emerges from this theory as the Hamilton operator of a new force whose agent is the Z-relating boson whose action results the stability of protons as well as their rather heavy weight, as compared to that of the electron. In the Standard Model, contrary to this, the stability is enforced by the barion numbers, which, however, do not yield any explanation neither for the heavy weight of protons nor for the parity violation. On Zeeman manifolds, the cooperative actions of the Stabler and the Decay operators govern all decays, such as alpha beta and gamma, and all above mentioned phenomena are spectrally explained there.

The Hamilton operator of the strong force is $\sqrt{-\Delta_Z}$, which appears in both Hamilton operators. Its action can completely be reduced to the Z-radial functions - the waves of the strong force. The strong force itself is defined by the gradient vector fields of these Z-radial functions. Since they vanish in the outside, the strong force can be infinitely strong at the boundary, thus it is appropriate to define the strong force Hamiltonian by $\sqrt{-\Delta_Z}$. In the inside of extended particles, it represents just a very weak force. At the boundary, however, it defines a distribution which corresponds to the infinitely large force acting at the boundary in order to keep the splinters in the inside of extended particles. When, due to the action of the strong force Hamiltonian, the waves go through quantum leaps, then the energy differences are balanced by corresponding emissions or absorptions of gluons.

It should be mentioned that Yukawa's strong force, whose agents are the pions, also appears in the theory. Namely, Yukawa's wave operator is inbuilt into the

Laplacian of the static Zeeman space-time (cf. Section 3.1). Its corresponding Hamilton operator is $-\Delta_Z$ and it appears together with $\partial^2/\partial t^2$. Due to its relation to the de Broglie waves, it becomes a major tool for establishing a new, spectral mass-assigning procedure which is analogous to the Higgs mechanism. On the accelerating Zeeman space-time, the $-\Delta_Z$ gets involved with the wave operator of dark energy and plays a major role to carry out the accelerating version of the new spectral mass assigning procedure. Term $\sqrt{-\Delta_Z}$, however, is engaged with the parabolic Schrödinger operator, expressed in terms of $\partial/\partial t$, which is the Wave operator of the non-expanding ordinary matter. According to these argument, the theory makes clear distinctions between the two forces and operator $\sqrt{-\Delta_Z}$ should really be considered as the Hamilton operator of Bjorken's strong force. It has an intimate relation to the graviton that is also described in the following discussions.

The arguments brought up sofar verify that the Stabler and Decay Hamiltonians describe nuclear processes in consent with the Standard Model. These Hamilton operators break down into part operators which not just reestablish the quantum world from Hamiltonian points of view but also imply such new features which actually are out of the scope of the Lagrangian theories. The most important among them is gravitation, appearing on the scene such that both the Stabler and Decay operators contain respective spool, $\odot|^{\mathbf{s}}_{\mathbf{v}}$, and winch, $\bigcirc|^{\mathbf{s}}_{\mathbf{w}}$ operators, binding any two splinter waves together produced by the repeated actions of the Monistic Hamilton Operator on a starting wave of 0 generation.

In order to understand this, imagine that one tries to determine eigenfunctions of Δ in a process, starting with the twisted Z-Fourier transform of a given function $\phi_{\mathbf{s}}(|X|,|K|)\Pi^{\mathbf{s}}_{K_u}(\psi^{(0)}(K_u,X))$. Index in $\psi^{(0)}$ indicate that it is a 0-generation function. If the Δ consisted only of the Stabler, then one would be able to find eigenfunctions in the form $\gamma_{\mathbf{s}}(|X|,|Z|)\Pi^{\mathbf{s}}_{Z_u}(\psi^{(0)}(Z_u,X))$, where double radial functions $\gamma_{\mathbf{s}}$ are derived from $\phi_{\mathbf{s}}$. Note that the generation index does not change in this case and the problem reduces to determining appropriate double radial functions $\gamma_{\mathbf{s}}(|X|,|Z|)$. However, due to the action of the Decay Hamiltonian, the problem can not be solved just by the 0-generation functions, but one also needs first generation function $\psi^{(1)}$ obtained from $\psi^{(0)}$ by the action of the Decay operator. Even these two generations will not be enough to determine eigenfunctions but one needs all generations, produced by repeated actions of the Decay operator. When all these splinter waves combined with double radial functions are produced, then all of them are needed to build up the eigenfunctions of Δ. Any two splinters of any two generations are connected by the spool and winch operators. They actually are Hamilton operators associated with very feeble pulling forces. The attraction is exerted in order to keep whole galaxies of particles together so as to form stable particle systems (galaxies) which appear to be as a single stable organization. If a particle in a stable galaxy is disturbed and tipped out of the equilibrium then the whole galaxy gets out of the eigen-state and this feeble force restores the balance by pulling the particle back to the stable position. This feeble force is nothing but the gravitation which manifests itself both on the microscopic and macroscopic levels. Its agents are the spool and winch gravitons which appear during quantum leaps caused by the actions of the operators.

Interactions due to spool and winch operators are just two out of three facets of gravitation. The third one emerges on most general Zeeman space-time models of the universe which are not just extensions of single H-type groups into the time direction but rather are Cartesian products of many of them in which a single H-type group is just an irreducible component. In such cases the spool and winch operators are not universal in the sense that they act just within the components of the Cartesian product. The missing universality of gravitational interaction is supplied by the universal gravitation operator emerging from the action of the "mammoth" Schrödinger wave operator, which, as explained later, acts in strong cooperation with the spool and winch gravitations in order to establish an all-embracing gravitation also acting among the distinct components of the Cartesian product. An other important feature is that this gravitation directly relates to the mass generated by the spectral mass-assignment procedure.

By an informal description, the Laplace operator acts on Zeeman manifolds like an operator which decomposes the non-stable de Broglie waves into splinter waves by which systems of stable particle-galaxies corresponding to the eigenfunctions of the Laplace operator can be built up. The process breaks down into alpha beta and gamma decays, resulting heavy protons, light electrons, and all particles the microscopic and macroscopic universes consist of.

This theory modifies Einstein's general relativity from several points of view. The gravitation manifesting according to the Zeeman manifold model escapes the most serious contradictions of Einstein's theory such as the collapse of large universes due to the overwhelming gravitational attraction. The theory also gives explanation for other phenomena. Explicit spectral computations reveal, for instance, as to how are the stable protons built up and why is the proton much heavier then the electron. All this information is derived from the very same operator. This is how the Zeeman manifold model realizes unification on many levels of the microscopic and the macroscopic universes as well.

3. Static and accelerating Zeeman space-times.

3.1. Static Zeeman space-times.

3.1.1. Static Monistic Wave Operator The static relativistic space-times are defined by metric Cartesian products, $N \times \mathbb{R}$, of Zeeman manifolds N with the time axis $\mathbb{R}$, where the latter is considered to be endowed with the indefinite inner product $< \partial_t, \partial_t >= -1$. Then, the tangent spaces $T(N)$ and $T(\mathbb{R})$ are perpendicular and, in the prototype case, the Laplacian Δ_{St} appears there in the form:

$$(\Delta_Z + \frac{2m\mathbf{i}}{\hbar}\partial_t - \partial^2_{tt}) + (\Delta_X + \frac{1}{4}|X|^2\Delta_Z + \sum \partial_\alpha D_\alpha \bullet - \frac{2m\mathbf{i}}{\hbar}\partial_t). \quad (12)$$

This formula represents the Δ_{St} as the sum of two operators which decomposition can be established by introducing the trivial term $0 = \frac{2m\mathbf{i}}{\hbar}\partial_t - \frac{2m\mathbf{i}}{\hbar}\partial_t$ into the undecomposed operator. The reason for doing so is that this decomposition results two original wave operators - the Yukawa and the Schrödinger operators- of classical quantum theory. Also notice that energy operator $-\frac{2m\mathbf{i}}{\hbar}\partial_t = \frac{2m\mathbf{i}}{\hbar}\partial_{-t}$ in the Schrödinger operator represents positive energies if it is considered regarding time direction $-t$.

3.1.2. The Yukawa operator and the de Broglie waves In order to properly exploit the apparent connections to the classical theory, one should recall some basic facts regarding Yukawa's strong force operator and its relation to de Broglie's waves:

$$\Psi_{deB}(Z,t) = \int\int\int A(K_1,K_2,K_3)e^{\mathbf{i}(\langle K,Z\rangle - \omega t)}dK_1dK_2dK_3, \quad (13)$$

$$\text{where} \quad \sqrt{|K|^2 + \frac{m^2c^2}{\hbar^2}} = \frac{\omega}{c}, \tag{14}$$

originally introduced for describing the ordinary matter having non-zero rest mass as matter-waves and the associated matter-particles as wave packets.

Waves $\Psi_{deB}(Z,t)$ satisfy the relativistic scalar wave equation:

$$(\Delta_Z - \frac{1}{c^2}\frac{\partial^2}{\partial t^2})\Psi_{deB}(Z,t) = \frac{m^2c^2}{h^2}\Psi_{deB}(Z,t) \tag{15}$$

(cf. [2.2] in [P], Vol. 5, pages 3), which also serves as scalar wave equation for the strong force, used by H. Yukawa to explore the Pion - the agent of the strong force. Most remarkably, this operator appears in this pure form in the undecomposed Laplacian of the static Zeeman space-time, hereby relating the theory both to de Broglie's waves and Yukawa's scalar wave equation of the strong force. These relations also clarify as to why should operator $\sqrt{-\Delta_Z}$ in the Stabler and Decay Hamilton operators be related to the strong force.

Wave $\tilde{\Psi}(Z,t) = \exp(\mathbf{i}mc^2t/h)\Psi(Z,t)$ satisfies:

$$(\Delta_Z + \mathbf{i}\frac{2m}{h}\frac{\partial}{\partial t} - \frac{1}{c^2}\frac{\partial^2}{\partial t^2})\tilde{\Psi}(Z,t) = 0 \tag{16}$$

(cf. [2.11] in [P], Vol. 5, pages 4), which, due to $\tilde{\Psi}\tilde{\Psi}^* = \Psi\Psi^*$, determines the same probabilistic density than Ψ. That is, Ψ and $\tilde{\Psi}$ describe the very same quantum state.

The transformation of Yukawa's original equation (15) to (16) allows to express the static Monistic Wave Operator (which originally contains only second order partial differentiations: ∂_{tt} regarding the time variable) as the sum of operator standing in (16) and the parabolic Schrödinger operator containing only first order differentiations regarding the time variable. The decomposition is simply implemented by introducing the trivial term $\frac{2m\mathbf{i}}{\hbar}\frac{\partial}{\partial t} - \frac{2m\mathbf{i}}{\hbar}\frac{\partial}{\partial t}$ into the static Monistic Wave Operator such that the first term is incorporated into Yukawa's and the second one into Schrödinger's operator. It is used below for establishing a new spectral mass-assigning procedure without using symmetry breaking. As an other important point, the mass is also related to the spectrum of the Hamilton operator appearing in the second part of the decomposed Monistic Wave Laplacian.

3.1.3. Spectral mass-assigning procedure In order to carry out this scheme, the static waves are introduced by:

$$\Psi_{\mathrm{st}}(X,Z,t) = \int_{\mathbb{R}^l} e^{\mathbf{i}(\langle Z,K\rangle - \frac{\omega}{c}t)}\phi_{\mathbf{s}}(|X|,|K|)\Pi^{\mathbf{s}}_{K_u}(\psi(X,K_u))dK, \tag{17}$$

where the ψ can be linear combination of generation functions. They are eigenfunctions of Yukawa's original operator which appears in the form $\Delta_Z - \partial^2_{tt}$ on the (Z,t)-space, over each X-vector. (As compared with (15), this one does not contains factor $1/c^2$ before ∂^2_{tt} but which difference is compensated by considering ω/c instead of ω in the above formula.) The mass appears in the eigenvalue m^2c^2/h^2 of this operator, which, however, has no relation to the spectrum of Hamilton operators also inbuilt into the Monistic wave Laplacian. To build up these connections, consider waves

$$\tilde{\Psi}_{st}(X,Z,t) = e^{\frac{\mathbf{i}mc^2}{\hbar}t}\Psi_{st}(X,Z,t), \tag{18}$$

which are harmonic regarding the so called exhausted Yukawa operator:

$$\Delta_Z + \frac{2m\mathbf{i}}{\hbar}\partial_t - \partial^2_{tt}. \tag{19}$$

The name (coined in this paper) is ment to indicate that the mass is exhausted regarding the Yukawa operator and the action of the whole Monistic Wave Laplacian on $\tilde{\Psi}_{st}(X, Z, t)$ is completely determined by the action of the Schrödinger operator. That is to say, exhausting waves (18) exhaust the mass just regarding the Yukawa operator which is furnished then into the Schrödinger operator in order to relate the mass to the spectrum of the Monistic Hamilton Operator. Due to the fact that the mass and time t appear both outside and inside of integration, the Schrödinger operator decomposes into the sum:

$$\Delta_X + \frac{1}{4}|X|^2\Delta_Z + \sum \partial_\alpha D_\alpha \bullet - \frac{2m\mathbf{i}}{h}\partial_t =$$

$$= \frac{2m^2c^2}{ch^2}\sqrt{1 - \frac{h^2}{m^2c^2}\Delta_Z} + \tag{20}$$

$$\Delta_X + \frac{1}{4}|X|^2\Delta_Z + \sum \partial_\alpha D_\alpha \bullet - \frac{2m\mathbf{i}}{h}\partial_t - \frac{2m^2c^2}{ch^2}\sqrt{1 - \frac{h^2}{m^2c^2}\Delta_Z}) \tag{21}$$

where the parts are called universal gravitation and actual Schrödinger operators, respectively. By this reason, the second in the decomposed Monistic Wave Laplacian (12) is called “mammoth” Schrödinger operator. Identity

$$\frac{1}{c}\frac{2m^2c^2}{h^2}\sqrt{1 - \frac{h^2}{m^2c^2}\Delta_Z}(\Psi_{\mathrm{st}})(X, Z, t) = \tag{22}$$

$$= -\frac{2m}{h}\int_{\mathbb{R}^l} \frac{\omega}{c} e^{\mathbf{i}(\langle Z,K\rangle - \frac{\omega}{c}t)}\phi_{\mathbf{s}}\Pi^{\mathbf{s}}_{K_u}(\psi)(X, K)dK = \tag{23}$$

$$= \frac{2m}{h}\Omega(\Psi_{\mathrm{st}})(X, Z, t) \tag{24}$$

regarding the universal gravitation operator is established in [Sz2, Sz6]. It primarily appears as an integral operator, but only its differential operator version indicates its relation to the strong, spool, and winch operators. It tells us as to how is the common mass m, defined for all particles participating in the system, establishes additional gravitational relation to those associated with the spool and winch operators, which can justifiably called "Mass-relating Gravitational Wave Operator". Its universality means that this kind of gravitation also exists among particles being in different components of the Cartesian product of H-type groups whose extension, into the time direction, defines the Zeeman space-time. Such universality, however, is not possessed by the spool and winch gravitations which do not involve mass-terms into their original definitions.

The harmonic solutions of (21) relate the mass to the eigenvalues of the actual Schrödinger operator according to the mass-energy equivalence. This spectral data depends on boundary conditions and refers to all waves produced by repeated actions of the Stabler and Decay operators. Only those mass values are physically appropriate which emerge from these spectral relations and include the masses of all constituent particles. Thus the possible masses are discrete and strongly relating to the boundary conditions. These computations actually determine the mass in relation to the sizes of particles. This mass-size relation allows the spectral establishments of the coupling constants defined insofar by Lagrangian means. There can also be explained as to why is the proton much heavier than the electron. This phenomenon is due to the parity violating spectrum and the extendedness of protons, as opposed to the electrons which are point particles.

The eigenfunctions of the actual Schrödinger operator are not eigenfunctions of the universal gravitation operator, whose action rather defines unbalanced waves which can be seen as being produced by the interaction existing between the commonly defined mass and the balanced system of particles associated with the eigenfunction-waves. During this interaction the boundary conditions are destroyed, after which effect the particles are compelled to seek their ways back to the eigenstate. There are two opposite tendencies inbuilt into the mammoth Schrödinger operator. One of them is mobilized by the actual Schrödinger operator which determines the possible eigenstates and compels the particles to move into those states. The other is orchestrated by the universal gravitation operator which tips out the system from the eigenstates. The cooperative actions of these two operators gives rise to motion – the most fundamental feature manifesting everywhere in Nature. This is why, the actual Schrödinger and the universal gravitation operators are called Dynamic Duo in the mammoth Schrödinger operator. Their interaction is resulted by that that they are two non-commutative operators. The duality is manifested also by the particular form how the time appears in the exhausting Yukawa waves. The exponential term outside of integration relates the wave to the actual Schrödinger operator, whereas the one standing inside relates it to the universal gravitation operator. This is called dynamical timing of waves, allowing multilateral exploration of the Monistic Wave Laplacian. This is contrary to Schrödinger's original waves, which only one-laterally explore his original equation.

3.1.4. Relations to symmetry breaking Although the spectral mass assignment does not use the idea of symmetry breaking, it can be related to it, particularly to the chiral symmetry breaking that was used by Nambu [N] to explain as to how do the protons and neutrons gain masses in the Yang-Mills theories. Chiral symmetry means no difference between right and lefthanded systems. In Nambu's theory the orientation of an object is relative to the observer and it depends on as to whether the observer is moving behind or ahead of the object. Thus the laws describing non-bypassable objects – because they are moving with speed of light – obey chiral symmetry. The laws controlling the movements of protons and neutrons in Yang-Mills theory are chiral symmetric, originally, thus they are massless in that stage. They gain mass by breaking the chiral symmetry, meaning slowing down to speed less than c.

The spectral mass assignment procedure evolves according to a very similar scenario. In the first step, the mass is generated as an eigenvalue of the time-symmetric Yukawa operator. For photons, the eigenvalue and therefore the mass is zero and they are moving with speed of light. They are the unbypassable objects whose tangents of their world-lines point into the direction of the light-cone of the (Z, t) space-time and the laws describing their motion obey chiral symmetry. If the eigenvalue is greater than zero, the tangents point into the inside of the light-cone, indicating speed less than c and the corresponding laws disobey chiral symmetry which is exhibited in relation to photons. This gives rise to a Lagrangian explanation as to why do these objects have positive masses. But they still are the products of the time-symmetric Yukawa operator and the mass does not have any relation to the spectrum of the rest part of the Monistic Wave Operator. This incompleteness is straiten out in the second step when the implementation of $0 = \frac{2m\mathbf{i}}{\hbar}\partial_t - \frac{2m\mathbf{i}}{\hbar}\partial_t$ into the undecomposed Operator breaks the time-symmetry of the Yukawa operator and provides the mass to the actual Schrödinger operator in order to bring it in connection with its spectrum. The implemented term also determines the time direction with respect to which the masses appear to be positive. By this reason, it is called Chiral Mass-Time Switch. The chiral and other symmetries are further explored in [Sz7].

3.2. Accelerating Zeeman space-times.

The prototype accelerating Zeeman space-times are carried out on the solvable extensions of H-type groups defined on the half-space $N \times \mathbb{R}_+$. The Lie algebra, $\mathcal{S} = \mathcal{N} \oplus \mathcal{T}$, is completely determined by:

$$[\partial_t, X] = \frac{1}{2}X;\ [\partial_t, Z] = Z;\ [T(N), T(N)]_{/SN} = [T(N), T(N)]_{/N}, \quad (25)$$

where $X \in \mathcal{X}$ and $Z \in \mathcal{Z}$. By the traditional interpretation, this Lie algebra is identified with the tangent space at the unity $(0,0,1)$. The indefinite metric tensor is introduced by the left-invariant extension of the indefinite inner product, $\langle\, , \rangle$, defined on the solvable Lie algebra S by $\langle \partial_t, \partial_t \rangle = -1$, $\langle \partial_t, T(N) \rangle = 0$, and $\langle T(N), T(N) \rangle = \langle T(N), T(N) \rangle_N$.

The left-invariant extensions $\mathbf{Y}_i; \mathbf{V}_\alpha; \mathbf{T}$ of unit tangent vectors $E_i = \partial_i; e_\alpha = \partial_\alpha; \epsilon = \partial_t$ picked up at $(0,0,1)$, are the vector fields:

$$\mathbf{Y}_i = t^{\frac{1}{2}}\mathbf{X}_i \quad ; \quad \mathbf{V}_\alpha = t\mathbf{Z}_\alpha \quad ; \quad \mathbf{T} = t\partial_t, \quad (26)$$

where $\mathbf{X}_i$ and $\mathbf{Z}_\alpha$ are invariant vector fields on the nilpotent subgroup N.

It follows that not t but T defined by $\partial_T = \mathbf{T}$ is the correct physical time-parameterization on the t parameter lines. Endowed with this parameterization, they become geodesics on SN. Transformation law $\partial_T = (dt/dT)\partial_t$ yields: $(dt/dT) = t$; $\ln t = T$; and $t = e^T$, thus the t-level sets are the same as the $T = \ln t$-level sets and subgroup N corresponds both to $t = 1$ and $T = 0$. The reversed time is $\tau = -T$.

Let $\mathbf{X}$ and $\mathbf{Z}$ be invariant vector fields on the subgroup N, furthermore, $c_x(s)$ and $c_z(s)$ be their integral curves of finite length denoted by $||c_x||$ and $||c_z||$, respectively. The flow generated by ∂_τ moves the curves to $c_x^\tau(s)$ resp. $c_z^\tau(s)$. By the above formulas, the tangent vectors $\dot{c}_x(s)$ and $\dot{c}_z(s)$ are of unit length just on N, defined as level set corresponding to $T = 0$, and they are changing according to the formulas $||c_x^\tau|| = ||c_x||e^{q\tau/2}$ resp. $||c_z^\tau|| = ||c_z||e^{q\tau}$. That is, by considering them as functions of the time-variable τ, the length is increasing such that the rate of change (derivative with respect to τ) is proportional to the length of the curves. In other words, this space-time represents an expanding universe where the distance between objects is growing according to a law similar to Hubble's.

But the law yielded on Zeeman space-time is very different from that of Hubble. Namely, on the Friedmann model, where the original Hubble law is mathematically established, the expansion is not accelerating. By contrast, the Zeeman space-time is expanding at an exponentially accelerating rate, meaning that the acceleration, together with the higher order ones, are also accelerating, and each of these higher order accelerations are proportional to the distance. That is, the farther is the galaxy, the bigger are the accelerations - of any order - by which it moves away from us. (To see the statement, consider the higher order derivatives of the arc-length formulas.)

3.3. The expanding Monistic Wave Laplacian.

The Monistic Wave Laplacian on accelerating Zeeman space-times is:

$$\Delta_E = \{(e^{-2\tau}\Delta_Z + \frac{2m\mathbf{i}}{\hbar}e^{-\tau}\partial_\tau - \partial^2_{\tau\tau}) - (\frac{k}{2} + l)\partial_\tau\} + \quad (27)$$

$$+e^T\{\Delta_X+\frac{1}{4}|X|^2\Delta_Z+\sum\partial_\alpha D_\alpha\bullet+\frac{2m\mathbf{i}}{\hbar}\partial_T\}. \tag{28}$$

Without $\frac{2m\mathbf{i}}{\hbar}e^{-\tau}\partial_\tau-\frac{2m\mathbf{i}}{\hbar}e^{-\tau}\partial_\tau=0$, one has the undecomposed Δ_E in which

$$EY=e^{-2\tau}\Delta_Z-\partial^2_{\tau\tau}-(\frac{k}{2}+l)\partial_\tau \tag{29}$$

is the expanding version of the original Yukawa operator. The first term in the decomposed operator (27) could be the candidate for the expanding exhausted Yukawa operator, however, it turns out that it is a much more complicated operator. The operator on the last line is the static mammoth Schrödinger operator multiplied by e^T on the left side. Thus, its action and waves can completely be described by the static operator. Since it is non-expanding even in the accelerating universe, this operator must be identified with the Monistic Wave Operator of ordinary matter.

3.4. Mass-assignment on accelerating Zeeman space-times

As compared with the static case, the mass-acquiring process evolves in a much more complicated way on expanding Zeeman space-times. The details of the following brief outline are established in [Sz2, Sz4].

The expanding wave functions are introduced by:

$$\Psi_{\rm ex}(X,Z,\tau)=\int_{\mathbb{R}^l}e^{\mathbf{i}(\langle Z,K\rangle-\omega e^{-\tau})}e^{Q\tau}\phi_{\mathbf{s}}\Pi^{\mathbf{s}}_{K_u}\psi^{(g)}(X,K_u)dK, \tag{30}$$

where the Q is determined such that $\Psi_{\rm ex}$ is an eigenfunction of (29). This is satisfied if and only if $Q=\frac{1}{2}(1-\frac{k}{2}-l)$, when the eigenvalue is:

$$\frac{m^2c^2}{h^2}-Q^2-(\frac{k}{2}+l)Q=\frac{m^2c^2}{h^2}-(\frac{1}{4}+\frac{1}{4}(\frac{k}{2}+l)^2)+\frac{1}{2}(\frac{k}{2}+l)^2. \tag{31}$$

Then, the expanding exhausting waves are defined by:

$$\hat\Psi_{\rm ex}(X,Z,\tau)=e^{\frac{\mathbf{i}mc^2e^{-\tau}}{\hbar}}\int_{\mathbb{R}^l}e^{\mathbf{i}(\langle Z,K\rangle-\frac{\omega e^{-\tau}}{c})}e^{Q\tau}\phi_{\mathbf{s}}\Pi^{\mathbf{s}}_{K_u}\psi^{(g)}dK, \tag{32}$$

on which, the action of the first operator:

$$(e^{-2\tau}\Delta_Z-\frac{2m\mathbf{i}}{\hbar}e^{-\tau}\partial_\tau-\partial^2_{\tau\tau})-(\frac{k}{2}+l)\partial_\tau. \tag{33}$$

appearing in the decomposed Monistic Wave Laplacian (27) results:

$$-(\frac{1}{4}+\frac{1}{4}(\frac{k}{2}+l)^2-\frac{1}{2}(\frac{k}{2}+l)^2+\frac{\mathbf{i}mc^2}{h}e^{-\tau})\hat\Psi_{\rm ex}. \tag{34}$$

That is, unlike in the static case, the candidate for the expanding exhausted Yukawa operator is not completely exhausted by $\hat\Psi_{\rm ex}$. They are not even eigenfunctions regarding this operator. The reason is that only the ordinary matter has been exhausted but the dark matter and dark energy are still there. This gives rise to the opportunity for finding their Wave Operators in (33) such that eigenvalue $\frac{1}{4}+\frac{1}{4}(\frac{k}{2}+l)^2$ is associated with the dark matter and the rest with the dark energy. The corresponding decomposition is:

$$\{(e^{2T}\Delta_Z+(\frac{2m\mathbf{i}}{h}e^T+\frac{\frac{k}{2}+l}{1-\frac{k}{2}-l})\partial_T-\partial^2_{TT}+\frac{\mathbf{i}(\frac{k}{2}+l)^2e^T}{1-\frac{k}{2}-l}\Omega\}+ \tag{35}$$

$$-\{(\frac{k}{2}+l-\frac{\frac{k}{2}+l}{1-\frac{k}{2}-l})\partial_\tau+\mathbf{i}(\frac{m}{h}e^{-\tau}+\frac{(\frac{k}{2}+l)^2}{1-\frac{k}{2}-l}e^{-\tau}\Omega)\}, \tag{36}$$

IOP Conf. Series: Journal of Physics: Conf. Series **804** (2017) 012041 doi:10.1088/1742-6596/804/1/012041

where (35) is the dark matter and (36) the dark energy operator. The action of the universal gravitation operator Ω on expanding waves is defined by:

$$-\frac{2m}{h}\int_{\mathrm{R}^l}\frac{\omega}{c}e^{\mathbf{i}(\langle Z,K\rangle-\frac{\omega}{c}e^{-\tau})}e^{Q\tau}(\phi_{\mathbf{s}}\Pi^{\mathbf{s}}_{K_u}(\psi))(X,K))dK= \tag{37}$$

$$=\frac{1}{c}\frac{2m^2c^2}{h^2}\sqrt{1-\frac{h^2}{m^2c^2}\Delta_Z(\Psi_{\mathrm{ex}})(X,Z,\tau)}=\frac{2m}{h}\Omega(\Psi_{\mathrm{ex}})(X,Z,\tau). \tag{38}$$

Terms including Ω and $p(k,l) = ((k/2)+l)/(1-(k/2)-l)$ terminate the relations of the dark matter to those represented by $-\frac{1}{2}(\frac{k}{2}+l)^2$ and $\frac{\mathbf{i}mc^2}{h}e^{-\tau}$ in (34). They entirely show up in the dark energy operator. The dark matter, dark energy, and the mammoth Schrödinger operators (the latter appears as the second long term on line (28)) fully establish the wave operators of the three matter-energy formations in terms of the expanding Monistic Wave Laplacian. In order to associate positive energy terms with them, the dark and ordinary matter operators should be considered regarding the shrinking time direction T, whereas, the dark energy regarding $\tau = -T$. The considered expanding waves are eigenfunctions of the dark matter operator with eigenvalue $-(\frac{1}{4}+\frac{1}{4}(\frac{k}{2}+l)^2)$. It represents positive energy level regarding the shrinking direction. The rest in (34) is counted for the dark energy operator. They form a Dynamic Duo in EY, displayed by the dynamical timing of expanding exhausted waves (32). The mammoth Schrödinger operator should be treated according to the static case. It acts on the static exhausting waves decomposes like a Dynamic Duo with components corresponding to the actual Schrödinger operator and the static universal gravitation operator. The orientations clearly indicate that the dark energy operator is associated with pushing away and the dark matter operator with pulling together force acting on the space-section, that is, on Zeeman manifolds. Namely, the accelerations are positive regarding τ and negative regarding T. Since the ordinary matter is described by static waves, this category also behaves according to the expectations.

3.5. Historical retrospect and outlook to the developments

The primary goal in this paper is to establish the relativistic wave operators of the dark energy, dark matter, and ordinary matter in the normal (non-anomalous) case by a new spectral mass-assignment procedure not using the idea of symmetry breaking, the main tool to establish the Higgs mechanism. Nonetheless, this new mechanism (purely based on Wave Operators) can be related to those evolved in SM and it can also be bridged to classical wave operators but which appear in new forms having contrasting features with the originals. Both relations need further clarifications.

3.5.1. Relations to pristine quantum operators It is known that Klein-Gordon's relativistic wave equation of the electron was also discovered by Schrödinger, who, however, was not able to handle the second order term ∂^2_{tt} present there which led to spectral computations not matching the experiments. He could restore the agreement with the reality by exchanging it for the first order differential operator $\mathbf{i}\hbar\partial_t$, which costed the price that the equation became non-relativistic valid only for small speeds $v << c$. The relativity was restored by Dirac, by considering appropriate first order partial derivatives regarding the space-variables, as well.

Operator called above as "static exhausted Yukawa operator" has also been well known in the quantum theory. The equation yielding the harmonic functions regarding

IOP Conf. Series: Journal of Physics: Conf. Series **804** (2017) 012041 doi:10.1088/1742-6596/804/1/012041

this operator is called "non-relativistic wave equation". It is noteworthy that Pauli determined solutions of this equation by means of de Broglie's waves such that he considered only lower order terms in the power series of ω (cf. [P], Vol. 5., pages 3-4). Such approximating solutions are really non-relativistic which describe physical reality just for small speeds $v << c$. This equation has always been associated with Yukawa's pion operator and never with the Schrödinger equation.

All these non-relativistic treatments are not to be confused with those applied on the static Zeeman space-time, where the two operators appear together after the relativistic Monistic Wave Laplacian is decomposed by introducing the trivial term $\frac{2m\mathbf{i}}{\hbar}\partial_t - \frac{2m\mathbf{i}}{\hbar}\partial_t$ there and establishing the exhausted operator by Yukawa's and with $\frac{2m\mathbf{i}}{\hbar}\partial_t$. The mammoth Schrödinger operator is equal to the rest of the Monistic Wave Laplacian. No operator exchange or any kind of negligence has been implemented, thus all these operators remain relativistic. They arise as two facets of the relativistic static Monistic Wave Laplacian, representing the unification of strong force interactions with the electromagnetic, weak, and gravitational. It should also be emphasized that they are written up in terms of a fixed coordinate system, and, in order to comply with relativity, in other coordinate system, they should be computed in accordance with Lorentz transformations. In this theory, term ∂^2_{tt} has not gotten rid of but remains there in the Yukawa operator as one of the contributors for establishing the new spectral mass assigning process. This program can be carried through just by the two operator together, which situation is present just on Zeeman space-time where the exterior and the common interior space appear as partners of each other. This tool was not available for Schrödinger who therefore was compelled to use non-relativistic tools in order to keep his equation close to physical reality.

There arise differences also between the ways how the harmonic solutions of the actual Schrödinger operator, defined on the static Zeeman space-time, and Schrödinger's original operator are obtained, respectively. To see this notice that the exhausting wave $\tilde{\Psi}$ - which is harmonic regarding the exhausted Yukawa operator, thus the result, due to the action of the complete Monistic Wave Laplacian on this wave, depends on the action of the mammoth Schrödinger operator which really contributes an idiosyncratic "mammoth "structure to the solution. Namely, the time is involved into two exponential functions standing both in the insides and outsides of integration. This feature actually results the decomposition into the actual Schrödinger operator and the universal mass-relating gravitation operator.

Also notice that the actual Schrödinger operator is related not to the linear mass m but to the squared mass, m^2. This is one of the reasons as to why should the Dirac operators apply square root in order to have linear masses for the fermions. An other reason why Dirac searched for quantum operators different from Schrödinger's was that the latter did not explain the observed double spectral lines which are due to the mutual spin of particles. In [Sz2, Sz3, Sz4, Sz5, Sz6], the proper spin operators are established by responding to all these requirements. To see the difference between the two approaches, it is enough to mention that, for seeking out the harmonic solutions of his equation, Schrödinger considered waves of the form $\Psi(x,t) = \psi(x)e^{-\mathbf{i}Et/\hbar}$, which neither respond to the universal gravitation nor have any relation to the exhausted Yukawa operator. Thus they are non-relativistic also on the Zeeman space-time. The anomalous case will also be different from that of Dirac whose operator has not been related to any kind of gravitational wave operator.

3.5.2. Densities of dark energy, dark matter, and ordinary matter The wave operators of dark energy, dark matter, and ordinary matter in the accelerating universe are completely unexplored in the literature, insofar. There are new arguments developed in [Sz2, Sz3, Sz4, Sz5, Sz6, Sz7] which demonstrate in consent with the experiments as to why are (28), (35), and (36) the appropriate wave operators for the three matter-energy formations. One of them is the explicit computations of the participation ratios which is carried out by relating these operators to the energy densities involved to the time-time component of the stress energy tensor $E(A, A^*) = Ri(A, A^*) - \frac{1}{2}\mathcal{R}\langle A, A^*\rangle$, where $A = X + Z + \mathbf{T}$ and $A^* = X^* + Z^* + \mathbf{T}$. These objects are considered both in the static and expanding cases, in [Sz2, Sz5, Sz6]. The point is that the above decomposition uniquely determines a corresponding decomposition of the energy densities appearing in the time-time component of the stress-energy tensor and their ratios determine the participation ratios of the three matter-energy formations on the Zeeman space-time model. Thus the model passes a very serious reality test if these participation ratios agree with those measured in Nature.

To carry out these computations, the density corresponding to the ordinary matter is the first to be determined. Explicit computations show that the static stress energy tensor $E_{St}(A, A^*)$ also appears in the expanding stress energy tensor with which the only non-expanding matter formation - the ordinary matter - must be associated. But there is a little confusion at this point, because a negative number, $-kl/8$, appears for the ordinary matter, whereas, the time-time component of the rest of the tensor, with which both the dark matter and dark energy is associated, is positive. But this enigmatic appearance only indicates that all matter-energy can not appear together with the ordinary matter such that everything is positively detected. Since the time directions for the operators are chosen such that they represent positive energies, the non-expanding ordinary matter must evidently be associated with density. $kl/8$. By the above arguments, the dark matter density is $\frac{1}{4}+\frac{1}{4}(\frac{k}{2}+l)^2$, directed to the shrinking direction. What is left in the total:

$$DE(k,l) = \frac{1}{4} + \frac{1}{2}(\frac{1}{2}(\frac{k}{2} + l)^2 + \frac{3k}{4} + 3l) \tag{39}$$

is the dark energy density which is counted regarding the expending direction.

These computations actually give rise to a new correspondence principle, associating energy densities, involved into the time-time component of the stress energy tensor, to the wave operators of the three energy-matter formations. This is the tool by which the participation ratios $(DER : DMR : OMR) = (70\% : 25\% : 5\%)$ on the dynamical Eternal Whizz model can be established. The computations are demonstrated on the most simple one particle model, defined by $k = 2$ and $l = 1$, where the corresponding densities are:

$$DE(2,1) = \frac{14}{4}, \quad DM(2,1) = \frac{5}{4}, \quad OM(2,1) = \frac{1}{4}. \tag{40}$$

Thus the total density is:

$$TOT(2,1) = \frac{14}{4} + \frac{5}{4} + \frac{1}{4} = \frac{20}{4}, \tag{41}$$

and the corresponding participation ratios are:

$$DER(2,1) = \frac{DE(2,1)}{TOT(2,1)}100\% = \frac{14}{20}100\% = 70\%, \tag{42}$$

$$DMR(2,1) = \frac{DM(2,1)}{TOT(2,1)}100\% = \frac{5}{20}100\% = 25\%, \tag{43}$$

$$OMR(2,1) = \frac{OM(2,1)}{TOT(2,1)}100\% = \frac{1}{20}100\% = 5\%. \tag{44}$$

That is, the participation ratios measured in Nature are surprisingly accurately represented by one particle systems and by all those models which are solvable extensions of Zeeman manifolds defined by Cartesian products of one particle Zeeman manifold models. They constitute very complex highly realistic physical models, that are made up by heavy protons, neutrons, electrons, and bosons associated with the mass assignment and the action of the Stabler. Although the spool gravitation is individualistic, holding together just the splinters derived from the individual particles, the Ω-gravity wraps them up into a physically connected system where two elements are related to each other at least by this gravitation. But one thing is still missing from this world, namely, there is no decay associated with higher dimensional Z-spaces and the actions of the $\mathbf{W}$-operators. The above correspondence principle is a working tool for computing the participation ratios when the system of independent 1-particles is mixed up with multiparticles having common higher dimensional Z-spaces. They represent a universe of such high complexity which can correspond to that observed today. All quantum physical laws are engraved into the Monistic Wave Laplacian. These ideas give rise to the Eternal Whizz model of the universe which started out infinitely long time ago and last forever. It is a dynamical model which represented independent 1-particles for an infinitely long time till the system got cool enough to admit multiparticles systems having common higher dimensional Z-spaces. The time when this took place is the Zeeman Big Bang, when the universe became of such high complexity as is today.

3.5.3. CERN's scalar boson The theory also explores as to how does CERN's scalar boson appear in this quantum universe and how are the particles of proper spin created there. The scalar boson emerges, in strong relation with the the participation ratios, as a compound Monistic Boson, having inner structure, which has components in each of the three matter-energy formations. Its mass overwhelmingly origins from dark matter and dark energy, which formations are completely made up by these two components of the scalar boson. The Z and W also appear as compound particles, originated from dark energy and ordinary matter only. Besides them, there are splinter W's of the same origin. The ordinary matter components of the scalar boson relating to these particles are called "Zolly" and "Willey", respectively, which together make up the complete amount of ordinary scalar matter from which all Bosonic and Fermionic ordinary matters are originated by the corresponding spin operators acting on spinors.

3.5.4. Zeeman space-time vs. Higgs field The Zeeman space-time theory does not contradict SM or the mass-assigning procedure established by the Higgs field. In case of Higgs mechanism, the particles - considered as tiny billiard balls - acquire mass through their interactions with an external field, the Higgs field. Whereas, on the Zeeman space-time, the particles are represented by waves, endowed with masses by the above described spectral procedure, which originates its name for relating the mass to the spectrum of the Hamilton operators involved into the Monistic Wave Laplacian. But, all these differences notwithstanding, the Higgs field and Zeeman space-time are not contradictory concepts, which point of view can be explained as follows.

IOP Conf. Series: Journal of Physics: Conf. Series **804** (2017) 012041 doi:10.1088/1742-6596/804/1/012041

The contradiction between general relativity and quantum theory origins from the different representations of matter. In general relativity, Einstein's field equation completely embeds it into the fabric of the curved space-time. While in quantum theory, the matter is released from there and is described, instead, by waves or wave-packets, living in a flat Minkowski space. Notice, for instance, that all $SU(n)$ models are defined over a Minkowski space, supporting these arguments. Additionally, there has also been created the Higgs field, a particular matter - a kind of plasma - that fills up the flat Minkowski space and provides masses to the released matter particles. Exactly this plasma matter brings up a question which helps us to build up connections between the SM and the Zeeman space-time theories. Namely, the question arises, as to whether there exists a curved space-time which realizes this plasma-matter according to the principles of general relativity.

The Zeeman space-time seems to be the relativistic space-time realization of this plasma-matter. But it should be pointed out that the theory breaks off both from general relativity and classical quantum theory and stands in the middle between them. The space-time does not absorbs into its fabric all the matter existing in the universe but only that much which presents the Monistic Wave Operator of the Zeeman space-time. The rest of the matter is represented by waves defined by Z-Fourier transforms. The Monistic Wave Operator governs all physical processes taking place there. In this theory, the only field which exists there is the pseudo Riemannian metric of Lorentz signature defined on the Zeeman space-time. This is it what corresponds to the Higgs plasma, which have been explored in SM by means available for Lagrangian theories.

Acknowledgments

Research partially supported by the Max Planck Institute in the Sciences, Leipzig, Germany, and Faculty Development Program 2012-13 of Lehman College of CUNY.

[ATLAS] ATLAS Collaboration: Observation of a new particle in the search for the Standard Model Higgs boson with the ATLAS detector at the LHC. *Physics Letters B*, Vol 716, Issue 1, 17 September 2012, Pages 1-29.

[B] D. Bohm: Quantum Theory *Dover Publications, INC., New York*, 1979

[CMS] CMS Collaboration: Observation of a new boson at a mass of 125 GeV with the CMS experiment at the LHC. *Physics Letters B*, Vol 716, Issue 1, 17 September 2012, Pages 30-61.

[D] P. A. M. Dirac: Lectures on Quantum Mechanics. *Belfer Graduate School of Science*, 1964; *Dover Publications, INC*, 2001.

[G] M. Gell-Mann: Quark Jaguar *W. H. Freeman and Company*, 1994.

[GN] M. Gell-Mann and Y. Ne'eman: The Eightfold Way *Benjamin*, 1964.

[H] S. Hawking: A Brief History of Time, Bantam Books, 1988.

[K] A. Kaplan: Riemannian nilmanifolds attached to Clifford modules. *Geom. Dedicata*, 11:127–136, 1981.

[KN] S. Kobayashi and K. Nomizu: Foundations of differential geometry II. *Interscience Publishers, New York*, 1969.

[LL] L. D. Landau, E. M. Lifshitz: Quantum Mechanics. Pergamon, 1958.

[N] Y. Nambu: Quarks. World Scientific, 1985.

[P] W. Pauli: Pauli Lectures on Physics. Dover, 2000.

[Pe1] S. Perlmutter et al: Discovery of a supernova explosion at half the age of the universe and its cosmological implications. (The Supernova Cosmology Project), *Nature*, 391, 51 (1998).

[Pe2] S. Perlmutter: Supernovae, Dark Energy, and the Accelerating Universe, *Physics Today*, April 2003.

[RS] A. G. Riess et al, including B. P. Schmidt: Observational evidence from supernovae for an accelerating universe and a cosmological constant. *The Astronomical Journal*, 116, 1009-1038, 1998 September.

[Sz1] Zoltán Imre Szabó: Ordinary matter, dark matter, and dark energy on normal Zeeman space-time. Preprint (This article).
[Sz2] Zoltán Imre Szabó: Zeeman manifolds and space-times: Comprehensive unification models of relativistic quantum theory; The spectral mass assignment and a preview, Preprint.
[Sz3] Zoltán Imre Szabó: Point particles on prototype Zeeman manifolds. Preprint.
[Sz4] Zoltán Imre Szabó: Extended particles on prototype Zeeman manifolds. Preprint.
[Sz5] Zoltán Imre Szabó: The prototype Zeeman space-time. Preprint.
[Sz6] Zoltán Imre Szabó: HyperKähler-Zeeman manifolds and space-times: General models of an all-comprehensive unification, Preprint.
[Sz7] Zoltán Imre Szabó: Symmetries, Symmetry Violations, and Isospectralities on Zeeman Manifolds and Zeeman Space-Times, Preprint.
[YM] C. N. Yang and R. L. Mills, *Phys. Rev.*, 96, 1954.
[V] M. Veltman: Facts and Mysteries in Elementary Particle Physics. *World Scientific,* 2003.

Action-angle approach to the geodesic motions in the homogeneous Sasaki-Einstein space $T^{1,1}$

Mihai Visinescu

Department of Theoretical Physics, National Institute for Physics and Nuclear Engineering, Magurele, P.O.Box M.G.-6, Romania

E-mail: mvisin@theory.nipne.ro

Abstract. The complete integrability of geodesics in the homogeneous Sasaki-Einstein space $T^{1,1}$ makes possible the explicit construction of the action-angle variables. This parametrization of the phase space represents a useful tool for developing perturbation theory. We find that two pairs of fundamental frequencies of the geodesic motions are resonant indicating a chaotic behavior when the integrable Hamiltonian is perturbed by a small non-integrable piece.

1. Introduction

Recently Sasaki geometry, as a natural odd-dimensional counterpart of the Kähler geometry, has become of significant interest in some modern studies in mathematics and physics [1, 2]. Sasaki-Einstein (SE) manifolds whose metric cones are Calabi-Yau manifolds find applications in string theory in connection with AdS/CFT correspondence which relates quantum gravity in a certain background to ordinary quantum field theory without gravity. A large class of examples consists of type IIB string theory on the background $AdS_5 \times Y_5$ with Y_5 a five-dimensional SE space. A particular interesting class of toric SE structures on $S^2 \times S^3$ have been studied by physicists [3, 4]. An extensively studied case is when Y_5 is $T^{1,1}$ which is homogeneous SE space with $SU(2)^2 \times U(1)$ isometry. The $AdS \times T^{1,1}$ is the first example of a supersymmetric holographic theory based on a compact manifold which is not locally S^5.

The purpose of this paper is to analyze the complete integrability of geodesics of the five-dimensional SE space $T^{1,1}$. We present the action-angle formulation of the geodesic motions in $T^{1,1}$ space. The description of the integrability of geodesics in $T^{1,1}$ in terms of action-angle variables give us a comprehensive description of the dynamics. We find that two pairs of frequencies of the geodesic motions are resonant giving way to chaotic behavior when the integrable Hamiltonian is perturbed by a small non-integrable piece.

The organization of the paper is as follows. In the next Section we present the functionally independent integrals of motions for geodesics in $T^{1,1}$ space. In Section 3 we describe the construction of action-angle variables and evaluate the fundamental frequencies of the geodesic motions. The paper ends with conclusions in Section 4.

2. $T^{1,1}$ space

Until recently, the only five-dimensional SE manifolds that were known explicitly were the round metric on S^5 and the homogeneous metric $T^{1,1}$ on $S^2 \times S^3$.

The metric on $T^{1,1}$ may be written down explicitly by utilizing the fact that it is a $U(1)$ bundle over $S^2 \times S^2$. We choose the coordinates (θ_i, ϕ_i) , $i = 1, 2$ to parametrize the two spheres S^2 in the standard way, while the angle $\psi \in [0, 4\pi)$ parametrizes the $U(1)$ fiber. The metric on $T^{1,1}$ may be written as [5]

$$ds^2(T^{1,1}) = \frac{1}{6}(d\theta_1^2 + \sin^2\theta_1 d\phi_1^2 + d\theta_2^2 + \sin^2\theta_2 d\phi_2^2) + \frac{1}{9}(d\psi + \cos\theta_1 d\phi_1 + \cos\theta_2 d\phi_2)^2 . \quad (1)$$

In what follows we introduce $\nu = \frac{1}{2}\psi$ so that ν has canonical period 2π.

On a Riemannian manifold (M, g) with local coordinates x^μ and metric $g_{\mu\nu}$ the geodesics are represented by the trajectories of test-particles with proper time Hamiltonian

$$H = \frac{1}{2} g^{\mu\nu} p_\mu p_\nu , \quad (2)$$

where p_μ are the canonical momenta conjugate to the coordinate x^μ, $p_\mu = g_{\mu\nu}\dot{x}^\nu$ with overdot denoting proper time derivative.

In the case of $T^{1,1}$ space, the conjugate momenta to the coordinates $(\theta_1, \theta_2, \phi_1, \phi_2, \nu)$ are

$$\begin{aligned}
p_{\theta_1} &= \frac{1}{6}\dot{\theta}_1 , \\
p_{\theta_2} &= \frac{1}{6}\dot{\theta}_2 , \\
p_{\phi_1} &= \frac{1}{6}\sin^2\theta_1\,\dot{\phi}_1 + \frac{1}{9}\cos^2\theta_1\,\dot{\phi}_1 + \frac{1}{9}\cos\theta_1\,\dot{\psi} + \frac{1}{9}\cos\theta_1\cos\theta_2\,\dot{\phi}_2 , \\
p_{\phi_2} &= \frac{1}{6}\sin^2\theta_2\,\dot{\phi}_2 + \frac{1}{9}\cos^2\theta_2\,\dot{\phi}_2 + \frac{1}{9}\cos\theta_2\,\dot{\psi} + \frac{1}{9}\cos\theta_1\cos\theta_2\,\dot{\phi}_1 , \\
p_\nu &= \frac{2}{9}\dot{\nu} + \frac{1}{9}\cos\theta_1\,\dot{\phi}_1 + \frac{1}{9}\cos\theta_2\,\dot{\phi}_2 ,
\end{aligned} \quad (3)$$

and the conserved Hamiltonian (2) takes the form:

$$H = 3\left[p_{\theta_1}^2 + p_{\theta_2}^2 + \frac{1}{4\sin^2\theta_1}(2p_{\phi_1} - \cos\theta_1 p_\nu)^2 + \frac{1}{4\sin^2\theta_2}(2p_{\phi_2} - \cos\theta_2 p_\nu)^2\right] + \frac{9}{8}p_\nu^2 . \quad (4)$$

Using the complete set of Killing vectors and Killing-Yano tensors of SE space $T^{1,1}$ we find a lot of conserved quantities [6, 7, 8, 9]. However the number of functionally independent integrals of motion is only *five* which means that the geodesic flow on $T^{1,1}$ space is completely integrable, but not superintegrable.

We can choose as functionally independent constants of motion the energy

$$E = H , \quad (5)$$

the momenta corresponding to the cyclic coordinates (ϕ_1 , ϕ_2 , ν)

$$\begin{aligned}
p_{\phi_1} &= c_{\phi_1} , \\
p_{\phi_2} &= c_{\phi_2} , \\
p_\nu &= c_\nu ,
\end{aligned} \quad (6)$$

where $(c_{\phi_1} , c_{\phi_2} , c_\nu)$ are some constants, and one of the two total $SU(2)$ angular momentum

$$\vec{J}_i^{\,2} = p_{\theta_i}^2 + \frac{1}{4\sin^2\theta_i}(2p_{\phi_i} - \cos\theta_i p_\nu)^2 + \frac{1}{4}p_\nu^2 , \quad i = 1, 2 . \quad (7)$$

3. Action-angle variables

The strategy to construct the action-angle variables consists in a few steps [10, 11]. First of all, we fix a level surface $\mathbf{F} = (H, p_{\phi_1}, p_{\phi_2}, p_\nu, \vec{J}_1^{\,2}) = \mathbf{c}$ of the mutually commuting constants of motion (5)–(7). After that, we introduce the generating function for the canonical transformation from the coordinates $(\mathbf{p}, \mathbf{q})$, where $\mathbf{p}$ are the conjugate momenta (3) to the coordinates $\mathbf{q} = (\theta_1, \theta_2, \phi_1, \phi_2, \nu)$, to the action-angle variables $(\mathbf{J}, \mathbf{w})$.

Taking into account that the Hamiltonian (4) has no explicit time dependence, we can write the Hamilton's principal function

$$S(\mathbf{q}, \mathbf{c}) = W(\mathbf{q}, \mathbf{c}) - Et \,, \tag{8}$$

where the Hamilton's characteristic function is

$$W = \sum_i \int p_i dq_i \,. \tag{9}$$

In the case of the geodesic motions in SE space $T^{1,1}$, the variables in the Hamilton-Jacobi equation are separable and we seek a solution with the characteristic function

$$W(\theta_1, \theta_2, \phi_1, \phi_2, \nu) = W_{\theta_1}(\theta_1) + W_{\theta_2}(\theta_2) + W_{\phi_1}(\phi_1) + W_{\phi_2}(\phi_2) + W_\nu(\nu) \,. \tag{10}$$

For completely separable Hamilton-Jacobi equation, considering only closed orbits, the *action variables* are defined as integrals over complete period in the (q_i, p_i) plane:

$$J_i = \oint p_i dq_i = \oint \frac{\partial W_i(q_i; c)}{\partial q_i} dq_i \qquad \text{(no summation)} \,. \tag{11}$$

J_i's form n independent functions of c_i's and can be taken as a set of new constant momenta.

For the cyclic variables $(\phi_1 \,,\, \phi_2 \,,\, \nu)$ from (6) and (9) we get straight

$$\begin{aligned} W_{\phi_1} &= p_{\phi_1}\phi_1 = c_{\phi_1}\phi_1 \,, \\ W_{\phi_2} &= p_{\phi_2}\phi_2 = c_{\phi_2}\phi_2 \,, \\ W_\nu &= p_\nu \nu = c_\nu \nu \,, \end{aligned} \tag{12}$$

and the corresponding action variables (11) are easily calculated

$$\begin{aligned} J_{\phi_1} &= 2\pi c_{\phi_1} \,, \\ J_{\phi_2} &= 2\pi c_{\phi_2} \,, \\ J_\nu &= 2\pi c_\nu \,. \end{aligned} \tag{13}$$

Conjugate *angle variables* w_i are given by

$$w_i = \frac{\partial W}{\partial J_i} = \sum_{j=1}^{n} \frac{\partial W_j(q_j; J_1, \cdots, J_n)}{\partial J_i} \,, \tag{14}$$

having a linear evolution in time

$$w_i = \omega_i t + \beta_i \,, \tag{15}$$

with β_i other constants of integration.

Having determined the characteristic functions corresponding to the cyclic variables, the Hamilton-Jacobi equation becomes

$$
\begin{aligned}
E = \; & 3\left[\left(\frac{\partial W_{\theta_1}}{\partial \theta_1}\right)^2 + \frac{1}{4\sin^2\theta_1}\left(2c_{\phi_1} - \cos\theta_1 c_\nu\right)^2\right] \\
& + 3\left[\left(\frac{\partial W_{\theta_2}}{\partial \theta_2}\right)^2 + \frac{1}{4\sin^2\theta_2}\left(2c_{\phi_2} - \cos\theta_2 c_\nu\right)^2\right] + \frac{9}{8}c_\nu^2 \,.
\end{aligned}
\tag{16}
$$

In this equation the dependencies on θ_1 and θ_2 are separated and we can set the quantities enter the two square brackets to be constants denoted by $c_{\theta_1}^2$ and respectively $c_{\theta_2}^2$. Consequently we have (16):

$$
\frac{\partial W_{\theta_i}}{\partial \theta_i} = \sqrt{c_{\theta_i}^2 - \frac{(2c_{\phi_i} - \cos\theta_i c_\nu)^2}{4\sin^2\theta_i}} \quad , \quad i = 1, 2 \,, \tag{17}
$$

and the corresponding action variables are

$$
J_{\theta_i} = \oint d\theta_i \sqrt{c_{\theta_i}^2 - \frac{(2c_{\phi_i} - \cos\theta_i c_\nu)^2}{4\sin^2\theta_i}} \quad , \quad i = 1, 2 \,. \tag{18}
$$

An efficient procedure to evaluate this integral is to put $\cos\theta_i = t_i$ and extend t_i to a complex variable z. The integral becomes a closed contour integral in the z-plane. The turning points of the t_i-motions are

$$
t_{i\pm} = 2\frac{c_{\phi_i} c_\nu \pm c_{\theta_i}\sqrt{4c_{\theta_i}^2 + c_\nu^2 - 4c_{\phi_i}^2}}{4c_{\theta_i}^2 + c_\nu^2} \quad , \quad i = 1, 2 \,, \tag{19}
$$

which are real for

$$
4c_{\theta_i}^2 + c_\nu^2 - 4c_{\phi_i}^2 \geq 0 \quad , \quad i = 1, 2 \,, \tag{20}
$$

and situated in the interval $(-1, +1)$. We cut the complex z-plane from t_- to t_+ and the closed contour integral is a loop enclosing the cut in a clockwise sence. The contour can be deformed to a large circular contour plus two contour integrals about the poles at $z = \pm 1$.

After the standard evaluation of the residues and the contribution of the large contour integral we finally get [12]

$$
J_{\theta_i} = 2\pi\left[\frac{1}{2}\sqrt{4c_{\theta_i}^2 + c_\nu^2} - c_{\phi_i}\right] \quad , \quad i = 1, 2 \,. \tag{21}
$$

We observe that the constants of motion $J_{\theta_i}, J_{\phi_i}, J_\nu, E$ are connected by the relation:

$$
H = E = \frac{3}{4\pi^2}\left[\left(J_{\theta_1} + J_{\phi_1}\right)^2 + \left(J_{\theta_2} + J_{\phi_2}\right)^2 - \frac{1}{8}J_\nu^2\right] . \tag{22}
$$

This result is characteristic for the complete integrable systems. The number of conserved quantities equals the number of degrees of freedom and the Hamiltonian depends only on the action variables [10, 11].

Moreover, we observe that the energy depends on J_{θ_i} and J_{ϕ_i} in the combination $J_{\theta_i} + J_{\phi_i}$ meaning that the frequencies of motions in θ_i and ϕ_i are identical:

$$
\omega_{\theta_i} = \omega_{\phi_i} = \frac{\partial H}{\partial J_{\theta_i}} = \frac{\partial H}{\partial J_{\phi_i}} = \frac{3}{2\pi^2}\left(J_{\theta_i} + J_{\phi_i}\right) \quad , \quad i = 1, 2 \,. \tag{23}
$$

IOP Conf. Series: Journal of Physics: Conf. Series **804** (2017) 012042 doi:10.1088/1742-6596/804/1/012042

These frequencies are *resonant* or *rational dependent*, the system is degenerate, and this fact has important consequences for developing perturbation theory.

Recall that in a geometric picture of an integrable Hamiltonian system, the whole phase space is foliated into an n-parameter family of invariant tori, on which the the flow is linear with constant frequencies. According to the classical Kolmogorov-Arnold-Moser (KAM) theorem [10], when an integrable Hamiltonian is perturbed by a small non-integrable piece, most KAM tori survive but suffer small alterations. In contrast with the customary case, the resonant tori which have rational ratios of frequencies get destroyed and motion on them becomes chaotic. The chaotic behavior of the perturbed Hamiltonian integrable system $T^{1,1}$ was observed in [13, 14] using numerical simulations in the study of certain classical string configurations in $AdS \times T^{1,1}$.

The angle variables (11) are evaluated from equations (13) and (21) [12]:

$$
\begin{aligned}
w_{\theta_i} &= \frac{\partial W}{\partial J_{\theta_i}} = \frac{\partial W_{\theta_i}}{\partial J_{\theta_i}} = -\frac{J_{\theta_i}+J_{\phi_i}}{2\pi} I_1(a_i,b_i,c_i;\cos\theta_i) \quad , \quad i=1,2\,,\\
w_{\phi_i} &= \frac{\partial W}{\partial J_{\phi_i}} = \frac{\partial W_{\theta_i}}{\partial J_{\phi_i}} + \frac{\partial W_{\phi_i}}{\partial J_{\phi_i}} = \frac{\partial W_{\theta_i}}{\partial J_{\phi_i}} + \frac{1}{2\pi}\phi_i\\
&= -\frac{J_{\theta_i}+J_{\phi_i}}{2\pi} I_1(a_i,b_i,c_i;\cos\theta_i) - \frac{2J_{\phi_i}-J_\nu}{8\pi} I_2(a_i+b_i+c_i, b_i+2c_i, c_i;\cos\theta_i-1)\\
&+ \frac{2J_{\phi_i}+J_\nu}{8\pi} I_2(a_i-b_i+c_i, b_i-2c_i, c_i;\cos\theta_i+1) + \frac{1}{2\pi}\phi_i \quad , \quad i=1,2\,,\\
w_\nu &= \frac{\partial W}{\partial J_\nu} = \frac{\partial W_{\theta_1}}{\partial J_\nu} + \frac{\partial W_{\theta_2}}{\partial J_\nu} + \frac{\partial W_\nu}{\partial J_\nu} = \frac{\partial W_{\theta_1}}{\partial J_\nu} + \frac{\partial W_{\theta_2}}{\partial J_\nu} + \frac{1}{2\pi}\nu\\
&= \sum_{i=1,2} \frac{2J_{\phi_i}-J_\nu}{16\pi} I_2(a_i+b_i+c_i, b_i+2c_i, c_i;\cos\theta_i-1)\\
&+ \sum_{i=1,2} \frac{2J_{\phi_i}+J_\nu}{16\pi} I_2(a_i-b_i+c_i, b_i-2c_i, c_i;\cos\theta_i+1) + \frac{1}{2\pi}\nu\,,
\end{aligned}
\tag{24}
$$

where:

$$
\begin{aligned}
a_i &= J_{\theta_i}^2 + 2J_{\theta_i}J_{\phi_i} - \frac{1}{4}J_\nu^2\,,\\
b_i &= J_{\phi_i}J_\nu\,,\\
c_i &= -\left(J_{\theta_i}+J_{\phi_i}\right)^2 \quad , \quad i=1,2\,.
\end{aligned}
\tag{25}
$$

In the above equations $I_1(a,b,c;t)$ and $I_2(a,b,c;t)$ are the integrals [15]

$$
\begin{aligned}
I_1(a,b,c;t) &= \int \frac{dt}{\sqrt{a+bt+ct^2}} = \frac{-1}{\sqrt{-c}} \arcsin\left(\frac{2ct+b}{\sqrt{-\Delta}}\right),\\
I_2(a,b,c;t) &= \int \frac{dt}{t\sqrt{a+bt+ct^2}} = \frac{1}{\sqrt{-a}} \arctan\left(\frac{2a+bt}{2\sqrt{-a}\sqrt{a+bt+ct^2}}\right),
\end{aligned}
\tag{26}
$$

evaluated for $c<0$, $\Delta = 4ac - b^2 < 0$ and for $a<0$ respectively. That is the case of the parameters (25) taking for granted (20).

Let us note that the degeneracy of the pairs of frequencies $(\omega_{\theta_i}, \omega_{\phi_i})$ may be removed by a canonical transformation from the variables

$$\{(w_{\theta_1}, J_{\theta_1}), (w_{\theta_2}, J_{\theta_2}), (w_{\phi_1}, J_{\phi_1}), (w_{\phi_2}, J_{\phi_2}), (w_\nu, J_\nu)\}$$

IOP Conf. Series: Journal of Physics: Conf. Series **804** (2017) 012042 doi:10.1088/1742-6596/804/1/012042

to new action-angle variables

$$\{(\Theta_1, J_{\Theta_1}), (\Theta_2, J_{\Theta_2}), (\Phi_1, J_{\Phi_1}), (\Phi_2, J_{\Phi_2}), (\Xi, J_\Xi)\}.$$

For this purpose we introduce the following generating function

$$F(w_{\theta_1}, w_{\phi_1}, w_{\theta_2}, w_{\phi_2}, w_\nu; J_{\Theta_1}, J_{\Phi_1}, J_{\Theta_2}, J_{\Phi_2}, J_\Xi) = \sum_{i=1,2}(w_{\phi_i} - w_{\theta_i})J_{\Phi_i} + \sum_{i=1,2} w_{\theta_i}J_{\Theta_i} + w_\nu J_\Xi\,, \quad (27)$$

which implies

$$\begin{aligned} \Phi_i &= \frac{\partial F}{\partial J_{\Phi_i}} = w_{\phi_i} - w_{\theta_i} \quad , \quad i = 1,2\,, \\ \Theta_i &= \frac{\partial F}{\partial J_{\Theta_i}} = w_{\theta_i} \quad , \quad i = 1,2\,, \\ \Xi &= \frac{\partial F}{\partial J_\Xi} = w_\nu\,, \end{aligned} \quad (28)$$

and

$$\begin{aligned} J_{\phi_i} &= \frac{\partial F}{\partial w_{\phi_i}} = J_{\Phi_i} \quad , \quad i = 1,2\,, \\ J_{\theta_i} &= \frac{\partial F}{\partial w_{\theta_i}} = J_{\Theta_i} - J_{\Phi_i} \quad , \quad i = 1,2\,, \\ J_\nu &= \frac{\partial F}{\partial w_\nu} = J_\Xi\,. \end{aligned} \quad (29)$$

The Hamiltonian (22) becomes

$$H = \frac{3}{4\pi^2}(J_{\Theta_1}^2 + J_{\Theta_2}^2 - \frac{1}{8}J_\Xi^2)\,, \quad (30)$$

with the fundamental frequencies

$$\omega_{\Phi_1} = \omega_{\Phi_2} = 0 \quad , \quad \omega_{\Theta_1} = \frac{3}{2\pi^2}J_{\Theta_1} \quad , \quad \omega_{\Theta_2} = \frac{3}{2\pi^2}J_{\Theta_2} \quad , \quad \omega_\Xi = -\frac{3}{16\pi^2}J_\Xi\,. \quad (31)$$

Therefore, in terms of the new transformed variables, the Hamiltonian (30) appears in a form involving only that action variables for which the corresponding frequencies are different from zero.

4. Conclusions

The action-angle approach to geodesic motions in $T^{1,1}$ space gives us a better understanding of the dynamics. The formulation of the integrable system in these variables represents a useful tool for developing perturbation theory.

We observe that for geodesics in $T^{1,1}$ the frequencies of the motions corresponding to θ_i and ϕ_i coordinates are equal. The resonant tori are in general destroyed by an arbitrary small perturbation to the integrable Hamiltonian giving way to chaotic behavior. A similar result has been obtained using numerical simulations [13] which shows that certain classical configurations in $AdS_5 \times T^{1,1}$ are chaotic.

It would be interesting to extend the action-angle formulations to other five-dimensional SE spaces as well to their higher dimensional generalizations relevant for the predictions of the AdS/CFT correspondence.

Acknowledgments
This work has been partly supported by the joint Romanian-LIT, JINR, Dubna Research Project, theme no. 05-6-1119-2014/2016 and partly by the program NUCLEU 16 42 01 01/2016, Romania.

References
[1] Boyer C and Galicki K 2008 *Sasakian geometry* (Oxford Mathematical Monographs, Oxford University Press, Oxford)
[2] Sparks J 2011 Sasaki-Einstein manifolds *Surv. Diff. Geom.* **16** 265 (*Preprint* 1004.2461)
[3] Gauntlett J P, Martelli D, Sparks J and Waldram D 2004 Sasaki-Einstein metrics on $S^2 \times S^3$ *Adv. Theor. Math. Phys.* **8** 711 (*Preprint* hep-th/0403002)
[4] Martelli D and Sparks J 2006 Toric geometry, Sasaki-Einstein manifolds and a new infinite class of AdS/CFT duals *Commun. Math. Phys.* **262** 51 (*Preprint* hep-th/0411238)
[5] Candelas P and de la Ossa X C 1990 Comments on conifolds *Nucl. Phys.* **B342** 246
[6] Babalic E M and Visinescu M 2015 Complete integrability of geodesic motion in Sasaki-Einstein toric $Y^{p,q}$ spaces *Mod. Phys. Lett. A* **30** 1550180 (*Preprint* 1505.03976)
[7] Slesar V, Visinescu M and Vîlcu G E 2015 Hidden symmetries on toric Sasaki-Einstein spaces *EPL* **110** 31001
[8] Slesar V, Visinescu M and Vîlcu G E 2015 Symplectic potential, complex coordinates and hidden symmetries on toric Sasaki-Einstein spaces *P. Romanian Acad. A* **16** 505
[9] Slesar V, Visinescu M and Vîlcu G E 2016 Toric data and Killing forms on homogeneous Sasaki-Einstein manifold $T^{1,1}$ *Rom. J. Phys.* **61** 260 (*Preprint* 1503.00443)
[10] Arnold V I 1989 *Mathematical Methods of Classical Mechanics* (Springer-Verlag, New York, 2nd edition)
[11] Goldstein H, Poole C and Safko J 2002 *Classical Mechanics* (Addison-Wesley, San Francisco, 3rd edition)
[12] Visinescu M 2016 Integrability of geodesics and action-angle variables in Sasaki-Einstein space $T^{1,1}$ *Eur. Phys. J. C* **76** 498 (*Preprint* 1604.03705)
[13] Basu P and Pando Zayas L A 2011 Chaos rules out integrability of strings on $AdS_5 \times T^{1,1}$ *Phys. Lett. B* **700**, 243 (*Preprint* 1103.4107)
[14] Basu P and Pando Zayas L A 2011 Analytic non-integrability in string theory *Phys. Rev. D* **84**, 046006 (*Preprint* 1105.2540)
[15] Gradshteyn I S and Ryzhik I M 2007 *Table of Integrals, Series, and Products* (Edited by A. Jeffrey and D. Zwillinger; Academic Press, New York)

Symmetry of String Scattering Amplitudes

Yi Yang

Abstract. We explore the hidden symmetry in string theory by studying string scattering amplitudes. We calculate 4-point open string scattering amplitudes with three tachyons and a massive higher spin string state. The result can be expressed as Type D Lauricella functions which are generalization of Gaussian Hypergeometric functions. In various high energy limits, the string amplitudes reduced to the expected results that we obtained previously. We find exact $SL(K+3,\mathbb{C})$ symmetry for the string amplitudes at general energy.

Department of Electrophysics, National Chiao-Tung University, Hsinchu, Taiwan

E-mail: yiyang@mail.nctu.edu.tw

1. Introduction

Quantum Field Theory (QFT) is a powerful theory in modern physics. Based on QFT, standard model of particle physics successfully describes our microcosmic world. All important predictions by standard model have been observed in various experiments under rather precise level. However, to solve the UV divergence problem in QFT, the key technical procedure, i.e. renormalization, is complicated and has not been fully understood. More seriously, the renormalization procedure does not work for gravity, which means that it is impossible to construct a consistent quantum gravity theory by using the conventional QFT. Most of people believe that the divergence in QFT comes from the fundamental topological structure of point-like particles, and it cannot be cured without modifying its topological structure. In string theory, one extends a point-like particle to a small piece of a string. This simple extension dramatically changes the topological structure of the theory. The new "Feynman diagram" now is a smooth world-sheet instead of a world-line with singularity at interacting points.

To understand the UV divergence problem better, let us briefly look at the high energy behavior in QFT by a simple power counting. In high energy hard limit, the tree amplitude by interchanging a spin-J particle behaves as $A^{(J)}_{tree} \sim E^{-2(1-J)}$, so that the one-loop amplitude behaves as

$$A^{(J)}_{1-loop} \sim \int d^4p \frac{\left(A^{(J)}_{tree}\right)^2}{(p^2)^2} \sim \int E^{-4(2-J)}\, d^4E, \tag{1}$$

which is finite for scalar particles ($J = 0$) and renormalizable for vector particles ($J = 1$), but is nonrenormalizable for particles with $J \geq 2$, including graviton ($J = 2$). However, there is a loophole to bypass this simple argument. If we sum over all tree amplitudes by interchanging states with different spins, the final amplitude will be

$$A_{tree} = \sum_J A^{(J)}_{tree} \sim \sum_J a_J E^{-2(1-J)}, \tag{2}$$

which could behave rather soft, so that loop amplitudes would be finite, if the following two conditions are satisfied simultaneously:

(i) there are infinite higher spin J particles
(ii) the coefficients a_J's are precisely related to each other.

In string theory, string scattering amplitudes are exponentially fall-off in high energy hard limit, which leads to string theory being a finite theory without UV divergence. We believe that the reason why the high energy behavior of string theory is so soft is that string theory satisfies the above two conditions.

The first condition is trivially satisfied in string theory because a string has infinite oscillation modes which correspond to infinite higher spin states. The second condition is highly nontrivial. We conjecture that it corresponds to a huge symmetry in string theory, which is complicated and not apparent so that we usually call it hidden symmetries. A useful way to investigate the hidden symmetry is to study the symmetry among the string scattering amplitudes. Gross has conjectured that the string scattering amplitudes are linearly related each other in the high energy, fixed scattering angle limit [1, 2, 3]. Using the methods of Ward identities of zero norm states [4, 5, 6], Virasoro algebra and direct calculation of scattering amplitudes, we are able to prove the Gross conjecture and compute the linear ratios among the different string amplitudes [7, 8, 9, 10, 11, 12, 13, 14, 15, 16, 17]. We also extend our study to the high energy, small angle limit, i.e. Regge scattering [18, 19, 20, 21, 22]. Recently, we calculated the four-point string amplitudes at arbitrary energy and found the amplitudes associate to $SL(K+3,\mathbb{C})$ algebra [23, 24]. In the next section, we will calculated the four-point string amplitudes, and study their relations in various high energy limits in section 3. In section 4, we show that how to get the $SL(K+3,\mathbb{C})$ algebra from string amplitudes. We conclude our result in section 5.

2. Four-Point String Amplitudes

To study the symmetry of string scattering amplitudes, we consider four-point open bosonic string scattering In the CM frame, the kinematics are

$$k_1 = \left(\sqrt{M_1^2+|\vec{k_1}|^2}, -|\vec{k_1}|, 0\right), \tag{3}$$

$$k_2 = \left(\sqrt{M_2^2+|\vec{k_1}|^2}, +|\vec{k_1}|, 0\right), \tag{4}$$

$$k_3 = \left(-\sqrt{M_3^2+|\vec{k_3}|^2}, -|\vec{k_3}|\cos\phi, -|\vec{k_3}|\sin\phi\right), \tag{5}$$

$$k_4 = \left(-\sqrt{M_4^2+|\vec{k_3}|^2}, +|\vec{k_3}|\cos\phi, +|\vec{k_3}|\sin\phi\right), \tag{6}$$

where ϕ is the scattering angle. The Mandelstam variables are defined as usual as

$$s = -\left(k_1+k_2\right)^2, t = -\left(k_2+k_3\right)^2, u = -\left(k_1+k_3\right)^2, \tag{7}$$

with $s+t+u = \sum M_i^2$.

On the 2-dimensional scattering plane, there are three independent polarizations which we can choose to be

$$e^T = (0,0,1), \tag{8}$$

$$e^L = \frac{1}{M_2}\left(|\vec{k_1}|, \sqrt{M_2+|\vec{k_1}|^2}, 0\right), \tag{9}$$

$$e^P = \frac{1}{M_2}\left(\sqrt{M_2+|\vec{k_1}|^2}, |\vec{k_1}|, 0\right). \tag{10}$$

IOP Conf. Series: Journal of Physics: Conf. Series **804** (2017) 012043 doi:10.1088/1742-6596/804/1/012043

Note that the string amplitude with polarizations orthogonal to the scattering plane vanish. For later use, we also define

$$k_i^X \equiv e^X \cdot k_i \quad \text{for} \quad X = (T, P, L). \tag{11}$$

The simplest four-point string amplitude is scattered by four tachyons with $M_i^2 = -2$, i.e. the Veneziano amplitudes. In (s,t) channel, the four-tachyon scattering amplitude can be easily calculated,

$$A_{st}^{(\text{4-tachyon})} = \left\langle e^{ik_1\cdot X(x_1)} e^{ik_2\cdot X(x_2)} e^{ik_3\cdot X(x_3)} e^{ik_4\cdot X(x_4)} \right\rangle = B\left(-\frac{s}{2}-1, -\frac{t}{2}-1\right), \tag{12}$$

where

$$B\left(-\frac{s}{2}-1, -\frac{t}{2}-1\right) = \frac{\Gamma\left(-\frac{s}{2}-1\right)\Gamma\left(-\frac{t}{2}-1\right)}{\Gamma\left(\frac{u}{2}+2\right)}, \tag{13}$$

is Beta function. It is easy to verify that the Veneziano amplitude behaves as exponentially fall-off $A_{st}^{(\text{4-tachyon})} \sim e^{-E}$ in the high energy limit ($s \sim E^2 \to \infty$, $t \to \infty$ and s/t fixed). This property holds for all four-point scattering amplitudes in string theory as we will show later.

Now let us consider a general 4-point string amplitudes with three tachyons and an arbitrary massive higher spin string state of the form,

$$|r_n^T, r_m^P, r_l^L\rangle = \prod_{n>0} \left(\alpha_{-n}^T\right)^{r_n^T} \prod_{m>0} \left(\alpha_{-m}^P\right)^{r_m^P} \prod_{l>0} \left(\alpha_{-l}^L\right)^{r_l^L} |0, k\rangle, \tag{14}$$

where $M_1^2 = M_3^2 = M_4^2 = -2$ are three tachyons and $M_2^2 \equiv M^2 = 2(N-1)$ is the higher spin string state with the mass level $N = \sum_{n,m,l>0} \left(nr_n^T + mr_m^P + lr_l^L\right)$.

The (s,t) and (t,u) channels string scattering amplitudes of states in Eq.(14) can be calculated to be [23, 24]

$$\begin{aligned} A_{st}^{(r_n^T, r_m^P, r_l^L)} &= B\left(-\frac{t}{2}-1, -\frac{s}{2}-1\right) F_D^{(K)}\left(-\frac{t}{2}-1; R_n^T, R_m^P, R_l^L; \frac{u}{2}+2-N; \tilde{Z}_n^T, \tilde{Z}_m^P, \tilde{Z}_l^L\right) \\ &\cdot \prod_{n=1} \left[-(n-1)!k_3^T\right]^{r_n^T} \prod_{m=1} \left[-(m-1)!k_3^P\right]^{r_m^P} \prod_{l=1} \left[-(l-1)!k_3^L\right]^{r_l^L}, \end{aligned} \tag{15}$$

$$\begin{aligned} A_{tu}^{(r_n^T, r_m^P, r_l^L)} &= B\left(-\frac{t}{2}-1, -\frac{u}{2}-1\right) F_D^{(K)}\left(-\frac{t}{2}-1; R_n^T, R_m^P, R_l^L; \frac{s}{2}+2-N; Z_n^T, Z_m^P, Z_l^L\right) \\ &\cdot \prod_{n=1} \left[-(n-1)!k_3^T\right]^{r_n^T} \prod_{m=1} \left[-(m-1)!k_3^P\right]^{r_m^P} \prod_{l=1} \left[-(l-1)!k_3^L\right]^{r_l^L}, \end{aligned} \tag{16}$$

where we have defined

$$R_k^X \equiv \left\{-r_1^X\right\}^1, \cdots, \left\{-r_k^X\right\}^k \quad \text{with} \quad \{a\}^n = \underbrace{a, a, \cdots, a}_{n}, \tag{17}$$

$$Z_k^X \equiv \left[z_1^X\right], \cdots, \left[z_k^X\right] \quad \text{with} \quad \left[z_k^X\right] = z_{k0}^X, \cdots, z_{k(k-1)}^X, \tag{18}$$

$$z_{kk'}^X = \left|\frac{k_1^X}{k_3^X}\right|^{\frac{1}{k}} e^{\frac{2\pi i k'}{k}} \quad \text{and} \quad \tilde{z}_{kk'}^X \equiv 1 - z_{kk'}^X,\ k' = 0, \cdots, k-1, \tag{19}$$

and the integer K is defined to be

$$K = \sum_{\substack{j=1 \\ \{\text{for all } r_j^T \neq 0\}}}^{n} j + \sum_{\substack{j=1 \\ \{\text{for all } r_j^P \neq 0\}}}^{m} j + \sum_{\substack{j=1 \\ \{\text{for all } r_j^L \neq 0\}}}^{l} j, \tag{20}$$

which is usually different from the mass level N.

Finally, the D-type Lauricella function $F_D^{(K)}$ is one of the four extensions of the Gauss hypergeometric function to K variables and is defined as

$$\begin{aligned}
&F_D^{(K)}\left(a;b_1,...,b_K;c;x_1,...,x_K\right)\\
&=\sum_{k_i}\frac{(a)_{k_1+\cdots+k_K}}{(c)_{k_1+\cdots+k_K}}\frac{(b_1)_{k_1}\cdots(b_n)_{k_K}}{k_1!\cdots k_K!}x_1^{k_1}\cdots x_K^{k_K}\\
&=\frac{\Gamma(c)}{\Gamma(a)\Gamma(c-a)}\int_0^1 dt\, t^{a-1}(1-t)^{c-a-1}\cdot(1-x_1t)^{-b_1}(1-x_2t)^{-b_2}...(1-x_Kt)^{-b_K},
\end{aligned} \tag{21}$$

where the integral representation of the Lauricella function $F_D^{(K)}$ in the last line was discovered by Appell and Kampe de Feriet (1926) [25].

By using the identity of Lauricella function for $b_i \in Z^-$

$$\begin{aligned}
&F_D^{(K)}\left(a;b_1,...,b_K;c;x_1,...,x_K\right)\\
&=\frac{\Gamma(c)\Gamma(c-a-\sum b_i)}{\Gamma(c-a)\Gamma(c-\sum b_i)}F_D^{(K)}\left(a;b_1,...,b_K;1+a+\sum b_i-c;1-x_1,...,1-x_K\right),
\end{aligned} \tag{22}$$

we can express the (s,t) channel amplitude (15) in the following form

$$\begin{aligned}
A_{st}^{(r_n^T,r_m^P,r_l^L)} &= B\left(-\frac{t}{2}-1,-\frac{s}{2}-1+N\right)F_D^{(K)}\left(-\frac{t}{2}-1;R_n^T,R_m^P,R_l^L;\frac{s}{2}+2-N;Z_n^T,Z_m^P,Z_l^L\right)\\
&\cdot\prod_{n=1}\left[-(n-1)!k_3^T\right]^{r_n^T}\cdot\prod_{m=1}\left[-(m-1)!k_3^P\right]^{r_m^P}\prod_{l=1}\left[-(l-1)!k_3^L\right]^{r_l^L}.
\end{aligned} \tag{23}$$

Now it is easy to see the string BCJ relation

$$\frac{A_{st}^{(r_n^T,r_m^P,r_l^L)}}{A_{tu}^{(r_n^T,r_m^P,r_l^L)}}=\frac{(-)^N\Gamma\left(-\frac{s}{2}-1\right)\Gamma\left(\frac{s}{2}+2\right)}{\Gamma\left(\frac{u}{2}+2-N\right)\Gamma\left(-\frac{u}{2}-1+N\right)}=\frac{\sin\left(\frac{\pi u}{2}\right)}{\sin\left(\frac{\pi s}{2}\right)}-\frac{\sin\left(\pi k_2\cdot k_4\right)}{\sin\left(\pi k_1\cdot k_2\right)}, \tag{24}$$

which was proved by monodromy of integration of string amplitudes [26, 27] and explicitly proved recently in [28].

3. Symmetry in High Energy Limits

To study the relations among the string scattering amplitudes, we consider two different high energy limits: hard scattering limit and Regge scattering limit. We will briefly describe the results in the following. Readers can find the detail in a current review paper [29].

3.1. Linear Relations in Hard Limit

Hard scattering limit is the fixed angle scattering with $s\sim E^2\to\infty$ and $\frac{t}{s}\sim\sin^2\frac{\phi}{2}=$ constant. The linear relations of string amplitudes in the hard scattering limit were conjectured by Gross [1, 2, 3] and proved in [7, 8, 9, 10, 11, 13]. In the hard scattering limit $e^P=e^L$ [7, 8], we can consider only the polarization e^L case. The relevant kinematics are

$$k_1^T=0,\quad k_3^T\simeq -E\sin\phi, \tag{25}$$

$$k_1^L\simeq-\frac{2p^2}{M_2}\simeq-\frac{2E^2}{M_2}, \tag{26}$$

$$k_3^L\simeq\frac{2E^2}{M_2}\sin^2\frac{\phi}{2}, \tag{27}$$

with $\tilde{z}^T_{kk'} = 1$, $\tilde{z}^L_{kk'} = 1 - \left(-\frac{s}{t}\right)^{1/k} e^{\frac{i2\pi k'}{k}} \sim O(1)$.

In the hard limit, the (s,t) channel string amplitude in Eq.(15) reduces to

$$A_{st}^{(r_n^T, r_l^L)} = B\left(-\frac{t}{2}-1, -\frac{s}{2}-1\right) F_D^{(K)}\left(-\frac{t}{2}-1; R_n^T, R_l^L; \frac{u}{2}+2-N; (1)_n, \tilde{Z}_l^L\right)$$
$$\cdot \prod_{n=1} \left[(n-1)!E\sin\phi\right]^{r_n^T} \prod_{l=1} \left[-(l-1)!\frac{2E^2}{M_2}\sin^2\frac{\phi}{2}\right]^{r_l^L}. \tag{28}$$

Next, we propose the following identity

$$\sum_{k_r=0}^{r_1^L} \frac{\left(-\frac{t}{2}-1\right)_{k_r}}{\left(\frac{u}{2}+2-N\right)_{k_r}} \frac{\left(-r_1^L\right)_{k_r}}{k_r!} \left(1+\frac{s}{t}\right)^{k_r}$$
$$= 0\cdot\left(\frac{tu}{s}\right)^0 + 0\cdot\left(\frac{tu}{s}\right)^{-1} + \cdots + 0\cdot\left(\frac{tu}{s}\right)^{-\left[\frac{r_1^L+1}{2}\right]-1} + C_{r_1^L}\left(\frac{tu}{s}\right)^{-\left[\frac{r_1^L+1}{2}\right]}$$
$$+ O\left\{\left(\frac{tu}{s}\right)^{-\left[\frac{r_1^L+1}{2}\right]+1}\right\}, \tag{29}$$

where $C_{r_1^L}$ is independent of energy E and depends on r_1^L and possibly scattering angle ϕ. For $r_1^L = 2m$ being an even number, we further propose that $C_{r_1^L} = \frac{(2m)!}{m!}$ and is ϕ independent. We have verified Eq.(29) for $r_1^L = 0, 1, 2, \cdots, 10$.

It should be noted that, taking Regge limit ($s \to \infty$ with t fixed) and setting $r_1^L = 2m$, Eq.(29) reduces to the Stirling number identity,

$$\sum_{k_r=0}^{2m} \frac{\left(-\frac{t}{2}-1\right)_{k_r}}{\left(-\frac{s}{2}\right)_{k_r}} \frac{(-2m)_{k_r}}{k_r!} \left(\frac{s}{t}\right)^{k_r} \simeq \sum_{k_r=0}^{2m} (-2m)_{k_r} \left(-\frac{t}{2}-1\right)_{k_r} \frac{(-2/t)^{k_r}}{k_r!}$$
$$= 0\cdot(-t)^0 + 0\cdot(-t)^{-1} + \cdots + 0\cdot(-t)^{-m+1} + \frac{(2m)!}{m!}(-t)^{-m} + O\left\{\left(\frac{1}{t}\right)^{m+1}\right\}, \tag{30}$$

which was proposed in [30] and proved in [31].

In Eq.(29), the 0 terms correspond to the naive leading energy orders of string amplitudes in the hard scattering limit. The true leading order of string amplitudes in the hard scattering limit can then be identified

$$A_{st}^{(r_n^T, r_l^L)} \simeq B\left(-\frac{t}{2}-1, -\frac{s}{2}-1\right) \cdot C_{r_1^L}\left(E\sin\phi\right)^{-2\left[\frac{r_1^L+1}{2}\right]} \cdot (\cdots)$$
$$\cdot \prod_{n=1} \left[(n-1)!E\sin\phi\right]^{r_n^T} \prod_{l=1} \left[-(l-1)!\frac{2E^2}{M_2}\sin^2\frac{\phi}{2}\right]^{r_l^L}$$
$$\sim E^{N-\sum_{n\geq 2} n r_n^T - \left(2\left[\frac{r_1^L+1}{2}\right] - r_1^L\right) - \sum_{l\geq 3} l r_l^L}, \tag{31}$$

which means that a string amplitude reaches its highest energy when $r_{n\geq 2}^T = r_{l\geq 3}^L = 0$ and $r_1^L = 2m$ being an even number. This is first observed in [7, 8, 9, 10, 11, 13]

Finally, the leading order string amplitudes in the hard scattering limit can be calculated to be

$$\begin{aligned}
A_{st}^{(r_1^T,2m,r_2^L)} &\simeq B\left(-\frac{s}{2}-1,-\frac{t}{2}-1\right)\cdot F_D^{(4)}\left(-\frac{t}{2}-1;R_1^T,R_2^L;\frac{u}{2}+2-N;1,Z_2^L\right)\\
&\cdot\left[E\sin\phi\right]^{r_1^T}\cdot\left[-\frac{2E^2}{M_2}\sin^2\frac{\phi}{2}\right]^{2m}\cdot\left[-\frac{2E^2}{M_2}\sin^2\frac{\phi}{2}\right]^{r_2^L}\\
&= B\left(-\frac{s}{2}-1,-\frac{t}{2}-1\right)(E\sin\phi)^N\cdot\left(r_1^L-1\right)!!\left(-\frac{1}{M_2}\right)^{2m+r_2^L}\left(\frac{1}{2}\right)^{m+r_2^L}\\
&= \left(r_1^L-1\right)!!\left(-\frac{1}{M_2}\right)^{2m+r_2^L}\left(\frac{1}{2}\right)^{m+r_2^L}\cdot A^{(N,0,0)},
\end{aligned}\tag{32}$$

which reproduces the ratios

$$\frac{A_{st}^{(r_1^T,2m,r_2^L)}}{A_{st}^{(N,0,0)}}=(2m-1)!!\left(-\frac{1}{M_2}\right)^{2m+r_2^L}\left(\frac{1}{2}\right)^{m+r_2^L},\tag{33}$$

which is consistent with the previous result [7, 8, 9, 10, 11, 13].

3.2. Recurrence Relations in Regge Limit

Regge scattering limit is the small angle scattering with $s\sim E^2\to\infty$ and $t\sim E^2\sin^2\frac{\phi}{2}=$ constant. The recurrence relations of string amplitudes in the Regge scattering limit have been studied in [30, 32, 33]. The relevant kinematics in Regge limit are

$$k_1^T=0,\quad k_3^T\simeq-\sqrt{-t},\tag{34}$$

$$k_1^P\simeq-\frac{s}{2M_2},\ k_3^P\simeq-\frac{\tilde{t}}{2M_2}=-\frac{t-M_2^2-M_3^2}{2M_2},\tag{35}$$

$$k_1^L\simeq-\frac{s}{2M_2},\ k_3^L\simeq-\frac{\tilde{t}'}{2M_2}=-\frac{t+M_2^2-M_3^2}{2M_2},\tag{36}$$

with $\tilde{z}_{kk'}^T=1$, $\tilde{z}_{kk'}^P=1-\left(-\frac{s}{t}\right)^{1/k}e^{\frac{i2\pi k'}{k}}\sim s^{1/k}$ and $\tilde{z}_{kk'}^L=1-\left(-\frac{s}{t'}\right)^{1/k}e^{\frac{i2\pi k'}{k}}\sim s^{1/k}$.

In the hard limit, the (s,t) channel string amplitude in Eq.(15) reduces to

$$\begin{aligned}
A_{st}^{(r_n^T,r_m^P,r_l^L)} &\simeq B\left(-\frac{t}{2}-1,-\frac{s}{2}-1\right)F_1\left(-\frac{t}{2}-1;-q_1,-r_1;-\frac{s}{2};\frac{s}{\tilde{t}},\frac{s}{\tilde{t}'}\right)\\
&\cdot\prod_{n=1}\left[(n-1)!\sqrt{-t}\right]^{r_n^T}\cdot\prod_{m=1}\left[(m-1)!\frac{\tilde{t}}{2M_2}\right]^{r_m^P}\prod_{l=1}\left[(l-1)!\frac{\tilde{t}'}{2M_2}\right]^{r_l^L},
\end{aligned}\tag{37}$$

where F_1 is the Appell function. Eq.(37) agrees with the result obtained in [33] previously.

The string amplitudes in Regge limit are much more complicated than that in hard limit and do not have linear relations. However, there are many recurrence relations of Appell functions F_1,

$$(a-b_1-b_2)F_1-aF_1(a+1)+b_1F_1(b_1+1)+b_2F_1(b_2+1)=0,\tag{38}$$

$$cF_1-(c-a)F_1(c+1)-aF_1(a+1;c+1)=0,\tag{39}$$

$$cF_1+c(x-1)F_1(b_1+1)-(c-a)xF_1(b_1+1;c+1)=0,\tag{40}$$

$$cF_1+c(y-1)F_1(b_2+1)-(c-a)yF_1(b_2+1;c+1)=0.\tag{41}$$

Using the above recurrence relations, we can obtain a lot of recurrence relations among the string amplitudes in Eq.(37). One can show that by solving the recurrence relations, all the string amplitudes at certain mass level can be expressed in term of a single amplitude [33].

4. Symmetry of Four-Point Amplitudes at General Energy

Let us recall the (s,t) channels string scattering amplitudes with three tachyons and a massive higher spin string state in Eq.(15),

$$A_{st}^{(r_n^T, r_m^P, r_l^L)} = B\left(-\frac{t}{2}-1, -\frac{s}{2}-1\right) F_D^{(K)}\left(-\frac{t}{2}-1; R_n^T, R_m^P, R_l^L; \frac{u}{2}+2-N; \tilde{Z}_n^T, \tilde{Z}_m^P, \tilde{Z}_l^L\right)$$
$$\cdot \prod_{n=1} \left[-(n-1)!k_3^T\right]^{r_n^T} \cdot \prod_{m=1} \left[-(m-1)!k_3^P\right]^{r_m^P} \prod_{l=1} \left[-(l-1)!k_3^L\right]^{r_l^L} . \quad (42)$$

To discover the symmetry of the above amplitudes, we need to understand their mathematical structure in a deeper way. To do it, we follow the mathematical construction in [34] and define the generating functions associated the D-type Lauricella function $F_D^{(K)}$ as

$$f_c^{a,b_j}(s, u_j, t, x_j) \equiv B(a, c-a) F_D^{(K)}(a; b_j; c; x_j) s^a u_1^{b_1} \cdots u_K^{b_K} t^c, \ j = 1, \cdots, K, \quad (43)$$

so that the (s,t) channels string scattering amplitudes can be expressed in term of the generating functions as

$$A_{st}^{(r_n^T, r_m^P, r_l^L)} \sim f_{\frac{u}{2}+2-N}^{-\frac{t}{2}-1, R_j^X}\left(1, k_3^X, 1, \tilde{Z}_j^X\right), \ X = T, P, L. \quad (44)$$

We next define the operators,

$$E^{\alpha} = s\left(\sum_j x_j \partial_{x_j} + s\partial_s\right), \ E^{\alpha\gamma} = st\left(\sum_j (1-x_j)\partial_{x_j} - s\partial_s\right),$$
$$E^{\beta_k} = u_k\left(x_k \partial_{x_k} + u_k \partial_{u_k}\right), \ E^{\beta_k\gamma} = u_k t\left(-(1-x_k)\partial_{x_k} + u_k \partial_{u_k}\right),$$
$$E^{\gamma} = t\left(\sum_j (1-x_j)\partial_{x_j} + t\partial_t - s\partial_s - \sum_j u_j \partial_{u_j}\right), \ E^{\alpha\beta_k\gamma} = su_k t\partial_{x_k},$$
$$J_{\alpha} = s\partial_s, \ J_{\beta_k} = u_k\partial_{u_k} - \frac{1}{2}t\partial_t + \frac{1}{2}\sum_{j\neq k} u_j\partial_{u_j}, \ J_{\gamma} = t\partial_t - \frac{1}{2}\left(s\partial_s + \sum_j u_j\partial_{u_j} + 1\right), \quad (45)$$

which act on the generating function gives,

$$E^{\alpha} f_c^{a,b_j} = (c-a-1) f_c^{a+1,b_j}, \quad E^{\alpha\gamma} f_c^{a,b_j} = \left(\sum_j b_j - 1\right) f_{c+1}^{a+1,b_j},$$
$$E^{\beta_k} f_c^{a,b_j} = b_k f_c^{a,b_k+1}, \quad E^{\beta_k\gamma} f_c^{a,b_j} = b_k f_{c+1}^{a,b_k+1},$$
$$E^{\gamma} f_c^{a,b_j} = \left(c - \sum_j b_j\right) f_{c+1}^{a,b_j}, \quad E^{\alpha\beta_k\gamma} f_c^{a,b_j} = b_k f_{c+1}^{a+1,b_k+1},$$
$$J_{\alpha} f_c^{a,b_j} = \left(a - \frac{c}{2}\right) f_c^{a,b_j}, \quad J_{\beta_k} f_c^{a,b_j} = \left(b_k - \frac{c}{2} + \frac{1}{2}\sum_{j\neq k} b_j\right) f_c^{a,b_j},$$
$$J_{\gamma} f_c^{a,b_j} = \left[c - \frac{1}{2}\left(a + \sum_j b_j + 1\right)\right] f_c^{a,b_j}. \quad (46)$$

Finally, we define a set of new operators in the following way,

$$\begin{aligned} &E^{\alpha} = \mathcal{E}_{12}, && E^{\alpha\gamma} = \mathcal{E}_{32}, && E^{\gamma} = \mathcal{E}_{31}, \\ &E^{\beta_k\gamma} = -\mathcal{E}_{k+3,1}, && E^{\alpha\beta_k\gamma} = -\mathcal{E}_{k+3,2}, && E^{\beta_k} = \mathcal{E}_{k+3,3}, \\ &J_\alpha = \frac{1}{2}\left(\mathcal{E}_{11} - \mathcal{E}_{22}\right), && J_\gamma = \frac{1}{2}\left(\mathcal{E}_{33} - \mathcal{E}_{11}\right), && J_{\beta_k} = \frac{1}{2}\left(\mathcal{E}_{k+3,k+3} - \mathcal{E}_{33}\right). \end{aligned} \tag{47}$$

The algebra satisfied by the new operators becomes

$$[\mathcal{E}_{ij}, \mathcal{E}_{kl}] = \delta_{jk}E_{il} - \delta_{li}E_{kj}, \tag{48}$$

which can be identified as $sl\,(K+3,\mathbb{C})$ algebra.

5. Conclusion

In this work, I briefly reviewed the hidden symmetry in string theory by studying the string scattering amplitudes. The four-point bosonic open string scattering amplitude with three tachyons and an arbitrary massive higher spin string state in both (s,t) and (t,u) channels have been explicitly calculated and expressed in term of D-type Lauricella function in Eqs.(15) and (16). The string BCJ relation can be verified easily. We also considered two high energy limits. In hard limit, the hidden symmetry reduces to the linear relations among the string amplitude. In Regge limit, the hidden symmetry exhibit to be the recurrence relations of the string amplitudes. At general energy, we mathematically showed that the hidden symmetry is associated to $sl\,(K+3,\mathbb{C})$ algebra. In the future, we will study this symmetry in more detail and try to understand its physical picture.

6. Acknowledgments

The main results in this talk are based on a series of works collaborated with S.H. Lai and Jen-Chi Lee. I would like to thank the organizers of The 24th International Conference on Integrable Systems and Quantum symmetries (ISQS24) for inviting me to present this work. This work is supported by the Ministry of Science and Technology (MoST), Taiwan.

[1] D. J. Gross and P. F. Mende, Phys. Lett. B **197**, 129 (1987); Nucl. Phys. B **303**, 407 (1988).
[2] D. J. Gross, Phys. Rev. Lett. **60**, 1229 (1988); Phil. Trans. R. Soc. Lond. A329, 401 (1989).
[3] D. J. Gross and J. L. Manes, Nucl. Phys. B **326**, 73 (1989). See section 6 for details.
[4] J. C. Lee, Phys. Lett. B **241**, 336 (1990); Phys. Rev. Lett. **64**, 1636 (1990); Prog. Theor. Phys.**91**, 353 (1994). J. C. Lee, Phys. Lett. B **326**, 79 (1994).
[5] T. D. Chung and J. C. Lee, Phys. Lett. B **350**, 22 (1995). Z. Phys. C **75**, 555 (1997). J. C. Lee, Eur. Phys. J. C **1**, 739 (1998).
[6] H. C. Kao and J. C. Lee, Phys. Rev. D **67**, 086003 (2003). C. T. Chan, J. C. Lee and Y. Yang, Phys. Rev. D **71**, 086005 (2005)
[7] C. T. Chan and J. C. Lee, Phys. Lett. B **611**, 193 (2005). J. C. Lee, [arXiv:hep-th/0303012].
[8] C. T. Chan and J. C. Lee, Nucl. Phys. B **690**, 3 (2004).
[9] C. T. Chan, P. M. Ho and J. C. Lee, Nucl. Phys. B **708**, 99 (2005).
[10] C. T. Chan, P. M. Ho, J. C. Lee, S. Teraguchi and Y. Yang, Nucl. Phys. B **725**, 352 (2005).
[11] C. T. Chan, P. M. Ho, J. C. Lee, S. Teraguchi and Y. Yang, Phys. Rev. Lett. 96 (2006) 171601.
[12] C. T. Chan, P. M. Ho, J. C. Lee, S. Teraguchi and Y. Yang, Nucl. Phys. B **749**, 266 (2006). "Comments on the high energy limit of bosonic open string theory," [arXiv:hep-th/0509009].
[13] C. T. Chan, J. C. Lee and Y. Yang, Nucl. Phys. B **738**, 93 (2006).
[14] C. T. Chan, J. C. Lee and Y. Yang, Nucl. Phys. B **749**, 280 (2006).
[15] Pei-Ming Ho, Xue-Yan Lin, Phys.Rev. D73 (2006) 126007.
[16] C. T. Chan, J. C. Lee and Y. Yang, " Scatterings of massive string states from D-brane and their linear relations at high energies", Nucl.Phys.B**764**, 1 (2007).
[17] J.C. Lee and Y. Yang, "Linear Relations of High Energy Absorption/Emission Amplitudes of D-brane", Phys.Lett. B646 (2007) 120, hep-th/0612059.

[18] Sheng-Lan Ko, Jen-Chi Lee and Yi Yang, "Patterns of High energy Massive String Scatterings in the Regge regime", JHEP 0906:028,(2009); "Kummer function and High energy String Scatterings", arXiv:0811.4502; "Stirling number Identities and High energy String Scatterings", arXiv:0909.3894 (published in the SLAC eConf series).

[19] Jen-Chi Lee and Yi Yang, "Regge Closed String Scattering and its Implication on Fixed angle Closed String Scattering", Phys.Lett.B687:84-88,2010.

[20] S. He, J.C. Lee, K. Takahashi and Y. Yang, "Massive Superstring Scatterings in the Regge Regime", arXiv:1001.5392. (accepted by PRD)

[21] J.C. Lee, Catherine H. Yan and Y. Yang, "High-energy String Scattering Amplitudes and Signless Stirling Number Identity", arXiv: 1012.5225.

[22] Jen-Chi Lee, Yoshihiro Mitsuka and Yi Yang, "Higher Spin String States Scattered from D-particle in the Regge Regime and Factorized Ratios of Fixed Angle Scatterings", arXiv: 1101.1228.

[23] Sheng-Hong Lai, Jen-Chi Lee and Y.Y., "The Exact $SL(K+3,\mathbb{C})$ Symmetry of String Scattering Amplitudes", arXiv:1603.00396.

[24] Sheng-Hong Lai, Jen-Chi Lee and Y.Y., "The Lauricella Functions and Exact String Scattering Amplitudes", arXiv:1609.06014, *JHEP* 11 (2016) 1-17.

[25] Joseph Kampe de Feriet and Paul Appell. Fonctions hypergeometriques et hyperspheriques 1926.

[26] N. E. J. Bjerrum-Bohr, P. H. Damgaard and P. Vanhove, Phys. Rev. Lett. 103, 161602 (2009) [hep-th/0907.1425].

[27] S. Stieberger, hep-th/0907.2211.

[28] S.H. Lai, J.C. Lee and Y. Yang, JHEP 1605 (2016) 186, arXiv: 1601.0381.

[29] J.C. Lee and Y. Yang, Review on High energy String Scattering Amplitudes and Symmetries of String Theory, arXiv: 1510.03297.

[30] S.L. Ko, J.C. Lee and Y. Yang, JHEP, 9060:028 (2009).

[31] J.C. Lee, C. H. Yan, and Y. Yang, "High energy string scattering amplitudes and signless Stirling number identity", SIGMA, 8:045, (2012).

[32] J.C. Lee and Y. Mitsuka, JHEP 1304:082 (2013).

[33] J.C. Lee and Y. Yang, Phys. Lett. B739, 370 (2014).

[34] Willard Miller. Jr., "Lie theory and generalizations of the hypergeometric functions", SIAM J. Appl. Math. Vol. 25 No. 2, 226 (1973).

Institute of Physics
Dirac House, Temple Back
Bristol BS1 6BE UK

ISSN: 1742-6588
ISBN 978-1-5108-3898-7

CURRAN ASSOCIATES INC.
proceedings.com